Texts and
Monographs
in Physics

K. Chadan and P. C. Sabatier

Inverse Problems in Quantum Scattering Theory

Second Edition

Revised and Expanded

With a Foreword by R. G. Newton

With 24 Illustrations

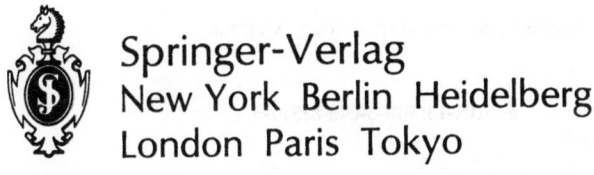

Springer-Verlag
New York Berlin Heidelberg
London Paris Tokyo

K. Chadan
Directeur de recherches CNRS
Laboratoire de Physique Théorique
Université de Paris-Sud
91405 Orsay
France

P. C. Sabatier
Professeur de physique mathématique
Université des Sciences et
  Techniques du Languedoc
34060 Montpellier Cedex
France

R. G. Newton
Professor of Mathematical Physics
Department of Physics
Indiana University
Bloomington, Indiana 47401
U.S.A.

Series Editors

Wolf Beiglböck
Institut für Angewandte
  Mathematik
Universität Heidelberg
D-6900 Heidelberg 1
Federal Republic of
  Germany

Joseph L. Birman
Department of Physics
The City College of the
  City University of
  New York
New York, NY 10031
U.S.A.

Robert P. Geroch
Enrico Fermi Institute
University of Chicago
Chicago, IL 60637
U.S.A.

Elliott H. Lieb
Department of Physics
Joseph Henry Laboratories
Princeton University
Princeton, NJ 08540
U.S.A.

Tullio Regge
Istituto de Fisica Teorica
Universita di Torino
I-10125 Torino
Italy

Walter Thirring
Institut für Theoretische Physik
  der Universität Wien
A-1090 Wien
Austria

Library of Congress Cataloging-in-Publication Data
Chadan, K. (Khosrow)
    Inverse problems in quantum scattering theory / K. Chadan and P. C.
Sabatier ; with a foreword by R. G. Newton. — 2nd ed.
    p.   cm. — (Texts and monographs in physics)
    Bibliography: p.
    Includes index.
    ISBN 0-387-18731-6
    1. Scattering (Physics) 2. Scattering (Mathematics) 3. Quantum
theory. 4. Inverse problems (Differential equations) I. Sabatier,
Pierre Célestin, 1935–   . II. Title. III. Series.
QC20.7.S3C45   1989
530.1'5—dc19   88-20147
                                                                CIP
Printed on acid-free paper.

Printed and bound by R. R. Donnelley & Sons, Harrisonburg, Virginia.

9 8 7 6 5 4 3 2 1

ISBN-13: 978-3-642-83319-9   e-ISBN-13: 978-3-642-83317-5
DOI: 10.1007/978-3-642-83317-5

Foreword

The normal business of physicists may be schematically thought of as predicting the motions of particles on the basis of known forces, or the propagation of radiation on the basis of a known constitution of matter. The inverse problem is to conclude what the forces or constitutions are on the basis of the observed motion. A large part of our sensory contact with the world around us depends on an intuitive solution of such an inverse problem: We infer the shape, size, and surface texture of external objects from their scattering and absorption of light as detected by our eyes. When we use scattering experiments to learn the size or shape of particles, or the forces they exert upon each other, the nature of the problem is similar, if more refined. The kinematics, the equations of motion, are usually assumed to be known. It is the forces that are sought, and how they vary from point to point.

As with so many other physical ideas, the first one we know of to have touched upon the kind of inverse problem discussed in this book was Lord Rayleigh (1877). In the course of describing the vibrations of strings of variable density he briefly discusses the possibility of inferring the density distribution from the frequencies of vibration. This passage may be regarded as a precursor of the mathematical study of the inverse spectral problem some seventy years later. Its modern analogue and generalization was given in a famous lecture by Marc Kac (1966), entitled "Can one hear the shape of a drum?"

With the invention of the Schrödinger equation the physical scope of mathematical notions connected with the spectra of differential equations with prescribed boundary conditions was enormously enlarged. The types of equations that previously had applications only to mechanical vibrations now were to be used for the description of atoms and molecules. It took only three years for the same type of question, that Rayleigh had asked half a century earlier, to be posed in the new context. In 1929 the inverse Schrödinger problem stimulated Ambarzumian to study a uniqueness question for eigenvalues of ordinary differential equations of second order. By then, there was enough confidence in kinematics for it to be used as a probe of the forces.

The specific problem of reconstructing the functions, or the boundary conditions, of a Sturm–Liouville system from its spectral data was first seriously attacked by Borg (1946). It was then pursued in detail and depth by Levinson (1949) and the Russian mathematicians Chudov (1949, 1956), Marchenko (1950, 1952, 1953, 1955), Krein (1951, 1953, 1954, 1955), Gel'fand and Levitan (1951), Levitan (1949, 1956, 1963, 1964), Berezanskii (1955, 1958), and Levitan and Gasymov (1964).

In the meantime, physicists had begun to use experimental information on the spectra of molecules to infer interatomic potential curves in a practical sort of way. Morse (1929) started such studies in the context of specific exactly solvable models, and this line of investigation was continued by Rosen and Morse (1932) and by Crawford and Jorgensen (1936). In a similar spirit Pöschl and Teller (1933) studied an anharmonic oscillator model, but this method of gaining information about the intramolecular forces did not turn out to be very fruitful.

A more promising approach was initiated by Rydberg (1931) who used the Bohr–Sommerfeld quantization rules to construct potentials from observed molecular spectra. This idea was extended by Klein (1932) to a systematic procedure using the semiclassical approximation to quantum mechanics, and Rydberg (1933) as well as Rees (1947) continued specific applications.

It was this semiclassical approach that spawned the first idea of using experimental information on the *continuous* spectrum to infer the forces: Hoyt (1939) used an Abel integral to obtain the interparticle potential from the scattering data. This development was picked up again later by Firsov (1953), Sanders and Mueller (1963), Mason and Monchik (1967), Miller (1968, 1969, 1971), Buck and Pauly (1969, 1971), Buck (1971), Boyle (1971), Klingbeil (1972), Pritchard (1972), and summarized by Buck (1974) and, in a lighter vein, by Wheeler (1976).

The Second World War interrupted most research after 1939. When it was over, Fröberg (1947, 1948, 1949, 1951) began an entirely new line of inquiry. The idea now was to start from the radial Schrödinger equation with a central potential, and to attempt to reconstruct it from a knowledge of the scattering phase shift of a given angular momentum without the intermediary of the JWKB approximation. A particular construction procedure was invented for this purpose by Hylleraas (1948, 1963, 1964). The method, however, was flawed; as Bargmann (1949) showed, the inference of a central potential from a phase shift of one angular momentum, given for all positive energies, is not necessarily unique. Levinson (1949) re-established uniqueness in case there are no bound states. It was at this point, by investigating the uniqueness question, that Marchenko (1950) initiated the earlier mentioned series of mathematical studies which culminated in the celebrated paper by Gel'fand and Levitan (1951).

Scattering theory had by this time become considerably more sophisticated than it had been before the war. Wheeler (1937) and Heisenberg (1943, 1944) had invented the S-matrix, and Heisenberg was promoting the conjecture that all physical information was contained in it. This meant, in particular, that the bound states of a system should be obtainable from the scattering. The additional information was to be gained by analytic continuation in the energy, and it was expected that the zeros of the eigenvalues of the S-matrix of a given angular momentum (or equivalently, its poles on another sheet of its Riemann surface) signified bound states of the same angular momentum. The discovery of "redundant zeros," which are not connected with bound states, seemed to deal a heavy blow to Heisenberg's program, and such zeros were discussed in a number of papers at that time: Ma (1946, 1947), Wildermuth (1949), and van Kampen (1951). Bargmann's (1949) proof of the independence of the bound states from the scattering phase shifts settled the question. The only connection that exists between a phase shift and the bound states of the same angular momentum is embodied in the now well-known Levinson theorem (1949). Families of phase-equivalent potentials, which also produced the same bound states, were explicitly exhibited by Jost and Kohn (1952) and by Holmberg (1952). It has been known since Faddeev (1956), however, that if the scattering amplitude is given for all angles and energies, then the potential is uniquely determined. Therefore the earlier arguments against Heisenberg's program were premature.

At this time Jost and Kohn (1952) produced a new systematic procedure for the construction of potentials from the phase shift and the bound states of one angular momentum, in the form of an infinite series. Moses (1956) and Prosser (1969, 1976) continued the development of this idea. It was overshadowed, however, by a much more elegant method based on the results of Gel'fand and Levitan (1951) and developed by Jost and Kohn (1953) and by Levinson (1953). Important alternatives and extensions were soon added by Krein (1953, 1955) and Marchenko (1955).

This approach was based on two alternative linear integral equations, one due to Gel'fand and Levitan and the other to Marchenko. Because of its power and elegance it remained the standard and became the model for the solution of most other types of inverse scattering problems until recently. Its general features were cast in a more abstract form that allowed application to the solution of a number of other problems in a series of papers by Kay and Moses (1955, 1956, 1957).

Two of the most interesting outcomes of these general inversion procedures were explicit constructions of an infinite family of potentials for which the radial Schrödinger equation of one angular momentum can be solved in closed form (Bargmann, 1949; Fulton and Newton, 1956; Theis, 1956), and of families of "transparent" potentials, which, for one angular momentum and all energies cause no scattering at all (Moses and Tuan, 1959; Chadan, 1967).

The powerful inversion method based on the Gel'fand–Levitan equation did not have to wait long to be generalized to coupled systems of differential equations (Newton and Jost, 1955; Agranovich and Marchenko, 1957) and thence to the inverse-scattering problem for particles with spin (Newton, 1955; Agranovich and Marchenko, 1958). After that time the method could be regarded as essentially complete, and it was summarized in two reviews, by Faddeev (1959) and by Agranovich and Marchenko (1963).

Later developments in this area concerned generalizations to coupled channels of given energy differences (Cox, 1962, 1964, 1966, 1967; Cox and Garcia, 1975), to energy-dependent potentials (Mal'cenko, 1966; Jaulent, 1972, 1975; and Jaulent and Jean, 1972, 1975, 1981a,b), and to relativistic equations by Corinaldesi (1954), Verde (1959), Prats and Toll (1959), Gasymov and Levitan (1966), Gasymov (1967, 1968), Weiss (1971), Weis et al. (1971, 1972). The results were also used to a limited extent for the interpretation of experimental data by Newton (1957), Fulton and Newton (1956), and Blazek (1962), and for approximations by Jost (1956), Swan and Pearce (1966), and Swan (1967).

In order to construct the potential from the phase shift by the methods based on the Gel'fand–Levitan equation or on the Marchenko equation, it is necessary to know the scattering for all energies. Since such knowledge is never available in practice, it was useful to investigate what partial information about the potential may be obtained from limited data, such as moments of a phase shift. Though such moments are integrals in which the phase shift appears as a function of the energy, from zero to infinity, different moments assign different weights to the high-energy region. They are therefore well suited to approximations. A fifteen-year development of studies concerned with this question began with Newton (1956) and Faddeev (1957). It was continued and gradually more and more generalized by Buslaev and Faddeev (1960), Buslaev (1962), Percival (1962), Percival and Roberts (1963), Roberts (1964, 1965), Calogero and Degasperis (1968), and finally summarized by Calogero (1971). These methods were then generalized to relativistic equations by Degasperis (1970) and Corbella (1970, 1971). Applications to approximations were discussed by Roberts (1963), and variational bounds on the potential were derived by similar means by Calogero et al. (1968).

It will be noted at this point that the idea of constructing the underlying potential from a single phase shift, given for all energies, is not only rather impractical and physically unsatisfactory, because at high energies the assumed nonrelativistic kinematics break down, but is also intrinsically unnatural. The scattering amplitude has to be fully known for all energies, but the information contained in all but one of the phase shifts is never used. A much more straightforward and practical procedure would utilize the full scattering amplitude, or all phase shifts, at one energy to construct the potential. Tentative steps in such a direction were made by Wheeler (1955) and Keller et al. (1956) on the basis of the JWKB approximation, by Regge (1959) on the basis of an analytic continuation in the angular-momentum variable, and by Martin and Targonsky (1961) by a procedure applicable to superpositions of Yukawa potentials only.

A more general construction method at fixed energy, based on an analogue of the Gel'fand–Levitan equation, was provided by Newton (1962), who also showed that the potentials could not be inferred uniquely. This procedure was significantly extended and deepened in a series of papers by Sabatier (1966, 1967, 1968), as well as by Newton (1967). A new method by Sabatier (1971) then led to a complete solution (Sabatier, 1972, 1974; Jean and Sabatier, 1973) of the fixed-energy inversion problem in the sense of producing, in a wide class, all the potentials that would lead to a given set of phase shifts. The ambiguities in the inversion were elucidated by Sabatier (1973) from the semiclassical point of view, and Regge's idea was studied in depth by Loeffel (1968). Sabatier (1968) and Hooshyar (1971, 1975) finally extended the fixed-energy method to particles with spin, and Hooshyar (1972, 1980) to potentials that depend on the angular momentum. Coudray (1977, 1979), Lipperheide and Fiedeldey (1978, 1981, 1982), Fiedeldey et al. (1982, 1984), and Lipperheide (1984) brought the fixed-energy inversion closer to application in nuclear physics, as did the modifications of the method by Munchow and Scheid (1980), May et al. (1984, 1987), and Ioannides and Mackintosh (1985).

All the methods mentioned so far apply only to scattering by central potentials or, in the case of particles with spin, by rotationally invariant spin-dependent interactions. To infer a general, noncentral interaction potential from the scattering amplitude presented much greater difficulties. In the electromagnetic case some results were obtained by Lax and Phillips (1967), but the generalization of the methods based on the Gel'fand–Levitan or Marchenko equation proved to be far from easy. The fact that the scattering amplitude in the noncentral case is a function of five variables (two initial, two final angles, and the energy), whereas the underlying potential is a function of only three, indicates that such amplitudes must be subject to strong restrictions and cannot be arbitrarily prescribed. What is more, the basis of the Gel'fand–Levitan and Marchenko procedures is the existence of an integral kernel with a "triangularity" property analogous to the form of a triangular matrix. This property was not easy to extend to more than one dimension. An attempt in that direction was made by Kay and Moses (1961), but it worked in a context of nonlocal potentials and failed to guarantee reproducibility of a local one.

The essential ingredient for a useful triangularity property in more than one dimension was provided by Faddeev (1965, 1966) and the full three-dimensional inverse scattering problem was thereupon attacked independently by Faddeev (1971, 1976) and Newton (1973, 1974, 1977). Nevertheless, there remained the important unanswered question whether the basic integral equation that uses Faddeev's Green's function had exceptional points, and if so, how to deal with them. As a result these methods were abandoned until recent results by Lavine and Nachman (1987a,b), and by Novikov and Henkin (1987a) and Henkin and Novikov (1988a,b) clarified this situation.

A different approach to the general three-dimensional inverse scattering problem for noncentral potentials was initiated by Newton in a series of papers (1980b,c, 1981, 1982b, 1983a, 1985c) and carried further by Cheney and Rose (1985), Rose et al. (1984, 1985c, 1988), Ramm (1987a), and Ramm and Weaver (1987). It incorporates a generalization both of the Gel'fand–Levitan equations and of the Marchenko equations.

The most recent new tool for attacking the multidimensional inverse scattering problem is the $\bar{\partial}$-method of the theory of several complex variables, which was introduced into this context by Beals and Coifman (1981, 1984a,b, 1985a,b, 1986, 1987) and used very effectively by Nachman and Ablowitz (1984a,b) as well as Novikov and Henkin (1987a,b) and Henkin and Novikov (1988a); see also Somersalo (1987).

The inverse scattering problem in two dimensions was solved by Cheney (1984, 1985); that in two dimensions at fixed-energy (which is of special interest because its data are not redundant) was solved for sufficiently weak scattering amplitudes by Novikov (1986a,b).

An entirely different, nonlinear, approach to the three-dimensional inverse problem, with data consisting of the backward scattering amplitude, was taken by Moses (1956) and Prosser (1969, 1976, 1980, 1982, 1983), and the whole subject of inverse scattering was formulated by Carroll (1979a,b, 1981, 1982a,b,c, 1983, 1984a,b,c,d, 1985) in terms of transmutations.

As a generalization beyond noncentral potentials, one may pose the inverse scattering problem for arbitrary nonlocal interactions. It is clear, however, from a simple counting of variables (an integral kernel in three dimensions is a function of six, the scattering amplitude, of only five) that without some restrictions this problem is ill-posed. It is the nature of these restrictions that makes the problem hard to formulate in a natural way. While local potentials certainly lose meaning at high energies, they are physically reasonable in a nonrelativistic context.

One class of nonlocal interactions (apart from velocity-dependent ones) has, in fact, been used extensively in nuclear physics, and these are the ones of finite rank. The reason is that they permit explicit solutions of the Schrödinger equation in terms of quadrature. In many applications and model calculations this is a considerable advantage. Their use, however, is not physically motivated. The inverse scattering problem for separable interactions can be solved in a much more straightforward manner than for local potentials, and this has been done by Gourdin and Martin (1957, 1958), and by Bolsterli and Mackenzie (1965). Generalizations were dealt with by Chadan (1958, 1967), and Mills and Reading (1969) and Tabakin (1969) solved the problem for general interactions of finite rank. An interesting new approach to the inverse scattering problem for general nonlocal potentials is due to Muzafarov (1986a,b, 1988).

It is a remarkable circumstance that for many years, indeed ever since the first attempt at inferring an interparticle potential from scattering data by Hoyt in 1939, the starting point of the inversion procedure was taken to be the scattering amplitude, in spite of the fact that what is experimentally observed is only the square of its modulus, the cross section. The question of systematically inferring the complex amplitude from its modulus, and of the possible ambiguities involved in this inference, was not seriously asked except in some notorious instances of particles with spin when insufficient polarization data made the problem urgent.

The idea of using the generalized optical theorem, or the unitarity of the S-matrix, as a means of determining the phase of the scattering amplitude from the differential cross section, was first put forward by Klepikov (1964) and then independently by Newton (1968) and Martin (1969). In these papers the generalized optical theorem is considered as a nonlinear integral equation for the phase of the amplitude, and conditions on the differential cross section which guarantee existence or uniqueness of a phase function are obtained from known fixed-point theorems. These first results were subsequently enlarged, refined, and generalized by Gerber and Karplus (1970), Tortorella (1972, 1974, 1975), Atkinson et al. (1972), and Sakmar (1978, 1979, 1980). Others added physically motivated analyticity conditions in the complex energy or momentum-transfer plane in order to remove ambiguities (Burkhardt, 1972, 1974; Martin, 1973; Itzykson and Martin, 1973; Atkinson et al., 1973, 1974, 1975; Burkhardt and Martin, 1975). The problem is still in a somewhat unsettled state because all the known conditions for existence or uniqueness of the phase are inapplicable in most cases of physical interest.

There are several related developments that deserve brief mention. The first is the one-dimensional version of the inverse scattering problem, namely, the inverse reflection problem. This has both an intrinsic interest and serves as an important mathematical proving ground for the three-dimensional version. It was studied and solved by Kay and Moses (1956), Faddeev (1958, 1964), and Deift and Trubowitz (1979); see also Melin (1985), Newton (1980a, 1984b), and Sabatier (1983b). The issue of nonuniqueness of the inversion for larger classes of potentials was studied by Abraham et al. (1981), Brownstein (1982), Moses (1983), Sabatier (1984, 1986), Aktosun and Newton (1985), Degasperis and Sabatier (1987), and Aktosun (1987). Marchenko (1968), Gray and Symes (1985), Carion (1986), and Aktosun (1987) investigated the stability of the inversion.

The application of the inversion procedures for the Schrödinger equation to the acoustic case, which is of great practical interest, encounters difficulties when the impedance contains discontinuities. These were largely overcome for one dimension by Sabatier (1983b, 1984) and Sabatier and Dolveck–Guilpart (1988), and for three dimensions by Sabatier (1987d, 1989). The one-dimensional inverse scattering problem for periodic potentials with "impurities" that vanish at infinity was solved by Firsova (1975a,b), and Newton (1983b, 1985d).

Another development is that of a discrete version of the inverse-scattering problem. Its solution by Case and Kac (1973) and Case (1973, 1974) was patterned after the continuous version. It has the virtue of a certain intuitiveness from the point of view of transport theory.

Finally, there is an important and altogether astonishing application of the results obtained in the inverse scattering problem to the solution of certain nonlinear wave equations, found by Gardner et al. (1967). The simplest one-parameter family of potential functions in the one-dimensional Schrödinger equation which leaves its spectrum invariant was discovered to satisfy the Korteweg–de Vries equation as a function of the parameter (considered as the "time" in the latter equation). Furthermore, the large-distance behavior of the regular solution of the Schrödinger equation, as a function of this parameter, is such that it causes a simple phase-change of the reflection coefficient. Consequently, the initial-value problem of the Korteweg–de Vries equation can be solved by the solution of the inverse reflection problem. The potential that underlies the given reflection co-efficient, as a function of the "time," solves the Korteweg–de Vries equation for all "times." The bound-state energies of the Schrödinger equation correspond to the propagation velocities of solitons, the solitary-wave solutions of the associated Korteweg–de Vries equation.

This unexpected use of the solution of the inverse-scattering problem has since been elaborated and generalized to other nonlinear wave equations, and it has important physical applications. The method and its results have been summarized by Bullough and Caudrey (1980), Ablowitz and Segur (1981), and Calogero and Degasperis (1982), but much work in this direction is still in progress.

Since the first edition was published eleven years ago, this book, by two important contributors to the field, has earned a well-deserved place on the shelf of every student and research worker in the area of inverse scattering. The new edition has been updated and enlarged to accommodate the many new results that have been obtained since then. I am sure that it will find an even larger and more appreciative audience of research scientists everywhere.

Bloomington, Indiana
April, 1988   R. G. Newton

Preface to the Second Edition

Between 1977 and 1988, the importance of inverse problems has grown considerably in many fields. Yet, my hope of 1977 (which goes back, in fact, to Sabatier (1972b) has been confirmed even more: general approaches, mathematical methods, and ways of modeling the problem, have so many points in common that one must consider it a distinct field—inverse problems. Meetings such as the annual one in Montpellier, and (since 1985) the journal *Inverse Problems* have progressively defined this field, which contains, on the one hand, the techniques of mathematical physics, applied mathematics, and information theory applied to inverse problems, together with computerized data inversion, theoretical imaging, diagnosis, and some problems of control and of stabilization. On the other hand, some mathematical techniques are themselves derived from the study of inverse problems. In particular, inverse methods have been used to solve exactly problems in nonlinear partial differential equations and in nonlinear signal processing. To give an indication of how rapidly this field is growing we may cite the following ten books, published in the West between July 1986 and December 1987 (and only in the theoretical domains *directly* connected with the inverse problems of this book): six textbooks or monographs by Craig and Brown (1986), Gladwell (1986), Poschel and Trubowitz (1986), Marchenko (1987), Hermann et al. (1987), and Tijhuis (1987); and four sets of proceedings edited by Cannon and Hornung (1986), Talenti (1986), Engl and Groetsch (1987), and Sabatier (1987). In addition, we should also cite the books on inverse methods and their applications to nonlinear partial differential equations, which are now a rich source of mathematical inverse problems and new methods of solution (Faddeev and Takhtadjan, 1987; Ablowitz et al., 1987). More than one hundred papers, in the same 18-month period, are evidence of the activity in theoretical inverse problems connected in some way to those of quantum scattering theory. As to the value of these new contributions, the $\bar{\partial}$-methods on the one hand, and the time-domain approaches on the other, have proved to be remarkably powerful in breaking down several difficulties of multi-dimensional inverse problems (never fear, however, many more remain). Furthermore, the first analytic results for inverse scattering and spectral problems with strongly discontinuous coefficients (corresponding to δ' distributions in a potential model) were published. If we extend this period back to 1977, we find of course many results that should be included in a new edition of this book. On the other hand, a general review of the unified field, as I mentioned in my preface to the first edition, has become impossible (unless we restrict ourselves to special points of view).

The word "unified" for the field of "inverse problems" certainly does not mean that there is a universal method of handling the problems. Although my statement is obvious, I must pause a little at this point. E. P. Wigner once remarked (1955) that when a model is much too general, it has no physical content. It is clear that all inverse problems can be put into a very general framework: let x be a parameter, let $\mathcal{M}$ be a mapping of the set of parameters $\mathscr{C}$ into the set of results $\mathscr{E}$, and let e be a result. Then any trio $(x, \mathcal{M}, e)$ has, intuitively, some "likelihood"—which can be defined as a conditional probability and which can be measured by a number after the probabilized spaces, etc., have been defined properly. Then a "measured result" is a probability distribution on the space of results and we can represent the set of possible mappings in the same way. If a "measured result" is given, we can calculate the *a posteriori* probability distribution of the parameters x. But this remark *is not* a general theory. It is either an obvious *preliminary* to a real theory or the beginning of a fraud: in the first case, the authors realize that, no matter how the "likelihood" has been defined as a probability, we must study the geometric relations between x, $\mathcal{M}$, and e, and there is no miraculous general method of obtaining them except case by case. In the second case, the author—or his readers—believe that it is sufficient to construct a "maximum likelihood" solution (the word "likelihood" is a matter of taste), minimizing a proper functional. Such a miraculous method simply ignores that even when a solution can be constructed without additional arbitrary constraints, it cannot be accepted as an *answer* to the problem *unless* all the others with an almost equal "likelihood" (and therefore acceptability) form a narrow enough set to be represented by *one* of its points (otherwise, you must handle the problem in other ways, see Sabatier (1987b)). Now since answering the "narrowness" question can be done only after we have completely disentangled the geometric relations between $\mathscr{C}$, $\mathcal{M}$, and $\mathscr{E}$, we again see that this last study is really the *body* of any theory of inverse problems. This is always the essential point in theories of "mathematical inverse problems," which start from classes of "exact" results, as well as in theories of problems that are ill-posed, whether because data is sparse, redundant, or contradictory; or indeed for even deeper reasons. Once this is done, probabilistic or statistical considerations are easy: consider, for instance, how much easier it is to use Shannon's remarks (on a linear inverse problem) *after* its geometry has been reduced by a "singular value decomposition." Thus, "inversion theory" is only a general guide to approaching this kind of problem: given a class of measured results, with errors, we try to analyze the deficiency in information, and ascertain the "*a priori* knowledge" or "constraints" that must be added to give a solution. We must never use the words "inverse theory" or "inversion theory" for a more or less sophisticated way of obtaining "in fine" "a unique physical solution" in a problem where the information is insufficient, without further qualification. Each "miraculous" method that pretends to do so is obviously incorrect, even if it works occasionally: with the development of computers, it is easier than ever to produce examples, and generalizing is such a huge temptation! A true "inversion theory" essentially yields a set of warnings against this temptation (Sabatier, 1987b).

Let us now return to this new edition. Fortunately, this book has always been understood as an *introduction* to the subject. Hence the chapters devoted to classical methods (for radial problems), which were essentially achieved several years before our first edition, need not be modified, except for minor points and in the last sections (Comments and References), which are meant to sketch more advanced developments. Admittedly, it would have been possible to simplify one or two proofs and to insert Gel'fand–Levitan methods into a $\bar{\partial}$-presentation, for instance; but this would have meant rewriting the book in a more abstract form, which would not necessarily meet the reader's needs. Hence we have only updated the Comments and References of chapters I to VIII (Chadan) and IX to XVI (Sabatier), without too many modifications inside the chapters, except for chapter XIV. Now, the most important recent developments of inverse quantum scattering problems could be divided into four classes: three-dimensional scattering problems, one-dimensional scattering problems on the line, spectral problems, and "numerical methods," which may also include most of the new theoretical developments in radial inverse problems, since most were directed towards applications. Since I have worked on most of these problems during the last few years, I was aware of new developments. Hence I decided to rewrite and expand chapter XIV (three dimensional), rewrite completely and expand considerably (three times!) chapter XVII on one-dimensional problems (which had been written by K. Chadan in the first edition), and to write two new chapters: XVIII (spectral) and XIX (numerical) (except sections XVIII.3 and XIX.4 which have been written by K. Chadan).

What should be the content of our new chapter XIV was very clear in my mind. The time-domain approaches, the Marchenko-type methods introduced by R. G. Newton since 1980, and the latest improvements of the Faddeev–Newton approach by means of $\bar{\partial}$-methods, obviously must be there. What was not at all obvious was how to put these very technical and conceptually difficult developments into a compact and understandable form. I do not know whether the final result is good, but if not, the fault is mine. I received the greatest help from B. De Facio for the time-domain approach, from A. Nachmann for the $\bar{\partial}$-method (they very kindly gave me drafts of the sections they were concerned with), and from G. Henkin and R. G. Newton, with whom I had several valuable discussions.

I have worked on several aspects of the one-dimensional problem (on the line) including "pedagogical" aspects, so I had no special problem in writing the new chapter XVII having once decided what to include in it. However this latter decision was more difficult. There have been three main aspects of recent studies of one-dimensional problems. First, there are applications of new methods to our old friend the one-dimensional Schrödinger problem, or extensions to new classes of parameters. Second, there may be restrictions to the Schrödinger problem of methods primarily designed for one-dimensional electromagnetic or elastic problems. Third, there are many one-dimensional problems whose main interest lies in the application of the inverse method to new classes of partial differential equations. For all these purposes, one-dimensional problems can be considered as a testing ground for various methods before they are applied to more complicated cases or to multidimensional problems. Whilst most problems of the first two classes cited above clearly belong to quantum scattering studies, this is only true for those of the third class in so much as either they produce extensions of the classes of potentials to be studied or they are closely related to the methods of chapter XIV. I have tried (perhaps arbitrarily) to keep along these lines. Finally, I decided to construct the chapter as follows. Section XVII.1 gives a detailed survey of the Faddeev approach to the direct problem, as in the first edition, but with a new presentation, new notation (more commonly used now), and several pedagogical devices that aim at increasing rigor and/or simplicity. In section XVII.2, the general inverse problem is successively studied through various approaches. In sections XVII.2.1 and XVII.2.2, the Faddeev–Marchenko equation is derived from the dispersion relations of the associated Jost solution $e^{-ikx}f_+(k, x)$; characterization conditions of scattering data are surveyed for potentials in L_2^1 and L_1^1 (without proofs), together with a detailed review of construction and reconstruction problems. In section XVII.2.3, I survey Newton's approaches to the one-dimensional problem that reduce it to a Riemann–Hilbert problem, itself reducible to solving a Marchenko-type equation. Section XVII.2 closes with an introduction to $\bar{\partial}$-approaches on the simple example of this one-dimensional Schrödinger problem. Inversions, where the potential belongs to a narrower class, are surveyed in section XVII.3 (reduced support potentials, rational reflection coefficients, Darboux transformations methods, generalized Born approximations, trace method, and finally time-domain approaches, including layer stripping and the invariant imbedding method). These approaches are included in this section because, from the point of view of Schrödinger problems, they are restricted to cases where strong causality exists and bound states have been ruled out. The one-dimensional problem that corresponds to the impedance equation is a generalization of the Schrödinger equation since it would correspond (if we may use this word) to potentials V including particular singularities like δ or δ'. It is studied in detail in sections XVII.4.1 and XVII.4.2, with several explicit examples written completely. First-order impedance systems and the difficult problem of potentials depending on k follow. In section XVII.5, we arrive at problems of the third class defined above, and we go from detailed reviews to brief ones, because my aim is to make the reader aware of certain problems rather than saving him reading the references. Thus the matrix Schrödinger equations, the Zakharov–Shabat problem, and the Beals and Coifman systems are sketched. Very good reviews on these subjects have been published, in connection with solvable nonlinear evolution equations, or are in preparation. I must take this opportunity to acknowledge the valuable help in preparing the "Zakharov–Shabat" section: I have read a few drafts of the second volume of the Calogero and Degasperis treatise on *Spectral Transforms and Solitons*, which will certainly be as good as the first one. I think that stopping our investigations of theoretical one-dimensional problems at this point is not arbitrary; in particular, the numerous other inverse problems connected with the inverse method definitely look farther from the one-dimensional Schrödinger one. But certainly it may happen that a method devised for one of these problems may become useful in quantum scattering. As for applications of a more applied or numerical character, they will be studied in chapter XIX.

After the first edition, several readers told me of their surprise at not finding any section on Sturm–Liouville problems, this having been the first object of the Gel'fand–Levitan study and the cornerstone of a large class of inversion methods. It also happens that, since 1977, two kinds of problem have been related to regular Sturm–Liouville problems, and in each case either Chadan or myself was involved. Hence we decided to add a chapter on inverse problems connected with discrete spectra and is of interest in quantum theory. The string and Sturm–Liouville problems are briefly reviewed in section XVII.1. I have given much more detailed reviews elsewhere (Sabatier, 1981, 1987c), and the excellent books by Levitan (1984) and Marchenko (1987) give a full mathematical study. In section XVII.2, further extensions are sketched and it is shown how problems of scattering by a finite range scatterer are related to spectral problems, and how this relation has helped in the understanding of some inverse mode problems. Section XVIII.3 is devoted to an inverse problem of the Schrödinger equation where the input is the spectrum $\{g_n\}$ in the coupling constant and where the potential is to be determined. Finally, in section XVIII.4, we give a short guide to the literature on inverse spectral problems. Again, studies of a more applied or numerical character are to be surveyed in chapter XIX.

If the last chapter, XIX, were devoted to a detailed survey of applied or numerical solutions of problems of quantum scattering theory, it would only have an *indirect* relation with the subject of this book, which is on *inverse problems*. As I have clearly stated many times, giving an approximate solution of a problem without discussing other possible solutions should not be called solving an inverse problem but only "getting a fit." Hence, we decided to first present questions of numerical analysis posed by the ill-posedness of our

inverse problems, and to continue the section as a long guide to the wide
literature of numerical studies, without giving a detailed technical survey of
any of them. Section XIX.1.1 is devoted to elements of inversion theory,
classification of different ill-posednesses, the regularization problem, etc. The
review is brief: we have given in a different textbook (Sabatier, 1987c) a
full pedagogical treatment of these subjects, with all the classical methods
of regularizing ill-posed problems and eventually solving them. In section
XVIII.1.2, we discuss the ill-posedness of inverse problems in quantum scattering theory, and we show how it can be understood more simply than by
means of a complete study; in particular, by an exploration method of $\mathscr{E}$
which combines the Born approximation and geometrical transformations.
As we have already announced, the long section XIX.2 (local potentials) and
the short section XIX.3 (nonlocal potentials) are essentially a guide to the
literature (essentially "Comments and References"). Since our primary purpose is the reader's convenience, I begin with a short survey of fitting procedures, when they are refined enough to discuss their validity. Next the radial
problems are surveyed (section XIX.2.3), together with the one-dimensional
problems. Preparing this last part was made easier by my reading of an unpublished report, prepared by Lesselier and Tabbara on the "Caractérisation
acoustique de milieux sous-marins stratifiés" (Ecole Supérieure d'Electricité,
1987). On the other hand, the references it gives on one-dimensional elastic
and electromagnetic inverse problems made necessary many difficult deci-
sions: nowhere in the present book is the border line between methods of
quantum theory and others more arbitrary. I can only say that I tried to
restrict myself to papers giving methods or results which are not too special-
ized in their domains of application. The chapter, and the book, finishes with
brief sections on numerical methods for inverse spectral problems and for
inverse scattering problems with nonlocal potentials.

As for the modifications to other chapters, the only noticeable point is the
way we include new results on generalized translation operators (now more
commonly called transmutations) and those on Darboux (also Bäcklund,
Crum, Krein, Marchenko, etc.) transformations. I am now used to considering
these objects as routes enabling joint exploration of the parameters set $\mathscr{C}$ and
the results set $\mathscr{E}$, and I have shown elsewhere how this point of view is useful
for understanding the geometrical structure of inverse problems (Sabatier,
1984, 1986) and in particular their ill-posedness, if any. But, in the first edition
the transformations were studied *with* their application, i.e., in chapters IV
and VI, and our review of transmutations, whose applications could have
been shared between chapters IX, XV, and XVII, was concentrated in chapter
XV. For the sake of simplicity, we did not try to modify these arrange-
ments. Section XV.2 on transmutations has been increased, in particular, by
citing those results of Carroll which may be of direct interest in quantum
theory (there are many other results connected with more mathematical
problems), although not necessarily of direct relevance to the title of chapter
XV. Darboux and other transformations are studied with their applications,
so can be found in chapters IV, VI, XVII, and XIX. Only the last two give

general methods of deriving them or how to apply them in ill-posedness studies.

I acknowledge the valuable help afforded by Calogero, De Facio, Degasperis, Henkin, Lesselier, Nachman, Newton, and Tabbara. I should also have cited all my colleagues on the editorial board of *Inverse Problems*, and more generally all the participants to our annual meeting "RCP 64: Etudes Inter-disciplinaires des problèmes inverses," and many faithful correspondents, because they all contributed by their preprints, questions, and remarks to our knowledge of inverse problems and of published or unpublished papers. By the way, although this new edition contains some 500 new references, to be added to the 432 old ones, we suspect that several important references are missing or have been overlooked, so we can only apologize in advance. The reader will find some additional references in the cited treatises. I also wish to express my appreciation to Professor A. Fordy for several stylistic improvements in this Preface. Finally, K. Chadan and myself are very grateful to our secretaries, in particular, Mrs. Cellier, Mrs. Croisard, and Mrs. Geurts, for their excellent work on this new edition, and to our editor and our families for their surprising patience.

Montpellier, 1988   P. C. SABATIER

Preface to the First Edition

The mathematical expression of a physical law is a rule which defines a mapping $\mathcal{M}$ of a set of functions $\mathscr{C}$, called the *parameters*, into a set of functions $\mathscr{E}$ called the *results*. This rule usually is a set E of equations, in which the parameters are elements of $\mathscr{C}$, the solutions are the corresponding elements of $\mathscr{E}$, so the definition of $\mathcal{M}$ is quite involved. However, from the very definition of a mapping, the solution of E must exist and be unique in $\mathscr{E}$ for any element of $\mathscr{C}$. This is the only constraint one has to put on E. We call "computed results" the elements of $\mathscr{E}$ which are thus obtained from those of $\mathscr{C}$. Deriving the "computed result" for a given element of $\mathscr{C}$ is called "solving the *direct problem*." Conversely, obtaining the subset of $\mathscr{C}$ that corresponds to a given element of $\mathscr{E}$ is called "*solving the inverse problem*."

To give a physical meaning to $\mathcal{M}$, $\mathscr{E}$ must be such that its elements can be *compared* to the *experimental results*. In the following, we assume that the result of any relevant measurement is an element of a subset, $\mathscr{E}_e$, of $\mathscr{E}$, called the set of experimental results. $\mathscr{E}$ therefore contains the union of the experimental and the computed results. It is also assumed that $\mathscr{E}$ can be given the structure of a metric space. The comparison of a given *computed result*, e_i, and a given *experimental result*, e_j, is then measured by the distance $d(e_i, e_j)$.

The set $\mathscr{C}$ was defined as the set of functions for which E can be solved. Stronger limitations usually come in when "physical properties" are taken into account. In other words, $\mathscr{C}$ could be the set $\mathscr{S}$ of all the functions for which E can be solved *and* which are consistent with all the "physical information" coming either from general principles or from previous measurements. However, the definition of $\mathscr{S}$ is, in most cases, indirect and difficult to make precise, and so one is led to choose for $\mathscr{C}$ a convenient subset of $\mathscr{S}$, with a clear definition. On the other hand, attempt to enlarge the definition of $\mathscr{C}$ sometimes gives access to new classes of "parameters" for which the direct and inverse problems can be solved. We shall see a recent illustration of this remark.

With these definitions, it may seem that all physical problems are inverse problems. Actually, one usually reserves this name for the problems where precise mathematical forms are sought for the generalized inverse mappings of $\mathscr{E}$ into $\mathscr{C}$. This excludes the so-called "fitting procedure" in which models depending on a few parameters and giving a good fit of the experimental results are obtained by trial and error or any other technique. The present book is devoted to the inverse problems of quantum scattering theory. Thus, the experimental results are the measured quantities in scattering experiments, e.g., cross sections or related quantities. Since these quantities are related to the asymptotic behavior of the wave functions, one will also consider problems where the set $\mathscr{E}$ consists of "theoretical measurements" of this asymptotic behavior, e.g., the scattering amplitude or the phase shifts. This naturally leads to the particular problem of constructing the scattering amplitude from the cross section. Leaving this aside, the equations E defining the mapping $\mathscr{M}$ consist of a wave equation (e.g., the Schrödinger equation, the Klein–Gordon equation, the Dirac equation together with appropriate conditions). The sets $\mathscr{C}$ of "parameters are sets of local or nonlocal "potentials," from which it is possible to predict scattering results.

We do not think there is any need to emphasize the interest of these problems. Since all the information we can obtain on nuclear, particle, and subparticle physics is obtained from scattering experiments, they are of obvious physical importance. Their intrinsic interest, as mathematical problems, appears in their connection with so many branches of analysis: advanced results in differential and integral equations, harmonic analysis, spectral theory of operators, holomorphic functions, asymptotic expansions, numerical analysis, are all necessary to study these improperly posed problems.

The book is made up of seventeen chapters of which the first two are introductory, surveying all useful results of potential scattering. The sixteenth chapter deals with approximate methods, the seventeenth with "one-dimensional inverse problems," and the remaining thirteen chapters form two parts. In them we deal with spherically symmetric potentials.

In the first part (Chapters III–VIII), the "result" to be analyzed is a phase shift $\delta_l(E)$, known as a function of E for all positive energies, and some additional information (e.g., bound states). The corresponding inverse problems are called "inverse problems at fixed l." In Chapter III, the so-called Gel'fand–Levitan–Jost–Kohn method is analyzed in great detail. We have chosen to work in a class of potentials which is larger than the one commonly used, and has the advantage of being in one-to-one correspondence with a large physically interesting class of phase shifts. We completely study the cases $l = 0$ and then any integer l for the Schrödinger equation with a spherical potential. In Chapter IV, we study some general applications of the Gel'fand–Levitan method: we construct classes of potentials which differ by the bound states only, and/or are phase equivalent, and also classes of potentials whose Jost functions differ by a rational function as a multiplicative factor (Bargmann potentials). We also derive the so-called Marchenko theorem, which relates a given asymptotic S-function $S(k)$ with several Schrödinger operators. The Marchenko method is studied in Chapter V, with full details again in the simple case of the Schrödinger operator with a spherically symmetric potential. Examples of results obtained by these spectral methods and that can be put in closed forms are exhibited in Chapter VI (Bargmann potentials, singular oscillating potentials). Chapter VII is devoted to complete or partial studies of the inverse problem at fixed l for special classes of potentials: Yukawa potentials (Martin's method and N/D methods) and holomorphic potentials (the so-called trace methods). In Chapter VIII, a nonlocal interaction is introduced in the Schrödinger equation, and the inverse problem is completely studied for this class of interactions, separable or not. Chapter IX gives a set of miscellaneous approaches to the inverse problems at fixed l. The large number of different methods which are introduced in this chapter yields a natural transition to the second part of the book. Meanwhile, it shows a lot of (as yet), unsolved problems, whose study is part of the current literature.

The second part of the book (Chapters X–XV), is devoted to problems in which the "experimental result" is the cross section, and this is, of course, the more realistic case. In Chapter X, we study the apparently simple, and actually very difficult, construction of the scattering amplitude from the cross section. Assuming then that this result has been obtained, at a given energy, in a complete way, we show how it is possible to construct the corresponding potential in the three following chapters: the first one (Chapter XI) contains general results and the general questions to be answered. The second one (Chapter XII) contains the matrix methods (e.g., the so-called Newton–Sabatier methods) in which the class of parameters $\mathscr{C}$ is defined in such a way that the construction of its elements from the "experimental" scattering amplitude is possible via a certain mapping which is the inverse of u. In other words, $\mathscr{C}$ is the image of $\mathscr{E}$ via this mapping and it is thus an ad hoc class for which the inverse problem has a simple solution which can be obtained with special matrices. The third part of this study (Chapter XIII) contains the operator methods, in which the class $\mathscr{C}$ of potentials is a priori defined on physical grounds. The method of Regge and Loeffel is a generalization of the Gel'fand–Levitan method applied to this problem, whereas the "complete solution" is the furthest extension of the matrix methods. The latter enables us to answer all the general questions posed in Chapter XI. In the Chapter XIV, we present an attempt to solve the three-dimensional inverse problem: how to construct the potential $V(\mathbf{r})$ from a knowledge of the scattering amplitude at all energies. In contrast with the previous inverse problems which were underdetermined, this problem is strongly overdetermined, and the consistency conditions which guarantee existence are not completely clear as yet. As an introduction to the chapter, a comparison is made with the procedures used in the simpler problems. It helps us enormously to understand the machinery of the problem, which, although complicated, is not essentially different from the others. In the fifteenth chapter a review of miscellaneous approaches to inverse problems at fixed energy enables us to unscramble the mechanisms of some approaches to more general problems.

It was difficult to decide exactly what to do in Chapter XVI on approximate methods. We have chosen to review most cases in which an approximate closed form has been obtained for a generalized inverse mapping from $\mathscr{E}$ to $\mathscr{C}$. Thus the computer numerical methods are not presented. On the other hand, this decision led us to review the semiclassical methods in a way which sometimes took us to the frontier of our subject. For instance, the reader will see that the RKR method is quoted, although the determination of a differential equation from several discrete spectra is obviously beyond the scope of this book. It is clear that the boundary of our subject is not very neatly defined. For instance, in Chapter XVII, we give an example of the inverse problem considered as a generalized transform and applied to solving nonlinear partial differential equations, but these problems are too wide to be treated here. Such applications, together with the determination of a differential operator from several discrete spectra, linear inverse problems, and classical inverse scattering problems, with all their applications in chemistry, geophysics, acoustics, and electromagnetic studies, deserve a book by themselves, and one of us (P.C.S) hopes to be able one day to undertake this task. The last chapter is devoted to the one-dimensional inverse problem, and some of its direct applications. It is also clear that the center of interest in the two parts of the book are somewhat different. The inverse problems at fixed l have known a much longer history than the inverse problems at fixed E. Thus many problems are now completely solved, and we have chosen to study them completely, and to survey their possible generalizations more rapidly, while putting the emphasis on the needs and interests of physicists. In particular, we have chosen proofs which should be familiar to physicists and we have often only given references to the more lengthy and more rigorous proofs preferred by mathematicians. The inverse problems at fixed E are younger, and classes for which results are neat and tidy have not always been obtained, so that more caution in the writing was necessary. On the other hand, since their physical applications are more direct, it is necessary to carefully compare the possible machineries giving solutions in order to select constructive methods. Hence the "organization" of the two parts of the book is slightly different; another source of differences is probably the fact that Chapters I–VIII and Chapter XVII were essentially written by K.C. and Chapters IX–XVI by P.C.S.

Three review papers have been of a great help to us. Faddeyev's review paper (1963), Martin's lectures (1973) on the construction of the phase shifts from the cross section and Newton's lectures (1966–1973), particularly the lecture on the three-dimensional inverse problem. It is no surprise. These three physicists are together responsible for most of the important work on inverse problems in the last 20 years. For helpful discussions, we would like to thank them, and with them all the physicists who improved our knowledge of inverse problems in quantum scattering theory by their personal comments on special points, in particular, Professors Atkinson, Cox, and Loeffel, and our friends Drs Jaulent and Miodek, who gave us personal contributions to Chapters IX and XVII. We do not forget as an important source the NASA report *Mathematics of Profile Inversion*, edited by L. Colin (1972), which has been of essential help in many interdisciplinary works on inverse problems. Besides, one of us (P.C.S.) is particularly indebted to Professor R. G. Newton who gave him all kinds of help to arrive at a better understanding of inverse problems, and who was kind enough to write the Foreword of this book.

The "Centre National de la Recherche Scientifique" has sponsored studies of inverse problems in the "Recherche coopérative sur programmes" "Etudes interdisciplinaires des Problèmes Inverses." Many meetings and discussions were made possible by this organization, which has been an important support for this work.

But in the preparation of such a book there are two essential tasks. One is to gather the references and to help with the preparation of the manuscript and of the drawings. This has been done by Mrs Albernhe, "documentaliste" at C.N.R.S.; the other is to correct the English language (in the second part of the book and in the introduction); this has been done by Dr Ilia Miodek. In both cases the competence was allied with the devotion to the task. If this book can be understood and if it can be useful to find a forgotten reference, it is thanks to them and thus we think that it is to them that the reader should be particularly grateful.

Dedication Je voudrais dédier ce livre à la mémoire de mon père, Célestin Sabatier, qui a suivi avec intérêt la préparation de l'ouvrage, et qu'un accident de la circulation a tué pendant les dernières corrections des épreuves.

Table of Contents

CHAPTER I

Some Results from Scattering Theory

I.1 The Reduced Radial Schrödinger Equation

Fundamental Assumptions

We consider the scattering of a particle of mass M by a central potential $V(r)$. The potential is supposed to be real and locally integrable (Lebesgue integrability), except perhaps at the origin, where it is supposed to satisfy the integrability condition

$$\int_0^a r\,|V(r)|\,dr < \infty, \qquad a < \infty. \tag{I.1.1}$$

At infinity, we shall assume that

$$\int_b^\infty |V(r)|\,dr < \infty, \qquad b > 0. \tag{I.1.2}$$

In most cases, we shall argue as if the potential were continuous. This is only for simplifying the presentation. At the end of this chapter, when we study a particular class of singular potentials, we shall see that we can also accommodate some mild generalized functions like the Dirac delta function, etc.

For potentials satisfying the above conditions, it can be shown that the usual scattering theory, and the connection between the asymptotic form of the wave function and the cross section, are valid. However, in order to avoid the occurrence of some unnecessary complications (infinitely many bound states, etc.) it is usually assumed that $rV(r)$ is absolutely integrable at infinity. Combining this with **(I.1.1)**, we obtain

$$\int_b^\infty |V(r)|r \, dr < \infty, \qquad b \geq 0. \tag{I.1.3}$$

This condition means, roughly, that the potential is less singular than r^{-2} at the origin, and that it decreases faster than r^{-2} at infinity. We shall call these potentials regular. The others are singular.

A more general scattering theory has been developed in recent years to deal with potentials violating **(I.1.1)** and or **(I.1.2)**. The necessity for violating **(I.1.2)** is fairly obvious (Coulomb potential). A general scattering theory for potentials violating **(I.1.1)** is also necessary because of the occurrence of singular potentials in elementary particle physics, and in many-body problems with strongly repulsive potentials. In short, there exists now a general rigorous scattering theory for potentials behaving at infinity like $r^{-\varepsilon}$, $\varepsilon > 0$, provided they are repulsive, and those which are more singular than r^{-2} at the origin. For some of these (singular oscillating potentials studied at the end of this chapter), everything is in fact similar to what we obtain for regular potentials satisfying **(I.1.3)**. For more general potentials, the usual theory must be modified in an appropriate way. It has been shown that a coherent and unique scattering theory can be developed for a wide class of singular potentials if the singularity (either at the origin, or at infinity, or both) is repulsive. When the potential is singular and attractive at the origin, it is known that the Hamiltonian is not essentially self-adjoint. There are an infinite number of self-adjoint extensions, and for each of these, a coherent scattering theory can be set up, and, surprisingly, contrary to the case of classical mechanics, there is no such a thing as the "collapse into the origin." The only problem is to find out which one is a good physical theory. Unfortunately, there does not exist any physical criterion which would allow us to single out one among these perfectly honorable theories satisfying all general postulates like asymptotic completeness, etc. On the other hand, if the singularity is at infinity, and is attractive, there does not exist any scattering theory.

At any rate, the inverse problems for these singular potentials, except those considered at the end of this chapter, have never been studied. The reason is that the direct problem itself is not well understood in all its details. This is why we shall not consider in this book the general case of singular potentials.

In order to simplify the writing of the formulae, we shall henceforth choose the units in such a way that

$$\hbar = 2M = 1. \tag{I.1.4}$$

The total energy of the particle is then given by

$$E = \frac{p^2}{2M} = \frac{\hbar^2 k^2}{2M} = k^2, \tag{I.1.5}$$

where p is its initial or final momentum, and k the corresponding wave number. The reduced radial Schrödinger equation for the partial wave of angular momentum l then reads

$$\frac{d^2}{dr^2}\psi_l(k,r) + \left(k^2 - \frac{l(l+1)}{r^2}\right)\psi_l(k,r) = V(r)\psi_l(k,r). \qquad \text{(I.1.6)}$$

The Physical Solution

For the time being, we assume that the parameter k is real. We shall see later how complex value of k can be introduced. The equation being of second order, and its coefficients generally singular at the origin (either $l \neq 0$, or a singular potential), it is well known that only one of its two linearly independent solutions can vanish at $r = 0$ while the other solution is generally singular there. In order to define this solution in a precise manner, we refer to the free case, i.e., when there is no interaction. We define the free solution to be

$$u_l(kr) = \left(\frac{\pi kr}{2}\right)^{1/2} J_{l+1/2}(kr), \qquad \text{(I.1.7)}$$

where J_ν is the Bessel function[1] of order ν. The behavior of this free solution near the origin and at infinity is given by

$$u_l \underset{r\to 0}{\simeq} \frac{(kr)^{l+1}}{(2l+1)!!} \qquad \text{(I.1.8)}$$

$$u_l \underset{r\to\infty}{\simeq} \sin(kr - \tfrac{1}{2}l\pi), \qquad \text{(I.1.9)}$$

where $(2l+1)!! = 1 \cdot 3 \cdot 5 \cdot \ldots \cdot (2l+1)$.

For the general case, when the interaction is present, we define the solution of (I.1.6), which we call regular, as opposed to the general solution, which is usually singular at $r = 0$, by the boundary condition

$$\lim_{r\to 0}(2l+1)!!\, r^{-l-1}\varphi_l(k,r) = 1. \qquad \text{(I.1.10)}$$

For "regular" potentials, we shall see that such a solution exists and it is the only one which vanishes at the origin. For physical reasons known in scattering theory, the physical wave function ψ_l also must vanish at the origin. It is therefore a multiple of the solution φ_l. The proportionality factor between the two solutions is determined by the physical requirement that ψ_l should have the precise asymptotic behavior

$$\psi_l \simeq \exp(i\delta_l)\sin(kr - \tfrac{1}{2}l\pi + \delta_l), \qquad r \to \infty, \qquad \text{(I.1.11)}$$

where δ_l, the phase-shift, is a real quantity which depends of course on energy, or momentum k.

---

[1] Throughout this book, we shall adopt the notations and definitions of Erdélyi et al. Higher Transcendental Functions (Erdélyi 1953–1955).

Jost Solutions and Jost Functions

In order to establish the precise connection between φ_l and ψ_l which, as we shall see, is extremely important for the inverse problem, we must introduce another well-defined solution of (**I.1.6**) by giving its asymptotic behavior at infinity. Here again, we refer to the free case, and choose the normalized free solution to be

$$w_l(kr) = i \exp(i\pi l)\left(\frac{\pi}{2} kr\right)^{1/2} H^{(1)}_{l+1/2}(kr), \qquad (\text{I.1.12})$$

where $H^{(1)}_\nu$ is the Hankel function of order ν and of the first kind. Near the origin, and at infinity, we have

$$w_l(kr) \underset{r\to 0}{\simeq} (-1)^l[(2l-1)!!](kr)^{-l} \qquad (\text{I.1.13})$$

and

$$w_l(kr) \underset{r\to\infty}{\simeq} e^{i\pi l/2} e^{ikr}. \qquad (\text{I.1.14})$$

We introduce now a solution of the Schrödinger equation (**I.1.6**), called the Jost solution, and denoted by $f_l(k, r)$, by demanding that

$$\lim_{r\to\infty}\{e^{-i\pi l/2} e^{-ikr} f_l(k, r)\} = 1, \qquad (\text{I.1.15})$$

so that f_l reduces to w_l in the free case. We shall see later that under condition (**I.1.3**), such a solution exists and satisfies also

$$\lim_{r\to\infty}\{f_l(k, r) - e^{i\pi l/2} e^{ikr}\} = 0 \qquad (\text{I.1.16})$$

and

$$\lim_{r\to\infty}\{f'_l(k, r) - ike^{i\pi l/2} e^{ikr}\} = 0. \qquad (\text{I.1.17})$$

From these, it follows that when $k \neq 0$, the two solutions $f_l(\pm k, r)$ are independent of each other since their Wronskian, according to (**I.1.16**) and (**I.1.17**) is

$$W\{f_l(k, r), f_l(-k, r)\} = (-1)^{l+1} 2ik. \qquad (\text{I.1.18})$$

Therefore, the regular solution can be written as a linear combination, with constant (r-independent) coefficients, of these two Jost solutions:

$$\varphi_l = \tfrac{1}{2}ik^{-l-1}\{F_l(k)f_l(-k, r) - (-1)^l F_l(-k)f_l(k, r)\}. \qquad (\text{I.1.19})$$

The fact that the same function F_l appears in front of both Jost solutions is due to the fact that the regular solution is an even function of k. This in turn is a consequence of the fact that the Schrödinger equation is even in k, and the boundary condition (**I.1.10**) is independent of it.

The function F_l, which we can define also by the Wronskian

$$F_l(k) = (-k)^l W\{f_l(k, r), \varphi_l(k, r)\} \qquad (\text{I.1.20})$$

is called the Jost function, and plays a very important role in scattering theory and the associated inverse problem. One of its important properties is that it satisfies the relation

$$F_l(-k) = \{F_l(k)\}^*, \qquad k \text{ real}, \tag{I.1.21}$$

where * means complex conjugate.

This property follows from

$$\{\varphi_l(k, r)\}^* = \varphi_l(k, r), \qquad k \text{ real}, \tag{I.1.22}$$

and

$$\{f_l(k, r)\}^* = (-1)^l f_l(-k, r), \qquad k \text{ real}, \tag{I.1.23}$$

which in turn follow by taking the complex conjugate of the Schrödinger equation, using the reality of the potential, and taking into account the boundary conditions (I.1.10) and (I.1.15). If we write now (definition of δ_l)

$$F_l(k) = |F_l(k)| \exp[-i\delta_l(k)] \tag{I.1.24}$$

and use the asymptotic estimate (I.1.16), we obtain immediately

$$\varphi_l \underset{r \to \infty}{\simeq} \frac{|F_l(k)|}{k^{l+1}} \sin(kr - \tfrac{1}{2}l\pi + \delta_l). \tag{I.1.25}$$

This shows that the physical solution is given by

$$\psi_l = \frac{k^{l+1}}{F_l(k)} \varphi_l \tag{I.1.26}$$

and that the phase shift is just equal to minus the phase of the Jost function. The connection between the phase shift and the phase of the Jost function is one of its remarkable properties. Another remarkable property of the Jost function is the intimate connection between some of its roots and the bound states. These are physical solutions of the Schrödinger equation which are square integrable on the whole real r-axis $[0, \infty)$. They occur for certain special values of the energy. We shall study them in detail in the next chapter. It can be shown that, under condition (I.1.2), they occur only for $E \leq 0$.

Inverse Scattering Problems

The above analysis, which we shall pursue in more detail in subsequent sections of this chapter, brings us now to the following: when we know the potential, we can solve the Schrödinger equation for a given l, and calculate its various solutions, from which we can then deduce the phase shift and its dependence on energy, as well as the energies and the wave function of the bound states. Conversely, would it be possible to determine the potential, in a more or less unique way, from the knowledge of the phase shift and bound states? This is the subject of the first part of this book, in which l is supposed to be fixed, the energy varying over its whole range $[0, \infty)$. The interest of this question is fairly obvious since the phase shift is a measurable quantity which can be obtained from scattering experiments (see Chapter X).

At this point, we may make some heuristic remarks about the inter-connection between these quantities. These remarks are by no means rigorous. However, the general conclusions we are going to reach here are true and will be shown in later chapters. We wish to determine the potential, which is a real integrable function of one real variable satisfying some conditions of the kind (**I.1.3**). Since there are no miracles in nature, and even less in mathematics and mathematical physics, it should be intuitively clear that in order to determine in a more or less unique way the potential from something else, that something else must have at least the same general-ity as the potential itself. That is to say, that something else must be at least a real function of a real variable. This is exactly the case of the phase shift $\delta_l(k)$ for real values of k. As we shall see, when the potential is repulsive, or not sufficiently attractive to admit bound states, the connection between the potential $V(r)$, and the phase shift $\delta_l(k)$, assumed to be known for all $k \geq 0$, is indeed one to one.

When bound states are present, we must know also their energies and the normalizations of their wave function. This is because, as it will become clear later, the fundamental ingredient in the solution of the inverse problem is the completeness of the set of solutions of the Schrödinger equation in the range $-\infty < E < \infty$. When there are no bound states, it is known that scattering states, which correspond to $E \geq 0$, are complete. In the presence of bound states, we must add their wave function to the set of scattering states in order to achieve completeness. This is the profound reason why bound states, when they are present, enter into the construction of the potential.

So far, we have not been assuming anything particular about the potential. Let's suppose we know *a priori* that the potential belongs to some particular class of functions. For instance, that it is an holomorphic function of r around the origin which does not reduce to a polynomial. In this case, the potential is equivalent to a countably infinite set of real numbers, which, for instance, we can take as the coefficients of its Taylor expansion. One may then ask whether the knowledge of another set of real numbers, such as the value of the phase shift at some infinite but discrete set of values of energy, would suffice to determine the potential. This, of course, could be the case if it can be shown that for such potentials the phase shift for all k is also determined by a set $\{\delta_l(k_i), i = 0, 1, \ldots\}$ of real numbers. Unfortunately, this does not seem to be the case if we keep l fixed. But if we keep the energy fixed, and vary $l(=0, 1, 2, \ldots)$, and if a certain function associated with the potential is an entire function of exponential type, then the set $\{\delta_l\}$ determines the potential up to some small ambiguities. This brings us to the inverse problem at fixed energy, which is the subject of the second part of this book. It will be shown there that the knowledge of all the phase shifts at one energy does not, in general, determine the potential. However, if one makes some special assumptions about the potential (holomorphic, or finite radius, or superpositions of Yukawa potentials, ..., etc.), one may then be able to obtain much more precise information, and reduce the arbitrariness of the inverse problem.

I.2   The Regular Solution: S-Wave ($l = 0$)

We are now going to study in detail the properties of the regular solution φ, which plays an important role in the Gel'fand–Levitan theory of the inverse problem. The choice of S-wave is motivated by the great simplification in the formula and the transparency of the results. The general case will be dealt with later in this chapter.

Combining the Schrödinger equation with the boundary condition (**I.1.10**), which reads here

$$\varphi(k, 0) = 0, \qquad \varphi'(k, 0) = 1, \tag{I.2.1}$$

and using the method of the "variation of the constants" of Lagrange, we obtain the Volterra integral equation

$$\varphi(k, r) = \frac{\sin kr}{k} + \int_0^r \frac{\sin k(r - r')}{k} V(r')\varphi(k, r')dr'. \tag{I.2.2}$$

In order to study the existence and unicity of the solution, we will use the well-known iterative method. Writing

$$\varphi^{(0)}(k, r) = \frac{\sin kr}{k} \tag{I.2.3}$$

and

$$\varphi^{(n)}(k, r) = \int_0^r \frac{\sin k(r - r')}{k} V(r')\varphi^{(n-1)}(k, r')dr', \tag{I.2.4}$$

and using the bound

$$\left| \frac{\sin Z}{Z} \right| \leq C \frac{e^{|\operatorname{Im} Z|}}{1 + |Z|}, \tag{I.2.5}$$

where C is an appropriate numerical constant, both for the inhomogeneous term $\varphi^{(0)}$ and the kernel, we get easily, for k arbitrary and complex,

$$|\varphi^{(1)}| \leq C \frac{r}{1 + |k|r} e^{|\operatorname{Im} k|r} \int_0^r \frac{r'}{1 + |k|r'} |V(r')| dr'$$

and, more generally,

$$|\varphi^{(n)}| \leq C \frac{r}{1 + |k|r} e^{|\operatorname{Im} k|r} \frac{1}{n!} \left\{ C \int_0^r \frac{r'}{1 + |k|r'} |V(r')| dr' \right\}^n. \tag{I.2.6}$$

From these estimates, it follows immediately that the series

$$\varphi = \sum_{n=0}^{\infty} \varphi^{(n)} \qquad \text{(I.2.7)}$$

is absolutely and uniformly convergent in any compact set of the complex k-plane provided that

$$\int_0^r \frac{r'|V(r')|}{1 + |k|r'} \, dr' < \infty$$

which is true if (**I.1.1**) holds.

It can now be easily seen that the series (**I.2.7**) is indeed the unique solution of the integral equation, and therefore satisfies (**I.1.6**) and (**I.2.1**), and that we have the estimates

$$|\varphi| \leq C \frac{r}{1 + |k|r} e^{|\text{Im } k|r} \exp\left\{ C \int_0^r \frac{r'|V(r')|}{1 + |k|r'} dr' \right\} \leq C_1 \frac{r}{1 + |k|r} e^{|\text{Im } k|r} \qquad \text{(I.2.8a)}$$

and

$$\left| \varphi - \frac{\sin kr}{k} \right| \leq C_2 \frac{r}{1 + |k|r} e^{|\text{Im } k|r} \int_0^r \frac{r'|V(r')|}{1 + |k|r'} dr', \qquad \text{(I.2.8b)}$$

where $C_1, C_2, \ldots$ are appropriate numerical constants.

From these estimates, and the fact that both $\varphi^{(0)}$ and the kernel of the integral equation are holomorphic entire functions of the variable k, it then follows that, for "regular potentials" satisfying (**I.1.1**):

1. φ is an entire function of k for every fixed finite r.
2. Its asymptotic behavior for large values of k is given by

$$\varphi(k, r) \underset{|k| \to \infty}{=} \frac{\sin kr}{k} + \frac{e^{|\text{Im } k|r}}{|k|} o(1); \qquad \text{(I.2.9a)}$$

and

$$\lim_{|k| \to \infty} \{\varphi'(k, r) - \cos kr\} = 0, \qquad \text{(I.2.9b)}$$

where the small term $o(1)$ is uniform in r for all $r \geq 0$. This entails that φ is in fact an entire function of exponential type r with respect to k (order 1, type r). For large values of k, φ is very close to the free solution $\varphi^{(0)}$.

As we saw before, φ is also an even function of k. And since it is real for real values of k, we have, by the Schwarz principle

$$\varphi(-k^*, r) = \varphi(k^*, r) = [\varphi(k, r)]^*. \qquad \text{(I.2.10)}$$

The asymptotic behavior of φ for large r, (**I.1.25**), and the properties of the phase shift δ as well as those of the Jost function will be established later, once we have studied the properties of the Jost solution (**I.1.15**). We mention here, for later use, that (**I.2.8b**) shows that

$$\int_0^\infty \left| \varphi(k, r) - \frac{\sin kr}{k} \right| dk < \infty. \qquad \text{(I.2.11)}$$

When the potential is integrable at the origin, (**I.2.8b**) shows in fact that, for large real values of k,

$$\varphi = \frac{\sin kr}{k} + O\left(\frac{1}{k^2}\right). \qquad \text{(I.2.9c)}$$

Similarly

$$\varphi' = \cos kr + O\left(\frac{1}{k}\right). \qquad \text{(I.2.9d)}$$

I.3   The Jost Solution: S-Wave ($l = 0$)

Here again, we restrict our study first to the S-wave for simplicity, and leave the general case for the end of this chapter. Using again the method of the "variation of constants" of Lagrange, we can combine the Schrödinger equation with the boundary condition (**I.1.15**) into the Volterra integral equation

$$f(k, r) = e^{ikr} + \int_r^\infty \frac{\sin k(r' - r)}{k} V(r') f(k, r') dr'. \qquad \text{(I.3.1)}$$

Iterating this integral equation, and using, as before, the bound (**I.2.5**) in the kernel, we are again led, in the half-plane Im $k \geq 0$, assuming (**I.1.3**), to the absolutely and uniformly convergent series $\sum f^{(n)}$,

$$f^{(0)} = e^{ikr}, \qquad f^{(n)} = \int_r^\infty \frac{\sin k(r - r')}{k} V(r') f^{(n-1)}(k, r') dr',$$

where[2]

$$|f^{(n)}| \leq C e^{-\operatorname{Im} kr} \frac{1}{n!} \left\{ \int_r^\alpha \frac{r' |V(r')|}{1 + |k| r'} dr' \right\}^n.$$

From this estimate, we find

$$|f(k, r)| \leq C_2 e^{-\operatorname{Im} kr}, \qquad \operatorname{Im} k \geq 0. \qquad \text{(I.3.2)}$$

---

[2] It is obvious here that when $k \neq 0$, and $r > 0$ (**I.1.2**) would be sufficient for the convergence of the series. We assume (**I.1.3**) in order to include $k = 0$.

Now using this in (**I.3.1**), we get successively

$$|f(k, r) - e^{ikr}| \leq Ke^{-\operatorname{Im} kr} \int_r^\infty r' |V(r')| dr', \tag{I.3.3}$$

$$|f(k, r) - e^{ikr}| \leq K \frac{e^{-\operatorname{Im} kr}}{|k|} \int_r^\infty |V(r')| dr', \qquad k \neq 0, \tag{I.3.4}$$

and

$$|f'(k, r) - ike^{ikr}| \leq Ke^{-\operatorname{Im} kr} \int_r^\infty |V(r')| dr'. \tag{I.3.5}$$

where the constant K is independent of r and k. The inequality (**I.3.4**) gives us the behavior of f when k is large (Im $k \geq 0$) and r fixed ($r > 0$), whereas (**I.3.3**) and (**I.3.5**) give the asymptotic behavior of f when r is large and k fixed in the closed upper half-plane.

For $r \to 0$, noticing that (**I.1.3**) implies that

$$\lim_{r \to 0} r \int_r^\infty |V(r')| dr' = 0$$

we obtain from (**I.3.5**) that

$$\lim_{r \to 0} rf'(k, r) = 0. \tag{I.3.6}$$

From this, we get, by using the definition (**I.1.20**), that

$$F(k) = f(k, 0). \tag{I.3.7}$$

We can also differentiate f with respect to k by using its perturbative series and the estimate previously given for $f^{(n)}$. It follows that f is continuously differentiable with respect to k in Im $k \geq 0$, with the possible exception of $k = 0$, if (**I.1.3**) holds. We also get, for all $r \geq 0$,

$$|\dot{f}(k, r) - ike^{ikr}| \leq K \frac{e^{-\operatorname{Im} kr}}{|k|}, \qquad \dot{f} = \frac{df}{dk}, \tag{I.3.8}$$

where K is independent of r and k.

The above estimates and asymptotic properties show that, for "regular potentials":

1. f is holomorphic in Im $k > 0$, and continuous and bounded in Im $k \geq 0$, for all values of $r \geq 0$.
2. The asymptotic estimates

$$f(k, r) \underset{r \to \infty}{=} e^{ikr} + e^{-\operatorname{Im} kr} o(1), \tag{I.3.9}$$

$$f'(k, r) \underset{r \to \infty}{=} ike^{ikr} + e^{-\operatorname{Im} kr} o(1), \tag{I.3.10}$$

and for $r > 0$,

$$f(k, r) \underset{\substack{k \to \infty \\ \operatorname{Im} k \geq 0}}{=} e^{ikr} + \frac{e^{-\operatorname{Im} kr}}{|k|} o(1) \tag{I.3.11}$$

are all uniform with respect to the "fixed" variable. If we wish to include $r = 0$, we must replace **(I.3.11)** by

$$f(k, r) \underset{k \to \infty}{=} e^{ikr} + e^{-\operatorname{Im} kr} o(1), \tag{I.3.12}$$

where, again, o(1) is uniform with respect to $r, r \geq 0$;

3. f is in fact continuously differentiable with respect to k in $\operatorname{Im} k \geq 0$, $k \neq 0$, and we have the uniform estimate **(I.3.8)**.
4. f has the following "symmetry" property:

$$f(-k^*, r) = [f(k, r)]^* \tag{I.3.13}$$

which can be checked on every term of its iterative series. It is real on the imaginary k-axis. Thus, it follows that the Jost function satisfies

$$F(-k^*) = [F(k)]^*. \tag{I.3.14}$$

This is the general form of **(I.1.21)** for $\operatorname{Im} k \geq 0$.

I.4 The Jost Function and the Phase Shift

From the definitions **(I.1.19)**[3] and **(I.1.24)**, and the asymptotic behaviors of f and f' given by **(I.3.9)** and **(I.3.10)**, we get

$$\varphi(k, r) \underset{r \to \infty}{=} \frac{|F(k)|}{|k|} \sin(kr + \delta) + o(1) \tag{I.4.1}$$

and

$$\varphi'(k, r) \underset{r \to \infty}{=} |F(k)| \cos(kr + \delta) + o(1) \tag{I.4.2}$$

both valid only for real values of k. From **(I.1.21)** and **(I.1.24)**, it is obvious that we can define the phase shift in such a way that, for $k \geq 0$,

$$\delta(-k) = -\delta(k). \tag{I.4.3}$$

In order to obtain the asymptotic behaviors of the Jost function and the phase shift for large values of energy. which are needed in the inverse problem, we start from the following integral representation

$$F(k) = 1 + \int_0^\infty e^{ikr} V(r) \varphi(k, r) dr \tag{I.4.4}$$

which is easily obtained from the definition **(I.1.20)** taken for large values of r, together with the integral equation **(I.2.2)** for φ, and the estimates **(I.3.9)** and **(I.3.10)**.

From the definition **(I.1.20)**, or from the above integral representation, it is obvious that the Jost function is a well-defined function in the upper

---

[3] Because of the simultaneous occurrence of both k and $-k$ in this formula, its validity is generally limited to the real k-axis.

half-plane Im $k \geq 0$, which is holomorphic in Im $k > 0$, and continuous in Im $k \geq 0$. It is also bounded there because of (**I.2.8**), which gives

$$|F(k) - 1| \leq C \int_0^\infty \frac{r|V(r)|}{1 + |k|r} \, dr. \qquad \text{(I.4.5)}$$

We, therefore, see that, when $k \to \infty$, Im $k \geq 0$, we have

$$\lim_{k \to \infty} F(k) = 1. \qquad \text{(I.4.6)}$$

It is also easily seen that F is continuously differentiable in the upper half-plane, with the possible exception of the origin. This follows from the definition (**I.3.7**) and what we have learnt about the Jost solution in the last section. In fact, from (**I.3.8**), it is obvious that $k\dot{F}(k)$ is continuous and bounded in the entire upper half-plane Im $k \geq 0$, and that

$$\dot{F}(k) \underset{k \to \infty}{=} o(1). \qquad \text{(I.4.7)}$$

From the asymptotic property (**I.4.6**), and the definition (**I.1.24**) of the phase shift, it is obvious that we can choose

$$\delta(+\infty) = 0. \qquad \text{(I.4.8)}$$

Starting from this, we can then obtain $\delta(k)$ by continuity for *all* finite values of the energy. In the future, we shall always keep this determination of the phase shift.

It is now easy to show that

$$\int^\infty k^{-1} |\delta(k)| \, dk < \infty \qquad \text{(I.4.9)}$$

and

$$\int^\infty |\text{Im } F(k)| k^{-1} \, dk < \infty. \qquad \text{(I.4.10)}$$

The two integrands being of the same order for large k because of (**I.1.24**), (**I.4.6**), and (**I.4.8**), it is sufficient to prove only (**I.4.10**). Now, using the definition (**I.4.4**), and the estimates (**I.2.5**) and (**I.2.8**), we have, for $k > 0$,

$$k^{-1} |\text{Im } F| \leq C \int_0^\infty |V(r)| \left(\frac{r}{1 + kr} \right)^2 dr$$

from which our assertion follows.

As we shall see, the integrability conditions (**I.4.9**) or (**I.4.10**) play an important role in the study of the inverse problem; they guarantee that the potential found from the phase shift is "regular," i.e., satisfies **(I.1.3)**. Introducing the S-matrix (k real)

$$e^{2i\delta} = S(k) = \frac{F(-k)}{F(k)} \qquad \text{(I.4.11)}$$

the integrability condition **(I.4.9)** can also be written in the form

$$\int^{\infty} |S - 1| k^{-1} \, dk < \infty. \qquad \text{(I.4.12)}$$

If the potential is integrable at the origin, **(I.2.9)** and **(I.4.4)** show that, as $k \to \infty$, one has at least

$$F(k) = 1 + O\left(\frac{1}{k}\right) \qquad \text{(I.4.6')}$$

and

$$\delta(k) = O\left(\frac{1}{k}\right). \qquad \text{(I.4.8')}$$

I.5   Higher Waves

So far, we have discussed the S-wave in detail. For higher waves, except for some obvious minor changes, the methods and the results are quite similar. We therefore give here only the results without going into the proofs. The most important difference between $l = 0$ and $l \geq 1$ is that, in the latter case, according to **(I.1.13)**, the free solution $w_l(z)$ has a pole of order l at the origin. This induces a pole of the same order in the Green's functions, and therefore leads to a singularity in the Jost solution f at the origin. We must therefore be careful in taking the limits $r \to 0$ or $k \to 0$.

The regular solutions of **(I.1.6)** are defined by the boundary conditions **(I.1.10)**. The integral equation satisfied by this solution is now

$$\varphi_l(k, r) = k^{-l-1} u_l(kr) + \int_0^r G_l(k, r, r') V(r') \varphi_l(k, r') dr', \qquad \text{(I.5.1)}$$

$$G_l = i(-1)^l (2k)^{-1} [w_l(kr') w_l(-kr) - w_l(kr) w_l(-kr')]. \qquad \text{(I.5.2)}$$

From the bounds

$$|u_l(z)| < C\left(\frac{|z|}{1 + |z|}\right)^{l+1} e^{|\operatorname{Im} z|} \qquad \text{(I.5.3)}$$

and ($r \geq r'$)

$$|G_l(k, r, r')| < C|k|^{-1} \left(\frac{|k|r}{1 + |k|r}\right)^{l+1} \left(\frac{|k|r'}{1 + |k|r'}\right)^{-l} e^{|\operatorname{Im} k|(r - r')} \qquad \text{(I.5.4)}$$

and proceeding along the same lines as in Section I.2, we obtain the following results:

1. φ_l, the unique solution of the Schrödinger equation satisfying the boundary conditions (**I.1.10**), is, for each fixed $r \geq 0$, an entire function of k of exponential type r, and we have

$$|\varphi_l(k, r)| \leq C\left(\frac{r}{1 + |k|r}\right)^{l+1} e^{|\text{Im }k|r}, \tag{I.5.5}$$

$$|\varphi_l - k^{-l-1}u_l(kr)| \leq C\left(\frac{r}{1 + |k|r}\right)^{l+1} e^{|\text{Im }k|r}o(1). \tag{I.5.6}$$

From the last inequality, it follows that

$$\varphi_l(k, r) \underset{k \to \infty}{=} \frac{\sin(kr - \frac{1}{2}l\pi)}{k^{l+1}} + \frac{e^{|\text{Im }k|r}}{k^{l+1}} o(1). \tag{I.5.7}$$

2. φ_l is a real even function of k. It therefore satisfies the symmetry relations

$$\varphi_l(-k, r) = \varphi_l(k, r) = [\varphi_l(k^*, r)]^*. \tag{I.5.8}$$

The Jost solution can be studied likewise. If we define it by the boundary condition (**I.1.15**), it satisfies the integral equation

$$f_l(k, r) = w_l(kr) - \int_r^\infty G_l(k, r, r')V(r')f_l(k, r')dr', \tag{I.5.9}$$

where G_l is again given by (**I.5.2**). From the bounds

$$|w_l(z)| \leq C\left(\frac{|z|}{1 + |z|}\right)^{-l} e^{-\text{Im }z} \tag{I.5.10}$$

and (**I.5.4**) in which we have to interchange r and r' (here $r \leq r'$), we obtain now

1. For each fixed $r > 0$, the Jost solution exists and is holomorphic in k in $\text{Im }k > 0$, and $k^l f_l$ is continuous in $\text{Im }k \geq 0$;
2. For each fixed $k \neq 0$ in $\text{Im }k \geq 0$, $r^l f_l$ is a continuous function of r for $r \geq 0$;
3. For k in $\text{Im }k \geq 0$, we have the bounds

$$|f_l(k, r)| < K\left(\frac{|k|r}{1 + |k|r}\right)^{-l} e^{-\text{Im }kr}, \tag{I.5.11}$$

$$|f_l(k, r) - w_l(kr)| < K\left(\frac{|k|r}{1 + |k|r}\right)^{-l} e^{-\text{Im }kr} \int_r^\infty r'|V(r')|dr' \tag{I.5.12}$$

and

$$|f_l - w_l| < \left(\frac{|k|r}{1 + |k|r}\right)^{-l} \frac{e^{-\text{Im }kr}}{|k|} \int_r^\infty |V(r')|dr'. \tag{I.5.13}$$

The bound **(I.5.11)** is useful when one of the variables is fixed $\neq 0$, while the other one becomes very large. Eq. **(I.5.12)** is good for $k \neq 0$ and $r \to \infty$. It shows that, for Im $k \geq 0$,

$$f_l(k, r) \underset{\substack{r \to \infty \\ k \neq 0}}{=} i^l e^{ikr} + e^{-\operatorname{Im} kr} o(1).$$ (I.5.14)

Finally, **(I.5.13)** gives us

$$f_l(k, r) \underset{\substack{k \to \infty \\ r \neq 0}}{=} i^l e^{ikr} + O\left(\frac{e^{-\operatorname{Im} kr}}{|k|}\right).$$ (I.5.15)

In order to calculate the Jost function via the Wronskian **(I.1.20)**, in which we take r very large, we must know the asymptotic behavior of f_l'. This is most easily found by differentiating **(I.5.9)** with respect to r, then letting $r \to \infty$, and using the asymptotic forms of the free solution u_l and w_l as well as the bound **(I.5.11)**. It is found that

$$f_l'(k, r) \underset{\substack{r \to \infty \\ k \neq 0}}{=} i^{l+1} k e^{ikr} + e^{-\operatorname{Im} kr} o(1).$$ (I.5.16)

Now, by taking **(I.1.20)**, replacing φ_l by the right-hand side of **(I.5.1)**, letting $r \to \infty$ and using the asymptotic properties of the Green's function G_l as well as **(I.5.15)** and **(I.5.16)**, we obtain the integral representation

$$F_l(k) = 1 + (-k)^l \int_0^\infty w_l(kr) V(r) \varphi_l(k, r) dr.$$ (I.5.17)

This integral representation, together with the bounds for w_l and φ_l, leads us to the following results:

1. The Jost function is holomorphic in Im $k \geq 0$;
2. It satisfies the symmetry relation

$$F_l(-k^*) = [F_l(k)]^*;$$ (I.5.18)

3.
$$F_l(k) \underset{k \to \infty}{=} 1 + o(1)$$ (I.5.19)

holds in Im $k \geq 0$.

From the definition **(I.1.24)**, and **(I.5.19)**, we see that we can choose the phase shift in such a way that

$$\delta_l(\infty) = 0.$$ (I.5.20)

We have also

$$\delta_l(-k) = -\delta_l(k), \qquad k \geq 0.$$ (I.5.21)

From the representation (**I.5.17**), together with the bound (**I.5.10**) for w_l and (**I.5.5**), we again obtain, as in the case of S-wave, the integrability conditions

$$\int^{\infty} |\delta_l(k)| \frac{dk}{k} < \infty, \tag{I.5.22}$$

$$\int^{\infty} |\operatorname{Im} F_l| \frac{dk}{k} < \infty, \tag{I.5.23}$$

$$\int^{\infty} |S_l - 1| \frac{dk}{k} < \infty, \qquad S_l = \frac{F_l(-k)}{F_l(k)}, \tag{I.5.24}$$

where S_l is the S-matrix. Also using the behavior of φ_l near $r = 0$ in (**I.1.20**), we find that

$$F_l(k) = \lim_{r \to 0} \frac{(-kr)^l}{(2l - 1)!!} f_l(k, r). \tag{I.5.25}$$

I.6   Singular Potentials

To our knowledge, the only singular cases for which the inverse scattering problem at fixed angular momentum has been studied are the following. First, let us define short range potential by $rV(r) \in L^1(b, \infty)$, $b > 0$. We then have

1. Regular potentials plus the Coulomb potential. Scattering theory with Coulomb potential plus a regular potential is now well understood, and the inverse problem has also been solved. We shall come back later to this case.
2. Short-range potentials which are singular and oscillating near the origin, in such a way that, if we define

$$W(r) = - \int_r^{\infty} V(t)dt \tag{I.6.1}$$

it satisfies [4]

$$W(r) \in L^1(0, \infty) \tag{I.6.2a}$$

and

$$\lim_{r \to 0} rW(r) = 0. \tag{I.6.2b}$$

---

[4] Here the main point is the integrability of W at the origin. Its integrability at infinity follows from (**I.1.3**) for $b > 0$. It follows also from this last condition that $\lim rW(r) = 0$ when $r \to \infty$.

In short, although the potential may be very singular at the origin, it oscillates there sufficiently rapidly to make $W(r)$ well behaved. Examples of this kind are numerous:

$$W(r) = r^{-a}e^{-\mu r}\sin\left(\exp\frac{1}{r}\right), \qquad a < 1, \tag{I.6.3}$$

etc. Differentiating W, one sees that V is indeed highly singular and very rapidly oscillating.

For this whole class of potentials, it can be shown that everything we said before for potentials satisfying (**I.1.3**) is valid again without any modification, and that the inverse problem can be formulated exactly in the same way. The main point of the analysis is the observation that, in order to be able to show the existence of a unique solution φ of (**I.2.2**) given by the series (**I.2.7**),[5] it is not necessary to require the absolute convergence of this series. As is well known, uniform convergence would be sufficient. And this can be shown to be true in the following way.

We replace first V by W' in (**I.2.2**), and integrate by parts, using (**I.6.2b**), and the fact that W is a continuous function for $r > 0$. We obtain in this way the coupled integral equations

$$\varphi(k, r) = \frac{\sin kr}{k} + \int_0^r dr'\,W(r')\left\{\cos k(r - r')\varphi(k, r') - \frac{\sin k(r - r')}{k}\varphi'(k, r')\right\} \tag{I.6.4a}$$

and

$$\varphi'(k, r') = \cos kr + W(r)\varphi(k, r)$$
$$- \int_0^r dr'\,W(r')\{k\sin k(r - r')\varphi(k, r') + \cos k(r - r')\varphi'(k, r')\}. \tag{I.6.4b}$$

We can now iterate these equations and use (**I.2.5**) and

$$|\cos Z| < e^{|\operatorname{Im} Z|}. \tag{I.6.5}$$

We obtain in this way absolutely and uniformly convergent series whose general terms satisfy

$$|\varphi^{(n)}(k, r)| \le C\left(\frac{r}{1 + |k|r}\right)e^{|\operatorname{Im} k|r}(n!)^{-1}\left\{\int_0^r (2C|W(t)| + C^2tW^2(t))dt\right\}^n \tag{I.6.6a}$$

$$|\varphi'^{(n)}| \le Ce^{|\operatorname{Im} k|r}(n!)^{-1}\{\cdots\}^n. \tag{I.6.6b}$$

The integrals exist because of (**I.6.2a**) and the fact that rW^2, the product of the bounded continuous function rW by the integrable function W is also integrable. It then follows, as before, that φ and φ' exist and satisfy

---

[5] For simplicity, we start again with the S-wave.

(I.1.6) and **(I.2.1)**, that they are unique, and, finally, that they have the same analyticity properties in k (entire functions). They also satisfy upper bounds similar to **(I.2.8)** and **(I.2.9)** the only change being that one must everywhere make the substitution

$$C \int_0^r r' |V(r')| dr' \to \int_0^r (2C|W| + C^2 t W^2) dt. \qquad (I.6.7)$$

This, of course, does not change the conclusions and again we obtain **(I.2.10)** and **(I.2.11)**.

For higher waves, the analysis is similar, and we reach again the conclusions of the previous section, namely **(I.5.5)**, **(I.5.6)**, **(I.5.7)** and **(I.5.8)**.

Since the existence and the properties of the Jost solution $f_l(k, r)$ for $r > 0$, do not depend on the singularity of the potential at the origin, and are guaranteed by the sole condition **(I.1.3)** with $b > 0$, it follows that we can define again the Jost function $F_l(k)$ through **(I.1.19)**, or **(I.1.20)**, and that, as before, it is analytic in the upper plane $\operatorname{Im} k > 0$, is continuous in $\operatorname{Im} k \geq 0$ and satisfies there **(I.5.18)** and **(I.5.19)**.

Also, the integral representation **(I.5.17)** is valid, and can be written, more conveniently (use integration by parts) as

$$F_l(k) = 1 - (-k)^l \int_0^\infty W(r) \{ w_l(kr) \varphi_l' + w_l'(kr) \varphi_l \} dr. \qquad (I.6.8)$$

Finally, the phase shift is again defined through **(I.1.24)** and we obtain all the properties shown in **(I.5.20)**–**(I.5.24)**.

The only difference between a regular potential satisfying **(I.1.3)** and a singular one satisfying **(I.6.2a)** and **(I.6.2b)** may be in the possible absence or the presence of infinitely many oscillations in the phase shift as $k \to \infty$. Of course, in both cases, the phase shift goes to zero as the energy goes to infinity. However, while in the regular case, the potential may keep a constant sign near the origin, which means that the phase shift may go to zero while keeping a constant sign (opposite to that of the potential near the origin), in the singular case the oscillations are necessary, and so the phase shift always oscillates indefinitely as $k \to \infty$.

For example, for

$$V(r) \simeq V_0 r^{-1-\beta}, \qquad r \to 0, \qquad \beta < 1 \qquad (I.6.9a)$$

we have, for the S-wave,

$$\delta(k) \underset{k \to \infty}{\simeq} - V_0 \beta^{-1} \cos\left(\frac{\pi \beta}{2}\right) \Gamma(1 - \beta)(2k)^{\beta - 1} \qquad (I.6.9b)$$

whereas, for

$$W(r) = 2(2r)^{-1/2} \sin\left(\frac{a}{2r}\right) e^{-\mu r} \qquad (I.6.10a)$$

we obtain, as $k \to \infty$,

$$\delta(k) \simeq -\frac{1}{2}\left(\frac{\pi}{2k}\right)^{1/2}\{\sin 2(ak)^{1/2} - \cos 2(ak)^{1/2}\} + \cdots. \quad \textbf{(I.6.10b)}$$

Here we see that there is a definite relationship between the oscillations of W and those of the phase shift. This is a general phenomenon. Of course, if we introduce an oscillating factor in **(I.6.9a)** we also obtain oscillations in **(I.6.9b)**. But then, because of these oscillations the phase shift goes to zero faster than is indicated in **(I.6.9b)**.

Besides the oscillating singularities at the origin we studied above, it is obvious in **(I.6.2a)** and **(I.6.2b)** that the theory can easily accommodate some mild (integrable) singularities of the potential at finite distances by working exclusively with $W(r)$. Likewise, mild generalized functions (Dirac δ-function, etc.) can be dealt with without using any special technique.

3. Potentials having a strong repulsive singularity at the origin. We shall study them at the end of chapters II and III.

I.7   Comments and References

This chapter gives a survey of all the classical and useful results on scattering theory with spherically symmetric potentials which are needed. The reader is referred to the books of De Alfaro and Regge (1965), Mott and Massey (1965), and Newton (1982), for further details and references on scattering theory.

Section 6 on the singular potentials is based on the paper by Baeteman and Chadan (1976). As was mentioned repeatedly, all the results of this chapter are true if the potential satisfies the conditions of section 6 : $rV(r) \in L^1(b, \infty)$, $b > 0$, $W(r) \in L^1(0, a)$, and $\lim rW(r) = 0$ as $r \to 0$. These conditions are weaker than the usual condition $rV(r) \in L^1(0, \infty)$. The reason we have first treated this restricted class of potentials is because of its wide spread use in all the textbooks on scattering theory. It seemed therefore appropriate, to give first the results for this restricted class, and then show that they are also true under weaker assumptions on the potential.

For the general theory of ordinary differential equations, the reader is referred to the books of Coddington and Levinson (1955), and Hille (1969). For the theory of integral equations, see, e.g., Tricomi (1957) and Pogorzelski (1966).

CHAPTER II

Bound States—
Eigenfunction Expansions

II.1   Bound States: The Levinson Theorem

Bound States

The bound states correspond to the solutions of **(I.1.6)** which satisfy the boundary conditions **(I.2.1)**, and are square-integrable on the whole positive r-axis. We are going to study them in detail because, as is well known, they are necessary in the completeness relation, and therefore, as was mentioned before, enter into the inverse problem. We again consider the S-wave first. The assumptions on the potential are those of the last section of chapter I: $rV(r) \in L^1(b, \infty)$, $b > 0$, $W(r) \in L^1(0, a)$, and $rW(r) \to 0$ as $r \to 0$, which are, evidently, weaker than the usual assumption $rV(r) \in L^1(0, \infty)$.

First of all, it is easy to show that every zero of the Jost function in $\mathrm{Im}\, k > 0$ corresponds to a bound state. Indeed, if

$$f(k_0, 0) \equiv F(k_0) = 0 \qquad \textbf{(II.1.1)}$$

it follows from **(I.1.20)** that $f(k_0, r)$ and $\varphi(k_0, r)$ are not independent of each other, i.e.,

$$f(k_0, r) = f'(k_0, 0)\varphi(k_0, r). \qquad \textbf{(II.1.2)}$$

Therefore, according to **(I.3.9)**, $\varphi(k_0, r)$ vanishes exponentially when r goes to infinity, and we have a bound state. Notice that φ is well behaved everywhere. Therefore, if $f'(k_0, 0) = 0$, this would contradict **(I.3.9)**.

For real values of k, it can be seen that the Jost function cannot vanish, except perhaps at $k = 0$. For, if $F(k_0) = 0$, $k_0 \neq 0$, and real, then, according to (I.1.21), we also have $F(-k_0) = 0$. It follows then from (I.1.19) that $\varphi(k_0, r) \equiv 0$, which contradicts (I.2.1). We shall come back in a moment to the case $k_0 = 0$. It is also easy to see that, in $\operatorname{Im} k > 0$, the zeros occur for purely imaginary values of k (real negative energies!). Indeed, writing the Schrödinger equation for $f(k_0, r)$ and $[f(k_0, r)]^*$, and combining them, we obtain

$$f'(k_0, 0)[f(k_0, 0)]^* - f(k_0, 0)[f'(k_0, 0)]^*$$

$$= 4i \operatorname{Re} k_0 \operatorname{Im} k_0 \int_0^\infty |f(k_0, r)|^2 \, dr. \quad \text{(II.1.3)}$$

From (II.1.1), it follows then that if k_0 corresponds to a zero, we must have $\operatorname{Re} k_0 = 0$, which proves our assertion. We shall see below that these zeros give all the bound states.

The case $F(0) = 0$ is somewhat exceptional. We consider the Schrödinger equation at zero energy

$$\varphi'' = V\varphi. \quad \text{(II.1.4)}$$

On the basis of the integral equation (I.2.2) taken at $k = 0$, and following the same line of reasoning as in section I.2, we find that the asymptotic form of φ for large r is given, if we assume either $rV(r) \in L^1(0, \infty)$, or that the potential belongs to the general class considered at the end of chapter I, by

$$\varphi = Ar + B + o(1), \quad \text{(II.1.5a)}$$
$$\qquad\qquad\qquad\qquad r \to \infty.$$
$$\varphi' = A + o(1), \quad \text{(II.1.5b)}$$

Likewise, it is also obvious that the asymptotic form of $f(0, r)$ is

$$f(0, r) \underset{r \to \infty}{=} 1 + o(1). \quad \text{(II.1.6)}$$

It is also obvious, on the basis of the results of section I.2 and I.3, that A is the Jost function at $k = 0$.

From (II.1.5a) and (II.1.6), we see that the asymptotic form of the general solution of (II.1.4) is $Ar + C$, i.e., it always blows up at infinity, except when $A = 0$. Therefore, even if $F(0) = 0$, we have no square integrability at infinity, and, consequently, no bound state at zero energy. This zero-energy singularity would therefore not enter into the eigenfunction expansions associated with the Schrödinger equation for the S-wave.

This analysis can be done also for $l \neq 0$. It can be shown that when the Jost function vanishes at $k = 0$, the asymptotic form of $\varphi_l(0, r)$ is just $\sim r^{-l}$. Therefore, we have square-integrability at infinity for all angular momenta $(l = 1, 2, \ldots)$, i.e., a root of the Jost function at the origin corresponds to a true bound state at zero energy. This distinguishes the higher waves from the S-wave.

Let us now look at the number and multiplicities of the zeros of the Jost function in $\operatorname{Im} k > 0$, which are all located, as we saw, on the imaginary axis.

On the basis of (**I.4.6**), it is obvious that they cannot be located too high on the axis, and that $+i\infty$ cannot be an accumulation point of the zeros. Likewise, it can be shown that the zeros cannot accumulate at $k = 0$. Their number is therefore finite.

We now proceed to show that the zeros are all simple. To see this, we consider the Schrödinger equation for f, and differentiate it with respect to k. Using the notation $\dot{f} = df/dk$, and combining the two equations, we obtain, for k in the upper half-plane,

$$\dot{f}'(k, 0)f(k, 0) - \dot{f}(k, 0)f'(k, 0) = 2k \int_0^\infty f^2(k, r)dr. \qquad \textbf{(II.1.7)}$$

From this relation, (**II.1.1**), and (**II.1.2**), we deduce that if $F(k_0) = 0$, we have, for the normalization of the bound state wave function, the relation

$$C_0^{-1} = \int_0^\infty \varphi^2(k_0, r)dr = \frac{-\dot{F}(k_0)}{2k_0 f'(k_0, 0)}. \qquad \textbf{(II.1.8)}$$

This implies, among other things, that $\dot{F}(k_0) \neq 0$ because the integral on the left-hand side is positive (remember (**I.2.10**)!). Consequently, the zero is simple. It can also be shown that if there is a zero at $k = 0$ it is also simple. More precisely $\dot{F}(k)$, when $k \to 0$ along a path in the upper half-plane, does not vanish. As we said before, we never have a bound state at zero energy for an S-wave.

The same general properties are valid for higher waves, $l = 1, 2, \ldots$, with the only exception that a zero at the origin would be double, and would correspond to a true bound state with zero energy.

Fredholm Integral Equation for the Physical Solution

So far, we have been considering the zeros of F in Im $k > 0$, and have shown that they correspond to bound states. We may now ask whether these zeros give us all the bound states. The answer is in the affirmative. Indeed, it is well known, and easily verified that the physical solution ψ discussed in section I.1, which satisfies the boundary conditions $\psi(k, 0) = 0$ and (**I.1.11**), is the solution of the Fredholm integral equation

$$\psi(k, r) = \sin kr - \int_0^\infty \frac{\sin kr_<}{k} e^{ikr_>} V(r')\psi(k, r')dr',$$

$$r_> = \max(r, r'), \qquad r_< = \min(r, r'). \qquad \textbf{(II.1.9)}$$

It has been shown by Jost and Pais (1951) that this equation is of the Fredholm type if the potential satisfies (**I.1.3**), and that its Fredholm determinant is exactly the Jost function $F(k)$. It then follows from the Fredholm alternative theorem that all the square-integrable solutions are given by the zeros of the Jost function. That their number is finite is the consequence of Bargmann inequality (1952) on the number of the bound states in the angular momentum state l:

$$n_l \leq \frac{1}{2l+1} \int_0^\infty r|V(r)|dr. \qquad \textbf{(II.1.10a)}$$

This bound cannot, in general, be improved if we have no other information than **(I.1.3)** on the potential. Moreover, there exist potentials which violate this condition, and which have effectively an infinite number of bound states.

For more general potentials of section I.6, the Jost function is again the Fredholm determinant of the integral equation **(II.1.9)**, which can also be written in the coupled form similar to **(I.6.4a, b)**. In this case, we have the bound

$$n_l \leq \frac{2}{2l+1} \int_0^\infty r^{4l+1} W_l^2(r)dr, \qquad W_l(r) = -\int_r^\infty s^{-2l}V(s)ds. \qquad \textbf{(II.1.10b)}$$

The Levinson Theorem

This theorem relates the variation of the phase shift along the real axis to the number of the bound states. It is merely the well-known relationship between the variation of the imaginary part of log $F(k)$ along the closed path made up of the whole real axis, plus the half-circle at infinity in the upper-plane, and the number of zeros of the Jost function inside the domain. Making use of **(I.4.6)** and **(I.4.3)**, we can write it[1]

$$l-0 \begin{cases} \delta(0) - \delta(\infty) = n\pi & \text{if } F(0) \neq 0, \\ \delta(0) - \delta(\infty) = (n + \frac{1}{2})\pi & \text{if } F(0) = 0. \end{cases} \qquad \textbf{(II.1.11)}$$

The validity of this theorem is much more general than stated here for local potentials satisfying **(I.1.3)**. It is, in fact, valid, as we shall see later, even for a large class of nonlocal potentials.

Higher Waves

For higher waves, the preceding analysis extends in a straightforward manner. Concerning the bound states, one can again show that the Jost function is the Fredholm determinant of the integral equation satisfied by the physical solution **(I.1.26)** which now reads, in general,

$$\psi_l(k, r) = u_l(kr) - \int_0^\infty u_l(kr_<)w_l(kr_>)V(r')\psi_l(k, r')dr'. \qquad \textbf{(II.1.9')}$$

---

[1] Here, we must assume that $r^2V(r)$ is absolutely integrable at infinity if $F(0) = 0$. The same remark applies to **(II.1.12)**.

Therefore, there is, in general, a one to one correspondence between the zeros of the Jost function in Im $k \geq 0$ and the bound states. For P-wave and higher waves, the wave function at zero energy has the asymptotic form

$$\varphi_l(0, r) = A_l r^{l+1} + B_l r^{-l} + o(r^{-l}), \tag{II.1.5c}$$

$$\varphi'_l(0, r) = A'_l r^l + B'_l r^{-l-1} + o(r^{-l-1}), \tag{II.1.5d}$$

which generalize (II.1.5a,b). Here, since $l \geq 1$, it follows that if $F_l(0) = 0$, i.e., $A_l = 0$, this corresponds to a true bound state at zero energy because now φ_l is square-integrable at infinity. This is the only difference with the S-wave.

Following the same method as for the S-wave, we find again that the zeros of the Jost function in the upper half-plane are simple, and situated, as they should, on the imaginary axis. If the Jost function vanishes at the origin, the zero is double for $l \geq 1$, whereas we noted that it was simple for $l = 0$.

The Levinson theorem now reads ($l \geq 1$)

$$\delta_l(0) - \delta_l(\infty) = n_l \pi, \tag{II.1.12}$$

where n_l is the number of bound states, including the one at zero energy, if it exists, with angular momentum l.

If $k = i\gamma_0$, $\gamma_0 > 0$, corresponds to a bound state, we have

$$f_l(k_0, r) = c_l \varphi_l(k_0, r) \tag{II.1.13a}$$

and

$$(-k_0)^{-l} \dot{F}_l(k_0) = -2c_l k_0 \int_0^\infty \varphi_l^2(k_0, r) dr \tag{II.1.13b}$$

which is the generalization of (II.1.8), and is obtained in the same way. Making $r \to 0$ in (II.1.13a), we obtain also

$$c_l = \lim_{r \to 0} (2l + 1)!! \, r^{-l-1} f_l(k_0, r). \tag{II.1.13c}$$

We shall see later that for potentials decreasing at infinity exponentially or faster ($V \sim e^{-\mu r}$), the Jost solution is holomorphic at least in Im $k > -\mu/2$. Therefore, there is a common domain of holomorphy for $f_l(\pm k, r)$, which is the strip

$$|\text{Im } k| < \frac{\mu}{2}.$$

It then follows also that the Jost functions $F_l(\pm k)$ have the same common domain of holomorphy. Thus, formula (I.1.19) is valid in this strip. Suppose now that there is a bound state $k_0 = i\gamma_0$ in this strip ($\gamma_0 < \mu/2$). From (I.1.19), (II.1.1) and (II.1.13a), we find

$$\varphi_l(i\gamma_0, r) = -\left(\frac{i^l}{2}\right) \gamma_0^{-l-1} F_l(-i\gamma_0) f_l(i\gamma_0, r) \tag{II.1.13d}$$

and

$$c_l = -2\gamma_0^{l+1} [i^l F_l(-i\gamma_0)]^{-1}. \tag{II.1.13e}$$

Substitution of this into **(II.1.13b)** then leads to

$$C_0^{-1} = \int_0^\infty \varphi_l^2(i\gamma_0, r)dr = (-1)^l \dot{F}_l(i\gamma_0) \frac{F_l(-i\gamma_0)}{4i\gamma_0^{2l+2}}. \qquad \text{(II.1.14)}$$

Since now the S-matrix is meromorphic in the strip, we can calculate its residue at the pole of the bound state. It is

$$s_0 = \frac{F(-i\gamma_0)}{F(+i\gamma_0)} = -i \frac{F^2(i\gamma_0/2)(-)^l C_0}{4\gamma_0^{2l+2}}. \qquad \text{(II.1.14')}$$

This shows that $i(-1)^l s_0$ is positive at the bound state poles of the S-matrix.

II.2 Integral Representation for the Jost Function

The analysis we have so far carried out shows the importance of the Jost function. It shows that this single function contains all the information about the scattering and the S-matrix, as well as the bound states (number and energies) save for the normalization of the bound state wave functions. It also plays a very important role in the formulation of the inverse problem because it provides the link between the experimental data (phase shifts and binding energies) and the so-called Povzner–Levitan kernel which determines the potential. This is shown in the next chapter. The first step is the integral representation

$$F_l(k) = \prod_1^{n_l} \left(1 - \frac{E_j}{E}\right) \exp\left[-\left(\frac{2}{\pi}\right) \int_0^\infty \frac{\delta_l(k')}{k'^2 - k^2} k' \, dk'\right] \qquad \text{(II.2.1)}$$

valid for $\text{Im } k > 0$, the E_j's being the binding energies of the bound states $(E_j = -\gamma_j^2)$.

This representation is obtained by noticing first that the function

$$\tilde{F}_l(k) = \prod_1^{n_l} \left(\frac{k + k_j}{k - k_j}\right) F_l(k) \qquad \text{(II.2.2)}$$

is holomorphic in $\text{Im } k > 0$, continuous in $\text{Im } k \geq 0$, and, according to **(I.5.9)**, also satisfies

$$\tilde{F}_l(k) = 1 + o(1), \qquad |k| \to \infty. \qquad \text{(II.2.3)}$$

We also notice that $\tilde{F}_l$ does not vanish any more in $\text{Im } k > 0$. It follows that $\text{Log } \tilde{F}_l$ is holomorphic in $\text{Im } k > 0$, and vanishes at infinity because of **(II.2.3)**. We can therefore apply the Hilbert integral transforms to the real and imaginary parts of $\text{Log } \tilde{F}_l$. Noticing that

$$\text{Im Log } \tilde{F}_l = -\delta_l(k) + 2\sum_j \text{tg}^{-1}(\gamma_j/k) \qquad \text{(II.2.4)}$$

we easily find **(II.2.1)**. When the energy becomes real, $k \to \text{Re } k + i0$, we see, as we expect, that the phase of the Jost function is indeed $-\delta_l(k)$, whereas its modulus is given by **(II.2.1)** with the principal value symbol in front of the integral:

$$|F_l(k)| = \prod \left(1 - \frac{E_j}{E}\right) \exp\left[-\left(\frac{2}{\pi}\right)P \int_0^\infty \frac{\delta_l(k')}{k'^2 - k^2} k' \, dk'\right], \qquad E = k^2 \geq 0.$$

$$\text{(II.2.1')}$$

II.3   Eigenfunction Expansion

The eigenfunction expansion associated with a second-order differential equation has been subject to extensive studies in the literature. We shall therefore give here only the fundamental results, and refer the reader to the literature on the subject. The differential equation we have to deal with is (take first $l = 0$):

$$-\varphi'' + V(r)\varphi = k^2\varphi \qquad \text{(II.3.1)}$$

on the interval $[0, \infty)$. If we assume that V satisfies the integrability condition **(I.1.1)**, or **(I.6.2a)** and **(I.6.2b)**, at the origin, it can be shown that the origin corresponds, in the terminology of Weyl and Titchmarsh, to a limit circle. This means that there are two independent solutions of **(II.3.1)** which are both $L^2(0, a)$ for any finite a. Therefore, every solution also belongs to $L^2(0, a)$.

In order to define a self-adjoint operator in $L^2(0, \infty)$, we need therefore to impose some boundary conditions. This is achieved by setting

$$\varphi(k, 0) = 0, \qquad \varphi'(k, 0) = 1. \qquad \text{(II.3.2)}$$

At infinity, we assume of course that $rV(r) \in L^1(b, \infty)$, i.e., we are dealing with short-range potentials. We are now able to obtain our self-adjoint operator by the extension of the symmetric operator **(II.3.1)** acting on the twice-continuously differentiable functions satisfying the boundary conditions and vanishing outside some finite interval. We denote this self-adjoint operator by D.

Consider now the kernel

$$R_E(r, t) = \frac{\varphi(k, r_<)f(k, r_>)}{F(k)}, \qquad \text{Im } k > 0, \qquad \text{(II.3.3)}$$

where φ and f, are, respectively, the regular solution and the Jost solution studied in chapter I, $F(k)$ is the Jost function, and $r_< = \min(r, t)$, $r_> = \max(r, t)$. We assume throughout this section that $F(0) \neq 0$. In the E-plane, R_E is an analytic function with a cut on the positive real axis, and simple poles on the negative real axis at points where the Jost function vanishes in the upper plane Im $k > 0$, i.e., at points $E_j = -\gamma_j^2$ corresponding to the energies of the bound states. Their number is finite. It is easily verified, using the definition (**I.1.20**) of the Jost function $F(k)$, that R_E satisfies the usual differential equation

$$\left(-\frac{d^2}{dr^2} + V(r) - E\right)R_E(r, t) = \delta(r - t) \tag{II.3.4}$$

and a similar one in t, and the boundary condition

$$R_E(0, t) = 0. \tag{II.3.5}$$

The boundedness character of R_E follows from the bounds (**I.2.8a**) and (**I.3.2**), which imply that, for k in Im > 0,

$$|R_E(r, t)| < Cr(1 + |k|r)^{-1}e^{-\operatorname{Im}k|r - t|}. \tag{II.3.6}$$

This shows clearly that R_E defines a bounded operator in $L^2(0, \infty)$, i.e., it is the resolvent operator

$$R_E = \frac{1}{D - \lambda I}. \tag{II.3.7}$$

The discontinuity across the cut $E \geq 0$ and the residues at the poles $E_j = -\gamma_j^2$ determine the spectral function of the operator D with the help of which we can now formulate the completeness relation for the solutions $\varphi(E, r)$, $E \geq 0$, and $\varphi_j(r) = \varphi(E_j, r)$, $j = 1, 2, \ldots, n$, associated with the differential operator (**II.3.1**) and the boundary conditions (**II.3.2**). This is formulated in the following theorem.

Theorem 1. *The functions* $\varphi(E, r)$, $E \geq 0$, *and* $\varphi_j(r)$ *form a complete orthogonal system, and the completeness relation for these can be written, symbolically,*

$$\frac{1}{\pi} \int_0^\infty \varphi(E, r)\varphi(E, t)|F(k)|^{-2}E^{1/2} \, dE + \sum_j C_j\varphi_j(r)\varphi_j(t) = \delta(r - t), \tag{II.3.8}$$

where the normalization constants are given by formula (**II.1.8**) *for each* $k_j = i\gamma_j$:

$$C_j = -\frac{2i\gamma_j f'(i\gamma_j, 0)}{F(i\gamma_j)}. \tag{II.3.9}$$

In the free case, when the potential vanishes, (**II.3.8**) reduces to the usual completeness relations for Fourier sine transforms

$$\frac{2}{\pi} \int_0^\infty \frac{\sin kr}{k}\frac{\sin kt}{k} k^2 \, dk = \delta(r - t). \tag{II.3.10}$$

The resolvent R_E considered above is of course identical to the resolvent of the Fredholm integral equation (**II.1.9**) for the physical solution defined by $\psi(k, 0) = 0$ and (**I.1.11**), and the Jost function $F(k)$ is, as we saw before, its

Fredholm determinant. Remembering (**I.1.26**), we see that (**II.3.8**) can be written in the form

$$\frac{2}{\pi} \int_0^\infty \psi(k, r)\psi(k, t)dt + \sum_j C_j' \psi_j(r)\psi_j(t) = \delta(r - t) \qquad \textbf{(II.3.11)}$$

which is the exact analog of (**II.3.10**), ψ_j being now the solution of the homogeneous integral equation associated with (**II.1.9**), and

$$C_j' \int_0^\infty \psi_j^2(r)dr = 1. \qquad \textbf{(II.3.12)}$$

Since we are dealing with continuous spectra, $\{\psi(k, r)\}$ are not elements of the Hilbert space, and therefore are not eigenfunctions in the usual sense. To attach meaning to them, we must consider them as kernels of integral transforms which diagonalize the self-adjoint operator D corresponding to (**II.3.1**), exactly as $(\sin kr/k)$ is the kernel of the Fourier sine transform corresponding to $-\varphi'' = k^2\varphi$ together with (**II.3.2**). We shall not enter into the details of such transformations since we shall not use them in this book.

The symbolic relations (**II.3.8**)–(**II.3.11**) are of course to be understood in the sense of L^2 space. Given any function $f(r)$ which is twice continuously differentiable and vanishes outside some finite interval not containing the origin, it can be shown with the use of the resolvent R_E and complex contour integration à la Titchmarsh (Titchmarsh, 1962; Jorgens, 1964) that

$$f(r) = \frac{2}{\pi} \int_0^\infty |F(k)|^{-2} k^2 \varphi(k, r) \left[\int_0^\infty \varphi(k, t) f(t)dt \right] dk$$

$$+ \sum_j C_j \varphi_j(r) \left[\int_0^\infty \varphi_j(t) f(t)dt \right]. \qquad \textbf{(II.3.13)}$$

Since these functions $f(r)$ are dense in $L^2(0, \infty)$, it follows that (**II.3.13**) holds in the L^2 sense, and we obtain (**II.3.8**). For later use, we shall write it in the condensed form

$$\int_{-\infty}^\infty \varphi(E, r)\varphi(E, t)d\rho(E) = \delta(r - t), \qquad \textbf{(II.3.14a)}$$

where

$$\frac{d\rho(E)}{dE} = \begin{cases} \dfrac{1}{\pi} \dfrac{E^{1/2}}{|F(E^{1/2})|^2}, & E \geq 0, \\[2ex] \sum_j C_j \delta(E - E_j), & E < 0. \end{cases} \qquad \textbf{(II.3.14b)}$$

For higher waves $(l \geq 1)$, everything is quite similar, except that now, because of the presence of the centrifugal potential in the Schrödinger equation,

$$-\varphi'' + \frac{l(l + 1)}{r^2} \varphi + V(r)\varphi = E\varphi \qquad \textbf{(II.3.15)}$$

the origin $r = 0$ is a limit point, i.e., there is only one solution which is L^2 there, namely the regular solution φ_l, whereas the Jost solution f_l is singular. Except for this point, everything else is quite similar to the case $l = 0$, as was seen in detail in chapter I, section 5. The normalization constants of the bound states are given now by (**II.1.13b**)

$$
\begin{aligned}
C_j^{-1} &= \int_0^\infty \varphi_j^2(r)dr \\
&= \frac{\dot{F}(i\gamma_j)}{2i\gamma_j^{l+1}} \left\{ (2l+1)!! \lim_{r\to 0} r^{-l-1} f_l(i\gamma_j, r) \right\}.
\end{aligned}
\qquad \text{(II.3.16)}
$$

To avoid unnecessary complications, which could easily be taken care of, we again assume $F_l(0) \neq 0$ (no bound state at zero energy). The completeness relation for the eigenfunctions reads now, exactly as before,

$$
\int_{-\infty}^{\infty} \varphi_l(E, r)\varphi_l(E, r')d\rho_l(E) = \delta(r - r'),
\qquad \text{(II.3.17a)}
$$

$$
\frac{d\rho_l(E)}{dE} = \begin{cases}
\dfrac{1}{\pi} (E)^{l+1/2} |F_l(E^{1/2})|^{-2}, & E \geq 0, \\[2ex]
\sum_j C_j \delta(E - E_j), & E < 0.
\end{cases}
\qquad \text{(II.3.17b)}
$$

In the free case, the analogue of (**II.3.10**) is the well-known formula

$$
\frac{2}{\pi} \int_0^\infty u_l(kr)u_l(kt)dk = \delta(r - t),
\qquad \text{(II.3.18)}
$$

where $u_l = (\pi kr/2)^{1/2} J_{l+1/2}(kr)$. If we use the physical solution ψ_l instead of φ_l, it is seen from (**I.1.26**) that (**II.3.11**) holds, as it is, for all l. All these results are valid under the general assumptions given at the beginning of this chapter (in particular, the absolute integrability of $rV(r)$ at the origin is not needed).

II.4   Miscellaneous Results

Some Integral Representations

For later use, we shall study here some integral representations for the Jost function F and the S-matrix, and some integral operators defined with the help of the spectral density $d\rho_l$ introduced in the previous section. First of all, we have, assuming either $rV(r) \in L^1(0, \infty)$, or (**I.6.2a**), (**I.6.2b**), and $rV(r) \in L^1(b, \infty)$, $b > 0$,

$$
F(k) = 1 + \int_0^\infty \Gamma(t)e^{ikt}\, dt,
\qquad \text{(II.4.1)}
$$

where $\Gamma(t) \in L^1(0, \infty)$. This representation can be shown most simply by using the Levin representation of the Jost solution, (V.1.5), which will be shown in chapter V, and which is the starting point of the Marchenko method for solving the inverse problem. Putting $r = 0$ in that representation, we find $\Gamma(t) = A(0, t)$. The upper bound (V.1.7) ensures then that Γ belongs to $L^1(0, \infty)$. From (II.4.1), we can now establish analogous representations for the S-matrix, the spectral density $d\rho_l(E)$, and the phase shift. These are all obtained with the help of the Wiener–Lévy theorem, which is as follows.

Theorem 2. *Let $G(z)$ be analytic in a domain D of the complex plane and let $F(k)$ be such that the curve $z = F(k)$, $-\infty \leq k \leq \infty$, lies inside D. If $F(k)$ is representable in the form*

$$F(k) = C + \int_{-\infty}^{\infty} f(t)e^{ikt}\, dt, \qquad k \text{ real,}$$

with $f(t) \in L^1(-\infty, \infty)$, then $G(F(k))$ also possesses a similar representation.

Notice that, because of the Riemann–Lebesgue theorem on Fourier transforms of integrable functions, $C = F(\infty)$.

This theorem applies directly to the S-matrix and to $|F(k)|^{-2}$, and leads to[2]

$$|F(k)F(-k)|^{-1} = |F(k)|^{-2} = 1 + \int_{-\infty}^{\infty} H(t)e^{ikt}\, dt$$

$$= 1 + 2\int_{0}^{\infty} H(t)\cos kt\, dt, \qquad k \text{ real,} \qquad \textbf{(II.4.2)}$$

and

$$S(k) = \frac{F(-k)}{F(k)} = 1 + \int_{-\infty}^{\infty} \Sigma(t)e^{-ikt}\, dt, \qquad \textbf{(II.4.3)}$$

where both $H(t)$ and $\Sigma(t)$ are integrable on the real line. In both these formulae, k is of course real since in general $|F|^{-2}$ and $S(k)$ have no analyticity properties in k. The reason for taking $\exp(-ikt)$ in (II.4.3) is because of later use of this formula in the Marchenko method (chapter V).

The theorem of Wiener–Lévy can also be applied to the phase shift itself. If there are no bound states, we can choose $\delta(-\infty) = \delta(\infty) = 0$ according to the Levinson theorem, and we have, from (II.4.3) and $\delta(-k) = -\delta(k)$, k positive,

$$\delta(k) = \frac{i}{2} \text{Log } S(k) = -\int_{0}^{\infty} \gamma(t)\sin kt\, dt, \qquad \textbf{(II.4.4)}$$

---

[2] We write $|F(k)|^{-2} - 1 = [F^{-1}(k) - 1][F^{-1}(-k) - 1] + [F^{-1}(k) - 1] + [F^{-1}(-k) - 1]$, and apply the theorem separately to each term, using the well-known fact (see Titchmarsh, 1948, chapter II, theorems 40 and 41) that the Fourier transform of the product of two functions whose Fourier transforms belong to $L^1(-\infty, +\infty)$ is also a function in $L^1(-\infty, +\infty)$. We proceed similarly with the S-matrix.

where again $\gamma(t)$ is in $L^1(0, \infty)$. From this, using the theory of reciprocal Hilbert transforms (see Titchmarsh, 1948, chapter V), we deduce that

$$|F(k)| = \exp\left[\int_0^\infty \gamma(t)\cos kt \, dt\right], \qquad k \text{ real}, \qquad \text{(II.4.5)}$$

and

$$F(k) = \exp\left[\int_0^\infty \gamma(t)e^{ikt} \, dt\right] \qquad \text{(II.4.6)}$$

which relate again the phase shift to the Jost function, and provide us an alternative for (II.2.1). When bound states are present, we can factor them out by using (remember (II.2.2) and (II.2.4))

$$\tilde{S}(k) = e^{2i\delta(k)} = S(k) \prod_j \left(\frac{k - i\gamma_j}{k + i\gamma_j}\right). \qquad \text{(II.4.7)}$$

To $\tilde{S}(k)$, we can apply the Wiener–Lévy theorem because now $\tilde{\delta}(-\infty) = \tilde{\delta}(0) = \tilde{\delta}(\infty) = 0$, and write $\tilde{\delta}(k)$ in the form (II.4.4). We can then calculate $\tilde{F}$ through (II.4.6), and obtain

$$F(k) = \tilde{F}(k) \prod_j \left(\frac{k - i\gamma_j}{k + i\gamma_j}\right). \qquad \text{(II.4.8)}$$

The Povzner–Levitan Kernel

For later use (chapter III, section 1), we study the properties of the function

$$\tilde{\psi}(r, t) = \int_0^\infty \cos kt \left[\varphi(k, r) - \frac{\sin kr}{k}\right] dk \qquad \text{(II.4.9)}$$

which plays an important role in the Gel'fand–Levitan theory. It is a well-defined continuous function of r and t because the integrand is, according to (I.2.11), an integrable function. We consider it in the domain $0 \le t \le r$. If we substitute (I.2.2) in the above definition, and change the orders of integrations, which is legitimate because we are dealing with absolutely convergent integrals, we find

$$\tilde{\psi}(r, t) = \int_0^r V(u)\Psi(r, t; u)du, \qquad \text{(II.4.10a)}$$

$$\Psi(r, t; u) = \int_0^\infty \cos kt \frac{\sin k(r - u)}{k} \frac{\sin ku}{k} dk$$

$$+ \int_0^\infty \cos kt \frac{\sin k(r - u)}{k} dk \left[\varphi(k, u) - \frac{\sin ku}{k}\right] dk = \Psi_1 + \Psi_2.$$

$$\text{(II.4.10b)}$$

We shall see that what is important is the derivative of $\tilde{\psi}(r, t)$ with respect to t. Let us therefore consider the derivative of (II.4.10b) with respect to t. It is easily seen, on the basis of (I.2.8b), that the second expression in the right-hand side of (II.4.10b) can be differentiated with respect to t under the integral sign, and leads to a finite result which is bounded by a constant multiplied by u. If this, in turn, is substituted in (II.4.10a), one obtains an absolutely convergent integral because, by assumption, $uV(u)$ is integrable at the origin. It remains therefore to study the first term in the right-hand side of (II.4.10b). This is an absolutely convergent integral whose derivative with respect to t is finite everywhere, and is given by

$$t < r: \qquad \frac{\partial}{\partial t} \Psi_1(r, t; u) = 0 \qquad \text{for} \quad 0 \le u < r - t,$$

$$t = r: \qquad \frac{\partial}{\partial t} \Psi_1(r, t; u) = \frac{\pi}{8} \qquad \text{if} \quad u = 0, \tag{II.4.11}$$

$$= \frac{\pi}{4} \qquad \text{if} \quad 0 < u < 2r.$$

Collecting all these results, we get the following

Lemma. *If the potential is integrable at the origin, $\tilde{\psi}(r, t)$ has a finite derivative with respect to t in $0 \le t \le r$. If, however, the potential is not integrable at the origin, $\partial_t \tilde{\psi}$ is finite for $0 \le t < r$, and infinite for $r = t$ in general.*

A related result, which is also useful in chapter III, concerns the function

$$G(r, t) = \frac{2}{\pi} \int_0^\infty \sin kr \, \sin kt [|F(k)|^{-2} - 1] dk$$

$$= H(r - t) - H(r + t), \tag{II.4.12}$$

where H is the function introduced in (II.4.2). From its definition, G is symmetric in its arguments.

On the basis of the asymptotic properties of the Jost function at large energies, it is easily seen, using the Abel theorem on the convergence of integrals whose integrand is oscillating, that G is finite for all $t < r$, but becomes in general infinite for $t = r$. However, when the potential is integrable at the origin, G is also finite at $t = r$. A rigorous proof of this fact can be given on the basis of (I.4.6'). In all cases, $G(r, t)$ is integrable in t everywhere, including $t = r$. This is obvious on (II.4.12) since we know that H is an integrable function.

To show that G is finite at $t = r$ when the potential is integrable at the origin, we use the fact that, in (I.4.4), the integral goes to zero when k goes to infinity, a fact which was shown before [see (I.4.6)]. It follows that, for large values of k,

$$|F(k)|^2 - 1 \simeq \int_0^\infty 2 \cos kr V(r) \varphi(k, r) dr. \tag{II.4.13}$$

Substitution of this in (**II.4.12**) leads to

$$G(r, t) \simeq 2 \int_0^\infty \sin kr \sin kt \, dk \int_0^\infty \cos ku V(u)\varphi(k, u)du. \qquad \textbf{(II.4.14)}$$

The important point here is the convergence of the k-integral at infinity. In the right-hand side of this expression, we therefore substitute φ by its asymptotic form $(\sin ku)/k$. Now introducing W, the integral of the potential (**I.6.1**), and integrating the u-integral by parts, we find (remember that $rW(r)$ vanishes at $r = 0$, and at infinity)

$$G(r, t) \simeq \int_0^\infty [\cos k(r - t) - \cos k(r + t)]dk \int_0^\infty W(u)\cos 2ku \, du$$

$$= W\left(\frac{r - t}{2}\right) - W\left(\frac{r + t}{2}\right). \qquad \textbf{(II.4.15)}$$

This shows several facts clearly. Since, by assumption, W exists and is continuous if its argument is strictly positive, we see that $G(r, t)$ is continuous for $t < r$. When the potential is integrable at the origin, $W(r)$ is also continuous at $r = 0$, so that $G(r, t)$ is finite even at $t = r$. Finally, it is clear from (**II.4.2**) and (**II.4.12**), that the dominant part of H is just W.

The above analysis was carried out on the dominant parts (in k) of the formulas (**II.4.12**) and (**II.4.14**). However, with some extra work, using (**I.2.9a,b,c, and d**), one can show that the contributions of the remainders do not alter the conclusions. It can be shown quite rigorously that if

$$V(r) \sim r^{-a}, \qquad 1 < a < 2, \qquad r \to 0 \qquad \textbf{(II.4.16a)}$$

then the singularity of $G(r, t)$ at $t = r$ is given by

$$G(r, t) = |r - t|^{1 - a} + \cdots. \qquad \textbf{(II.4.16b)}$$

We shall see in the next chapter that the kernel (Povzner–Levitan)

$$K(r, t) \equiv \partial_t \tilde\psi(r, t)$$

$$= -\frac{1}{\pi} \int_0^\infty \left[\varphi(k, r) - \frac{\sin kr}{k}\right] k \sin kt \, dk \qquad \textbf{(II.4.17)}$$

plays a very important role in the Gel'fand–Levitan theory.

A last formula which will be used later (chapter V) is the relation between the normalized bound state wave function (of energy $-\gamma^2$, S-wave)

$$\psi_0(r) = (C_0)^{1/2}\varphi(i\gamma, r),$$

where C_0 is defined by (**II.1.8**), and the Jost solution $f(i\gamma, r)$. According to (**II.1.2**) and (**II.1.8**), we have

$$\psi_0(r) = s_0 f(i\gamma; r), \qquad \textbf{(II.4.18a)}$$

where

$$s_0 = \left[\frac{2i\gamma}{-\dot F(i\gamma)f'(i\gamma, 0)}\right]^{1/2}. \qquad \textbf{(II.4.18b)}$$

For higher waves there is an exact analogue of theorem 2 using the Hankel transform, i.e.,

$$e^{ikx} \rightarrow e^{-i\pi l/2} w_l(kx). \tag{II.4.19}$$

It can be obtained either directly or by using the integral representation of the exponential in terms of Hankel functions, which are well known. This leads to integral representations quite similar to **(II.4.1)–(II.4.6)**, where we now have to use $e^{-i\pi l/2} w_l$, $-e^{i\pi l/2} v_l$, and $e^{i\pi l/2} u_l$, respectively, instead of exponential, cosine, and sine. The function $v_l(z)$ is $\sqrt{(\pi z/2)} N_{l+1/2}(z)$, so that

$$w_l(z) = e^{i\pi(l+1)} [w_l(z) - i u_l(z)]. \tag{II.4.20}$$

We have to remember that, although w_l and v_l are singular like z^{-l} at the origin, there is no problem in defining the above integral representations for the S-matrix, Jost function, etc., for short-range potentials which decrease at infinity faster than any power of r. Indeed, in this case, we have, for all l, the well-known expansion

$$k^{2l+1} \cot g\delta_l = -\frac{1}{a_l} + \cdots, \tag{II.4.21}$$

which can be obtained either directly, or from **(I.5.17)**, using the real part of w_l. This takes care of the singularities of w_l and v_l. If the potential decreases like $r^{-3-\alpha}$, then we again have **(II.4.21)** for $2l < \alpha$. For higher l, we have to make the necessary subtractions, i.e., to deal with generalized functions, etc.

Concerning the Povzner–Levitan kernel, **(II.4.9)**, similar remarks apply for higher l. We have to deal with the integral

$$\partial_t \int_0^\infty [\varphi_l - k^{-l-1} u_l(kr)] u_l(kt) dk \tag{II.4.22}$$

from which we reach similar conclusions.

II.5   Singular Potentials

We consider here the case of short-range potentials which have a truly strong singularity at the origin (outside the W-class studied at the end of chapter I). The singularities at the origin can be classified into three classes. The first class is a singularity which is repulsive. Examples are

$$V(r) \underset{r\to 0}{\sim} gr^{-n}, \quad n > 2, g > 0; \qquad V(r) \underset{r\to 0}{\sim} ge^{ar^{-n}}, \quad n, a, g > 0, \tag{II.5.1}$$

etc. The second class contains potentials which are attractive ($g < 0$). Finally, the third class is made up of very nasty singular and violently oscillating potentials, like the one devised by Pearson (1975). For the second and third classes, it is well known that there are several difficulties either with the definition of the Hamiltonian as a unique self-adjoint operator, or with the asymptotic completeness of wave operators. We shall therefore consider only the first class, i.e., singular repulsive potentials, for which the scattering theory is well defined. We shall assume that the potential is of the form

$$V(r) = V_s(r) + V_1(r), \qquad \lim_{r \to 0} r^2 V_s(r) = \infty, \qquad V_1(r) \in W\text{-class}, \quad \textbf{(II.5.2)}$$

where V_s is the singular part similar to **(II.5.1)**.

It is well known that for singular repulsive potentials the S-matrix exists and is unitary, which means that the phase shift $\delta(k)$ is a real continuous function of the momentum k. Moreover, it can be shown that the S-matrix is obtained as the limit of the S-matrix for the regularized potential. We think here of regularizing procedures like

$$gr^{-n} = \lim_{\varepsilon \downarrow 0} g(r + \varepsilon)^{-n}; \qquad gr^{-n} = \lim_{\varepsilon \downarrow 0} gr^{-n}\theta(r - \varepsilon), \qquad \textbf{(II.5.3)}$$

or any other regularizing procedure. When the potential is regularized at the origin, we have of course the usual decomposition $S_\varepsilon(k) = F_\varepsilon(-k)F_\varepsilon^{-1}(k)$, where F_ε is the Jost function. When we let $\varepsilon \downarrow 0$, we recover the S-matrix of the singular potential. The difference between the regular (W-class) and the truly singular repulsive potentials is that, at the limit $\varepsilon = 0$, F_ε tends uniformly to the Jost function of the limiting potential in the case of potentials of W-class, whereas F_ε tends to infinity in the case of singular potentials. In other words, it is only the ratio as a whole which has a finite limit in the case of singular potentials.

Another difference is that, in the (regular) case of the W-class, the phase shift tends to zero at $k = \infty$ (at high energies the potential is negligible compared to the kinetic energy), whereas for singular repulsive potentials the phase shift tends to $-\infty$. As an example, we have, for the potential gr^{-n}, the high-energy behavior

$$\delta(k)_{k \to \infty} = -\alpha_n k^{(n-2)/n} + \cdots, \qquad \alpha_n = -\tfrac{1}{2}g^{2/n} B\left(\frac{1}{2}, \frac{n-1}{n}\right), \quad \textbf{(II.5.4)}$$

where B is the beta function.

This fact, and the fact that $F_\varepsilon(k)$ does not have a finite limit for such potentials when $k = \infty$, are intimately related, as will be seen below. As before, let us notice that the most rapid growth of the phase shift at large energies is linear, corresponding to $n = \infty$ in **(II.5.1)**, i.e., a hard core, which is the most singular case.

For regular potentials belonging to the W-class, we saw previously the integral representation

$$F(k) = \exp\left[\int_0^\infty \gamma(t)e^{ikt}\,dt\right], \qquad \gamma(t) \in L^1(0, \infty). \qquad \textbf{(II.5.5)}$$

The phase shift being minus the phase of the Jost function, we have (k real!)

$$\delta(k) = -\int_0^\infty \gamma(t)\sin kt\,dt, \qquad |F(k)| = \exp\left[\int_0^\infty \gamma(t)\cos kt\,dt\right]. \quad \textbf{(II.5.6)}$$

These representations show that when $k = \infty$, the phase shift is zero, whereas the modulus of the Jost function is 1 (by the Riemann–Lebesgue theorem).

Now consider the singular case. When we regularize the potential at the origin we obtain, for its Jost function $F_\varepsilon(k)$, a representation of the form (II.5.5), with $\gamma_\varepsilon(t) \in L^1(0, \infty)$. Now suppose we let $\varepsilon \downarrow 0$. As was said before, F_ε does not have a limit. However, the S-matrix has a limit:

$$\lim_{\varepsilon \downarrow 0} S_\varepsilon(k) = \lim \exp[2i\, \delta_\varepsilon(k)] = e^{2i\delta(k)}.$$

According to formula (II.5.6), this means that the limiting function $\gamma(t)$ is such that

$$\Gamma(t) = -\int_t^\infty \gamma(u)du \tag{II.5.7}$$

is negative near $t = 0$, and is singular there (weaker than t^{-1}). Indeed, before taking the limit $\varepsilon = 0$, we can integrate by parts, and write (II.5.6) as

$$\delta_\varepsilon(k) = k \int_0^\infty \Gamma_\varepsilon(t)\cos kt\, dt. \tag{II.5.8}$$

This shows, according to standard theorems on the asymptotic behavior of Fourier sine and cosine transforms (Titchmarsh (1948), see "Comments and References" at the end of this chapter), that the limiting function $\Gamma(t)$ must satisfy the above conditions if we demand that the high-energy behavior of the phase shift should be of the type (II.5.4), i.e., a negative main term with a behavior less singular than k. Notice that if $\Gamma(0)$ is finite, the phase shift does not become infinite when k becomes very large. Therefore, $F_\varepsilon(0) = \exp[\int_0^\infty \gamma_\varepsilon(t)dt]$ becomes infinite when $\varepsilon \downarrow 0$. However,

$$|\tilde{F}_\varepsilon(k)| \equiv |F_\varepsilon(k)| F_\varepsilon^{-1}(0) = \exp\left[\int_0^\infty \gamma_\varepsilon(t)(\cos kt - 1)dt\right]$$

$$= \exp\left[k \int_0^\infty \Gamma_\varepsilon(t)\sin kt\, dt\right] \tag{II.5.9}$$

has a finite limit, $|\tilde{F}(k)|$, when $\varepsilon \downarrow 0$. Again, using the same asymptotic theorems as before, we can now show that this limiting function has the asymptotic form

$$|\tilde{F}(k)| \simeq \exp[-\alpha k^{1-\eta}] \tag{II.5.10}$$

when k is large, with $\alpha > 0$ and $0 < \eta < 1$.

We call this function the modified Jost function. It is given by

$$\tilde{F}(k) = \exp\left[-ik \int_0^\infty \Gamma(t)e^{ikt}\, dt\right]. \tag{II.5.11}$$

As in the regular case, the phase shift is now given by minus the phase of the modified Jost function:

$$\delta(k) = k \int_0^\infty \Gamma(t)\cos kt\, dt. \tag{II.5.12}$$

For the regularized potential, the regular solution of the Schrödinger equation satisfying $\varphi(0) = 0$ and $\varphi'(0) = 1$ is given by

$$\varphi_\varepsilon(k, r) = (2ik)^{-1}[F_\varepsilon(-k)f_\varepsilon(k, r) - F_\varepsilon(k)f_\varepsilon(-k, r)], \qquad \text{(II.5.13)}$$

where f_ε is the Jost solution. In fact, since the Jost solution is defined only by its asymptotic behavior at $r = \infty$, it does not depend on ε except at very short distances if we use $V_\varepsilon = V\theta(r - \varepsilon)$. For all finite values of r, the Jost solution of the true potential is the same as that of the regularized potential. Therefore, according to our analysis, the limit of the regularized wave function $\tilde{\varphi}_\varepsilon(k, r) = F_\varepsilon^{-1}(0)\varphi_\varepsilon(k, r)$ exists for all $r > 0$ as $\varepsilon \downarrow 0$, and defines our regular solution for the true potential:

$$\tilde{\varphi}(k, r) = (2ik)^{-1}[\tilde{F}(-k)f(k, r) - \tilde{F}(k)f(-k, r)] \qquad \text{(II.5.14)}$$

exactly as in the regular case. However, here, obviously

$$\tilde{\varphi}'(k, 0) = \lim_{\varepsilon \downarrow 0} F_\varepsilon^{-1}(0)\varphi_\varepsilon'(k, 0) = 0$$

as expected, since for singular repulsive potentials we have $\varphi(0) = \varphi'(0) = \cdots = 0$. To see exactly how $\tilde{\varphi}$ is normalized, we have to look at a point $r \neq 0$. We choose $r = \infty$. For r large enough, we have

$$\tilde{\varphi} \simeq k^{-1}|\tilde{F}(k)|\sin(kr + \delta). \qquad \text{(II.5.15)}$$

Now using the fact that, because of (II.5.9), $\tilde{F}(0) = \lim_{\varepsilon \downarrow 0} \tilde{F}_\varepsilon(0) \equiv 1$ we have, for large values of r,

$$\tilde{\varphi}(0, r) \simeq r - a_0, \qquad \text{(II.5.16)}$$

where a_0 is the scattering length. This gives the exact normalization of the regular solution of the radial equation at zero energy. Once this is given, we have, for other values of energy,

$$\tilde{\varphi}(k, r) = \tilde{\varphi}(0, r) - k^2 \int_0^r G_v(r, r')\tilde{\varphi}(k, r')dr', \qquad \text{(II.5.17)}$$

where G_v is the appropriate total "Green's function" at zero energy.

Now consider the completeness of the solutions of the radial equation. For the regularized potential, we have, as before, assuming for simplicity that there are no bound states,

$$\frac{2}{\pi} \int_0^\infty \varphi_\varepsilon(k, r)\varphi_\varepsilon(k, r')|F_\varepsilon(k)|^{-2}k^2\, dk = \delta(r - r'). \qquad \text{(II.5.18)}$$

Dividing φ_ε and F_ε by $F_\varepsilon(0)$, and taking the limit, the above relation becomes

$$\frac{2}{\pi} \int_0^\infty \tilde{\varphi}(k, r)\tilde{\varphi}(k, r')|\tilde{F}(k)|^{-2}k^2\, dk = \delta(r - r'). \qquad \text{(II.5.19)}$$

There is therefore no formal change in going from (II.2.14a) to (II.5.19), except that all quantities should be replaced by modified ones.

We shall use these results in chapter III for the inverse problem with two potentials. As a final remark, let us note that, proceeding as above, the Volterra integral equation (**I.2.2**) now becomes homogeneous:

$$\tilde{\varphi}(k, r) = \int_0^r \frac{\sin k(r - r')}{k} V(r')\tilde{\varphi}(k, r')dr'. \qquad \text{(II.5.20)}$$

We can now insert (**II.5.17**) into the right-hand side of this last equation, and obtain a new inhomogeneous equation which reads

$$\tilde{\varphi}(k, r) = \tilde{\varphi}_0(k, r) - k^2 \int_0^r \tilde{G}_V(k; r, r')\tilde{\varphi}(k, r')dr', \qquad \text{(II.5.21)}$$

where

$$\tilde{\varphi}_0(k, r) = \int_0^r \frac{\sin k(r - r')}{k} \tilde{\varphi}(0, r')V(r')dr', \qquad \text{(II.5.22)}$$

and

$$\tilde{G}_V(k; r, r') = \int_{r'}^r \frac{\sin k(r - r'')}{k} V(r'')G_V(r', r'')dr''. \qquad \text{(II.5.23)}$$

In this last equation, we have used the symmetry of G_V with respect to r' and r''. Equation (**II.5.21**) can now be iterated, as was done for (**II.2.2**), and we obtain a convergent series which can be used for the study of the properties of $\tilde{\varphi}$ at high energies, etc.

II.6   Comments and References

The results of this chapter concerning scattering theory can be found in the books by De Alfaro and Regge (1965) and Newton (1982), already referred to in chapter I. For the theory of eigenfunction expansions associated with second-order differential equations, see Coddington and Levinson (1955), Hille (1969), Titchmarsh (1962), Jorgens (1964), and Dunford and Schwartz (1963).

The integral representations of the Jost function, the S-matrix, and the phase shift (**II.4.1**)–(**II.4.4**), as Fourier integrals of absolutely integrable functions, are due to Krein (1955, 1958). See also the review paper by Faddeev (1963).

Again, all these results for $l \geq 0$ are valid under more general assumptions than the absolute integrability of rV at the origin, for the W-class of chapter I, provided the potential is short range, as was mentioned in section II.5. This generalization to W-class is due to Baeteman and Chadan (1975, 1976). See also Chadan and Martin (1979). The bound (**II.1.10b**) is due to Chadan and Grosse (1983).

For potentials having a strong repulsive singularity at the origin, see Chadan (1978), and also Rofe–Beketov and Hristov (1969). For a derivation of (**II.5.4**), see Brander (1981), which contains many references to earlier works. A general study of scattering theory with singular potentials can be found in

Frank et al. (1971), where many soluble examples are given. See also Ivanov et al. (1985), which contains references to earlier work.

The theorem of Titchmarsh, which we have used in connection with (**II.5.8**) and (**II.5.9**), reads as follows: let $f(x)$ and $f'(x)$ be integrable over any finite interval not ending at $x = 0$; let $x^{1+\alpha} f'(x)$ be bounded for all x, and let $f(x) \sim x^{-\alpha}$, $\alpha < 1$, as $x \to \infty$ ($x \to 0$). Then $F_c(x) \sim \sqrt{(2/\pi)} \, \Gamma(1 - \alpha) \sin(\tfrac{1}{2}\pi\alpha) x^{\alpha-1}$ as $x \to 0$ ($x \to \infty$), where F_c is the Fourier cosine transform of f. Similarly for F_s, the Fourier sine transform, with $\sin \tfrac{1}{2}\pi\alpha$ replaced by $\cos \tfrac{1}{2}\pi\alpha$. See Titchmarsh (1948). The Hilbert integral transforms used in deriving (**II.2.1**) are also given in chapter V of this book.

CHAPTER III

The Gel'fand–Levitan–Jost–Kohn Method

III.1   The Povzner–Levitan Representation

Generally speaking, there are two methods for solving the inverse problem at fixed angular momentum. The first one, which uses the regular solution of the radial Schrödinger equation, is that of Gel'fand and Levitan, and Jost and Kohn, and is the subject of the present chapter. The second method is that of Marchenko, which uses the Jost solution. It will be studied in the following chapters. Each one has its own advantages, and permits us to clarify some particular aspect of the problem. The two methods are of course equivalent to each other.

The key to the Gel'fand–Levitan method is an integral representation of the regular solution of the radial Schrödinger equation in terms of its free solution. This is based on the Paley–Wiener theorem (or variants of it, like the Wiener–Boas theorem, see below) for the Fourier transforms of entire functions of exponential type which are L^2 (or L^1) on the real axis.

A function $f(z)$ of the complex variable z is said to be entire if it is holomorphic in the entire finite z-plane. If its growth at infinity can be compared to the growth of $\exp(\alpha|z|)$ where α is positive, then $f(z)$ is said to be of exponential type α. More precisely, if we define

$$M(r) = \max_{0 \le \theta \le 2\pi} |f(re^{i\theta})|, \qquad \text{(III.1.1)}$$

f is a function of exponential type α if and only if (Boas, 1954)

$$\varlimsup_{r \to \infty} [(\text{Log } r)^{-1} \, \text{Log Log } M(r)] = 1 \qquad \text{(III.1.2)}$$

and

$$\overline{\lim_{r \to \infty}} \left[r^{-1} \operatorname{Log} M(r) \right] = \alpha. \qquad \text{(III.1.3)}$$

For these functions, we have the following two theorems (Boas, 1954):

Theorem 1 (Paley–Wiener). $f(z)$ is an entire function of exponential type α and its restriction on the real axis belongs to $L^2(-\infty, \infty)$, if and only if it is given by

$$f(z) = \int_{-\alpha}^{\alpha} \tilde{f}(t) e^{itz} \, dt, \qquad \text{(III.1.4a)}$$

where $\tilde{f}(t)$ is a function of $L^2(-\alpha, \alpha)$. In words, the Fourier transform of the restriction of f to the real axis is an L^2-function of finite support $(-\alpha, \alpha)$.

Theorem 2 (Wiener–Boas). $f(z)$ is an entire function of exponential type α and belongs to $L^1(-\infty, \infty)$ on the real axis if and only if

$$f(z) = \int_{-\alpha}^{\alpha} \tilde{f}(t) e^{itz} \, dt, \qquad \text{(III.1.4b)}$$

where $\tilde{f}(\alpha) = \tilde{f}(-\alpha) = 0$, and the function obtained by extending $\tilde{f}(t)$ to be 0 outside $(-\alpha, \alpha)$ has an absolutely convergent Fourier series on any larger interval $(-\alpha - \varepsilon, \alpha + \varepsilon)$, $\varepsilon > 0$. Notice that here $\tilde{f}(t)$, if we define it by the inverse F.T. of f, is continuous because $f(x)$ belongs to $L^1(-\infty, \infty)$.

Consider now, again for simplicity, the S-wave. The regular solution φ, for each fixed r, has the asymptotic properties (**I.2.10**). It is therefore an entire function of exponential type r in the variable k, and its restriction on the real k-axis is L^2. But we know more. In fact, we know that if the potential satisfies the conditions (**I.6.2a**) and (**I.6.2b**), or condition (**I.1.1**), $\psi = \varphi - \sin kr/k$, according to (**I.2.11**), is L^1 on the real k-axis. We can therefore apply to it the Wiener–Boas theorem, and get (remember that φ is even in k)

$$\psi(k, r) \equiv \varphi(k, r) - \frac{\sin kr}{k} = \int_{-r}^{r} \tilde{\psi}(r, t) e^{ikt} \, dt$$

$$= 2 \int_{0}^{r} \tilde{\psi}(r, t) \cos kt \, dt, \qquad \text{(III.1.5a)}$$

where $\tilde{\psi}(r, t)$ is continuous and satisfies

$$\tilde{\psi}(r, \pm r) = 0. \qquad \text{(III.1.5b)}$$

Inverting the Fourier transform, which is manifestly legitimate, we obtain

$$\tilde{\psi}(r, t) = \frac{1}{\pi} \int_{0}^{\infty} \psi(k, r) \cos kt \, dk. \qquad \text{(III.1.6)}$$

This representation shows that $\tilde{\psi}$ can be differentiated at least once with respect to r or t for $t < r$. Indeed, taking the derivative inside the integral in **(III.1.6)** leads to

$$\frac{\partial}{\partial r} \tilde{\psi}(r, t) = \frac{1}{\pi} \int_0^\infty \frac{\partial}{\partial r} \psi(k, r)\cos kt \, dk \tag{III.1.7a}$$

and

$$\frac{\partial}{\partial t} \tilde{\psi}(r, t) = -\frac{1}{\pi} \int_0^\infty k\psi(k, r)\sin kt \, dk. \tag{III.1.7b}$$

Both integrals exist because of the Abel theorem since the integrands oscillate and vanish in the limit of large k, the first one because of **(I.2.9b)**, and the second because of **(I.2.8b)**. It follows that taking the derivative signs under the integral is justified, and that $\tilde{\psi}$ has first derivatives with respect to t and r which are finite everywhere for $t < r$. The reason for which this property fails in general for $t = r$ is that $k\psi$ and $\partial\psi/\partial r$ contain both sin kr and cos kr in their expansions, so that, in the limit of large k, and for $r = t$, parts of the integrals in **(III.1.7)** and **(III.1.8)** may keep constant signs. And since, in general, if we assume only **(I.6.2a)** and **(I.6.2b)**, or **(I.1.1)**, both factors[1] are only o(1) for large k, the integrals do not converge at the upper limit. At any rate, it was shown at the end of chapter II that $\partial_t\tilde{\psi}(r, t)$ is finite for all $t < r$. It is also finite at $t = r$ if the potential is integrable at the origin, and is generally infinite if we have only **(I.1.1)**.

To obtain the Povzner–Levitan representation for φ, it remains only to integrate **(III.1.5a)** by parts, and use **(III.1.5b)**. The final result is

$$\varphi(k, r) = \frac{\sin kr}{k} + \int_0^r K(r, t) \frac{\sin kt}{k} \, dt, \tag{III.1.8a}$$

where

$$K(r, t) \equiv \frac{\partial}{\partial t} \tilde{\psi}(r, t) = -\frac{1}{\pi} \int_0^\infty \psi(k, r)\sin kt \, k \, dk. \tag{III.1.8b}$$

This representation is the key to the solution of the inverse problem. Notice that the kernel $K(r, t)$ is independent of k, and **(III.1.8a)** should not be confused with the integral equation **(I.2.2)**.

For the validity of the representation **(III.1.8a)**, we must secure the convergence of the integral at both ends. At the lower limit, there is no problem. $K = (\partial/\partial t)\tilde{\psi}|_{t=0}$ is not only finite but actually vanishes. This is obvious from **(III.1.7b)** in which we can take the limit $t = 0$ to obtain

$$K(r, 0) = 0. \tag{III.1.9}$$

On the other hand, at the upper limit $t = r$, K is generally infinite since, as we saw before, the integral in **(III.1.7b)** may not converge at its upper limit

---

[1] According to **(I.2.8b)**, ψ is k^{-1} o(1) for large k. However, differentiating it, we get an extra k factor, and so $\partial\psi/\partial r$ is only o(1) in general. For more details, see the end of chapter II.

when $t = r$. However, what we need is only the integrability of $K(r, t)$ in t at $t = r$. And this is obvious since the integral of K with respect to t is just $\tilde{\psi}(r, t)$, which is finite everywhere in $0 \le t \le r$. Alternatively, we can argue that, starting from (III.1.5a) and proceeding through legitimate steps, we end up at (III.1.8a). The integral in this equation must therefore be well defined.

So far, the representation (III.1.8a) has nothing to do with the Schrödinger equation. To go further, we introduce it into both sides of (I.2.2) and take the Fourier sine transform, assuming $t \le r$ (the domain of definition of K), changing at will the orders of integrations. The result is

$$K(r, t) = \frac{1}{2} \int_{(r-t)/2}^{(r+t)/2} V(s)ds$$

$$+ \int_{(r-t)/2}^{(r+t)/2} ds \int_0^{(r-t)/2} V(s + u)K(s + u, s - u)du. \quad \text{(III.1.10)}$$

We see immediately that both the inhomogeneous term and the kernel of this integral equation are meaningful if $t < r$. We can therefore try to solve it by iteration, as is usually done, and see whether we can justify a posteriori the formal operations we have performed for its obtention.

Let us assume first that the potential is integrable at the origin. It is then easily seen that the iteration leads to a convergent series as the unique solution of (III.1.10), and that

$$|K(r, t)| < \frac{1}{2} \int_{(r-t)/2}^{(r+t)/2} |V(s)|ds \exp\left[\int_0^{(r+t)/2} u|V(u)|du\right]. \quad \text{(III.1.11)}$$

This result shows that $K(r, t)$ is finite everywhere in $0 \le t \le r$; and permits us, with some lengthy but trivial extra work which we shall not reproduce here, to justify the formal steps for going from (I.2.2) and (III.1.8a) to (III.1.10). Putting $t = r$ in this equation, we now deduce

$$K(r, r) = \frac{1}{2} \int_0^r V(t)dt. \quad \text{(III.1.12)}$$

This result is, so far, the most important conclusion we obtain from the Povzner–Levitan representation. It gives a direct method for calculating the potential from the kernel K. Notice that, with our assumption about the integrability of the potential itself at the origin, we can deduce the representation (III.1.8a) directly from the Paley–Wiener theorem (theorem 1). Indeed, we have now (I.2.9c). Therefore, $k\psi(k, r)$ belongs to $L^2(-\infty, \infty)$, and admits a representation of the form (III.1.4). Taking into account the fact that it is an odd function of k, we obtain (III.1.8a). Here, for each fixed r, K belongs to $L^2(0, r)$ in the variable t, but we saw that, in fact, K is finite everywhere in $0 \le t \le r$ according to (III.1.11), a result which could also be shown directly.

Now let us go back to the general case where we have only (I.6.2a) and (I.6.2b), or (I.1.1). In this case, $K(r, r)$ is generally infinite, and (III.1.12) obviously breaks down. Nevertheless, we can still solve (**III.1.10**) by iteration for $t < r$. To see this, let us make the change of variables

$$x = \tfrac{1}{2}(r + t), \qquad y = \tfrac{1}{2}(r - t). \tag{III.1.13}$$

This leads to

$$K(r, t) \equiv L(x, y) = \tfrac{1}{2}(W(x) - W(y))$$

$$+ \int_y^x ds \int_0^y du \; W'(s + u)L(s, u), \tag{III.1.14}$$

where W is an integral of V, which we can take conveniently to be

$$W(r) = - \int_r^\infty V(t)dt. \tag{III.1.15}$$

Now, for $t < r$, we have $y > 0$, so that the inhomogeneous term in the above integral equation is finite, and the kernel W' is integrable because $s > 0$. However, we have to integrate $L(s, u)$ down to $u = 0$, which is a singular point. If we could show that $L(s, u)$ is integrable at $u = 0$, then everything would be meaningful in (**III.1.14**). This can be seen in each term of the series one obtains by iterating this equation, and can be shown to hold for the sum with some extra work. We shall not give here the details since this concerns only the direct problem. As for the integrability of $K(r, t)$ at $t = r$, this was shown on general grounds at the beginning of this section, in the discussion after (**III.1.9**).

When the potential is an ordinary continuous function, we deduce from (**III.1.12**) that

$$V(r) = 2 \frac{d}{dr} K(r, r). \tag{III.1.16}$$

When it is only an integrable function, down to the origin, we have (**III.1.12**), and this is sufficient for scattering theory, as was shown at the end of chapter I, where we saw that what is really important is the integral of the potential, $W(r)$, and not the potential itself. Since we assume the potential to be integrable at infinity, (**III.1.12**) can be written

$$K(R, R) - K(r, r) = \frac{1}{2} \int_r^R V(t)dt$$

$$= \tfrac{1}{2}[W(R) - W(r)]. \tag{III.1.17}$$

In the right-hand side of this equation, no reference is made to the origin $r = 0$, and so one may expect this equation to be more general than (**III.1.12**). One can imagine introducing a regularization parameter ε into the potential, so that $V_\varepsilon(r)$ becomes integrable at the origin when ε is, say, strictly positive. Equation (**III.1.17**) is now certainly valid with subscript ε everywhere. We let now ε go to zero in order to recover the original potential. Since the right-hand side has a finite limit, the same must be true for the left-hand side.

We shall see at the end of the next section that it is easy to introduce this cut-off from the beginning into the scattering data.

We end up this section by mentioning the differential equation satisfied by $K(r, t)$. We assume K to be differentiable twice with respect to r and t, and we calculate φ'' by twice differentiating the formula (III.1.8a). We also calculate $E\varphi$ by integrating the same formula twice by parts. Combining now the results, and using (III.1.9), we find that the Schrödinger equation is satisfied if

$$\left(\frac{\partial^2}{\partial r^2} - \frac{\partial^2}{\partial t^2}\right)K(r, t) = V(r)K(r, t), \qquad 0 \le t \le r \qquad \text{(III.1.18)}$$

and if the potential is given by (III.1.16). This differential equation is of course equivalent to the integral equation (III.1.10) as can be seen by differentiating the latter, which is most easily done using (III.1.14). One can now start a discussion of (III.1.18) by first extending the domain of definition of K to $-r \le t \le r$ with

$$K(r, -t) = -K(r, t) \qquad \text{(III.1.19)}$$

which is the natural choice according to (III.1.8b) and (III.1.7b), and then

$$K(r, -r) = -\int_0^r V(t)dt. \qquad \text{(III.1.20)}$$

The differential equation becomes now a standard hyperbolic equation with boundary values given by (III.1.12) and (III.1.20) on the two characteristics $t = \pm r$. Therefore, it has a unique solution under some assumptions on the potential which would ensure the differentiability of K. In this manner, we see that we can replace the Schrödinger equation by (III.1.18), which is independent of energy, and (III.1.8a). We shall not pursue this procedure here any further, and refer the reader to the existing literature (Gel'fand–Levitan, 1951).

A last equation which will be useful later in the next section is the inverse of (III.1.8a):

$$\frac{\sin kr}{k} = \varphi(k, r) + \int_0^r \tilde{K}(r, t)\varphi(k, t)dt. \qquad \text{(III.1.21)}$$

It is obtained in the following way. We know that $K(r, t)$ is finite in $0 \le t < r$, satisfies (III.1.9) and is integrable in t at $t = r$. We can now consider (III.1.8a) as a Volterra integral equation for $\sin kr/k$, with the inhomogeneous term given by φ. It can be solved again by the method of successive approximations, and leads to the unique solution (III.1.21), as is the case for Volterra equations with the resolvent kernel $\tilde{K}$ finite for all $t < r$, and integrable in t at $t = r$. If we substitute (III.1.8a) in (III.1.21) and take the Fourier sine transform of both sides, we obtain

$$K(r, t) + \tilde{K}(r, t) + \int_t^r \tilde{K}(r, s)K(s, t)ds = 0 \qquad \text{(III.1.22)}$$

which is just the equation for the resolvent kernel $\tilde{K}$ of (III.1.8a) considered as an integral equation for $\sin kr$ (Tricomi, 1957; Pogorzelski, 1964).

III.2   The Gel'fand–Levitan Integral Equation

The Gel'fand–Levitan equation is obtained most simply by combining the completeness relation (**II.3.14a,b**)

$$\int_{-\infty}^{\infty} \varphi(E, r)\varphi(E, t)d\rho(E) = \delta(r - t), \qquad \textbf{(III.2.1a)}$$

$$\frac{d\rho(E)}{dE} = \begin{cases} \pi^{-1}|F(E^{1/2})|^{-2}E^{1/2}, & E \geq 0, \\ \sum_j C_j\delta(E - E_j), & E < 0, \end{cases} \qquad \textbf{(III.2.1b)}$$

with the two reciprocal equations (**III.1.8a**) and (**III.1.21**). From (**III.2.1a**) we see that

$$\int_{-\infty}^{\infty} \varphi(E, r)\varphi(E, r')d\rho(E) = 0, \qquad r \neq r'. \qquad \textbf{(III.2.2)}$$

We now multiply both sides of (**III.1.21**) by $\varphi(E, r_1)$, and integrate with $d\rho(E)$. If $0 \leq r < r_1$, recalling (**III.2.2**), we obtain

$$\int_{-\infty}^{\infty} \frac{\sin kr}{k} \varphi(E, r_1)d\rho(E) = 0. \qquad \textbf{(III.2.3)}$$

Now we multiply both sides of (**III.1.4a**) by φ, and integrate with $d\rho(E)$. Using (**III.2.3**), we find, with a slight change of notations, and for $t < r$,

$$\int_{-\infty}^{\infty} \frac{\sin kr}{k}\frac{\sin kt}{k} d\rho(E) + \int_0^r K(r, s)ds \int_{-\infty}^{\infty} \frac{\sin ks}{k}\frac{\sin kt}{k} d\rho(E) = 0. \quad \textbf{(III.2.4)}$$

If we now use (definition of $d\sigma$)

$$d\sigma \equiv d\rho - d\rho_0 = \begin{cases} d\rho(E) - d\left(\dfrac{2E^{3/2}}{3\pi}\right), & E \geq 0, \\ d\rho(E), & E < 0, \end{cases} \qquad \textbf{(III.2.5)}$$

in the last equation, and remember that

$$\frac{2}{\pi} \int_0^{\infty} \frac{\sin kr}{k}\frac{\sin kt}{k} k^2\, dk = \delta(r - t), \qquad \textbf{(III.2.6)}$$

we obtain the Gel'fand–Levitan equation

$$K(r, t) + G(r, t) + \int_0^r K(r, s)G(s, t)ds = 0, \qquad \textbf{(III.2.7a)}$$

where

$$G(r, t) = \int_{-\infty}^{\infty} \frac{\sin kr}{k}\frac{\sin kt}{k} d\sigma(E). \qquad \textbf{(III.2.7b)}$$

Remember that σ is the difference between the spectral measure ρ and the free spectral measure ρ_0. The kernel G can also be written

$$G(r, t) = \frac{2}{\pi} \int_0^\infty \frac{\sin kr}{k} \frac{\sin kt}{k} \left[\frac{1}{|F(k)|^2} - 1 \right] k^2 \, dk + \sum_j \frac{C_j}{4\gamma_j^2} \sinh(\gamma_j r)\sinh(\gamma_j t)$$

$$= H(r - t) - H(r + t) + \sum_j \cdots, \qquad \text{(III.2.8)}$$

where H is the function introduced in (II.4.2). We studied at the end of chapter II the properties of G. It was shown there that $G(r, t)$ is finite for $t < r$, and integrable at $t = r$. If the potential is integrable at the origin, G is also finite at $t = r$.

Equation (III.2.7a), for each fixed r, is a Fredholm equation for $K(r, t)$ in its second variable t. We have therefore to show that it has a unique solution under the restrictions on $d\rho(E)$ we found in chapter II. Once the solution is found, we have to see whether the potential we obtain has the desired properties. So there are two problems: the unicity of K, and therefore V, and the properties of V. The last problem will be dealt with in Section III.6 below. The unicity of the solution is shown as follows.

We suppose that $d\rho$, and therefore $d\sigma$, are given. For the Fredholm equation (III.2.7a) for the unknown function $K(r, t)$ in the variable t to have a unique solution, we must show that the homogeneous equation

$$\overline{K}(r, t) + \int_0^r \overline{K}(r, s)G(s, t)ds = 0 \qquad \text{(III.2.9)}$$

has no nontrivial L^2 solution. To this end, we multiply it by $\overline{K}(r, t)$, and integrate:

$$J = \int_0^r \overline{K}^2(r, t)dt + \int_0^r ds \int_0^r G(s, t)\overline{K}(r, s)\overline{K}(r, t)dt = 0. \qquad \text{(III.2.10)}$$

On the other hand, we can write (III.2.7b) as

$$G(r, t) = \int_{-\infty}^\infty \frac{\sin kr}{k} \frac{\sin kt}{k} \, d\rho(E) - \delta(r - t)$$

$$= \overline{G}(r, t) - \delta(r - t) \qquad \text{(III.2.11)}$$

so that J becomes

$$J = \int_0^r ds \int_0^r \overline{K}(r, t)\overline{K}(r, s)\overline{G}(s, t)dt$$

$$= \int_{-\infty}^\infty d\rho(E)\left[\int_0^r \overline{K}(r, t)\frac{\sin kt}{k} \, dt \right]^2 = 0. \qquad \text{(III.2.12)}$$

Now the integral inside the brackets is an entire function of k (and of E, since the integrand is even) by virtue of the Paley–Wiener theorem. It is well defined for all values of E, and is real for E real. Since the spectral density $d\rho$ is positive, it follows that

$$\int_0^r \overline{K}(r, t) \sin kt \, dt = 0, \qquad \forall \, k, \qquad \text{(III.2.13)}$$

and therefore that $\overline{K}(r, t) \equiv 0$ in $t \in (0, r)$ for each value of r, which is the desired result. Therefore, the Gel'fand–Levitan equation has a unique solution for every fixed r.

Let us go back to the discussion of the "ε-trick" mentioned in the discussion after (III.1.17). It is now obvious, on the basis of equations (III.2.7a,b), that the differentiability properties of K are related to the differentiability of G, and, therefore, to the decrease of $d\sigma$ when E becomes very large. Given $d\sigma$, we can improve its decrease by multiplying it, for instance, by a factor $1/(1 + \varepsilon k^2)$, ε strictly positive. This, at the same time, guarantees that $G(r, r)$ and $K(r, r)$ with subscript ε exist and are finite. Equation (III.1.17) now becomes meaningful, as long as ε is positive. If we now let ε go to zero, (III.1.17) will remain valid because its right-hand side has a finite limit. We shall see in chapter VI examples with singular potentials.

III.3   Krein's Equation

In order to complete the proof that the potential we obtain from Gel'fand–Levitan theory leads indeed to the Jost function we started from, we must show that the asymptotic form of the wave function φ calculated via (III.1.8a) and (III.2.7a) is given by (I.4.1)

$$\varphi(k, r) = \frac{|F(k)|}{k} \sin(kr + \delta) + o(1), \qquad r \to \infty, \qquad \text{(III.3.1)}$$

where F and δ are related by (II.2.1). This cannot be done easily on (III.1.8a) because of its complicated structure. However, it is possible to rewrite it in an equivalent form, due to Krein, from which (III.3.1) easily follows. The starting point is the Gel'fand–Levitan equation in which we assume, for simplicity, that there are no bound states present. Its kernel is given by (III.2.8) and (II.4.2). If we now introduce the function Γ, solution of Krein's equation

$$\Gamma_{2r}(u) + H(u) + \int_0^{2r} \Gamma_{2r}(s) H(s - u) ds = 0 \qquad \text{(III.3.2)}$$

where H is defined by (II.4.12), it can be seen that if this equation has a solution, then

$$K(r, t) = \Gamma_{2r}(r - t) - \Gamma_{2r}(r + t) \qquad \text{(III.3.3)}$$

satisfies the Gel'fand–Levitan equation. Obviously, if the solution of (III.3.2) is unique, then this equation is equivalent to the Gel'fand–Levitan equation.

This has been shown by Krein in a series of papers (1954, 1958). The advantage of the Krein's equation is that the solution φ can be expressed directly in terms of Γ, and has a form which permits us to take easily the limit $r \to \infty$. Indeed, substituting (**III.3.2**) into (**III.1.8a**), we find

$$\varphi(k, r) = \frac{\sin kr}{k} + \int_0^r \Gamma_{2r}(r - t) \frac{\sin kt}{k} dt$$

$$- \int_0^r \Gamma_{2r}(r + t) \frac{\sin kt}{k} dt$$

$$= \frac{\sin kr}{k} + \int_0^{2r} \Gamma_{2r}(t) \frac{\sin k(r - t)}{k} dt$$

$$= k^{-1} \operatorname{Im}\left[e^{ikr}\left(1 + \int_0^{2r} \Gamma_{2r}(t)e^{-ikt} dt \right)\right]. \qquad \textbf{(III.3.4)}$$

Taking now the limit, we find

$$\varphi = k^{-1} \operatorname{Im}\left[e^{ikr}\left(1 + \int_0^\infty \Gamma(t)e^{-ikt} dt \right)\right] + o(1), \qquad \textbf{(III.3.5)}$$

where $\Gamma(t) = \Gamma_\infty(t)$ is the solution of the limit of (**III.3.2**), i.e.,

$$\Gamma(u) + H(u) + \int_0^\infty \Gamma(s)H(s - u)ds = 0. \qquad \textbf{(III.3.6)}$$

To solve this equation, we notice that, for real values of k,

$$(|F(k)|^{-2} - 1)(F(k) - 1) + (F(k) - 1) + (|F(k)|^{-2} - 1) = (F(-k))^{-1} - 1.$$
$$\textbf{(III.3.7)}$$

We know by assumption (no bound states) that $F(-k)$ is free of zeros in the lower half-plane. We can then take the Fourier transform of both sides of (**III.3.7**), and close the contour by a large circle in the lower half-plane. If we remember now the definition (**II.4.2**) and (**II.4.1**), we obtain (**III.3.6**). It follows that its solution is given by

$$\Gamma(t) = (2\pi)^{-1} \int_{-\infty}^\infty (F(k) - 1)e^{-ikt} dk. \qquad \textbf{(III.3.8)}$$

It then follows immediately from (**III.3.5**) that, as r becomes very large,

$$\varphi \to k^{-1} \operatorname{Im}[F(-k)e^{ikr}] \qquad \textbf{(III.3.9)}$$

which is identical to (**III.3.1**) (remember (**I.1.21**) and (**I.1.24**)).

III.4   Higher Waves

We have already seen in detail in chapters I and II the properties of the various solutions of the Schrödinger equation (**I.1.6**), as well as the eigenfunction expansion associated with the regular solution φ_l. In short, the

Jost function and the S-matrix have essentially the same analyticity and asymptotic properties as for $l = 0$. Of course, for $l > 0$, the free solutions (Bessel and Hankel functions, etc.) are a little more complicated than sine and cosine. However, their asymptotic properties are closely related to sine and cosine, and this was what really mattered in the previous sections. Now, instead of dealing with Fourier sine, cosine, and exponential transforms, we have to deal with integral transforms defined by means of Bessel and Hankel functions. These are all well known, and, except for some algebraic changes, there is no extra work to be done in order to go from $l = 0$ to $l > 0$.

The Povzner–Levitan representation reads again, in the notations of **(I.1.7)** and **(I.1.10)**,

$$\varphi_l(k, r) = \varphi_l^{(0)}(k, r) + \int_0^r K_l(r, t)\varphi_l^{(0)}(k, t)dt, \qquad \textbf{(III.4.1a)}$$

$$\varphi_l^{(0)}(k, r) = k^{-l-1}u_l(kr). \qquad \textbf{(III.4.1b)}$$

Combining the above representation with the completeness relation **(II.3.17a)**, we find

$$K_l(r, t) + G_l(r, t) + \int_0^r K_l(r, s)G_l(s, t)ds = 0, \qquad \textbf{(III.4.2a)}$$

$$G_l(r, t) = \frac{2}{\pi} \int_0^\infty \varphi_l^{(0)}(k, r)\varphi_l^{(0)}(k, t)k^{2l+2}[\,|F_l(k)|^{-2} - 1]dk$$

$$+ \sum_j C_j \varphi_j(i\gamma_j, r)\varphi_l(i\gamma_j, t). \qquad \textbf{(III.4.2b)}$$

The proof of the uniqueness of the solution of this equation is exactly the same as for the case $l = 0$.

One can also obtain differential equations similar to **(III.1.18)**, which would now read

$$\left[\left(\frac{\partial^2}{dr^2} - \frac{l(l+1)}{r^2}\right) - \left(\frac{\partial^2}{dt^2} - \frac{l(l+1)}{t^2}\right)\right]K_l(r, t) = V(r)K_l(r, t). \quad \textbf{(III.4.3)}$$

The potential is again given by

$$V(r) = 2\frac{d}{dr} K_l(r, r) \qquad \textbf{(III.4.4)}$$

as can easily be verified. We see that there is really no essential difference between the case $l = 0$ and $l > 0$. We shall see in fact, in the next chapter, together with some applications of the Gel'fand–Levitan theory, a theorem due to Krein and Marchenko which states essentially that a function $S_l(k) = \exp 2i\delta_l$ associated with some angular momentum l and a potential $V(r)$ can also be the S-matrix of any angular momentum $l' = 0, 1, 2, \ldots$, and a potential $V_{l'}(r)$, which behaves at $r = 0$ and $r = \infty$ exactly as does $V(r)$.

III.5   More General Equations

So far, our starting point has always been the Povzner–Levitan representation of the solution φ in terms of the free solution. The more general case, where one starts from a given potential V_1 for which everything is known and would like to solve the inverse problem with respect to this reference potential, can be dealt with as well. We assume again that V_1, and the unknown potential V_2, both satisfy the condition (I.1.3), i.e., $rV_{1,2} \in L^1(0, \infty)$, and consider, for simplicity, the S-wave. According to our previous results, we have

$$\varphi_1(k, r) = \frac{\sin kr}{k} + \int_0^r K_1(r, t) \frac{\sin kt}{k} \, dt \qquad \text{(III.5.1)}$$

and

$$\varphi_2(k, r) = \frac{\sin kr}{k} + \int_0^r K_2(r, t) \frac{\sin kt}{k} \, dt. \qquad \text{(III.5.2)}$$

Solving the first equation for $(\sin kr)/k$, we obtain (III.1.21). If we now substitute this result into (III.5.2), we obtain

$$\varphi_2(k, r) = \varphi_1(k, r) + \int_0^r K_{12}(r, t)\varphi_1(k, t)dt, \qquad \text{(III.5.3a)}$$

where the kernel is given by

$$K_{12}(r, t) = \tilde{K}_1(r, t) + K_2(r, t) + \int_t^r K_2(r, s)\tilde{K}_1(s, t)ds. \quad \text{(III.5.3b)}$$

The equation (III.5.3a) is the Povzner–Levitan representation in its most general form. Notice that, according to our previous results, everything is well defined in (III.5.3b), and the integral is meaningful for all values of r and $t, r \geq t$.

We can develop now, along the same lines as done previously, the complete theory of the inverse problem. Everything being quite similar to the previous case, we shall content ourselves by enumerating just the essential results. First of all, according to (III.5.3b), (III.1.22), and (III.1.9), and in complete analogy with the latter, we have

$$K_{12}(r, 0) = 0. \qquad \text{(III.5.4)}$$

We do not write down the integral equation similar to (III.1.10), and relating K_{12} to V_1, V_2, and K_1. It can be treated as before, and leads to similar results concerning the properties of K_{12}. Making $t = r$ in (III.5.3b) and (III.1.22), we find

$$K_{12}(r, r) = \tilde{K}_1(r, r) + K_2(r, r) = -K_1(r, r) + K_2(r, r). \quad \text{(III.5.5)}$$

It follows that

$$K_{12}(r,r) = \frac{1}{2} \int_0^r [V_2(t) - V_1(t)]dt \qquad \text{(III.5.6)}$$

which is the generalization of (III.1.12). From this, we obtain again, whenever meaningful, that

$$V_2(r) - V_1(r) = 2 \frac{d}{dr} K_{12}(r,r). \qquad \text{(III.5.7)}$$

Here too, when V_1 and V_2 are not integrable at the origin, $K_{12}(r,r)$ is infinite, but we can use the ε-trick discussed at the end of section 2, and use the more general form (III.1.17) to get the integral of the potential.

Proceeding, we can easily establish the general form of the Gel'fand–Levitan integral equation, which now reads

$$K_{12}(r,t) + G_{12}(r,t) + \int_0^r K_{12}(r,s)G_{12}(s,t)ds = 0, \qquad \text{(III.5.8a)}$$

where

$$G_{12}(r,t) = \int_{-\infty}^{\infty} [d\rho_2(E) - d\rho_1(E)]\varphi_1(E,r)\varphi_1(E,t), \qquad \text{(III.5.8b)}$$

$$\frac{d\rho_2(E) - d\rho_1(E)}{dE} = \begin{cases} \dfrac{1}{\pi} E^{1/2}[|F_2(k)|^{-2} - |F_1(k)|^{-2}], & E \geq 0, \\[2mm] \displaystyle\sum_{j_2} C_{j_2}\delta(E - E_{j_2}) - \sum_{j_1} C_{j_1}\delta(E - E_{j_1}), & E < 0. \end{cases}$$

$$\text{(III.5.8c)}$$

This general form of the equations is useful in many instances, as we shall see in the next chapter. Notice that, in the last formula, the two summations may refer to different numbers of bound states for each potential. In particular, V_1 may have no bound states, as in the free case studied before. It can again be shown that (III.5.8a) has a unique solution under our general assumptions on the scattering data. Having shown this, one can proceed, and show that the asymptotic form of the wave function φ_2 is indeed

$$\varphi_2(k,r) = k^{-1}|F_2(k)|\sin(kr + \delta_2) + o(1), \qquad r \to \infty. \qquad \text{(III.5.9)}$$

Again, generalization to higher waves is straightforward. Equations (III.5.8a,b, and c) are valid for any angular momentum if we use the appropriate wave functions φ_l, etc. Finally, the differential equation (III.4.3) becomes

$$\left[\left(\frac{\partial^2}{\partial r^2} - \frac{l(l+1)}{r^2}\right) - \left(\frac{\partial^2}{\partial r'^2} - \frac{l(l+1)}{r'^2}\right)\right]K_{12}(r,r') = [V_2(r) - V_1(r')]K_{12}(r,r').$$

$$\text{(III.5.10)}$$

Singular Potentials. As an application of these two-potential results we now consider the case of a singular potential of the form (**II.5.2**)

$$V(r) = V_s(r) + V_1(r), \qquad \lim_{r \to 0} r^2 V_s(r) = \infty, \qquad V_1 \in W\text{-class}, \qquad \textbf{(III.5.11)}.$$

and would like to generalize the Gel'fand–Levitan theory with two potentials to this case. Let us regularize V_s, as we did at the end of chapter II. We then have, in obvious notations, the Povzner–Levitan representation

$$\varphi_\varepsilon(k, r) = \varphi_{s\varepsilon}(k, r) + \int_0^r K_\varepsilon(r, r')\varphi_{s\varepsilon}(k, r')dr'. \qquad \textbf{(III.5.12)}$$

Dividing both sides by $F_{s\varepsilon}(0)$, where $F_{s\varepsilon}$ is the Jost function of the singular potential $V_{s\varepsilon}$, and letting $\varepsilon \downarrow 0$, we obtain, according to what we saw at the end of chapter II,

$$\tilde{\varphi}(k, r) = \tilde{\varphi}_s(k, r) + \int_0^r K(r, r')\tilde{\varphi}_s(k, r')dr'. \qquad \textbf{(III.5.13)}$$

Indeed, $\varphi/F_{s\varepsilon}(0)$ and $\varphi_s/F_{s\varepsilon}(0)$ have well-defined limits $\tilde{\varphi}$ and $\tilde{\varphi}_s$. Therefore, K_ε should also have a limit. From this, and the completeness relations both for $\tilde{\varphi}$ and $\tilde{\varphi}_s$, (**II.5.19**), we find the integral equation

$$K(r, t) + G(r, t) = \int_0^r K(r, s)G(s, t)ds, \qquad \textbf{(III.5.14)}$$

$$G(r, t) = \frac{2}{\pi} \int_0^\infty \tilde{\varphi}_s(k, r)\tilde{\varphi}_s(k, t)[|\tilde{F}(k)|^{-2} - |\tilde{F}_s(k)|^{-2}]k^2 \, dk. \qquad \textbf{(III.5.15)}$$

To see the convergence of the integral here, we have, from the fact that the large k behavior of the Jost solution is given by $f \sim e^{ikr}$, that

$$\tilde{\varphi}(k, r) \simeq k^{-1}|\tilde{F}(k)|\sin(kr - \alpha k^\beta) \qquad \textbf{(III.5.16)}$$

with $\beta < 1$. Since the large k behavior of F and F_s are the same (remember that V_1 is regular, and therefore it can be neglected compared to the kinetic energy plus the singular part of the potential when k is large), we can show the convergence of the integral in the same way as for the regular case. When bound states are present, (**III.5.15**) has to be modified in the well-known way to include their contributions.

To solve the inverse problem, we assume the phase shift to be given for all $k \geq 0$, and that it can be written as

$$\delta(k) = \delta_s(k) + \delta_1(k), \qquad \textbf{(III.5.17)}$$

where $\delta_s(k)$, the singular part, has an asymptotic behavior similar to (**II.5.4**), whereas $\delta_1(\infty) = 0$. We assume the corresponding singular potential V_s and its solution $\tilde{\varphi}_s$, as well as its Jost function $\tilde{F}_s$, to be explicitly known. From the total phase shift δ, we also calculate $\tilde{F}$ via (**II.5.11**) and (**II.5.12**). From these, we can now calculate the kernel G of the generalized Gel'fand–Levitan equation (III.5.15). The potential V_1 is now given by

$$V_1(r) = 2(d/dr)K(r, r). \qquad \textbf{(III.5.18)}$$

In résumé, the inverse problem with two potentials, when the known part is repulsive and singular, is formally identical to the case where both potentials are regular, except that we must use the modified solutions $\tilde{\varphi}$ and the modified Jost functions $\tilde{F}$, etc., everywhere.

III.6   Concluding Summary of the Method

Assumptions. We are given the phase shift $\delta_l(k)$ of a partial-wave of angular momentum l, and assume that it is known for all positive energies $E = k^2 \geq 0$. We assume that it is a continuous function of k, and satisfies the conditions (**I.5.20**) and (**I.5.22**):

$$\delta_l(\infty) = 0, \qquad \int^{\infty} k^{-1} |\delta_l(k)| dk < \infty, \qquad \textbf{(III.6.1)}$$

and that the function $H(t)$ defined by (**II.4.2**) or (**II.4.12**) belongs to $L^1(0, \infty)$ and satisfies

$$\lim_{r \to 0} rH(r) = 0. \qquad \textbf{(III.6.2)}$$

We are also given a finite number of bound states with binding energies $0 \geq E_1 > E_2 > \cdots > E_n$, and assume that $\delta_l(k)$ is consistent with this number, i.e., the Levinson theorem, (**II.1.11**) and (**II.1.12**), is satisfied. We assume for simplicity that $E_1 \neq 0$, i.e., $F_l(0) \neq 0$. This means that there is no resonance at zero energy in the S-wave, or a bound state with zero binding energy for higher waves. The case where $F_l(0) = 0$ can be obtained as a limit when $E_1 \to 0$. Remember that $|F(k)|$, which enters into the definition of H, is given by (**II.2.1**).

Problem. We wish to construct all the potentials that reproduce $\{\delta_l(k),\ k \geq 0,$ and $E_j, j = 1, 2, \ldots, n\}$, and belong to the class defined by

$$rV(r) \in L^1(b, \infty), \qquad b > 0, \qquad \textbf{(III.6.3)}$$

and

$$W \in L^1(0, \infty), \qquad \textbf{(III.6.4)}$$

$$\lim_{r \to 0} rW(r) = 0, \qquad \textbf{(III.6.5)}$$

where

$$W(r) = - \int_r^{\infty} V(t)dt.$$

Solution. We first construct the Jost function $F_l(k)$ with the help of **(II.2.1)**. Its modulus is given by **(II.2.1')**. Then, with n arbitrary positive parameters C_j, we construct the spectral function $d\rho_l(E)/dE$ by means of **(II.3.14b)** for $l = 0$, or **(II.3.17b)** for general l. We calculate then the difference between this spectral function and the free one. We then use the result in **(III.2.8)** for $l = 0$, or **(III.4.2b)** for general l, to calculate the kernel $G(r, t)$. This symmetric kernel is then used in the Gel'fand–Levitan integral equation **(III.2.7a)** or **(III.4.2a)** to calculate the transformation kernel K. The potential is then readily obtained from the kernel K by means of **(III.1.16)** or **(III.1.17)** for $l = 0$, and **(III.4.4)** or its integral analogue for general l.

We shall see in chapter V, with the help of the Marchenko method, that the additional conditions to be imposed on the phase shift in order to have $rV \in L^1(0, \infty)$ are as follows.

Let $S(k) = e^{2i\delta(k)}$, and

$$A_0(t) = \frac{1}{2\pi} \int_{-\infty}^{\infty} [S(k) - 1]e^{ikt} \, dk, \qquad \text{(III.6.6)}$$

where $\delta(k)$ for $k < 0$ is defined by **(I.5.21)**. Then

$$\int_0^{\infty} r |V(r)| dr < \infty \qquad \text{(III.6.7)}$$

and

$$\int_0^{\infty} t |A_0'(t)| dt < \infty \qquad \text{(III.6.8)}$$

are completely equivalent, i.e. $rV(r)$ and $rA_0'(r)$ behave identically as far as their absolute integrability at $r = 0$ and at $r = \infty$ are concerned.

In the general case treated here, since we demand **(III.6.3)** to be satisfied, we must of course assume $rA_0'(r) \in L^1(b, \infty)$. This, together with **(III.6.1)** and **(III.6.2)**, is now sufficient to entail **(III.6.4)** and **(III.6.5)**. Indeed, it is easily seen on the Neumann series obtained by iterating the Gel'fand–Levitan equation, by using **(III.1.17)** and (see **(III.2.8)**)

$$G(R, R) - G(r, r) = \frac{1}{\pi} \int_0^{\infty} \left(\frac{1}{|F(k)|^2} - 1 \right)(\cos 2kr - \cos 2kR)dk$$

$$+ \Sigma \text{ (bound states)} = H(2r) - H(2R) + \Sigma \text{ (bound states)},$$
$$\text{(III.6.9)}$$

that the successive approximations to the potential satisfy all **(III.6.5)** and $W \in L^1(0, a)$ for every finite a. This last condition, together with **(III.6.3)**, entail now **(III.6.4)**. The convergence of the series being absolute and uniform, it is obvious that the above properties hold for the exact potential. In short, we have:

Theorem. *Let the phase shift be a continuous function of k, satisfying the conditions* **(III.6.1)**, **(III.6.2)**, *the Levinson theorem (in accordance with the number of bound states), and be such that $rA_0'(r) \in L^1(b, \infty)$. Then, the potential* *obtained from the Gel'fand–Levitan equation satisfies* (**III.6.3**), (**III.6.4**), *and* (**III.6.5**). *The converse of this theorem, except for the absolute integrability of $rA_0'(r)$ at infinity, which will be seen in chapter V, was shown in chapters I and II when we studied the direct problem.*

In the more general case treated in the fifth section, where one starts from a given potential V_1, for which everything is known, instead of starting from the free case ($V_1 \equiv 0$), the procedure and the results are quite similar.

Remark. Suppose that complete scattering data (or spectral function) are given for several l. Each one would determine a unique potential. In general, these potentials are all different from each other. The compatibility conditions to be imposed among these different scattering data for different l in order to obtain the same potential are not known.

For the generalization of the Gel'fand–Levitan method to coupled equations, many channel problems, and relativistic equations, see chapter IX.

III.7   Comments and References

The inverse problem has been the subject of very extensive studies from two points of view. The first one is that of mathematicians, who have been among the forerunners in the field, and have been primarily interested in constructing a differential operator either from one or several discrete spectra (different boundary conditions), or from the complete spectral measure corresponding to a given boundary condition. The first problem is beyond the scope of this book, although it is probably the oldest spectral inverse problem which has been treated by physicists (Rayleigh 1877). It is the second one which is important for our purpose here. In the case of the Schrödinger equation $\varphi'' + E\varphi = V\varphi$, it means finding the potential V from the knowledge of the spectral measure corresponding to specified boundary conditions, and whose support determines the spectrum completely.

Among mathematicians whose names are attached to the solution of the above problems, we mention Ambarzumian (1929), Povzner (1948), Borg (1945, 1949), Chudov (1949, 1956), Gel'fand and Levitan (1951), Berezanskii (1955, 1958), Levitan (1956, 1963, 1964), Levitan and Gasymov (1964), Agranovich and Marchenko (1957), Leibenzon (1968), Ljance (1966), Friedman (1957), Krein (several references below), Marchenko (several references below).

Marchenko (1950) was the first who showed that the spectral function uniquely determines the potential. In the fundamental paper of Gel'fand and Levitan (1951), it is shown that the potential can be determined from the spectral measure by solving a linear Fredholm equation, which now bears their names.

The second aspect of the problem, in which physicists of scattering theory are primarily interested, is that of finding the potential from the S-matrix and the bound states. The first attempts at solving this problem were made by Froberg (1947a,b, 1948) and Hylleraas (1948), who found formal procedures using series expansions whose convergence and generality seemed plausible.

The first one who showed that, in the presence of bound states, the potential is not uniquely determined by the phase shift and the binding energies was Bargmann (1949a,b), who constructed several examples leading to the same S-matrix and the same binding energies. Levinson (1949a,b) showed that a regular potential such that $V(r) + [l(l + 1)/r^2] \geq 0$ is uniquely determined from the knowledge of the phase shift $\delta_l(k)$ for all $k \geq 0$. The precise mathematical reason for this was given by Marchenko (1952, 1953) who showed that the S-matrix determines only the continuous part of the spectral measure. To construct the spectral function when bound states are present, one must know not only their locations but also an additional real parameter for each of them. This parameter may be taken, for instance, to be the value of the derivative at the origin of the corresponding normalized eigenfunction. This was also shown at about the same time by Borg (1949). Following Levinson's method, Jost and Kohn (1952a) came to the same conclusion concerning the lack of uniqueness of the potential in the presence of bound states. They derived a procedure for finding the potential by a series expansion. They also derived explicit formulae for constructing all the phase-equivalent potentials and their wave functions (1952b). Analogous formulae for phase-equivalent potentials were obtained at the same time by Holmberg (1952).

The procedure for explicitly constructing the potential from its spectral function, as presented in this chapter, is due to Gel'fand and Levitan (1951) as already mentioned. Their general procedure was immediately applied by Jost and Kohn (1953) and Levinson (1953) to the physical problem of finding the potential from scattering data. These authors gave the complete solution of the problem. Simple examples of phase equivalent potentials were worked out by Jost and Kohn in their paper. Necessary and sufficient conditions on the spectral function were obtained by Krein (1953a). Krein's method and equations used in this chapter are to be found in his papers (1953b, 1954, 1955, 1956).

For further references on the applications of Gel'fand–Levitan equations, see chapters IV and VI, and references therein.

The extensions of the Gel'fand–Levitan theory to higher waves ($l > 0$) are due to Stashevskaya (1953) and Volk (1953). See also Jost and Kohn (1953), and Newton (1956).

A more general and more abstract approach, based on the general theory of transformation operator of Friedrichs, can be found in Kay and Moses (1955, 1956a).

For other references, see chapters IV, VI, IX, and XVII.

In this chapter, we have treated the inverse problem in the general case where the potential may be as singular as those studied in the last section of chapter I, and again in chapter II about the bound states, eigenfunction expansions, and other useful results. It is shown in those chapters, and in the present one, that the natural frame for scattering theory and the Gel'fand–Levitan method, with short range potentials, i.e., those satisfying

$$rV \in L^1(b, \infty), \qquad b > 0$$

is the one provided by the conditions on the integral of the potential, as given in the summary of the present chapter. The more restrictive condition that $rV \in L^1(0, \infty)$ imposes more restrictions on the scattering data, given in the summary. These extra restrictions are found in chapter V with the help of Marchenko method. They provide necessary and sufficient conditions on the scattering data to have $rV \in L^1(0, \infty)$. The more general case treated in the present chapter is due to Baeteman and Chadan (1975, 1976).

For the application of the Gel'fand–Levitan equation with two potentials to the singular case, with a strongly repulsive singularity at the origin, see Rofe–Beketov and Hristov (1969) and Chadan (1978). For higher waves, the Paley–Wiener theorem and related theorems have been generalized by Levin (1956), by using the appropriate Hankel functions, instead of the exponential, in (**III.1.4a,b**). They lead directly to (**III.4.1a,b**). See also Sohin (1972).

Let us mention here the symmetrical Tauberian theorem into which the theorem at the end of section 6 can be transformed, and which reads: The integrability properties of W, (**III.6.4**), and the Fourier–Sine transform of the phase shift, (**II.4.4**), are the same at the origin, and at infinity, respectively (Brander and Chadan, 1980).

The Gel'fand–Levitan equation, and the Marchenko equation of chapter V, have been generalized to higher order differential equations. We refer the reader to Beals (1985) and Zachary (1986), where complete references to earlier works of Sakhnovich, Butler, Leibenzon, and others, are given. For the three-body problem, see Zakhariev (1986, 1987).

Other generalizations appear in chapters XIV, XVII, and XVIII.

Methods Competing with the Gel'fand–Levitan Method. The most important alternatives to the Gel'fand–Levitan–Jost–Kohn method will be studied in chapters V (Marchenko method) and VII (methods for special classes of potential). Let us quote here two other historical alternatives. The first one is an approach introduced by Jost and Kohn (1952a) and developed by Moses (1956). They obtained an algorithm which reproduced the potential essentially by inverting the Born series for the scattering matrix. The Jost–Kohn algorithm is less efficient (the results are series whose convergence is not generally proved) and less attractive than its more famous competitor, but it is simple in conception. It has been applied by Prosser (1969, 1976) to inverse problems for scattering from a potential, a variable index of refraction, and a soft boundary. The other historical approach is due to Hylleraas (1948, 1963, 1964). By using systems of biorthonormal functions, he was able to give a very simple derivation of a dissymmetric expansion of the potential that is essentially equivalent, when no bound state is present, to the formula (**V.2.27**) obtained later in Marchenko's method. The potential is then obtained by an obvious iteration process, which, however, has no reason to converge unless the potential is small enough.

CHAPTER IV

Applications of the Gel'fand–Levitan Equation

IV.1 Introduction of New Bound States

The Gel'fand–Levitan equation being a Fredholm equation, it is obvious that it cannot be solved in general in closed form. Except in a very few exceptional cases, the most general case solvable in closed form is when the kernel is degenerate. In this case, it is known that one can solve the equation algebraically. The most interesting instances of this situation are the following: In the first one, we start from a potential V_1 which has no bound states, and for which everything (wave function φ_1, spectral density $d\rho_1(E)$) is known, and we would like to find all the potentials V which have the same spectral density for positive energies (the same continuum) but have, in addition, some bound states. The second case is when we start from a given Jost function, assuming again that all quantities related to it, including the potential, are known, and would like to find all the potentials whose Jost functions differ from the first one by a rational factor of k. We leave this problem for the third section, and now study the first problem.

We must consider the general form of the Gel'fand–Levitan equation, discussed at the end of the last chapter, in which we start from a given potential V_1 instead of the free case. Since now, by assumption, $d\rho(E) = d\rho_1(E)$ for $E \geq 0$, equation (**III.5.8a**) reduces to

$$K(r, t) + G(r, t) + \int_0^r K(r, s)G(s, t)ds = 0 \qquad \textbf{(IV.1.1)}$$

with G given by the sum over the finite number of bound states of $V(r)$:

$$G(r, t) = \sum_j C_j \varphi_1(i\gamma_j, r)\varphi_1(i\gamma_j, t), \qquad \textbf{(IV.1.2)}$$

where $\varphi_1(k, r)$ is the regular solution for potential V_1. Notice that $\varphi_1(i\gamma_j, r)$ are not bound state wave functions, and therefore do not remain bounded when r becomes very large. Since the kernel G is now a degenerate kernel, we can easily solve the equation. In vector notations, if we write

$$G(r, t) = (\psi(r), \chi(t)), \qquad \textbf{(IV.1.3)}$$

where ψ and χ are vectors with components $C_j\varphi_1(i\gamma_j, r)$ and $\varphi_1(i\gamma_j, t)$, respectively, it is obvious that we must seek a solution of **(IV.1.1)** in the form

$$K(r, t) = (K(r), \chi(t)) \qquad \textbf{(IV.1.4)}$$

$K(r)$ being a vector with n components. Equation **(IV.1.1)** becomes now

$$(K(r), \chi(t)) + (\psi(r), \chi(t)) + \left\{ \left[\int_0^r R(s)ds \right] K(r), \chi(t) \right\} = 0, \qquad \textbf{(IV.1.5)}$$

where $R(s)$ is the tensor product of ψ and χ, i.e., the matrix with elements

$$R_{jk}(s) = C_j\varphi_1(i\gamma_j, s)\varphi_1(i\gamma_k, s). \qquad \textbf{(IV.1.6)}$$

The general results of chapter III about the existence of a unique solution of the Gel'fand–Levitan equation ensure that the matrix

$$C(r) = I + \int_0^r R(s)ds \qquad \textbf{(IV.1.7)}$$

has an inverse for all r. This is also obvious since from its definition and **(IV.1.6)**, one sees that $C(r)$ is positive definite.

The solution of **(IV.1.5)** is given by applying C^{-1} to the inhomogeneous term

$$K(r, t) = -(C^{-1}(r)\psi(r), \chi(t)). \qquad \textbf{(IV.1.8)}$$

It then follows that

$$K(r, r) = -(C^{-1}(r)\psi(r), \chi(r))$$

$$= -\text{Tr}[C^{-1}(r)R(r)] = -\text{Tr}\left[C^{-1}(r)\frac{d}{dr}C(r) \right]$$

$$= -\frac{d}{dr} \text{Log det } C(r). \qquad \textbf{(IV.1.9)}$$

The potential V is now given by

$$V(r) - V_1(r) = 2\frac{d}{dr}K(r, r) = -2\frac{d^2}{dr^2} \text{Log det } C(r) \qquad \textbf{(IV.1.10)}$$

and the solution φ by

$$\varphi(k, r) = \det \begin{vmatrix} C(r) & \psi(r) \\ \beta(k, r) & \varphi_1(k, r) \end{vmatrix} [\det C(r)]^{-1}. \qquad \textbf{(IV.1.11)}$$

In this last equation, the matrix in the numerator is obtained by adding to $C(r)$ the column ψ and the row β and the last diagonal element φ_1, where

$$\beta(k, r) = \int_0^r \chi(t)\varphi_1(k, t)dt. \qquad \textbf{(IV.1.12)}$$

From these equations, it is possible to find the Jost function $F(k)$, and the asymptotic properties of φ and the potential increment $V - V_1$. Now, it will be shown that, as expected,

$$F(k) = F_1(k) \prod_j \left(\frac{k - i\gamma_j}{k + i\gamma_j}\right) \qquad \textbf{(IV.1.13)}$$

and that the potential increment ΔV satisfies

$$\int_0^\infty r|\Delta V(r)|dr < \infty \qquad \textbf{(IV.1.14)}$$

In fact, we shall find that

$$\Delta V(r) = -4 \sum_j C_j r + r \, o(1), \qquad r \to 0, \qquad \textbf{(IV.1.15)}$$

and

$$\Delta V(r) = -\frac{2}{C_0} (2\gamma_0)^5 e^{-2\gamma_0 r}[1 + o(1)], \qquad r \to \infty, \qquad \textbf{(IV.1.16)}$$

where γ_0 is the smallest of γ_j's, and C_0 its corresponding normalization constant.

In order to verify the above assertions, we begin with the case of one bound state of energy $-\gamma^2$ and normalization constant C. The matrices we have to deal with are simple two by two matrices, and it is easily verified that (IV.1.11) gives

$$\varphi(k, r) = \varphi_1(k, r) - C\varphi_1(i\gamma, r)$$

$$\times \left[1 + C \int_0^r \varphi_1^2(i\gamma, t)dt\right]^{-1} \int_0^r \varphi_1(i\gamma, t)\varphi_1(k, t)dt. \quad \textbf{(IV.1.17)}$$

From this expression, we have to find the Jost solution and the Jost function. For this purpose, we use the identity

$$(k^2 + \gamma^2) \int_0^r \varphi_1(i\gamma, t)\varphi_1(k, t)dt = W[\varphi_1(k, r), \varphi_1(i\gamma, r)], \quad \textbf{(IV.1.18)}$$

which is obtained by integrating the Schrödinger equation and by using the boundary conditions. If we now substitute this expression into **(IV.1.17)**, and make the decomposition

$$\varphi_1(k, r) = (2ik)^{-1}[g_1(k, r) - g_1(-k, r)], \qquad \text{(IV.1.19a)}$$

$$g_1(k, r) = F_1(-k)f_1(k, r), \qquad \text{(IV.1.19b)}$$

and a similar one for $\varphi(k, r)$, we find, in a unique way

$$F(-k)f(k, r) = F_1(-k)\left[f_1(k, r) - C(k^2 + \gamma^2)^{-1}\varphi_1(i\gamma, r) \right.$$

$$\left. \times \left(1 + \int_0^r \varphi_1^2(i\gamma, t)dt\right)^{-1} W(f_1(k, r), \varphi_1(i\gamma, r)) \right].$$

$$\text{(IV.1.20)}$$

This decomposition is unique is easily verified on the basis of **(I.2.8b and c)**, which show that $\varphi_1(i\gamma, r)$ is exponentially increasing as $r \to \infty$, **(I.3.9)**, and **(I.3.10)**. These formulas show that the asymptotic behavior of the right-hand side for large values of r is indeed given, up to a constant factor, by $\exp(ikr)$, and that any other decomposition would lead to terms containing also $\exp(-ikr)$. In fact, calculating the asymptotic behavior of the right-hand side from

$$\varphi_1(i\gamma, r) = (2\gamma)^{-1}e^{\gamma r}(1 + o(1)) \qquad \text{(IV.1.21)}$$

we find

$$F(k) = F_1(k)\left(\frac{k - i\gamma}{k + i\gamma}\right). \qquad \text{(IV.1.22)}$$

This shows clearly that the potential $V(r)$ has indeed a bound state of energy $-\gamma^2$ since its Jost function vanishes at $k = i\gamma$. Notice that, making $r = 0$ in **(IV.1.20)**, we find, in agreement with **(IV.1.22)**,

$$|F(k)|^2 = |F_1(k)|^2, \qquad k \text{ real}, \qquad \text{(IV.1.23)}$$

which guarantees that the spectral densities $d\rho$ and $d\rho_1$ are identical for positive energies, i.e., on the continuous part of the spectrum.

Making $k = i\gamma$ in **(IV.1.17)**, we find, for the bound-state wave function

$$\varphi(i\gamma, r) = \frac{\varphi_1(i\gamma, r)}{1 + C\int_0^r \varphi_1^2(i\gamma, t)dt} \qquad \text{(IV.1.24)}$$

from which we easily verify that C is indeed the normalization constant defined by **(II.3.16)**.

The Jost solution can also be calculated by using **(IV.1.21)** in **(IV.1.20)**. Thus

$$f(k, r) = \frac{k - i\gamma}{k + i\gamma}\left[f_1(k, r) - C\frac{\varphi_1(i\gamma, r)}{1 + C\int_0^r \varphi_1^2(i\gamma, t)dt} \right.$$

$$\left. \times \frac{1}{k^2 + \gamma^2} W(f_1(k, r), \varphi_1(i\gamma, r)) \right]. \qquad \text{(IV.1.25)}$$

This procedure can be repeated any number of times we wish, step by step. At each step we add one new bound state, keeping the normalization constants of the previous bound states the same. The difference between the old and the new spectral densities reduces again to one term—that of the new bound state—and we obtain simple formulas quite similar to (IV.1.7), (IV.1.20) and (IV.1.22) to calculate all the quantities of interest. It is obvious from this last formula that, at the end, we could get (IV.1.13).

To see the behavior of the potential increment ΔV near the origin, and at infinity, we use formula (IV.1.10) and assume again that there is only one bound state. It is obvious from (IV.1.6) and (IV.1.7) that $C(r)$ is twice continuously differentiable, even when the potential V_1 is a δ-function (see the discussion at the end of the last section of chapter I). Therefore, the potential increment is an ordinary continuous function. From

$$\varphi(k, r) = r + o(r), \qquad r \to 0, \tag{IV.1.26}$$

and (IV.1.6), we see that, near the origin,

$$C(r) = 1 + C \int_0^r \varphi_1^2(i\gamma, t)dt$$

$$= 1 + \frac{C}{3} r^3 + o(r^3) \tag{IV.1.27}$$

where C is the bound state normalization constant. Thus,

$$\Delta V = -2 \frac{d^2}{dr^2} \operatorname{Log} C(r) = -4Cr[1 + o(1)]. \tag{IV.1.28}$$

We can again reach the general case of n bound states step by step by introducing the bound states one by one. Since the potential increments add up, we finally get (IV.1.15).

For the behavior of ΔV at infinity, we use (IV.1.21) in (IV.1.10). This leads, for large values of r, to

$$C(r) = 1 + \frac{C}{8\gamma^3} e^{2\gamma r} + \cdots \tag{IV.1.29}$$

and related estimates for $C'(r)$ and $C''(r)$, so that

$$\Delta V = -2 \frac{C''(r) - C'^2(r)}{C^2(r)} = -\frac{2}{C} (2\gamma)^5 e^{-2\gamma r} + \cdots. \tag{IV.1.30}$$

The potential increment behaves therefore, at infinity, like the square of the wave function of the bound state. Introducing now other bound states one by one, as we did before, we add each time to the potential a term which has a behavior similar to (IV.1.30). It is then obvious that, at the end, the tail of the total potential increment would have the behavior given by (IV.1.16). It is also obvious that we have, in general,

$$\int_0^\infty |\Delta V(r)| r \, dr < \infty. \tag{IV.1.31}$$

IV.2 Phase Equivalent Potentials

The general formula given above for the potential V having n bound states in terms of an initial potential V_1 with no bound state but with the same spectral density on the continuous part of the spectrum gives us also the dependence of V on the normalization constants C_j. In the general expression for V, we can now keep all γ_j's fixed, and vary only C_j's. We obtain in this way an n-parameter family of potentials with the same phase shift and the same binding energies. Indeed, since the spectral density for positive energies, and hence $|F(k)|$ for $k \geq 0$, is the same for all members of the family, and since γ_j's are also the same, it follows from the second section of chapter II that the phase shift is also the same. Because of this, this family is called phase-equivalent, but we have to keep in mind that the binding energies are also fixed. If we keep the phase shift fixed, and vary the binding energies and the corresponding normalizations constants, it follows from (**II.2.1**) that the continuous part of the spectral density changes also, so that the solution of the problem looks more difficult than before. Nevertheless, it can be solved analytically, as we shall see in the next section. We give here a more detailed study of the phase-equivalent potentials.

The main problem in the phase-equivalent family of potentials, i.e., when we vary only C_j's in the general formula (**IV.1.10**) for the potential, is the relation among different members of the family. This is of course contained in the above formula, but in simple cases, one can obtain more explicit results. We shall consider first the case where one of the parameters, which has the initial value $\bar{C}$, is changed to C, the others remaining fixed. We may ask then how to calculate the wave function and the potential for C in terms of related quantities for $\bar{C}$, which we distinguish also by a bar above. The answer is again provided by the general form of the Gel'fand–Levitan equation discussed at the end of the last chapter. In the case at hand, we have

$$d\rho(E) - d\bar{\rho}(E) = (C - \bar{C})\delta(E + \gamma^2) \qquad \textbf{(IV.2.1)}$$

in obvious notations. This is just the simplest case leading to a degenerate kernel with one term similar to (**IV.1.2**) with the fundamental difference that now the initial wave function $\bar{\varphi}(i\gamma, r)$ corresponds to a bound state, i.e., it vanishes like $\exp(-\gamma r)$ for large values of r. Everything else being the same, we can read off the answer from the results of the last section. From (**IV.1.17**), we have

$$\varphi(k, r) = \bar{\varphi}(k, r) - (C - \bar{C})\bar{\varphi}(i\gamma, r)$$

$$\times \left[1 + (C - \bar{C}) \int_0^r \bar{\varphi}^2(i\gamma, t)dt \right]^{-1} \int_0^r \bar{\varphi}(i\gamma, t)\bar{\varphi}(k, t)dt. \quad \textbf{(IV.2.2)}$$

In order to avoid singularities in φ, we must verify that the denominator never vanishes. This is the case if C is positive, as it should, because

$$1 + (C - \bar{C}) \int_0^r \bar{\varphi}^2(i\gamma, t)dt = C \int_0^r \bar{\varphi}^2(i\gamma, t)dt + \bar{C} \int_r^\infty \bar{\varphi}^2(i\gamma, t)dt \quad \textbf{(IV.2.3)}$$

and we know that $\bar{C}$ is positive too. Making r very large in (**IV.2.2**), we see also that, for k real, both φ and $\bar{\varphi}$ have the same asymptotic behavior, i.e., the phase shift is the same for both, as expected. Remembering also that now

$$F(k) \equiv \bar{F}(k) \qquad \text{(IV.2.4)}$$

we get from (**IV.1.20**)

$$f(k, r) = \bar{f}(k, r) - \frac{C - \bar{C}}{k^2 + \gamma^2} \frac{\bar{\varphi}(i\gamma, r)}{1 + (C - \bar{C}) \int_0^r \bar{\varphi}^2(i\gamma, t)dt} W[\bar{f}(k, r), \bar{\varphi}(i\gamma, r)]$$

$$\text{(IV.2.5)}$$

and we can again check that the asymptotic behavior of both sides coincides because the second term in the right-hand side vanishes exponentially for large values of r. As for the wave function of the bound state, it is obtained from (**IV.2.2**):

$$\varphi(i\gamma, r) = \frac{\bar{\varphi}(i\gamma, r)}{1 + (C - \bar{C}) \int_0^r \bar{\varphi}^2(i\gamma, t)dt}. \qquad \text{(IV.2.6)}$$

We end this section with a somewhat different way of writing the above formulas for phase-equivalent potentials. We assume that we know one potential of the family, $V_0(r)$, and its Jost solution $f_0(k, r)$. Consider first, for simplicity, the case of one bound state with binding energy $-\gamma^2$. Denoting by $\psi_0(r)$ the wave function of the bound state:

$$\psi_0(r) = f_0(i\gamma, r) \qquad \text{(IV.2.7)}$$

it is easy to verify that for other members of the family, we have

$$\psi_0(\lambda, r) = \frac{\psi_0(r)}{N(\lambda, r)} \qquad \text{(IV.2.8)}$$

$$V(\lambda, r) = V_0(r) + \lambda N^{-1}(\lambda, r)\psi_0(r)\psi_0'(r) + \frac{\lambda^2}{8} N^{-2}(\lambda, r)\psi_0^4(r)$$

$$= V_0(r) - 2\frac{d^2}{dr^2} \text{Log}\left[1 + \frac{\lambda}{4} \int_r^\infty \psi_0^2(t)dt\right], \qquad \text{(IV.2.9)}$$

where

$$N(\lambda, r) = 1 + \frac{\lambda}{4} \int_r^\infty \psi_0^2(t)dt. \qquad \text{(IV.2.10)}$$

The allowed domain of variation of the parameter λ is

$$-4\left[\int_0^\infty \psi_0^2(t)dt\right]^{-1} < \lambda < \infty. \qquad \text{(IV.2.11)}$$

When λ is more negative than the left-hand side, $N(\lambda, r)$ vanishes at some strictly positive value of r, and the potential becomes singular with a double pole. Such potentials are meaningless from the point of view of scattering theory. From **(IV.2.8)**, we can calculate easily the normalization constant of the wave function of the bound state for arbitrary λ. The result is

$$\tilde{C}^{-1}(\lambda) \equiv \int_0^\infty \psi_0^2(\lambda, r)dr = \frac{\tilde{C}^{-1}(0)}{1 + \lambda \tilde{C}^{-1}(0)/4} \qquad \textbf{(IV.2.12a)}$$

which can also be written in the form

$$\lambda = 4[\tilde{C}(\lambda) - \tilde{C}(0)]. \qquad \textbf{(IV.2.12b)}$$

When λ varies over its admissible range $(-4\tilde{C}(0), \infty)$, $\tilde{C}(\lambda)$ takes on all the values in $(0, \infty)$.

The Jost solution for arbitrary λ is given by

$$f(\lambda; k, r) = f_0(k, r) - \frac{\lambda}{4N(\lambda, r)} \psi_0(r) \int_r^\infty \psi_0(t)f_0(k, t)dt \quad \textbf{(IV.2.13a)}$$

which can be easily verified. Since all the potentials of the family have the same Jost function $F(k)$, which is obvious on **(IV.2.5)** if we make $r = 0$, the regular solution φ is given by

$$\varphi(\lambda; k, r) = \varphi_0(k, r) - \frac{\lambda}{4} N^{-1}(\lambda, r)\psi_0(r) \int_r^\infty \psi_0(t)\varphi_0(k, t)dt. \quad \textbf{(IV.2.13b)}$$

It is easily seen on these formulae that the additional terms due to λ all vanish exponentially in the limit of large r.

If more than one bound state is present, we can solve the problem step by step, starting from one bound state and adding the others one by one. We can generate in this way the whole family of phase-equivalent potentials. We can also solve the problem globally by a method quite analogous to that of the first section. Indeed, since all the potentials of the family have the same Jost function, the kernel of the general form of the Gel'fand–Levitan equation **(III.5.8a)** in which we start from one particular member of the family corresponding to normalization constants C_j^0 reads

$$G(r, t) = \sum_j (C_j - C_j^0)\varphi_0(i\gamma_j, r)\varphi_0(i\gamma_j, t), \qquad \textbf{(IV.2.14a)}$$

where $\varphi_0(i\gamma_j, r)$ is the wave function of the jth bound state. This is similar to **(IV.1.2)** with the exception that there $\varphi_1(i\gamma_j, r)$ is not a bound-state wave function. Anyway, the integral equation can again be solved in closed form, with a result quite similar to **(IV.1.10)**:

$$V(r) = V_0(r) - 2\frac{d^2}{dr^2} \text{Log det} \|M_{jk}(r)\|, \qquad \textbf{(IV.2.14b)}$$

where

$$M_{jk}(r) = -\delta_{jk} - (C_j - C_j^0) \int_0^r \varphi_0(i\gamma_j, t)\varphi_0(i\gamma_k, t)dt. \qquad \textbf{(IV.2.14c)}$$

It can be shown that the matrix M_{jk} has an inverse, so that (**IV.2.14c**) is well defined. The potential increment is again $O(r)$ at the origin, and $O(e^{-\gamma_0 r})$ at large distances, where γ_0 is the smallest of the γ_j's. We shall see examples in chapter VI.

IV.3 Bargmann Potentials

We studied in the first section the case where the initial potential had no bound state, and showed how one can construct a potential having the same spectral density for positive energies, but with n bound states. The problem could be solved explicitly because the kernel of the Gel'fand–Levitan equation in its general form was then degenerate. Another case where one can solve the problem explicitly is when we start from a given Jost function F_1 for which everything, including the potential itself, is known, and we should like to construct the potential whose Jost function F_2 differs from the previous one by multiplication by a rational function of k. The case treated previously is in fact a particular instance of this general case, as it is apparent from (**IV.1.13**). A Bargmann potential attached to a given potential V_1 is a potential V_2 such that F_2 differs from F_1 by a rational factor.

Because of the fact that the Jost function must tend to 1 when k becomes very large, it is obvious that we should have in general

$$F_2(k) = \prod_j \left(\frac{k + ia_j}{k + ib_j} \right) F_1(k). \qquad (\textbf{IV.3.1})$$

In order for F_2 to be holomorphic in the upper half-plane, we must assume $\mathrm{Re}\, b_j > 0$. Also, we must have either $a_j < 0$, i.e., introduction of a new bound state, or $\mathrm{Re}\, a_j > 0$. Because of the symmetry property (**I.1.21**), it is obvious that to the factor $k + ib_j$ there corresponds also to the factor $k + ib_j^*$, and if $\mathrm{Re}\, a_j > 0$, we should also have the term $k + ia_j^*$ in the numerator. If a_j is real and negative (new bound state), we can always write

$$\frac{k - i|a|}{k + ib} = \left(\frac{k - i|a|}{k + i|a|} \right)\left(\frac{k + i|a|}{k + ib} \right) \qquad (\textbf{IV.3.2})$$

and treat first the second factor (no bound state), as will be done below, and then the first factor, according to the method of the first section. We shall therefore assume that the rational factor does not introduce new bound states.

The simplest case is when the initial potential has no bound states, and when we assume also that both a_j and b_j satisfy

$$a_j > 0, \qquad b_j > 0, \qquad (\textbf{IV.3.3})$$

and are all distinct. This is already general enough to show the important features of the results. The general case where some of the a's or b's are complex can be dealt with, as well, with the method we shall develop. Because of the symmetry property (**I.1.21**), it is obvious that such complex a's or b's appear always in pairs (a and a^*, b and b^*) and we must treat each pair together. This slightly complicates the algebra, but does not change the fundamental results. We therefore keep (**IV.3.3**). The kernel of Gel'fand–Levitan then becomes

$$G(r, t) = \frac{2}{\pi} \int_0^\infty \varphi_1(k, r)[|F_2(k)|^{-2} - |F_1(k)|^{-2}]\varphi_1(k, t)k^2 \, dk$$

$$= \frac{2}{\pi} \int_0^\infty \varphi_1(k, r)|F_1(k)|^{-2}\left[\prod_j \left(\frac{k^2 + b_j^2}{k^2 + a_j^2} \right) - 1 \right]\varphi_1(k, t)k^2 \, dk. \quad \textbf{(IV.3.4)}$$

Using now the identity

$$\prod_j \left(\frac{k^2 + b_j^2}{k^2 + a_j^2} \right) - 1 = \sum_j \frac{A_j}{k^2 + a_j^2}, \quad \textbf{(IV.3.5a)}$$

$$A_j = \frac{\prod_l (b_l^2 - a_j^2)}{\prod_{l \neq j} (a_l^2 - a_j^2)}, \quad \textbf{(IV.3.5b)}$$

the kernel becomes

$$G(r, t) = \sum_j \frac{2}{\pi} A_j \int_0^\infty \varphi_1(k, r)\varphi_1(k, t)(k^2 + a_j^2)^{-1}|F_1(k)|^{-2}k^2 \, dk. \quad \textbf{(IV.3.6)}$$

It is easy to recognize that each term of the above sum is the resolvent kernel (**II.3.4**) of the Schrödinger equation for the potential V_1 and energy $-a_j^2$. Indeed, if we apply the differential operator $-(d^2/dr^2) + V_1 + a_j^2$ to each integral, and use the completeness relation (**II.3.8**), we obtain (**II.3.4**). This, together with symmetry in r and t, and the boundary condition (**II.3.5**), both trivially satisfied here, proves the assertion. If we use now the expression (**II.3.3**) for the resolvent, we can write (**IV.3.6**), for $r < t$

$$G(r, t) = \sum_j A_j R_{-a_j^2}(r, t)$$

$$= \sum_j A_j F_1^{-1}(ia_j) f_1(ia_j, t)\varphi_1(ia_j, r) \quad \textbf{(IV.3.7)}$$

and its symmetric for $t < r$. These have to be used now in

$$K(r, t) + G(r, t) + \int_0^r K(r, s)G(s, t)ds = 0. \quad \textbf{(IV.3.8)}$$

Since the kernel is again degenerate, we shall seek a solution in the form

$$K(r, t) = \sum_j \omega_j(r)\varphi_1(ib_j, t). \quad \textbf{(IV.3.9)}$$

The integral equation now reduces to

$$\sum_j \omega_j(r)\varphi_1(ib_j, t) + \sum_j A_j \varphi_1(ia_j, t)F_1^{-1}(ia_j)f_1(ia_j, r)$$

$$+ \sum_j \sum_l A_l \omega_j(r) \int_0^r \varphi_1(ib_j, s)R_{-a_j^2}(s, t)ds = 0 \quad \textbf{(IV.3.10)}$$

The last integral, when use is made of the identity

$$(a^2 + k^2) \int_{r_1}^{r_2} R_{-a^2}(r, t)\psi(t)dt = \psi(r) + W[R_{-a^2}(r, t), \psi(t)]\Big|_{r_1}^{r_2}, \quad \text{(IV.3.11a)}$$

where ψ is any solution of the Schrödinger equation

$$\psi'' + k^2\psi = V_1\psi \qquad \text{(IV.3.11b)}$$

becomes

$$\sum_j \sum_l \omega_j(r)A_l(a_l^2 - b_j^2)^{-1}\varphi_1(ib_j, t)$$

$$+ \sum_j \sum_l \omega_j(r)W[R_{-a^2}(r, t), \varphi_1(ib_j, r)]A_l(a_l^2 - b_j^2)^{-1}. \quad \text{(IV.3.12)}$$

If we make use now of (IV.3.5b), we see that the first term in the above expression cancels the first term in (IV.3.10), so that we are left with

$$\sum_l A_l\varphi_1(ia_l, t)F_1^{-1}(ia_l)[f_1(ia_l, r) + \sum_j W_{lj}(r)\omega_j(r)] = 0, \quad \text{(IV.3.13a)}$$

where

$$W_{lj}(r) = (a_l^2 - b_j^2)^{-1}W[f_1(ia_l, r), \varphi_1(ib_j, r)]. \qquad \text{(IV.3.13b)}$$

In vector notation, using the fact that $\varphi_1(ia_l, t)$ are linearly independent of each other, we can write (IV.3.13a)

$$f(r) + W(r)\omega(r) = 0. \qquad \text{(IV.3.14)}$$

Thus

$$\omega(r) = -W^{-1}(r)f(r) \qquad \text{(IV.3.15)}$$

and, denoting by (,) the scalar product of two vectors,

$$K(r, t) = (\omega(r), \varphi(t)) = -(W^{-1}(r)f(r), \varphi(t)) \qquad \text{(IV.3.16)}$$

We now notice that in analogy with (IV.1.8) and (IV.1.10), we have

$$\frac{d}{dr}W_{lj}(r) = f_1(ia_l, r)\varphi(ib_j, r). \qquad \text{(IV.3.17)}$$

Therefore, exactly as for (IV.1.9) and (IV.1.10), we obtain

$$\Delta V(r) = V_2(r) = -2\frac{d^2}{dr^2}\text{Log det } W(r) \qquad \text{(IV.3.18)}$$

and

$$\varphi_2(k, r) = \frac{\det\begin{vmatrix} W(r) & f(r) \\ \beta(k, r) & \varphi_1(k, r) \end{vmatrix}}{\det W(r)}, \qquad \text{(IV.3.19a)}$$

where the vector β is given by its components

$$
\begin{aligned}
\beta_j(k, r) &= \frac{W[\varphi_1(ib_j, r), \varphi(k, r)]}{k^2 + b_j^2} \\
&= \int_0^r \varphi_1(ib_j, t)\varphi_1(k, t)dt.
\end{aligned} \qquad \textbf{(IV.3.19b)}
$$

We shall see in the next section that φ_2 is indeed the solution of the Schrödinger equation with potential V_2 (Krein). Just as was done in the first section to introduce bound states, we can calculate the asymptotic behavior of φ_2 for large values of r. We find

$$
\varphi_2(k, r) = (2ik)^{-1}[F_2(-k)e^{ikr} - F_2(k)e^{-ikr}] + o(1), \qquad \textbf{(IV.3.20)}
$$

where

$$
F_2(k) = F_1(k) \prod_j \left(\frac{k + ia_j}{k + ib_j} \right)
$$

as expected.

The behavior of the potential increment at the origin and at infinity can be analyzed as before. At infinity it decreases exponentially like $\exp(-2br)$, where b is the smallest of the b_j's. At the origin, unlike the case of section 1, the potential increment does not vanish anymore, but is finite.

IV.4   Transformations of the Schrödinger Equation

In this last section, we shall prove two theorems stated earlier. The first one, the theorem of Marchenko, mentioned at the end of section 4 of the last chapter is as follows:

Theorem (Marchenko). *If $S(k)$ is the "S-matrix" associated with the potential $V(r)$ and angular momentum l, then it is also the "S-matrix," for any other angular momentum l' and a potential $V_{l'}(r)$, which behaves exactly as $V(r)$ near the origin, and at infinity.*

The second theorem (Krein) is that the function φ_2 defined by (IV.3.19a) is the solution of the Schrödinger equation with the potential V_2 given by (IV.3.18).

The proof of these theorems is based on some general transformations of second-order differential equations. Let φ_0 be a solution of

$$
\varphi''(E, r) + E\varphi(E, r) = V(r)\varphi(E, r) \qquad \textbf{(IV.4.1)}
$$

for $E = E_0$, which does not vanish in an interval containing r_0. Consider now

$$
\varphi_1(E, r) = W[\varphi(E, r), \varphi_0(E_0, r)][(E - E_0)\varphi_0(E_0, r)]^{-1}, \qquad \textbf{(IV.4.2)}
$$

where φ is an arbitrary solution of **(IV.4.1)**, and W is the Wronskian. The function φ_1 is a solution of the same equation with a potential $V_1 = V + \Delta V$, where

$$\Delta V(r) = V_1(r) - V(r) = -2 \frac{d}{dr} \frac{\varphi_0'(E_0, r)}{\varphi_0(E_0, r)}. \qquad \text{(IV.4.3)}$$

For the proof of this lemma we use the identity

$$\frac{d}{dr} \frac{W[\varphi(E_1, r), \varphi(E_2, r)]}{E_1 - E_2} = \varphi(E_1, r)\varphi(E_2, r) \qquad \text{(IV.4.4)}$$

satisfied by any two solutions of **(IV.4.1)** with the same potential. It follows now from **(IV.4.2)**, that

$$\varphi_1'(E, r) = \varphi(E, r) - \frac{W[\varphi(E, r), \varphi_0(E_0, r)]}{(E - E_0)\varphi_0^2(E_0, r)} \varphi_0'(E_0, r). \qquad \text{(IV.4.5)}$$

One more differentiation leads then to the desired result.

The solution defined in **(IV.4.2)** is of course meaningful only if $E \neq E_0$. When $E = E_0$, one of the solutions of the transformed equation, i.e., with the potential $V_1(r)$, is

$$\psi_1(E_0, r) = \frac{1}{\varphi_0(E_0, r)} \qquad \text{(IV.4.6)}$$

as can be verified by direct computation. To obtain a second linearly independent solution of the transformed equation, we take, as is well known,

$$\varphi_1(E_0, r) = \psi_1(E_0, r) \int^r \frac{dt}{\psi_1^2(E_0, r)} = \varphi_0^{-1}(E_0, r) \int^r \varphi_0^2(E_0, t)dt. \quad \text{(IV.4.7)}$$

It is easily verified that the Wronskian of ψ_1 and φ_1 is indeed 1. Formulae **(IV.4.2)**, **(IV.4.6)** and **(IV.4.7)** give the solution of the Schrödinger equation with the potential V_1 in terms of the solutions with the potential V, V_1 and V being related by **(IV.4.3)**. Conversely, the solution $\varphi(E, r)$ of the initial equation can be expressed in terms of the solutions of the transformed equation. Indeed, **(IV.4.6)** yields

$$\frac{\varphi_0'(E_0, r)}{\varphi_0(E_0, r)} = -\frac{\psi_1'(E_0, r)}{\psi_1(E_0, r)}. \qquad \text{(IV.4.8)}$$

It follows then that **(IV.4.5)** can be written

$$\varphi(E, r) = \frac{W[\varphi_1(E, r), \psi_1(E_0, r)]}{\psi_1(E_0, r)}. \qquad \text{(IV.4.9)}$$

This is the inverse of formula **(IV.4.2)**.

We have so far assumed that $\varphi(E_0, r)$ does not vanish in the neighborhood of the point $r = r_0$, and considered everything in that neighborhood.

We shall assume now that $r_0 = 0$, and $\varphi_0(E_0, r)$ does not vanish at the origin, and that the potential $V(r)$ has the behavior

$$V = l(l + 1)r^{-2} + \bar{V}(r) \underset{r \to 0}{=} l(l + 1)r^{-2} + O\left(\frac{1}{r^{2-\varepsilon}}\right), \qquad \varepsilon > 0. \quad \textbf{(IV.4.10)}$$

The form of the second term is slightly less general than $r|\bar{V}(r)|$ integrable at the origin. However, it simplifies the calculations. According to what we know, we have two types of solutions of the Schrödinger equation with the potential V:

$$\varphi(E, r) = Cr^{l+1}[1 + O(r^\varepsilon)] \qquad \textbf{(IV.4.11a)}$$

and

$$\psi(E, r) = Dr^{-l}[1 + O(r^\varepsilon)]. \qquad \textbf{(IV.4.11b)}$$

If we use the regular solution $\varphi(E_0, r)$ in the transformation studied above, the potential increment becomes

$$\Delta V(r) = -2\frac{d^2}{dr^2} \text{Log } r^{l+1}[1 + O(r^\varepsilon)]$$

$$= 2(l + 1)r^{-2} + O(r^{\varepsilon-2}), \qquad r \to 0. \quad \textbf{(IV.4.12)}$$

The new potential V_1 near the origin corresponds therefore to the angular momentum $l' = l + 1$. This can also be verified on the solutions φ and ψ. From **(IV.4.4)**, we deduce

$$\frac{W[\varphi(E, r), \varphi(E_0, r)]}{E - E_0} = \int_0^r \varphi(E, t)\varphi(E_0, t)dt. \qquad \textbf{(IV.4.13)}$$

Now using this in the definition **(IV.4.2)** we find

$$\varphi_1(E, r) = \frac{C}{2l + 3} r^{l+2}[1 + O(r^\varepsilon)] \qquad \textbf{(IV.4.14)}$$

and

$$\psi_1(E, r) = \frac{(2l + 1)D}{r^{l+1}}[1 + O(r^\varepsilon)]. \qquad \textbf{(IV.4.15)}$$

Therefore, under a transformation with the regular solution, the singularity of the potential at the origin goes from l to $l + 1$, and regular and singular solutions go, respectively, to regular and singular solutions of the new potential. One can also see easily that a transformation using the irregular solution lowers the singularity of the potential at the origin from l to $l - 1$, the characters of the solution being again preserved.

We now come to the proof (Krein) that **(IV.3.19a)** is the solution of the Schrödinger equation with the potential V_2 of **(IV.3.18)**. For this, we perform first on the Schrödinger equation with the potential V_1 the transformation defined above, using its regular solution $\varphi_1(ib, r)$.[1] We distinguish all quantities

---

[1] Notice the change of notations. We shall have two steps now, $V_1 \to \bar{V}$ and $\bar{V} \to V_2$.

related to this new equation by a bar, $\bar{\varphi}, \ldots$ etc. Then we perform our transformation on this equation, using this time its singular solution $\bar{f}(ia, r)$ (obtained also from $f_1(ib, r)$ by the transformation with $\varphi_1(ib, r)$, as previously noted). Combining now the successive formulae

$$\bar{\varphi}(k, r) = \frac{W[\varphi_1(k, r), \varphi_1(ib, r)]}{(k^2 + b^2)\varphi_1(ib, r)} \tag{IV.4.16}$$

$$= \frac{1}{\varphi_1(ib, r)} \int_0^r \varphi_1(k, t)\varphi_1(ib, t)dt,$$

$$\bar{f}(ia, r) = \frac{W[f_1(ia, r), \varphi_1(ib, r)]}{(b^2 - a^2)\varphi_1(ib, r)} \equiv \frac{W(r)}{\varphi_1(ib, r)}, \tag{IV.4.17}$$

$$\varphi_2(k, r) = \frac{W[\bar{\varphi}(k, r), \bar{f}(ia, r)]}{\bar{f}(ia, r)}, \tag{IV.4.18}$$

$$V_2 - \bar{V} = -2\frac{d^2}{dr^2}\text{Log}[\varphi_1(ib, r)\bar{f}(ia, r)], \tag{IV.4.19}$$

we find that

$$\varphi_2(k, r) = \varphi_1(k, r) - \frac{f_1(ia, r)}{W(r)}\frac{W[\varphi_1(k, r), \varphi_1(ib, r)]}{k^2 + b^2} \tag{IV.4.20}$$

is the solution of the Schrödinger equation with the potential

$$V_2(r) = V_1(r) - 2\frac{d^2}{dr^2}\text{Log } W(r)$$

$$= V_1(r) - 2\frac{d^2}{dr^2}\text{Log}\frac{W[f_1(ia, r), \varphi_1(ib, r)]}{b^2 - a^2} \tag{IV.4.21}$$

which is the result we were looking for.

We now come to the proof of the theorem of Marchenko. Transformations studied so far and using regular solutions φ change the behavior of the potential near the origin only, increasing the singularity from l to $l + 1$. At infinity, the behavior of the potential increment depends on E_0. In all the cases considered above, the potential increment decreases exponentially. In order to get a real potential, it is obvious that we must assume E_0 real in order to secure the reality of ΔV in (IV.4.3), where now φ_0 is the regular solution vanishing at the origin. However, as is obvious from (IV.4.3), for E_0 positive, ΔV would have singularities at finite points on the r-axis because the regular solution vanishes infinitely many times for positive energies. Taking E_0 negative leads to an exponential decrease at infinity. It is then obvious that in order to have a polynomial decrease of ΔV at infinity, we must take $E_0 = 0$. We are going now to see that in this case, if the initial potential contains the centrifugal term $l(l + 1)r^{-2}$, the increment ΔV behaves like $2(l + 1)r^{-2}$ at infinity. Since it was shown before that it had also the same behavior near the origin, it is now obvious that the final total potential contains the centrifugal term $(l + 1)(l + 2)r^{-2}$, i.e., we have really added one unit to the angular momentum in the Schrödinger equation.

The proof of this statement goes as before. Since we start from angular momentum l, it is obvious that we have, for r very large,

$$\varphi_l(0, r) = Cr^{l+1}[1 + O(r^{-\delta})] \qquad (\text{IV.4.22})$$

and we can find, among the irregular solutions, a solution ψ such that

$$\psi_l(0, r) = Dr^{-l}[1 + O(r^{-\delta})]. \qquad (\text{IV.4.23})$$

If we use φ in (IV.4.2) and (IV.4.3), we get, for r very large,

$$\Delta V = 2(l + 1)r^{-2} + O(r^{-2-\delta}) \qquad (\text{IV.4.24})$$

whereas, using ψ in (IV.4.3), we find, again for large values of r,

$$\Delta V = -2lr^{-2} + O(r^{-2-\delta}). \qquad (\text{IV.4.25})$$

Thus, by using φ or ψ in transforming the Schrödinger equation, we increase or decrease, respectively, the angular momentum by one unit. We have to see now what is the "S-matrix" for the transformed equation. It is easily seen first that the Jost solutions are

$$f_{l+1}(k, r) = -\frac{W[f_l(k, r), \varphi_l(0, r)]}{ik\varphi_l(0, r)}, \qquad (\text{IV.4.26})$$

$$f_{l-1}(k, r) = -\frac{W[f_l(k, r), \psi_l(0, r)]}{ik\psi_l(0, r)}, \qquad (\text{IV.4.27})$$

and have the asymptotic property

$$f_{l\pm1}(k, r) = i^{l\pm1}e^{ikr} + o(1), \qquad r \to \infty. \qquad (\text{IV.4.28})$$

According to (I.5.25) the Jost solution is defined as

$$f_l(k, r) = (2l - 1)!!(kr)^{-l}F_l(k)[1 + O(r^\varepsilon)]. \qquad (\text{IV.4.29})$$

Now taking the same limit in (IV.4.26) and (IV.4.27), we find, for r going to zero,

$$f_{l+1}(k, r) = (2l + 1)!!(kr)^{-l-1}F_l(k)[1 + O(r^\varepsilon)] \qquad (\text{IV.4.30})$$

and

$$f_{l-1}(k, r) = (2l - 3)!!(kr)^{-l+1}F_l(k)[1 + O(r)]. \qquad (\text{IV.4.31})$$

It follows that the Jost function and therefore the "S-matrix" are invariant under the above transformations, whereas the angular momentum goes from l to $l + 1$ or $l - 1$, respectively. This completes the proof of the theorem of Marchenko.

IV.5   Comments and References

We have given, in the last chapter, the general references for the Gel'fand–Levitan method of solving the inverse problem. In the first section of the present chapter, the results are mainly due to Jost and Kohn (1953) who give the general formula (**IV.1.10**) for the potential increment. Formula (**IV.1.11**) giving the wave function, in the case of one bound state, was given previously by Krein (1953a, 1954). In section 2 on phase-equivalent potentials, the results are also mainly due to Jost and Kohn (1952b, 1953) and Holmberg (1952). In section 3 on Bargmann potentials, we follow the procedure developed by Theiss (1956), and Fulton and Newton (1956), which is based on ideas and methods of Bargmann (unpublished), who was the first to solve the inverse problem for rational Jost functions (see also Bargmann (1949a,b) and Blažek (1962b)). In section 4, the theorem stated at the beginning is due to Marchenko (1956). The general transformations of this section were first used by Crum (1955) to lessen the number of eigenvalues by one. Krein (1956) extended Crum's method and applied the results to get a complete characterization of the spectral function of a potential with a $l(l + 1)/r^2$ singularity at the origin. Marchenko (1956) made use of an analogous transformation to prove his theorem. Our presentation follows that of the review article of Faddeev (1963).

Bargmann potentials in the case where spin-orbit and tensor forces are present are treated thoroughly in Fulton and Newton (1956).

For other references, see chapters V, VI, VII, and IX.

The transformations of section IV.4 are not isolated to the study of inverse problems. Others are found in XVII.3.2, XVII.4.2, XVII.4.3, and XVII.4.4. Darboux and Backlund transformations, which are of first importance in the inverse method for solving nonlinear evolution equations, are found in section XIX.1.2 where it is shown how transformations can be used to understand ill-posedness problems, and in section XV.2 where we cite the transmutation theory, which is (also) a powerful method for constructing transformations. Comments on these sections will give further references on this subject.

For the application of the transformations of section 4 to the ordering of the levels of confining potentials, see Baumgartner et al. (1985) and Common and Martin (1987).

CHAPTER V

The Marchenko Method

V.1   The Levin Representation

The Levin representation is the analogue of the representation **(III.1.8a)**, for the Jost solution $f(k, r)$. We know from chapter I that this function is holomorphic in Im $k > 0$, and satisfies there, for $r \neq 0$, the bound **(I.3.4)**. It follows from this that $h = f - e^{ikr}$ is an L^2 function of k on any line parallel to the real axis, and that

$$\int_{-\infty}^{\infty} |h(\sigma + i\tau)|^2 \, d\sigma = O(e^{-2\tau r}).$$

(V.1.1)

We know also that $f(k, r)$, and therefore $h(k, r)$, are continuous on the real k-axis. To such functions, we can apply the following theorem:

Theorem (Titchmarsh).[1]   *A necessary and sufficient condition that* $F(x) \in L^2(-\infty, \infty)$ *should be the limit as* $y \to 0$ *of an analytic function* $F(z = x + iy)$ *in* $y > 0$ *such that*

$$\int_{-\infty}^{\infty} |F(x + iy)|^2 \, dx = O(e^{2\lambda y})$$

(V.1.2)

is that

$$\int_{-\infty}^{\infty} F(x)e^{-itx} \, dx = 0$$

(V.1.3)

for all values of $x < -\lambda$

[1] Titmarsh, 1948, theorem 96, p. 129.

This theorem applies without any modification to $h(k, r)$, and gives us

$$A(r, t) \equiv \tilde{h}(t, r)$$

$$= (2\pi)^{-1} \int_{-\infty}^{\infty} [f(k, r) - e^{ikr}] e^{-ikt} \, dk = 0, \qquad t < r. \quad \textbf{(V.1.4)}$$

The inversion of this Fourier transform leads then to

$$f(k, r) = e^{ikr} + \int_{r}^{\infty} A(r, t) e^{ikt} \, dt \qquad \textbf{(V.1.5)}$$

valid everywhere in Im $k \geq 0$, $A(r, t)$ being an L^2-function in t for all $r > 0$, $t \geq r$.

The representation (**V.1.5**), which is the analogue of (**III.1.8a**), can now be substituted into the integral equation (**I.3.1**), and gives, after unfolding the Fourier transform, the integeral equation

$$A(r, t) = \frac{1}{2} \int_{(r+t)/2}^{\infty} V(s) ds$$

$$- \int_{(r+t)/2}^{\infty} ds \int_{0}^{(t-r)/2} V(s - u) A(s - u, s + u) du, \qquad t \geq r. \quad \textbf{(V.1.6)}$$

This equation, the analogue of (**III.1.10**), can now be solved by the method of successive approximations. It admits a unique solution satisfying

$$|A(r, t)| < \frac{1}{2} \int_{(r+t)/2}^{\infty} |V(s)| ds \, \exp\left[\int_{r}^{\infty} u|V(u)| du\right] \qquad \textbf{(V.1.7)}$$

under the condition $rV(r) \in L^1(b, \infty)$, $b > 0$. It is also easily seen from (**V.1.6**) that

$$\left|\frac{\partial}{\partial r} A(r, t) + \frac{1}{4} V\left(\frac{r+t}{2}\right)\right| \leq K \int_{r}^{\infty} |V(s)| ds \int_{(r+t)/2}^{\infty} |V(u)| du \quad \textbf{(V.1.8)}$$

and a similar bound for $(\partial/\partial t)A$, K being an appropriate constant.

All the formal steps in going from (**V.1.5**) to (**V.1.6**) are easily justified since we are dealing now, everywhere, with L^2 functions. Moreover, for $0 < r \leq t$, the integral equation (**V.1.6**) is meaningful under the sole condition of integrability of the potential $rV(r) \in L(b, \infty)$, $b > 0$. From (**V.1.6**), making $r = t$, we also find, according to (**I.6.1**),

$$A(r, r) = \frac{1}{2} \int_{r}^{\infty} V(t) dt = -\tfrac{1}{2} W(r). \qquad \textbf{(V.1.9)}$$

If the potential exists as a continuous function, we have

$$V(r) = -2 \frac{dA(r, r)}{dr}. \qquad \textbf{(V.1.10)}$$

These last two equations are analogues of (**III.1.12**) and (**III.1.16**). However, they have the advantage of being meaningful without any assumption about the integrability of the potential at the origin. This is the reason why the Levin representation (**V.1.5**), and the integral equation (**V.1.6**) are more general and more useful when dealing with potentials which are singular at the origin. We shall see in the next section that one can get from the Marchenko equation precise statements about the behavior of the potential near the origin.

We can again go further, and assume that A is twice differentiable in r and t. Then we get from (**V.1.5**) or (**V.1.6**), using (**V.1.9**), the differential equation (Marchenko)

$$\left(\frac{\partial^2}{\partial r^2} - \frac{\partial^2}{\partial t^2}\right) A(r, t) = V(r) A(r, t). \qquad (\mathbf{V.1.11})$$

This equation, the analogue of (**III.1.18**), can now be studied per se, as an alternative for the Schrödinger equation. Once the energy-independent solution $A(r, t)$ is found, we recover the solution of the Schrödinger equation through (**V.1.5**).

V.2   The Marchenko Integral Equation

We start here from the expansion theorem (**II.3.14a**), assuming for simplicity that there are no bound states. Using (**I.1.19**) and (**I.4.11**), it can be written

$$(2\pi)^{-1} \int_{-\infty}^{\infty} f(k, r)[f(-k, t) - S(k)f(k, t)]dk = \delta(r - t). \qquad (\mathbf{V.2.1})$$

On the other hand, exactly in the same way we found the representation (**V.1.5**), we can derive also

$$e^{ikr} = f(k, r) + \int_{r}^{\infty} \tilde{A}(r, t) f(k, t) dt. \qquad (\mathbf{V.2.2})$$

This can be done either by using the general theory of integral transforms with L^2 kernel $A(r, t)$, or, more simply, by considering (**V.1.5**) as a Volterra integral equation for e^{ikr} with the kernel A and the inhomogeneous term f. Because of (**V.1.7**), we can solve this equation, and obtain (**V.2.2**) with $\tilde{A}(r, t) \in L^2(r, \infty)$ in t for all $r > 0$.

Now combining (**V.2.1**) and (**V.2.2**), we get

$$\int_{-\infty}^{+\infty} f(k, r)[e^{-ikt} - S(k)e^{ikt}]dk = 0, \qquad r > t. \qquad (\mathbf{V.2.3})$$

If we substitute now in this equation the Jost solution f by its representation (**V.1.5**), we obtain

$$A(r, t) = A_0(r + t) + \int_{r}^{\infty} A(r, s) A_0(s + t) ds, \qquad t > r, \qquad (\mathbf{V.2.4a})$$

where

$$A_0(t) = (2\pi)^{-1} \int_{-\infty}^{\infty} [S(k) - 1] e^{ikt} \, dk. \qquad \text{(V.2.4b)}$$

When bound states are present, we obviously have

$$A_0(t) = (2\pi)^{-1} \int_{-\infty}^{\infty} [S(k) - 1] e^{ikt} \, dk + \sum_j s_j e^{-\gamma_j t}, \qquad \text{(V.2.4c)}$$

where the s_j's are the normalization constants of the bound states, (II.4.18b).

Equation (V.2.4a) with the kernel (V.2.4b or c) is the Marchenko equation. It is somewhat simpler than the Gel'fand–Levitan equation in that here the kernel is directly related to $S(k) = \exp(2i\delta)$, whereas with the Gel'fand–Levitan equation we have first to calculate the Jost function through (II.2.1), and use it to calculate the spectral density $d\rho(E)$ by (II.3.14b), which in turn is used in (III.2.5) and (III.2.7b).

We now have two problems to study, namely, the unicity of the solution of the Marchenko equation, and the verification that the potential we obtain has the desired properties. We shall not study the first problem. It can be done along the same lines as for the Gel'fand–Levitan equation (see section 2 of chapter III) with some extra work. Instead, we shall show that the Marchenko equation is equivalent to the Gel'fand–Levitan equation. Then the properties of the potential and their relation with those of $S(k)$ are obtained with the help of the Marchenko theory.

The proof of the equivalence of the two formulations of the inverse problem goes as follows. Let us consider the function

$$\bar\varphi(k, r) = \frac{\varphi(k, r)}{F(-k)F(k)} = (2ik)^{-1}\left[\frac{f(k, r)}{F(k)} - \frac{f(-k, r)}{F(-k)} \right] \qquad \text{(V.2.5)}$$

and the related kernel (see chapter III, formulas (III.1.5a) and (III.1.6))

$$\begin{aligned}
\bar K(r, t) &= \frac{2}{\pi} \int_0^{\infty} \left[\frac{\varphi(k, r)}{|F(k)|^2} - \frac{\sin kr}{k} \right] \frac{\sin kt}{k} k^2 \, dk \\
&= \frac{1}{2\pi} \int_{-\infty}^{\infty} \left[\frac{f(k, r)}{F(k)} - e^{ikr} \right] e^{-ikt} \, dk
\end{aligned} \qquad \text{(V.2.6)}$$

which is meaningful for $r \neq t$ according to various asymptotic estimates established in chapter II. We assume first that there are no bound states. It then follows that

$$\frac{e^{-ikr} f(k, r)}{F(k)}$$

is holomorphic and bounded in the upper half-plane (see chapter II). Therefore, for $r > t$, we can close the path of integration in the upper half-plane, and use the residue theorem. Thus,

$$\bar K(r, t) = 0, \qquad r > t. \qquad \text{(V.2.7)}$$

If we now invert (**V.2.6**), we find

$$\frac{f(k, r)}{F(k)} = e^{ikr} + \int_r^\infty \overline{K}(r, t)e^{ikt} \, dt. \tag{V.2.8}$$

Now consider

$$\Pi(t) = \frac{1}{2\pi} \int_{-\infty}^\infty \left(\frac{1}{F(k)} - 1 \right) e^{-ikt} \, dk. \tag{V.2.9}$$

Since the function F^{-1} does not vanish in the upper half-plane, it follows that

$$\Pi(t) = 0, \qquad t < 0. \tag{V.2.10}$$

Recalling now the Levin representation (**V.1.5**), and using the convolution theorem, we find from (**V.2.8**) that

$$A(r, t) + \Pi(t - r) + \int_r^t A(r, s)\Pi(s - t)ds = \overline{K}(r, t). \tag{V.2.11}$$

It follows from this equation that

$$A(r, r) = \overline{K}(r, r) - \Pi(0) \tag{V.2.12}$$

and

$$V(r) = -2\frac{dA(r, r)}{dr} = -2\frac{d\overline{K}(r, r)}{dr}. \tag{V.2.13}$$

We now have to relate the kernel $\overline{K}$ to the kernel K of the Gel'fand–Levitan equation. This is most easily done in the following way. In the representation (**V.2.6**), we substitute φ by the right-hand side of (**III.1.8a**), take $r > t$, and use the middle expression in (**III.2.8a**). Recalling (**V.2.7**), formula (**V.2.6**) becomes

$$\overline{K}(r, t) = G(r, t) + K(r, t) + \int_0^r K(r, s)G(s, t)ds = 0 \tag{V.2.14}$$

which is the Gel'fand–Levitan integral equation. When $r < t$, $\overline{K}(r, t)$ is no longer zero. In fact, it is easy to see that, in this case, $\overline{K}(r, t)$ is identical to the resolvent kernel $\tilde{K}(t, r)$ introduced in chapter III, formulae (**III.1.21**) and (**III.1.22**). Indeed, we have, for $r < t$,

$$\overline{K}(r, t) = \int_0^\infty \varphi(k, r)\frac{\sin kt}{k} \, d\rho(k^2) - \delta(r - t). \tag{V.2.15}$$

It is now easy to calculate the integral by multiplying (**III.1.21**) by φ and integrating with $d\rho$. Taking into account the completeness relation (**II.3.14a**), we find

$$\overline{K}(r, t) \equiv \tilde{K}(t, r), \qquad r < t, \tag{V.2.16}$$

where $\tilde{K}$ is the resolvent kernel of the Gel'fand–Levitan equation. However, we know that $\tilde{K}$ satisfies the integral equation (III.1.22)

$$K(r, t) + \tilde{K}(r, t) + \int_t^r \tilde{K}(r, s)K(s, t)ds = 0. \qquad \text{(V.2.17)}$$

It follows that

$$\overline{K}(r, r) = -K(r, r). \qquad \text{(V.2.18)}$$

Therefore, the potential calculated via (V.2.13) is identical to the potential calculated via (III.1.16)

$$V(r) = 2\frac{dK(r, r)}{dr}. \qquad \text{(V.2.19)}$$

This proves the equivalence of the Marchenko method with the Gel'fand–Levitan approach.

We assumed so far that no bound state is present. If there are bound states, we have only to take care of the corresponding poles when calculating $\overline{K}$ in (V.2.6) and Π in (V.2.9) by contour integration in the upper half-plane. The extra terms we get from the residues cancel each other appropriately, and we again end up with (V.2.11), (V.2.13), and (V.2.19).

The last problem we study in this chapter is the integrability property of the potential. We shall see that $V(r)$ behaves in many respects like the derivative of $A_0(t)$. If we introduce

$$s(r) = \int_r^\infty |V(t)|dt, \qquad \text{(V.2.20a)}$$

$$s_1(r) = \int_r^\infty t|V(t)|dt, \qquad \text{(V.2.20b)}$$

the bounds (V.1.7) and (V.1.8) can be written

$$|A(r, t)| < Cs\left(\frac{r + t}{2}\right), \qquad \text{(V.2.21a)}$$

$$\left|\frac{\partial A}{\partial r} + \frac{1}{4}V\left(\frac{r + t}{2}\right)\right| < Cs(r)s\left(\frac{r + t}{2}\right), \qquad \text{(V.2.21b)}$$

where C is an appropriate constant containing $\exp s_1(0)$.

From the Marchenko integral equation, we have

$$A(r, r) = A_0(2r) + \int_r^\infty A(r, t)A_0(r + t)dt \qquad \text{(V.2.22a)}$$

which can be written

$$A_0(2r) = A(r, r) + \int_r^\infty A(r, 2t - r)A_0(2t)dt. \qquad \text{(V.2.22b)}$$

This equation may now be solved for A_0 by iteration, and one easily finds

$$|A_0(2r)| < Cs(r). \tag{V.2.23}$$

If we now differentiate (**V.2.22b**) with respect to r, and use (**V.2.20b**) and (**V.2.23**), we obtain

$$|A_0'(2r) + \tfrac{1}{4}V(r)| < Cs^2(r). \tag{V.2.24}$$

Along similar lines, it is not difficult to find estimates similar to (**V.2.23**), namely

$$|A_0'(2r) + \tfrac{1}{4}V(r)| < C\tau^2(2r), \tag{V.2.25a}$$

where

$$\tau(r) = \int_r^\infty |A_0'(t)| \, dt. \tag{V.2.25b}$$

The bounds (**V.2.24**) and (**V.2.25a**) show the strong similarity between $F'(2r)$ and the potential. In particular, it is easily seen that they entail the complete equivalence between

$$s_1(0) = \int_0^\infty r|V(r)| \, dr < \infty \tag{V.2.26a}$$

and

$$\tau_1(0) = \int_0^\infty t|A_0'(2t)| \, dt < \infty. \tag{V.2.26b}$$

For other relations between the properties of the potential and those of the S-matrix or the Jost function, the reader is referred to the papers quoted in the references. It is clear from the analysis of this chapter, and of chapter III, that the conclusions here are the same as those of that chapter as regards the relation between the phase shift and the potential, i.e., the most general conditions under which there is a one-to-one correspondence between a short range potential and the phase shift are those given in the sixth section of chapter III.

It is interesting to notice that $A(r, r')$ (or $V(r)$) can be expanded in terms of products of the perturbed and unperturbed wave functions. From (**V.2.4a**), (**V.2.4b**), and (**V.1.5**), we easily derive the expansion (valid if no bound states):

$$A(r, t) = (2i\pi)^{-1} \int_{-\infty}^{+\infty} [S(k) - 1] f(k, r)e^{ikt} \, dk \tag{V.2.27}$$

with trivial additions in presence of bound states. Similar results could be derived in the Gel'fand–Levitan method and in the inverse problem at fixed energy (see e.g. (**XII.2.3**)). Such formulas (for $t = r$, or for $V(r)$) were used in certain iterative methods.

Concluding Summary of the Marchenko Method. The physicist who wants to derive the potential from the scattering data should first derive $A_0(t)$ from $S(k)$ and the information available on bound states,[2] using (**V.2.4b**) or (**V.2.4c**). He should then solve the Marchenko equation (**V.2.4a**) and thus obtain the potential from $A(r, r)$ via (**V.1.10**).

Remarks. Our derivations of the Gel'fand–Levitan and Marchenko methods are the most commonly used in theoretical physics. A general scheme for these derivations is given in Section (XIII.1), and one can find many examples in this book (e.g. in chapter IX)—one can call these derivations the standard derivations. From the point of view of a mathematician, certain steps in these derivations look sketchy. The reader who is interested in very complete and rigorous mathematical derivations is referred to the following papers:

1. For the step spectral function → potential (Gel'fand–Levitan method), see Gel'fand–Levitan (1951) or Levitan–Gasymov (1964).
2. For the step $\delta_0(k) \to$ spectral function, and the consistency of the Gel'fand–Levitan method, see Faddeyev (1959).
3. For Marchenko's method, see Agranovich and Marchenko (1963).

However, it should be noticed that many steps which have been written here according to the standards of physicists can easily be rewritten to meet the standards of mathematicians. As an example, we can avoid using δ-functions in deriving Marchenko's equation:

We assume for the sake of simplicity that there is no bound state. Studies of the direct problem have given the representation (**V.1.5**), with $t \to A(r, t) \in L_1 \cap L_2$. Besides, $S(k) - 1$ is the Fourier transform of a function A_0 of $L_1(R)$:

$$S(k) = 1 + \int_{-\infty}^{+\infty} A_0(t)e^{-ikt}\, dt. \tag{V.2.28}$$

From (**I.1.19**) (with $l = 0$), (**V.1.5**), (**I.4.11**), and (**V.2.28**), we obtain the equation

$$2ik\left[\varphi(k, r)\left(\frac{1}{\Gamma(k)} - 1\right) + \left(\varphi(k, r) - \frac{\sin kr}{k}\right)\right] = \int_{-\infty}^{+\infty} B(r, t)e^{-ikt}\, dt, \tag{V.2.29}$$

where

$$B(r, t) = A_0(r + t) + \int_r^{\infty} A_0(t + u)A(r, u)du - \theta(t - r)A(r, t)$$

$$+ \theta[-(t + r)]A(r, -t) \tag{V.2.30}$$

belongs to L_1, and θ is the step function.

---

[2] The consistency conditions are those of section III.6. In particular, $S(k)$ must be of the form $\exp[2i\delta(k)]$, where $\delta(k)$ is real for real k (unitarity) and satisfies the Levinson formula (**II.1.11**).

The Fourier transform $t \to \tilde{B}(r, t)$ of the right-hand side is almost everywhere equal to $t \to B(r, t)$ for $t \in R$, and, in particular, for $t \geq r \geq 0$. We can evaluate $\tilde{B}(r, t)$ by the formula

$$\tilde{B}(r, t) = \frac{1}{2\pi} \lim_{R \to \infty} \int_{-R}^{R}$$

$$\times 2ik \left[\varphi(k, r)\left(\frac{1}{\Gamma(k)} - 1\right) + \left(\varphi(k, r) - \frac{\sin kr}{k}\right) \right] e^{ikt} \, dk. \quad \text{(V.2.31)}$$

The integrand in (V.2.31) is an analytic function of k in the half-plane Im $k > 0$, and, because of the bound (I.2.8), it goes to zero exponentially (for any $t > r \geq 0$), as $|k| \to \infty$, Im $k > 0$. Besides, it goes to zero as $k \to \pm\infty$. A contour integration readily shows that $\tilde{B}(r, t)$ is zero for $t > r \geq 0$ so that $B(r, t) = 0$ almost everywhere for $t > r \geq 0$. Thus we obtain almost everywhere the Marchenko equality:

$$A(r, t) \overset{\text{a.e.}}{=} A_0(r + t) + \int_{r}^{\infty} A_0(t + u)A(r, u)du \qquad (t > r \geq 0) \qquad \text{(V.2.32)}$$

and this can be extended everywhere (for $t \geq r \geq 0$) by simple continuity arguments.

For higher l, we can proceed as before, using integral transforms with Hankel functions w_l instead of the Fourier transform. It is known that these transforms have essentially the same properties as Fourier transforms. Most importantly, we have the Titchmarch theorem given earlier, where we have to replace e^{-itx} in (V.1.3) by $w_l(-tx)$. In this way, we get

$$f_l(k, r) = w_l(kr) + \int_{r}^{\infty} A_l(r, t)w_l(kt)dt, \qquad \text{(V.2.33)}$$

and the differential equations

$$\left[\left(\frac{\partial^2}{\partial r^2} - \frac{l(l + 1)}{r^2}\right) - \left(\frac{\partial^2}{\partial t^2} - \frac{l(l + 1)}{t^2}\right) \right] A_l(r, t) = V(r)A_l(r, t). \quad \text{(V.2.34)}$$

In order for f_l to be the right solution of the Schrödinger equation, we must have

$$V(r) = -2\frac{d}{dr} A_l(r, r), \qquad \text{(V.2.35)}$$

and

$$\lim_{t \to \infty} A_l(r, t) = \lim_{t \to \infty} \frac{\partial}{\partial t} A_l(r, t) = 0. \qquad \text{(V.2.36)}$$

The rest of the analysis follows as before, and we finally get the Marchenko integral equation

$$A_l(r, s) + \int_{r}^{\infty} A_l(r, t)A_l^0(s + t)dt + A_l^0(s + r) = 0, \qquad \text{(V.2.37)}$$

where

$$A_l(r + s) = \frac{1}{2\pi} \int_{-\infty}^{\infty} [S_l(k) - 1] w_l(kr) w_l(ks) k^2 \, dk + \sum_j s_{lj} w_l(i\gamma_j r) w_l(i\gamma_j s).$$

$$\text{(V.2.38)}$$

Again, we can follow Marchenko's analysis step by step, and prove the existence and unicity of $A_l(r, s)$, and of the potential. Notice here that, because of **(II.4.21)** which means $S_l(k) - 1 = O(k^{2l+1})$, the integrand in **(V.2.38)** is well defined at $k = 0$. Also, we have, according to **(II.1.14′)** and **(II.1.13b)**,

$$(-)^l s_{lj} = -C_{lj}^{-1} \frac{4i\gamma_j^{2l+2}}{[\dot{F}(i\gamma_j)]^2}.$$

$$\text{(V.2.39)}$$

V.3   Comments and References

The representation **(V.1.5)** is due to Levin (1956). The integral equation **(V.2.4a)** is due to Marchenko (1955). Other results and conclusions of this chapter, especially the intimate relation between the absolute integrability of $rV(r)$ and $tA_0'(t)$ are due to Agranovich and Marchenko (1957, 1958, 1963). Other works on the relation between the behavior of the potential and the behavior of the Jost function or the S-matrix can be found in Neuhaus (1955), Friedman (1957), Jost (1956), Newton (1956). A complete study of the consistency of the Marchenko method in radial problems appears, for instance, in Ramm (1988a).

The book by Agranovich and Marchenko (1963) treats the general case of coupled equations in matrix notations, and gives a good list of references.

The relationship between the Gel'fand–Levitan kernel K and the kernel A, **(V.2.11)** and **(V.2.16)**, is due to Faddeev (1963).

The stability of Marchenko's method has been studied by Marchenko (1968). See also Aktosun (1987b).

Generalization to the class of singular potentials of chapter I is due to Baeteman and Chadan (1975, 1976).

Conditions for which the potential is of finite range were given by Ramm (1965), and Chadan and Montes (1968).

The generalization of the Marchenko equation to higher waves was given by Blazek (1966). See also Trlifaj (1981).

Marchenko equations have also been obtained in many "nonradial" inverse problems. In the Schrödinger inverse problem "on the line," studied in chapter XVII, the Faddeev method quite naturally leads us to a Marchenko equation very similar to the one of chapter V (see sections XVII.2.1 and XVII.2.2). Besides, the method given in section XVII.2.1 could have been used above. More generally, several inverse problems can be reduced to "Riemann–Hilbert" problems whose solution is obtained by a Marchenko-type method, so that a Marchenko equation can appear in problems with no symmetry (section XIV.5).

Comments on these sections will give references to these generalizations of the Marchenko equation.

CHAPTER VI

Examples

VI.1   Bargmann Potentials

We consider here some simple examples of Bargmann potentials. These are, in general, potentials for which the Jost function is a rational function of k. In fact, in the third section of chapter IV we studied an even more general case consisting of an initial arbitrary potential for which everything is known, and an increment which has the effect of multiplying the initial Jost function by a rational function of k.

Here, we consider simply the case of S-wave, where the initial potential is zero. The simplest such case is when the Jost function has one zero and one pole. Because of the symmetry property (**I.3.14**), we have then

$$F(k) = \frac{k + ia}{k + ib},$$ (**VI.1.1**)

where a and b are both real, and we have necessarily $b > 0$. When $a > 0$, we have no bound state. When $a < 0$, we have a bound state of energy $-a^2$. Finally, for $a = 0$, the bound state goes into a "resonance" at zero energy.

No Bound State. The method developed in the third section of Chapter IV takes now its simplest form. Because of

$$F_1(k) = 1,$$
$$f_1(k, r) = e^{ikr},$$
$$\varphi_1(k, r) = \frac{\sin kr}{k},$$

and

$$|F(k)|^{-2} - 1 = \frac{b^2 - a^2}{k^2 + a^2}, \qquad k \text{ real}, \qquad \text{(VI.1.2)}$$

we have, in accordance with (IV.3.7),

$$G(r, t) = -(a^2 - b^2)e^{-ar}\frac{\sinh at}{a}, \qquad r \geq t, \qquad \text{(VI.1.3)}$$

and the symmetric expression for $r < t$. Solving now the Gel'fand–Levitan equation with this kernel, we obtain, for $r \geq t$,

$$K(r, t) = -2\frac{(b - a)e^{-br}}{1 + \beta e^{-2br}}\sinh bt, \qquad \text{(VI.1.4)}$$

$$V(r) = -8b^2\beta\frac{e^{-2br}}{(1 + \beta e^{-2br})^2}, \qquad \text{(VI.1.5)}$$

where $\beta = (b - a)/(b + a)$. This potential is one of the Eckart potentials (Bargmann, 1949a, b).

It is easily seen that the above potential corresponds to

$$k \cot \delta = \frac{ab}{b - a} + \frac{1}{b - a}k^2 = -\frac{1}{a_0} + \frac{1}{2}r_0 k^2. \qquad \text{(VI.1.6)}$$

It corresponds to the case where the effective range approximation is exact at all energies. It is a two-parameter family of potentials for which the two parameters may conveniently be taken as the scattering length a_0 and the effective range r_0, so that

$$a = r_0^{-1}[-1 + (1 - 2r_0 a_0^{-1})^{1/2}], \qquad b = r_0^{-1}[1 + (1 - 2r_0 a_0^{-1})^{1/2}].$$
$$\text{(VI.1.7)}$$

If $b > a > 0$, we have $r_0 > 0$, $a_0 < 0$. The potential, although attractive, is not strong enough to produce a bound state. When $a > b > 0$, we have $r_0 < 0$, $a_0 > 0$, and the potential is everywhere repulsive. The effective range theory of low energy neutron–proton interaction in the singlet state corresponds to the case $b > a > 0$.

The normalized regular wave function is given by

$$\varphi(k, r) = \frac{\sin kr}{k} + \frac{b^2 - a^2}{k(k^2 + b^2)}\frac{k \sinh(br)\cos(kr) - b\cosh(br)\sin(kr)}{b\cosh(br) + a\sinh(br)} \qquad \text{(VI.1.8)}$$

and the Jost solution by

$$f(k, r) = e^{ikr}\left[1 - \left(\frac{2ib}{k + ib}\right)\frac{\beta e^{-2br}}{1 + \beta e^{-2br}}\right]. \qquad \text{(VI.1.9)}$$

We notice that the asymptotic tail of the potential is given, in general, by e^{-2br}, except when $b = 0$. In this case, we have

$$V(r) = \frac{2a^2}{(1 + ar)^2}. \qquad \text{(VI.1.10)}$$

This is a long-range potential violating the condition (**I.1.3**). The phase shift, normalized to zero at infinity. is $-\pi/2$ at zero energy, and the Levinson theorem is violated.

When $a = 0$, we have a "resonance" at zero energy. The phase shift is $\pi/2$ there, and we have an infinite cross section. The corresponding potential is

$$V(r) = \frac{-2b^2}{\cosh^2 br}. \tag{VI.1.11}$$

One Bound State. Here, we have $a < 0$, and therefore a bound state of energy $-a^2$. There is now a whole family of phase-equivalent potentials differing from each other by the normalization C of the bound state. We can proceed here in two ways: either solve the problem directly for the whole class, which now contains three parameters, a, b, and the normalization constant C, or write the Jost function as

$$F(k) = \frac{k - ia}{k + ib}\frac{k + ia}{k - ia} = F_1(k)F_2(k) \tag{VI.1.12}$$

solve the problem for F_1, which has no bound state, and then introduce F_2 and the bound state by the method developed in the first section of chapter IV. Both methods are again elementary and lead to simple algebraic calculations. We shall adopt here the notations of (**IV.2.8, 2.9, and 2.10**). For $\lambda = 0$, we take everything given by the same formulae as in the case with no bound state:

$$V_0(r) = -8b^2 \frac{\beta e^{-2br}}{(1 + \beta e^{-2br})^2} \tag{VI.1.13}$$

with $\beta = (b - a)/(b + a) = (b + \gamma)/(b - \gamma)$, and, again,

$$k \cot \delta = \frac{-b\gamma}{b + \gamma} + \frac{1}{b + \gamma} k^2. \tag{VI.1.14}$$

We have also the same relations (**VI.1.7a,b**) between the scattering length, the effective range, a, and b. The regular solution is again (**VI.1.8**), and the Jost solution (**VI.1.9**). Remember that, here, we have $a = -\gamma$. Finally, the bound state wave function is obtained by making $k = -ia$ in (**VI.1.8**) or in (**VI.1.9**). The result is, according to (**I.1.19**),

$$\varphi_0(r) = \frac{-1}{2\gamma} F(-i\gamma)f(i\gamma, r) = -\frac{1}{b - \gamma} f(i\gamma, r), \tag{VI.1.15a}$$

$$\psi_0(r) \equiv f(i\gamma, r) = e^{-\gamma r} \frac{(b - \gamma)\sinh br}{b \cosh br - \gamma \sinh br}. \tag{VI.1.15b}$$

The normalization of bound state wave function is given, according to (**II.1.14**), by

$$C_0^{-1} = \int_0^\infty \varphi_0^2(r)dr = \frac{1}{2\gamma(b^2 - \gamma^2)}. \tag{VI.1.16}$$

We see that it is a given function of a and b.

The fundamental difference between this and the previous case is that now a cannot be too large (negative). Indeed, if $a < -b$, the potential V_0 has a double pole somewhere on the positive r-axis, in which case it has no longer any physical meaning. One also sees on (VI.1.16) that C_0 becomes negative.

From the above formulae, we can now construct the whole phase-equivalent family and their wave functions, for any value of λ via (IV.2.7)–(IV.2.13b). Their complete expressions being too cumbersome, we give only the essential results, from which the reader can calculate easily the wave functions.

As is obvious from (VI.1.13), V_0 has the asymptotic tail e^{-2br}, whereas all other members of the family, as is clear from (IV.2.9), have the longer tail $e^{-2\gamma r}$ because now $\gamma < b$. V_0 is therefore the unique member of the family which has the shortest tail, or the most rapid decrease at infinity. The function $N(\lambda, r)$ of chapter IV is given by

$$N(\lambda, r) = 1 + \frac{\lambda}{8\gamma} \frac{b - \gamma}{b + \gamma} \frac{b \cosh br + \gamma \sinh br}{b \cosh br - \gamma \sinh br} e^{-2\gamma r}. \qquad \text{(VI.1.17)}$$

The integrals in (IV.2.13a,b) can be easily calculated by the Wronskian method (IV.4.4) valid for any two solutions of the Schrödinger equation at different energies, and one would therefore get explicit analytical expressions for the regular solution and the Jost solution. Thus

$$f(\lambda; k, r) = f_0(k, r) + \frac{\lambda \psi_0(r)}{4N(\lambda, r)} \frac{1}{k^2 + \gamma^2} W[f_0(k, r), \psi_0(r)] \qquad \text{(VI.1.18)}$$

and

$$\varphi(\lambda; k, r) = \varphi_0(k, r) + \frac{\lambda \psi_0(r)}{4N(\lambda, r)} \frac{1}{k^2 + \gamma^2} W[\varphi_0(k, r), \psi_0(r)]. \qquad \text{(VI.1.19)}$$

A more compact form of writing the potential is

$$V(\mu, r) = -4\gamma \frac{d}{dr} \left[\sinh br \, \frac{g(\mu; \gamma, r)}{g(\mu; b + \gamma, r) - g(\mu; \gamma - b, r)} \right] \qquad \text{(VI.1.20a)}$$

where

$$g(\mu; x, r) = x^{-1}(e^{-xr} + \mu \sinh xr). \qquad \text{(VI.1.20b)}$$

The normalized bound state wave function becomes then

$$(C_0)^{1/2} \varphi_0(\mu, r) = 2 \left(\frac{\mu\gamma}{b^2 - \gamma^2} \right)^{1/2} \frac{\sinh br}{g(\mu, \gamma + b, r) - g(\mu, \gamma - b, r)}. \qquad \text{(VI.1.21)}$$

The relation between λ and μ is

$$\mu = 2 \left[1 + \frac{\lambda}{8\gamma} \frac{b - \gamma}{b + \gamma} \right]^{-1}, \qquad \lambda \in \left[-\frac{8\gamma(b + \gamma)}{b - \gamma}, \infty \right] \qquad \text{(VI.1.22)}$$

and one sees that, as λ varies over its admissible range, μ varies between 0 and ∞. It is also seen that, for $\lambda = 0$, we have $\mu = 2$, and the potential has the shortest range.

As a special case, we may let $\gamma \to b$. V_0 is now identically zero because $\beta = \infty$, and so are $\varphi_0, \ldots$ etc. The whole phase-equivalent family therefore vanishes if λ is finite. However, if at the same time as we make $\gamma \to b$, we let also $\mu \to \infty$ such that

$$2\mu(b - \gamma) = A^{-1}. \tag{VI.1.23}$$

It can be seen on (**VI.1.20**) that the limiting potential is

$$V(r) = 2\frac{\sinh(\gamma r)[2\gamma(\tfrac{1}{2}r - A)\cosh(\gamma r) - \sinh(\gamma r)]}{[A + \tfrac{1}{2}((\sinh 2\gamma r/2\gamma) - r)]^2}, \tag{VI.1.24}$$

where $\gamma^2 = a^2 = b^2$ is the energy of the bound state. From the Jost function

$$F(k) = \frac{k - i\gamma}{k + i\gamma} \tag{VI.1.25}$$

we find

$$\cos \delta = \frac{k^2 - \gamma^2}{k^2 + \gamma^2}, \qquad \sin \delta = \frac{2k\gamma}{k^2 + \gamma^2}. \tag{VI.1.26}$$

The wave functions are given by

$$\varphi(k, r) = \frac{\sin kr}{kr} - \frac{\sinh(\gamma r)[\gamma \sin(kr)\cosh(\gamma r) - k\cos(kr)\sinh(\gamma r)]}{k(k^2 + \gamma^2)[A + \tfrac{1}{2}((\sinh(2\gamma r)/2\gamma) - r)]^2} \tag{VI.1.27}$$

and

$$\varphi_0(r) = \frac{A \sinh \gamma r}{\gamma[A + \tfrac{1}{2}((\sinh 2\gamma r/2\gamma) - r)]}. \tag{VI.1.28}$$

It is easily seen that the normalization constant of the bound state is

$$C_0^{-1} = \frac{A}{\gamma^2}. \tag{VI.1.29}$$

As a limiting case, we can now make $\gamma \to 0$. The limiting potential is, with $(2\gamma^2/A) = \mu$ finite,

$$V(r) = \frac{6r(r^3 - 2\mu)}{(r^3 + \mu)^2}. \tag{VI.1.30}$$

This is a long-range potential which violates (**I.1.3**). Here, we have a genuine bound state at zero energy, whose wave function is

$$\varphi_0(r) = \frac{\mu r}{\mu + r^3}. \tag{VI.1.31}$$

The Jost function being identically 1, the phase shift is identically zero.[1] This is also seen immediately on the wave function

$$\varphi(k, r) = k^{-1}\sin(kr) + 3r^2(r^3 + \mu)^{-1}k^{-2}\cos(kr) - 3r(r^3 + \mu)^{-1}\frac{\sin kr}{k^3}. \tag{VI.1.32}$$

---

[1] Because of this property, these potentials are called transparent (see chapter VIII for more examples).

A More General Case. In the two previous cases, we had basically two parameters in the scattering data, the scattering length a_0, and the effective range r_0. The binding energy $-a^2$ was a function of these two parameters. In the present case, we consider again the effective range formula (VI.1.14) and assume that it is exact for all energies. However, we assume now that the binding energy $-\gamma^2$ is independent of a_0 and r_0, i.e., the Jost function has the two zeros $k = \pm i\gamma$, and two poles:

$$F(k) = \frac{k^2 + \gamma^2}{[k + i(\beta - \sigma)][k + i(\beta + \sigma)]},$$

$$\beta = r_0^{-1}, \qquad \sigma = r_0^{-1}(1 - 2r_0 a_0^{-1})^{1/2}, \tag{VI.1.33}$$

and we assume of course $(2r_0/a_0) < 1$, $a_0 > 0$, $r_0 > 0$.

We can use here the method of the first section of chapter IV by considering first the auxiliary potential V_1 which has no bound state and is such that $|F_1(k)| = |F(k)|$. Its Jost function is given by

$$F_1(k) = \frac{(k + i\gamma)^2}{[k + i(\beta - \sigma)][k + i(\beta + \sigma)]}. \tag{VI.1.34}$$

The kernel G of the Gel'fand–Levitan equation, which permits to calculate V in terms of V_1 has now only one term, and so everything is trivial. If we write

$$\gamma = (\beta + \sigma)\frac{1 + \lambda}{1 - \lambda}, \qquad -1 < \lambda < 1, \tag{VI.1.35}$$

then $V_1(r)$ calculated according to the third section of chapter IV, can be written as

$$V_1(r) = 2\beta\sigma\{(\beta + \lambda\sigma)^2(\sigma + \lambda\beta)^2(\beta - \sigma)^2[e^{1/2(\beta + \sigma)r} - \lambda^2 e^{-1/2(\beta + \sigma)r}]^2$$
$$- \lambda^2(\beta + \sigma)^2[(\beta + \lambda\sigma)^2 e^{1/2(\beta - \sigma)r} - (\sigma + \lambda\beta)^2 e^{-1/2(\beta - \sigma)r}]^2\}$$
$$\times \{(\beta + \lambda\sigma)^2[(\sigma e^{\beta r} + \beta\lambda^2 e^{-\sigma r})] - (\sigma + \lambda\beta)^2[\beta e^{\sigma r} + \sigma\lambda^2 e^{-\beta r}]\}^{-2}$$

$$\tag{VI.1.36}$$

and the corresponding regular function at $E = k^2 = -\gamma^2$ is

$$\varphi_1(r) = \varphi_1(i\gamma, r)$$

$$= e^{\gamma r}(1 - \lambda^2)\frac{(\beta + \lambda\sigma)[\sigma e^{\beta r} + \beta\lambda e^{-\sigma r}] - (\sigma + \lambda\beta)[\beta e^{\sigma r} + \lambda\sigma e^{-\beta r}]}{(\beta + \lambda\sigma)^2[\sigma e^{\beta r} + \beta\lambda^2 e^{-\sigma r}] - (\sigma + \lambda\beta)^2[\beta e^{\sigma r} + \lambda^2 \sigma e^{-\beta r}]}. \tag{VI.1.37}$$

Remember that this does not correspond to a bound state.

The integral equation for $K(r, t)$ is

$$K(r, t) + C\varphi_1(r)\varphi_1(t) + C\varphi_1(t)\int_0^r K(r, s)\varphi_1(s)ds = 0$$

with the solution

$$K(r, t) = -\frac{C\varphi_1(r)\varphi_1(t)}{1 + C\int_0^r \varphi_1^2(t)dt} \qquad \text{(VI.1.38)}$$

and the potential

$$V(r) = V_1(r) - 2C\partial_r \frac{\varphi_1^2(r)}{1 + C\int_0^r \varphi_1^2(t)dt}. \qquad \text{(VI.1.39)}$$

The bound state function is

$$\varphi(i\gamma, r) = \varphi_1(i\gamma, r) + \int_0^r K(r, t)\varphi_1(i\gamma, t)dt$$

$$= \frac{\varphi_1(i\gamma, r)}{1 + C\int_0^r \varphi_1^2(i\gamma, t)^2 \, dt}. \qquad \text{(VI.1.40)}$$

VI.2   Singular Potentials

As a last example, we consider the case of singular oscillating potentials treated at the end of chapter I. We again use the Gel'fand–Levitan equation, and assume for simplicity that there are no bound states. Giving the Jost function, or the spectral density, amounts to giving $H(t)$ of formulae (**II.4.12**) and (**III.2.8**), which determines the kernel $G(r, t)$ of the integral equation. It was shown before (see the formula (**II.4.15**)) that at $t = r$, which is usually a singular point for G and K, the dominant part of H is

$$H(r - t) \simeq W(r - t), \qquad \text{(VI.2.1)}$$

where W is the integral of the potential, (**I.6.1**). To have an oscillating singular potential, we must therefore take $H(t)$ to be oscillating and singular at the origin.

An example of such kind is

$$H(t) = t^{-\alpha} \sin \frac{a}{t} e^{-\mu t}, \qquad \alpha < 1. \qquad \text{(VI.2.2)}$$

It corresponds to a short-range potential, and it is seen that

$$G(r, t) = \frac{\bar{G}(r, t)}{|r - t|^\alpha}, \qquad \text{(VI.2.3)}$$

where $\bar{G}$ is bounded.

According to the general theory of Fredholm equations with singular but integrable kernels, if one iterates the equation a sufficient number of times, one ends up with a finite kernel. The singularity is then only in the inhomogeneous term. With our kernel (**VI.2.3**), when $\alpha \neq \frac{1}{2}$, one has to iterate n times, $n \geq (1 - \alpha)^{-1}$ (for $\alpha = \frac{1}{2}$, one has to iterate three times because of the presence of the logarithm in the second iterate). In any case, after having performed the necessary amount of iterations, we end up with

$$K + G_1 + \int KG_2 \, dt = 0,$$

where G_1 is the sum over the iterates of G, and G_2 is the last iterate, $G^{(n)}$, of G. One can then show that the singular part of V is obtained by taking $K = G_1$. For instance, for $\alpha = \frac{1}{2}$, one obtains that

$$V(r) \underset{r \to 0}{\simeq} 2e^{-2\mu r} \left[2a(2r)^{-5/2} \cos \frac{a}{2r} + (2r)^{-3/2} \sin \frac{a}{2r} + 2\mu r^{-1/2} \sin \frac{a}{2r} \right] + \cdots .$$

$$(VI.2.4)$$

The high-energy behavior of the phase shift is given by

$$\delta(k) \underset{k \to \infty}{\simeq} \frac{1}{2} \left(\frac{\pi}{2} \right)^{1/2} \text{Im}[(1 - i)(k + i\mu)^{-1/2}(e^{2ia^{1/2}(k+i\mu)^{1/2}} - e^{-2a^{1/2}(k+i\mu)^{1/2}})].$$

$$(VI.2.5)$$

One sees that it decreases like $k^{-1/2}$ while oscillating indefinitely.

Another example can be obtained by giving $\gamma(t)$ of equation (II.4.4). The Jost function is given now by (II.4.5) and (II.4.6). Taking, for instance,

$$\gamma(t) = t^{-1/2} \sin \frac{a}{t} \qquad (VI.2.6a)$$

we obtain

$$\delta(k) = -\frac{1}{2} \left(\frac{\pi}{2k} \right)^{1/2} [\sin 2(ak)^{1/2} - \cos 2(ak)^{1/2} - e^{-2(ak)^{1/2}}] \quad (VI.2.6b)$$

which is again oscillating. For $t \to 0$, one has

$$H(t) \approx \gamma(t) \qquad (VI.2.7)$$

and the integral equation leads to

$$V(r) = 2\partial_r \left[(2r)^{-1/2} \sin \frac{a}{2r} \right] + O(r^{-1}). \qquad (VI.2.8)$$

VI.3   Comments and References

Many examples of Bargmann potentials were first given by Bargmann himself (1949a,b). The examples (VI.1.5), (VI.1.10), (VI.1.11), (VI.1.13), and (VI.1.24) are due to him. The general forms (VI.1.17)–(VI.1.19) for phase-equivalent potentials are due to Jost and Kohn (1952b). The potentials (VI.1.5), (VI.1.13), for which the effective range formula is exact at all energies, were also treated by the Gel'fand–Levitan method by Jost and Kohn (1953). The compact forms (VI.1.20)–(VI.1.21) are due to Bargmann (unpublished), as quoted by Newton (1957a). The potential (**VI.1.24**) and its wave function (**VI.1.27**) were also given by Moses and Tuan (1959), who give also the limiting case (**VI.1.30**) together with (**VI.1.31**) and (**VI.1.32**).

The more general case corresponding to (**VI.1.33**) can be found in Bargmann (1949b), and also in Jost and Kohn (1953).

The simple Bargmann potentials for which the effective range theory is exact at all energies have been used by Chadan (1955, 1956), and Newton (1957a,b), for representing the low energy phenomenological neutron–proton interaction without tensor force. See also Blažek (1962a).

We have already mentioned in chapter III the work of Fulton and Newton (1956) on Bargmann potential for coupled equations in the presence of spin-orbit and tensor force. For applications to phenomenological neutron–proton interactions, see Newton and Fulton (1957). For other methods or further numerical calculations, see Swan and Pearce (1966), Swan (1967), and Benn and Scharf (1972).

Singular potentials studied in the second section of this chapter in the framework of Gel'fand–Levitan and Marchenko methods have been introduced by Baeteman and Chadan (1975, 1976).

The two-body problem at low energies in which the effective range expansion is first replaced by Padé approximants, and then used in connection with the Marchenko method of chapter V, is treated in detail in Lambert et al. (1975). They show that, in the absence of bound states, $\delta(0) - \delta(\infty) = 0$ is sufficient in order that a given Padé approximant $[M, N]$ of $\cot \delta$, with $M \leq N - 1$, be exactly reproducible by a regular Bargmann potential.

Attempts to apply the general exact methods given in chapters III and VI to problems which are not relevant to quantum mechanics appear for instance in geophysics; see, e.g., Weidelt (1972) and Sabatier (1974b,c). For general references on these interdisciplinary aspects, see Newton (1970) and Sabatier (1972b, 1975).

For Bargmann potentials which have bound states imbedded in the continuum, see Piovarchik et al. (1986) which contains references to previous works. See also the book by Eastham and Kalf: *Schrödinger-Type Operators with Continuous Spectra*, Pitman, London, 1982.

Multichannel and multidimensional Bargmann potentials are studied in Plekhanov et al. (1982).

Bargmann-type potentials also appear in inverse problems that are not radial problems at fixed l. They are then defined by their correspondence to a scattering function, which is a rational fraction with only a finite number of poles. Hence, when the inverse problem is used in an inverse method, they are also related with "multisolitons" solutions of nonlinear partial differential equations. We will find information on these generalized Bargmann potentials in sections XIX.1.3 and XIX.2.2 below, and further references in the Comments on these sections.

CHAPTER VII

Special Classes of Potentials

VII.1 Yukawa Potentials and the Direct Problem

Yukawa potentials have played an important role in particle physics and have been the subject of extensive studies in the past. Although the inverse problem for them can be dealt with either by the Gel'fand–Levitan or by the Marchenko method, a special technique has been found by Martin which permits solving the direct as well as the inverse problem by an iterative method based on special properties of the Laplace transform. We follow his method.

Our purpose here is to show that the inverse scattering problem for Yukawa potentials can be solved in the following way. We calculate first the S-matrix for real values of momentum k from the phase shift. If the scattering data correspond to Yukawa potentials, the S-matrix must necessarily be holomorphic in the k-plane with the two cuts (Martin, 1959, 1965; De Alfaro and Regge, 1965)

$$ k = \pm i\chi, \qquad \chi > \frac{\mu_0}{2}, $$

along the imaginary axis. In the upper half-plane we know that its poles correspond to the bound states, and are located on the imaginary axis. They are simple. All these singularities (cuts and poles) can in principle be obtained by analytic continuation from the real axis. What we are going to show is that the discontinuity of the S-matrix along its cut in the upper half-plane determines in a simple way the potential.

We consider again the S-wave for simplicity, and assume the potential to be given by

$$V(r) = \int_{\mu_0}^{\infty} C(\alpha)e^{-\alpha r}\, d\alpha, \qquad \mu_0 > 0, \qquad \text{(VII.1.1)}$$

where $C(\alpha)$ is a bounded function with some integrability properties at infinity that we shall specify later. Notice that the single Yukawa potential $e^{-\mu r}/r$ is obtained by taking $C(\alpha) = 1$. The Jost solution $f(k, r)$ can be written

$$f(k, r) = e^{ikr}g(k, r), \qquad \text{(VII.1.2)}$$

where g satisfies the differential equation

$$g'' + 2ikg' = Vg. \qquad \text{(VII.1.3)}$$

The boundary condition (**I.1.15**) now becomes

$$\lim_{r \to \infty} g(k, r) = 1. \qquad \text{(VII.1.4)}$$

This suggests to use the "Ansatz"

$$g(k, r) = 1 + \int_0^{\infty} \rho(k, \alpha)e^{-\alpha r}\, d\alpha \qquad \text{(VII.1.5)}$$

which is quite natural because the potential is a Laplace transform, and also because of (**VII.1.4**). Substitution of (**VII.1.5**) and (**VII.1.1**) into (**VII.1.3**) leads then to

$$\alpha(\alpha - 2ik)\rho(k, \alpha) = C(\alpha) + \int_0^{\alpha} C(\alpha - \beta)\rho(k, \beta)d\beta. \qquad \text{(VII.1.6)}$$

This is a Volterra integral equation in which k enters as a parameter. As long as k is real, it is a regular equation, and shows immediately that

$$\rho(k, \alpha) = 0, \qquad \alpha < \mu_0, \qquad \text{(VII.1.7)}$$

because of (**VII.1.1**). Therefore, the integration in (**VII.1.5**) runs from μ_0 to infinity, and (**VII.1.6**) can be written, using the step function θ,

$$\alpha(\alpha - 2ik)\rho(k, \alpha) = C(\alpha) + \theta(\alpha - 2\mu_0) \int_{\mu_0}^{\alpha - \mu_0} C(\alpha - \beta)\rho(k, \beta)d\beta. \quad \text{(VII.1.8)}$$

This equation has the remarkable property of being exactly soluble, step by step, in successive intervals $[n\mu_0, (n + 1)\mu_0)$, that is, knowing $\rho(k, \alpha)$ in the interval $[\mu_0, n\mu_0)$, we can calculate it exactly in the next interval. Indeed, from (**VII.1.8**), we have, to begin with,

$$\rho(k, \alpha) = \frac{C(\alpha)}{\alpha(\alpha - 2ik)}, \qquad \mu_0 \le \alpha < 2\mu_0. \qquad \text{(VII.1.9)}$$

Suppose therefore that $\rho(k, \alpha)$ is known in the interval $[\mu_0, n\mu_0)$. It is clear that for $\alpha \in [n\mu_0, (n + 1)\mu_0)$, the integration in (**VII.1.8**) concerns only

$[\mu_0, n\mu_0)$, in which we know $\rho(k, \alpha)$ by hypothesis. Therefore, we obtain $\rho(k, \alpha)$ in the new interval by simple integration in the previous interval. Starting from (VII.1.9), we can, in this step by step method, find exactly $\rho(k, \alpha)$ in any finite interval after a finite number of steps. Thus, from (VII.1.9), we find

$$\rho(k, \alpha) = \frac{1}{\alpha(\alpha - 2ik)} \left[C(\alpha) + \int_{\mu_0}^{\alpha - \mu_0} \frac{C(\alpha - \beta)C(\beta)}{\beta(\beta - 2ik)} d\beta \right], \qquad 2\mu_0 \leq \alpha < 3\mu_0,$$

(VII.1.10)

In order to complete the proof that the solution we obtain in this way is indeed the solution of (VII.1.3), we have to show that the integral in (VII.1.5) converges at its upper limit. For this purpose, we have to impose some restrictions on $C(\alpha)$. We shall not enter into the details, and only give the results. We notice first that, in all the successive intervals, there is no need to keep k real. As long as k does not belong to $[-i\mu_0/2, -i\infty)$, all the successive integrands are meaningful. Consider therefore k outside this cut, i.e.,

$$\left| \arg k - \frac{3\pi}{2} \right| > \varepsilon.$$

(VII.1.11)

In this cut plane, we have

$$|\alpha - 2ik| > \alpha \sin \varepsilon,$$

(VII.1.12a)

$$|\alpha - 2ik| > 2|k| \sin \varepsilon.$$

(VII.1.12b)

These inequalities are obtained by projecting $\alpha - 2ik$ on the line $(0, 2ik)$, and $-2ik$ and $\alpha - 2ik$ on the real axis, respectively. Introducing

$$\tilde{\rho}(k, \alpha) = \alpha(\alpha - 2ik)\rho(k, \alpha)$$

(VII.1.13)

the integral equation becomes

$$\tilde{\rho}(k, \alpha) = C(\alpha) + \theta(\alpha - 2\mu_0) \int_{\mu_0}^{\alpha - \mu_0} \frac{C(\alpha - \beta)}{\beta(\beta - 2ik)} \tilde{\rho}(k, \beta) d\beta.$$

(VII.1.14)

Suppose now that there exists a positive increasing function $B(\alpha)$ such that, everywhere,

$$|C(\alpha)| \leq B(\alpha), \qquad B'(\alpha) \geq 0.$$

(VII.1.15)

We have, according to (VII.1.12a),

$$\left| \frac{C(\alpha - \beta)}{\beta(\beta - 2ik)} \right| < \frac{B(\alpha)}{\beta^2 \sin \varepsilon}.$$

(VII.1.16)

It is now obvious that the solution of

$$\bar{\rho}(\alpha) = B(\alpha) \left[1 + \int_{\mu_0}^{\alpha} \frac{\bar{\rho}(\beta)}{\beta^2 \sin \varepsilon} d\beta \right],$$

(VII.1.17)

that is

$$\bar{\rho}(\alpha) = B(\alpha)\exp\left[\int_{\mu_0}^{\alpha} \frac{B(\beta)}{\beta^2 \sin \varepsilon} d\beta\right] \tag{VII.1.18}$$

provides a majorant for $\tilde{\rho}(k, \alpha)$. Indeed, in going from (VII.1.14) to (VII.1.17), we have taken absolute values, have majorized all the factors, and have also increased the interval of integration. Thus

$$|\rho(k, \alpha)| < \frac{B(\alpha)}{\alpha^2 \sin \varepsilon} \exp\left[\int_{\mu_0}^{\alpha} \frac{B(\beta)}{\beta^2 \sin \varepsilon} d\beta\right]. \tag{VII.1.19}$$

It follows that $\rho(k, \alpha)$ exists for all values of α and k:

$$\alpha \in [\mu_0, \infty), \tag{VII.1.20a}$$

$$|\arg k - \tfrac{3}{2}\pi| > \varepsilon, \tag{VII.1.20b}$$

with $\varepsilon > 0$ as small as we wish, provided that

$$\int_{\mu_0}^{\infty} \beta^{-2} B(\beta) d\beta < \infty. \tag{VII.1.21}$$

Notice that, because of (VII.1.15), we have $rV(r) \in L^1(0, \infty)$. Moreover, ρ, for all values of α in its domain of definition, is holomorphic in k in the plane cut along the lower part of the imaginary axis, from $-i\mu_0/2$ to $-i\infty$. We have also, because of (VII.1.19)

$$\int_{\mu_0}^{\infty} |\rho(k, \alpha)| d\alpha < C_\varepsilon < \infty, \tag{VII.1.22}$$

where $C_\varepsilon \to \infty$ as $\varepsilon \to 0$. This guarantees the absolute and uniform convergence of the integral in (VII.1.5) in any compact set of the k-plane, even for $r = 0$. We can also differentiate both sides of (VII.1.6) with respect to k, and using again (VII.1.19), we find

$$\int_{\mu_0}^{\infty} |\partial_k \rho(k, \alpha)| d\alpha < C_\varepsilon < \infty \tag{VII.1.23}$$

in the cut k-plane. It follows that

$$\partial_k g(k, \alpha) = \int_0^{\infty} \partial_k \rho(k, \alpha) e^{-\alpha r} d\alpha \tag{VII.1.24}$$

exists also in the same cut-plane for all $r \geq 0$. This entails that $g(k, r)$, and therefore $f(k, r)$ are holomorphic in k outside the cut

$$k = -i\chi, \quad \chi \geq \frac{\mu_0}{2},$$

that g is indeed the solution of **(VII.1.3)**, and that it satisfies the boundary condition **(VII.1.4)**, for all k outside the cut. It is also easy to show on the basis of **(VII.1.8)**, **(VII.1.12b)**, and **(VII.1.19)**, that

$$\lim_{k \to \infty} g(k, r) = 1, \qquad |\arg k - \tfrac{3}{2}\pi| > \varepsilon, \qquad \text{(VII.1.25)}$$

uniformly in $r \geq 0$.

Since the Jost function is given by

$$F(k) = f(k, 0) = g(k, 0) \qquad \text{(VII.1.26)}$$

we see that it is also holomorphic in the same cut plane, and that

$$\lim_{k \to \infty} F(k) = 1, \qquad |\arg k - \tfrac{3}{2}\pi| > \varepsilon, \qquad \text{(VII.1.27)}$$

with ε as small as we wish, but positive. Also, **(I.3.14)** generalizes to this whole cut plane:

$$F(-k^*) = [F(k)]^*. \qquad \text{(VII.1.28)}$$

Consider now the regular solution

$$\varphi(k, r) = (2ik)^{-1}[F(-k)f(k, r) - F(k)f(-k, r)]. \qquad \text{(VII.1.29)}$$

We know that it is an entire function of k. This means that the cuts which are present in the right-hand side must cancel each other. At the upper cut (remember that φ is even in k)

$$k = i\chi \pm 0, \qquad \chi \geq \frac{\mu_0}{2},$$

which appears in $F(-k)$ and $f(-k, r)$, we get, from **(VII.1.28)** and

$$\varphi(i\chi + 0, r) = \varphi(i\chi - 0, r)$$

using **(VII.1.5)** and **(VII.1.8)**, that the discontinuity of the Jost function is given by

$$2i\Delta(\chi) = F(-i\chi + 0) - F(-i\chi - 0)$$

$$= 2i\pi \frac{F(i\chi)}{2\chi}\left[C(2\chi) + \theta(2\chi - \mu_0) \int_{\mu_0}^{2\chi - \mu_0} C(2\chi - \beta)\rho(-i\chi, \beta)d\beta \right].$$

$$\text{(VII.1.30)}$$

Notice that, because of **(VII.1.28)**, $\Delta(\chi)$ is real.

Suppose now that the bound-state poles (zeros of $F(i\chi)$, $\chi > 0$) are outside the upper cut, i.e., are in the interval $(0, i\mu_0/2)$. The discontinuity of the S-matrix

$$S(k) = \frac{F(-k)}{F(k)} \qquad \text{(VII.1.31)}$$

along the upper cut being (definition of D!)

$$S(i\chi + 0) - S(i\chi - 0) = 2iD(\chi) = \frac{2i\Delta(\chi)}{F(i\chi)}, \qquad \textbf{(VII.1.32)}$$

we obtain from (**VII.1.30**)

$$D(\chi) = -\pi \frac{1}{2\chi} \left[C(2\chi) + \theta(2\chi - \mu_0) \int_{\mu_0}^{2\chi - \mu_0} C(2\chi - \beta)\rho(-i\chi, \beta)d\beta \right].$$

$$\textbf{(VII.1.33)}$$

This equation enables us to calculate $D(\chi)$ step by step in successive intervals $[n\mu_0/2, (n + 1)\mu_0/2, n = 1, 2, 3, \ldots]$ from $C(\alpha)$ according to the procedure described before. If $C(\alpha)$ is continuous, $D(\alpha)$ is continuous. Since $F(i\chi)$ is continuous, it follows that $\Delta(\chi)$ is also continuous. This shows, according to (**VII.1.30**), and $F(i\chi) \neq 0$ on the cut, that $g(-i\chi + 0, r) - g(-i\chi - 0, r)$ is also continuous on the cut for all $r \geq 0$.

VII.2   Yukawa Potentials and the Inverse Problem

To solve the inverse problem, i.e., to calculate $C(\alpha)$ from $D(\chi)$, as claimed at the beginning of this chapter, is an easy matter from (**VII.1.33**). It is obvious in this equation that

$$C(\alpha) = -\frac{\alpha}{\pi} D\left(\frac{\alpha}{2}\right), \qquad \mu_0 \leq \alpha < 2\mu_0. \qquad \textbf{(VII.2.1)}$$

From this $C(\alpha)$, we can calculate, as was shown previously, $\rho(k, \alpha)$ in $\alpha \in [\mu_0, 2\mu_0)$. It is given by (**VII.1.9**). We now go back again to (**VII.1.33**), with $\mu_0 \leq \chi < 3\mu_0/2$. The range of integration in β is now included in $[\mu_0, 2\mu_0)$, where we know the integrand. Inverting (**VII.1.33**), we find

$$C(\alpha) = -\frac{\alpha}{\pi} D\left(\frac{\alpha}{2}\right) - \int_{\mu_0}^{\alpha - \mu_0} C(\alpha - \beta)\rho\left(-i\frac{\alpha}{2}, \beta\right)d\beta$$

$$= -\frac{\alpha}{\pi} D\left(\frac{\alpha}{2}\right) - \frac{1}{\pi^2} \int_{\mu_0}^{\alpha - \mu_0} D\left(\frac{\alpha - \beta}{2}\right)D\left(\frac{\beta}{2}\right)d\beta. \qquad \textbf{(VII.2.2)}$$

It is obvious that this procedure can be continued indefinitely, and enables us to calculate $C(\alpha)$ in larger and larger intervals from the discontinuity $D(\chi)$ of the S-matrix. As a by-product, we have also $\rho(k, \alpha)$, and we can check now whether

$$F(i\chi) = 1 + \int_{\mu_0}^{\infty} \rho(i\chi, \alpha)d\alpha \qquad \textbf{(VII.2.3)}$$

does vanish at the bound state poles of the S-matrix in the interval $[0, \mu_0/2)$. If that is the case, then we have found the solution of our inverse problem, which is essentially unique. Of course, from the beginning, we must verify

that the residues of the S-matrix of the bound state poles are, according to
(II.1.14'), pure imaginary, with a negative imaginary part. If this imaginary
part is positive, we have a bound state with a negative norm, which is not
allowed for a real potential.

Consider now the case where a bound state is situated on the upper cut of
the S-matrix, i.e.,

$$F(i\chi_0) = 0, \qquad \chi_0 \geq \frac{\mu_0}{2}. \qquad \text{(VII.2.4)}$$

This root being simple, we have a simple pole on the cut. If we write

$$\tilde{S}(k) = \frac{k - i\chi_0}{i} S(k) \cdot \qquad \text{(VII.2.5)}$$

$\tilde{S}$ has no pole on the cut, and we can calculate its discontinuity

$$\tilde{D}(\chi) = \frac{\chi - \chi_0}{F(i\chi_0)} (2i)^{-1}[F(-i\chi - 0) - F(-i\chi_0 + 0) = (\chi - \chi_0)D(\chi) \quad \text{(VII.2.6)}$$

which is a continuous function when $C(\alpha)$ is continuous, as we shall see in
a moment. Before showing this, let us show that $\tilde{D}(\chi)$ vanishes at $\chi = \chi_0$.
Indeed, writing as before, that φ given by (VII.1.29) has the same value on
both sides of the upper cut, and using (VII.2.3), we immediately find

$$F(-i\chi_0 + 0) = F(-i\chi_0 - 0) \qquad \text{(VII.2.7)}$$

which shows our assertion. Consequently, $D(\chi)$ is finite at χ_0. Proceeding
now as before, we find

$$D(\chi) = \frac{\tilde{D}(\chi)}{\chi - \chi_0} - \frac{\pi}{2\chi}\left[C(2\chi) + \int_{\mu_0}^{2\chi - \mu_0} C(2\chi - \beta)\rho(i\chi, \beta)d\beta \right] \quad \text{(VII.2.8)}$$

which is the analogue of (VII.1.33).

From the above relation, it is again possible to solve the inverse problem,
i.e., to calculate $C(\alpha)$ from $D(\chi)$. One is led again to relations identical to
(VII.2.1), (VII.2.2), etc. And it is immediately seen that D is continuous if C
is so, and conversely. The case of several bound states on the cut can be
treated along the same lines, and the conclusions are the same. After having
found $C(\alpha)$, we have of course then to check that (VII.2.3) vanishes at the
bound states.

The last point to study is the asymptotic behavior of $C(\alpha)$ for large values
of α. This is important, for the behavior of the potential near the origin is
closely related to it. Consider first the direct problem. From (VII.1.15)
and (VII.1.21), it is easy to find, via (VII.1.8) and (VII.1.33), that if

$$|C(\alpha)| < B(\alpha) = A_1\alpha^{1-\eta}, \qquad \eta > 0, \qquad \text{(VII.2.9a)}$$

then

$$|D(\chi)| < A_2\chi^{-\eta}. \qquad \text{(VII.2.9b)}$$

In fact, one can show that the dominant contribution to $D(\chi)$ comes from the first term in the right-hand side of (VII.1.33) when $\eta > 0$, i.e.,

$$D(\chi) \simeq -\frac{\pi}{2\chi} C(2\chi), \qquad \chi \to \infty. \qquad \text{(VII.2.10)}$$

The converse of the above statements does not hold in general. In order to obtain (VII.2.9a) from (VII.2.9b), one has to assume something more on D. To see this, let us notice that from the definition

$$D(\chi) = \frac{\Delta(\chi)}{F(i\chi)} \qquad \text{(VII.2.11)}$$

and

$$F(i\chi) \to 1 \qquad \text{as} \qquad \chi \to \infty \qquad \text{(VII.2.12)}$$

if (VII.2.9a) holds, then $\Delta(\chi)$ is also bounded by $\chi^{-\eta}$ when χ is large.

The analyticity properties of F and (VII.1.27) entail now that F satisfies the dispersion relation

$$F(k) = 1 + \frac{i}{\pi} \int_{\mu_0/2}^{\infty} \frac{\Delta(\chi)}{k + i\chi} \, d\chi. \qquad \text{(VII.2.13)}$$

Thus (Martin, 1961)

$$\Delta(\chi) = -D(\chi) \left[1 + \frac{1}{\pi} \int_{\mu_0/2}^{\infty} \frac{\Delta(\chi')}{\chi' + \chi} \, d\chi' \right]. \qquad \text{(VII.2.14)}$$

For a given D, this is a Fredholm integral equation in Δ, whose kernel is

$$K(\chi, \chi') = \frac{D(\chi)}{\chi + \chi'}.$$

In general, when this kernel is not too large, i.e., when all its eigenvalues are larger than 1, it is obvious that we have a unique solution for Δ, which is obtained, as usual, by the method of successive approximations. It can be shown that such is the case when

$$\int_{\mu_0/2}^{\infty} \chi^{-1} |D(\chi)| \, d\chi < 2\pi \, \text{Log} \, 2. \qquad \text{(VII.2.15)}$$

If this condition is satisfied, then we can show that (VII.2.9b) leads to (VII.2.9a).

The above analysis is independent of the sign of D or $C(\alpha)$. We can go further if we know the sign of $C(\alpha)$. Indeed, suppose that $C(\alpha)$ is everywhere negative. Then, from the equations relating $C(\alpha)$ to $D(\chi)$ one sees immediately that $D(\chi)$ is positive everywhere. If this holds, then it can be shown that the Fredholm determinant of (VII.2.14) is positive, and that there is a unique solution, as long as

$$\int_{\mu_0/2}^{\infty} \chi^{-1} D(\chi) d\chi < \infty. \qquad \text{(VII.2.16)}$$

When $C(\alpha)$ is positive, no conclusion can be drawn on the sign of $D(\chi)$, and we have only (**VII.2.15**). Suppose now that the input $D(\chi)$ is negative. Then it is obvious that it cannot be too large. Indeed, doing the change of function

$$\tilde{\Delta}(\chi) = \frac{\Delta(\chi)}{(-D(\chi))^{1/2}}$$

the equation (**VII.2.14**) becomes a Fredholm equation with a positive symmetric kernel, and it is known that such equations have always eigenvalues. Therefore, if the kernel is too large, there are eigenvalues less than 1.

A careful study of equation (**VII.2.14**), and of its generalization for the Jost solution $f(k, r)$ has been made by Cornille. Here we shall give the essential results because of their bearings on the relativistic case where one has sometimes to deal with the so-called N/D equations similar to (**VII.2.14**).

We start from a potential of the form (**VII.1.1**) for which $\alpha^{-2}C(\alpha) \in L^1(\mu_0, \infty)$. This guarantees that $rV(r) \in L^1(0, \infty)$. According to (**VII.1.32**), the integral equation (**VII.2.14**) can be written in the form

$$F(i\chi) \equiv \bar{F}(x) = 1 + \lambda \int_{\mu_0/2}^{\infty} \frac{\bar{F}(y)D(y)}{x + y} \, dy, \qquad \text{(VII.2.17)}$$

where D is the discontinuity of the S-matrix across its upper cut in the k-plane. We have added here the parameter λ representing the strength of the discontinuity D, and would like to study the solutions of the equation when we make λ vary. We saw before that $x^{-1}D(x) \in L^1(\mu_0, \infty)$. We assume now that we have

$$\int_{\mu_0/2}^{\infty} x^{-1}|D(x)|dx < \infty \qquad \text{(VII.2.18a)}$$

and

$$\int_{\mu_0/2}^{\infty} \int_{\mu_0/2}^{\infty} \left(\frac{D(x)}{x + y}\right)^2 dx \, dy < \infty \qquad \text{(VII.2.18b)}$$

so that the integral equation becomes of Hilbert–Schmidt type.

When λ is small, i.e., when, for instance, λD satisfies (**VII.2.15**), we know that we have a unique solution, and therefore a good potential of the type (**VII.1.1**) with (**VII.2.9a**). The problem is to know what happens to the potential when λ becomes too large, i.e., when we cross an eigenvalue λ_0 of the kernel.

The answer to this question is given by the Marchenko equation for the inverse problem studied in chapter V. According to the results of that chapter, and neglecting the eventual bound states which are of no consequence, the potential is obtained by solving the integral equation

$$A(r, t) = A_0(r + t) + \int_r^{\infty} A(r, s)A_0(s + t)ds, \qquad \text{(VII.2.19a)}$$

where

$$A_0(t) = (2\pi)^{-1} \int_{-x}^{\infty} [1 - S(k)]e^{ikt} \, dk. \qquad \text{(VII.2.19b)}$$

The potential and the Jost solution are then obtained by

$$V(r) = -2 \frac{d}{dr} A(r, r) \qquad \text{(VII.2.20)}$$

and

$$\bar{F}(x, r) = f(ix, r) = e^{-xr} + \int_{r}^{\infty} A(r, t)e^{-xt} \, dt. \qquad \text{(VII.2.21)}$$

For Yukawa potentials, we can calculate the integral in (VII.2.19b) by contour integration, using the fact that $S(k)$ is now holomorphic in the upper half-plane cut from $i\mu_0/2$ to $i\infty$, and goes to 1 when k goes to infinity in all directions not parallel to the imaginary axis. The result is (again with the introduction of λ)

$$A_0(t) = \lambda \int_{\mu_0/2}^{\infty} e^{-xt} D(x) dx. \qquad \text{(VII.2.22)}$$

Substituting now this equation in (VII.2.19a), and using the result in (VII.2.21), we obtain (Petras, 1962; Cornille, 1967)

$$\bar{F}(x, r) = e^{-xr} + \lambda \int_{\mu_0/2}^{\infty} \frac{D(y)}{x + y} e^{-r(x+y)} \bar{F}(y, r) dy. \qquad \text{(VII.2.23)}$$

Making $r = 0$ in this equation, we obtain (VII.2.17). The above integral equation is obviously of the Hilbert–Schmidt type if (VII.2.18a and b) are satisfied.

If we denote by $d(\lambda, r)$ the Fredholm determinant of (VII.2.23), it can be shown that the potential corresponding to the discontinuity λD is given by

$$V(\lambda, r) = -2 \frac{\partial^2}{\partial r^2} \text{Log } d(\lambda, r). \qquad \text{(VII.2.24)}$$

It is then obvious in this formula that when $d(\lambda, r)$ vanishes for some value of r, a pole of second order appears in general in the potential. To see exactly how this singularity moves when we vary λ, one has to study the roots of

$$d(\lambda, r) = 0 \qquad \text{(VII.2.25)}$$

and follow continuously the eigenvalues $\lambda_j^{\pm}(r)$ (positive or negative) as functions of r. It is then found that, as soon as we cross the first eigenvalue, and have $\lambda > \lambda_1^+(0)$, or $\lambda < \lambda_1^-(0)$, the potential becomes first singular like r^{-2} at the origin, and then the singularity moves, in general, to $r > 0$, or sometimes to Re $r > 0$. Making λ still larger, new singularities appear as we cross $\lambda_2^{\pm}(0), \ldots$. These singularities are often double poles, and we also have often the appearance of ghosts (states with square-integrable wave functions with negative, and even complex, norms). At any rate, when singularities appear in Re $r > 0$, we have, in general, no longer a potential of the form (**VII.1.1**) with a decent $C(\alpha)$ because decent Yukawa potentials are holomorphic in Re $r > 0$. Also, when the double poles of the potential are on the real r-axis, the two regions to the left and to the right of the poles are completely disconnected from each other from the point of view of quantum mechanics (we can no longer define the Hamiltonian as a unique self-adjoint operator) and therefore are disconnected for scattering theory. There is therefore no meaning to talk about ghosts and other features when we have such wild singularities. The models no longer have any physical content, and become rather mathematical curiosities. Any analogy with relativistic dispersion relations and the so-called N/D or F/F equations becomes therefore meaningless.

Comparison with the Gel'fand–Levitan Method

After this digression, we now go back again to a well-behaved potential and a physically meaningful discontinuity D. Suppose also that there are some bound states, but none of them on the cut. We obtain then a well-defined and unique potential V_0 by the method of Martin, where the normalizations of the bound states wave functions are given by (**II.1.14**):

$$C_n^{-1} = \int_0^\infty \varphi_n^2 \, dr = \frac{2i\gamma_n F(-i\gamma_n)}{\dot{F}(i\gamma_n)}. \qquad (\text{VII.2.26})$$

On the other hand, from the integral equation (**VII.2.14**), we can calculate a unique Δ, the discontinuity of the Jost function, and therefore the Jost function itself through (**VII.2.13**). We know of course also the location of the bound states. From all these, we can now calculate, via Gel'fand–Levitan equations of chapters III and IV, all the phase-equivalent potentials in which the normalization constants of the bound states are arbitrary. Now one may ask how to characterize the unique potential obtained by Martin among these phase-equivalent potentials. The answer is given by the results of chapter IV, especially (**IV.2.9**) and (**IV.2.10**) in which we take the initial potential to be precisely the potential of Martin. According to (**IV.2.9**) (see also (**IV.1.30**)), we have, for any potential of the family,

$$V(r) = V_0(r) + a_1 e^{-2\gamma_1 r} + \cdots, \qquad r \to \infty, \qquad (\text{VII.2.27})$$

where $-\gamma_1^2$ is the weakest bound state. Since, by assumption, $\gamma_1 < \gamma_2 < \cdots < \gamma_n < (\mu_0/2)$, it is immediately seen that the Martin potential $V_0(r)$ is the unique member of the family which decreases the most rapidly at large distances. All others have longer ranges related to the energies of the bound states.

The above analysis can also be done when there is a bound state $i\gamma_0$ on the cut. The residue of the S-matrix, i.e., the normalization of the bound state, is again determined by

$$\text{Res } S(i\gamma_0) = \frac{F(-i\gamma_0)}{F(i\gamma_0)} \qquad (\text{VII.2.28})$$

which is again well defined. Indeed, it was seen before that even though $i\gamma_0$ is on the cut of $F(-k)$, we have (VII.2.23) if γ_0 corresponds to a bound state. We obtain therefore, again, a unique Martin potential with a well-defined normalization constant of the wave function. However, this time, the Martin potential cannot be singled out among phase-equivalent potentials for having the shortest range. This is clear in (VII.2.27) which shows that now all the potentials differing from Martin potential by variations in the sole normalization constant C_0 have the same range, i.e., the same asymptotic behavior $e^{-\mu_0 r}$. The case of several bound states on the cut is similar.

In general, whether or not one has bound states on the cut, it is obvious on (VII.2.27) that if the Martin potential corresponds to $C(\alpha)$ which is an ordinary integrable function, all other members of the phase-equivalent family correspond to generalized functions

$$C(\alpha) + a_1 \delta(\alpha - 2\gamma_1) + \cdots.$$

In this respect, in the presence of bound states, the Martin potential is the only potential of the family that corresponds to $C(\alpha)$ being an ordinary function.

VII.3 Higher Waves—Coulomb Potential

For higher waves ($l \geq 1$), there are two methods to include the centrifugal term $l(l + 1)r^{-2}$. The first method is again due to Martin. It consists in writing

$$r^{-2} = \int_0^\infty \alpha e^{-\alpha r} \, d\alpha \qquad \text{(VII.3.1)}$$

and use again the "ansatz" (VII.1.5). The equation (VII.1.6) becomes now

$$\alpha(\alpha - 2ik)\rho_l(k, \alpha) = C(\alpha) + l(l + 1)\alpha$$

$$+ \int_0^\alpha [C(\alpha - \beta) + l(l + 1)(\alpha - \beta)]\rho_l(k, \beta)d\beta. \quad \text{(VII.3.2)}$$

This cannot be transformed again into an equation similar to (VII.1.8), and therefore be solved step by step. However, taking advantage of the fact that, for $0 \leq \alpha < \mu_0$, we have

$$\alpha(\alpha - 2ik)\rho_l(k, \alpha) = l(l + 1)\left[\alpha + \int_0^\alpha (\alpha - \beta)\rho_l(k, \beta)d\beta\right] \quad \text{(VII.3.3)}$$

and writing

$$Y_l(k, \alpha) = 1 + \int_0^\alpha \rho_l(k, \beta)d\beta$$

we obtain from (**VII.3.2**)

$$\alpha(\alpha - 2ik)\rho_l(k, \alpha) - l(l + 1)\left[\alpha + \int_0^\alpha (\alpha - \beta)\rho_l(k, \beta)d\beta\right] = \xi(\alpha), \quad \textbf{(VII.3.4a)}$$

$$\xi(\alpha) \equiv C(\alpha) + \int_0^{\alpha - \mu_0} C(\alpha - \beta)Y_l'(k, \beta)d\beta, \quad \textbf{(VII.3.4b)}$$

and

$$Y_l(k, \alpha) = P_l\left(1 - \frac{\alpha}{ik}\right) + \int_0^\alpha \left[P_l\left(1 - \frac{\alpha}{ik}\right)Q_l\left(1 - \frac{\beta}{ik}\right)\right.$$

$$\left. - P_l\left(1 - \frac{\beta}{ik}\right)Q_l\left(1 - \frac{\alpha}{ik}\right)\right]\frac{d\xi}{d\beta} d\beta, \quad \textbf{(VII.3.5)}$$

where P_l and Q_l are Legendre functions of the first and second kind,

$$Q_l(z) = \frac{1}{2}\int_{-1}^{+1} \frac{P_l(t)}{z - t} dt. \quad \textbf{(VII.3.6)}$$

It is now easily seen that (**VII.3.4b**) and (**VII.3.5**) can again be solved by iteration, as was done previously for $l = 0$. Indeed, if we know Y_l for $\alpha < n\mu_0$, we can replace it in (**VII.3.4b**), and calculate $\xi(\alpha)$ for $\alpha < (n + 1)\mu_0$. Now inserting these values in (**VII.3.5**), we obtain Y_l in $\alpha < (n + 1)\mu_0$, and we can continue the procedure. Since, for $0 \le \alpha < \mu_0$, we have

$$Y_l(k, \alpha) \equiv P_l\left(1 - \frac{\alpha}{ik}\right) \quad \textbf{(VII.3.7)}$$

the assertion is proved.

From these results, one reaches exactly the same conclusions as before concerning the analyticity of $f_l(k, r)$ and $F_l(k)$ in the k-plane cut from $-i\mu_0/2$ to $-i\infty$, and about their asymptotic properties in this cut plane. Notice that here, f_l, like the free solution $w_l(kr)$, has a pole of order l at $r = 0$.

Concerning the inverse problem, one obtains again the potential step by step from the discontinuity D_l of the S-matrix, as was done for $l = 0$. We shall not enter into the details because of lack of space, but the conclusions are quite identical to those of the previous section.

Notice that here, because the integral representation is essentially a Laplace transform, the same for all waves, it can be shown that the procedure can easily be generalized to problems in which one has to deal with coupled equations for different angular momenta, for instance when dealing with tensor force, etc.

The second method, due to De Alfaro and Rossetti, uses, instead of the Laplace transform, the integral transform defined by $w_l(i\alpha r)$, where w_l is the free Jost solution. This choice is quite natural, and permits elimination of the centrifugal potential. Instead of (**VII.1.5**) one now writes

$$f_l(k, r) = w_l(kr) + \int_{\mu_0}^\infty \rho_l(k, \alpha)w_l[(k + i\alpha)r]d\alpha. \quad \textbf{(VII.3.8)}$$

The difficulty with this integral transform is that it does not have simple convolution properties as does the Laplace transform, so that it is not obvious that one can write the transform of the product $V(r)f_l(k, r)$ as a convolution of the transforms of V and f_l. This difficulty can be overcome, but the resulting equations are slightly more complicated than (**VII.1.8**). Nevertheless, they can again be solved exactly step by step, and the results one obtains are of course identical to the results obtained via (**VII.3.2**).

As a last generalization, we mention that the same techniques can be used to include the Coulomb potentials (Cornille and Martin, 1962).

A variant of the method of Martin has been found by Calogero and Cox, which is simpler for solving the inverse problem. In Martin's method, we have, basically, two equations, (**VII.1.8**) and (**VII.1.33**), which may be written, symbolically,

$$\rho = C + \int C\rho,$$

$$D = C + \int C\rho.$$

These two equations have to be solved together to obtain D from C, or to do the inverse operation. At any stage of the step by step operation, ρ is a necessary intermediary.

Calogero and Cox have been able, using Marchenko equation (**VII.2.23**), to obtain equations which can be written, symbolically,

$$\rho = D + \int D\rho,$$

$$C = D + \int D\rho.$$

These equations have the same structure as those of Martin, so that they can also be solved step by step. However, while the equations of Martin are simpler and give in a straightforward manner D from C, those of Calogero and Cox are manifestly simpler for doing the inverse problem, i.e., for going from D to C. Written in detail, they read

$$\rho(ik, \alpha) = \frac{\theta(\alpha - \mu_0)}{\alpha + \beta/2} D\left(\frac{\alpha}{2}\right)$$

$$+ \frac{\theta(\alpha - 2\mu_0)}{2} \int_{\mu_0}^{\alpha - \mu_0} \left(\gamma + \frac{\beta}{2}\right)^{-1} D\left(\frac{\gamma}{2}\right) \rho\left(\frac{i\gamma}{2}, \alpha - \gamma\right) d\alpha \qquad \text{(VII.3.9)}$$

and

$$C(\alpha) = \alpha\left[\theta(\alpha - \mu_0)D\left(\frac{\alpha}{2}\right) + \theta(\alpha - 2\mu_0) \int_{\alpha}^{\alpha - \mu_0} D\left(\frac{\gamma}{2}\right) \rho\left(\frac{i\gamma}{2}, \alpha - \frac{\gamma}{2}\right) d\frac{\gamma}{2}\right].$$

$$\text{(VII.3.10)}$$

The results one obtains from these equations are of course identical to those of Martin. With these equations, it is also easy to include bound states as part of the input of the inverse problem by making the substitution

$$D(\chi) \to D(\chi) + \sum_j \lambda_j \delta(\chi - \gamma_j),$$

where λ_j are the residues of the S-matrix, and are related to the renormalization of the bound states, as we know. This permits us to generate the whole family of phase-equivalent potentials of chapters IV and VI.

VII.4   Holomorphic Potentials

We shall consider here the potentials which, besides satisfying (I.1.3) at infinity, are holomorphic in r in a neighborhood of the origin. They admit therefore the Taylor expansion

$$V(r) = \sum_n v_n \frac{r^n}{n!} \tag{VII.4.1}$$

with a finite radius of convergence. The constants v_n are the derivatives of the potential at the origin, and the problem is to find these constants from the scattering data (phase shift and binding energies for one angular momentum). This problem has been solved progressively by the successive contributions of Newton, Faddeev, Buslaev, Percival, Roberts, and finally, Calogero and Degasperis, who gave the complete solution of the problem in its most general form. It is not possible to give here a detailed account of the methods and general formulas that have been obtained. We shall content ourselves with giving just some of the simplest results.

The general form of v_0, the value of the potential at the origin, is given by

$$v_0 = \frac{8}{(2l + 1)\pi} \int_0^\infty \left[-k \frac{d}{dk} k\delta_l(k) \right] dk + \frac{4}{(2l + 1)} \sum_n E_{nl}, \tag{VII.4.2}$$

where l is the angular momentum, and the sum extends over all the corresponding bound states.

The next term, v_1, the value of the first derivative of the potential at the origin, in the case of S-wave, is given by

$$v_1 = -4 \left[\sum_j C_j - \frac{2}{3!\pi} \int_0^\infty k^3 \frac{d^3}{dk^3} (k^2 |F(k)|^{-2}) dk \right], \tag{VII.4.3}$$

where C_j's are the normalization constants of the bound states, (II.1.8), and F is the Jost function, which is also given in terms of scattering data by (II.2.1'). For higher terms, one has more complicated formulas containing higher derivatives of the phase shift and the binding energies of bound states. In the very special case where the potential is even, i.e.,

$$v_1 = v_3 = \cdots = v_{2n+1} = \cdots = 0$$

it is obvious that one should obtain relations among various parameters of the scattering data. For instance, if only one bound state is present (necessarily an S-wave), its binding energy and normalization constant are given by

$$C_0 = \frac{1}{3\pi} \int_0^\infty k^3 \frac{d^3}{dk^3} [k^6(k^2 - E_0)^{-2} e^{-2\Delta(k)}]dk \qquad \textbf{(VII.4.4a)}$$

and

$$E_0 = \frac{1}{20} \left\{ \int_0^\infty k^5 \frac{d^5}{dk^5} [k^8(k^2 - E_0)^{-2} e^{-2\Delta(k)}]dk \right\}$$

$$\times \left\{ \int_0^\infty k^3 \frac{d^3}{dk^3} k^6(k^2 - E_0)^{-2} e^{-2\Delta(k)} dk \right\}^{-1}, \qquad \textbf{(VII.4.4b)}$$

where

$$\Delta(k) = -\frac{2}{\pi} P \int_0^\infty \frac{q\delta_0(q)}{q^2 - k^2} dq. \qquad \textbf{(VII.4.4c)}$$

We end this chapter by the relation

$$E_0 = \frac{2}{\pi} \int_0^\infty k \frac{d}{dk} [k(\delta_0 - \tfrac{1}{3}\delta_1)]dk \qquad \textbf{(VII.4.5)}$$

which gives the binding energy of the bound state (S-wave) if the potential does not have any other bound state. δ_0 and δ_1 are the S-wave and P-wave phase shifts.

VII.5   Comments and References

The special techniques and the results of the first two sections are due to Martin (1959, 1961a, 1965). See especially the last paper which gives a complete review of the problem. The case where some of the bound states are located on the cut was first studied by Chadan (1962). The fact that the Martin potential is the only one that has the shortest range among the phase equivalent family, when no bound states are located on the cut, was shown there. It is based on a theorem of Newton (1956) about the phase-equivalent potentials. An alternative method to that of Martin (1961a) for the case $l > 0$, is given by De Alfaro, Regge, and Rossetti (1962), who use (VII.3.8).

A detailed analysis of the integral equations (VII.2.14) is given in De Alfaro and Regge (1961). The results (VII.2.15) and (VII.2.16) are due to them. The integral equation (VII.2.23) appears first in Petras (1962). It is rederived by Cornille and its full content is analyzed in two long papers by him (1967, 1970a,b). Formula (VII.2.24) is due to him. See also Cornille and Rubinstein (1968), Weiss J. (1969, 1970, 1971a,b).

We have given above the references of Martin, and De Alfaro, Regge, and Rossetti for higher waves. The case where a Coulomb potential is also present is treated in Cornille and Martin (1962). The alternative method of Calogero and Cox for Yukawa potentials is given in their paper (1968).

For holomorphic potentials, the results of section 3 are due to the successive contributions of Newton (1956), Faddeev (1957), Buslaev and Faddeev (1960), Buslaev (1962, 1967, 1971), Percival (1962), Percival and Roberts (1963), Roberts (1963, 1964a,b; 1965). The solution of the problem, in its most general form for an S-wave was given by Calogero and Degasperis (1968, 1972). Formula (**VII.4.2**) is due to Newton (1956) and Faddeev (1957). For related mathematical work see Dikii (1958). Notice that these methods and results are known in the literature under the denomination "Trace Methods."

For the extension of the results to the case of Klein–Gordon and Dirac equations, see, respectively, Degasperis (1970) and Corbella (1970). The extension to higher waves is due to Corbella (1971). Other direct relations between V and δ, e.g., variational bounds, were obtained by Calogero et al. (1968).

CHAPTER VIII

Nonlocal Separable Interactions

VIII.1   The Direct Problem

Nonlocal separable two-body interactions have often been used in nuclear physics and many-body problems in the past because of the simple fact that the two-body Schrödinger equation is easily solvable for them, and leads to closed expressions for a large class of such interactions. They have also been used very systematically with Faddeev equations for the three-body problem. Their main feature is that the partial t-matrix has a very simple form, and can be continued off the energy-shell in a straightforward manner, a feature which is most important, as is well known, in nuclear physics, and in the Faddeev equations.

The three-dimensional Schrödinger equation for the scattering of a particle by a general nonlocal interaction reads

$$(\Delta + E)\Psi(\vec{k}, \vec{r}) = \int U(\vec{r}, \vec{r}')\psi(\vec{k}, \vec{r}')d\vec{r}'. \qquad \textbf{(VIII.1.1)}$$

Separable interactions are those for which

$$U(\vec{r}, \vec{r}') = \sum_{l=0}^{\infty} \sum_{n=1}^{N_l} \varepsilon_{nl} u_{nl}(r)u_{nl}(r')P_l(\cos \theta), \qquad \textbf{(VIII.1.2)}$$

$$r = |\vec{r}|, \qquad r' = |\vec{r}'|, \qquad \cos \theta = \frac{\vec{r} \cdot \vec{r}'}{rr'}, \qquad \varepsilon_{nl} = \pm 1.$$

For the time being, we shall consider the case where only one term is present for each l, i.e., the summation over n contains only one term. We shall come back later to the general case.

Making use of the partial wave decomposition

$$\Psi(\vec{k}, \vec{r}) = \sum_{l=0}^{\infty} (2l + 1)i^l \frac{\psi_l(k, r)}{kr} P_l(\cos \theta), \qquad \text{(VIII.1.3)}$$

$$\theta = (\vec{k}, \vec{r}),$$

we obtain

$$\left[\frac{d^2}{dr^2} + k^2 - \frac{l(l + 1)}{r^2}\right]\psi_l(k, r) = \varepsilon_l U_l(r) \int_0^{\infty} U_l(r')\psi_l(k, r')dr', \quad \text{(VIII.1.4a)}$$

$$U_l(r) = (4\pi)^{1/2}ru_l(r). \qquad \text{(VIII.1.4b)}$$

For simplicity, we consider again the S-wave ($l = 0$). We shall see later how one can generalize everything to higher waves. We have therefore to solve

$$\psi'' + k^2\psi = \varepsilon N(k)U(r), \qquad \varepsilon = \pm 1, \qquad \text{(VIII.1.5a)}$$

$$N(k) = \int_0^{\infty} U(r)\psi(k, r)dr \qquad \text{(VIII.1.5b)}$$

with the usual boundary condition

$$\psi(k, 0) = 0. \qquad \text{(VIII.1.6)}$$

Equations (VIII.1.5a,b) can be solved exactly by two methods: Fourier transform, or the Lagrange method. Both are, of course, equivalent. We choose the first method. Because of the boundary condition (VIII.1.6), we have to use Fourier sine transform, which we write, according to the notations of chapter II, third section,

$$\tilde{\psi}(k, p) = \int_0^{\infty} \psi(k, r) \frac{\sin pr}{p} dr, \qquad \text{(VIII.1.7a)}$$

$$\psi(k, r) = \int_0^{\infty} \tilde{\psi}(k, p) \frac{\sin pr}{p} d\rho_0(p^2), \qquad \text{(VIII.1.7b)}$$

$$U(r) = \int_0^{\infty} \tilde{U}(p) \frac{\sin pr}{p} d\rho_0(p^2), \qquad \text{(VIII.1.7c)}$$

$$\tilde{U}(p) = \int_0^{\infty} U(r) \frac{\sin pr}{p} dr, \qquad \text{(VIII.1.7d)}$$

where

$$d\rho_0(p^2) = \frac{2}{\pi} p^2 \, dp, \qquad \text{(VIII.1.7e)}$$

so that, for instance, (**VIII.1.7b**) reads also

$$\psi(k, r) = \frac{2}{\pi} \int_0^\infty \tilde{\psi}(k, p) \frac{\sin p}{p} p^2 \, dp. \qquad \text{(VIII.1.7b$'$)}$$

Obviously, $\tilde{\psi}$ is not an ordinary function, but rather a distribution. However, one knows well how to deal with such objects. Using (**VIII.1.7b,c**) in (**VIII.1.5a**), we obtain

$$(k^2 - p^2)\tilde{\psi}(k, p) = \varepsilon N(k)\tilde{U}(p), \qquad \text{(VIII.1.8a)}$$

$$N(k) = \int_0^\infty \psi(k, r)U(r)dr = \frac{2}{\pi} \int_0^\infty \tilde{\psi}(k, p)\tilde{U}(p)p^2 \, dp. \qquad \text{(VIII.1.8b)}$$

In order to be able to show the existence of the solution of (**VIII.1.8a**), we must impose some restrictions on $U(r)$. We assume that it is locally integrable, and satisfies, at the origin and at infinity, the following conditions:

$$\int_0^\infty r^{1/2-\eta}|U(r)|dr < \infty, \qquad \eta > 0, \qquad \text{(VIII.1.9a)}$$

$$\int_0^\infty r^{1+\eta'}|U(r)| \, dr < \infty, \qquad \eta' > 0, \qquad \text{(VIII.1.9b)}$$

where η, η' may be very small, but positive. It will be seen below that these conditions are sufficient for the existence and uniqueness of the solution. It is easily checked that (**VIII.1.9a**) ensures that the boundary condition (**VIII.1.6**) is satisfied, and (**VIII.1.9a,b**) together that

$$rU(r) \in L^1(0, \infty). \qquad \text{(VIII.1.9c)}$$

This in turn implies that $\tilde{U}(p)$ is continuous everywhere, and that $p\tilde{U}(p)$ is differentiable for all $p \geq 0$. Also, a careful analysis shows that

$$p\tilde{U}(p) \underset{p \to \infty}{=} O(1), \qquad \text{(VIII.1.10a)}$$

$$\tilde{U}(p) \underset{p \to 0}{=} O(1). \qquad \text{(VIII.1.10b)}$$

Scattering States. Here $E = k^2 \geq 0$, so that the solution of (**VIII.1.8a**) is

$$\tilde{\psi}(k, p) = \frac{\delta(k^2 - p^2)}{d\rho_0(k)} + \varepsilon N(k)P \frac{\tilde{U}(p)}{k^2 - p^2}, \qquad \text{(VIII.1.11)}$$

where P means the principal value. The normalization of the wave function, i.e., the factor in front of δ-function, has been chosen in such a way that, when $\varepsilon = 0$ (no interaction), (**VIII.1.7b**) gives $\psi(k, r) = \sin kr/k$. Substitution of (**VIII.1.11**) into (**VIII.1.8b**) shows then immediately that

$$N(k) = \tilde{U}(k)\left[1 + P \int_0^\infty \frac{\varepsilon\tilde{U}^2(p)}{p^2 - k^2} \, d\rho_0(p^2)\right]^{-1} \qquad \text{(VIII.1.12)}$$

and

$$\psi(k, r) = \frac{\sin kr}{k} + \varepsilon N(k) P \int_0^\infty \frac{\tilde{U}(p)}{k^2 - p^2} \frac{\sin pr}{p} \, d\rho_0(p^2). \quad \text{(VIII.1.13)}$$

Since $\tilde{U}(p)$ is differentiable, the principal value integrals exist. Also, because of (VIII.1.10a,b), all integrals are convergent at both extremities.

The asymptotic behavior of ψ for large values of r is now obtained quite easily by writing

$$P \frac{1}{p^2 - k^2} = \frac{1}{2} \frac{1}{p^2 - k^2 - i\varepsilon} + \frac{1}{p^2 - k^2 + i\varepsilon}.$$

The result is

$$\psi(k, r) \underset{r \to \infty}{=} \frac{\sin kr}{k} + \frac{1}{k} \tan \delta \sin(kr + \delta) + o(1), \quad \text{(VIII.1.14)}$$

where

$$-\frac{2k}{\pi} \tan \delta = A(k) \left[1 + P \int_0^\infty \frac{A(p)}{p^2 - k^2} \, dp \right]^{-1}, \quad \text{(VIII.1.15a)}$$

$$A(k) = \frac{2\varepsilon}{\pi} k^2 \tilde{U}^2(k). \quad \text{(VIII.1.15b)}$$

Notice that, because of (VIII.1.10a), the integral is absolutely convergent. Also, because of (VIII.1.10b), the phase shift is well defined for all k, $k \geq 0$. It vanishes as $O(k)$ at $k = 0$. Since $k\tilde{U}(k)$ is differentiable, so is $A(k)$. From well-known theorems on Hilbert transforms of Hölder-continuous functions, it follows that in fact the integral is Hölder-continuous of index $1 - \varepsilon$, $\varepsilon > 0$ as small as we wish.

Bound States. Contrary to the case of local potentials of short range studied previously, we are going to see that we may have here bound states of either positive or negative energy. We consider first the latter case. If there exists a bound state of energy $-X(X > 0)$, formula (VIII.1.8a) gives

$$\tilde{\psi}(-X, p) = \frac{-\varepsilon N \tilde{U}(p)}{p^2 + X}. \quad \text{(VIII.1.16)}$$

Substitution of this expression into (VIII.1.8b) leads to the eigenvalue equation

$$\Phi(X) \equiv \varepsilon + \frac{2}{\pi} \int \frac{\tilde{U}^2(p)}{p^2 + X} p^2 \, dp = 0. \quad \text{(VIII.1.17)}$$

The integral in this expression being a decreasing function of X for $X \geq 0$, in order to have a negative-energy bound state, we must have

$$\varepsilon = -1, \quad \text{(VIII.1.18a)}$$

$$\frac{2}{\pi} \int_0^\infty \tilde{U}^2(p) dp > 1. \quad \text{(VIII.1.18b)}$$

When these conditions are satisfied, the unique bound state with negative energy is given by

$$\frac{2}{\pi} \int_0^\infty \frac{\tilde{U}^2(p)}{p^2 + X_0} p^2 \, dp = 1. \tag{VIII.1.19}$$

Having found X_0, we obtain the wave function via (VIII.1.16), and it is not difficult to see that we have (VIII.1.6) and

$$\psi(-X_0, r) \simeq e^{(-X_0^{1/2})r}, \qquad r \to \infty. \tag{VIII.1.20}$$

Consider now the eventual bound states with positive energy. These are square-integrable solutions of the Schrödinger equation, and are therefore solutions of (VIII.1.11) where the first term in the right-hand side (the δ-function) is missing. This leads again to the eigenvalue equation

$$\varepsilon + \frac{2}{\pi} P \int_0^\infty \frac{\tilde{U}^2(p)}{p^2 - X_1} p^2 \, dp = 0, \qquad X_1 > 0. \tag{VIII.1.21}$$

Here, we may have of course solutions for both signs of ε. Suppose we have such a solution. The asymptotic behavior of the wave function is now given by

$$\psi(X_1, r) = -\frac{2}{\pi} P \int_0^\infty \frac{\tilde{U}(p)}{p^2 - X_1} \frac{\sin pr}{p} p^2 \, dp$$

$$= \frac{i\pi}{2} \tilde{U}(X_1^{1/2}) \cos X_1^{1/2} r + o(1), \qquad r \to \infty. \tag{VIII.1.22}$$

It follows that ψ has a vanishing asymptotic value if and only if

$$\tilde{U}(X_1^{1/2}) = \int_0^\infty U(r) \frac{\sin \sqrt{X_1} r}{\sqrt{X_1}} \, dr = 0. \tag{VIII.1.23}$$

If this is true, a more careful analysis shows that, in fact, ψ is square integrable on $r \geq 0$. Let us write $k_1 = \sqrt{X_1}$. Since $\tilde{U}(k_1) = 0$, there is no need to keep the symbol P in front of the integral in (VIII.1.22). This is now a regular absolutely convergent Fourier sine transform. It is well known that when $r \to \infty$, the behavior of the integral is governed by that of the integrand near $p = 0$, which is

$$p\tilde{U}(p) \underset{p \to 0}{=} p \int_0^\infty t U(t) dt = O(p). \tag{VIII.1.24}$$

From this result, one easily obtains that

$$\psi(X_1, r) \underset{r \to \infty}{=} O(r^{-2/3}). \tag{VIII.1.25}$$

Since (VIII.1.6) is also satisfied, our assertion follows, and X_1 corresponds to a bound state. In conclusion, positive-energy bound states are given by the simultaneous roots of (VIII.1.21) and (VIII.1.23).

At these energies, it follows from (**VIII.1.15a**) that the phase shift crosses a value $n\pi$, n being an integer, downward. Indeed, at $k_1 = \sqrt{X_1}$, say, both the numerator and the denominator vanish. However, as is clear from (**VIII.1.15b**), A has at least a double zero at k_1, whereas the denominator has only a simple zero because

$$\frac{\partial}{\partial k}\left[1 + P \int \frac{A(p) - A(k_1)}{p^2 - k^2}\, dp\right]\bigg|_{k=k_1} = +2k_1 \int \frac{A(p)}{(p^2 - k_1^2)^2}\, dp > 0.$$

$$(VIII.1.26)$$

From these results, we immediately deduce that δ vanishes at k_1, and changes sign:

$$\delta(k_1) = n\pi, \qquad (VIII.1.27a)$$

$$\frac{d\delta}{dk}\bigg|_{k_1} < 0. \qquad (VIII.1.27b)$$

Conversely, suppose that $\tan \delta$ vanishes at $k = k_1$. Since the denominator of (**VIII.1.15b**) is never infinite, we must have a zero in the numerator. But since $\tilde{U}$ is differentiable, the zero in the numerator is necessarily double. Two cases have to be considered. If the denominator does not vanish at $k = k_1$, the phase shift touches the line $n\pi$ either from above, or from below, without crossing it. It is tangent to this line. If the denominator vanishes also, we saw that it has only a simple zero, with a positive first derivative, so that the phase shift crosses $n\pi$ downward. Moreover, now, both (**VIII.1.21**) and (**VIII.1.23**) are satisfied, and, according to our previous analysis, we have a bound state. We can summarize all these results in the following:

Theorem. (a) *Positive energy bound states are given by the simultaneous roots of* (**VIII.1.21**) *and* (**VIII.1.23**); (b) *Equivalently, positive energy bound states correspond to those energies at which the phase shift crosses a value $n\pi$, $n = 0, \pm 1, \ldots$ downward, with a negative slope, and conversely*; (c) *The phase shift never crosses $n\pi$ upward*; (d) *The phase shift may become tangent to $n\pi$ either from below, or from above.*

Let us now look at the high-energy behavior of the phase shift. On the basis of (**VIII.1.10a**), one sees from (**VIII.1.15a**) that $\tan \delta(\infty) = 0$. We can therefore choose $\delta(k)$ in such a way that

$$\delta(\infty) = 0. \qquad (VIII.1.28)$$

We then define by continuity the phase shift for finite values of k starting from infinity. This is obviously possible because, according to (**VIII.1.15a**), $\tan \delta$ is a continuous function of k, except when the denominator vanishes. However, we know that the denominator is a Hölder-continuous function, so that there is no problem for defining δ by continuity for all k starting from (**VIII.1.28**).

The Levinson theorem, generalized by Martin and others to the general case of nonlocal interactions, reads now

$$\delta(0) - \delta(\infty) = \delta(0) = (v + n)\pi, \qquad \text{(VIII.1.29)}$$

where v is the number of positive-energy bound states, and $n(0$ or $1)$ is the number of those with negative energy.

As a final remark, let us notice that, for large values of k, the sign of the phase shift is $-\varepsilon$. Therefore, by simple inspection of the phase shift at high energies, one obtains ε. Also, since, according to (VIII.1.10a) and (VIII.1.10b), $A(p)$ belongs to some $L^p(0, \infty)$, its Hilbert transforms belongs to some $L^q(0, \infty)$. Since this Hilbert transform is also Hölder-continuous, with some positive index, it follows that the phase shift has also the same property, and that

$$\delta(k) \underset{k \to \infty}{=} O(k^{-\eta}), \qquad \eta > 0. \qquad \text{(VIII.1.30)}$$

VIII.2   The Inverse Problem

To determine $U(r)$ from the phase shift, we must first assume that the latter meets all the requirements for the problem to have a solution. For instance, if the phase shift crosses upward the value $n\pi$, n an integer, when we increase k, we are sure that our problem with one separable potential has no solution.

We consider first the inverse problem for the case of $\varepsilon = 1$. As we saw before, $\delta(k)$ at large energies must be small but negative. The eventual bound states with positive energy are then determined by just looking when δ crosses (upward!) the values $n\pi$, $n = 0, 1, 2, \ldots$, as we decrease k from infinity. At $k = 0$, we must have $\delta(0) = v\pi$. To find the potential from the phase shift, we have to solve the integral equation (VIII.1.15a) for $A(k)$. Once this quantity is found, there is no problem for finding $\tilde{U}(k)$ via (VIII.1.15a), and then undoing the Fourier transform (VIII.1.7c). For simplicity, we consider the case $v = 1$. We may write (VIII.1.15a) in the form

$$\varphi(x) = 1 + \frac{1}{\pi} \int_0^\infty \frac{h^*(y)\varphi(y)}{y - x - i0} \, dy, \qquad \text{(VIII.2.1a)}$$

where $x = k^2$, and

$$\tilde{\delta}(x) \equiv \delta(\sqrt{x}), \qquad \text{(VIII.2.1b)}$$

$$g(x) = -\frac{2\sqrt{x}}{\pi} \tan \tilde{\delta}(x), \qquad \text{(VIII.2.1c)}$$

$$\varphi(x) = \frac{A(\sqrt{x})}{g(x)} \left[1 + i\pi \frac{g(x)}{2\sqrt{x}} \right], \qquad \text{(VIII.2.1d)}$$

$$h(x) = \frac{\pi g(x)}{2\sqrt{x}} \left[1 - i\pi \frac{g(x)}{2\sqrt{x}} \right]^{-1} = -\sin \tilde{\delta} e^{-i\tilde{\delta}}. \qquad \text{(VIII.2.1e)}$$

We assume now, according to our previous analysis, that $\delta(k)$ is Hölder-continuous with some positive index and that it behaves like $k^{-\eta}$, η some positive quantity, when $k \to \infty$. These conditions are necessary if we wish to find a potential satisfying Equations (**VIII.1.9a and b**). They are also sufficient according to what follows.

To solve the integral equation (**VIII.2.1a**), we consider first the function

$$H(z) = 1 + \frac{1}{\pi} \int_0^\infty \frac{h^*(y)\varphi(y)}{y - z} \, dy. \qquad \text{(VIII.2.2)}$$

Assuming *a priori* the Hölder-continuity of φ and the proper convergence of the integral, we see that $H(z)$ is analytic in the z plane cut from 0 to $+\infty$. Moreover,

$$\lim H(z) = 1, \qquad z \to \infty, \qquad \text{(VIII.2.3)}$$

in all directions.

From the integral equation (**VIII.2.1a**) we see that its solution is necessarily of the form

$$\varphi(x) = H(x_+) \equiv \lim_{\varepsilon \to +0} H(x + i\varepsilon), \qquad 0 \le x \le \infty. \qquad \text{(VIII.2.4)}$$

On the other hand, the discontinuity of $H(z)$ on the cut is given by

$$H(x_+) - H(x_-) = -2i \sin \delta(x) e^{i\delta(x)}\varphi(x). \qquad \text{(VIII.2.5)}$$

Upon substituting φ by its value, we find

$$\exp[2i\delta(x)]H(x_+) - H(x_-) = 0, \qquad 0 \le x \le \infty, \qquad \text{(VIII.2.6)}$$

which is a homogeneous Riemann-Hilbert type equation. It is now easily verified that a particular solution of this equation, satisfying (**VIII.2.3**), is given by

$$H(z) = \exp[\omega(z)], \qquad \text{(VIII.2.7a)}$$

$$\omega(z) = -\frac{1}{\pi} \int_0^\infty \frac{\delta(y)}{y - z} \, dy. \qquad \text{(VIII.2.7b)}$$

Indeed, from our assumptions on the phase shift, we have

$$\lim_{|z| \to \infty} \omega(z) = 0 \qquad \text{(VIII.2.8)}$$

in all directions. Moreover, $\omega(z)$ is a well-defined function on the cut, except perhaps at $z = 0$, where its behavior is given by

$$\omega(z) = \frac{\delta(0)}{\pi} \log z + \tilde{\omega}(z) \qquad \text{(VIII.2.9)}$$

$\tilde{\omega}(z)$ being finite when $z \to 0$, and $\delta(0) = \delta(0) = \pi$ (one positive energy-bound state). Therefore, the function $\exp[\omega(z)]$ has only a simple zero at $z = 0$.

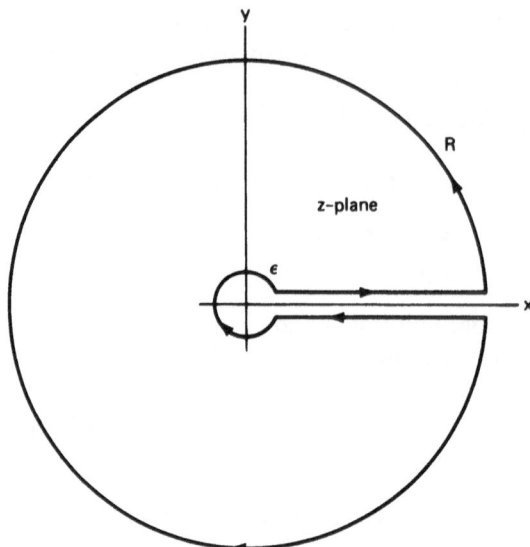

Figure VIII.1 The Contour Γ

According to (**VIII.2.7a** and **b**), $\varphi(x)$ is given by

$$\varphi(x) = e^{\omega(x+)} = e^{\alpha(x)}e^{-i\tilde\delta(x)}, \qquad (\text{VIII.2.10})$$

$$\alpha(x) = -\frac{1}{\pi} P \int_0^\infty \frac{\tilde\delta(y)}{y - x}\, dy. \qquad (\text{VIII.2.11})$$

We have to verify now that this indeed satisfies the integral equation (**VIII.2.1a**). This is most easily done since by Cauchy's theorem we have

$$-\frac{1}{\pi}\int_0^\infty \frac{e^{i\tilde\delta(y)} \sin \tilde\delta(y)}{y - x - i0} e^{\alpha(y)}e^{-i\tilde\delta(y)}\, dy = \lim_{\varepsilon \to 0}\left[\frac{1}{2i\pi}\left(\int_\Gamma \frac{e^{\omega(z)}}{z - x - i\varepsilon}\, dz - \int_R (\cdots)\right)\right]$$

$$= e^{\omega(x+)} - 1 = \varphi(x) - 1, \qquad (\text{VIII.2.12})$$

where the contour Γ is shown on figure 1, R being the circle with a very large radius. (With $H(z)$ regular at $z = 0$, it is obvious that the contribution of the small circle enclosing the origin vanishes when its radius r tends to zero). Note that $\varphi(x)$ is regular at $x = 0$. In fact it has a simple zero there. It is also obvious from (**VIII.2.10** and **2.11**) that $\varphi(x)$ is Hölder-continuous with the same index as the phase shift, and remains bounded when $x \to \infty$. Our *a priori* hypothesis on φ for solving the integral equation is therefore justified. Let us consider now the homogeneous equation

$$\varphi_0(x) = \frac{1}{\pi}\int_0^\infty \frac{h^*(y)}{y - x - i0} \varphi_0(y)dy \qquad (\text{VIII.2.13})$$

to which we associate the function

$$H_0(y) = \frac{1}{\pi} \int_0^\infty \frac{h^*(y)}{y - z} \varphi_0(y) dy. \tag{VIII.2.14}$$

Assuming *a priori* that φ_0 is Hölder-continuous, ... etc., we get

$$\lim_{|z| \to \infty} H_0(z) = 0 \tag{VIII.2.15}$$

in all directions. Proceeding as for H, we obtain once more

$$e^{2i\delta(x)} H_0(x_+) - H_0(x_-) = 0. \tag{VIII.2.16}$$

Using the Ansatz

$$H_0(z) = \sum_s A_s \frac{e^{\omega(z)}}{z^s}, \tag{VIII.2.17}$$

where $\omega(z)$ is again defined by (VIII.2.7b) and A_s are arbitrary constants, we find

$$\varphi_0(z) = \frac{A_0}{x} e^{\omega(x_+)}. \tag{VIII.2.18}$$

Again it can be checked by contour integration that (VIII.2.18) is indeed a solution of (VIII.2.13) and that it has all the required properties.

The general solution of our integral equation is therefore

$$\varphi(x) = \left(1 + \frac{A_0}{x}\right) e^{\omega(x_+)}. \tag{VIII.2.19}$$

Going back to our earlier notations, we obtain

$$A(k) = -\frac{2k}{\pi} \sin \delta(k) e^{\alpha(k^2)} \left(1 + \frac{A_0}{k^2}\right), \tag{VIII.2.20}$$

$$\alpha(k^2) = -\frac{2}{\pi} P \int_0^\infty \frac{\delta(p)}{p^2 - k^2} p \, dp. \tag{VIII.2.21}$$

This solution depends on the arbitrary constant A_0. However, from its definition, A must keep a constant sign for all values of k. Here A should be positive since $\varepsilon = 1$. On the other hand, we know that δ changes sign at the energy of the bound state, which occurs at $E = k_0^2$, k_0 being the point where the phase shift crosses the value zero. It follows that we must have

$$A_0 = -k_0^2 \tag{VIII.2.22}$$

in order for A given by (VIII.2.20) to keep a constant (positive) sign. We therefore see that the solution is completely determined by the phase shift alone, since this also determines k_0. Once A is known, we obtain

$$\left(\frac{2}{\pi}\right)^{1/2} k \tilde{U}(k) = [A(k)]^{1/2} \tag{VIII.2.23}$$

which gives in turn $U(r)$ by (VIII.1.7c). It can then be checked that U satisfies (VIII.1.9a and b) since A is Hölder-continuous and decreases like $k^{-\eta}$, $\eta > 0$, when $k \to \infty$. From (VIII.2.20) and (VIII.2.21) it is clear that these properties hold for A if they hold for the phase shift.

The case when several positive energy bound states occur at $E_\alpha = k_\alpha^2$, $\alpha = 1, 2, \ldots, \nu$, and $\varepsilon = 1$, is quite similar, and leads to

$$A(k) = -\frac{2k}{\pi} \sin \delta(k) e^{\alpha(k^2)} \prod_\alpha \left(1 - \frac{k_\alpha^2}{k^2}\right). \qquad \textbf{(VIII.2.24)}$$

We consider now the case where $\varepsilon = -1$, and where there are two bound states, one with positive energy $E_0 = k_0^2$, and the other a true bound state with negative energy $E = -\gamma^2$. Since $\delta(0) = 2\pi$, we get that $H(z)$ has, according to (VIII.2.9), a double zero at the origin. Proceeding as before, and taking into account the eigenvalue equation with $X = \gamma^2$, and the fact that the solution $A(k)$ must keep a constant (negative) sign for all k, we obtain

$$A(k) = -\frac{2k}{\pi} \sin \delta(k) e^{\alpha(k^2)} \left(1 - \frac{k_0^2}{k_2}\right)\left(1 + \frac{\gamma^2}{k^2}\right). \qquad \textbf{(VIII.2.25)}$$

In this case, we see that the potential is determined unambiguously from the phase shift and the energy of the true bound state. The generalization of the above method to higher partial waves is now obvious. We only have to use the Hankel integral transform

$$\tilde{\psi}(k, p) = \int_0^\infty \psi(k, r) \frac{(pr)^{1/2} J_{l+1/2}(pr)}{p^{l+1}} dr \qquad \textbf{(VIII.2.26)}$$

instead of the sine transform (VIII.1.7a). This permits us to eliminate the centrifugal potential and to obtain equations quite similar to those obtained for the S-wave. The conclusions are then quite similar.

VIII.3   More General Interactions

The method we have been discussing can be used with more general nonlocal interactions. The first generalization consists in considering the interaction

$$U(\mathbf{r}, \mathbf{r}') = V(r)\delta(\mathbf{r} - \mathbf{r}') + \sum_l \varepsilon_l u_l(r) u_l(r') P_l(\cos\theta). \qquad \textbf{(VIII.3.1)}$$

where the local potential V is assumed to be known. The radial equation now reads

$$\psi_l'' + \left(E - \frac{l(l+1)}{r^2}\right)\psi_l = V\psi_l + \varepsilon_l U_l(r) \int_0^\infty U_l(r')\psi_l(k, r')dr'. \qquad \textbf{(VIII.3.2)}$$

Here, the inverse problem consists of finding U_l when the phase shift $\delta_l(k)$ and the local potential V are both known. For this purpose, we assume that the local potential satisfies **(I.1.3)** with $b = 0$. Suppose we can solve the radial equation with V alone, and find ψ_l and the spectral density $d\rho_l(E)$. We can then define the integral transforms

$$\tilde{U}_l(E) = \int_0^\infty U_l(r)\varphi_l(E, r)dr, \qquad \text{(VIII.3.3a)}$$

$$U_l(r) = \int_{-\infty}^\infty \tilde{U}_l(E)\varphi_l(E, r)d\rho_l(E). \qquad \text{(VIII.3.3b)}$$

These integral transforms can be shown to have properties quite similar to those of Hankel transforms used earlier. They reduce, of course, to the latter when $V = 0$. If we write the phase shift as (we drop the subscript l)

$$\delta = \delta_v + \delta_u^v, \qquad \text{(VIII.3.4)}$$

where δ_v is the known phase shift of V alone, we obtain for the increment δ_u^v an equation identical to **(VIII.1.15a)**, i.e.,

$$-\frac{2k}{\pi} \tan \delta_u^v(k) = \frac{A(k)}{1 + P \int_0^\infty \dfrac{A(\sqrt{s})}{s - k^2} d\sqrt{s}}, \qquad \text{(VIII.3.5a)}$$

$$A(k) = \frac{2\varepsilon}{\pi} k^2 \tilde{U}^2(k)|F(k)|^{-2}, \qquad \text{(VIII.3.5b)}$$

where $\tilde{U}$ is defined now by **(VIII.3.3a)**, and $F(k)$ is the Jost function of V. Assuming again **(VIII.1.9a and b)**, we find, by studying **(VIII.3.5a)**, that δ_u^v behaves exactly as the phase shift of one pure separable term. In particular, the theorem of the last section, and its relation to the positive energy bound states applies in the present case to the phase shift increment. Therefore, by inspecting its variation with energy, we find immediately the positive energies at which there are bound states. Similarly, the sign of the phase shift increment at high energies determines ε.

Concerning negative-energy bound states, let us consider the general case where the local potential has n bound states. Then, the total potential admits $N = n$, $n + 1$, or $n - 1$ bound states, depending on the sign of ε, and the strength of the separable potential. If $\varepsilon = +1$, and the separable interaction is not too strong, we have $N = n$. If this interaction is strong enough, we find $N = n - 1$, i.e., the repulsive separable potential, if it is strong enough, may remove one of the bound states (the weakest) of the local potential. If $\varepsilon = -1$, we have, similarly, $N = n$ or $n + 1$, depending on the strength of the additional separable potential. The Levinson theorem applies of course to the total phase shift, and reads now

$$\delta(0) - \delta(\infty) = (N + v)\pi, \qquad N = n - 1, n, \text{ or } n + 1. \quad \text{(VIII.3.6)}$$

The inverse problem can be solved here in exactly the same manner as in the previous case. We have again to solve equations of the Riemann–Hilbert–Mushkhelishvili type. If we assume again Hölder-continuity of δ_u^v

and a behavior similar to (**VIII.1.30**), we can solve the inverse problem in a unique way (knowing V, φ, $d\rho$, δ_V) if and only if δ_u^v has the desired properties, i.e., if and only if the theorem of the preceding section applies to it. The results one obtains are then quite similar to (**VIII.2.24**), (**VII.2.25**), etc.

We shall not enter into the details, and refer the reader to the literature on the subject (Chadan 1958a, 1967; Mills and Reading 1969).

Another generalization consists of having several separable potentials in each angular momentum state:

$$U(\vec{r}, \vec{r}') = V(r)\delta(\vec{r} - \vec{r}') + \sum_l \sum_s \varepsilon_l u_{ls}(r)u_{ls}(r')\, P_l(\cos\theta). \quad \text{(VIII.3.7)}$$

The right-hand side of the radial equation (**VIII.1.4a**) then reads

$$V\psi_l + \sum_s \varepsilon_{ls}\, U_{ls}(r) \int_0^\infty U_{ls}(r')\psi_l(r')dr'. \quad \text{(VIII.3.8)}$$

There are two ways of solving this equation: either we solve it step by step, using the integral transforms defined by the wave function of the previous step, exactly as for (**VIII.3.2**); or we use the integral transform defined by φ_l throughout, obtaining a set of algebraic equations to determine the relation between the potentials U_{ls} and the phase shift. Once more, one is led to singular integral equations of the Hilbert–Mushkhelishvili type, and finally to conclusions quite similar to those reached before for the inverse problem. In this case, whatever δ_l and the negative energy bound states may be, we can always find an appropriate set of U_{ls} that solves the problem. In other words, δ_l may have any shape if we choose the set $\{s\}$ large enough.

VIII.4 Applications

The inverse problem with separable potentials being exactly soluble, one can find many cases where the solution can be obtained in closed form in terms of elementary functions. A simple case is when the $\tan\delta$ is given by a rational function of k. These have been worked out in detail in the literature. The simplest such a case is when the effective-range approximation is assumed to be valid at all energies for the S-wave.

$$k \cot\delta = -\frac{1}{a} + \frac{1}{2}r_0 k^2. \quad \text{(VIII.4.1)}$$

For the singlet case ($a < 0$), there is no bound state of any kind. If we write

$$\tan\delta = \frac{2}{r_0}\frac{k}{k^2 + \alpha^2}, \qquad \alpha^2 = \frac{-2}{ar_0}, \quad \text{(VIII.4.2)}$$

we find $\varepsilon = -1$ (the interaction is attractive, but not sufficiently strong to have a bound state), and

$$\alpha(k^2) = \frac{1}{2} \log \frac{k^2 + \beta_2^2}{k^2 - \beta_1^2}, \qquad \text{(VIII.4.3)}$$

where β_1 and β_2 are the roots of

$$\beta^2 - \frac{2}{r_0} \beta - \alpha^2 = 0. \qquad \text{(VIII.4.4)}$$

This equation has always two real roots, one of each sign, and we have chosen

$$\beta_2 > |\beta_1| \qquad \text{(VIII.4.5)}$$

in order to satisfy the conditions of regularity of the potential. From the above expressions, it is easy to calculate the potential via Fourier sine transform, and one finds

$$U(r) = \sqrt{4\pi} r u(r) = \frac{2\sqrt{2}\beta_2}{\pi\sqrt{r_0}} K_1(\beta_2 r), \qquad \text{(VIII.4.6)}$$

where K_1 is the modified Hankel function of the first kind of order 1. Notice that, here, contrary to the local case (see chapter VI), the interaction is attractive. Since the potential U behaves like r^{-1} near the origin, it is easily verified that the wave function behaves like $r \, \text{Log} \, r$.

In the triplet case ($a > 0$), we write

$$\tan \delta = \frac{2k}{r_0(k^2 - \gamma^2)}, \qquad \gamma^2 = \frac{-2}{ar_0}. \qquad \text{(VIII.4.7)}$$

Here, we have again $\varepsilon = -1$, but also a bound state of negative energy. The calculations are the same as in the previous case, and we obtain

$$\alpha(k^2) = \frac{1}{2} \text{Log} \frac{k^2}{(k^2 + \beta_1^2)(k^2 + \beta_2^2)}, \qquad \text{(VIII.4.8)}$$

where β_1 and β_2 are the real roots of

$$\beta^2 - \frac{2}{r_0} \beta + \gamma^2 = 0. \qquad \text{(VIII.4.9)}$$

The solution fitting the energy of the bound state, according to (VIII.2.25), with $k_0 = 0$, is

$$A(k) = \frac{-4\pi}{r_0} \frac{k^2(k^2 + \gamma^2)}{(k^2 + \beta_1^2)(k^2 + \beta_2^2)}. \qquad \text{(VIII.4.10)}$$

Unfortunately, the expression for $U(r)$ cannot be given in terms of elementary functions.

For more complicated rational expressions for $\tan \delta$, one obtains similar formulae with more terms. Similarly, the analysis is the same for higher waves if one uses the Hankel transforms (**VIII.2.26**), etc., and one gets quite analogous formulas.

As a final application, let us mention that one can use the general form (**VIII.3.7**), with a good local potential $V(r)$, and choose the separable terms $\{U_{ls}(r)\}$ in a such a way that the overall interaction is entirely equivalent to zero for all energies and for all l, i.e., a completely transparent interaction at all energies. It turns out that, for each V, there is a whole class of such interactions. Since the potential V is also arbitrary, one gets really a very large class of completely transparent potentials. Surprisingly, all these interactions can be made well behaved and finite everywhere, and have even exponentially decreasing tails in all directions if V has such a tail (Chadan, 1967).

VIII.5   Comments and References

The inverse problem for one separable potential was first studied, and then solved completely by Gourdin and Martin (1957a,b, 1958). Several years later, the same problem was rediscovered and solved, although in a less exhaustive way (in particular ignoring positive energy bound states) by Bolsterli and Mackenzie (1965)—and then Tabakin (1969).

The generalization of the Levinson theorem is due to Martin (1958). See also Jauch (1957), Tabakin (1969), (see some other references in Newton (1966)).

The more general case of the inverse problem where a local potential is also present was first treated by Chadan (1958a). The case of several separable potentials was considered and solved by Chadan (1967), and Mills and Reading (1969).

For a complete study of singular integral equations, see Mushkhelishvili (1953), where Hölder-continuity of principal value integrals are treated in detail.

For the general theory of Fourier integrals, see Titchmarsh (1959) where integrability properties of Fourier transforms as well as their Hölder-continuity and asymptotic properties are studied thoroughly. Hölder-continuity and asymptotic properties of P.V. integrals are also studied there.

The inverse problem for spin-orbit and tensor force was studied first by Chadan (1958b). They are also studied in Bolsterli and MacKenzie (1965) and Tabakin (1969).

A large number of papers have been devoted to the applications of separable potentials in nucleon–nucleon scattering, nuclear physics, and the properties of nuclear matter, as well as in connection with the applications of Faddeev equations. The first use of separable potentials in the nucleon–nucleon problem, including the tensor force, is due to Yamaguchi and Yamaguchi (1954). For more recent use of these potentials in various domains mentioned above, see the current literature in nuclear physics, particularly Oryu (1983) and Plessas (1986). The last paper contains full references to earlier works. For a general study of the inverse problem for nonlocal potentials in connection with half-off-shell two-body t-matrices needed in the Faddeev equations, see Muzafarov (1986a,b, 1988). For an explicit example of a nonlocal potential for which the Hamiltonian is a unitary equivalent to that of a local potential, see Abraham et al. (1980). For the inverse problem for Frahn–Lemmer-type nonlocal potentials, see Fiedeldey et al. (1986).

CHAPTER IX

Miscellaneous Approaches to the Inverse Problems at Fixed l

This chapter is a survey of further generalizations of the methods which have been described in previous chapters, new approaches, and approaches to more complicated problems. The study is limited to radial problems (for "one dimensional" problems see Chapter XVII). Throughout this chapter, most results are stated without proof and the reader who is interested in these proofs should refer to the literature. The notations are either the ones that are used elsewhere in this book or are specially given in each section.

IX.1   Generalization to Other Central Potentials

In the chapter III, we met a simple generalization of the Gel'fand–Levitan method to the case where one starts from a given potential instead of starting from the zero potential. The corresponding generalization of the Marchenko's method is straightforward. It suffices

1. To everywhere replace the solutions of the Schrödinger equation corresponding to the potential 0 by the ones corresponding to the reference potential V_0 which has been chosen.
2. To replace the free wave Schrödinger operator in the differential and partial differential equations by the perturbed Schrödinger operator corresponding to V_0.
3. To replace $S(k)$ by the difference $S(k) - S_0(k)$, where the subscript 0 refers to V_0.

The method is consistent where V_0 satisfies precisely the conditions for which Marchenko's method works. Thus it is not consistent with V_0 singular.

The Gel'fand–Levitan method for $l > 0$ partial waves was given in chapter III. The Marchenko method for an $l > 0$ partial wave was given by Marchenko (1963), who used an uneasy recursive method. Blazek (1963, 1966) obtained a more simple generalization by using Hankel transforms in place of Fourier transforms; his derivation extends the one which is valid for $l = 0$ and his results can be summarized as follows:

Under the assumptions that the first and second absolute moments $\int_0^\infty r^{1,2}|V(r)|dr$ of the potential are finite, one can determine the potential $V(r)$ as

$$V(r) = V_l(r) = -2\frac{dK_l(r, r)}{dr}, \qquad \text{(IX.1.1)}$$

where $K_l(r, \rho)$ is the solution of the linear integral equation ($r \leq \rho$)

$$K_l(r, \rho) + \int_r^\infty K_l(r, t)F_l(t, \rho)dt + F_l(r, \rho) = 0. \qquad \text{(IX.1.2)}$$

If there is no bound state (we give this case for the sake of simplicity), the kernel of (IX.1.2) is given by

$$F_l(r, \rho) = -(2\pi)^{-1} \int_{-\infty}^{+\infty} dk[1 - S_l(k)][-krh_l^{(2)}(-kr)][-k\rho h_l^{(2)}(-k\rho)]$$

$$\text{(IX.1.3)}$$

provided that

$$\lim_{k\to 0} \frac{1 - S_l(k)}{k^{2l+1}} < \infty. \qquad \text{(IX.1.4)}$$

The $h_l^{(2)}$'s are the spherical Hankel functions of the second kind. Clearly the condition (IX.1.4) is imposed to guarantee the convergence of (IX.1.3) at $k = 0$. If several $S_l(k)$'s are known, (IX.1.1) appears as a consistency condition because $V_l(r)$ should turn out to be independent of l.

From the partial wave equation

$$u_l'' + (k^2 - l(l + 1)r^{-2} - V(r))u_l = 0 \qquad \text{(IX.1.5)}$$

it is possible to define the Jost solution $f_l(k, r)$ as the one which goes to $\exp[-ikr]$ as r goes to infinity. The kernel $K_l(r, \rho)$ is a transformation kernel for this solution:

$$f_l(k, r) = i^{-l-1}\left[krh_l^{(2)}(kr) + \int_r^\infty K_l(r, t)kth_l^{(2)}(kt)dt\right]. \qquad \text{(IX.1.6)}$$

Thus the method is quite similar to the Marchenko method, to which it reduces when $l = 0$.

Singular potentials, i.e. those for which the usual formalism of scattering theory fails, have never to our knowledge been studied by exact methods from the point of view of inverse problems. The only exceptional case is that of the Coulomb potential. Some results on this case, including an uniqueness theorem, were already obtained in 1950 by Jauho and 1951 by Fröberg. To go further, one may think in terms of approaches enabling one to directly construct the transformation kernel in the presence of an extra potential. A convenient way to do it uses the Riemann–Green function of the partial differential equation satisfied by Marchenko's kernel (Gugushvili and Mentkovski 1972a,b—see also Coudray and Coz 1976). These authors studied the equation (**IX.1.5**) with a potential V of the form:

$$V(r) = ar^{-1} + u(r), \tag{IX.1.7}$$

where $u(r)$ is assumed to be bounded by a decreasing exponential. They construct a Marchenko method like above but starting from the pure Coulomb potential ar^{-1}, so that the Hankel functions should be replaced by Coulomb Jost solutions, the free wave Schrödinger operator should be replaced by the perturbed one, and $(1 - S_l(k))$ should be replaced by the difference between $S_l(k)$ and the Coulomb partial wave scattering amplitude $S_l^{(0)}(k)$. The transformation kernel $K_l^c(r, r')$ should then be a solution of the partial differential equation

$$\left[\left(\frac{\partial^2}{\partial r^2} - \frac{l(l+1)}{r^2} - \frac{a}{r}\right) - \left(\frac{\partial^2}{\partial \rho^2} - \frac{l(l+1)}{\rho^2} - \frac{a}{\rho}\right)\right] K_l^C(r, \rho) = u(r) K_l^C(r, \rho).$$

$$\tag{IX.1.8}$$

Using the standard method to solve a hyperbolic equation with inhomogeneous term (here the right-hand side of (**IX.1.8**)) and taking into account the boundary conditions, one easily transforms (**IX.1.8**) into an integral equation:

$$K_l^C(R, P) = K_l^B(R, P) - \frac{1}{2} \iint_{\mathscr{D}} \mathscr{R}_l(R, P, r, \rho) u(r) K_l^C(r, \rho) d^2 S, \tag{IX.1.9}$$

where $\mathscr{R}_l$ is Riemann's function, K_l^B is the Born approximation of K_l^C, and $d^2 S$ refers to the integration in a domain $\mathscr{D}(r, r')$ which is bounded by characteristic lines of the hyperbolic equation. A study of this integral equation gives the properties of the transformation kernel. Obviously this method gives a general way to study inverse problems in which the transformation kernels should be solutions of partial differential equations. We shall meet it again in the inverse problem at fixed energy, where it was used by Sabatier (1971). In the present problem, the method has made possible the study of potentials including a Coulomb term and has extended the Marchenko method to nonintegral values of l. For the same purpose, generalized translation operators (Coz and Coudray 1973) have also been proposed.

IX.2   Krein's Approach

An alternative and very interesting approach to the $l = 0$ inverse problem was given by Krein (1954, 1955, 1956). It also applies to matrix generalizations. Only certain of Krein's results are given here. The starting point is to replace the Schrödinger equation

$$-\frac{d^2 y}{dr^2} + V(r)y = k^2 y \qquad\qquad \textbf{(IX.2.1)}$$

by the equivalent system of first-order differential equations:

$$\frac{dy}{dr} + A(r)y = kz,$$

$$-\frac{dz}{dr} + A(r)z = ky, \qquad\qquad \textbf{(IX.2.2)}$$

where $A(r)$ and $V(r)$ are connected through the Ricatti equation

$$V(r) = -\frac{dA}{dr} + A^2(r) \qquad\qquad \textbf{(IX.2.3)}$$

[note that $(d^2/dr^2) - V(r) = ((d/dr) - A(r))(d/dr) + A(r))$]. The analytical properties of the regular solutions $\varphi(k, r)$ of **(IX.2.1)** and the Paley–Wiener theorem enable then to show the following representation:

$$\varphi(k, r) = k^{-1} \, \text{Im}\left[e^{ikr}\left(1 + \int_0^{2r} \Gamma_{2r}(t)e^{-ikt}\, dt\right)\right] \qquad\qquad \textbf{(IX.2.4)}$$

with

$$A(r) = 2\Gamma_{2r}(2r). \qquad\qquad \textbf{(IX.2.5)}$$

Besides, for each fixed r, the function $\Gamma_{2r}(t)$ is the solution of the Fredholm integral equation

$$\Gamma_{2r}(t) + H(t) + \int_0^{2r} \Gamma_{2r}(s)H(s - t)ds = 0 \qquad\qquad \textbf{(IX.2.6)}$$

where (in the no bound state case), $H(t)$ is obtained from $W(k) = |F(k)|^{-2}$ by the formula

$$H(t) = \pi^{-1} \int_0^{\infty} [W(k) - 1]\cos kt \, dk. \qquad\qquad \textbf{(IX.2.7)}$$

The procedure for solving the inverse problem clearly appears: construct H from (7), solve (6), obtain $A(r)$ from (5) and $V(r)$ from (3).

Finally, let us recall that Krein's equation (6) was introduced in section III.3 from the Gel'fand–Levitan equation.

IX.3   Systems of Equations

For a real $N \times N$ matrix potential $V_{\alpha\beta}(r)$, one considers a system of differential equations of the type

$$y''_\alpha(E, r) + Ey_\alpha(E, r) = \sum_{\beta=1}^{N} y_\beta(E, r)V_{\beta\alpha}(r) \qquad \text{(IX.3.1)}$$

and one tries to recover $V_{\alpha\beta}(r)$ from a knowledge of the S-matrix $S(E)$ for all E and of the bound states.

Newton and Jost (1955) gave a first generalization of the Gel'fand–Levitan method to this problem in a nonphysical case ($V_{\alpha\beta}$ regular at the origin and symmetric) (see also Krein 1956). Extension to more realistic cases, including centrifugal terms, spin-orbit and tensor force, was then given by Newton (1955), with the usual assumptions on the potentials. In particular, he was able to show that if a short-range potential reproducing the S-matrix does exist, it is uniquely determined by the S-matrix, the binding energies, and as many real, symmetric, positive, semidefinite matrices as there are bound states. The systems (IX.3.1), with centrifugal potential, tensor and spin orbit term, have also been studied by a generalized Marchenko method (Agranovich and Marchenko 1957, 1958). An authoritative exposition of these results, which are fairly complete, is given in the book of Agranovich and Marchenko (1963). We state here a typical result. A Hermitian potential matrix $V(x) = \{V_{\alpha\beta}(x)\}$ satisfying the first moment condition:

$$\int_0^\infty x |V_{\alpha\beta}(x)| dx < \infty \qquad \text{(IX.3.2)}$$

can be determined from the scattering matrix $S(k)$, the negative numbers k_n^2 ($n = 1, 2, \ldots, p$), and Hermitian matrices M_n ($n = 1, 2, \ldots, p$) provided these scattering data satisfy five necessary and sufficient conditions:

The matrix $I - S(\lambda)$ is the Fourier transform of a Hermitian matrix $F_s(t)$; each element of the matrix

$$F_s(t) = \frac{1}{2\pi} \int_{-\infty}^{+\infty} [I - S(\lambda)]e^{i\lambda t} \, d\lambda \qquad \text{(IX.3.3)}$$

is summable over $(0, \infty)$ and can be expressed in $(-\infty, 0)$ as the sum of two functions, one of which is summable and the other square summable and bounded.

The derivative $F'_s(t)$ exists for all $t > 0$ and

$$\int_0^\infty t |F'_s(t)| dt < \infty, \qquad \text{(IX.3.4)}$$

where $|A| = \sup_j \sum_k |A_{jk}|$.

The equation

$$-\mathbf{x}(t) + \int_{-\infty}^{0} \mathbf{x}(\xi)\mathbf{F}_s(t + \xi)d\xi = 0 \qquad (-\infty < t \leq 0) \qquad \textbf{(IX.3.5)}$$

has no nontrivial vector solutions with components that are square summable over $(-\infty, 0)$.

The equation

$$\mathbf{x}(t) + \int_{0}^{\infty} \mathbf{x}(\xi)\mathbf{F}(t + \xi)d\xi = 0 \qquad (0 \leq t < \infty) \qquad \textbf{(IX.3.6)}$$

where

$$\mathbf{F}(t) = \sum_{k=1}^{p} \mathbf{M}_n^2 e^{-|k_n|t} + \frac{1}{2\pi} \int_{-\infty}^{+\infty} [\mathbf{I} - \mathbf{S}(\lambda)]e^{i\lambda t}\, d\lambda \qquad \textbf{(IX.3.7)}$$

has no nontrivial vector solutions with components summable over $(0, \infty)$.

The number of linearly independent solutions of

$$\mathbf{x}(t) + \int_{0}^{\infty} \mathbf{x}(\xi)\mathbf{F}_s(t + \xi)d\xi = 0 \qquad (0 \leq t < \infty) \qquad \textbf{(IX.3.8)}$$

is equal to the sum of the ranks of the normalization matrices $\mathbf{M}_1, \mathbf{M}_2, \ldots, \mathbf{M}_n$. Where these conditions hold, it is possible to construct $\mathbf{V}(x)$ in three steps

1. Construct $\mathbf{F}(t)$ from the scattering data and **(IX.3.7)**.
2. From $\mathbf{F}$ obtain the transformation kernel via the integral equation:

$$\mathbf{F}(x + y) + \mathbf{K}(x, y) + \int_{x}^{\infty} \mathbf{K}(x, t)\mathbf{F}(t + y)dt = 0, \qquad 0 < x \leq y.$$

$$\textbf{(IX.3.9)}$$

3. The potential matrix is readily derived from $K(x, x)$:

$$\mathbf{V}(x) = -2\frac{d}{dx}\mathbf{K}(x, x) \qquad (x > 0). \qquad \textbf{(IX.3.10)}$$

All this is a formally straightforward generalization of Marchenko's method. The derivation is also very similar and the important step is again to obtain for a solution of the Schrödinger equation the representation **(V.1.3)** with a triangular kernel.

We now state Newton's results for a potential of the form

$$\mathscr{V}(r) = v_d(r) + v_\sigma(r)\boldsymbol{\sigma}_1\boldsymbol{\sigma}_2 + v_t(r)S_{1,2} + v^0(r)\mathbf{S}\cdot\mathbf{L} \qquad \textbf{(IX.3.11)}$$

which includes central, spin–spin, tensor and spin orbit interactions. It is applied to the triplet state wave function of two spin $\frac{1}{2}$ particles, whose two relevant radial components of the lth and $(l + 2)$nd angular momentum, may be combined in a row-matrix. The Schrödinger equation for the radial part of the wave function containing the components can then be written as follows:

$$\Psi''(E, r) + E\Psi(E, r) = [r^{-2}l(l + 1)(1 - P)$$
$$+ r^{-2}(l + 2)(l + 3)P + V(r)]\Psi(E, r) \qquad (IX.3.12)$$

where $P = \begin{pmatrix} 0 & 0 \\ 0 & 1 \end{pmatrix}$ and $V(r)$ is a real symmetric 2×2 matrix function of r; $\Psi(E, r)$ is a 2×2 matrix whose rows are individually vector solutions of (IX.3.1).

Again it is possible in this problem to define following quantities:

1. The regular solution, $G(k, r)$, which vanishes at $r = 0$ and which, for a zero potential reduces to the free solution

$$G_0(k, r) = \begin{pmatrix} k^{-(l+1)}u_l(k, r) & 0 \\ 0 & k^{-(l+3)}u_{l+2}(k, r) \end{pmatrix}. \qquad (IX.3.13)$$

2. The Jost solution, $F(k, r)$, which satisfies the boundary condition

$$\lim_{r \to \infty} \exp[i(kr - \tfrac{1}{2}\pi l)]\mathscr{E}F(k, r) = 1 \qquad (IX.3.14)$$

where $\mathscr{E} = \begin{pmatrix} 1 & 0 \\ 0 & -1 \end{pmatrix}$.

3. The Jost function, or matrix, is defined as the wronskian of G and F, i.e.

$$F(k) = [F(k, r)\tilde{G}'(k, r) - F'(k, r)\tilde{G}(k, r)]\mathscr{K}_l(k), \qquad (IX.3.15)$$

where

$$\mathscr{K}_l(k) = k^l(1 - P) + k^{l+2}P \qquad (IX.3.16)$$

and the tilde stands for transpose. $F(k)$ is analytic in Im $k < 0$, continuous for Im $k \le 0$, except possibly at $k = 0$, where

$$F_e(k) \equiv \mathscr{K}_l(k)F(k)\mathscr{K}_l(k^{-1}) \qquad (IX.3.17)$$

however, is continuous.

4. The S-matrix is equal to

$$S(k) = F(k)F^{-1}(-k). \qquad (IX.3.18)$$

5. The nonzero eigenvalues of (IX.3.17) correspond exactly to the points in the lower half plane where $F(k)$ is a singular matrix, (or det$[F(k)]$ is zero). Thus $F^{-1}(k)$ is analytic for Im $k < 0$ except at a finite number of points $k = -iK_i$, $K_i > 0$, where $F^{-1}(k)$ has simple poles. These poles correspond to the bound states. Some difficulties arise at $k = 0$ which are particular to this problem and we refer, for them, to Newton (1955). The phase of det $S(k)$ is also related to the number of bound states in a way which is similar to the Levinson theorem.

6. The regular solutions form a complete set. The completeness relation reads:

$$\int \tilde{G}(\sqrt{E}, r)G(\sqrt{E}, r')dP(E) = \delta(r - r') \qquad (IX.3.19)$$

where the spectral function $P(E)$ is defined by

$$P(-\infty) = 0, \tag{IX.3.20}$$

$$\frac{d\mathbf{P}(E)}{dE} = \begin{cases} \pi^{-1} E^{l+(1/2)} \mathscr{K}_0(\sqrt{E})[\mathbf{F}^T(\sqrt{E})\mathbf{F}(-\sqrt{E})]^{-1} \mathscr{K}_0(\sqrt{E}), \\ \qquad\qquad\qquad\qquad\qquad\qquad\qquad\qquad E > 0 \\ \displaystyle\sum_{n=0}^{L} \mathbf{C}_n \delta(E - E_n) \qquad\qquad\quad E_0 = 0, E \le 0. \end{cases}$$

The $\mathbf{C}_h$'s are normalization constants for the bound states at energies E_n. Thus the spectral function is a real, symmetric, positive semidefinite matrix function of E.

7. The connection between the spectral function and the potential via the generalized Gel'fand–Levitan equation is obtained from the completeness relations in the standard way. If the subscript 0 denotes quantities belonging to a suitable comparison potential V_0 and

$$\mathscr{F}(s, r) = -\int \tilde{\mathbf{G}}_0(\sqrt{E}, s) d[\mathbf{P}_0(E) - \mathbf{P}(E)] \mathbf{G}_0(\sqrt{E}, r), \tag{IX.3.21}$$

then there exists a matrix function $\mathscr{K}(s, r)$ which is the unique solution of the integral equation

$$\mathscr{K}(s, r) + \mathscr{F}(s, r) + \int_0^r \mathscr{F}(s, t) \mathscr{K}(t, r) dt = 0, \qquad s \le r, \tag{IX.3.22}$$

and satisfies

$$2 \frac{d}{dr} \mathscr{K}(r, r) = \mathbf{V}(r) - \mathbf{V}_0(r). \tag{IX.3.23}$$

It is also possible to construct the spectral function from $\mathbf{S}(k)$ (Newton 1955).

IX.4 Coupled Channels

We consider the set of coupled radial equations

$$-\Psi_\alpha'' + \sum_\beta V_{\alpha\beta} \Psi_\beta = K_\alpha^2 \Psi_\alpha \tag{IX.4.1}$$

where the Ψ_α's are the channel components of the time-independent wavefunction. $V_{\alpha\beta} = V_{\beta\alpha}$ is the potential matrix, which is symmetric because of the time reversal invariance, K_α is the channel wave number. For the sake of simplicity, we assume a common channel-reduced-mass μ. Thus we write the conservation of energy $(E = \hbar^2 K_1^2/2\mu)$ between channels as

$$K_1^2 = K_i^2 + \Delta_i^2 \tag{IX.4.2}$$

where $\hbar^2\Delta_i^2/2\mu$ is the constant threshold energy of the ith channel. The threshold energy of the first channel (entrance channel) is, of course, zero. It is evident from **(IX.4.2)** that K_i is real for an open channel, imaginary for a closed channel.

Clearly, the solution Ψ_α can be written as a column vector. It is convenient to combine n of these column solutions into a single square matrix. Each column solution is distinguished from the others by its boundary condition. Thus we write Ψ, respectively V, for the square $n \times n$ matrix $\Psi_{\alpha\beta}$, respectively $V_{\alpha\beta}$ and K for the diagonal matrix $K_\alpha\delta_{\alpha\beta}$. In matrix form, **(IX.4.1)** is

$$-\Psi'' + V\Psi = K^2\Psi. \tag{IX.4.3}$$

Assuming that V depends only on r, and that its first and second moments $\int_0^\infty r^{1,2}\|V(r)\|dr$ are finite, it is possible to generalize some results of the s-wave potential scattering theory, and in particular:

1. The regular solution $\Phi(K, r)$, which is defined by **(IX.4.3)** and the boundary conditions $\Phi(K, 0) = 0$, $\Phi'(K, 0) = 1$, is the unique solution of a Volterra integral equation which is quite similar to **(I.2.2)**. It follows that Φ, as a function of $K_1, \ldots, K_n$, is an analytic function of all the K_α's regular in their entire complex-plane. It can further be shown that Φ has the following asymptotic behavior for large K. As $|K_1|, |K_2|, \ldots, |K_n| \to \infty$,

$$\Phi(K, r) \to K^{-1}\sin[K, r] + o(|K|^{-1})\exp[|\operatorname{Im}K|r]. \tag{IX.4.4}$$

2. The Jost solution, $F(K, r)$, which is defined by **(IX.4.3)** and the boundary condition $\lim_{r \to \infty}\exp[iKr]F(K, r) = 1$, is the unique solution of a Volterra integral equation which is quite similar to **(I.3.1)**. It follows that $F_{\alpha\beta}(K_1, \ldots, K_n, r)$ is an analytic function of K_β regular in the lower half of the complex plane. However, it appears that $F_{\alpha\beta}$ has no further regularity properties as a function of the other K_α's unless much stronger assumptions are made concerning the potential. Increasing exponentials, which in general prevent convergence, appear in the integrands as soon as E goes below the threshold of any of the channel.

3. The Jost matrix, $F(K)$, is defined as the wronskian of $F(K, r)$ and $\Phi(K, r)$:

$$\tilde{F}(K) = \tilde{F}(K, r)\Phi'(K, r) - \tilde{F}'(K, r)\Phi(K, r) \tag{IX.4.5}$$

where the tilda denote the transposition. Thus $F(K)$ is nothing but $F(K, 0)$. Although $F(K)$ seems in general to have no regularity at all as soon as any of the K_α's is no longer real, Newton (1961) has shown that the determinant of $F(K)$ is an analytic function of all the K_α's regular in the whole upper plane. This regularity of det $F(K)$ will be exploited later.

4. The complete Green's function $\mathscr{Y}(K, r, r')$ is equal to

$$\mathscr{Y}(K, r, r') = \begin{cases} -\Phi(K, r)\tilde{F}^{-1}(-K)\tilde{F}(-K, r'), & r < r', \\ -F(-K, r)F^{-1}(-K)\tilde{\Phi}(K, r), & r' < r. \end{cases} \tag{IX.4.6}$$

Clearly it reduces for $\mathbf{V} = 0$ to

$$\mathscr{Y}_0(\mathbf{K}, r, r') = -\mathbf{K}^{-1} \sin[\mathbf{K}, r]_< \exp[i\mathbf{K}r']_> \qquad \text{(IX.4.7)}$$

which is also the asymptotic limit of $\mathscr{Y}(\mathbf{K}, r, r')$ as $|\mathbf{K}|$ goes to infinity. It has been possible to show that $\mathscr{Y}(\mathbf{K}, r, r')$ is multiple valued as a function of the energy with threshold branch points on the real axis. Besides, it is a meromorphic function of the energy in the region characterized by Im $K_1, K_2, \ldots, K_n \geq 0$, with simple poles. The significance of these poles is that they correspond to the energies of the bound states of the system (Newton 1961). Since such bound state energies must be real it follows that the poles of $\mathscr{Y}$ in Im $\mathbf{K} > 0$ must always lie on the imaginary axis of all the K_α's. It is also possible for $\mathscr{Y}$ to have poles on the real K_1 axis for real energies below the highest threshold. These poles must occur in pairs symmetric about the imaginary K_1 axis. They describe bound states "embedded in the continuum" (Fonda and Newton 1960). The residue of $\mathscr{Y}$ at a bound state pole is equal to $i\Phi(\mathbf{K}^{(j)}, r)\mathbf{C}_j\tilde{\Phi}(\mathbf{K}^{(j)}, r')$, where $\mathbf{C}_j$ is a real symmetric constant matrix and $E_j = \hbar^2[K_1^{(j)}]^2/2\mu$ indicates the energy of the jth bound state pole.

5. The functions $\Phi(\mathbf{K}, r)$ $(E > 0)$ and $\Phi(\mathbf{K}^{(j)}, r)$, $j = 1, \ldots, N$, form a complete set. The completeness relation is given by:

$$\int_{-\infty}^{+\infty} \Phi(\mathbf{K}, r)d\rho(E)\tilde{\Phi}(\mathbf{K}, r') = 1\delta(r - r') \qquad \text{(IX.4.8)}$$

where

$$\frac{d\rho(E)}{dE} = \begin{cases} \dfrac{2\mu}{\pi\hbar^2} \tilde{\mathbf{F}}^{-1}(-\mathbf{K})\mathbf{K}\Theta\mathbf{F}^{*-1}(-\mathbf{K}), & E > 0, \\[2ex] \displaystyle\sum_{j=1}^N \delta(E - E_j)\mathbf{C}_j(-2i\mathbf{K}_1^{(j)}), & E \leq 0, \end{cases} \qquad \text{(IX.4.9)}$$

and Θ is the diagonal matrix $|\mathbf{K}|^{-1}$ Re $\mathbf{K}$ which projects onto the open channels.

6. The S-matrix is equal to

$$\mathbf{S}(\mathbf{K}) = \mathbf{K}^{1/2}\mathbf{F}^{-1}(-\mathbf{K})\mathbf{F}(\mathbf{K})\mathbf{K}^{-1/2}. \qquad \text{(IX.4.10)}$$

7. det $\mathbf{F}(\mathbf{K})$ may have zeros on the imaginary K_1 axis. They correspond to bound states. It may also have zeros on the real K_1 axis in the region $-\Delta_n \leq K_1 \leq \Delta_n$, which correspond to "bound states embedded in the continuum." In the latter case, the zeros lie in pairs symmetric about the imaginary K_1 axis. Besides, det $\mathbf{F}^*(-\mathbf{K}^*)$ is equal to det $\mathbf{F}(\mathbf{K})$, and det $\mathbf{F}(\mathbf{K})$ goes to 1 as $|K_1| \to \infty$ in Im $K_1, K_2, \ldots, K_n \geq 0$.

Now, three inverse problems have been solved in this framework:

1. Constructing $V(r)$ from its spectral functions $d\rho(E)/dE$. This can be done thanks to a generalization of the Gel'fand–Levitan method which

is obtained from the completeness relation in the standard way (Cox 1967). Let $\boldsymbol{\Phi}^{(0)}(\mathbf{K}, r)$ and $\rho^{(0)}(E)$ be the wave function and the spectral function corresponding to the zero matrix potential. The Gel'fand–Levitan symmetric kernel is:

$$\mathscr{F}(r, r') = \int_{-\infty}^{+\infty} \boldsymbol{\Phi}^{(0)}(\mathbf{K}, r) d[\boldsymbol{\rho}^{(0)}(E) - \boldsymbol{\rho}(E)] \boldsymbol{\Phi}^{(0)}(\mathbf{K}, r'). \quad \textbf{(IX.4.11)}$$

The transformation kernel is the solution of the Gel'fand–Levitan integral equation

$$\mathscr{K}(r, r') = \mathscr{F}(r, r') + \int_0^r \mathscr{K}(r, \rho) \mathscr{F}(\rho, r') d\rho \qquad \textbf{(IX.4.12)}$$

and the potential

$$\mathbf{V}(r) = 2 \frac{d}{dr} \mathscr{K}(r, r). \qquad \textbf{(IX.4.13)}$$

There is no known method to construct the spectral function from the open-channel submatrix **S**, the only accessible to experiment.

2. Constructing $\mathbf{V}(r)$ from a knowledge of the S-matrix for all positive energy and from the bound states. This can be done thanks to a generalization of the Marchenko method which again is obtained from the completeness relation (Cox 1967). Defining a symmetric kernel

$$\mathbf{F}(r, r') = \mathbf{1}\delta(r - r') - \int_{-\infty}^{+\infty} \boldsymbol{\Phi}_\infty(\mathbf{K}, r) d\boldsymbol{\rho}(E) \tilde{\boldsymbol{\Phi}}_\infty(\mathbf{K}, r') \quad \textbf{(IX.4.14)}$$

it is first shown that there exists a transformation kernel $\mathbf{A}(r, r')$ such that

$$\boldsymbol{\Phi}(\mathbf{K}, r) = \boldsymbol{\Phi}_\infty(\mathbf{K}, r) + \int_r^\infty \mathbf{A}(r, r') \boldsymbol{\Phi}_\infty(\mathbf{K}, r') dr', \qquad \textbf{(IX.4.15)}$$

where $\boldsymbol{\Phi}_\infty(\mathbf{K}, r)$ means the asymptotic expression for $\boldsymbol{\Phi}(\mathbf{K}, r)$ as $r \to \infty$. This transformation kernel is the solution of Marchenko's equation

$$\mathbf{A}(r, r') = \mathbf{F}(r, r') + \int_r^\infty \mathbf{A}(r, \rho) \mathbf{F}(\rho, r') d\rho \qquad (r < r') \quad \textbf{(IX.4.16)}$$

$\mathbf{F}(r, r')$ is then related to the S-matrix and the bound state energies by means of Green's function. More precisely, the asymptotic behavior of $\mathscr{Y}(\mathbf{K}, r, r')$ when both r and r' are large is inserted in (**IX.4.14**). Thus, taking (**IX.4.9**) into account, one obtains

$$\mathbf{F}(r, r') = \frac{1}{2\pi} \int_{-\infty}^{+\infty} K_1 dK_1 e^{i\mathbf{K}r} \mathbf{K}^{-1/2} [S(\mathbf{K}) - \mathbf{1}] \mathbf{K}^{-1/2} e^{i\mathbf{K}r'}$$

$$- \sum_{j=1}^N e^{-|\mathbf{K}^{(j)}|r} \mathbf{A}_j e^{-|\mathbf{K}^{(j)}|r'}, \qquad \textbf{(IX.4.17)}$$

where we have assumed no continuum bound states for simplicity. $\mathbf{K}^{(j)}$ is the imaginary value corresponding to an ordinary bound state, and $\mathbf{A}_j$ a symmetric matrix easily related to $\mathbf{C}_j$. The potential $\mathbf{V}(r)$ is easily obtained from $\mathbf{A}(r, r')$:

$$\mathbf{V}(r) = -2\frac{d}{dr}\mathbf{A}(r, r). \qquad (IX.4.18)$$

This method is certainly very interesting, but only the open-channel submatrix is accessible. Although it is possible in principle to continue analytically all elements of this submatrix to energies where all channels are closed and find $\mathbf{S}(k)$, no general method for doing this is known to us.

3. Constructing $\mathbf{V}(r)$ from a knowledge of det $\mathbf{F}(\mathbf{K})$. This determinant actually is also the Fredholm determinant of the Lipmann–Schwinger equation. It most conveniently characterizes the scattering, bound states and resonances of the coupled-channel system. When it is known in analytic form, all elements of the open-channel $\mathbf{S}$ matrix can easily be constructed from it. It has been proved by Cox (1967) that $\mathbf{V}(r)$ can also be constructed from it.

Again, this very nice theoretical result does not solve a physical inverse problem.

Apart from these remarkable but not sufficient results, only a few others are known. Cox (1967, 1975), obtained a formal generalization of the many-channel Marchenko equations to $l > 0$ waves. Besides, he was able to construct Bargmann potentials both in the $l = 0$ and $l = 1$ case (Cox and Garcia 1975).

There is in the recent literature some confusion between these very difficult "inelastic" problems, and the "elastic" ones described in (IX.3). The elastic case is comparatively trivial, and much less important from the physical point of view.

IX.5   Relativistic Problems

IX.5.1   General Remarks

These problems, are probably much more "academic" than the previous ones, and yet, they have been the object of many more studies.

When one tries to generalize the Gel'fand–Levitan formalism to relativistic equations such as the Dirac or Klein–Gordon equation, one meets several difficulties:

1. The spectral function is difficult to define properly because either the eigenfunctions are not a complete orthonormal set in the usual sense or one has to deal with a system of equations.

2. The phase shift does not tend to zero as the energy goes to infinity (notice that in the Klein–Gordon equation, the potential appears multiplied by the energy E).

3. Because of the occurrence of the two-valued quantity $E = \sqrt{k^2 + m^2}$, the solutions of a relativistic equation must be represented as functions of k on a two-sheeted Riemann surface. This has the effect that the potential can be constructed only if the phase-shift is given for each branch of E. Moreover, the binding energies, which lie now in the interval $(-m, m)$, have to be given. The phase shift for the negative branch of E, and the negative eigenvalues, must be interpreted physically as describing the scattering and bound states of antiparticles.

The first generalization of the Gel'fand–Levitan procedure to the S-wave Klein–Gordon equation was given by Corinaldesi (1954). Simpler derivations are due to De Alfaro (1958), who used the Jost and Kohn procedure, Verde (1958), who used dispersion relations (see also Malčenko (1966)). Verde's method can also be applied to the Dirac equation, which was also studied by Prats and Toll (1959). Further results, and more mathematical elaboration, have been given for Dirac systems by Gasimov and Levitan (1966), and Gasimov (1968). See also Levitan and Sargsjan (1965). Generalizations of the Marchenko method were given for the Klein–Gordon equation by Weiss and Scharf (1971), for the Dirac equation by Gasimov (1967), and by Weiss, Stahel and Scharf (1972). Also, Cornille (1970), reconstructed the Klein–Gordon operator with exponentially decreasing potentials from the S-matrix discontinuity in the complex k-plane. The results are complicated. For somewhat related work, see sections VII.5 and IX.10. We shall give here only the results obtained by the generalization of the Marchenko formalism (Gasimov and Levitan, 1966; Weiss, Stahel and Scharf, 1972; Weiss and Scharf, 1971). They illustrate sufficiently well the kinds of results one obtains in relativistic problems.

IX.5.2   The Dirac Equation

The radial dependent parts $F(r)/r$ and $G(r)/r$ of the Dirac wave function for a particle of mass m in a central potential $V(r)$, for a given parity and a given angular momentum $j (= 1/2, 3/2, \ldots)$ satisfy the coupled differential equations

$$\omega \frac{d\varphi}{dr} + \sigma_1 \frac{\lambda}{r} \varphi - \sigma_3 m\varphi + V\varphi = E\varphi, \qquad \text{(IX.5.1a)}$$

$$\omega = \begin{pmatrix} 0 & 1 \\ -1 & 0 \end{pmatrix}, \qquad \sigma_1 = \begin{pmatrix} 0 & 1 \\ 1 & 0 \end{pmatrix}, \qquad \sigma_3 = \begin{pmatrix} 1 & 0 \\ 0 & -1 \end{pmatrix},$$

$$\varphi = \begin{pmatrix} \varphi_1 \\ \varphi_2 \end{pmatrix} = \begin{pmatrix} F \\ G \end{pmatrix}, \qquad \text{(IX.5.1b)}$$

where the units are chosen so that $\hbar = c = 1$, and $\lambda = (j + \frac{1}{2})$.

We shall assume that

$$\int_0^\infty r^n |V(r)| dr < \infty, \qquad n = 0, 1, 2. \qquad \text{(IX.5.2)}$$

Under these assumptions, one can show, in complete analogy with what we have seen in previous chapters, that for each real E, $|E| > m$, there exists a solution (the Jost solution) $f(k, r)$ such that[1]

$$f(k, r) \underset{r \to \infty}{\sim} (-ik)^\lambda \begin{pmatrix} i(E - m)/k \\ 1 \end{pmatrix} e^{ikr}, \qquad \text{(IX.5.3)}$$

where $k = (E^2 - m^2)^{1/2}$. In words, this solution is asymptotic to an outgoing free spherical wave. Similarly, one can define the solution

$$f(-k, r) = f(-k, r) = [f(k, r)]^* \qquad \text{(IX.5.4)}$$

which is asymptotic to an incoming free spherical wave. Both solutions are finite for each k and r ($r > 0$), and have a singularity of the type $r^{-\lambda}$ at the origin. They form a fundamental system of solutions of (IX.5.1). This is obvious since the determinant (remember that we are dealing here with a system of coupled first-order differential equations!)

$$\Delta(k) = \det\{f(k, r), f(-k, r)\} = 2ik^{2\lambda - 1}(E - m) \qquad \text{(IX.5.5)}$$

is different from zero. It follows that the regular solution $\varphi(E, r)$, defined by the boundary condition

$$r^{-\lambda}\varphi(E, r)\Big|_{r=0} = \begin{pmatrix} [(2\lambda - 1)!!]^{-1} \\ 0 \end{pmatrix} \qquad \text{(IX.5.6)}$$

can be written

$$\varphi(E, r) = \Delta^{-1}(k)[F(k)f(-k, r) - F(-k)f(k, r)], \qquad \text{(IX.5.7)}$$

where we have introduced the "Jost function"

$$F(k) = [(2\lambda - 1)!!]^{-1} r^\lambda f_2(k, r)\Big|_{r=0} \qquad \text{(IX.5.8)}$$

f_2 being the second component of f. As it is seen, the above formulae are quite analogous to those of the first chapter for the Schrödinger equation. From the asymptotic behavior of $f(k, r)$ given by (IX.5.3), it follows that

$$\varphi(E, r) \underset{r \to \infty}{\sim} k^{-\lambda} |F(k)| \begin{pmatrix} \cos[kr - \frac{1}{2}\pi + \delta(k)] \\ [k/(E - m)]\sin[kr - \frac{1}{2}\pi + \delta(k)] \end{pmatrix}, \qquad \text{(IX.5.9)}$$

---

[1] To simplify the formulae, we shall not write explicitly the dependence of the wave functions and other related quantities on λ since we are dealing with the inverse problem for fixed λ.

where $\delta(k)$, the phase-shift, is given by

$$\delta(k) = -\operatorname{Arg} F(k). \tag{IX.5.10}$$

All quantities with $E \geq m$ correspond to the scattering of the particle, and those with $E \leq -m$ to the scattering of the antiparticle. Likewise, the bound states, which always lie in the interval $(-m, m)$, correspond to the particle, or the antiparticle, according to the sign of their energy. We shall see below, that in order to solve the inverse problem, we must know the scattering data (phase shifts, binding energies, and their normalization constants) both for particle and for antiparticle.

On the basis of the above assumptions on the potential, one can study the properties of various solutions of the Dirac equation (the regular solution, the "Jost solution" $f(k, r)$), quite in analogy with the nonrelativistic case. In order to study the bound states and their connection with the Jost function, we must first extend the definition of $f(k, r)$ and of $F(k)$ to the complex k-plane Since $E = \pm(k^2 + m^2)^{1/2}$, we introduce the Riemann surface of this double-valued function by cutting the k-plane from im to $i\infty$, and from $-im$ to $-i\infty$. This gives us a single-valued function on the two-sheeted Riemann surface, the two sheets being distinguished by the sign of the real part of E, which is constant through each sheet. We distinguish therefore the two sheets by the subscript $\sigma = \pm$ according to the sign of Re E. In complete analogy with the case of the Schrödinger equation, it can be seen that $f(k, r)$ can be extended to the complex k-plane, and becomes an analytic function in the upper half-planes of both sheets (Im $k > 0$, $\sigma = \pm$). It is also a continuous function on Im $k = 0$, except perhaps at $k = 0$.

One can then show that there is a one-to-one correspondence between the zeros of the Jost function in the upper planes, which all lie on imaginary axis between 0 and im, and the bound states. The eigenfunction expansions can be again dealt with by contour integration in the complex plane, and one is led to formulae similar to the nonrelativistic case, containing now contributions from both sheets, i.e. both for particle and for antiparticle.

Likewise, one can prove the existence of integral representations analogous to the Povzner–Levitan and Marchenko representations of chapters III and V. Combining them now with the completeness relations, one obtains integral equations of the Gel'fand–Levitan and Marchenko types, which permit a solution of the inverse problem in terms of the scattering data. In the Marchenko formalism, the kernel $A(x, y)$ is now a 2×2 matrix

$$A(x, y) = \begin{pmatrix} A_{11}(x, y) & A_{12}(x, y) \\ A_{21}(x, y) & A_{22}(x, y) \end{pmatrix}, \tag{IX.5.11}$$

which satisfies the integral equation

$$A(x, y) + F(x, y) + \int_x^\infty dt\, A(x, t)F(t, y) = 0, \tag{IX.5.12}$$

where the kernel F is given by

$$F(x, y) = \int_{-\infty}^{+\infty} dk \left\{ \sum_\sigma \frac{E_\sigma + m}{2\pi E_\sigma} (1 - S_\sigma(k)) f_\sigma^0(k, x) \tilde{f}_\sigma^0(k, y) \right.$$

$$+ \frac{1}{\pi} \begin{pmatrix} 1 - 2\cos\delta_\alpha & \sin 2\delta_\alpha \\ \sin 2\delta_\alpha & -1 + \cos 2\delta_\alpha \end{pmatrix} e^{ik(x+y) - i\pi\lambda} \right\}$$

$$+ \sum_{n,\sigma} C_{\sigma n}^{-1} f^0(i\gamma_{\sigma n}, x) \tilde{f}^0(i\gamma_{\sigma n}, y). \qquad \text{(IX.5.13)}$$

Here $i\gamma_{\sigma n}$ are the bound state zeros of the Jost function, with the bound state energies

$$E_{\sigma n} = \sigma(m^2 - \gamma_{\sigma n}^2)^{1/2} \qquad \text{(IX.5.14)}$$

and

$$S_\sigma(k) = e^{2i\delta_\sigma(k)}, \qquad \text{(IX.5.15)}$$

$$\delta_\alpha = \lim_{E \to \infty} \delta_+(E) = - \int_0^\infty V(r)dr, \qquad \text{(IX.5.16)}$$

is the S-matrix. By $f^0(k, r)$, we have denoted the free radial wave function

$$f_\sigma(\pm k, x) = \begin{pmatrix} (E_\sigma - m)h_{\lambda-1}^\pm(kx) \\ kxh_\lambda^\pm(kx) \end{pmatrix} \qquad \text{(IX.5.17)}$$

where h_λ are the Hankel functions.

Solving the above integral equation, we obtain the potential by the following formulae

$$V(r) = - \frac{1}{2} \frac{d}{dr} \arcsin 2 \frac{(\lambda/r)A_{12}(r, r) - mA_{11}(r, r)}{(\lambda^2/r^2) + m^2} \qquad \text{(IX.5.18)}$$

or

$$V(r) = - \frac{d}{dr} \operatorname{arctg} \frac{A_{11}(r, r)}{A_{12}(r, r)} - \frac{d}{dr} \operatorname{arctg}\left(\frac{mr}{\lambda}\right). \qquad \text{(IX.5.19)}$$

We see therefore that whenever the integral equation has a unique solution, the potential is uniquely determined from the scattering data. The uniqueness of the solution has been treated in detail by Gasimov and Levitan. Unfortunately, the results are too complicated to be given here, and so we refer the reader to their work.

IX.5.3   The Klein–Gordon Equation

The solution of the inverse scattering problem here is quite similar to the previous case. One has again an integral equation of the Marchenko type similar to (IX.5.12) whose kernel is given by the Fourier transform of the elements of the S-matrix for both particle and antiparticle plus a sum of terms for the contributions of bound states of both kinds. Everything being quite similar to the case of the Dirac equation, we shall not give the formulae here, and refer the reader to the work of Weiss and Scharf (1971) which contains references to the earlier works of Corinaldesi and Verde.

IX.6   Discrete Forms of the Methods

A discrete version of the inverse scattering problems was considered by Case and Kac (1973), as regards the Gel'fand–Levitan method. Instead of the Schrödinger equation

$$\frac{1}{2}\frac{d^2\psi(E, x)}{dx^2} - q(x)\psi(E, x) = -E\psi(E, x), \tag{IX.6.1}$$

$$\psi(E, 0) = 0; \qquad \psi'(E, 0) = 1, \tag{IX.6.2}$$

they study the difference equation

$$\tfrac{1}{2}\psi(E; (n + 1)\Delta) + \tfrac{1}{2}\psi(E; (n - 1)\Delta) = \lambda \exp[v(n)]\psi(E; n\Delta) \tag{IX.6.3}$$

where

$$\lambda = 1 - E\Delta^2; \qquad v(n) = \Delta^2 q(n\Delta). \tag{IX.6.4}$$

The equation (IX.6.3) goes over into the Schrödinger equation (IX.6.1) in the limit $\Delta \to 0$. Setting

$$\varphi(\lambda, n) = \exp[\tfrac{1}{2}v(n)]\psi(E, n\Delta), \tag{IX.6.5}$$

$$a_{m,n} = \tfrac{1}{2}\exp[-\tfrac{1}{2}(v(m) + v(n))][\delta_{m,n+1} + \delta_{m,n-1}], \tag{IX.6.6}$$

we see that (IX.6.3) can be rewritten in the equivalent form

$$A\varphi = \lambda\varphi \tag{IX.6.7}$$

where the symmetric matrix $A = \{a_{m,n}\}$ is tridiagonal, and φ is the sequence $\{\varphi(\lambda, n)\}$ $(m, n \geq 1$, so that the boundary condition $\varphi(\lambda, 0) = 0$ is automatic). It is convenient to first take $\varphi(\lambda, 1) = 1$ rather than $\varphi(\lambda, 1) = \Delta$ as the other boundary condition. For $v(n) \equiv 0$, the function $\varphi(\lambda, n)$ is equal to

$$\mathring{\varphi}(\lambda, n) = \frac{[\lambda + (\lambda^2 - 1)^{1/2}]^n - [\lambda - (\lambda^2 - 1)^{1/2}]^n}{2(\lambda^2 - 1)^{1/2}} \tag{IX.6.8}$$

and the spectral distribution $\sigma(\lambda)$ is given by the well-known formula

$$\sigma(\lambda) = \begin{cases} 0, & \lambda < -1, \\ \dfrac{2}{\pi}\displaystyle\int_{-1}^{\lambda} (1 - \mu^2)^{1/2}\, d\mu, & -1 < \lambda < 1, \\ 1, & \lambda > 1. \end{cases} \tag{IX.6.9}$$

Now let the spectral distribution of A be $\rho(\lambda)$, so that

$$\int \varphi(\lambda; m)\varphi(\lambda; n)d\rho(\lambda) = \delta_{mn}. \qquad \textbf{(IX.6.10)}$$

$$\int \varphi(\lambda; m)\varphi(\lambda; n)\lambda \, d\rho(\lambda) = a_{m.n}. \qquad \textbf{(IX.6.11)}$$

Since the $\varphi(\lambda, n)$ are polynomials, they are *the* orthogonal polynomials (properly normalized) with respect to the weight $\rho(\lambda)$.

Now a first inverse problem is to determine $v(n)$ from the spectral distribution $\rho(\lambda)$. It is almost trivial because given $\rho(\lambda)$ we can determine the orthonormal polynomials with respect to ρ and then calculate $a_{n,n+1}$. There are two important points which have to be fixed. The first is that since $a_{n,n+1}$ is positive, one must be sure that the orthonormal polynomials are such that

$$\int \lambda\varphi(\lambda; n)\varphi(\lambda; n + 1)d\rho(\lambda) > 0. \qquad \textbf{(IX.6.12)}$$

This condition is fulfilled because of the result **(IX.6.22)** below. The second point is that ρ cannot be prescribed arbitrarily, since we must have

$$a_{nn} = \int \lambda\varphi^2(\lambda; n)d\rho(\lambda) = 0. \qquad \textbf{(IX.6.13)}$$

We must therefore impose an *a priori* condition on ρ to ensure **(IX.6.13)**. The simplest such condition is that

$$\rho(-\lambda) = 1 - \rho(\lambda) \qquad \textbf{(IX.6.14)}$$

which ensures that the orthogonal polynomials are either even or odd and hence **(IX.6.13)** follows. From now on **(IX.6.14)** will be assumed.

Given ρ, we seek the orthonormal polynomials in the form

$$\varphi(\lambda; n) = K(n, n)\dot\phi(\lambda, n) + \sum_{m=1}^{n=1} K(n, m)\dot\phi(\lambda, m). \qquad \textbf{(IX.6.15)}$$

The polynomial $\varphi(\lambda, n)$ which is of degree $n - 1$, is orthogonal to every polynomial of degree lower than $n - 1$ and hence to every $\dot\phi(\lambda; m)$ for $m < n$. Thus, orthogonality conditions are equivalent to

$$\int \varphi(\lambda, n)\dot\phi(\lambda, m)d\rho(\lambda) = 0, \qquad m < n, \qquad \textbf{(IX.6.16)}$$

and setting

$$q(m, l) = \int \dot\phi(\lambda; m)\dot\phi(\lambda, l)d[\rho(\lambda) - \sigma(\lambda)], \qquad \textbf{(IX.6.17)}$$

we see that (**IX.6.16**) can be rewritten in the form

$$0 = K(n, n)q(n, m) + K(n, m) + \sum_{l=1}^{n-1} K(n, l)q(l, m), \qquad n > m. \qquad \textbf{(IX.6.18)}$$

Clearly, the kernel $K(n, m)$ acts in (**IX.6.15**) as a transformation kernel generating $\varphi(\lambda, n)$ from $\hat{\varphi}(\lambda, n)$. The equation (**IX.6.18**), which relates this transformation kernel with the symmetric kernel $q(n, m)$ is strongly reminiscent of the Gel'fand–Levitan equation.

So as to calculate $K(n, m)$ and $K(n, n)$, it is convenient to write down the equations for

$$\mathcal{K}(n, m) = \frac{K(n, m)}{[K(n, n)]}, \qquad m < n. \qquad \textbf{(IX.6.19)}$$

We see that (**IX.6.18**) can be written in the form

$$0 = q(n, m) + \mathcal{K}(n, m) + \sum_{l=1}^{n-1} \mathcal{K}(n, l)q(l, m). \qquad \textbf{(IX.6.20)}$$

Then the normalization condition $[\int \varphi^2(\lambda, n)d\rho(\lambda) = 1]$ readily yields $K(n, n)$ as

$$K(n, n) = \left[1 + q(n, n) + \sum_{l=1}^{n-1} \mathcal{K}(n, l)q(l, n) \right]^{-1/2} \qquad \textbf{(IX.6.21)}$$

It remains to determine $v(n)$. Now

$$a_{n, n+1} = \int \lambda \varphi(\lambda; n)\varphi(\lambda; n + 1)d\rho(\lambda)$$

$$= \tfrac{1}{2}K(n, n) \int \hat{\varphi}(\lambda; n + 1)\varphi(\lambda; n + 1)d\rho(\lambda)$$

$$= \tfrac{1}{2}K(n, n)[K(n + 1, n + 1)]^{-1} \qquad \textbf{(IX.6.22)}$$

or, equivalently

$$\tfrac{1}{2}[v(n) + v(n + 1)] = \log K(n + 1, n + 1) - \log K(n, n). \quad \textbf{(IX.6.23)}$$

Equation (**IX.6.23**) does not in general determine $v(n)$ but if, e.g. it is known that $\lim_{n \to \infty} v(n) = 0$, then, in fact, v becomes uniquely determined, since:

$$\tfrac{1}{2}[v(1) + (-1)^{n+1}v(n + 1)] = \sum_{r=1}^{n}(-1)^{r+1} \log \frac{K(r + 1, r + 1)}{K(r, r)}. \quad \textbf{(IX.6.24)}$$

Thus the inverse problem $\rho \to v$ is completely solved. The method of solution is obtained in a straightforward way and closely follows the Gel'fand–Levitan method. It is instructive to see the limiting form of the equations as $\Delta \to 0$. We first replace φ, $\hat{\varphi}$ by $\varphi' = \Delta\varphi$, $\varphi'_0 = \Delta\hat{\varphi}$, and require that

$$\int_{-\infty}^{+\infty} \varphi'(\lambda, n)\varphi'(\lambda, m)d\rho'(\lambda) = \int_{-\infty}^{+\infty} \varphi'_0(\lambda, n)\varphi'_0(\lambda, m)d\sigma' = \Delta^{-1}\delta(m, n). \quad \textbf{(IX.6.25)}$$

Thus $\rho' = \rho\Delta^{-3}, \sigma' = \sigma\Delta^{-3}$. Hence we obtain the Gel'fand–Levitan method, and, in particular, the Gel'fand–Levitan equation:

$$0 = q'(x, y) + \mathscr{K}'(x, y) + \int_0^x \mathscr{K}'(x, t)q'(t, y)dt \qquad \textbf{(IX.6.26)}$$

with

$$q'(x, y) = \int_{-\infty}^{+\infty} \varphi_0'(E, x)\varphi_0'(E, y)d[\rho' - \sigma'] \qquad \textbf{(IX.6.27)}$$

and the expression for $\varphi(E, x)$:

$$\varphi(E, x) = \frac{\sin \sqrt{2E}\,x}{\sqrt{2E}} + \int_0^x \mathscr{K}'(x, y)\frac{\sin \sqrt{2E}\,y}{\sqrt{2E}}\, dy. \qquad \textbf{(IX.6.28)}$$

The step $\delta \to \rho$ can also be solved in the discrete problem, but the method is not simpler than in the continuous one.

Discrete versions of the inverse scattering problem also have been considered with respect to the Marcenko method (Case 1973). Besides, applications to orthogonal polynomials and related problems (in particular, the transport equation) were given (Case, 1974a,b; Sharpe, 1977). These discrete forms have the advantage of getting rid of certain analytical details so that the results are more transparent than in the continuous case. However, we think that this pedagogical interest is nowhere more apparent than in the case we have treated above.

IX.7   Dispersion Relation Approach

The Marchenko Gel'fand–Levitan methods are both based on the very special analyticity properties in the k-plane of the various solutions (each specified by a particular boundary condition) of the underlying differential equation which itself depends analytically on k^2. A very elegant and instructive way to highlight the central importance of these analyticity properties is to derive the above methods from dispersion relations for the solutions. These dispersion relations being themselves just an expression of the analytic properties. Karlson (1974) (1976) has done this for the radial Schrödinger equation. The fixed (r, l) dispersion relation for the Jost solutions

$$f_{l\pm}(k; t) \underset{|k|t \to \infty}{\sim} e^{\pm ikt}; \qquad (\pm \text{Im } k \geq 0)$$

yields the Marchenko equation while that for the regular solution

$$\varphi_l(k^2; r) \underset{r \to 0}{\sim} r^{l+1}$$

yields the Gel'fand–Levitan equation. The relevant dispersion relations are obtained in two easy steps. First the analytic properties are translated into a Cauchy-type dispersion relation which tells us how to extend the functions from $k \in \mathbb{R}$ to Im $k \geq 0$. Second the S-matrix is introduced into the Cauchy-type dispersion relation by using the usual expansion of the regular solution in terms of the Jost solutions which, for $k \in \mathbb{R}$, satisfy the symmetry property $f_{l-}(k; r) = f_{l+}(-k; r) = f_{l+}(k; r)^*$. We will sketch the details of Karlson's results for the case of the Jost solution.

Dispersion relations have also been used as the basis for solving the inverse problem for equations other than the radial Schrödinger equation. An important example is furnished by Ablowitz et al (1973) who solved the inverse problem for the 2×2 matrix equation

$$\left(i \mathcal{O}_3 \frac{d}{dx} - \begin{bmatrix} 0 & iq(x) \\ -ir(x) & 0 \end{bmatrix} \right) \Psi = k \Psi, \qquad \mathcal{O}_3 = \begin{bmatrix} 1 & 0 \\ 0 & -1 \end{bmatrix}, \qquad x \in \mathbb{R},$$

(IX.7.1)

with $q(x)$ and $r(x)$ tending to zero fast enough as $|x| \to \infty$ and satisfying some other conditions (which are sufficient but not necessary):

$$r(x) = \mp q(x)^*, \qquad \int_{-\infty}^{\infty} dx |q(x)| < 0.523.$$

Karlson's Results. The analytic properties of $f_{l+}(k'; r)$ (Im $k' \geq 0$) are summarized by saying that $\{f_{l+}(k'; r)e^{-ik'r} - 1\}$ is analytic for Im $k' > 0$, is continuous in k' for Im $k' \geq 0$ and that it tends to zero as $|k'| \to \infty$ (Im $k' > 0$). This is the same as saying that we have the following Cauchy integral representations for Im $k' \geq 0$:

$$\{f_{l+}(k'; r)e^{-ik'r} - 1\} = \frac{1}{2\pi i} \int_C dk \frac{1}{k - k' - io} \{f_{l+}(k; r)e^{-ikr} - 1\}$$

$$= \frac{1}{2\pi i} \int_{-\infty}^{+\infty} dk \frac{1}{k - k' - io} \{f_{l+}(k; r)e^{-ikr} - 1\}$$

$$= \frac{1}{2\pi i} \int_{-\infty}^{+\infty} dk'' \frac{-1}{k + k' + io} \{f_{l+}(-k''; r)e^{ik''r} - 1\},$$

(IX.7.2)

where C is any counterclockwise oriented contour in the upper half of the k-plane which encloses the pole at $k = k' + io$. The second integral is a special case of the first which exploits the property $\{ \ \} \to 0$ as $|k| \to \infty$ (Im $k \geq 0$). It is a dispersion relation in the sense that it gives us $f_{l+}(k'; r)$ (Im $k' \geq 0$) from $f_{l+}(k; r)$ (Im $k = 0$). [The limit Im $k' \to 0_+$ can be obtained by using the recipe $\int dk(k - k' - io)_0^{-1} \equiv \int dk$ P.V.$(k - k')_0^{-1} + i\pi \int dk \, \delta(k - k')_0$.] The third integral is then obtained by changing the real integration variable to $k'' = -k$. This is useful because it introduces $f_{l-}(+k''; r) = f_{l+}(-k''; r) = f_{l+}(kr; r)^*$ ($k'' \in \mathbb{R}$), which in turn leads us to the second step, the introduction of the S-matrix. The S-matrix enters via the usual expansion of the regular solution in terms of the Jost solutions for $k \in \mathbb{R}$:

$$2ik(-ik)^l \varphi_l(k^2; r)/(2l+1)!! = (-1)^l \mathcal{L}_{l-}(k) f_{l+}(k; r) - \mathcal{L}_{l+}(k) f_{l-}(k; r),$$

where $S_l(k) \equiv \mathcal{L}_{l-}(k)/\mathcal{L}_{l+}(k)$ defines the S-matrix and

$$\mathcal{L}_{l\pm}(k) \equiv \lim_{r \to 0} (-ikr)^l f_{l\pm}(k; r)/(2l+1)!!$$

defines the Jost functions. We recall that (a) $\mathcal{L}_{l\pm}(k)^{-1}$ is meromorphic for $\text{Im } k > 0$, that it tends to 1 as $|k| \to \infty$, and that it has poles at the Schrödinger bound states as its only singularities; (b) $\varphi_l(k^2; r)$ is an entire function of k^2 for fixed (r, l) and asymptotically behaves like

$$\varphi_l(k^2; r) = (2l+1)!! k^{-l-1} \sin(kr - \tfrac{1}{2} l\pi) + o(|k|^{-l-1} \exp(r|\text{Im } k|)) \tag{IX.7.3}$$

as $|k|r \to \infty$. Thus for $\text{Im } k > 0$

$$2ik(-ik)^l e^{ikr} \varphi_l(k^2; r)/(2l+1)!! \to -1 \tag{IX.7.4}$$

as $|k| \to \infty$.

For the sake of simplicity we now proceed with the $l = 0$ case. We can substitute

$$f_{0+}(-k''; r) = f_{0-}(k''; r) = S_0(k) f_{0+}(k; r) - 2ik\varphi_0(k^2; r)/\mathcal{L}_{0+}(k)$$

into the third integral and rearrange the result to obtain the following dispersion-like equation for the Jost solution $f_{0+}(k'; r)$ $(\text{Im } k' \geq 0)$ in terms of $[S_0(k) - S_0^{BS}(k; r)] f_{0+}(k; r)$ $(k \in \mathbb{R})$:

$$f_{0+}(k'; r) = e^{ik'r} - \int_{-\infty}^{\infty} dk \, \frac{\exp(i(k'+k)r)}{k'+k+io}$$

$$\times [S_0(k) - S_0^{BS}(k; r) - \tilde{S}_0(k; k')] f_{0+}(k; r), \tag{IX.7.5}$$

where $S_0^{BS}(k; r)$ $(\text{Im } k \geq 0)$ has the appropriate bound-state pole contributions and tends to zero as $|k| \to \infty$, while $\tilde{S}_0(k; k')$ $(\text{Im } k \geq 0)$ is an arbitrary analytic function of k which satisfies the rather weak r-dependent boundedness condition $\lim_{k \to \infty} e^{2ikr} \tilde{S}_0(k; k') = 0$. This boundedness condition is just what is needed to assure that $\tilde{S}_0$ doesn't contribute to the integral since then

$$\int_{-\infty}^{\infty} dk'' \left[\frac{\exp(ik''r)}{k''+k'+io} f_{0+}(k''; r) \right] = \int_C dk[\cdots] = 0. \tag{IX.7.6}$$

Let us call equation (IX.7.5) Karlson's dispersion relation. For $r > 0$ the choices $\tilde{S}_0(k''; k') = 1$ and $\tilde{S}_0(k''; k') = S_0(+k') - S_0^{BS}(k'; r)$ have been found useful by Karlson. Different choices of $\tilde{S}_0$ lead to different equations for solving the inverse problem when the limit $\text{Im } k' \to 0+$ is taken.

The Marchenko equation can now be obtained by choosing $\tilde{S}(k; k') = 1$ and replacing $f_{0+}(k'; r)$ (Im $k' \geq 0$) with its "triangular kernel" integral representation:

$$f_{0+}(k'; r) = e^{ik'r} + \int_r^\infty dr' \, A(r, r') e^{ikr'}. \tag{IX.7.7}$$

This representation is nothing but the Fourier representation of $\{f_{0+}(k'; r) e^{-ik'r} - 1\}$:

$$\{f_{0+}(k'; r) e^{ik'r} - 1\} = \int_{-\infty}^{+\infty} dr' \, A(r, r + r') e^{ik'r'}$$

$$= \int_0^\infty dr' \, A(r, r + r') e^{ik'r'}. \tag{IX.7.8}$$

The triangular property, $A(r, r') \equiv 0$ for $r' < r$, is necessary and sufficient for $\{f_0(k'; r) e^{-ik'r} - 1\} \to 0$ as $|k'| \to \infty$ Im $k' \geq 0$ and $A(r, r + r')$ is thus the Fourier transform of $\{f_{0+}(k'; r) e^{-ik'r} - 1\}$ with respect to k'. Thus the Marchenko equation is essentially the k'-Fourier transform of Karlson's dispersion relation when $\tilde{S}(k; k') = 1$.

The Gel'fand–Levitan equation can be obtained in a similar and related way via the following dispersion relation for the regular solution which is, in turn, obtained by considering analogous Cauchy integral representations of $[1 - f_{0+}(k; r) e^{-ikr} / \mathcal{L}_{0+}(k)]$:

$$k' \varphi_0(k'^2; r) = \sin k'r + \frac{1}{\pi} \int_{-\infty}^\infty dk \, \frac{\sin(k + k')r}{k + k'} [|\mathcal{L}_{0+}(k)|^{-2} - 1] k \varphi_0(k^2; r).$$

$$\tag{IX.7.9}$$

Momentum-Space Versions. Choices of $\tilde{S}(k; k')$ other than $\tilde{S} = 1$ have also been found useful by Karlson. In particular for $\tilde{S}(k; k') = S(+k')^*$ he has considered Fourier transformations of the dispersion relation with respect to r. These yield momentum-space versions of the relation. In this way, for $l = 0$ at least, he was able to develop practical momentum-space methods for solving the inverse problem. One of these is a Fredholm series solution which converges for a large class of potentials even if there are bound states. For the simplest Bargmann potential, for which $\mathcal{L}_{0+}(k) = (k + i\alpha)/(k + i\beta)$, this Fredholm series converges after two terms giving the exact result.

As a by-product Karlson obtained the result that originally motivated him, a simple procedure for continuing the two-body transition amplitude off the energy shell. The Fourier transform with respect to r of the wave function is simply related to the half-off-shell transition amplitude which he wanted to use as input to Faddeev's formulation of the three-body problem.

IX.8   Energy Dependent Potentials

Except for some remarks by Malčenko, the only problem which has been treated is the inverse s-wave scattering problem for potentials depending linearly on $\sqrt{E}$. The results are almost complete (Jaulent and Jean 1972; Jaulent 1972, 1973). They use a generalization of the Marchenko method which is similar to the generalization of the Regge–Newton procedure for spin-orbit potentials (chapter XV below) (actually, it was first built in analogy to it). Let us give an outline of the method, in the case of real potentials. We consider the differential equation

$$\frac{d^2 y^\pm}{dx^2} + [k^2 - V^\pm(k, x)]y^\pm = 0, \qquad x \geq 0, \qquad \text{(IX.8.1)}$$

with

$$V^\pm(k, x) = U(x) \pm 2kQ(x), \qquad x \geq 0, k \in C, \qquad \text{(IX.8.2)}$$

(k is related to the energy E by $k = |E|^{1/2} \exp[\frac{1}{2} \operatorname{Arg} E]$ for $0 < \operatorname{Arg} E \leq 2\pi$). $V(x)$ and $Q(x)$ are two "potentials." They satisfy the usual boundedness conditions and, in addition, a convenient differentiability condition (Jaulent 1972, 1973). The scattering amplitudes $S^\pm(k)$ correspond to the appropriate superscript in the differential equations (IX.8.1). If $S^+(k)$ is only known, for for $k > 0$, $S^-(k)$ appears in a certain way as a free parameter (see below). The inverse problem is then to recover U and Q from S^+ and $S^-(k)$. It is shown that Marchenko's representation of the Jost solution is generalized

$$f^\pm(k, x) = F^\pm(x)e^{-ikx} + \int_x^\infty A^\pm(x, t)e^{-ikt}\, dt, \qquad \operatorname{Im} k \leq 0, x \geq 0, \quad \text{(IX.8.3)}$$

where

$$F^\pm(x) = \exp\left[\mp i \int_x^\infty Q(t)dt\right] \qquad \text{(IX.8.4)}$$

and where $A^\pm(x, t)$ is a function such that

$$F^\pm(x)U(x) = \frac{d^2 F^\pm(x)}{dx^2} - 2\frac{d}{dx} A^\pm(x, x) + 2\frac{(d/dx)F^\pm(x)}{F^\pm(x)} A^\pm(x, x), \qquad x \geq 0.$$

$$\text{(IX.8.5)}$$

The Marchenko equation is also generalized to

$$A^\pm(x, t) = F^\mp(x)s^\pm(x + t) + \int_x^\infty s^\pm(u + t)A^\mp(x, u)du \qquad \text{(IX.8.6)}$$

where

$$s^\pm(t) = \lim_{R \to \infty} \frac{1}{2\pi} \int_{-R}^R [S^\pm(k) - (F^\pm(0))^2]e^{ikt}\, dk \qquad (t > 0) \quad \text{(IX.8.7)}$$

when there is no bound state. Because the potentials are real, $F^\pm(x)$, $A^\pm(x, t)$, are conjugate quantities, and the coupled integral equations (**IX.8.6**) can be reduced to one:

$$A^+(x, t) = \overline{F^+(x)}s^+(x + t) + \int_x^\infty s^+(u + t)\overline{A^+(x, u)}du. \quad \text{(IX.8.8)}$$

The inverse problem can be solved if $S^\pm(k)$ satisfies a number of conditions, coming either from unitarity, or from asymptotic properties, or consequences of the reality of potentials:

1. $$|S^\pm(k)| = S^\pm(0) = 1. \qquad\qquad\qquad \text{(IX.8.9)}$$

2. $$S^\pm(\infty) = S^\pm(-\infty) = (F^\pm(0))^2. \qquad \text{(IX.8.10)}$$

3. $$S^\pm(-k) = \overline{S^\mp(k)} = [S^\pm(k)]^{-1}. \qquad \text{(IX.8.11)}$$

4. $S^\pm(k)$ is continuous for $k \in \mathbb{R}$.

5. $S^\pm(k) - (F_0^\pm)^2$ is the Fourier transform of an integrable function $s^\pm(t)$.

6. $\arg S^\pm(k)]_{-\infty}^{+\infty} = 0$.

Because of (**IX.8.11**), $S^-(k)$ is not a free parameter if we know both $S^+(k)$ and $S^+(-k)$. Clearly, it is convenient to take $S^+(-k)$ as the free parameter. Once $S^+(k)$ is given, and $S^+(-k)$ is chosen, $[F^+(0)]^2$ is known from (**IX.8.10**), and $s^+(t)$ can be determined from (**IX.8.7**). The equation (**IX.8.8**) can be completed with some properties of F^+ and A^+, (the most obvious one being unimodularity of F^+ which follows from (**IX.8.4**)), and the system which is thus obtained can be solved. The potential is then obtained from (**IX.8.5**). The problem can also be solved in the complex case (actually the spin orbit problem of chapter XV is similar with the case U real and Q imaginary).

   Notice that the essential difference between this energy dependent problem and the energy independent one is that when k goes to $-k$, the equation with V^+ becomes the equation with V^- so that the analytical properties in the k-plane are not quite as simple.

IX.9   Miscellaneous Results

Along with the various generalizations of the Gel'fand–Levitan or Marchenko methods, there is room for studies in which these methods are restricted to particular classes of potentials, (like the classes which were studied by special methods in chapter VII).

   The Marchenko equation was studied in detail with potentials that are finite or infinite combinations of exponentials. They correspond respectively to Jost functions with N poles and to Jost functions with a discontinuity along a cut, and the potentials can be reconstructed from their singularities (Petras 1962, Blazek 1962).

Methods of reconstruction of the interaction from the scattering data have been formulated in the framework of R-matrix theory, in which the potential is determined by the positions of the resonances E_λ and their reduced widths γ_λ^2 (Chadan and Montes, 1968; Zachariev, 1976).

The former approach gives necessary and sufficient conditions on the parameters of the R-matrix in order for the potential to have a finite radius. The potential can then be constructed.

The latter approach has been derived in a finite difference approximation to make the logic more transparent. It was also applied to matrix problems of the type studied in section 9.3.

A method for reconstructing the Schrödinger inversion formalisms at fixed l and at fixed k, with potentials depending or not depending on k or l, from the properties of Fredholm determinants, has been derived by Cornille (1976). It is worth noticing that the Gel'fand–Levitan equations not only appear in scattering problems connected with second-order differential equations but also sometimes in more unexpected problems, e.g., the design of optical systems compensating in real time the effects of fluctuating atmospheric perturbation of the incoming light (Dyson, 1975). Besides, although we have seen in this book many derivations of the Gel'fand–Levitan equation, we still have to quote two interesting ones: a variational method (Dyson, 1976a), and, more unexpected, a random walk method (M. Kac, unpublished lecture in the Summer School on Inverse Problems, Los Angeles, 1974).

To close this chapter, let us notice that it is sometimes possible, and of course very interesting, to make the inverse scattering problem at fixed l a well-posed inverse problem. A sufficient condition is that no discrete spectrum (bound states) is possible. There exist several inequalities on the potential that guarantee the nonexistence of bound states (see, e.g., Bargmann, 1952; Glaser et al., 1976). Also, when the scattering problem is equivalent to certain problems, the absence of any discrete spectrum may follow from very natural physical constraints, see, e.g., Sabatier (1974a,b, 1977) and Martin and Sabatier (1977).

Inverse problems of a complicated nature, either corresponding to systems of first- or second-order differential equations or to higher order differential equations, can sometimes be analyzed by methods using a Povzner–Levitan representation of the functions of interest. Zachary made use of a transformation due to M. K. Fage which generalizes the Povzner–Levitan transformation for the Sturm–Liouville operator, and was able to present an inverse scattering formalism for certain linear differential operators of even order n that generalizes the Gel'fand–Levitan theory (Zachary, 1980, 1984, 1986). The transmutation or generalized translation theory (see section XV.2.3) has also been partly able to generalize Gel'fand–Levitan and Marchenko formalisms. It should be noticed however at this point that the most physical way to seek generalization of these theories is by considering the problem in a form where it is hyperbolic (to the price of ruling out, for example, bound states) and using causality: a beautiful example of the power of this method has been given by Burridge (1980). Other methods of generalization appear, for instance, in

Moses (1978) and Suzuki (1986), whereas Chadan (1978), Coz (1980), and Göckeler (1981) introduce singularities in the potential which can be derived. A maximum entropy method has also been related with the exact methods for inverse scattering theory (Inguva and Baker-Jarvis, 1986).

For the Levinson theorem for the Dirac equation, see Z.-Q. Ma: The Levinson theorem for Dirac particles, *Phys. Rev. D* 32 (1985), 2213–2215. This paper contains all the references to earlier works (Barthélémy, etc.)

CHAPTER X

Scattering Amplitudes from Elastic Cross Sections

X.1   Introduction

The problem of constructing the scattering amplitude from the cross section could theoretically be avoided by performing measurements of intensity correlations on scattered particles at different space-time points. The interest of such measurements has been pointed out by Goldberger et al. (1963, 1964, 1965, 1966), following Hanbury Brown and Twiss (1954, 1956). They proved that, with this technique, it is possible to explicitly measure the phase of a scattering amplitude (with particular relevancy to X-ray scattering). However, obtaining the correlated counting rates of two detectors with the beam intensities nowadays available is hopeless in particle physics. Besides, it is not obvious that overcoming this difficulty will readily yield a solution of the problem. In typical situations, the scattering phase remains deeply involved in the formulas. Therefore, mathematical ways of constructing the scattering amplitude from the cross sections will be of interest for years.

On the other hand, it is fair to say that studies of this inverse problem have not yet been used in the phase-shift analyses. In most cases, interactions are represented through a model including a few parameters, and cross sections are calculated for several values of the parameters, until a satisfactory fit is obtained. With this technique, uniqueness problems (and, to some extent, existence problems) are simply ignored. In other cases, it is assumed that only a few phase shifts are not negligible. Fits are obtained by optimisation techniques. With this method, certain existence and uniqueness problems may be fixed, but not necessarily within the class that is physically interesting.

We quote Martin (1973): "What is incredible but true is that even if we see hundreds of phase shift analyses performed and published, this problem is not really solved, even 'in principle' i.e. even if you start from extremely accurate measurements."

It has been shown long ago (Bessis and Martin (1967), Alvarez–Estrada (1971)), that if $|F|$ is known in the whole physical region, F is uniquely determined. (See also Burkhardt and Martin (1975)). This determination is rather instable (Martin 1973). In the following, we do consider only results at fixed energy. For the sake of simplicity we first study the case of spin-zero elastic scattering ($A + B \to A + B$, at an energy which is below the first inelastic threshold). We assume that we know the differential cross section exactly at one fixed energy and for all physical scattering angles. Hence we know, for $0 \leq \theta \leq \pi$, the function:

$$\sigma(\theta) = |F(\cos \theta)|^2 \qquad \text{(X.1.1)}$$

where the "scattering amplitude" $F(\cos \theta)$ is related to the "phase shifts" δ_l by the Legendre polynomial expansion:

$$F(x) = \sum_0^\infty (2l + 1)F_l P_l(x), \qquad \text{(X.1.2)}$$

where

$$F_l = \exp[i\delta_l]\sin[\delta_l]. \qquad \text{(X.1.3)}$$

According to the unitarity condition, when applied to elastic scattering, the δ_l's are *real*, and conversely this condition, when expansion (2) converges, is sufficient to give us a unitary amplitude $F(x)$. The condition δ_l real is equivalent to the equality

$$\text{Im } F_l = |F_l|^2. \qquad \text{(X.1.4)}$$

Now, let $F(1, 2)$ be the scattering amplitude, in which 1 and 2 denote unit vectors in the initial and final directions of the particles ($x = \cos[\widehat{1, 2}]$). The unitarity condition can also be written as:

$$\text{Im } F(1, 2) = (4\pi)^{-1} \int F(1, 3)F^*(2, 3)d\Omega_3, \qquad \text{(X.1.5)}$$

where the integration extends over all values of the solid angle Ω_3.

Let $y = \cos[\widehat{1, 3}]$, $z = \cos[\widehat{2, 3}]$, the formula (X.1.5) can then be written as

$$\text{Im } F(x) = \int_{-1}^{+1} \int_{-1}^{+1} F(y)F^*(z)S(x, y, z)dy\, dz, \qquad \text{(X.1.6)}$$

where

$$S(x, y, z) = \frac{1}{2\pi}[1 - x^2 - y^2 - z^2 + 2xyz]^{-1/2}$$

$$\times \theta[1 - x^2 - y^2 - z^2 + 2xyz]. \qquad \text{(X.1.7)}$$

θ is the Heaviside function. One can also derive (**X.1.6**) directly from (**X.1.2**) and assuming that the δ_l's are real, (notice that $S(x, y, z)$ is equal to $\frac{1}{4} \sum_0^\infty (2l + 1)P_l(x)P_l(y)P_l(z)$). Now let $\varphi(x)$ (resp. $\varphi(1, 2)$) be the phase of $F(x)$ (resp. $F(1, 2)$) (**X.1.5**) and (**X.1.6**) can be used to reduce this inverse problem to solving the following integral equation for $\varphi(x)$ in a suitable class of functions

$$\sin \varphi(x) = \iint Q(x, y, z)\cos[\varphi(y) - \varphi(z)]dy\, dz. \qquad \text{(X.1.8)}$$

The kernel $Q(x, y, z)$ is defined as the nonnegative function

$$Q(x, y, z) = |F(x)|^{-1}|F(y)||F(z)|S(x, y, z). \qquad \text{(X.1.9)}$$

With a convenient definition of the sets of functions, one may reduce this problem to the search for fixed points of the operator T defined by

$$\sin[(T\varphi)_x] = \iint Q(x, y, z)\cos[\varphi(y) - \varphi(z)]dy\, dz. \qquad \text{(X.1.10)}$$

For whatever formulation of the inverse problem, a complete investigation would involve:

1. the choice of the set $\mathscr{E}$ where σ is given, or the set $\mathscr{F}$ where $|F|$ is given, and the set Φ where $\varphi(x)$ is sought, or the set $\mathscr{C}$ where $F(x)$ is sought;
2. proofs of existence, uniqueness or nonuniqueness, constructability and stability of the solutions.

Two choices for $\mathscr{E}$ and $\mathscr{C}$ are interesting for physical applications:

1. $\mathscr{E}$ and $\mathscr{C}$ can be set of functions, say $\mathscr{E}_Q$ and $\mathscr{C}_Q$, which are analytic in certain ellipses with foci -1 and $+1$, which agree with fundamental results obtained in axiomatic quantum theory.
2. $\mathscr{E}$ and $\mathscr{C}$ can be set of functions that are simply continuous for $-1 \le x \le 1$, say $\mathscr{E}_c$ and $\mathscr{C}_c$ (or Φ_c if we deal with the phase). Such sets are larger than the previous ones, and the question of stability with regards to random experimental errors is easier to deal with.

A third choice can help current phase-shifts analyses, the choice of sets of polynomials which are consistent with each other: $\mathscr{C}$ is the set $\mathscr{C}_N$ of *finite* Legendre polynomial expansions like (**X.1.2**) of degree $\le N$: Thus we call $\mathscr{C}_N$ the set of unitary polynomials of degree $\le N$. $\mathscr{E}$ is the set $\mathscr{E}_{2N}$ of moduli-square of all unitary polynomials with degree $\le N$—it is therefore a subset of the set of polynomials of degree $\le 2N$.

The results which have been obtained up to now are not satisfactory, and, except for the small subset $\mathscr{D}$ of $\mathscr{E}_c$ to be described below, never answer all questions simultaneously.

Let us first notice that there is an overall ambiguity corresponding to the complex conjugation

$$F(x) \to -F^*(x). \qquad \text{(X.1.11)}$$

In other words, if $\varphi(x)$, $-1 \leq x \leq 1$, is a solution of (**X.1.8**), so is $\pi - \varphi(x)$. This ambiguity, called the trivial ambiguity, is always present in the problem and can be taken into account by conveniently choosing the subclass of $\mathscr{C}_c$ or Φ_c in which the solution is sought. We shall forget it in the following and all our statements must be understood modulo this trivial ambiguity.

As for the *existence* of a solution, the optical theorem:

$$\sin \varphi(1) = \frac{1}{2} \int_{-1}^{+1} \frac{|F(x)|^2}{|F(1)|} \, dx \qquad (\mathbf{X.1.12})$$

to which (**X.1.8**) reduces for $x = 1$, shows an immediate restriction. We must have

$$|F(1)| \geq \frac{1}{2} \int_{-1}^{+1} |F(x)|^2 \, dx. \qquad (\mathbf{X.1.13})$$

Otherwise there can be no (real) solution $\varphi(x)$ of (**X.1.8**).

Are there any other restrictions on the differential cross section implied by unitarity? That there must be such restrictions can be made plausible as follows (Newton 1968): Suppose that for a given $|F(x)|$, the function $\varphi(x)$ solves (**X.1.8**). Let us now decrease $|F(x)|$ only in the interval $(x \in \Delta || x - x_0| \leq \varepsilon)$. Clearly we can make ε so small that, no matter how $|F(x)|$ is made in Δ, the right-hand side of (**X.1.8**) changes by as little as we like. Hence the original solution $\varphi(x)$ needs to be changed only "infinitesimally" in order to solve the altered equation (**X.1.8**) everywhere except inside Δ. But inside Δ we need only make $|F(x)|$ small enough in order to prevent $|\sin \varphi(x)| \leq 1$ from being able to solve (**X.1.8**).

The foregoing argument shows that at every point x, $-1 \leq x \leq 1$, (**X.1.8**) implies a lower bound on $|F|$ relative to its values everywhere else in the interval. This however, does not imply that the positive function

$$M(x) = \iint Q(x, y, z) dy \, dz \qquad (\mathbf{X.1.14})$$

must not be too large. Consider the special case in which F consists of a single l partial wave. It is easy to see that $M(x)$ has infinities at the zeros of $P_l(x)$.

After several partial studies that belong to the prehistory of this subject (Fröberg 1947, 1948, Puzikov et al. (1957), Klepikov (1962)), it was proved (Newton 1968, Martin 1969, Gersten 1969, Gerber and Karplus (1970), Atkinson et al. 1972, Tortorella 1972, 1974a,b, 1975a,b see also Eftimiu 1972, 1973) that when $|F(x)|$ is such that the positive function $M(x)$ is uniformly bounded by 0.79 (*domain $\mathscr{D}$*), T is a contracting mapping in Φ_c. This yields existence, uniqueness, and a constructive method to get the solution in $\mathscr{D}$; but since no resonant amplitude can hence be obtained, this domain is not very interesting. Conjectures of existence have been studied by the same authors in the *domain $\mathscr{D}'(M(x) < 1)$*. Local existence results have been obtained (Atkinson et al. 1973a, 1975a), for $\sigma \in \mathscr{E}_Q$, $F \in \mathscr{C}_Q$: it has been proved that if a solution $F_0(x)$ is known for a cross section σ, there exists an open ball in $\mathscr{E}_c$, centered at σ_0, in which the solution or solutions can be constructed. Finally an existence and uniqueness theorem has been proved (Martin 1973) for unitary polynomials if the total cross section $\int_{-1}^{+1} \sigma(x)dx$ is small enough (*domain* Γ).

Apart from the domain $\mathcal{D}$ quoted above, results on uniqueness have been obtained only for special classes of functions. For $\sigma \in \mathcal{E}_Q$, $F \in \mathcal{C}_Q$, it has been proved (Atkinson et al. 1973a, 1975a) that the number of branches at a bifurcation point is finite, but nothing has been given concerning the partition of $\mathcal{C}_Q$ by the branches of solution, so that the result which has been proved is purely local. Global results have been obtained (Itzykson and Martin (1973)) in sets $\mathcal{C}_E$ of genuine entire functions of finite order: it has been proved that there are no more than two solutions (viz. 4 if one takes into account the trivial ambiguity). The same has been proved (Crichton 1966, Atkinson, Johnson, Mehta, de Roo, (1973), Berends and Ruijsenaars (1973a,b), Cornille and Drouffe (1975)) for the sets $\mathcal{C}_N$ of unitary polynomials of degree $N \leq 4$. Examples of Crichton ambiguities that are not polynomials have also been exhibited (Atkinson et al. 1977). Finally, we recall the uniqueness results given above for polynomials in the domain Γ.

This ends our knowledge of the subject! In particular, few results have been given on the problem of stability (needless to say, it is trivial in $\mathcal{D}$). Besides, the results on existence (local and global) and uniqueness, which have been studied by different authors, are in most cases concerned with rather disconnected sets!

To illustrate this situation, we shall prove in section X.5 the following remark (Sabatier 1974a): If there were three different solutions in the class $\mathcal{C}_c$, one could find three different sequences of "unitary" entire functions F of finite order, such that the corresponding sequences of functions σ have the same limit-point. Thus, if the stability could be proved in a class of entire functions, one could use this result to generalize the no-more-than-two theorem from $\mathcal{C}_F$ to $\mathcal{C}_Q$ or $\mathcal{C}_c$. But, in the present state of our knowledge, one cannot say more than: from the experimentalist's point of view, the no-more-than-two-solutions proofs that have been given in *special classes* do not prove anything of much use, even in those classes.

Our aim here is not to give a complete study of the results which have been summarized above, but to give details and proofs of those results which we choose either because they are general or because they are simple.

In section X.2 we present the only two constructive methods that have been given—the iterative method, which converges in the domain $\mathcal{D}$, and the descending construction, which yields a unique solution in the domain Γ.

In sections X.3, we present two uniqueness studies—the one in the class of entire functions of order $\rho \in]0, 1[$, and the one for unitary polynomials of degree ≤ 2.

In section X.4, we briefly present local existence and uniqueness proofs that define F as an implicit function of σ in the neighborhood of a couple of related points σ_0, F_0.

Section X.5 is a set of remarks on the uniqueness and stability problems trying to understand the relations between the different questions to be dealt with.

In section X.6, we survey the more general inverse problems which are related to the present one.

X.2   Constructive Methods

X.2.1   The Successive Approximations Method

We will try to obtain a fixed point of (**X.1.8**) through an iterative procedure.

Definition of the Sets $\mathscr{E}$ and Φ. Let us assume that $M(x)$, as defined by (**X.1.10**), is everywhere smaller than 1, and let us set

$$\sin \mu = \sup_{x \in [-1, +1]} M(x) < 1. \tag{X.2.1}$$

This condition obviously is sufficient to guarantee that the right-hand side of (**X.1.8**) belongs to $]-1, +1[$ for any function of Φ. Since φ is defined up to a 2π translation, and since the "trivial" ambiguity means that if φ is a solution, so is $\pi - \varphi$, we can choose a solution whose value at a given point belongs to $[-(\pi/2), (\pi/2)]$. If this solution is continuous, it follows from (**X.2.1**) that its values necessarily belong to $[-\mu, \mu]$. Now let us assume that $\sup \varphi(x)$ belongs to $[0, \mu]$. Then, we claim that $\varphi(x)$ belongs to $[0, \mu]$.

Proof. Either

(a) $$0 \leq \sup \varphi - \inf \varphi \leq \frac{\pi}{2},$$

or

(b) $$\frac{\pi}{2} < \sup \varphi - \inf \varphi \leq 2\mu < \pi.$$

In case (a), $\cos[\varphi(y) - \varphi(z)] \geq 0$, and hence $\sin \varphi \geq 0$. In case (b), we get successively:

$$\sin[\inf \varphi] > \sin \mu \cos[\sup \varphi - \inf \varphi],$$

$$\text{tg}[\inf \varphi] > \frac{\sin \mu \cos[\sup \varphi]}{1 - \sin \mu \sin[\sup \varphi]} > 0,$$

and this last inequality contradicts the inequality (b), so that the case (b) cannot hold.

If we assume that $\sup \varphi(x)$ belongs to $[-\mu, 0]$, $\sup \varphi - \inf \varphi$ is smaller than $\pi/2$, so that φ is positive, which contradicts the assumption.

Hence, we are led to define $\mathscr{E}$ as the set of real continuous functions on $[-1, +1]$ that are consistent with (**X.2.1**), and Φ as the set Φ_μ^+ of continuous functions on $[-1, +1]$ with values on $[0, \mu]$. For every solution of (**X.1.8**) that is contained in Φ_μ^+, there exists a solution which is associated with it by the trivial ambiguity, and no other continuous solution can exist as long as (**X.2.1**) does hold.

Algorithm. The successive approximations sequence is defined by

$$\varphi_0(x) = 0$$

$$\varphi_{n+1}(x) = \sin^{-1}\left[\iint Q(x, y, z)\cos[\varphi_n(y) - \varphi_n(z)]dy \, dz\right]. \quad \text{(X.2.2)}$$

Every φ_n clearly belongs to Φ_μ^+. We get for two elements φ and ψ of Φ_μ^+ the following inequalities:

$$|(T\varphi)_x - (T\psi)_x| < (\cos \mu)^{-1}\left|\iint Q(x, y, z)\{\cos[\varphi(y) - \varphi(z)]\right.$$

$$\left. - \cos[\psi(y) - \psi(z)]\}dy \, dz\right|$$

$$< 2 \tan \mu \iint Q(x, y, z)\left|\sin\left[\frac{\varphi(y) - \psi(y)}{2} - \frac{\varphi(z) - \psi(z)}{2}\right]\right|$$

$$< 2 \tan \mu \iint Q(x, y, z)|\varphi(y) - \psi(y)|dy \, dz, \quad \text{(X.2.3)}$$

where we have taken into account the symmetry $Q(x, y, z) = Q(x, z, y)$. Now let us introduce in Φ_μ^+ the norm:

$$\|\varphi\| = \sup_{x\in[-1, +1]} |\varphi(x)|. \quad \text{(X.2.4)}$$

Φ_μ^+ is now a Banach space, and it follows readily from (X.2.3) that T is a contraction mapping in Φ_μ^+ whose unique fixed point is given by (X.2.2), when

$$0 \le 2 \sin \mu \tan \mu < 1 \quad (\Leftrightarrow \quad 0 \le \sin \mu < 0.62\cdots). \quad \text{(X.2.5)}$$

Using a more refined analysis, it has been possible to prove that the algorithm converges if

$$\sin \mu < 0.86\cdots. \quad \text{(X.2.6)}$$

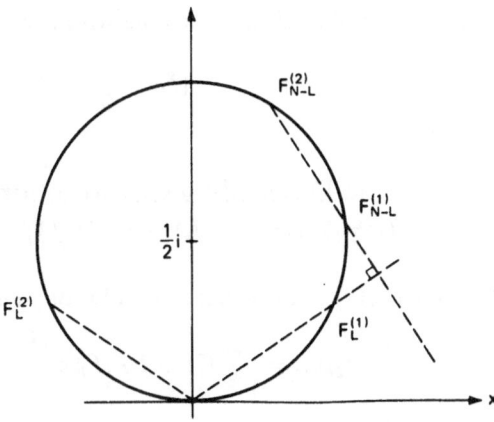

Figure X.1a

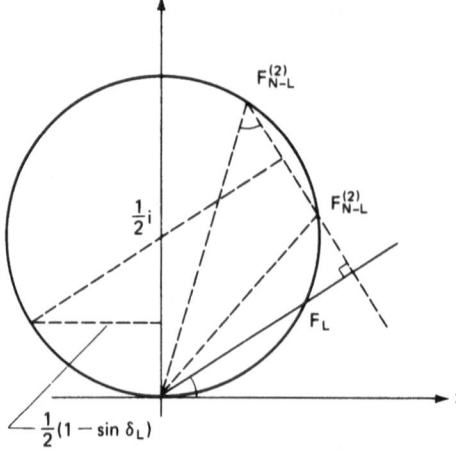

$$-\tfrac{1}{2}(1 - \sin \delta_L)$$

Figure X.1b

Remark. It is a common guess that the solution exists and is unique for any $\sin \mu < 1$. This condition (see, e.g., Martin 1973) is realized at energies sufficiently close to the threshold (if amplitudes have a normal threshold behavior and if the scattering length is not zero). It should be noticed, however, that, even if the conjecture is proved, the result corresponds to very particular scattering amplitudes. Indeed, for $\varphi \in [0, \mu]$, Re F and Im F are positive. Hence, for $l \geq 1$, we get:

$$\text{Re } F_0 \pm \text{Re } F_l = \frac{1}{2} \int_{-1}^{+1} \text{Re } F(x)[1 \pm P_l(x)]dx > 0$$

and, similarly Im $F_0 \pm$ Im $F_l > 0$. But F_0 and F_l lie on the "unitarity" circle $B(\tfrac{1}{2}i, \tfrac{1}{2})$ (figure X.1). For Im $F_0 >$ Im $F_l > \tfrac{1}{2}$, the condition $|\text{Re } F_0|\text{Re } F_l|$ implies $|\text{Im } F_0| < |\text{Im } F_l|$, which contradicts the second inequality. Hence Im $F_l < \tfrac{1}{2}$. If we make the convention $-(\pi/2) < \delta_l < \pi/2$ (the phase shifts are defined modulo 2π), we see that the inequality $\sin \mu < 1$ implies

$$|\delta_l| < \frac{\pi}{4} \qquad (l \geq 1). \tag{X.2.7}$$

A more refined analysis yields the stronger inequality

$$|\delta_l| < \frac{\pi}{6} \qquad (l \geq 1). \tag{X.2.8}$$

X.2.2   The Descending Construction
(Itzykson and Martin 1973)

Let now $\sigma_{2L}(x)$ be the square-modulus of an unitary polynomial of degree L:

$$\sigma_{2L}(x) = \left| \sum_0^L (2l + 1)F_l P_l(x) \right|^2 = \sum_0^{2L} (2l + 1)\sigma_l P_l(x) \tag{X.2.9}$$

and let us try to reconstruct the F_l's from the σ_l's. The Legendre expansion of a product $P_l P_n$ only contains Legendre polynomials of degree $\in [|l - n|, l + n]$. Hence the Legendre coefficients of σ_{2L} with order $\geq L$ yield a triangular system for $F_0, F_1, \ldots, F_L$:

$$\sigma_N = \sum_{l = E[(1/2)(N+1)]}^{L} \sum_{l' = N-l}^{l} A_N^{ll'} \, \text{Re}[F_l^* F_{l'}], \qquad \text{(X.2.10)}$$

$$(N = L, L + 1, \ldots, 2L),$$

where $A_N^{ll'}$ is zero unless $l + l' \geq N$. This system can be solved step by step:

$$\sigma_{2L} = A_{2L}^{LL}[F_L^* F_L],$$
$$\vdots \qquad\qquad\qquad\qquad\qquad\qquad \text{(X.2.11)}$$
$$\sigma_N = A_N^{L(N-L)} \, \text{Re}[F_L^* F_{N-L}] + [\text{terms previously obtained}],$$

$$(N = L, L + 1, \ldots, 2L - 1).$$

At the first step, the modulus of F_L is given. The circle $B(0, |F_L|)$ intersects the unitarity circle $B(\frac{1}{2}i, \frac{1}{2})$ at two points $F_L^{(1)}$ and $F_L^{(2)}$. Choosing one of them — say $F_L^{(1)}$ $(0 \leq \delta_L \leq \pi/2)$ — suppresses the trivial ambiguity. On following steps, $\text{Re}(F_L^* F_{N-L})$ is given, and defines a straight line in the complex plane which intersects the unitarity circle at two points. This seems to yield 2^L possible unitarity amplitudes, but one still has to compute $\sigma_0, \sigma_1, \ldots, \sigma_{L-1}$ with these various sets and seek the set which agrees with these values.

There are two cases in which the amplitude that may thus be obtained is unique:

$$\boxed{\sin \mu < 1.}$$

Suppose there were two different solutions, with the same F_l's for $l = L, L - 1, \ldots, N - L + 1$. For $l = N - L$, the two solutions correspond to $F_{N-L}^{(1)}$ and $F_{N-L}^{(2)}$, as given in figure X.1. It is clear from the geometrical constructions that

$$\delta_{N-L}^1 + \delta_{N-L}^2 = \delta_L + \tfrac{1}{2}\pi. \qquad \text{(X.2.12)}$$

Clearly the condition (X.2.8) cannot be simultaneously satisfied by δ_{N-L}^1 and δ_{N-L}^2, so that one of them only must be kept. This argument fails for δ_0, but it is easy to see that $|F_0|$ can then be unambiguously given from the equality:

$$\sigma_T = \frac{1}{2} \int_{-1}^{+1} \sigma(x) dx = |F_0|^2 + \sum_{1}^{N} (2l + 1)|F_l|^2. \qquad \text{(X.2.13)}$$

The circle $B(0, |F_0|)$ intersects the unitarity circle at two points. One of them only can coincide with one of the points $F_0^{(1)}$ and $F_0^{(2)}$, since δ_L is smaller than $\pi/2$. Hence the solution is unique. In fact, it has been possible to show that if Re $F > 0$ and Im $F > 0$, (which is weaker than sin $\mu < 1$) *and* if the number of nonzero F_l is *finite*, the solution is unique (Martin 1973).

$$\boxed{\sigma_T \text{ small}}$$

The total cross section σ_T being defined by (X.2.13), we see that if there are two different amplitudes, such that F_n is unique for $n = N - L + 1, \ldots, L$, and not for $n = N - L$ we can write

$$\sigma_T > (2L + 1)\sin^2 \delta_L + [2(N - L) + 1]\left[\frac{|F_{N-L}^{(1)}|^2 + |F_{N-L}^{(2)}|^2}{2}\right]. \quad \text{(X.2.14)}$$

From the geometrical construction, we see that $\frac{1}{2}[\text{Im } F_{N-L}^{(1)} + \text{Im } F_{N-L}^{(2)}]$ is certainly larger than $\frac{1}{2}[1 - \sin \delta_L]$. Using (X.1.3) we therefore obtain, if there are two different amplitudes

$$\sigma_T > (2L + 1)\sin^2 \delta_L + \frac{1}{2}[2(N - L) + 1](1 - \sin \delta_L). \quad \text{(X.2.15)}$$

The smallest value of the right-hand side occurs for $L = 2$, $N = 3$, sin $\delta_L = 3/20$, so that $\sigma_T > 1.38$. Hence the condition

$$\sigma_T < 1.38 \quad \text{(X.2.16)}$$

guarantees that the solution is unique. This condition is very different from the condition sin $\mu < 1$.

X.3   Other Uniqueness Studies

X.3.1   Polynomials

Let $F_L(z)$ be a polynomial amplitude of degree L. In order to remove the trivial ambiguity, we assume $0 \leq \delta_L \leq \frac{1}{2}\pi$. $F_L(z)$ can be written as a product over its zeros:

$$F_L(z) = \alpha(z - z_1)(z - z_2) \cdots (z - z_L), \quad \text{(X.3.1)}$$

where α is equal to

$$\alpha = \exp[i\delta_L]\sin[\delta_L] \times 2^{L+1} \frac{\Gamma(L + 3/2)}{\Gamma(L + 1)}. \quad \text{(X.3.2)}$$

Suppose now that $|F_L(z)|^2$ is a know polynomial, of degree $2L$, with zeros associated by pairs (zeros and their conjugate, so that there may be 2^L polynomials of the form (X.3.1)). Among this set, we have to select the unitary polynomials of the form (X.1.3).

Let us first study the case $L = 2$. The four polynomials with the same α and modulus are:

$$F_2^{++} = \alpha(z - z_1)(z - z_2),$$
$$F_2^{+-} = \alpha(z - z_1)(z - z_2^*),$$
$$F_2^{-+} = \alpha(z - z_1^*)(z - z_2), \qquad \text{(X.3.3)}$$
$$F_2^{--} = \alpha(z - z_1^*)(z - z_2^*).$$

The Legendre coefficients of F_2^{++} and F_2^{--} have the same modulus. It is easy to see that they cannot be simultaneously unitary. Similarly F_2^{+-} and F_2^{-+} cannot be simultaneously unitary. Hence, there are at most two unitary polynomials of degree 2 with the same modulus. What is most remarkable is that these two solutions really can exist under certain conditions, which are physically interesting. The first example of this ambiguity was given by Crichton (1966). Since then, it has been fairly well studied, and the range in which this ambiguity appears has been quite well explored (figure X.2). The function

$$F(\cos \theta) = F(1) \prod_{i=1}^{2} \frac{\cos \theta - F_i}{1 - F_i}$$

has two determinations when $|F(\cos \theta)|^2$ is given. In the first figure, the values of the phase shifts versus the parameter $|F_2|^2$ are given, showing that the ambiguity is not isolated (from Berends and Ruijsenaars (1973)). In the second figure, $F(\cos \theta)$ is given versus θ for a certain function $|F(\cos \theta)|^2$ (from Crichton 1966). Besides, the case of polynomials of degree 3 (Berends and Ruijsenaars (1973)) and 4 (Cornille and Drouffe (1975)) have been studied and it has also been proved in these cases that the number of nontrivial ambiguities cannot exceed the number of two solutions. However, the proof becomes increasingly complicated—for example, in the case $L = 4$, it was necessary to check inequalities by means of a computer—and the methods which are used cannot be generalized for arbitrary L.

For $L > 4$, an heuristic argument is known, which supports the result obtained for $L \le 4$. A priori there may be 2^L possible amplitudes. Suppose

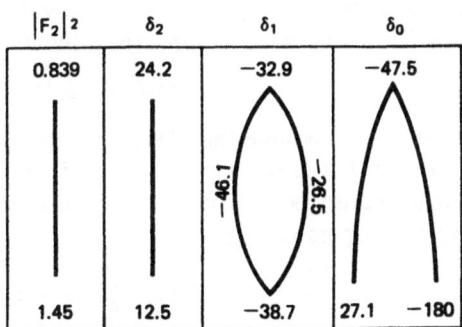

$\lvert F_2 \rvert^2$	δ_2	δ_1	δ_0
0.839	24.2	−32.9	−47.5
1.45	12.5	−38.7	27.1 −180

Figure X.2a Crichton's ambiguity

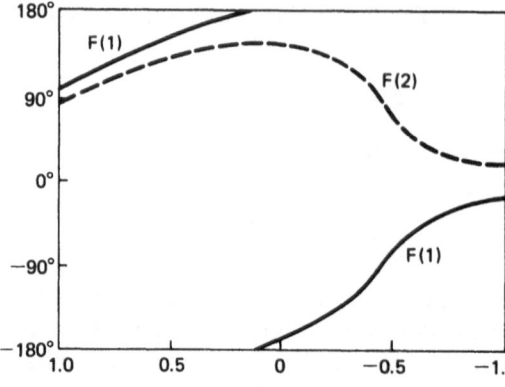

Figure X.2b $F(1)$ and $F(2)$ denote the two equivalent solutions.

there are N unitary amplitudes F^k. They depend on $(2L + 1)$ parameters: the real parts and moduli of the imaginary parts of the L zeros, as well as α. The unitarity conditions read:

$$\text{Im } F_l^k = |F_l^k|^2, \qquad k = 1, 2, \ldots, N, \qquad l = 0, 1, 2, \ldots, L - 1. \quad \textbf{(X.3.4)}$$

Equations **(X.3.4)** constitute a system of NL nonlinear algebraic equations. *If they are independent,* we must have

$$NL \le 2L + 1, \quad \text{i.e.,} \quad N \le 2. \qquad \textbf{(X.3.5)}$$

If $N = 2$, there is one parameter left. Hence a twofold ambiguity may arise on a one-dimensional variety of differential cross sections. This really happens in the cases $L = 2, L = 3, L = 4$, which have been studied. Needless to say, an essential part of this argument (equations "independent") is not proven.

X.3.2   Entire Functions

In the case of polynomials, the partial wave series stops abruptly. A case which seems closer to physical reality is that in which there are infinitely many partial waves. In that case, the problem of uniqueness can be considered through the asymptotic behavior of the coefficients F_l: to guarantee the convergence of **(X.1.3)**, it is necessary that F_l should go to zero as l goes to infinity. It follows that, for large l, the unitarity relations yield

$$\text{Im } F_l = |F_l|^2 \sim (\text{Re } F_l)^2. \qquad \textbf{(X.3.6)}$$

Now, let us assume that $F(x)$ is a "genuine" entire function, i.e., an entire function that is *not* a polynomial. $F(x)$ is continued in the whole complex plane: we use the notation $F(z)$. The behavior of $F(z)$ for large $|z|$ is dominated by its "dispersive" part:

$$D(z) = \sum_{0}^{\infty} (2l + 1)\text{Re } F_l P_l(z). \qquad \textbf{(X.3.7)}$$

Hence, if $\sigma(z)$ is continued in the complex plane

$$\sigma(z) = F(z)F^*(z^*) \tag{X.3.8}$$

its behavior for large $|z|$ is dominated by

$$[\sum (2l + 1)\text{Re } F_l P_l(z)]^2. \tag{X.3.9}$$

This should simplify the uniqueness problem, since deriving the asymptotic behavior of F from that of σ can be done with only a sign ambiguity.

To go further, let us introduce the "absorptive part" A of the scattering amplitude, which is given by (X.1.5) as

$$A(1, 2) = (4\pi)^{-1} \int F(1, 3)F^*(2, 3)d\Omega_3 \tag{X.3.10}$$

or

$$A(\cos[1, 2]) = \sum_0^\infty (2l + 1)\text{Im } F_l P_l(\cos[\widehat{1, 2}]). \tag{X.3.11}$$

Let F be entire *and* unitary. Then, according to (X.3.6) and (X.3.11), $A(y)$ can be continued in the complex plane as an entire function:

$$A(y) = \sum_0^\infty (2l + 1)\text{Im } F_l P_l(y). \tag{X.3.12}$$

We can also continue it by another way: choosing as z axis the bisector of $\widehat{1, 2}$, we write (X.3.10) as:

$$A(\cos[\widehat{1, 2}]) = A(\cos[2\theta_0])$$

$$= (4\pi)^{-1} \int d\varphi \, d(\cos \theta)F(\cos \theta_0 \cos \theta + \sin \theta_0 \sin \theta \cos \varphi)$$

$$\times F^*(\cos^* \theta_0 \cos \theta - \sin^* \theta_0 \sin \theta \cos \varphi). \tag{X.3.13}$$

Let E_{θ_0} be an ellipse in the complex plane, with foci $-1, +1$, going through $\cos \theta_0$. We see that the argument of F in (X.3.13) is on the segment connecting the points $\cos(\theta_0 + \theta)$ and $\cos(\theta_0 - \theta)$, both on E_θ. The same is true for the argument of F^*, so that (X.3.13) yields a continuation of A. In particular, for real ψ:

$$A(\cos h2\psi) = (4\pi)^{-1} \int d\varphi \, d(\cos \theta)|F(\cos \theta \cos h\psi + i \sin \theta \sin \psi \cos \varphi)|^2. \tag{X.3.14}$$

For an analytic function f, let $M_f(x)$ be the supremum of $|f|$ inside the ellipse of semi-major axis x, foci $-1, +1$. According to (X.3.6), A is a sum of Legendre polynomials with positive coefficients, and therefore is maximum at the right extremity of the ellipse. Hence, we get from (X.3.14) the inequality

$$M_A(2x^2 - 1) \le |M_F(x)|^2. \tag{X.3.15}$$

It follows from (**X.3.15**) that if F is an entire function of order ρ, this inequality, used for $x \to \infty$, yields for any $\varepsilon > 0$:

$$M_A(y) \le \exp\left[\left(\frac{1 + y}{2}\right)^{(1/2)\rho + \varepsilon}\right]$$ (X.3.16)

so that A is an entire function of order $\tfrac{1}{2}\rho$.

Now, suppose that two entire unitary amplitudes F and F' give the same cross section:

$$F = D + iA,$$

$$F' = D' + iA',$$ (X.3.17)

$$D^2(z) + A^2(z) \equiv D'^2(z) + A'^2(z).$$

We exclude in the following the case $A^2 - A'^2 = 0$, which admits as a solution the trivial ambiguity $D' = -D$. If F is of order $\rho > 0$, A is of order $\tfrac{1}{2}\rho$, and D is of order ρ. This implies that D' is of order ρ and A' of order $\tfrac{1}{2}\rho$. Hence F and F' must have the same order. We can write (**X.3.17**) in the form

$$(D + D')(D - D') = -(A + A')(A - A')$$ (X.3.18)

Clearly, both sides must be of order $\tfrac{1}{2}\rho$.

Assume now that $\rho < 1$. $(D - D')$ and $(D + D')$ are separately entire functions of order $\rho < 1$ and can therefore be written as absolutely convergent products over their zeros. The zeros of $D - D'$ are a subset of the zeros of $(A - A')(A + A')$. Hence $(D - D')$ is necessarily of order $\tfrac{1}{2}\rho$. So is $(D + D')$. Hence D and D' are separately of order $\tfrac{1}{2}\rho$, which contradicts the assumption unless $\rho = 0$. Hence there is at most one amplitude of order $0 < \rho < 1$ reproducing a given cross section.

This result has been generalized (Itzykson and Martin 1973) for arbitrary finite noninteger order > 0. If the order is an integer, there can be two solutions at most. It has also been extended to the case of genuine entire functions of order 0, where, again, there may be two solutions at most. An interesting physical consequence of this study is that if we *know* that the scattering amplitude is given by the Schrödinger equation with a *truncated* potential, so that it is an entire function of order $\tfrac{1}{2}$, there is only one solution to our problem.

X.4   Local Results

The local problem has been formulated and studied in the following terms: Suppose that we know an unitary amplitude $F_0(z)$ that is analytic in the small Martin ellipse $\varepsilon(z_0)$ (i.e., the ellipse of foci -1, $+1$, passing through z_0). Let $\sigma_0(z)$ be the corresponding cross section, defined in such a way that it can be continued in the complex plane:

$$\sigma_0(z) = F_0(z)F_0^*(z^*) = D_0^2(z) + A_0^2(z),$$ (X.4.1)

where D_0 and A_0 are, respectively, the dispersive and absorptive parts of F_0. What sufficient conditions can permit us to change $\sigma_0(z)$ into $\sigma_0(z) + \delta\sigma(z)$, and find a corresponding unitary amplitude $F_0(z) + \delta F(z)$?

To solve this problem:

1. We shall define a Hilbert space of functions that are analytic in a convenient ellipse with foci -1 and $+1$.
2. We shall define an operator yielding the solution $F(z)$ from $\sigma(z)$ and to which some form of the "implicit-function theorem" can be applied in some "neighborhood" of $\sigma_0 \to F_0$.

1. We first define a set of real Hilbert spaces, $\mathscr{H}(\zeta)$, parametrized by the real number $\zeta > 1$, and having the inner product

$$\langle f, g \rangle_\zeta = \sum_{l=0}^{\infty} (2l + 1) f_l g_l |Q_l(\zeta)|^{-2} \qquad \text{(X.4.2)}$$

defined on the set of functions that are real analytic within $\varepsilon(\zeta + \varepsilon)$ $(\varepsilon > 0)$. f_l and g_l are the Legendre coefficients of f and g:

$$f_l = \frac{1}{2} \int_{-1}^{+1} f(z) P_l(z) dz. \qquad \text{(X.4.3)}$$

P_l and Q_l are the Legendre functions of 1st and 2nd kind. It is possible to define numbers $\zeta_0'', \zeta_0', \zeta_0, \zeta_1$ in such a way that

$$\zeta_0 < \zeta_0' < \zeta_0'' < z_0 < \zeta_1 = 2\zeta_0^2 - 1 \qquad \text{(X.4.4)}$$

and that (a) $A(z)$ is real analytic within $\varepsilon(\zeta_1)$, (b) $\sigma_0(z)$ (which is real-analytic within $\varepsilon(z_0)$) belongs to $\varepsilon(\zeta_0'')$ and so is $D_0(z)$. (c) $D_0(z)$ is bounded below by $n_0 > 0$ on the boundary $\partial\varepsilon(\zeta_0')$ of $\varepsilon(\zeta_0')$:

$$n_0 = \inf_{z \in \partial\varepsilon(\zeta_0')} |D_0(z)|. \qquad \text{(X.4.5)}$$

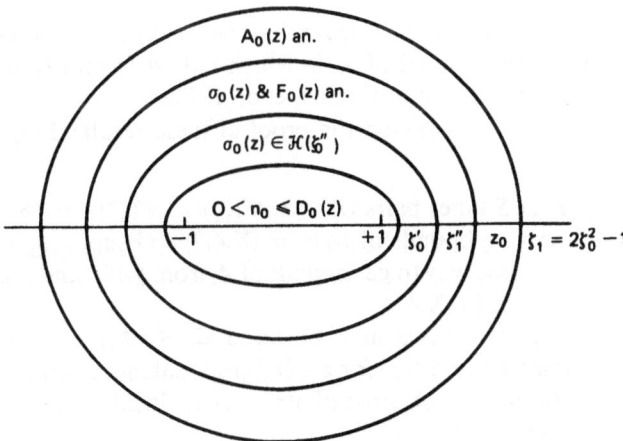

Figure X.3

2. Now, for each pair

$$(A, \sigma) \in \mathscr{H}(\zeta_1) \times \mathscr{H}(\zeta_0'') \tag{X.4.6}$$

we can define the following two sequences

$$D_l = \frac{1}{2} \int_{-1}^{+1} [\sigma(x) - A^2(x)]^{1/2} P_l(x) dx,$$

$$\bar{D}_l = \frac{1}{2\pi i} \int_{\partial\varepsilon(\zeta_0')} [\sigma(z) - A^2(z)]^{1/2} Q_l(z) dz. \tag{X.4.7}$$

Odd-order zeros of $\sigma(z) - A^2(z)$ give rise to branch points of the integrand of (X.4.7). If their number in $\varepsilon(\zeta_0')$ is even, suitable cuts may be drawn such that the integrand is continuous around $\partial\varepsilon(\zeta_0')$. If not, we draw a cut along the real axis and the contour is open. In both cases, it is easy to see that $\bar{D}_l^2$ is well defined and real. D_l and $\bar{D}_l$ coincide if and only if there are no odd-order zeros in $\varepsilon(\zeta_0')$. It follows from the assumptions that $D_0(z)$ is analytic in $\varepsilon(\zeta_0')$, and equal to $[\sigma_0(z) - A_0^2(z)]^{1/2}$, so that certainly

$$D_{0l} = \bar{D}_{0l}. \tag{X.4.8}$$

Thus it is possible to define the operator

$$S(A, \sigma; z) = A(z) - \sum_{l=0}^{\infty} (2l + 1)(A_l^2 + \bar{D}_l^2) P_l(z). \tag{X.4.9}$$

Because $F_0(z)$ is unitary and because the equality (X.4.8) we know that

$$S(A_0, \sigma_0) = 0. \tag{X.4.10}$$

It is possible to use the Hildebrandt–Graves theorem to show that

$$S(A, \sigma) = 0 \tag{X.4.11}$$

defines $A(\sigma)$ as an implicit function of σ, for σ in some neighborhood of σ_0 such that

$$A(\sigma_0) = A_0. \tag{X.4.12}$$

Besides, it is possible to prove that there exists an infinite-dimensional subset of the neighborhood of σ_0 in which indeed $\bar{D}_l$ and D_l are the same, so that unitarity is satisfied (Atkinson et al. 1974a).

Here, we only sketch the proof of these results. For this we successively prove.

1. That S takes pairs $(A, \sigma) \in \mathscr{H}(\zeta_1) \times \mathscr{H}(\zeta_0'')$ into $\mathscr{H}(\zeta_1)$. This is done by deriving bounds of $\bar{D}_l$ from (X.4.7) and $\sup_{z \in \partial\varepsilon(\zeta_0')} |\sigma(z) - A^2(z)|$, using the same way to get bounds of A_l from $\|A\|_\zeta$, and then substituting these bounds in (X.4.9).
2. That there is a neighborhood Ξ, of (A_0, σ_0), such that $\sigma(z) - A^2(z)$ remains zero-free for $z \in \partial\varepsilon(\zeta_0')$ so that, according to Rouché's theorem, the number of zeros of $\sigma(z) - A^2(z)$ inside $\partial\varepsilon(\zeta_0')$ is the same as that of $\sigma_0(z) - A_0^2(z)$.

For this, we notice that for any function $f(z)$ of $\mathcal{H}(\zeta_0'')$, we can write for $z \in \varepsilon(\zeta_0')$:

$$|f(z)|^2 = \left| \sum_l (2l + 1)P_l(z)f_l \right|^2$$

$$\leq \left\{ \sum_l (2l + 1) \frac{|f_l|^2}{|Q_l(\zeta_0'')|^2} \right\} \left\{ \sum_l (2l + 1)[P_l(\zeta_0')Q_l(\zeta_0'')]^2 \right\}$$

$$\text{(X.4.13)}$$

by the Schwarz inequality. This yields a bound of $\sup_{z \in \varepsilon(\zeta_0')} |\sigma(z) - \sigma_0(z)|$ by $C\|\sigma - \sigma_0\|_{\zeta_0''}$, and similarly for $A(z) - A_0(z)$. Hence it can be proved that making $\|\sigma - \sigma_0\|_{\zeta_0''}$ and $\|A - A_0\|_{\zeta_1}$ small enough, we can make $\sup_{z \in \varepsilon(\zeta_0')} |[\sigma(z) - A^2(z)] - [\sigma_0(z) - A_0^2(z)]|$ smaller than $n_0^2 - n^2$ $(0 < n < n_0)$ and that suffices to complete the proof.

3. That the Fréchet derivative $S_A(A, \sigma)$ of S with respect to A exists as a bounded linear operator on $\mathcal{H}(\zeta_1)$ and that $S(A, \sigma)$ and $S_A(A, \sigma)$ are continuous with respect of (A, σ) in the neighborhood of A_σ, σ_0.

For this one writes down

$$S_A(A, \sigma) = 1 - M_A(A, \sigma), \qquad \text{(X.4.14)}$$

$$M_A(A, \sigma)\delta A = \sum_{l=0}^{\infty} (2l + 1)P_l(z)\varphi_l, \qquad \text{(X.4.15)}$$

with

$$\varphi_l = \frac{1}{i\pi} \oint_{\partial_0(\zeta_0')} Q_l(t) \left\{ A_l - \frac{\bar{D}_l A(t)}{[\sigma(t) - A^2(t)]^{1/2}} \right\} \delta A(t)dt, \qquad \text{(X.4.16)}$$

and then get from (X.4.15) bounds for $|\varphi_l|$, which prove the convergence of (X.4.14), so that $M_A(A, \sigma)$ is a bounded linear operator. The other results can be proved by similar, but somewhat longer, calculations.

4. That $M_A(A_0, \sigma_0)$ is compact. For this it is shown that if $B(0, r)$ is the ball of center 0, radius r in $\mathcal{H}(\zeta_1)$, it is transformed by $M_A(A_0, \sigma_0)$ into a precompact subset. Indeed, given any $\varepsilon > 0$, one finds L such that $\sum_{l=L}^{\infty} (2l + 1)|\varphi_l|^2|Q_l(\zeta_1)|^{-2} < \varepsilon$. Hence, one can first cover the set in R^L by a finite ε-net which also serves as a finite net for $M_A(A_0, \sigma_0)$ $B(0, r)$.

5. Hence, unless unity belongs to the (point) spectrum of $M_A(A_0, \sigma_0)$, $S_A(A_0, \sigma_0)$ has a bounded inverse and the Hildebrandt–Graves theorem applies, proving the desired result. The problem of the splitting of zeros (which might prevent $\bar{D}_l$ being equal to D_l) can also be circumvented.

Bifurcation Case. Now suppose that 1 is an eigenvalue of $M_A(A_0, \sigma_0)$; we define

$$U(\delta A, \delta \sigma) = S(A, \sigma) - S_A(A_0, \sigma_0)\delta A, \qquad \text{(X.4.17)}$$

where

$$\left.\begin{array}{l} \delta A = A - A_0, \\ \delta\sigma = \sigma - \sigma_0. \end{array}\right\} \tag{X.4.18}$$

If there is a solution $A(\sigma)$ of (X.4.11), at such a solution

$$S_A(A_0, \sigma_0)\delta A = -U(\delta A, \delta\sigma). \tag{X.4.19}$$

$S_A(A_0, \sigma_0)$ has no inverse. Nevertheless we can show that there exists a so-called generalized inverse. Indeed, since $M_A(A_0, \sigma_0)$ is compact, (a) the null space N of $S_A(A_0, \sigma_0)$ is a linear subspace of $\mathcal{H}(\zeta_1)$ of finite dimension— say n; (b) the range R of $S_A(A_0, \sigma_0)$ is (strongly) closed. We can then define the quotient space $\mathcal{H}_N(\zeta_1) = \mathcal{H}(\zeta_1)/N$ and make it a Banach space with the norm

$$\|\delta A_N\|_N = \inf_{\alpha \in N} \|\delta A + \alpha\|_{\zeta_1}, \tag{X.4.20}$$

where δA_N is the equivalence class, modulo N, that contains δA. There exists a continuous, linear one-to-one mapping, from $\mathcal{H}_N(\zeta_1)$ to R, say S_N, such that

$$S_N \delta A_N = S_A(A_0, \sigma_0)\delta A. \tag{X.4.21}$$

Then the inverse mapping theorem tells us that S_N has a continuous inverse S_N^{-1} as a linear mapping from R to $\mathcal{H}_N(\zeta_1)$. Let Π_N and Π_K be the orthogonal projection operators (figure X.4) from $\mathcal{H}(\zeta_1)$ onto N and R, respectively. The generalized inverse of $S_A(A_0, \sigma_0)$ is defined, as a bounded linear mapping from $\mathcal{H}(\zeta_1)$ to $N^\perp$, the subspace of $\mathcal{H}(\zeta_1)$ orthogonal to N, by

$$S_A^{-1} = (I - \Pi_N)_N S_N^{-1}\Pi_R, \tag{X.4.22}$$

where $(I - \Pi_N)_N$ is the linear mapping from $\mathcal{H}(\zeta_1)$ to N defined by

$$(I - \Pi_N)_N \delta A_N = (I - \Pi_N)\delta A. \tag{X.4.23}$$

This generalized inverse can then be used to solve (X.2.48) provided that U belongs to R. To introduce this constraint, let us write

$$\delta A = -\tilde{S}_A^{-1} U(\delta A, \delta\sigma) + u, \tag{X.4.24}$$

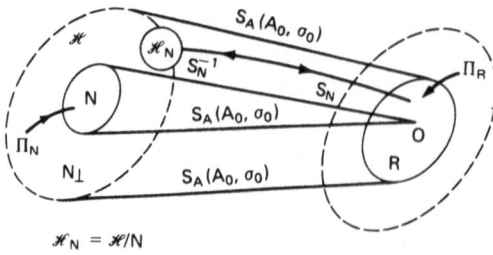

$$\mathcal{H}_N = \mathcal{H}/N$$

Figure X.4

where $u \in N$. Let $\{u_m\}$ be a real basis for N, λ_m a set of real numbers, we write

$$u = \sum_{m=1}^{n} \lambda_m u_m. \tag{X.4.25}$$

If $\|\delta\sigma\|_{\zeta_0''}$ and $\|u\|_{\zeta_1}$ are small enough, it is possible to determine the λ_m in such a way that U belongs to R, and that (X.4.24) defines a contraction mapping so that a locally unique solution, say $\delta A(\delta\sigma, u)$, exists. The argument is that (X.4.24) and (X.4.22) imply that

$$S_A(A_0, \sigma_0)\delta A = -\Pi_R U(\delta A, \delta\sigma) \tag{X.4.26}$$

which reduces to (X.4.19) if

$$(I - \Pi_R)U(\delta A(\delta\sigma, u), \delta\sigma) = 0 \tag{X.4.27}$$

$I - \Pi_R$ is the projection operator onto the null space of the adjoint operator $S_A^+(A_0, \sigma_0)$. u depends on the n real numbers $\lambda_1, \lambda_2, \ldots, \lambda_n$. Hence we may regard (X.4.27) as a system of n nonlinear algebraic equations for the n variables λ_m, with $\delta\sigma$ as an infinite dimensional parameters. Clearly, only real solutions are of interest; complex ones are to be ignored. For each real solution $\{\lambda_m\}$, there corresponds a solution $u(\delta\sigma)$ of the bifurcation equation (X.4.27); and for this solution, the function $\delta A(\delta\sigma, u(\delta\sigma))$ solves not only (X.4.24) (with $u = u(\delta\sigma)$), but also the original equation (X.4.19).

Like in the nonsingular case, it is possible to go from the linear theory to the nonlinear one by the way of a contraction mapping, this being proved after some simple but somewhat tedious majorations. Besides, the problem with the shift of the zeros can be managed like in the nonsingular case.

Hence we see that the existence of a solution $(\sigma_0 \rightarrow F_0)$ implies the existence of one or a finite number of solutions $(\sigma \rightarrow F)$ provided that σ is close enough to σ_0 in $\mathcal{H}(\zeta_0'')$. Unfortunately there does not yet seem to be in the literature any result available concerning the maximum dimension of N, nor global results on the bifurcation lines in the Banach space $\mathcal{H}$.

Besides, it has been given a numerical method (like the Newton–Kantorovich one) to get the new amplitude, and to prove that the Crichton ambiguity is not isolated (Atkinson et al. 1974a,b).

X.5   Uniqueness and Stability: A Reassessment

Again, in all the following, we forget about the trivial ambiguity. It follows from the review given above that the no-more-than-two-solutions conjecture (n-m-t-t-s) conjecture holds in large classes of functions, so that a physicist might produce the following remarks:

1. The n-m-t-t-s conjecture holds if I assume that F belongs to a class of entire functions of finite order.
2. Given any $\varepsilon > 0$, and any function that is continuous on $[-1, +1]$, I can find an entire function of finite order that approaches $f(x)$ within ε for the norm $\|f\| = \sup_{[-1, +1]} |f|$.

3. The only solutions which really are of physical interest are those which are stable with respect to experimental errors, viz., the mappings $\sigma \to F$ that are continuous for certain norms in the space $\mathscr{E}(\sigma \in \mathscr{E})$ and $\mathscr{C}(F \in \mathscr{C})$.

The physicist may then ask:

Question 0. Is it not possible to deduce from these remarks that the n-m-t-s conjecture holds in all cases of physical interest?

Both these remarks and this question are neither precise or really relevant to the physical centers of interest: (a) From (1) and (2) it simply follows that, given σ_0 in $\mathscr{E}$ that is equal to $|F_0|^2$ ($F_0 \in \mathscr{C}$), and a ball $B(F_0, \varepsilon)$ in $\mathscr{C}$, there exists at least a point F in this ball such that the problem $|F|^2 \to F$ has at most two solutions. (b) It is necessary to make the remark (3) more precise to give it a meaning at all. In particular the concept of stability with respect to experimental error has a meaning only if we define properly the metric spaces $\mathscr{E}$ and $\mathscr{C}$. Besides, the structure of $\mathscr{E}$ should preferably be defined "by itself"—for example, we may decide that $\mathscr{E}$ is the space of entire functions, or of polynomials, or of analytic functions, or of continuous functions, with the norm so and so. It should not be defined with a special reference at the properties of the space $\mathscr{C}$ in which the solution F *is to be* found. Similarly this metric space $\mathscr{C}$ should also be defined "by itself." Then only the concept of stability with experimental errors would really be equivalent with the assessment "$\sigma \to F$ is a continuous mapping." Unfortunately, this makes it necessary to define both $\mathscr{C}$ and $\mathscr{E}$ by weak conditions—"continuous functions spaces" are, maybe, the narrowest ones in which the stability problem can be *usefully* studied. Indeed, it is very easy to see that certain polynomials σ cannot correspond to a polynomial unitary amplitude (take for example an odd-order polynomial σ!). The same is true for genuine entire functions (take for example an unitary entire function F_0 of order 1, corresponding to $\sigma_0 = |F_0|^2$, and to D_0 of order 1, A_0 of order $\frac{1}{2}$, and try to solve within the class $\mathscr{C}$ of entire functions of order 1 the inverse problem $\sigma \to F$, with $\sigma = D_0^2 + A^2$, and A of order $\frac{3}{4}$). Similar examples can be built in the class of analytic functions by putting an absorptive perturbation to σ which does not modify the inclusion of σ in a given space $\mathscr{H}$, but is not itself analytic in a sufficiently large ellipse.

Now we claim that the question which is really of physical interest is the following one:

Let $\mathscr{E}$ and $\mathscr{C}$ be defined as metric spaces, with definitions that are *convenient* for physical studies. Let σ be a function of $\mathscr{E}$, and $B(\sigma, \delta)$ a ball centered at σ, of radius δ positive (arbitrarily small). We consider the images of the points of B through all the possible mappings ($\sigma \to$ unitary amplitude) of $\mathscr{E}$ into $\mathscr{C}$. The question is:

Question 1. In this image made of one, or two, or a finite number of disjunct sets, and what do these sets become as δ goes to zero?

Completing the n-m-t-s conjecture, the answer we *would like* to hear would run as follows.

Answer 1. "*Except maybe* for isolated points, without any physical interest, the image of B is contained in 1 or 2 (no more than 2) disjunct balls. As $\delta \to 0$, the distance of these balls from each other does not vanish, whereas the diameter of each ball goes to zero, and each ball never fails to contain at least a point solution."

Only such an answer would be plainly satisfactory for the physicist who, for any given σ, thus would be able to find one or two unitary amplitudes, $F_{1,2}$ (two amplitudes really "different" from each other), such that $|F_{1,2}|^2 - \sigma$ is arbitrarily small. We claim that, unfortunately, there is no evidence to support this statement as long as the uniqueness problem is treated from the global point of view in certain spaces and from the local one in others.

To prove our point, let us assume that three different solutions may exist near a bifurcation point in the space $\mathscr{C}_D$ of the functions that are twice continuously differentiable on $[-1, +1]$. Hence there exist three different unitary amplitudes, $F_1(x)$, $F_2(x)$, $F_3(x)$, with the same modulus $\sigma(x)$, which belongs itself to the space $\mathscr{E}_c$ of the functions that are continuous on $[-1, +1]$:

$$\sigma(x) = |F_1(x)|^2 = |F_2(x)|^2 = |F_3(x)|^2, \tag{X.5.1}$$

$$F_j(x) = \sum_0^\infty (2l + 1)(\exp[2i\delta_l^j] - 1)P_l(x), \qquad j = 1, 2, 3. \tag{X.5.2}$$

Let us introduce both in $\mathscr{C}_D$ and $\mathscr{E}_D$ the norm

$$\|f\| = \sup_{x \in [-1, +1]} |f(x)|. \tag{X.5.3}$$

Now, let $F_0(x)$ be an unitary amplitude which is a genuine entire function of order ρ, type τ

$$F_0(x) = \sum_0^\infty (2l + 1)(\exp[2i\delta_l^0] - 1)P_l(x). \tag{X.5.4}$$

The differentiability assumptions enable us to get for $j = 0, 1, 2, 3$ the inequality

$$|\delta_l^j| < c(l + \tfrac{1}{2})^{-5/2} \sup_{x \in [-1, +1]} [|F_j(x)|, |F_j'(x)|, |F_j''(x)|] \tag{X.5.5}$$

where c is a number. Let ε be any positive number smaller than

$$\tfrac{1}{2} \inf_{i,j=1,2,3} \|F_i - F_j\|.$$

It follows from (X.5.5) that it exists a set of numbers N_j $(j = 0, 1, 2, 3)$ such that

$$\left\| F_j - \sum_0^{N_j} (2l + 1)(\exp[2i\delta_l^j] - 1)P_l \right\| < \tfrac{1}{2}\varepsilon, \qquad j = 0, 1, 2, 3. \tag{X.5.6}$$

Now let us define, for $j = 1, 2, 3$, the functions F_j^ε:

$$F_j^\varepsilon(x) = \sum_0^{N_j} (2l + 1)(\exp[2i\delta_l^\varepsilon] - 1)P_l(x)$$

$$+ \sum_{l = \sup(N_j, N_0)}^{\infty} (2l + 1)(\exp[2i\delta_l^0] - 1)P_l(x). \qquad (\text{X.5.7})$$

$F_j^\varepsilon(x)$ differs from F_0 by a polynomial and therefore is a genuine entire function of order ρ type τ. Besides, it follows from (X.5.6) that

$$\|F_j^\varepsilon - F_j\| < \varepsilon. \qquad (\text{X.5.8})$$

From (X.5.1) and (X.5.8) we derive the inequality

$$\|\,|F_j^\varepsilon|^2 - \sigma\| < \varepsilon^2 + 2\varepsilon\|F_j\|. \qquad (\text{X.5.9})$$

For any positive number δ, it is obviously possible to define a positive number, say $\varepsilon(\delta)$, such that

$$\varepsilon^2(\delta) + 2\varepsilon(\delta) \sup_{j=1, 2, 3} \|F_j\| = \delta. \qquad (\text{X.5.10})$$

Hence, if δ is a positive number (which can be made arbitrarily small) there exist in the image of the ball $B(\sigma, \delta)$ at least three points $F_1^\varepsilon, F_2^\varepsilon, F_3^\varepsilon$, which are entire functions of order ρ, type τ, and which are "really different," since their distance does not vanish as ε goes to zero:

$$\|F_i - F_j\| - 2\varepsilon \leq \|F_i^\varepsilon - F_j^\varepsilon\| \leq \|F_i - F_j\| + 2\varepsilon \qquad (j = 1, 2, 3). \quad (\text{X.5.11})$$

This result proves that global uniqueness results, when they are obtained alone in a class of functions, are generally irrelevant for a physicist. In particular this remark would apply to an n-m-t-t-s theorem obtained in the class of functions analytic in an ellipse, or in the class of functions continuous on $[-1, +1]$, or in any class $\mathscr{C}$, as long as there may exist more than two solutions in a space $\mathscr{C}'$ in which $\mathscr{C}$ is dense for a certain norm, that makes use of values of x only on $[-1, +1]$! Thus we are led to look how should a global uniqueness theorem be completed to yield an answer like the answer 1. There are several possibilities:

(1) **Uniqueness + Stability.** Suppose that we can prove:

    (a) that there are at most N different solutions, i.e., N mappings, of a ball $B(\sigma_0, \delta)$ contained in a certain normed space $\mathscr{E}'$ into a normed space $\mathscr{C}'$;

    (b) that a class $\mathscr{C}$ does exist, which is dense in $\mathscr{C}'$, and whose image in $\mathscr{E}'$ is, say $\mathscr{E}$;

    (c) that the n-m-t-t-s conjecture holds for the mappings $\mathscr{E} \to \mathscr{C}$;

    (d) that the two mappings $\mathscr{E} \to \mathscr{C}$ are continuous for the norms $\|\cdot\|_{\mathscr{E}'}$ and $\|\cdot\|_{\mathscr{C}'}$.

Then necessarily $N \leq 2$.

Proof. According to (a), there are, at most, N unitary amplitudes $F_N(x)$ fitting $\sigma(x)$. According to (b), given a ball $B(\sigma, \delta)$, we can construct, like above, N functions F_j^ε that belong to $\mathscr{C}$ and correspond to N functions σ_j that belong to $\mathscr{E}$ *and* to the ball $B(\sigma, \delta)$. It follows from (c) and (d) that the functions F_j^ε can be put into two balls of $\mathscr{C}$, at most, whose diameters go to zero as $\delta \to 0$. This contradicts (a) unless $N \leq 2$.   Q.E.D.

Hence it would be *really interesting for physicists* to prove in a class of functions *both* uniqueness (or the n-m-t-t-s conjecture) *and* stability.

(2) **Uniqueness + "S-stability"**. Let us use again the assumption (a), (b), (c) of our previous study, and replace (d) by the following one:

(e) Let M_1 and M_2 be two mappings $\mathscr{E} \to \mathscr{C}$. We shall say that M_1 and M_2 are S stable (S for shift) for the norms in $\mathscr{E}$ and $\mathscr{C}$ if for any two unitary amplitudes F_1^ε and F_2^ε such that $\|\sigma_1^\varepsilon - \sigma_2^\varepsilon\| < \delta$, it exists in $\mathscr{C}$, for δ small enough, two unitary amplitudes $\tilde{F}_1^\varepsilon$ and $\tilde{F}_2^\varepsilon$ such that

$$|\tilde{F}_1^\varepsilon|^2 = |\tilde{F}_2^\varepsilon|,$$
$$\|F_1^\varepsilon - \tilde{F}_1^\varepsilon\| < \varepsilon, \qquad \|F_2^\varepsilon - \tilde{F}\| < \varepsilon. \tag{X.5.12}$$

Clearly, if this property holds, it is possible to construct from N functions F_j^ε N functions $\tilde{F}_j$ that are in $\mathscr{C}$ and correspond to a same "cross section σ." This contradicts the assumption (c) unless $N \leq 2$. Clearly also such a property holds if the stability property does. The interest of the S stability is that we have an heuristic argument in favor of it even in the narrow class of "polynomials" (Sabatier 1974a), so that we may hope to find a class of functions in which it can be proved.

X.6   Generalizations

The studies which have been done for the elastic scattering of spinless particles below the first inelastic threshold have been extended towards three directions;

X.6.1   Inelastic Region

The unitarity relation between the imaginary part and the modulus of the partial scattering amplitudes is replaced by a weaker one, which confines F_l to the interior of the Argand circle, or which is expressed by including an arbitrary "inelastic parameter":

$$|F_l|^2 - \operatorname{Im} F_l + I_l = 0, \tag{X.6.1}$$

where

$$I_l = \tfrac{1}{4}[1 - |\eta_l|^2] \in [0, \tfrac{1}{4}] \tag{X.6.2}$$

and the I_l's are constrained by the optical theorem:

$$\frac{1}{4} \sum_{l=0}^{\infty} (2l + 1)I_l = \frac{q^2}{4\pi} (\sigma_{\text{tot}} - \sigma_{el}). \tag{X.6.3}$$

Thus, a continuous ambiguity appears. To understand how it comes in, we refer ourselves to the notations of section X.4, replacing (**X.4.9**) by:

$$S(A, I) = \sum_{l=0}^{\infty} (2l + 1)S_l P_l(z) \qquad (\textbf{X.6.4})$$

with

$$S_l = A_l - A_l^2 - D_l^2 - I_l. \qquad (\textbf{X.6.5})$$

Now, let A^0 and I^0 be a first set (giving a first fit) for which $S(A^0, I^0) = 0$ and let $\{p_j^{(0)}\}$ $(j = 1, 2, \ldots, N)$ be the zeros of

$$R(A^0; z) = \sigma(z) - [A^0(z)]^2. \qquad (\textbf{X.6.6})$$

Since $R^{1/2}(A^0; z)$, the dispersive part of the amplitude, is analytic in the ellipse $\varepsilon(\zeta_0)$, these zeros must be of even order, and for simplicity we suppose that they are double. We choose ζ_0 such that the boundary of the ellipse, $\partial \varepsilon(\zeta_0)$, is zero-free. We know then that

$$R(A^0; p_j^{(0)}) = R'(A^0; p_j^{(0)}) = 0 \qquad (j = 1, 2, \ldots, N). \qquad (\textbf{X.6.7})$$

Starting from A^0, I^0, and $p_j^{(0)}$, we wish to generate a nearby solution, A, I, p_j, of the equations.

$$S(A, I) = P,$$
$$R(A, p_j) = R'(A, p_j) = 0, \qquad (\textbf{X.6.8})$$

with $\|A - A^0\|$, $\|I - I^0\|$, and $|p_j - p_j^0|$ all small. This is a typical implicit function problem. The study has been done by Atkinson et al. (1974) who showed how to iterate the solutions. They have applied the method to $\alpha - \alpha$ elastic scattering at 35 Mev. The results are illustrated for two waves in figure X.5, where only a few points are given. We should stress they are merely

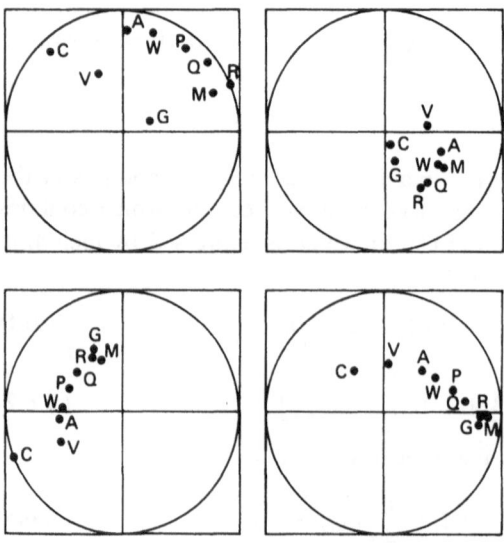

Figure X.5

points in a continuum for the particular values of the δl_i chosen. We see at once that the ambiguity is large—perhaps a third of the circle for S-wave, with a tendency to become more narrow in the higher waves. Other studies have been given by the same authors (Atkinson et al., 1973, 1973a,b, 1975) and also by Burkhardt (1972), Bart et al. (1973a,b), and Pietarinen (1974). For a good review and a survey of the ways to restore uniqueness in principle, see Burkhardt (1974) and Burkhardt and Martin (1975).

X.6.2   Higher Spin Waves

The case of spin and isospin (spin 0 − spin $\frac{1}{2}$) has been studied both in the elastic and the inelastic region (Atkinson et al., 1973b), see also Alvarez–Estrada and Carreras (1973), and Berends-Ruijsenaars (1973b).) At energies below the inelastic threshold, the problem is overdetermined for a complete set of measurements (differential cross sections, polarizations and Wolfenstein parameters for all charge states). Under favourable circumstances, a knowledge of a suitable subset of the experimentally independent observables suffices to determine all the amplitudes and therefore all the remaining observables. On the other hand, for energies above the inelastic threshold, the continuum ambiguity that was discussed for scalar particles persists in general, despite the spin and isospin complications. Thus the amplitudes, and so the phase shifts, cannot in general be determined uniquely simply from a knowledge of the observable at a fixed energy.

X.6.3   Nonspherical Case

The unitarity equation (**X.1.5**) holds in the absence of spherical symmetry (we still assume invariance under time reversal). Thus

$$F(1, 2) - F^*(1, 2) = \frac{i}{2\pi} \int F(1, 3)F^*(2, 3)d\Omega_3. \qquad \textbf{(X.6.9)}$$

It is easy to see that the trivial ambiguity

$$F(1, 2) \to -F^*(1, 2) \qquad \textbf{(X.6.10)}$$

still holds. So as to study existence and uniqueness, let us introduce the quantities

$$H(1, 2, 3) = \frac{|F(1, 3)||F(2, 3)|}{4\pi|F(1, 2)|}, \qquad \textbf{(X.6.11)}$$

$$Q(1, 2) = \int_S H(1, 2, 3)d\Omega(3), \qquad \textbf{(X.6.12)}$$

$$M = \sup_{1, 2} Q(1, 2). \qquad \textbf{(X.6.13)}$$

The inverse problem consists of obtaining $F(1, 2)$ from $|F(1, 2)|$. Thus H, Q and M are known quantities. $M < 1$ is a very strong condition, but which is very convenient to study the problem, as in the spherically symmetric case.

Assume this condition holds, the following algorithm is well defined:

$$F^{(0)}(1, 2) = |F(1, 2)|,$$

$$\text{Im } F^{(n+1)}(1, 2) = \frac{1}{4\pi} \int_S F^{(n)}(1, 3)F^{(n)*}(2, 3)d\Omega(3),$$

$$F^{(n+1)}(1, 2) = \{|F(1, 2)|^2 - [\text{Im } F^{(n+1)}(1, 2)]^2\}^{1/2} + i \text{ Im } F^{(n+1)}(1, 2),$$

$$(\text{X.6.14})$$

where the determination of the square root is fixed so as to suppress the trivial ambiguity. The convergence of this algorithm can be studied by using fixed point theorems. A sufficient condition is $M < 0.6248$ (Tortorella 1974) It guarantees existence and uniqueness of the solution, but certainly implies strong restrictions on its physical interest.

X.7   Comments and References

The study given here has been limited to a fixed-energy analysis. Thus the only constraint on the scattering amplitude was the unitarity condition, which has been written for elastic scattering at energies below any open inelastic channel. This case corresponds to most phase-shift analyses performed in nuclear physics, and to most theoretical analyses published between 1946 and 1976 (the references were given by the way). Since 1975, no important new result was obtained for this fully elastic fixed-energy problem, but an increasing number of new analyses dealt with the phase reconstruction from data supposedly known over a range of energies in (relativistic) particle scattering. Thus, in addition to the (general) unitarity property, our *a priori* knowledge on the problem includes several new constraints:

(a) the scattering amplitude $F(s, z)$ (where $s = k^2$, $z = \cos \theta$) must be an analytic function at least in an "axiomatic domain" of C^2;
(b) it has the crossing property versus the three processes that are consistent with a given scattering scheme;
(c) it is polynomially bounded in one variable, at most finite fixed values of other variables.

These properties, together with group invariances, may reduce or suppress the nonuniqueness. In particular, they emphasize the importance of following the zeros' trajectories as a way of reducing the discrete ambiguities (Barrelet, 1972; Stefanescu, 1987). On the other hand, the leading importance of analytic properties in the process of defining reconstructions suppresses ambiguities (Stefanescu, 1981, 1982, 1985; Atkinson and Stefanescu, 1985), but only at the price of instabilities so that regularizations are necessary (Stefanescu, 1980, 1986). For a complete review of these problems, which are beyond the scope of the present introduction, see Stefanescu (1987). Related phase problems appear, for instance, in Moses and Prosser (1983), and Barakat and Newson (1984), with references to former papers. Another one was studied very recently, that of reconstructing a unitary $N \times N$ matrix from the matrix of moduli. Continuous ambiguities appear for $N \geq 4$ (Auberson, 1988).

CHAPTER XI

Potentials from the Scattering Amplitude at Fixed Energy: General Equation and Mathematical Tools

XI.1 Introduction

In the following three chapters we only deal with problems in which spherical symmetry holds. Now, let us assume that the scattering amplitude $F(k, \cos \theta)$ has been constructed from the cross section. The next step in studying the interaction is to introduce a mathematical model in which the interaction appears as a parameter, and to determine that parameter from $F(k, \cos \theta)$. Such is the Schrödinger equation with a spherically symmetric potential $V(r)$ in the case of two colliding particles, below the first inelastic threshold, in the nonrelativistic range. In detail we have

$$\nabla^2 \Psi + [k^2 - V(r)] \Psi = 0, \qquad \text{(XI.1.1)}$$

where r is the distance between the two particles, k^2 and $V(r)$ are, respectively, proportional to the energy and the potential energy in terms of the reduced mass, $k^2 = 2mE/h^2$, $V(r) = 2m\mathbf{V}(r)/h^2$. The well-known partial wave decomposition leads one to the partial wave equations:

$$\frac{d^2\varphi_l}{dr^2} + [k^2 - V(r) - l(l+1)r^{-2}]\varphi_l = 0, \qquad \text{(XI.1.2)}$$

where φ_l is the solution which, as $r \to 0$, behaves like

$$\varphi_l(r) \sim u_l(r) = \left(\frac{\pi}{2} kr\right)^{1/2} J_{l+1/2}(kr) \sim \frac{\pi^{1/2}(\tfrac{1}{2}kr)^{l+1}}{\Gamma(l + \tfrac{3}{2})}. \qquad \text{(XI.1.3)}$$

It is assumed that $V(r)$ is taken in the largest class of potentials—say—$\mathcal{P}$ for which φ_l and $d\varphi_l/dr$ exist and behave asymptotically as follows:

$$\left.\begin{array}{l} \varphi_l(r) = A_l \sin\left(kr - l\dfrac{\pi}{2} + \delta_l \right) + o(1), \\[2mm] \dfrac{d\varphi_l(r)}{dr} = kA_l \cos\left(kr - l\dfrac{\pi}{2} + \delta_l \right) + o(1), \end{array}\right\} \qquad r \to \infty. \qquad \textbf{(XI.1.4)}$$

This mathematical model is able to represent the elastic scattering of two particles. The scattering amplitude is related to the phase shifts by the formula:

$$F(k, \cos \theta) = (k)^{-1} \sum_{l=0}^{\infty} (2l + 1)e^{i\delta_l} \sin \delta_l \, P_l(\cos \theta). \qquad \textbf{(XI.1.5)}$$

Hence, giving $F(k, \cos \theta)$ is equivalent to giving the phase shifts for all l. Now, assume that $F(k, \cos \theta)$ is known at a certain energy E_0. In the following we choose the corresponding values of k^{-1}, say k_0^{-1}, as the unit of length. The formulas **(X.1.2)** to **(X.1.4)** then reduce to the simple form:

$$\frac{d^2\varphi_l}{dr^2} + [1 - V(r) - l(l + 1)r^{-2}]\varphi_l = 0, \qquad \textbf{(XI.1.6)}$$

$$u_l(r) = \left(\frac{\pi}{2}r\right)^{1/2} J_{l+1/2}(r), \qquad \textbf{(XI.1.7)}$$

$$\left.\begin{array}{l} \varphi_l = A_l \sin\left(r - l\dfrac{\pi}{2} + \delta_l \right) + o(1), \\[2mm] \dfrac{d\varphi_l}{dr} = A_l \cos\left(r - l\dfrac{\pi}{2} + \delta_l \right) + o(1), \end{array}\right\} \qquad r \to \infty. \qquad \textbf{(XI.1.8)}$$

We now have to use them to derive $V(r)$ from the sequence $\{\delta_l\}$. Throughout the successive studies of this problem, like in most inverse problems, the questions of interest have changed. In the first studies, authors used trial and error methods to obtain approximate solutions of the problem. Their research and the later development of exact methods showed that solutions were not unique and this suggested further questions. Let us write down the six main questions that are basic to the inverse problem at fixed energy:

A. Existence. Being given a sequence $\{\delta_l\}$, does a potential V exist in $\mathcal{P}$, which generates $\{\delta_l\}$, at the given energy, through the Schrödinger equation? Such a potential is called a "solution" of the inverse problem.

B. Uniqueness. Let us be given a class $\mathcal{W}$ of functions ($\mathcal{W} \subseteq \mathcal{P}$), that contains a solution V. Is V the only solution in $\mathcal{W}$?

The answer to A being in general affirmative, the answer to B being in general negative, we are led to the following questions.

C. Constructibility. Give methods to construct some solutions of the problem.

D. Complete Constructibility in $\mathscr{W}$. Let $\{v\}$ be the set of all potentials in $\mathscr{W}$ that correspond to a sequence $\{\delta_l\}$ (at the given energy). Construct $\{v\}$. In the following we call two potentials "equivalent" if they produce the same scattering amplitude.

E. Approximation Theory. Measure the deviation from each other of all the equivalent potentials in $\mathscr{W}$. Clearly, if $\mathscr{W}$ is a metric space, such a measure is the diameter of $\{v\}$. If this measure is small, the phase shifts give the physical potential with a good approximation.

F. Stability Theory. Describe the evolution of the set of equivalent potentials when the scattering amplitude is submitted to small perturbations.

Clearly, the answer to questions A, B, C, D, E, F is a prerequisite to any numerical study of the problem, and it enlightens all those mathematical aspects which are important for the physicist. *But the physicist needs more.* If the problem has an infinity of solutions, he wants to know what additional information would be sufficient to discard all but one of these solutions, and what experiments would yield this additional information. In other words, he wants to give a *physical meaning* to the ambiguities. Such a goal is almost reached in the available solutions of the first ("fixed l") inverse problem, where the ambiguities correspond to the possible existence of bound states. We would like to obtain a similar result here. As we shall see, this has not been done. Another interesting physical question would be "are there some physical properties shared by all the inverse 'solutions'?" (like the number of holes in a drum in the relevant inverse problem).

The historical development of the subject has been similar to that of other inverse problems. The early studies tried to solve questions A and C by trial-and-error, or by approximate methods. General methods started to appear in 1959, but they remained formal until 1966, when they were systematically developed, giving definite answers to questions A, B, C. The solutions, or attempts at solution, to questions D, E, F are more recent, and the subject is not exhausted. A number of tools in this problem have been constructed by analogy with the inverse problem at fixed angular momentum. Others have been found independently and it might be interesting to look for their relevance in other problems.

In the present chapter, we introduce in a simple way a set of equations which turns out to be of general use in the problem.

To fully understand this machinery, it is convenient for the reader to refer himself at each step to a diagram showing the connection between all the tools which are introduced (figure XI.1).

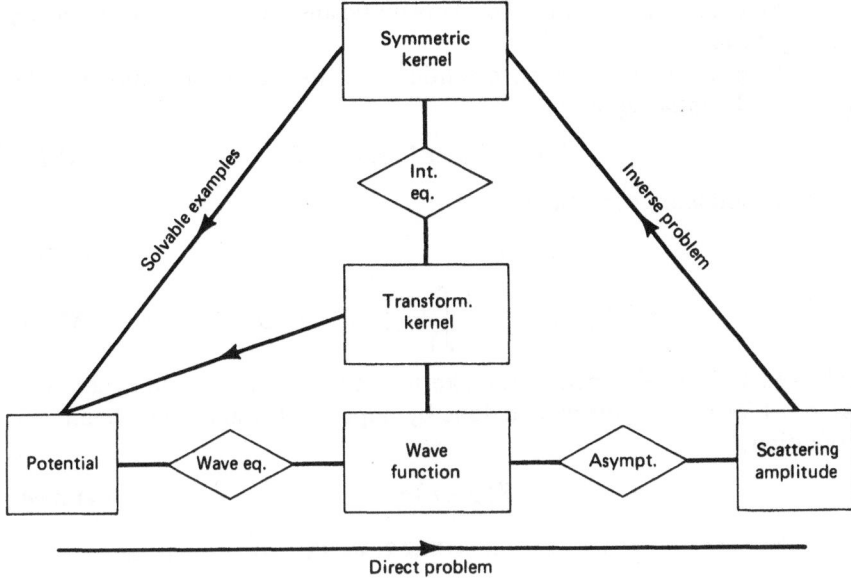

Figure XI.1

XI.2   The Transformation Kernel

Let D_r^V be the differential operator

$$D_r^V = r^2 \left[\frac{\partial^2}{\partial r^2} + 1 - V(r) \right], \qquad \text{(XI.2.1)}$$

where $V \in \mathscr{P}$, and let $\varphi_l^V(r)$ be the solution of the equation

$$D_r^V \varphi_l^V(r) = l(l+1)\varphi_l^V(r) \qquad \text{(XI.2.2)}$$

such that

$$\frac{\varphi_l^V(r)}{u_l(r)} \to 1 \qquad (r \to 0). \qquad \text{(XI.2.3)}$$

Clearly, **(XI.2.2)** is equivalent to **(XI.1.6)**. We introduce in the same way the operator $D_r^{V_0}(r)$ and the function $\varphi_l^{V_0}(r)$, where V_0 is a "well-known" potential, for which we know the solutions of **(XI.2.2)**. We define a "V_0 to V transformation kernel" as a continuous l-independent function, $K_{V_0}^V(r, r')$, which transforms $\varphi_l^{V_0}(r)$ into $\varphi_l^V(r)$ for any l and r as follows:

$$\varphi_l^V(r) = \varphi_l^{V_0}(r) - \int_0^r K_{V_0}^V(r, \rho)\varphi_l^{V_0}(\rho)\rho^{-2}\,d\rho. \qquad \text{(XI.2.4)}$$

Hence, knowing $\varphi_l^{V_0}$ and $\mathbf{K}_{V_0}^V$ yields the whole solution of the Schrödinger equation. The operator which is defined by the transformation kernel in **(XI.2.4)** is an example of the so-called "generalized translation operators."

They have been fully studied by mathematicians, but here we will not use their general theory.

Let us prove now that a twice continuously differentiable solution of the partial differential equation

$$[D_r^V - D_{r'}^{V_0}]\mathbf{K}_{V_0}^V(r, r') = 0 \tag{XI.2.5}$$

with the boundary conditions:

$$\mathbf{K}_{V_0}^V(r, 0) = 0, \tag{XI.2.6a}$$

$$r^{-1}\mathbf{K}_{V_0}^V(r, r) = -\frac{1}{2}\int_0^r \rho[V(\rho) - V_0(\rho)]d\rho, \tag{XI.2.7}$$

is indeed a " V_0 to V transformation kernel." Later, we shall need to continue $\mathbf{K}_{V_0}^V(r, r')$ for $r' \geq r$. This will be done by imposing the additional boundary condition:

$$\mathbf{K}_{V_0}^V(0, r') = 0. \tag{XI.2.6b}$$

Proof of the above statement. From (**XI.2.2**) and (**XI.2.4**) we readily get

$$D_r^V\varphi_l^V(r) = D_r^{V_0}\varphi_l^{V_0}(\rho) - r^2[V(r) - V_0(r)]\varphi_l^{V_0}(r)$$

$$- r^2\frac{d}{dr}[\mathbf{K}_{V_0}^V(r, r)r^{-2}\varphi_l^{V_0}(r)] - \varphi_l^{V_0}(r)\left[\frac{\partial}{\partial r}\mathbf{K}_{V_0}^V(r, \rho)\right]_{\rho = r}$$

$$- \int_0^r D_r^V\mathbf{K}_{V_0}^V(r, \rho)\varphi_l^{V_0}(\rho)\rho^{-2}\, d\rho. \tag{XI.2.8}$$

Using (**XI.2.5**) in the last term of (**XI.2.8**), we integrate twice by parts, and taking into account (**XI.2.6**) and (**XI.2.7**), we verify (**XI.2.2**). The condition (**XI.2.3**) follows readily from (**XI.2.4**), the boundary condition on $\varphi_l^{V_0}(r)$ and the finiteness of $\mathbf{K}_{V_0}^V(r, \rho)$. Since the solution of (**XI.2.2**) with (**XI.2.3**) is unique, $\mathbf{K}_{V_0}^V$ is a transformation kernel from V_0 to V.   Q.E.D.

Clearly, the partial differential equation (**XI.2.5**) is not easier to study than the Schrödinger equation, because it still depends explicitly on V. To go further, we need to introduce another function, the "symmetric kernel."

XI.3   The Symmetric Kernel and the Integral Equation

Let $f_{V_0}^V(r, r')$ be for any $r, r' \geq 0$ a three times continuously differentiable symmetric solution of the partial differential equation.

$$[D_r^{V_0} - D_{r'}^{V_0}]f_{V_0}^V(r, r') = 0 \tag{XI.3.1}$$

with the boundary condition

$$f_{V_0}^V(r, 0) = f_{V_0}^V(0, r') = 0. \tag{XI.3.2}$$

The way in which $f_{V_0}^V$ depends on V will soon be made clear. Let us now introduce, by analogy with the Gel'fand–Levitan equation, the following integral equation:

$$K_{V_0}^V(r, r') = f_{V_0}^V(r, r') - \int_0^r K_{V_0}^V(r, \rho) f_{V_0}^V(\rho, r') \rho^{-2} \, d\rho \qquad \text{(XI.3.3)}$$

and let us prove that the solution of (XI.3.3), when it exists and is unique, is indeed a V_0 to V transformation kernel (V_0 is introduced in (XI.3.1) and $(V - V_0)$ is defined by (XI.2.7) from the solution).

Proof. We first notice that (XI.3.3) is, for any fixed r, a linear Fredholm integral equation, which is more conveniently written as follows:

$$k_{V_0}^V(r, r') = \phi_{V_0}^V(r, r') - \int_0^r k_{V_0}^V(r, \rho) \phi_{V_0}^V(\rho, r') d\rho, \qquad \text{(XI.3.4)}$$

where k and ϕ are defined from K and f by:

$$k_{V_0}^V(r, r') = (rr')^{-1} K_{V_0}^V(r, r'), \qquad \text{(XI.3.5)}$$

$$\phi_{V_0}^V(r, r') = (rr')^{-1} f_{V_0}^V(r, r'). \qquad \text{(XI.3.6)}$$

Since $f_{V_0}^V(r, r')$ is three times continuously differentiable, it follows from (XI.3.2) and (XI.3.6) that $\phi_{V_0}^V(r, r')$ is twice continuously differentiable (three times for $rr' \neq 0$). Hence the kernel of (XI.3.4) is continuous and the Fredholm alternative holds: either the solution of (XI.3.4) exists and is unique or the homogeneous equation

$$\xi_{V_0}^V(r, r') = - \int_0^r \xi_{V_0}^V(r, \rho) \phi_{V_0}^V(\rho, r') d\rho \qquad \text{(XI.3.7)}$$

has a nontrivial solution. Let Ω be the set of values of r for which (XI.3.7) has a nontrivial solution. Ω cannot cover the whole axis $r \geq 0$: since $\phi_{V_0}^V(r, r')$ is continuous, the norm of the kernel in (XI.3.4) goes to zero as r goes to zero, so that there is a value of r under which the Neumann series converges. For any r that does not belong to Ω, we can construct the solution of (XI.3.4) by the method of Fredholm determinants:

$$k_{V_0}^V(r, r') = [\mathcal{D}(r)]^{-1} \mathcal{D}_r(r, r'), \qquad \text{(XI.3.8)}$$

where

$$\mathcal{D}(r) = 1 + \sum_{m=1}^{\infty} (m!) \int_0^r \cdots \int_0^r \Phi \begin{pmatrix} r_1, r_2, \ldots, r_m \\ r_1, r_2, \ldots, r_m \end{pmatrix} dr_1 \cdots dr_m, \qquad \text{(XI.3.9)}$$

$$\mathcal{D}_r(r, r') = - \sum_{m=0}^{\infty} (m!)^{-1} \int_0^r \cdots \int_0^r \Phi \begin{pmatrix} r, r_1, r_2, \ldots, r_m \\ r', r_1, r_2, \ldots, r_m \end{pmatrix} dr_1 \cdots dr_m, \qquad \text{(XI.3.10)}$$

(where $\Phi(r, r')$ is the term with $m = 0$ and reduces to $\phi(r, r')$).
The functions Φ stand for the well-known Fredholm determinants:

$$\Phi \begin{pmatrix} x_1, x_2, \ldots, x_m \\ y_1, y_2, \ldots, y_m \end{pmatrix} = \begin{vmatrix} \phi(x_1, y_1) & \phi(x_1, y_2) & \cdots & \phi(x_1, y_m) \\ \phi(x_2, y_1) & \phi(x_2, y_2) & \cdots & \phi(x_2, y_m) \\ \vdots & & & \\ \phi(x_m, y_1) & \phi(x_m, y_2) & \cdots & \phi(x_m, y_m) \end{vmatrix}. \qquad \text{(XI.3.11)}$$

The set Ω is the set of zeros of $\mathscr{D}(r)$. For $r \notin \Omega$, the continuity of the first and second order partial derivatives of $\mathbf{k}_{V_0}^V(r, r')$ is easily shown from (XI.3.8) by means of the Hadamard's theorem to bound the series, just as it is usually done to prove the convergence of (XI.3.9) and (XI.3.10) (according to this theorem, $|\Phi|$ in (XI.3.11) is bounded by $|\sup \phi|^m m^{(1/2)m}$). Hence, for $r \notin \Omega$, $\mathbf{K}_{V_0}^V(r, r')$ is twice continuously differentiable and (XI.3.7) has only the solution 0.

Now for $r \notin \Omega$, let us use (XI.3.3) to calculate the function

$$\chi_{V_0}^V(r, r') = (rr')^{-1}(D_r^V - D_{r'}^{V_0})\mathbf{K}_{V_0}^V(r, r'). \qquad \text{(XI.3.12)}$$

We obtain

$$D_r^V \mathbf{K}_{V_0}^V(r, r') = D_r^{V_0} f_{V_0}^V(r, r') - \mathbf{K}_{V_0}^V(r, r') \frac{\partial}{\partial r} f_{V_0}^V(r, r')$$

$$- \int_0^r [D_r^V \mathbf{K}_{V_0}^V(r, \rho)] f_{V_0}^V(\rho, r') \rho^{-2} \, d\rho$$

$$- f_{V_0}^V(r, r') \left\{ r^2(V(r) - V_0(r)) + r^2 \frac{d}{dr}(r^{-2}\mathbf{K}_{V_0}^V(r, r)) \right.$$

$$\left. + \left[\frac{\partial}{\partial r} \mathbf{K}_{V_0}^V(r, \rho) \right]_{\rho = r}^0 \right\}, \qquad \text{(XI.3.13)}$$

$$D_{r'}^{V_0} \mathbf{K}_{V_0}^V(r, r') = D_{r'}^{V_0} f_{V_0}^V(r, r') - \mathbf{K}_{V_0}^V(r, r) \frac{\partial}{\partial r} f(r, r')$$

$$- \int_0^r [D_\rho^{V_0}(\rho) \mathbf{K}_{V_0}^V(r, \rho)] f_{V_0}^V(\rho, r') \rho^{-2} \, d\rho$$

$$- f_{V_0}^V(r, r') \left[\frac{\partial}{\partial \rho} \mathbf{K}_{V_0}^V(r, \rho) \right]_{\rho = r}^0, \qquad \text{(XI.3.14)}$$

where we have integrated twice by parts and used (XI.3.1) subtracting (XI.3.14) from (XI.3.13) and *defining* $V(r)$ from $\mathbf{K}_{V_0}^V(r, r')$ by (XI.2.7). We find that $\chi_{V_0}^V(r, r')$ is a solution of (XI.3.7) and therefore is zero. Hence $\mathbf{K}_{V_0}^V(r, r')$ is indeed a solution of (XI.2.5) with the boundary conditions (XI.2.6) and (XI.2.7).   Q.E.D.

The physical meaning of the set Ω, where $\mathbf{K}_{V_0}^V(r, r')$ does not exist, as a solution of (XI.3.3), can be made more clear. It follows from (XI.3.9) and (XI.3.10) that

$$\mathscr{D}_r(r, r) = -\mathscr{D}'(r) \qquad \text{(XI.3.15)}$$

and from (XI.2.7)

$$r\mathscr{D}_r(r, r) = -\tfrac{1}{2}\mathscr{D}(r) \int_0^r \rho[V(\rho) - V_0(\rho)] d\rho. \qquad \text{(XI.3.16)}$$

From (XI.3.16), we see that $V(r) - V_0(r)$ is not integrable for $r \in \Omega$. If $\mathscr{D}(r)$ has isolated zeros, it follows from (XI.3.15) and (XI.3.16) that $V(r)$ has poles of order 2 at these points. In any case, the point $r \in \Omega$ corresponds to non-integrable singularities of $rV(r) - rV_0(r)$.

XI.4   The General Machinery

The mathematical tools that have been introduced above can be used in several ways, which are schematized in the diagram reproduced in figure XI.1. We can use this machinery in several ways.

(1) Construction of Examples. Let us first notice that any product $\varphi_\mu^{V_0}(r)\varphi_\mu^{V_0}(r')$ is a solution of (**XI.3.1**). Since $V_0 \in \mathscr{P}$, such a product satisfies the boundary condition (**XI.3.2**) provided that $\mathrm{Re}\,\mu > -\frac{1}{2}$.

Now, let $\mathscr{F}$ be a set of curves, in the half plane $\mathrm{Re}\,\mu > -\frac{1}{2}$, which is symmetric with respect to the real axis. Let $\rho'(\mu)$ the derivative of a real piecewise differentiable function defined on $\mathscr{F}$ and taking equal values on symmetric parts. Let $f_{V_0}^V(r, r')$ be the real function defined by

$$f_{V_0}^V(r, r') = \int_{\mu \in \mathscr{F}} \varphi_\mu^{V_0}(r)\varphi_\mu^{V_0}(r')d\rho(\mu). \tag{XI.4.1}$$

We assume that $d\rho(\mu)$ fulfills conditions that are sufficient to guarantee the convergence of (**XI.4.1**) and of the series obtained by differentiating (**XI.4.1**) up to three times. Then $f_{V_0}^V(r, r')$ can be taken as an input function in the machinery. If we know how to exactly solve the integral equation (**XI.3.3**), we can then construct potentials $(V - V_0)$ for which the Schrödinger equation is exactly solvable. Such is the case if $d\rho(\mu)$ reduces to a finite set of δ functions. The kernel $f_{V_0}^V(r, r')$ of (**XI.3.3**) is then a degenerate kernel and the solution of (**XI.3.3**) is readily obtained as a quotient of two determinants. A particularly simple case is for real μ:

$$f_{V_0}^V(r, r') - c_\mu \varphi_\mu^{V_0}(r)\varphi_\mu^{V_0}(r'). \tag{XI.4.2}$$

We readily obtain from (**XI.3.3**) and (**XI.2.4**)

$$K_{V_0}^V(r, r') = c_\mu \varphi_\mu^V(r)\varphi_\mu^{V_0}(r'), \tag{XI.4.3}$$

$$\left.\begin{array}{l} \varphi_\mu^V(r) = \varphi_\mu^{V_0}(r)[1 + c_\mu L_{\mu\mu}^{V_0}(r)]^{-1}, \\[2mm] V(r) = V_0(r) - 2c_\mu r^{-1}\dfrac{d}{dr}\{r^{-1}[\varphi_\mu^{V_0}(r)]^2(1 + c_\mu L_{\mu\mu}^{V_0}(r))^{-1}\}, \end{array}\right\} \tag{XI.4.4}$$

where

$$L_{ll'}^{V_0}(r) = \int_0^r \varphi_l^{V_0}(\rho)\varphi_{l'}^{V_0}(\rho)\rho^{-2}\,d\rho \tag{XI.4.5}$$

and

$$\varphi_l^V(r) = \varphi_l^{V_0}(r) - c_\mu \varphi_\mu^V(r)L_{\mu l}^V(r). \tag{XI.4.6}$$

Let us introduce generalized notations in (XI.1.8):

$$
\left.\begin{aligned}
\varphi_l^V(r) &\sim A_l^V \sin\left(r - l\frac{\pi}{2} + \delta_l^V\right), \\
\frac{d}{dr}\,\varphi_l^V(r) &\sim A_l^V \cos\left(r - l\frac{\pi}{2} + \delta_l^V\right).
\end{aligned}\right\}
\tag{XI.4.7}
$$

From (XI.4.4) and (XI.4.6), it is easy to derive the values of A_l^V and δ_l^V that correspond to our special example:

$$
\begin{aligned}
A_l^V e^{i\delta_l^V} &= A_l^{V_0} e^{i\delta_l^{V_0}} - L_{l\mu}^{V_0}(\infty)c_\mu A_\mu^V e^{i[\delta_\mu^V + (l-\mu)\pi/2]}A_\mu^V e^{i\delta_\mu^V} \\
&= [1 + L_{\mu\mu}^{V_0}(\infty)c_\mu]^{-1} A_\mu^{V_0} e^{i\delta_\mu^{V_0}}.
\end{aligned}
\tag{XI.4.8}
$$

The "unphysical meaning" of the set Ω can be illustrated with the formulas (XI.4.4) and (XI.4.8). For positive c_μ, this set is complex. But suppose c_μ is negative, and smaller than $-[L_{\mu\mu}^{V_0}(\infty)]^{-1}$. There exists a value R of r for which $L_{\mu\mu}^{V_0}(R)c_\mu = -1$. For this value, $V(r) - V_0(r)$ has a repulsive double pole, since it follows from (XI.4.4) that, for $r \to R$:

$$
V(r) - V_0(r) \sim 2(c_\mu)^2 r^{-4}[\varphi_\mu(V_0)]^4[1 + c_\mu L_{\mu\mu}^{V_0}(r)]^{-2}. \tag{XI.4.9}
$$

Now assume that $V_0(r)$ is a "good potential," so that $A_l^{V_0} e^{i\delta_l^{V_0}} \to 1$ as $l \to \infty$, and that $c_\mu L_{\mu\mu}^{V_0}(\infty)$ is precisely -1. From (XI.4.8) we see that A_l^V is no longer finite. Simultaneously, we deduce from (XI.4.4) that $V(r)$ asymptotically behaves like $16r^{-1} \sin[2r - 2\delta_\mu^{V_0}]$, and hence is no longer a "regular potential." It is remarkable that this singularity has disastrous consequences for the asymptotic amplitude but not on the phase shifts, which remain bounded.

(2) Study of the Inverse Problem. As soon as we know how to construct $f_{V_0}^V(r, r')$ from $\{\delta_l^V\}$, we will know how to solve the inverse problem. Now, the first idea is to choose $f_{V_0}^V(r, r')$ in a class of functions that depend on a sequence of parameters $\{c_\mu^V\}$ which can be bijectively associated to the sequence $\{\delta_l^V\}$. This is the basis of the matrix methods of chapter XII. To go further, one must show what class of functions $f_{V_0}^V(r, r')$ correspond to a given class of potentials. It turns out that $f_{V_0}^V(r, r')$ can be written as the sum of a function depending on a sequence $\{c_l\}$ that can be determined from $\{\delta_l^V\}$ and a function that can be arbitrarily chosen in a certain class. This yields the complete matrix method of chapter XIII. One can also associate $f_{V_0}^V(r, r')$ with the spectrum of the differential operator D_r^V. This is the main result of the spectral methods to be described in chapter XIII. But this spectrum does not include the physical (half-integer) values of l. Hence $f_{V_0}^V(r, r')$ is actually related not to $\{\delta_l\}$, but to a certain interpolation of the phase shifts which should be determined indirectly.

In each problem where an interpolation in the l complex-plane takes place, it is useful to introduce new notations, in which the main variable is no longer l but $l + \frac{1}{2}$:

$$u_l \rightarrow u_{l+1/2}, \qquad \varphi^V_{\lambda-1/2} \rightarrow \varphi^V_\lambda,$$

$$\delta_{\lambda-1/2} \rightarrow \delta(\lambda), \qquad c_{\mu-1/2} \rightarrow \gamma_\mu. \tag{XI.4.10}$$

Now, if $f^V_{V_0}(r, r')$ is of the form:

$$f^V_{V_0}(r, r') = \int_{\lambda \in \mathscr{F}} \varphi^{V_0}_\lambda(r) \varphi^{V_0}_\lambda(r') d\rho^V_{V_0}(\lambda) \tag{XI.4.11}$$

one easily derives from (XI.2.4) and (XI.3.3) the corresponding expansions for $K^V_{V_0}(r, r')$ and $\varphi^V_\mu(r)$:

$$K^V_{V_0}(r, r') = \int_{\lambda \in \mathscr{F}} \varphi^V_\lambda(r) \varphi^{V_0}_\lambda(r') d\rho^V_{V_0}(\lambda), \tag{XI.4.12}$$

$$\varphi^V_\lambda(r) = \varphi^{V_0}_\lambda(r) - \int_{\lambda \in \mathscr{F}} \Lambda^{V_0}_{\lambda\mu}(r) \varphi^V_\mu(r) d\rho^V_{V_0}(\mu), \tag{XI.4.13}$$

where

$$\Lambda^{V_0}_{\lambda\mu}(r) = \int_0^r \varphi^{V_0}_\lambda(\rho) \varphi^{V_0}_\mu(\rho) \rho^{-2} \, d\rho$$

$$= (\mu^2 - \lambda^2)^{-1} \left[\varphi^{V_0}_\lambda(r) \frac{d}{dr} \varphi^{V_0}_\mu(r) - \varphi^{V_0}_\mu(r) \frac{d}{dr} \varphi^{V_0}_\lambda(r) \right], \quad \text{(XI.4.14)}$$

and solving the inverse problem will always consist in finding a way to go from the asymptotic behaviour of $\varphi^V_\lambda(r)$ to $d\rho^V_{V_0}(\mu)$, the most obvious one making use of (XI.4.13), if that is possible.

XI.5   Further Study of the Integral Equation

The resolvent of the integral equation (XI.3.3) is the solution of the following integral equation:

$$S^V_{V_0 r}(r_1, r_2) = f^V_{V_0}(r_1, r_2) - \int_0^r S^V_{V_0 r}(r_1, \rho) f^V_{V_0}(\rho, r_2) \rho^{-2} \, d\rho. \tag{XI.5.1}$$

For $r \notin \Omega$, $S^V_{V_0 r}(r_1, r_2)$ exists. It can be constructed by Fredholm determinants, which can easily be used to prove that, for $r \notin \Omega$, $S^V_{V_0 r}(r_1, r_2)$ is a continuously differentiable function of r. For $r_1 = r$, the equation (XI.5.1) reduces to equation (XI.3.3), whose $K^V_{V_0}(r, r')$ is the unique solution. Hence

$$S^V_{V_0 r}(r, r') = K^V_{V_0}(r, r'). \tag{XI.5.2}$$

Now, let us prove the following formula for any r such that $[0, r] \cap \Omega = \varnothing$:

$$S^V_{V_0 r}(r_1, r_2) = f^V_{V_0}(r_1, r_2) - \int_0^r K^V_{V_0}(\rho, r_1) K^V_{V_0}(\rho, r_2) \rho^{-2} \, d\rho, \tag{XI.5.3}$$

where $\mathbf{K}_{V_0}^V(r, r')$ is defined for $r' \geq r$ by solving (XI.2.5) with the boundary conditions (XI.2.6b) and (XI.2.7).

Proof. Let us introduce the function

$$\chi_r(r_1, r_2) = S_{V_{0r}}^V(r_1, r_2) - f_{V_0}^V(r_1, r_2) + \int_0^r \mathbf{K}_{V_0}^V(\rho, r_1)\mathbf{K}_{V_0}^V(\rho, r_2)\frac{d\rho}{\rho^2}, \qquad \text{(XI.5.4)}$$

where $S_{V_{0r}}^V$ is defined from (XI.5.1). The function $r \to \chi_r(r_1, r_2)$ is differentiable, and reduces to zero as r goes to zero. Using (XI.5.1), it is easy to see that

$$\frac{d\chi_r(r_1, r_2)}{dr} = [\mathbf{K}_{V_0}^V(r, r_1)\mathbf{K}_{V_0}^V(r, r_2) - \mathbf{K}_{V_0}^V(r, r_1)f_{V_0}^V(r, r_2)]r^{-2}$$

$$- \int_0^r \frac{d}{dr}[\chi_r(r_1, \rho)]f_{V_0}^V(\rho, r_2)\rho^{-2}\,d\rho$$

$$+ r^{-2}\int_0^r \mathbf{K}_{V_0}^V(r, r_1)\mathbf{K}_{V_0}^V(r, \rho)f_{V_0}^V(\rho, r_2)\rho^{-2}\,d\rho. \qquad \text{(XI.5.5)}$$

Comparing to (XI.3.3), we see that the right-hand side is equal to its third term, so that the homogeneous forms of (XI.5.1) holds for $[(d/dr)\chi_r(r_1, r_2)]$. Hence $(d/dr)\chi_r(r_1, r_2)$ is zero for any $r \notin \Omega$. Integrating from 0 to r yields the same result for $\chi_r(r_1, r_2)$. Q.E.D.

From (XI.5.3) and (XI.5.2) it follows that $f_{V_0}^V(r, r')$ is uniquely determined from $\mathbf{K}_{V_0}^V(r, r')$:

$$f_{V_0}^V(r, r') = \mathbf{K}_{V_0}^V(r, r') + \int_0^r \mathbf{K}_{V_0}^V(\rho, r)\mathbf{K}_{V_0}^V(\rho, r')\rho^{-2}\,d\rho. \qquad \text{(XI.5.6)}$$

Modifications of the Reference Potential. We consider $d\rho_{V_0}^V(\lambda)$ and $d\rho_V^W(\lambda)$, and construct from (XI.4.11) and (XI.3.3) $\mathbf{K}_{V_0}^V(r, r')$ and $\mathbf{K}_V^W(r, r')$. We call $\mathscr{F}$ the union of the sets, in the right half λ plane, on which $\rho_{V_0}^V$ and ρ_V^W are defined. As for (XI.2.4), we can write

$$\varphi_\lambda^V(r) = \varphi_\lambda^{V_0}(r) - \int_0^r \mathbf{K}_{V_0}^V(r, \rho)\varphi_\lambda^{V_0}(\rho)\rho^{-2}\,d\rho, \qquad \text{(XI.5.7)}$$

$$\varphi_\lambda^W(r) = \varphi_\lambda^V(r) - \int_0^r \mathbf{K}_V^W(r, \rho)\varphi_\lambda^V(\rho)\rho^{-2}\,d\rho. \qquad \text{(XI.5.8)}$$

Substituting (XI.5.7) into (XI.5.8), we obtain the following after some algebra.

$$\varphi_\lambda^W(r) = \varphi_\lambda^{V_0}(r) - \int_0^r \mathbf{K}_{V_0}^W(r, \rho)\,\varphi_\lambda^{V_0}(\rho)\rho^{-2}\,d\rho, \qquad \text{(XI.5.9)}$$

where we have *defined* $\mathbf{K}_{V_0}^W(r, r')$ by

$$\mathbf{K}_{V_0}^W(r, r') = \mathbf{K}_{V_0}^V(r, r') + \mathbf{K}_V^W(r, r') - \int_{r'}^r \mathbf{K}_V^W(r, \tau)\mathbf{K}_{V_0}^V(\tau, r')\tau^{-2}\,d\tau. \qquad \text{(XI.5.10)}$$

This kernel is a V_0 to W transformation kernel. It is also valid for $V_0 = W$. Now, let $\mathscr{F}'$ be the set of all continuous kernels $\mathbf{K}_V^W(r, r')$ that can be obtained from functions $\rho(\mu)$ defined on $\mathscr{F}$. For $V_0 = W$, the only transformation kernel $\mathbf{K}_{V_0}^W$ in $\mathscr{F}'$ is zero (proof: multiplying both sides of (**XI.5.9**) by $d\rho_{V_0}^W(\lambda)$ and integrating on $\mathscr{F}$ yields $0 = \int_0^r [K_{V_0}^{V_0}(r, \rho)]^2 \rho^{-2} \, d\rho$, and K is continuous!). In this proof, and in the result, it is assumed that φ_λ can be constructed by means of the transformation kernel for all λ in $\mathscr{F}$. It follows from the result that a kernel $\mathbf{K}_{V_0}^V(r, r')$ belonging to a class $\mathscr{F}'$ is unique. Thus, from (**XI.2.5**), (**XI.2.6**), and (**XI.2.7**), we obtain

$$\mathbf{K}_{V_0}^V(r, r') = -\mathbf{K}_V^{V_0}(r', r). \tag{XI.5.11}$$

Applying (**XI.4.12**) and (**XI.5.8**) successively we easily show that

$$\mathbf{K}_{V_0}^V(r, r') - \int_0^r \mathbf{K}_V^W(r, \tau)\mathbf{K}_{V_0}^V(\tau, r')\tau^{-2} \, d\tau$$

$$= \int_{\mathscr{F}} \varphi_\lambda^{V_0}(r')\left[\varphi_\lambda^V(r) - \int_0^r \mathbf{K}_V^W(r, \tau)\varphi_\lambda^V(\tau)\tau^{-2} \, d\tau\right]d\rho_{V_0}^V(\lambda)$$

$$= \int_{\mathscr{F}} \varphi_\lambda^{V_0}(r')\varphi_\lambda^W(r)d\rho_{V_0}^V(\lambda). \tag{XI.5.12}$$

In the same way, also taking (**XI.5.11**) into account:

$$\mathbf{K}_V^W(r, r') + \int_0^{r'} \mathbf{K}_V^W(r, \tau)\mathbf{K}_{V_0}^V(\tau, r')\tau^{-2} \, d\tau = \int_{\mathscr{F}} \varphi_\lambda^W(r)\varphi_\lambda^{V_0}(r')d\rho_V^W(\lambda). \tag{XI.5.13}$$

From (**XI.5.12**) and (**XI.5.13**), and comparing (**XI.4.12**), one obtains the remarkable additive formula:

$$d\rho_{V_0}^V(\lambda) + d\rho_V^W(\lambda) = d\rho_{V_0}^W(\lambda). \tag{XI.5.14}$$

In particular

$$d\rho_V^W(\lambda) = -d\rho_W^V(\lambda). \tag{XI.5.15}$$

XI.6   Remarks on This Chapter

1. So as to be simple, we have introduced the equations of this chapter by imposing a strong differentiability assumption to $f_{V_0}^V(r, r')$. In the next chapters, particularly chapter XIII, this assumption is violated and the machinery still holds, although we shall not prove it.

2. The continuation of $\mathbf{K}_{V_0}^V(r, r')$ for $r' \geq r$ that we have introduced in section XI.5 is not necessary nor is it always used. $\mathbf{K}_{V_0}^V(r, r')$ being useful only for $r' \leq r$, it is sometimes convenient to complete its definition by setting it equal to zero for $r' > r$ (e.g., in Loeffel's method). Then, our $\mathbf{K}_{V_0}^V(r, r')$ for $r' \geq r$ is replaced by a certain "transposed kernel," which is conveniently defined.

XI.7   Connection Between the Problem at Fixed E and the Problem at Fixed l

The Schrödinger equation for the $l = 0$ partial wave reads:

$$\frac{d^2}{d\rho^2} \varphi(k, \rho) - V(\rho)\varphi(k, \rho) = -k^2\varphi(k, \rho), \qquad \textbf{(XI.7.1)}$$

$$\varphi(k, 0) = 0, \qquad \textbf{(XI.7.2)}$$

where we are interested in the dependence on k. The Schrödinger equation for the λ partial wave reads

$$\left\{ r^2\left[\frac{d^2}{dr^2} + 1 - V(r) \right] + \frac{1}{4} \right\} \varphi(\lambda, r) = \lambda^2\varphi(\lambda, r), \qquad \textbf{(XI.7.3)}$$

$$\varphi(\lambda, 0) = 0, \qquad \textbf{(XI.7.4)}$$

where we are interested in the dependence on λ. Now, in **(XI.7.4)**, let us set, (following Langer (1937)):

$$r = \exp[\rho]; \qquad \varphi(\lambda, r) = \exp[\tfrac{1}{2}\rho]\varphi(\lambda, \rho). \qquad \textbf{(XI.7.5)}$$

We readily obtain

$$\frac{d^2}{d\rho^2} \varphi(\lambda, \rho) + \exp[2\rho][1 - V(\exp[\rho])]\varphi(\lambda, \rho) = \lambda^2\varphi(\lambda, \rho). \quad \textbf{(XI.7.6)}$$

Clearly, the equation **(XI.7.6)** is similar to **(XI.7.1)**, and we can say that the inverse problem at fixed E is similar to the inverse problem at $l = 0$ if the variable is $\log r$, and k is replaced by $i\lambda$. But the boundary condition **(XI.7.4)** at $r = 0$ yields a boundary condition at $\rho = -\infty$, and not at $\rho = 0$. Difficulties come from this substitution in the boundary condition, and the complex substitution in eigenvalues. Thus, the analogy given here is helpful to derive formal structures in the problem, but the analysis has to be made separately.

Historical Notes. The general machinery for solving this inverse problem has progressively grown out of several works. It is given here in a very general form, which is partly new, and encompasses most approaches. Details concerning the history of these approaches will be given in the following chapters. Here we only point out a few dates to trace the evolution of the general machinery, giving what seems to be the first place where results appeared (although usually in a simpler form). The Regge–Newton equation **(XI.3.3)** and the machinery introduced in sections XI.2 and XI.3 were introduced by Regge (1959) rather formally and were used for the first time in a constructive method by Newton (1962), who also gave the first solvable example (Eqs. **XI.4.1** to **XI.4.8**). The study of the integral equation was carried further by Sabatier (1967b) (Eqs. **XI.3.8** to **XI.3.6**, and section XI.5, in particular the additive formula **(XI.5.14)**). The check list of section XI.1 and the diagram appear in Sabatier (1971).

CHAPTER XII

Potentials from the Scattering Amplitude at Fixed Energy: Matrix Methods

XII.1 Introduction

The simplest way to introduce a function $f_{V_0}^V(r, r')$ which is a solution of (**XI.3.1**) and which depends on a sequence of parameters $\{c_\mu\}$ is to construct a linear combination of products $\varphi_\mu^{V_0}(r)\varphi_\mu^{V_0}(r')$ with coefficients c_μ. An alternative method might introduce products $\varphi_\mu^{V_0}\varphi_\mu^{V_0}$, with strong consistency conditions. In all cases, a relation between $\{c_\mu\}$ and $\{\delta_l\}$ must be found by investigating the asymptotic behavior of $\varphi_l^V(r)$. This aim is achieved in the matrix methods we present in this chapter. They yield potentials in special classes that are defined by the nature of $\{\mu\}$. We strictly limit our study to real potentials.

Throughout this chapter, the reference potential V_0 is 0. Hence the regular wave function $\varphi_l^{V_0}(r)$ reduces to $u_l(r)$, which was defined by (**XI.1.7**). We shall suppress the indices V_0 and V in most of the formulas, and the boldface letters, so that the fundamental relation (**XI.2.4**) reads:

$$\varphi_l(r) = u_l(r) - \int_0^r K(r, \rho)u_l(\rho)\rho^{-2}\, d\rho \qquad \text{(XII.1.1)}$$

and the integral equation (**XI.3.3**)

$$K(r, r') = f(r, r') - \int_0^r K(r, \rho)f(\rho, r')\rho^{-2}\, d\rho. \qquad \text{(XII.1.2)}$$

XII.2   A Method in Which the Index μ Runs Through Integers

Let us introduce functions $f(r, r')$ of the following form (Newton 1962):

$$f(r, r') = \sum_0^\infty c_l u_l(r) u_l(r'),$$ (XII.2.1)

where $\{c_l\}$ is a sequence of real numbers. To be able to obtain the asymptotic behavior of the ψ_l's in the framework of this method, we have to impose a strong condition on the c_l's (Sabatier 1966b), not weaker than

$$\sum_1^\infty l^{-2} |c_l| < \infty.$$ (XII.2.2)

Inserting (XII.2.1) into (XII.1.2) and taking into account (XII.1.1) yields

$$K(r, r') = \sum_0^\infty c_l \varphi_l(r) u_l(r').$$ (XII.2.3)

Inserting (XII.2.3) into (XII.1.2) yields

$$\varphi_l(r) = u_l(r) - \sum_{l'} L_{ll'}(r) c_{l'} \varphi_{l'}(r),$$ (XII.2.4)

where $L_{ll'}(r)$ is defined by (XI.4.5) (with $V_0 = 0$). Letting $r \to \infty$ in (XII.2.4), and using (XI.4.7), we obtain the infinite system:

$$A_l e^{i\delta_l} = 1 - \frac{1}{2} \pi \frac{c_l A_l e^{i\delta_l}}{2l + 1} + i \sum_{l'} M_{ll'} c_{l'} A_{l'} e^{i\delta_{l'}}$$ (XII.2.5a)

or

$$A_l \left(1 + \frac{1}{2} \pi \frac{c_l}{2l + 1} \right) = e^{-i\delta_l} + i \sum_{l'} M_{ll'} c_{l'} A_{l'} e^{i(\delta_{l'} - \delta_l)},$$ (XII.2.5b)

where

$$M_{ll'} = \begin{cases} [(l' + \tfrac{1}{2})^2 - (l + \tfrac{1}{2})^2]^{-1} & \text{if } |l - l'| \text{ is odd,} \\ 0 & \text{if } |l - l'| \text{ is even or } 0. \end{cases}$$ (XII.2.6)

Considering separately the real and imaginary parts of (XII.2.5b) and setting $a_l = c_l A_l \cos \delta_l$, we obtain the systems:

$$\tan \delta_l = \sum_{l'} M_{ll'} a_{l'} (1 + \tan \delta_l \tan \delta_{l'}),$$ (XII.2.7)

$$c_l = a_l (1 + \tan^2 \delta_l) \left\{ 1 - \frac{1}{2} \pi \frac{a_l}{2l + 1} (1 + \tan^2 \delta_l) \right.$$

$$\left. - \sum_{l'} M_{ll'} a_{l'} (\tan \delta_{l'} - \tan \delta_l) \right\}^{-1}.$$ (XII.2.8)

In matrix notations, the system (**XII.2.7**) reads (Newton 1962)

$$\tan \Delta \mathbf{e} = M(1 + R)\mathbf{a}, \qquad\qquad \text{(XII.2.9)}$$

where we have assumed that M can be inversed, and where $\mathbf{a}$ is the vector $\{a_i\}$, $\mathbf{e}$ is the vector $\{1\}$, $\tan \Delta$ is the diagonal matrix $\{\tan \delta_i\}$, R is the matrix

$$R = M^{-1} \tan \Delta M \tan \Delta. \qquad\qquad \text{(XII.2.10)}$$

This ends the formal inversion procedure. It is clear that if M^{-1} and $(1 + R)^{-1}$ can be obtained, a set of coefficients c_l can be obtained from (**XII.2.7**) and (**XII.2.8**). Thus $f(r, r')$ can be constructed, and then $K(r, r')$ can be constructed from $f(r, r')$ by solving (**XII.1.2**). $\varphi_l(r)$ is given by (**XII.1.1**) and $V(r)$ by (**XI.2.7**) which here reads simply:

$$V(r) = -2r^{-1} \frac{d}{dr} r^{-1} K(r, r). \qquad\qquad \text{(XII.2.11)}$$

XII.3 Inversion of the Matrix M and Other Properties

We first prove that the matrix M, which has been defined by (**XII.2.6**), is the two-sided inverse of a matrix M^{-1} defined by

$$M^{-1} = -\mu M \mu, \qquad\qquad \text{(XII.3.1)}$$

where μ is a diagonal matrix with the elements

$$\mu_{2n} = \frac{2}{\pi} (4n + 1) \left[\frac{\Gamma(n + \frac{1}{2})}{\Gamma(n + 1)} \right]^2,$$

$$\mu_{2n+1} = \frac{2}{\pi} (4n + 3) \left[\frac{\Gamma(n + \frac{3}{2})}{\Gamma(n + 1)} \right]^2. \qquad\qquad \text{(XII.3.2)}$$

Then we show that there exists a sequence (column matrix) which is annihilated by M. Its elements are

$$v_{2n} = \mu_{2n}; \qquad v_{2n+1} = 0. \qquad\qquad \text{(XII.3.3)}$$

These results originally were obtained by an inductive method, which yields similar results for similar matrices (Sabatier 1966b), and may be adapted to larger classes of matrices. The method we give here is less general but saves time.

Proof.

1. Let us introduce the matrices **M** and m:

$$\mathbf{M} = i\mu^{1/2} M \mu^{1/2} = im. \qquad\qquad \text{(XII.3.4)}$$

We have to prove

$$\mathbf{MM} = -mm = 1. \qquad\qquad \text{(XII.3.5)}$$

Using the well-known formula[1]

$$\int_0^1 P_{2n}(x)P_{2q+1}(x)dx = \frac{2}{\pi}(-1)^{q+n}[(2q+\tfrac{3}{2})^2$$

$$-(2n+\tfrac{1}{2})^2]^{-1}\frac{\Gamma(\tfrac{1}{2}+n)\Gamma(\tfrac{3}{2}+q)}{\Gamma(1+n)\Gamma(1+q)}. \qquad \text{(XII.3.6)}$$

We can write

$$\mathbf{M}_{ll'} = -\mathbf{M}_{l'l} = i^{l'-l}\int_{-1}^{+1}\text{sgn}(x)p_l(x)p_{l'}(x)dx, \qquad \text{(XII.3.7)}$$

where sgn(x) is the function "sign of x," $p_l(x)$ is the normalized Legendre polynomial:

$$p_l(x) = (l+\tfrac{1}{2})^{1/2}P_l(x). \qquad \text{(XII.3.8)}$$

Clearly, the terms with even $|l - l'|$ vanish because of the parity properties of the Legendre polynomials. Using the Bessel–Parseval theorem, one straightforwardly shows that $\mathbf{M}$ is its own inverse:

$$\sum_{l''=0}^{\infty}\mathbf{M}_{ll''}\mathbf{M}_{l''l'} = i^{l'-l}\sum_{l''=0}^{\infty}\int_{-1}^{+1}[p_l(x)\text{sgn }x]p_{l''}(x)dx\int_{-1}^{+1}[p_{l'}(y)\text{sgn }y],$$

$$p_{l''}(y)dy = i^{l'-l}\int_{-1}^{+1}p_l(x)p_{l'}(x)(\text{sgn }x)^2\,dx = \delta_{ll'}, \qquad \text{(XII.3.9)}$$

where $\delta_{ll'}$ is the Kronecker's symbol.

2. Let $\mathbf{v}$ be a nonnull vector such that

$$\mathbf{Mv} = 0 = i^{-l}\int_{-1}^{+1}p_l(x)\text{sgn }x\sum_0^{\infty}i^{l'}v_{l'}\,p_{l'}(x). \qquad \text{(XII.3.10)}$$

$\{v_l\}$ cannot belong to l_2. If it did, $\sum_{l=0}^{\infty}i^lv_lp_l(x)$ would belong to $L_2(-1, +1)$ and its product by sgn x too. According to the Bessel–Parseval theorem, this product would be zero, and, hence, so would be $\{v_l\}$. Let us then assume that $\{(l+\tfrac{1}{2})^{-2}v_l\}$ belongs to l_2 and consider the sequence

$$\mathbf{v}_0 = \{v_{2l}\}. \qquad \text{(XII.3.11)}$$

By subtracting $\sum_{q=0}^{\infty}\mathbf{M}_{2n+1}^{2q}v_{2q}$ from $\sum_{q=0}^{\infty}\mathbf{M}_1^{2q}v_{2q}$, it follows that

$$\sum_0^{\infty}[(2q+\tfrac{1}{2})^2 - (2n+\tfrac{3}{2})^2]^{-1}\frac{\Gamma(q+\tfrac{1}{2})}{\Gamma(q+1)}\sqrt{2q+\tfrac{1}{2}}$$

$$\times [(2q+\tfrac{1}{2})^2 - (\tfrac{3}{2})^2]^{-1}v_{2q} = 0 \qquad \text{(XII.3.12)}$$

must hold for any n but $n = 0$. This equality can also be written as

$$\int_{-1}^{+1}P_{2n+1}(x)\sum_{q=0}^{\infty}\text{sgn}(x)p_{2q}(x)\frac{v_{2q}(-1)^q}{(2q+\tfrac{1}{2})^2 - (\tfrac{3}{2})^2} = 0. \qquad \text{(XII.3.13)}$$

---

[1] *Higher Transcendental Functions*. McGraw-Hill, 1953, Formula 3.2. (15).

Since the sequence $(-1)^q v_{2q}[(2q + \frac{1}{2})^2 - (\frac{3}{2})^2]^{-1}$ belongs to l_2, it follows from (**XII.3.13**) that

$$\sum_{q=0}^{\infty} p_{2q}(x) \frac{v_{2q}(-1)^q}{(2q + \frac{1}{2})^2 - (\frac{3}{2})^2} = C \operatorname{sgn} x p_1(x). \qquad \text{(XII.3.14)}$$

Setting $\mathbf{v} = \mu^{1/2}\mathbf{v}$, (**XII.3.14**) readily yields (**XII.3.3**). Besides, if we apply the same token to vectors with odd-order components, and to more divergent sequences, we easily show that the only vector $\mathbf{v}$ for which $M\mathbf{v}$ converges and is equal to zero is the one that is defined by (**XII.3.3**) (notice that a *necessary* convergence condition is $v_{2q} = o(q)$, $v_{2q+1} = o(1)$ as $q \to \infty$).   Q.E.D.

Mathematical Identities Involving M and Asymptotic Behavior of Certain Sums*.[2] Let us now give two ways of transforming the matrix M that are useful for the study of asymptotic properties. We introduce the matrices:

$$P_{ll'} = \tfrac{1}{2}\{\delta_{ll'} + (-1)^l\delta_{ll'}\}, \qquad \text{(XII.3.15)}$$

$$K_{ll'} = \begin{cases} 1 & \text{for odd } |l - l'|, \\ 0 & \text{for even } |l - l'|, \end{cases} \qquad \text{(XII.3.16)}$$

$$L_{ll'} = (l + \tfrac{1}{2})^{-2}\delta_{ll'}. \qquad \text{(XII.3.17)}$$

Now, from the identity

$$\frac{1}{(l' + \frac{1}{2})^2 - (l + \frac{1}{2})^2} = \left\{\frac{1}{(l' + \frac{1}{2})^2} + \frac{1}{(l' + \frac{1}{2})^2} \frac{(l + \frac{1}{2})^2}{[(l' + \frac{1}{2})^2 - (l + \frac{1}{2})^2]}\right\}$$

$$\text{(XII.3.18)}$$

we get

$$M = KL + L^{-1}ML. \qquad \text{(XII.3.19)}$$

From the identity

$$\frac{1}{(l' + \frac{1}{2})^2 - (l + \frac{1}{2})^{1/2}} = \left\{-\frac{1}{(l + \frac{1}{2})^2} + \frac{(l' + \frac{1}{2})^2}{(l + \frac{1}{2})^2[(l' + \frac{1}{2})^2 - (l + \frac{1}{2})^2]}\right\}$$

$$\text{(XII.3.20)}$$

we get

$$M = -LK + LML^{-1}. \qquad \text{(XII.3.21)}$$

Obviously, these identities can be iterated, and/or multiplied by the matrices P or $(1 - P)$. To use them in an example, let us calculate the asymptotic behavior of

$$g_l = \sum_{l'=0}^{\infty} m_{ll'} \tan \delta_{l'} f_{l'}, \qquad \text{(XII.3.22)}$$

---

[2] The asterisked paragraphs can be skipped in a first reading.

where[3]

$$|f_{2l}| = O(1), |f_{(2l+1)}| = O(l); (l \to \infty) \quad \text{and} \quad |\tan \delta_l| \le Cl^{-3+\varepsilon}. \quad \textbf{(XI.3.23)}$$

We use the inequality, valid for $\alpha \ne \gamma, -1 < \beta \le 1$,

$$\sum_{0}^{\infty} \frac{(q + \frac{1}{2})^{-\beta}}{(r + \alpha)^2 - (q + \gamma)^2} = O[r^{-\beta-1} \text{Log } r + r^{-2}] \quad (r \to \infty) \quad \textbf{(XII.3.24)}$$

which can be proved by comparison with integrals (Sabatier 1966b), and we derive from **(XII.3.21)** the identity:

$$\mu^{1/2} M \mu^{1/2} \tan \Delta f = -\mu^{1/2} L K \mu^{1/2} \tan \Delta P f - \mu^{1/2} L K \mu^{1/2} \tan \Delta (1 - P) f$$
$$+ \mu^{1/2} L M L^{-1} \mu^{1/2} \tan \Delta (1 - P) f$$
$$+ \mu^{1/2} L^2 K L^{-1} \mu^{1/2} \tan \Delta P f$$
$$+ \mu^{1/2} L^2 M L^{-2} \mu^{1/2} \tan \Delta P f. \quad \textbf{(XII.3.25)}$$

According to **(XII.3.2)**, $(\mu_l)^{1/2}$ is $O(1)$ for even l and $O(l)$ for odd l as $l \to \infty$. Using **(XII.2.6)** and **(XII.3.24)** in **(XII.2.5)**, we readily obtain:

$$g_{2l} = -(\mu_{2l})^{1/2}(2l + \tfrac{1}{2})^{-2} \sum_{p=0}^{\infty} (\mu_{2p+1})^{1/2} \tan \delta_{2p+1} f_{2p+1}$$
$$+ O(l^{-3-\varepsilon} \text{Log } l) \quad (l \to \infty), \quad \textbf{(XII.3.26)}$$

$$g_{2l+1} = -(\mu_{2l+1})^{1/2}(2l + \tfrac{1}{2})^{-2} \sum_{p=0}^{\infty} (\mu_{2p})^{1/2} \tan \delta_{2p} f_{2p}$$
$$- (\mu_{2l+1})^{1/2}(2l + \tfrac{1}{2})^{-4} \sum_{p=0}^{\infty} (2p + \tfrac{1}{2})^2 (\mu_{2p})^{1/2} \tan \delta_{2p} f_{2p}$$
$$+ O(l^{-3-\varepsilon} \text{Log } l) \quad (l \to \infty), \quad \textbf{(XII.3.27)}$$

which are the sought-after asymptotic properties for this example.

XII.4   Construction of $\{c_l\}$ from $\{\tan \delta_l\}$

From **(XII.3.4)** and **(XII.2.9)**, using the results of section XII.3, we obtain

$$\tfrac{1}{2}\alpha v - m \tan \Delta f = [1 - m \tan \Delta][1 + m \tan \Delta]b, \quad \textbf{(XII.4.1)}$$

where

$$\left.\begin{array}{l} f = \mu^{1/2} e, \\ v = \mu^{-1/2} v, \\ b = \mu^{-1/2} a. \end{array}\right\} \quad \textbf{(XII.4.2)}$$

Now, let us study the equation

$$w - u = \pm m \tan \Delta w. \quad \textbf{(XII.4.3)}$$

---

[3] Except if stated otherwise, C and ε are the following positive numbers whose exact value either is not given or can be chosen arbitrarily.

Using (**XII.3.4**) and (**XII.3.6**), we can write (**XII.4.3**) in the form:

$$\sum_0^\infty i^l[w_l - u_l]p_l(x) = \pm \int_{-1}^{+1} T(x, y)\left[\sum_0^\infty i^l w_l p_l(y)\right] dy, \qquad \text{(XII.4.4)}$$

where

$$T(x, y) = i \operatorname{sgn} x \sum_0^\infty \tan \delta_{l'} \, p_{l'}(x)p_{l'}(y). \qquad \text{(XII.4.5)}$$

Let us assume that $\{u_l\}$ is an l_2-sequence, so that $\sum i^l u_l p_l(x)$ belongs to $L_2 (-1, +1)$. Besides, we assume in the following that positive numbers C and ε exist, for which the following inequality uniformly holds.

$$|\tan \delta_l| < Cl^{-3-\varepsilon}. \qquad \text{(XII.4.6)}$$

Hence $T(x, y)$ is a piecewise continuous function in the square $[-1, +1] \times [-1, +1]$, and is uniformly bounded by

$$|T(x, y)| \le \sum_0^\infty |\tan \delta_l|(l + \tfrac{1}{2}). \qquad \text{(XII.4.7)}$$

Exceptional Sets. The equations (**XII.4.4**) are Fredholm equations. They can be solved unless the homogeneous equation has a nontrivial solution. We call "exceptional" these sets of phase shifts for which at least one of the two homogeneous equations

$$\mathbf{w} = \pm m \tan \Delta \mathbf{w} \qquad \text{(XII.4.8)}$$

has a nontrivial solution. For these sets, the Fredholm determinants $\mathscr{D}^\pm$ of the equations (**XII.4.4**) vanish. It follows from (**XII.4.5**) that the L^2 norm of $T(x, y)$ is nothing but $(\sum_0^\infty \tan^2 \delta_l)^{1/2}$. Hence the condition

$$\sum_0^\infty \tan^2 \delta_l < 1 \qquad \text{(XII.4.9)}$$

ensures that a set of phase shift is *not* exceptional, since it ensures that the Neuman series for the resolvent of (**XII.4.4**) converges. Exceptional sets can be constructed by a convenient choice of w in the following formula, which is readily derived from (**XII.4.8**):

$$\tan \Delta \mathbf{w} = \pm m\mathbf{w}. \qquad \text{(XII.4.10)}$$

As an exercise, the reader may check that a set in which all phase shifts but δ_0 and δ_1 are zero is exceptional if

$$\tan \delta_0 \tan \delta_1 \, m_0^1 m_1^0 = 1 \qquad \text{(XII.4.11)}$$

the corresponding $\mathbf{w}$ has even components proportional to $m_1^0 m_{2p}^1$ and odd ones proportional to $m_{2p+1}^0 \cotg \Delta_1$.

Properties * of $\{a_l\}$. When $\{\tan \delta_l\}$ is not exceptional, equation (**XII.4.3**) can be solved by the Fredholm method:

$$\sum_0^\infty i^l[w_l - u_l]p_l(x) = \int_{-1}^{+1} \frac{\mathscr{D}^\pm(x, y)}{\mathscr{D}^\pm} \left(\sum_0^\infty i^l u_l p_l(y) \right) dy, \quad \textbf{(XII.4.12)}$$

where

$$\mathscr{D}^\pm = 1 + \sum_{m=1}^\infty \frac{(\mp 1)^m}{m!} \int_{-1}^{+1} \cdots \int_{-1}^{+1} \begin{bmatrix} T(\xi_1, \xi_1) & \cdots & T(\xi_1, \xi_m) \\ \vdots & & \\ T(\xi_m, \xi_1) & \cdots & T(\xi_m, \xi_m) \end{bmatrix} d\xi_1 \cdots d\xi_m$$

$$\textbf{(XII.4.13)}$$

and

$$\mathscr{D}^\pm(x, y) = \sum_{m=0}^\infty \frac{(\mp 1)^m}{m!} \int_{-1}^{+1} \cdots \int_{-1}^{+1}$$

$$\times \begin{bmatrix} T(x, y) & T(x, \xi_1) & \cdots & T(x, \xi_m) \\ T(\xi_1, y) & & \cdots & \\ \vdots & & & \\ T(\xi_m, y) & T(\xi_m, \xi_1) & \cdots & T(\xi_m, \xi_m) \end{bmatrix} d\xi_1 \cdots d\xi_m. \quad \textbf{(XII.4.14)}$$

By expanding $T(x, y)$ and $T(x, \xi_i)$ in (**XII.4.14**), and performing the y integration in **XII.4.12**, we obtain:

$$\sum_0^\infty i^l[w_l - u_l]\dot{p}_l(x) = \mp i \, \mathrm{sgn} \, x \sum_0^\infty \tan \delta_{l'} \, p_{l'}(x) i^{l'} \, d_{l'}^\pm, \quad \textbf{(XII.4.15)}$$

where

$$\mathscr{D}^\pm d_l^\pm = -i^{-l} \sum_{m=0}^\infty \frac{(\mp 1)^m}{m!} \int_{-1}^{+1} \cdots \int_{-1}^{+1}$$

$$\times \begin{bmatrix} i^l u_l & p_l(\xi_1) & \cdots & p_l(\xi_m) \\ T(\xi_1) & T(\xi_1, \xi_1) & \cdots & T(\xi_1, \xi_m) \\ \vdots & \vdots & & \\ T(\xi_m) & T(\xi_m, \xi_1) & \cdots & T(\xi_m, \xi_m) \end{bmatrix} d\xi_1 \cdots d\xi_m. \quad \textbf{(XII.4.16)}$$

In (**XII.4.16**), $T(\xi)$ stands for $i \, \mathrm{sgn}(\xi) \sum_0^\infty \tan \delta_l i^l u_l p_l(\xi)$. With the assumption (**XII.4.6**) $T(\xi)$ is absolutely and uniformly bounded. Let T be $\sup_{x, y, \xi}(|T(\xi)|, |T(x, y)|)$. We expand the determinants along their first line, derive bounds for the cofactors by Hadamard's theorem, and notice that if $|F(\xi)|$ is smaller than F, then $|\int_{-1}^{+1} p_l(\xi)F(\xi)d\xi|$ is smaller than $F\sqrt{2}$. Hence we obtain

$$|d_l^\pm| \le |\mathscr{D}^\pm|^{-1} \sum_0^\infty (m!)^{-1}(m\sqrt{2} + |u_l|)m^{(1/2)m}(2T)^m. \quad \textbf{(XII.4.17)}$$

This inequality shows that the sequence $|d_l|$ is uniformly bounded if the sequence $|u_l|$ is. Hence, from (**XII.4.15**) and (**XII.3.7**), we get

$$\mathbf{w} - \mathbf{u} = m \tan \Delta \mathbf{d}^{\pm} \qquad \text{(XII.4.18)}$$

which gives the solution of (**XII.4.3**) for any set of phase shifts that is not exceptional. Besides, if $\mathbf{u}$ is a bounded sequence, it follows from (**XII.3.26**) and (**XII.3.27**) that $m \tan \Delta \mathbf{u}$ belongs to l_2. Then writing (**XII.4.3**) in the equivalent form:

$$(\mathbf{w} - \mathbf{u}) - (\pm m \tan \Delta \mathbf{u}) = \pm m \tan \Delta(\mathbf{w} - \mathbf{u}) \qquad \text{(XII.4.19)}$$

we see that the result (**XII.4.18**), with a convenient definition of $\mathbf{d}^{\pm}$, holds for a bounded sequence $\mathbf{u}$. Such is the sequence on the left-hand side of (**XII.4.1**).

Applying (**XII.4.18**) twice readily yields

$$b_l = \tfrac{1}{2}\alpha v_{2l} - \sum_{l'=0}^{\infty} m_{ll'} \tan \delta_{l'} h_{l'}, \qquad \text{(XII.4.20)}$$

where h_l is $O(1)$ for even l, $O(l)$ for odd l, as $l \to \infty$. From (**XII.3.26**), (**XII.3.27**), (**XII.3.2**) and (**XII.4.2**), we derive the asymptotic behaviour of a_l as $l \to \infty$:

$$\begin{cases} a_{2l} = \tfrac{1}{2}\alpha\mu_{2l} + \beta_1(2l + \tfrac{1}{2})^{-2} + O(l^{-3-\varepsilon} \operatorname{Log} l), \\ a_{2l+1} = \tfrac{1}{2}\beta(l + \tfrac{3}{4})^{-2}\mu_{2l+1} + \beta_2(2l + \tfrac{3}{2})^{-2} + O(l^{-2-\varepsilon} \operatorname{Log} l), \end{cases} \qquad \text{(XII.4.21)}$$

where β, β_1, β_2, are certain known convergent series of $\tan \delta_l$ (Sabatier 1966b). We notice that $\tfrac{1}{2}\mu_{2l}$ and $\tfrac{1}{2}(l + \tfrac{3}{4})^{-2}\mu_{2l+1}$ are both equal to $4\pi^{-1} + O(l^{-2})$ as $l \to \infty$.

Asymptotic Behavior of the c_l's. From (**XII.2.8**) and (**XII.4.6**), we see that the asymptotic behavior of c_l is that of $a_l(1 - (\pi/2)a_l/2l + 1)^{-1}$, up to terms of order l^{-2}:

$$c_l = a_l\left(1 - \frac{\pi}{2}\frac{a_l}{2l + 1}\right)^{-1} + O(l^{-2}) = 4\pi^{-1} + O(l^{-2}). \qquad \text{(XII.4.22)}$$

Thus c_l is, in particular, bounded.

XII.5 Construction of $V(r)$—Consistency of the Method

The free regular wave function, $u_l(r)$, satisfies the inequality

$$|u_l(r)| < C(\tfrac{1}{2}|r|)^{l+1} \exp \frac{|r|^2}{4l + 1}(l!)^{-1} \qquad \text{(XII.5.1)}$$

for any r in the complex-plane, where C is a constant that is independent of l and r. Thus, because of (**XII.2.2**), $\sum_{l=0}^{\infty} c_l u_l(r)u_l(r')$ and its partial derivatives of any finite order converge towards entire functions of r and r'. Actually,

it would suffice that, for a certain p, $|c_l| = O(l^p)(l \to \infty)$. Thus, $f(r, r')$ satisfies
(XI.3.1) and (XI.3.2). The kernel $K(r, r')$, which is defined by (XI.3.3), can be
obtained from (XI.3.8). Bounds for (XI.3.11) can be obtained from Hadamard's
theorem in any finite part of the r and r' complex planes, so that $\mathscr{D}_r(r, r')$ and
$\mathscr{D}(r)$ can be continued as entire functions of r and r'. From (XI.3.16) it follows
that $V(r)$ is obtained from the c_l's as a meromorphic function of r, with
isolated double poles, in any finite region of the r complex plane. $V(r)$ is
real on the real axis. Besides, there always exists a nonvanishing disk, centered
at the origin, and in which $V(r)$ is analytic. Obtaining the asymptotic behavior of $V(r)$ is complicated, and we only sketch the argument. We first
notice that the asymptotic behaviors of $f(r, r)$ and $r^{-1}(d/dr)[r^{-1}f(r, r)]$ as r
goes to ∞ are dominated, for $\alpha \neq \beta$, by the contributions of the leading
terms in (XII.4.21):

$$f(r, r) = \tfrac{1}{2}(\alpha + \beta)r \int_0^{2r} J_0(t) + \beta r J_1(2r) + O(1), \qquad \text{(XII.5.2)}$$

$$r^{-1}\frac{d}{dr}r^{-1}f(r, r) = (\alpha - \beta)r^{-1}J_0(2r) + O(r^{-2}). \qquad \text{(XII.5.3)}$$

Hence, if anything like a linear approximation holds in certain cases,
yielding $K(r, r) \simeq f(r, r)$, we cannot expect that the potential goes to zero
more rapidly than $(\alpha - \beta)r^{-3/2}$ as r goes to ∞. On the other hand, if α and β
were both zero, $f(r, r')$ and its derivatives would be uniformly bounded;
$\mathscr{D}(r)$ obviously would go to a definite limit $\mathscr{D}(\infty)$ as r goes to ∞, and, provided
that $\mathscr{D}(\infty) \neq 0$, the solution of (XI.3.4) would uniformly approach the
(uniformly bounded) solution of the equation:

$$\gamma(r, r') = \phi(r, r') - \int_0^\infty \gamma(r, \rho)\phi(\rho, r')d\rho. \qquad \text{(XII.5.4)}$$

Thus, one could easily show in that case that $V(r)$ is $O(r^{-2})$ as r goes to ∞.
    In the general case, one first has to prove the relevance of (XII.5.4). This is
done by exactly solving (XII.5.4) for a part of $\phi(r, r')$, then taking into account
the remainder by a standard technique of integral equation theory (this
technique works only if 1 does not happen to be an eigenvalue of the kernel
$\phi(r, r')$—a condition obviously equivalent to $\mathscr{D}(\infty) \neq 0$).
    Equation (XII.5.4) is then compared to (XI.3.4) and the remainders are
appraised, giving the asymptotic behavior of $K(r, r')$. Similar methods are
then applied to obtain the asymptotic behavior of the derivatives of $K(r, r')$.
The result confirms what has been seen above in the linear approximation.
It has been proved (Sabatier 1966a) that for any positive value of ε':

$$V(r) = -2\pi^{-1/2}(\alpha - \beta)r^{-3/2} \cos\left(2r - \frac{\pi}{4}\right) + O(r^{\varepsilon'-2}). \qquad \text{(XII.5.5)}$$

For such a potential, the wave function has the asymptotic behavior
(XI.4.7) which is required to define a phase shift. Besides, the asymptotic
amplitudes A'_l are bounded by a number which is independent of l, and δ'_l goes

to zero as l goes to ∞. Thus the equations (**XII.2.5**)–(**XII.2.7**) hold, and, from (**XII.2.5**), we get

$$A_l' - 1 = O(l^{-1}), \qquad \delta_l' = O(l^{-1}). \tag{XII.5.6}$$

It remains to prove that the A_l' and δ_l' are equal to A_l and δ_l.

Proof. If A_l' and δ_l' are not A_l and δ_l, it follows from (**XII.2.5**) that:

$$\Gamma_l e^{i\gamma_l}\left(1 + \frac{1}{2}\frac{\pi c_l}{2l+1}\right) = i\sum_{l'} M_{ll'} c_{l'} \Gamma_{l'} e^{i\gamma_{l'}}, \tag{XII.5.7}$$

where

$$\Gamma_l e^{i\gamma_l} = A_l e^{i\delta_l} - A_l' e^{i\delta_l'}. \tag{XII.5.8}$$

Since (**XII.5.6**) also holds for unprimed quantities, Γ_l and γ_l are $O(l^{-1})$ as $l \to \infty$. From (**XII.5.7**) it follows that

$$\sum_{l'} M_{ll'} c_{l'} \Gamma_{l'} \cos\gamma_{l'}(1 + \tan\gamma_l \tan\gamma_{l'}) = 0. \tag{XII.5.9}$$

According to the analysis of section XII.3, this is impossible unless $\Gamma_l = 0$. Hence $A_l = A_l'$ and $\delta_l = \delta_l'$. Q.E.D.

Thus, the method is consistent provided that $\{\delta_l\}$ is not "exceptional" and "$\mathscr{D}(\infty)$ is not zero." Since the latter condition can be achieved only for "exceptional sets $\{c_l\}$" (it is somewhat similar to the condition for phase-shifts), *the method is consistent for almost every set of phase shifts satisfying the condition* (**XII.4.6**). It yields one (only one) potential which decreases faster than $r^{-3/2}$ (the one with $\alpha = \beta$).

XII.6   Generalized Matrix Methods

Direct Generalizations. Let $\mathscr{M}$ be any set of numbers in the half plane $\mathrm{Re}\,\mu > -\frac{1}{2}$, symmetric with respect to the real axis (so as to get real functions on the real axis). Provided that "everything converges," the function $\sum_{\mu \in \mathscr{M}} c_\mu u_\mu(r) u_\mu(r')$ can be used as an input function $f(r, r')$ (it is understood that the summation may of course be replaced in certain subsets by an integration). A system like (**XII.2.4**) is obtained. The functions again are replaced by their asymptotic behavior:

$$\varphi_\mu(r) \sim (2i)^{-1}[f_{1,\mu} e^{ir} - f_{2,\mu} e^{-ir}], \tag{XII.6.1}$$

$$L_{\lambda,\mu}(r) \to L_{\lambda,\mu}(\infty) = \frac{\sin[(\lambda - \mu)\pi/2]}{(\lambda + \frac{1}{2})^2 - (\mu + \frac{1}{2})^2}. \tag{XII.6.2}$$

If $\mathscr{M}$ contains an infinite real subsequence $\{\mu_n\}$, a "matrix method," is obtained by choosing c_μ for $\mu \in \mathscr{M}$, $\mu \notin \{\mu_n\}$, and trying to solve the infinite system which is obtained:

$$A_l e^{i\delta_l} = 1 - \sum_{\substack{\mu \in \mathscr{M} \\ \mu \notin \{\mu_n\}}} L_{l,\mu} c_\mu f_{1,\mu} e^{il(\pi/2)} - \sum_{n=0}^{\infty} L_{l,\mu_n} c_{\mu_n} A_{\mu_n} e^{i[\delta_{\mu_n} + (l-\mu_n)\pi/2]}. \tag{XII.6.3}$$

Two of these generalizations (Sabatier 1967a) have some interest. In the first one, $\mathcal{M}$ involves integers and half integers. Setting $\mu_n = n$, multiplying both sides of (**XII.6.3**) by $e^{-i\delta_l}$, and taking the imaginary part of the result yields a system of matrix equations which can be solved like the one appearing in section XII.2. In the second generalization, $\mathcal{M}$ involves only half-integers, and we set $\mu_n = n + \frac{1}{2}$. The systems can be solved if the phase shifts are small enough. To see the interest of these generalization, let us assume that $rV(r)$ is analytic in the disc $|r| \leq R$. Using the Frobenius method to derive bounds for the Taylor expansion of $r^{-l-1}\varphi_l(r)$, it is possible to show that $K(r, r)$ has the following *unique* expansion, which converges for $|r| < R$:

$$K(r, r) = \sum_0^\infty c_l \varphi_l(r)u_l(r) + \sum_0^\infty c_{l+1/2} \varphi_{l+1/2}(r)u_{l+1/2}(r). \qquad \textbf{(XII.6.4)}$$

Besides, if $V(r)$ is even, c_l is zero for all l, and the function $r^{-1/2}\varphi_{l+1/2}(r)$ are even, like $r^{-1/2}u_{l+1/2}(r)$. Thus, the matrix method described in sections XII.2–XII.5 cannot give even potentials, whereas the method with only the $c_{l+1/2}$'s gives *only* even potentials. This does not mean, however, that a matrix method using both c_l's and $c_{l+1/2}$'s would yield *all* meromorphic potentials: the condition "$rV(r)$ analytic" does not imply in any way that the condition (**XII.2.2**) is fulfilled and that the asymptotic system has a meaning at all (see section XII.9 below). Thus we see there is an infinite number of equivalent potentials. Generalizations giving access to potentials that are analytic functions of r^α are studied in a quite similar way.

Further Generalizations. From the results of section XI.3, we know that the essential property of an "input function" is that it solves (**XI.3.1**) and (**XI.3.2**). Cox and Thompson (1969, 1970a,b) proposed including terms of the form

$$g(r, r') = \sum_{l \in S} \gamma_l u_l(r_<)v_l(r_>) \qquad \textbf{(XII.6.5)}$$

in $f(r, r')$, where $v_l(r)$ is the irregular spherical Bessel function $(\frac{1}{2}\pi r)^{1/2} Y_{l+1/2}(r)$, and the notation $><$ has been defined in (**II.1.9**). S is a finite set of real numbers $> -\frac{1}{2}$, and the γ_l's are real numbers. One readily sees that $g(r, r')$ satisfies all the required properties except maybe for $r = r'$, where there can be a derivative discontinuity and the equations need a particular study. Cox and Thompson made it in the most difficult case where the input function is $g(r, r')$ alone. They were able to show that equation (**XII.1.2**) still has a solution $K(r, r')$, with the expansion

$$K(r, r') = \sum_{L \in T} A_L(r)u_L(r'), \qquad \textbf{(XII.6.6)}$$

where T is a set of real numbers $(T \cap S = \emptyset)$, T and $\{\gamma_l\}$ satisfy the consistency condition:

$$\sum_{l \in S} [l(l + 1) - L(L + 1)]^{-1}\gamma_l = 1 \qquad (L \in T) \qquad \textbf{(XII.6.7)}$$

and $A_L(r)$ is obtained by solving the system:

$$\sum_{L \in T} \frac{W[u_L(r), v_l(r)]}{l(l + 1) - L(L + 1)} = v_l(r) \qquad (l \in S), \qquad \textbf{(XII.6.8)}$$

where W is the Wronskian.

It is possible to show that the function $\varphi_l(r)$ which is constructed from $K(r, r')$ by (XII.1.1) is actually a regular solution of the Schrödinger equation, but with a particular normalization.

$g(r, r')$ can be used to construct examples of potentials for which the Schrödinger equation (at fixed energy) is exactly solved. It can also be used to augment $f(r, r')$. An interesting property is that $\int_0^\infty \rho V(\rho) d\rho$ can then be different from zero even if the number of terms in the input function is *finite*, (a property which does not hold for the direct generalizations we gave above). However, this is paid for by the greater complication. The complete solution given in chapter XIII is in some way a generalized matrix method.

XII.7   Miscellaneous Results

Transparent Potentials. For $\{\delta_l\} = \{0\}$, the inversion methods we have studied in the preceding sections yield a potential V equal to zero and other potentials which are not. They are called "transparent potentials" (Sabatier 1966b). Suppose for instance we make $\tan \Delta = 0$ in (XII.4.1). Solving the matrix equations readily yields $c_{2l+1} = 0$ and

$$c_{2l} = \tfrac{1}{2}\alpha v_{2l}\left[1 - \frac{1}{4}\pi\alpha\frac{v_{2l}}{4l + 1}\right]^{-1}. \qquad \textbf{(XII.7.1)}$$

According to (XII.5.5), the corresponding "transparent potential" has an asymptotic tail decreasing like $r^{-3/2}$.

One-Term Examples. In section XI.4, we have seen how solvable examples can be constructed from a function $f(r, r')$ consisting of one term $c_\mu u_\mu(r) u_\mu(r')$ (Newton 1962, Sabatier 1966a,c, 1967a). A similar study can be made with an input function $\gamma_\nu u_\nu(r_<) v_\nu(r_>)$ (Cox and Thompson 1969, 1970a). It is remarkable that, in both cases (Sabatier 1966c, Cox and Thompson 1969, 1970a), the scattering amplitudes can be calculated in a closed form. They yield somewhat similar but different results (see Cox and Thompson 1970a). Here we give only the simple example $f(r, r') = c_L u_L(r) u_L(r')$ with L being an even integer

$$f(\theta) = \frac{a\pi}{2\cos\pi b}[P_{b-1/2}(\cos\theta) - P_{b-1/2}(-\cos\theta)], \qquad \textbf{(XII.7.2)}$$

where

$$a = c_L[1 + \pi c_L(4L + 2)^{-1}]^{-1},$$
$$b = [(L + \tfrac{1}{2})^2 - ia]^{1/2}. \qquad \textbf{(XII.7.3)}$$

First Moment of the Potential. In the method which has been described in sections XII.2–XII.5, the leading asymptotic behavior of $K(r, r)$ as r goes to ∞ is that of $f(r, r)$, which is given by (**XII.5.2**). Using (**XI.2.7**) we obtain the formula (Sabatier 1967a)

$$\int_0^\infty rV(r)dr = -(\alpha + \beta). \qquad \text{(XII.7.4)}$$

The first moment is therefore zero if α and β are. Now, α and β appear in the asymptotic behavior (**XII.4.22**) of the c_i's. More generally, Newton (1967) proved that the first moment is zero if the coefficients c_μ in the generalized expansion $\sum c_\mu u_\mu(r)u_\mu(r')$ go to zero too rapidly when μ goes to ∞. In numerical computations, it is a good idea to take care of this property—for instance by using closed form for a part (index zero) of the infinite series, with $c_{2l}^{(0)}$ or $c_{2l+1}^{(0)}$ going to the definite constants $\lim c_{2l}$ or $\lim c_{2l+1}$, and making a cutoff only in the remainder.

Expansions of the Jost Functions. We use the notations (**XI.4.10**) and introduce the Jost functions $f_1(\lambda)$ and $f_2(\lambda)$ which are normalized in such a way that

$$\varphi_\lambda^V(r) \sim (2i)^{-1}[f_1(\lambda)e^{ir} - f_2(\lambda)e^{-ir}]. \qquad \text{(XII.7.5)}$$

Now let S be a sequence of real positive numbers and suppose $f(r, r') = \sum_{\mu \in S} \gamma_\mu(v, 0)u_\mu(r)u_\mu(r')$. It is easy to derive from (**XII.6.3**) the following expansion of the Jost function (Sabatier 1966c, 1967a, Newton 1967):

$$f_{1,2}(\lambda) = \exp[\pm i\pi(\lambda - \tfrac{1}{2})]$$

$$- \sum_{\mu \in S} \frac{\sin[(\lambda - \mu)\pi/2]}{\lambda^2 - \mu^2} \gamma_\mu(V, 0)f_{1,2}(\mu). \qquad \text{(XII.7.6)}$$

Newton (1967) showed that the nature of S, and the singularity of this expansion can be related to the set of singularities of the Mellin transform of the potential ($\int_0^1 x^{\sigma-1}V(x)dx$). Besides, he interpreted (**XII.7.6**) as a special dispersion formula, and gave conditions that are sufficient to guarantee its existence. The Jost functions given by the Cox and Thompson generalizations have similar properties (Cox and Thompson 1970a).

Methods Starting with a Potential $V_0 \neq 0$. We use the notations (**XI.4.10**), a reference potential $V_0 \neq 0$, and we define the input function by a series:

$$f_{V_0}^V(r, r') = \sum_{\mu \in S} \gamma_\mu(V, V_0)\varphi_\mu^{V_0}(r)\varphi_\mu^{V_0}(r'). \qquad \text{(XII.7.7)}$$

If the coefficients $\gamma_\mu(V, V_0)$ are bounded as in (**XII.2.2**):

$$\gamma_\mu(V, V_0) = o(\mu) \qquad (\mu \to \infty) \qquad \text{(XII.7.8)}$$

and if the wave functions corresponding to the reference potential have the asymptotic behavior (**XI.4.7**), with $\delta_\mu^{V_0} \to 0$ and $|A_\mu^{V_0}| = O(1)$ when μ goes to infinity, then it is possible to make $r \to \infty$ in (**XII.7.11**), again obtaining an infinite set of coupled equations (Newton 1967). Solving this set in terms of $(\delta_\mu^V - \delta_\mu^{V_0})$ reduces to inverting M and some other infinite matrices, which depend on the phase shifts of the reference potential. Clearly there cannot be convergence difficulties when $\delta_\mu^{V_0}$ and δ_μ^V are $O(\mu^{-3-\varepsilon})$ since the problem could also be solved directly by the methods given above in two steps (thus obtaining a potential that is *equivalent* to $V_0 + V$). The most interesting case would be the one where V_0 is a pure Coulomb potential. As we see below (equation **XII.8.13**) such a potential corresponds to γ_μ's that are defined for integer μ and $\mu - \frac{1}{2}$, so that the matrix methods could be introduced very naturally. Unfortunately, these γ_μ's are not bounded (they are $O(\mu)$ as $\mu \to \infty$). and the $\delta_l^{V_0}$'s go to zero slowly as l goes to infinity (they are $O(l^{-1})$), so that the convergence questions in the inversions must be carefully checked. The arguments given by Coudray and Coz (1970) are not convincing.

XII.8   Interpolation Properties

The interpolation formulas introduced by Sabatier (1966d, 1967b) give expansions of the wave functions or of the Jost functions for any λ in terms of their values for $\lambda \in S$, where S is a certain sequence $\{s_n\}$ of real numbers. It turns out that the coefficients are trivially related to the γ_μ's. The formulas can be derived for any potential for which $rV(r)$ is an analytic function of r^α (α real positive) in a disc centered at the origin.

Fine Analytic Properties of the Wave Functions in the λ-Plane. We study here the example $\alpha = 1$, using the notations (**XI.4.10**). We assume that $rV(r)$ and $rV_0(r)$ are analytic in the disc $|r| \leq R$. The expansion (**XII.6.4**) reads:

$$K_V^W(r, r) = \sum_{\mu \in S} \gamma_\mu(V, V_0)\varphi_\mu^V(r)\varphi_\mu^{V_0}(r), \qquad \text{(XII.8.1)}$$

where S contains positive integers and half integers. Using the Frobenius method to solve the Schrödinger equation, it is possible to show that for any $r < R$, both for V and V_0, $r^{-\mu-1/2}\varphi_\mu(r)$ is analytic, and its Taylor coefficients φ_μ^p are bounded in the following way (Sabatier 1967a)

$$|\varphi_\mu^p/\varphi_\mu^0| \leq M(\mu + p)^{-1}R^{-p}, \qquad \text{(XII.8.2)}$$

where M depends only on R. Hence

$$|(\tfrac{1}{2}r)^{-\mu-1/2}\varphi_\mu(r)\Gamma(1 + \mu) - \pi^{1/2}| = o(1), \qquad \mu \to \infty. \qquad \text{(XII.8.3)}$$

The coefficient $\gamma_\mu(V, V_0)$ are obtained by comparing the Taylor expansions of both sides of (**XII.8.1**). Hence they are unique (incidentally, one readily sees that $\gamma_\mu(V, V_0) = -\gamma_\mu(V_0, V)$, see eq. (**XI.5.15**)), and it follows from (**XII.8.2**) that

$$|\gamma_\mu(V, V_0)| = O(|(\Gamma(1 + \mu))^2(\tfrac{1}{2}R')^{-2\mu}|) \qquad (\mu \to \infty), \qquad \text{(XII.8.4)}$$

where R' can be chosen arbitrarily in $[0, R[$.

The formula (**XI.4.13**) takes the form:

$$\varphi_\lambda^V(r) = \varphi_\lambda^{V_0}(r) - \sum_{\mu \in S} \Lambda_{\lambda\mu}^{V_0}(r)\varphi_\mu^V(r)\gamma_\mu(V, V_0). \qquad \textbf{(XII.8.5)}$$

Now, suppose that $V_0 = 1$. From (**XI.1.6**) and (**XI.1.7**), we get

$$\varphi_\lambda^1(r) = (\tfrac{1}{2}\pi r)^{1/2}[\Gamma(1 + \lambda)]^{-1}(\tfrac{1}{2}r)^\lambda. \qquad \textbf{(XII.8.6)}$$

Using (**XI.4.14**) to derive $\Lambda_{\lambda\mu}^1(r)$, and substituting into (**XII.8.5**) yields the formula:

$$[\Gamma(-\lambda)]^{-1}\left(\frac{r}{2}\right)^{-\lambda}(\tfrac{1}{2}\pi r)^{-1/2}\varphi_\lambda^V(r)$$

$$= \frac{\sin \pi\lambda}{\pi}\left\{-1 + \sum_{\mu \in S} (\tfrac{1}{2}\pi)^{1/2}(\tfrac{1}{2}r)^\mu[\Gamma(1 + \mu)]^{-1}(\lambda + \mu)^{-1}\gamma_\mu(V, 1)\varphi_\mu^V(r)\right\}.$$

$$\textbf{(XII.8.7)}$$

It follows from (**XII.8.3**) and (**XII.8.4**) that (**XII.8.7**) can be used to continue $\varphi_\lambda^V(r)$ is the λ plane, as a meromorphic function with simple poles at the point $-s_n$ except for integer s_n. Besides, from the value of $\Lambda_{-q, q}^1(r)$ (which is equal to $-(\pi/2q)(-1)^q$) and that of $\Lambda_{-q, q}^0(r)$, using (**XII.8.5**), it is easy to derive the "reflection" formulas:

$$\varphi_{-q}^V(r) = (-1)^q \frac{\pi}{2q} \gamma_q(V, 1)\varphi_q^V(r)$$

$$= (-1)^q\left[1 + \frac{\pi}{2q} \gamma_q(V, 0)\right]\varphi_q^V(r). \qquad \textbf{(XII.8.8)}$$

Interpolation Formulas. Now the keys for constructing interpolation formulas are

(a) the Lagrange–Valiron theorem: (Boas 1954)

   Theorem. *Any entire function $f(z)$ of order 1 and type τ which is bounded on the real axis has the representation*:

$$f(z) = \left\{\frac{f'(0)}{\tau} + \frac{f(0)}{\tau z} + \tau z \sum_{\substack{n \neq 0 \\ n = -\infty}}^{n = +\infty} \frac{(-1)^n f[n\pi/\tau]}{n\pi(\tau z - n\pi)}\right\}\sin[\tau z]. \qquad \textbf{(XII.8.9)}$$

(b) a remark which follows from the results given above: if $V(r)$ is analytic in the disc $|r| < R$, then $\bar{\varphi}_\lambda^V = \cos[\pi\lambda]\varphi_\lambda^V(r)$ is entire and satisfies the reflection properties:

$$\bar{\varphi}_{-q}^V = (-1)^q\gamma_q(V, 1)\bar{\varphi}_q^V(r),$$

$$\bar{\varphi}_{q+1/2}^V(r) = 0, \qquad \textbf{(XII.8.10)}$$

$$\bar{\varphi}_{-q-1/2}^V(r) = -[\tfrac{1}{2}\pi(q + \tfrac{1}{2})^{-1}]\gamma_{q+1/2}(V, 0)\varphi_{q+1/2}^V(r).$$

From **(XII.8.7)** and the bounds **(XII.8.3–XII.8.4)**, it is easy to see that $\lambda \cos(\pi\lambda)\varphi_\lambda^V(r)\varphi_{-\lambda}^{V_0}(r)$ is an entire function of order 1, type 2π, bounded on the real azis. Applying then **(XII.8.8)** and using **(XII.8.10)**, we obtain the interpolation formula:

$$\frac{\pi\lambda}{\sin\pi\lambda}\,\varphi_\lambda^V(r)\varphi_{-\lambda}^{V_0}(r) = \varphi_0^V(r)\varphi_0^{V_0}(r)$$

$$+ \sum_1^\infty \varphi_n^V(r)\varphi_n^{V_0}(r)\frac{\pi}{2n}\left[\frac{\lambda}{\lambda-n}\gamma_n(V_0,1) + \frac{\lambda}{\lambda+n}\gamma_n(V,1)\right]$$

$$+ \sum_1^\infty \varphi_{n-1/2}^V(r)\varphi_{n-1/2}^{V_0}(r)\frac{\pi}{2n}\left[\frac{\lambda}{\lambda-n+\frac{1}{2}}\gamma_{n-1/2}(V_0,1)\right.$$

$$\left.+ \frac{\lambda}{\lambda+n-\frac{1}{2}}\gamma_{n-1/2}(V,1)\right]. \tag{XII.8.11}$$

Interesting particular cases of this formula are all those for which the wave functions are exactly known (i.e., $V = 0, 1, 1 - k^2, 2\eta r^{-1}$). Together with the additivity formula **(XI.5.14)**, which was in fact originally proved from **(XII.8.11)** (Sabatier 1967b), it gives us a way to calculate, for instance

$$\gamma_\mu(0,1) = \begin{cases} \dfrac{2\mu}{\pi} & \text{for positive integer } \mu, \\[2mm] 0 & \text{otherwise.} \end{cases} \tag{XII.8.12}$$

(Thus, notice in passing that $\gamma_{n-1/2}(V_0,1) = \gamma_{n-1/2}(V_0,0)$). Also

$$1 + \frac{\pi}{2p}\gamma_p(2\eta r^{-1},0) = \frac{\Gamma(p+\frac{1}{2}-i\eta)\Gamma(p+\frac{1}{2}+i\eta)}{\Gamma(p+\frac{1}{2})\Gamma(p+\frac{1}{2})}\cosh(\pi\eta);$$

$$\frac{\pi}{2p-1}\gamma_{p-1/2}(2\eta r^{-1},0) = -\frac{\Gamma(p+i\eta)\Gamma(p-i\eta)}{\Gamma(p+\frac{1}{2})\Gamma(p+\frac{1}{2})}\sinh(\pi\eta). \tag{XII.8.13}$$

From **(XII.8.2)** and **(XII.8.3)** we see that the coefficients c_μ which in a "matrix method" would give either the constant or the Coulomb potential go to ∞ like l when l goes to infinity. Byproducts of **(XII.8.11)** are interpolation formulas for Bessel and Laguerre functions, e.g.,

$$J_\lambda(r)J_{-\lambda}(r) = \frac{\sin\pi\lambda}{\pi\lambda}\left\{J_0^2(r) + \sum_1^\infty J_n^2(r)(\lambda^2-n^2)^{-1}\lambda^2\right\} \tag{XII.8.14}$$

and their generalizations, and also the values, in particular cases, of $\mathbf{K}_1^V(r,r')$, which appears as a generating function of the φ_l^V's. **(XII.8.7)** can itself be obtained as a byproduct.

Assume now not only that the potential is analytic in a disc but also that the coefficients $\gamma_\mu(V,0)$ satisfy bounds such that the series $\sum \mu^{-2}|\gamma_\mu|$ converges. Then the Jost function is given by the formula **(XII.7.6)**. From this formula, it follows that $\bar{f}_1(\lambda) = \cos[\pi\lambda]f_1(\lambda)$ is an entire function of order 1, and type $\frac{3}{2}\pi$, bounded on the real axis, which satisfies the additional properties:

(a) Let $t(\lambda)$ be the ratio $\bar{f}_1(-\lambda)/f_1(\lambda)$. For any positive integer, $t(q) - 1$ and $t(q + \frac{1}{2})$ are real, and go to zero as q goes to infinity in such a way that the series $\sum q^{-1}|t(q) - 1|$ and $\sum q^{-1}|t(q + \frac{1}{2})|$ converge.

(b)

$$\lim_{|\mathrm{Im}\ \lambda| \to \infty} \frac{f_1(\lambda) + f_1(-\lambda)}{2\cos(\lambda\pi/2)} = i \lim_{|\mathrm{Im}\ \lambda| \to \infty} \frac{f_1(\lambda) - f_1(-\lambda)}{2\sin(\lambda\pi/2)} = e^{i\pi/4}.$$

The Lagrange–Valiron theorem and some algebra enable us to show that, conversely, a function $f_1(\lambda)$ which satisfies the above property necessary satisfies a formula like (XII.7.6). Since the condition on the γ_μ's is essentially the condition (XII.2.2) which makes possible the matrix methods, we arrive at this very important conclusion:

In the class of analytic potentials, the matrix methods are defined for those potentials for which the (dynamical) interpolation of the Jost function is an entire function of exponential type.

All the interpolation formulas are easily extended to potentials which are analytic functions of r^α (with rational α). A JWKB study, and an elementary derivation of the interpolation formulas (see chapter XV) are also available (Sabatier 1967b).

XII.9   Limitation of the Matrix Methods

The matrix methods have the remarkable property of providing a way to easily construct a potential which fits the cross section exactly. Their limitations have been described in several places in this chapter. Let us recapitulate them:

1. For all matrix methods, the c_μ's must satisfy bounds equivalent to (XII.2.2). Hence, $f(r, r)$ is holomorphic at any finite point of the complex plane except for a possible branch point at $r = 0$. Furthermore $|f(r, r)|$ cannot have an arbitrary asymptotic behavior in the r plane. For $|r| \to \infty$, the functions $u_\mu(r)$ increase at most like $\exp|r|$. Using recurrence relations (like $J_{\nu-1} + J_{\nu+1} = 2\nu r^{-1}J_\nu$), it is easy to see that if $\sum \mu^{-2}|c_\mu|$ is finite, $|f(r, r)|$ is $O(|r|^2 \exp[2|r|])$. Thus, when the allowed values of μ are the ones for which 2μ is an integer, $f(r, r)$ is an entire function whose exponential type is 2—obviously a special behavior.

2. The Jost functions have special dynamical interpolations. As we have seen, in the case of analytic potential, this interpolation involves an entire function of exponential type, divided by $\cos \pi\lambda$. It is possible to show (Sabatier 1967b) that such an interpolation actually excludes

potentials of the Yukawa class $(\int_{\mu_0}^{\infty} C(\mu)\exp[-\mu r]d\mu)$, while these potentials are very interesting for physical reasons.

3. If we restrict ourselves to the method in which only integer l come in, we have shown that this method cannot even lead to even analytic potentials.

Obviously, the reason for which these methods are limited is that they have been introduced *to construct* potentials, and the classes in which these potentials can be constructed have been discovered afterwards. This leads us to the studies of the next chapter.

Historical Notes. The methods using expansions in products $c_\mu u_\mu(r)u_\mu(r')$ are sometimes called in the literature "Newton–Sabatier" methods after Newton (1962) who introduced the fundamental inversion procedure of section XII.2 and showed that M has a nonunique inverse M^{-1}, and Sabatier (1966a,b,c 1967a), who constructed M^{-1} and **v** (section XII.3), derived the properties of the potentials and of the Jost functions (sections XII.4, XII.5), proved the consistency of the method (section XII.5), and gave certain generalizations (section XII.6). Cox and Thompson (1970a,b) gave alternative generalizations. The miscellaneous properties of sections XII.7, XII.8, are essentially due to these four authors (they appear in the text with their references). A mathematical adjunction to the study of M has also been given by Redmond (1964).

Modifications of the matrix methods have been devised for numerical applications and will be studied in chapter XIX.

CHAPTER XIII

Potentials from the Scattering Amplitude at Fixed Energy: Operator Methods

XIII.1 Introduction

In matrix methods, the goal was to construct potentials in a large param-etrized class of functions, then to identify this class. A different point of view is to define a certain large class of functions *a priori* and then try to answer the questions of section XI.1 (existence, uniqueness, construction ...) in this class. In the first kind of approach, one risks obtaining potentials that violate certain fine physical requirements. In the second kind, one risks obtaining poor or very partial answers to the question C and the following questions (constructibility, stability ...). This remark is verified in the two first approaches given in this chapter. In the third approach, the combination of the two points of view enables one to answer all questions, but the scattering equivalent classes are large, and identifying special physical classes is still a problem.

All these approaches have analogous mathematical foundations: one introduces a function (1st and 3rd approach) or a set (2nd and 3rd approach), whose relation to the Schrödinger operator and to the scattering amplitude are both well described. Thus, founding the method is choosing and describ-ing this mathematical object for a Schrödinger operator in which V belongs to the chosen class.

XIII.2   Method for Potentials of the Yukawa Class

A. Martin and Gy. Targonski (1961) looked for a solution of the problem in the Yukawa class of potentials:

$$V(r) = r^{-1} \int_{\mu}^{\infty} C(\alpha)\exp[-\alpha r]d\alpha, \qquad \text{(XIII.2.1)}$$

where $\mu > 0$, and $C(\alpha) \in L_1(\mu, \infty)$. They proved that if V is in this class, the scattering amplitude has a unique analytic continuation for complex angles, that can be constructed from the discontinuity across the cut in the $\cos \theta$ plane, and it is possible to derive from it the potential. Let us now give some details.

The mathematical tool is the analytic continuation of the scattering amplitude $T(t)$ in the complex plane of momentum transfer:

$$t = -2(1 - \cos \theta). \qquad \text{(XIII.2.2)}$$

(Remember that units are chosen so that $k = 1$). The physical values of t belongs to $[-4, 0]$. For a potential satisfying (XIII.2.1), $T(t)$ is analytic in the whole complex plane, except along a cut on the real axis from $t = \mu^2$ to $t = +\infty$ (figure XIII.1). Thus $T(t)$ can be continued in a unique way in the whole complex t plane. In particular, μ itself and the discontinuity of $T(t)$ across the cut are *uniquely* defined when $T(t)$ is known in the physical region. A way of obtaining this discontinuity $2i\pi D(t)$ is to try to solve the following Fredholm equations of the first kind:

$$T(t) = \int_{\mu^2}^{\infty} \frac{D(t')dt'}{t' - t}, \qquad \text{(XIII.2.3)}$$

$$T(t) = a_0 + a_1 t + \cdots + a_{n-1}t^{n-1} + t^n \int_{\mu^2}^{\infty} \frac{D(t')dt'}{(t' - t)t'^n}, \quad \text{with} \quad -4 \le t \le 0,$$

and where n has to be finite according to Regge's work. If one of these equations has a solution, those with larger number of subtractions give exactly

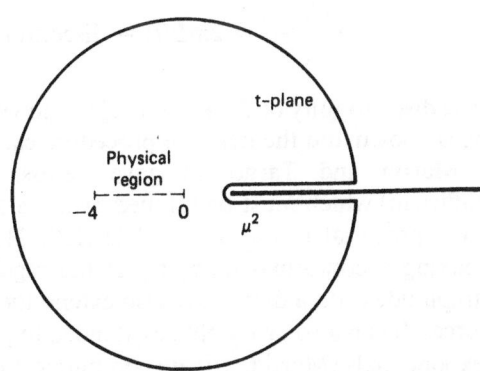

Figure XIII.1

the same solution. Here it is obvious that one needs to know $T(t)$ with infinite accuracy to draw such a conclusion. Otherwise a fit by a polynomial in t, with $D(t) = 0$, is always possible when only n values of $T(t)$ are known.

Once the discontinuity has been obtained, the potential can be reconstructed. Writing the scattering amplitude as

$$T = T_1 + \cdots + T_n + R_n, \qquad \text{(XIII.2.4)}$$

where T_n is the nth Born term, it is possible to show that the discontinuity across the cut in the region $\mu^2 < t < (n + 1)^2\mu^2$ comes from the n first Born terms only. For a given n let us split the potential into the parts:

$$V = \sum_{l=1}^{n} V_l + V_{n+1}, \qquad \text{(XIII.2.5)}$$

where

$$V_l = r^{-1} \int_{l\mu}^{(l+1)\mu} C(\alpha)\exp[-\alpha r]d\alpha, \qquad l \le n,$$

$$\text{(XIII.2.6)}$$

$$V_{n+1} = r^{-1} \int_{(n+1)\mu}^{\infty} C(\alpha)\exp[-\alpha r]d\alpha.$$

In the region $\mu^2 < (n + 1)^2\mu^2$, it is possible to show that:

1. V_{n+1} does not contribute to the discontinuity.
2. V_n contributes to the discontinuity through the first Born term T_1 and nowhere else; the discontinuity of T_1 is most easily related to $C(\alpha)$ since:

$$T_1(t) = \int \frac{C(\alpha)d\alpha}{t - \alpha^2}; \qquad T_1(t + i\varepsilon) - T_1(t - i\varepsilon) = i\pi \frac{C(\sqrt{t})}{\sqrt{t}}. \qquad \text{(XIII.2.7)}$$

Thus, $C(\alpha)$ can be constructed by an iterative procedure: assume $C(\alpha)$ is known from $\alpha = \mu$ to $\alpha = n\mu$, then $V_1, \ldots, V_{n-1}$ are known, and, in the region $n^2\mu^2 < t < (n + 1)^2\mu^2$, we have

$$\pi i \frac{C(\sqrt{t})}{\sqrt{t}} = 2\pi i D(t) - \text{discontinuity of } (T_2 + \cdots + T_n).$$

The discontinuity of $T_2 + \cdots + T_n$ is known in terms of $V_1, \ldots, V_{n-1}$. Hence V_n is known and the iterative procedure can go on.

Martin and Targonski gave necessary conditions (not necessarily sufficient) which must be fulfilled by a scattering amplitude to be produced by a potential in the family (**XIII.2.1**). These conditions are obtained by making a conformal mapping of the regularity domain of the scattering amplitude onto a disc. They also extend the method to the case of exchange forces. It can also probably be extended to potentials bounded by decreasing exponentials (Martin, private communication to PCS).

Clearly, numerical computations are not feasible. Besides, since the method involves an analytic continuation (or solving **XIII.2.3**), it is in general not stable with respect to experimental or round-off errors (Viano 1969, 1976).

In connection with the Martin and Targonski approach, it is interesting to quote the works by Charap and Fubini (1959, 1960).

XIII.3   Methods Using the Spectrum of the Schrödinger Operator

XIII.3.1   Regge–Loeffel Approach

Introduced by Regge (1959), and thoroughly studied by Loeffel (1968), this method uses the machinery described in chapter XI, with $V_0 = 1$. This means that the unperturbed equation reads:

$$r^2 \frac{d^2}{dr^2} \varphi_l^1(r) = l(l + 1)\varphi_l^1(r) \qquad \text{(XIII.3.1)}$$

and φ_l^1 is simply constant times r^{l+1}. Now, let $r\phi_V(l + \frac{1}{2}, r)$ be the solution of **(XI.1.2)** which behaves like r^{l+1} at $r = 0$

$$\phi_V(l + \tfrac{1}{2}, r) = \pi^{-1/2}\Gamma(l + \tfrac{3}{2})2^{l+1}r^{-1}\varphi_{l+1/2}^V(r). \qquad \text{(XIII.3.2)}$$

Let $\mathcal{O}$ be the class of locally integrable functions such that there exists a positive number ε for which:

$$\|V\|_{\mathcal{O}} = \int_0^1 r^{1-\varepsilon}|V(r)|dr + \int_1^\infty |V(r)|dr < \infty. \qquad \text{(XIII.3.3)}$$

Transformation Kernel. Using methods of functional analysis, Loeffel (1968) was able to derive analytic properties and bounds for ϕ such that the functions $f_{ro}(\lambda + \varepsilon)$ and $f_r(\lambda + \varepsilon)$, corresponding to a potential V in φ

$$f_{ro}(\lambda + \varepsilon) = (2\lambda)^{-1}\int_0^r u[V(u) - 1]\left[1 - \left(\frac{u}{r}\right)^{2\lambda}\right]du,$$

$$f_r(\lambda + \varepsilon) = r^{-\lambda+1/2}\phi_V(\lambda, r) - 1 - f_{ro}(\lambda + \varepsilon) \qquad (r > 0, \operatorname{Re}\lambda \geq 0, \varepsilon > 0)$$

$$\text{(XIII.3.4)}$$

belong to the class $\mathcal{H}$, the set of all functions f which are holomorphic for $\operatorname{Re}\lambda > 0$ and such that there exists a number M (depending on f), with

$$\int_{-\infty}^{+\infty} |f(\sigma + i\tau)|^2\, d\tau \leq M \qquad \text{for all } \sigma > 0. \qquad \text{(XIII.3.5)}$$

According to a theorem of R. Paley and N. Wiener (1934), $\mathcal{H}$ is identical with the set of the Laplace transforms $\int_0^\infty e^{-\rho\lambda}g(\rho)d\rho$ of the functions g of

$L_2(0, \infty)$. Hence, if g_{ro} and g_r are the functions associated with f_{ro} and f_r by this Laplace transform, and if

$$h_r(\rho) = g_{ro}(\rho) + g_r(\rho) \qquad \text{(XIII.3.6)}$$

one obtains, for Re $\lambda > -\varepsilon$:

$$\phi_V(\lambda, r) = r^{\lambda - 1/2} + r^{\lambda - 1/2} \int_0^\infty h_r(\rho) e^{-\rho(\lambda + \varepsilon)} \, d\rho \qquad \text{(XIII.3.7)}$$

or with

$$k_1^V(r, r') = -(rr')^{-1/2} \left(\frac{r'}{r}\right)^\varepsilon h_r \left[\text{Log} \left(\frac{r}{r'}\right) \right] \qquad \text{(XIII.3.8)}$$

for $0 < r' \le r$

$$\phi_V(\lambda, r) = \phi_1(\lambda, r) - \int_0^r k_1^V(r, \rho) \phi_1(\lambda, \rho) d\rho. \qquad \text{(XIII.3.9)}$$

It is possible to show that $(rr')^{-1/2} |k_1^V(r, r')|$ is bounded by $(rr')^\varepsilon C(r)$, where $C(r)$ is bounded for finite r, so that

(a)   for any $\sigma > -\varepsilon$, $\displaystyle\int_0^r |k_1^V(r, \rho)| \rho^{\sigma - 1/2} \, d\rho < \infty$,   (XIII.3.10)

(b)   for any $a > 0$, $k_1^V(r, \rho) \in L_2[(0, a) \times (0, a)]$.   (XIII.3.11)

Clearly, $k_1^V(r, r')$ is the λ-independent transformation kernel which was sought.

Symmetric Kernel. The formula **(XIII.3.9)** can be considered as a Volterra equation for $\phi_1(\lambda, \rho)$. Resolving it yields the "inverse" of the transformation kernel, say, $k_V^1(r, r')$, which is such that

$$\phi_1(\lambda, r) = \phi_V(\lambda, r) - \int_0^r k_V^1(r, \rho) \phi_V(\lambda, \rho) dp. \qquad \text{(XIII.3.12)}$$

From **(XIII.3.11)** it readily follows (see, e.g., Tricomi 1957) that $k_V^1(r, \rho)$ belongs itself to $L_2[(0, a) \times (0, a)]$ for any positive a, and that $k_V^1(r, r')$ and $k_1^V(r, r')$ are associated by

$$k_V^1(r, r') + k_1^V(r, r') = \int_{r'}^r k_1^V(r, \rho) k_V^1(\rho, r') d\rho \qquad (0 < r' \le r). \quad \text{(XIII.3.13)}$$

Now, let us *define* the symmetric operator $\phi_1^V(r, r')$ by:

$$\phi_1^V(r, r') = \begin{cases} -k_V^1(r, r') + \displaystyle\int_0^{r'} k_V^1(r, \rho) k_V^1(r', \rho) d\rho, & 0 < r' \le r, \\[4mm] \phi_1^V(r', r), & 0 < r \le r'. \end{cases} \qquad \text{(XIII.3.14)}$$

From (XIII.3.10), (XIII.3.11), (XIII.3.13) and (XIII.3.14), it is easy to prove that $\phi_1^V(r, r')$ belongs to $L_2[(0, a) \times (0, a)]$ for any positive a and that, for $r \geq r'$

$$(rr')^{1/2}|\phi_1^V(r, r')| \leq (rr')^\varepsilon k(r)[1 + (rr')^\varepsilon k(r')], \qquad \text{(XIII.3.15)}$$

where $k(r)$ is bounded for any finite r.

Fundamental Equation. It is a matter of simple algebraic manipulations, using (XIII.3.13) (and Fubini's theorem to interchange integrations), to prove from (XII.3.14) the following equation $(0 < r' \leq r)$:

$$k_1^V(r, r') = \phi_1^V(r, r') - \int_0^r k_1^V(r, \rho)\phi_1^V(\rho, r')d\rho. \qquad \text{(XIII.3.16)}$$

Clearly, (XIII.3.16) is nothing but the Regge–Newton equation (XIII.3.4). On the other hand, considering the way it has been obtained from the formally self-adjoint differential operator

$$\tau = -\frac{d}{dr}\left(r^2 \frac{d}{dr}\right) - \frac{1}{4} + r^2(V(r) - 1) \qquad \text{(XIII.3.17)}$$

we can also call it a Gel'fand–Levitan equation.

Relation of ϕ_1^V to the Asymptotic Behavior of $\phi(\lambda, r)$. Let $\psi_V^+(\lambda, r)$ and $\psi_V^-(\lambda, r)$ the (Jost) solutions of (XI.1.2), such that

$$\lim_{r \to \infty} \{r \exp[\mp ir]\psi_V^\pm(\lambda, r)\} = 1 \qquad \text{(XIII.3.18)}$$

and let $\psi_V^-(\lambda)$ and $\psi_V^+(\lambda)$ be the Jost functions, defined in such a way that, for $r \to \infty$:

$$\begin{cases} \phi_V(\lambda, r) - \psi_V^-(\lambda)\psi_V^+(\lambda, r) + \psi_V^+(\lambda)\psi_V^+(\lambda, r) + o(r^{-1}), \\ \dfrac{d}{dr}\phi_V(\lambda, r) = \psi_V^-(\lambda)\dfrac{d}{dr}\psi_V^+(\lambda, r) + \psi_V^+(\lambda)\dfrac{d}{dr}\psi_V^-(\lambda, r) + o(r^{-1}). \end{cases}$$

$$\text{(XIII.3.19)}$$

Let $\mathcal{D}_1$ denote the space of the functions f such that

$$f \in C_2(0, \infty) \cap L_2(0, \infty); \qquad \tau(f) \in L_2(0, \infty) \qquad \text{(XIII.3.20)}$$

and let $\mathcal{D} = \mathcal{D}(\theta)$ be the subset of those f in $\mathcal{D}_1$ such that, for a certain real number θ:

$$\lim_{r \to \infty} W(f, r^{-1} \cos(r + \theta)) = 0. \qquad \text{(XIII.3.21)}$$

Let $T(= T(\theta))$ the operator defined on $\mathcal{D}$ by $T(f) = \tau(f)$.

For any potential V belonging to the class $\mathcal{O}$ defined in (XIII.3.3), Loeffel (1968) was able to prove that T is self-adjoint, with a countable set of negative eigenvalues. Then applying a method of Krein, he was able to obtain an expansion theorem for $\phi_1^V(r, r')$ in terms of the regular solutions of **(XI.1.2)**:

Expansion Theorem. For any positive a, and any function $f \in L_2(0, a)$,

$$\int_0^a f(\rho)\phi_1^V(r, \rho)d\rho = \int_0^a f(\rho) \sum_{k=1}^\infty d_k^V \phi_1(\lambda_k^V, r)\phi_1(\lambda_k^V, \rho)d\rho - (2\pi)^{-1} \text{ l.i.m.}_{\omega \to \infty}$$

$$\times \int_{-\omega}^\omega d\tau \gamma_V(i\tau)\phi_1(i\tau, r) \text{ l.i.m.}_{b \to 0} \int_b^a f(\rho)\phi_1(i\tau, \rho)d\rho,$$

(XIII.3.22)

where[1]

(a) $\quad d_k^V = \left(\int_0^\infty |\phi_V(\lambda_k^V, \rho)|^2 \, d\rho \right)^{-1},$ (XIII.3.23)

(b) $\quad \gamma_V(\lambda) = \dfrac{\psi_V^-(-\lambda)\exp[-i\theta] - \psi_V^+(-\lambda)\exp[i\theta]}{\psi_V^-(\lambda)\exp[-i\theta] - \psi_V^+(\lambda)\exp[i\theta]}.$ (XIII.3.24)

The series converges uniformly for (r, r') in any compact subset of $\mathbb{R}^+ \times \mathbb{R}^+$. The continuous function $\gamma_V(i\tau)$ and the sequences d_k^V and λ_k^V, which are readily associated with the spectrum of T, are called "the spectral data." It should be understood that they correspond to the "spectral function" of the Gel'fand–Levitan method which has been studied in the first part of this book. Like the spectral function they completely, and uniquely, characterize the potential in $\mathcal{O}$. Thus, they can constitute this essential link we were seeking between V and the asymptotic behavior of ϕ.

Uniqueness Theorems. Using the above characterization of potentials in $\mathcal{O}$, Loeffel (1968) was able to prove by function-theoretic methods the two following uniqueness theorems:

1. Let V_1 and V_2 be in $\mathcal{O}$. Let ψ_1 and ψ_2 be the corresponding Jost functions. If $\psi_1^-(\lambda) = \psi_2^-(\lambda)$ for Re $\lambda \geq 0$ then $V_1(r) = V_2(r)$ for almost all positive r.
2. Let $s_l = \exp[2i\delta_l]$, and let $\sigma(\lambda)$ be the interpolation of $s_l(s_l = \sigma(l + \frac{1}{2}))$ obtained through the Jost functions (the so-called Regge interpolation). Let now V_1 and V_2 be in the class $\mathcal{O}$. If the corresponding Regge interpolations σ_1 and σ_2 satisfy $\sigma_1(\lambda) = \sigma_2(\lambda)$ for all λ with Re $\lambda > 0$ where both are holomorphic, then $V_1(x) = V_2(x)$ for almost all positive x.

The problem of uniqueness reduces therefore to the step $s_l \to \sigma(l + \frac{1}{2})$. At this point, there can be uniqueness only for particular classes of potentials, those allowing the interpolation to be unique. This holds if the analytic properties of the Jost functions enable one to apply a uniqueness theorem such as Carlson's theorem (Yukawa class) or a form of Lagrange–Valiron's theorem (classes used in the matrix methods).

---

[1] l.i.m. means "limit in mean"

Existence and Construction Problems. If a subclass of $\mathcal{O}$ is chosen, such that the solution is unique in this subclass, and if the Regge interpolation is known, the question of existence can be answered positively in certain cases (Loeffel 1968), and formal answers can be given to the question C.

The best known example is the one of potentials that are truncated at $r = b > 0$:

$$V \in \mathcal{O}, \qquad V(r) = 0 \quad \text{for} \quad r > b > 0. \qquad \text{(XIII.3.25)}$$

For each V, let $\xi(\lambda)$ be the logarithmic derivative of the regular solution at $r = h$:

$$\xi_V(\lambda) = [\phi_V(\lambda, b)]^{-1}\left[\frac{d}{dr}\,\phi_V(\lambda, r)\right]_{r=b}. \qquad \text{(XIII.3.26)}$$

Clearly, $\xi_V(\lambda)$ is bijectively related to the Regge interpolation $\sigma(\lambda)$. Besides, it is easy to see that $\xi_V(\lambda)$ belongs to the class of functions that are meromorphic in Re $\lambda \geq 0$, and of the form:

$$\xi(\lambda) = (\lambda - \tfrac{1}{2})b^{-1} - \sum_0^m b_j(\lambda - \lambda_j)^{-1} + \int_0^1 m(r)r^{\lambda - 1/2}\,dr \qquad \text{(XIII.3.27)}$$

with $b_j \geq 0$ ($j = 0, 1, \ldots, m$), $0 = \lambda_0 < \lambda_1 < \cdots < \lambda_m$, and Im $\xi(i\tau) > 0$ for $\tau > 0$. Let $\mathcal{B}_b$ be this class of functions. A subclass is the set of functions $\xi(\lambda)$ of the form:

$$\xi(\lambda) = (\lambda - \tfrac{1}{2})b^{-1} + \frac{P_{n-1}(\lambda)}{Q_n(\lambda)}, \qquad \text{(XIII.3.28)}$$

where P_{n-1} is a polynomial of degree $\leq n - 1$, Q_n a polynomial of degree n, P_{n-1} and Q_n have no common divisor. Now, it is not difficult, for these truncated potentials, to deduce from (**XIII.3.23**), (**XIII.3.24**), (**XIII.3.19**), (**XIII.3.26**) that the spectral data are homographic functions of $\xi(\lambda)$ with known coefficients, so that they are explicitly given, if $\xi(\lambda)$ is in the class (**XIII.3.28**), by the zeros and the poles of $\lambda \to \xi(\lambda)$ and $\tau \to$ Im $\xi(i\tau)$. Besides, in this case, the Regge–Newton equation then becomes separable.

Heiniger and Loeffel (1973) recalled these results and suggested algorithms using rational approximants $\xi^{(n)}$ of ξ. The convergence of these algorithms is not proved.

XIII.3.2   Approach Using a Complete Set of Jost Solutions

The Regge–Loeffel approach was very similar to those used in the first part of this book for the inverse scattering problem at fixed l. Still more similar, and still more formal, is the following approach, which was described in the paper by Burdet, Giffon, and Predazzi (1965).

For a potential in the class $\mathcal{O}_1$ of functions whose two first absolute moments on $[0, \infty]$ are finite, let $G(\lambda, r, r')$ be the Green's function, which enables us to go from the free radial wave to the actual radial wave. In terms of the functions defined in (section XIII.3.1) $G(\lambda, r, r')$ is given by

$$(rr')^{-1}G(\lambda, r, r') = - \frac{\phi_V(\lambda, r_<)\psi_V^+(\lambda, r_>)}{\psi_V^+(\lambda)}. \qquad \text{(XIII.3.29)}$$

Let $h(r)$ be any function of $L_2(0, \infty)$. For $V \in \mathcal{O}_1$, we know that $\psi_V^+(\lambda, r)$ is an entire function of λ^2, and that $\phi_V(\lambda, r)$ and $\psi_V^+(\lambda)$ are holomorphic functions of λ for Re $\lambda \geq 0$. These analytic properties give us two ways to evaluate the integral

$$I(r) = \int \lambda \, d\lambda \int_0^\infty h(\rho)[-(r\rho)^{-1}G(\lambda, r, \rho)]d\rho \qquad \text{(XIII.3.30)}$$

finally obtaining:

$$h(r) = \int_0^\infty h(r')dr'\Bigg\{\sum_i \psi_V^+(\alpha_i, r)\psi_V^-(\alpha_i, r)M^{-2}(\alpha_i)$$

$$+ \frac{2i}{\pi}\int_0^{i\infty} \frac{\psi_V^+(\lambda, r)\psi_V^-(\lambda, r)}{\psi_V^+(\lambda)\psi_V^+(-\lambda)} \lambda^2 \, d\lambda\Bigg\}, \qquad \text{(XIII.3.31)}$$

where $\{\alpha_i\}$ is the sequence of zeros of $\psi_V^+(\lambda)$ in the right half plane Re $\lambda \geq 0$, and

$$M^2(\alpha_i) = \int_0^\infty [\psi_V^+(\alpha_i, r)]^2 \, dr. \qquad \text{(XIII.3.32)}$$

Clearly, the completeness relation (XIII.3.31) yields the "spectral function" $\rho_V(\rho)$ such that $d\rho_V(\lambda)$ is the distribution multiplying $\psi_V^+(\lambda, r)\psi_V^-(\lambda, r)$ in (XIII.3.31). Using it as was done in the Marchenko method, one finds a transformation formula for the Jost solution:

$$\psi_V^+(\lambda, r) = \psi_{V_0}^+(\lambda, r) + \int_r^\infty \psi_{V_0}^+(\lambda, \rho)B(r, \rho)d\rho, \qquad \text{(XIII.3.33)}$$

where the transformation kernel $B(r, r')$ is λ-independent:

$$B(r, r') = \int \psi_{V_0}^+(\mu, r)\psi_V^+(\mu, r')d[\rho_{V_0}(\mu) - \rho_V(\mu)]. \qquad \text{(XIII.3.34)}$$

It is not difficult to show that $B(r, r')$ obeys a partial differential equation similar to those given in chapter XI and that $V(r)$ is related to $B(r, r)$ by

$$[V_0(r) - V(r)] = 2r^{-1} \frac{d}{dr}[rB(r, r)]. \qquad \text{(XIII.3.35)}$$

Besides, it is possible to introduce a symmetric kernel

$$j(r, r') = \int \psi_{V_0}^+(\mu, r)\psi_{V_0}^+(\mu, r')d[\rho_{V_0}(\mu) - \rho_V(\mu)] \qquad \text{(XIII.3.36)}$$

and to relate it to the transformation kernel by a "Marchenko" equation:

$$B(r, r') = j(r, r') + \int_r^\infty B(r, \rho)j(\rho, r')d\rho. \qquad \text{(XIII.3.37)}$$

Hence the knowledge of the spectral function determines $B(r, r')$ through **(XIII.3.36)** and **(XIII.3.37)**. The spectral function is determined in principle, from the knowledge of the Jost function, as a holomorphic function in the half-plane Re $\lambda \geq 0$. Thus this approach gives us ways to study uniqueness problems and, when an interpolation of the Jost function has been given, existence problems. It is not a constructive method. In the inverse problem at fixed l, the Marchenko method had the very seductive feature of giving the symmetric kernel readily from the scattering amplitude. Here, this interesting property is completely lost because the spectral function depends on the scattering amplitude in a very complicated way.

This treatment has been extended to the case of singular potentials behaving more singularly that the centrifugal term near the origin (Burdet Giffon and Goldberg (1966)).

XIII.4   Complete Solution

XIII.4.1   Scheme of the Complete Solution

Like for the Regge-Loeffel method, the foundation is based on certain properties of $K(r, r')$ and $f(r, r')$, which hold when V belongs to a certain class of potentials. The representation of $f(r, r')$ leads us to make two parts out of it, $f_I(r, r')$ and $f_E(r, r')$. $f_I(r, r')$ is of the form $\sum_{l=0}^\infty c_l u_l(r)u_l(r')$, and $f_E(r, r')$ belongs to a certain class of functions. Thus $f(r, r')$ is used as the input function in the general machinery of chapter XI. To go to the asymptotic limit $r \to \infty$, and to obtain a relation between $\{c_l\}$, f_E, and $\{\delta_l\}$, it is necessary to make a particular study of the asymptotic behavior. This is possible because of particular representations of $K(r, r')$ and $f(r, r')$ when r goes to ∞. Thus, one obtains an infinite system which relates $\{c_l\}$, $\{\tan \delta_l\}$, and asymptotic quantities coming from $f_E(r, r')$. To solve the inverse problem, one can choose $f_E(r, r')$ arbitrarily in its class, calculate the corresponding quantities, insert them into the infinite system, and finally solve the problem as in the first matrix method of chapter XII. Thus the method is constructive. Besides, since the function which can be arbitrarily chosen explicitly appears, it is easy to answer all the questions posed in chapter XI.

In the present section, we closely follow this scheme. It is to be understood that we start from a potential 0, and, for the sake of simplicity, we drop the indices 0, V, for K_0^V or similar quantities when there is no confusion.

XIII.4.2   Foundation of the Complete Solution

Historically, the complete solution was based on the properties of $K(r, r')$ that can be derived (Sabatier 1971), by integrating the partial differential equation (**XI.2.5**), with the boundary conditions (**XI.2.6**) and (**XI.2.7**). By using Riemann's representation of the solution, it was possible to transform these equations into the equivalent integral equation:

$$K(r, r') = K_0(r, r') \pm \iint_{D^\pm} N(r, r', \rho, \rho')\rho^2 V(\rho)K(\rho, \rho')d\rho\, d\rho', \quad \textbf{(XIII.4.1)}$$

where

$$K_0(r, r') = -\tfrac{1}{2}(rr')^{1/2} \int_0^{(rr')^{1/2}} J_0[(r - r')(1 - \rho^2/rr')^{1/2}]\rho V(\rho)d\rho, \quad \textbf{(XIII.4.2)}$$

$$N(r, r', \rho, \rho') = \tfrac{1}{2}(rr')^{1/2}(\rho\rho')^{-3/2} J_0(\sqrt{(rr' - \rho\rho')(r/r' + r'/r - \rho/\rho' - \rho'/\rho)}),$$

$$\textbf{(XIII.4.3)}$$

and the domains D^+, D^-, the signs $+$ and $-$, are to be used, respectively, for $r \geq r'$ and $r' \leq r$ (figure XIII.2). Thus, $K_0(r, r')$ is a well-defined linear transform of the potential; the kernel of (**XIII.4.1**) contains $V(\rho)$ in a simple way; the equation (**XIII.4.1**) is of the Volterra form, and it can be solved by successive approximations.

Now let $\mathscr{E}$ be the class of twice differentiable functions $V(r)(0 < r < \infty)$ such that for any positive ε, $|V(r)|$, $r|V'(r)|$, and $r^2|V''(r)|$ are $O(r^{\varepsilon-1})$ as $r \to 0$ and $O(r^{-\varepsilon-3})$ as $r \to \infty$. Parenthetically, this differentiability restriction is convenient to establish formulas like (**XIII.4.4**) below but is probably not necessary if one is ready to accept very weak convergences. For $V \in \mathscr{E}$, the properties of the solutions of (**XIII.4.1**) were thoroughly investigated (Sabatier 1971), with the following results:

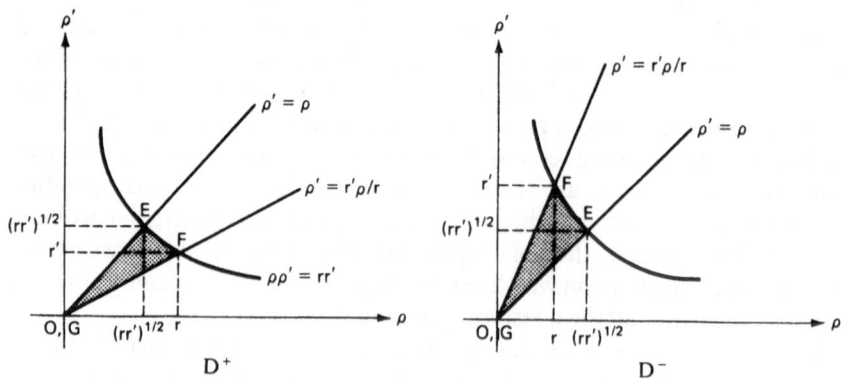

Figure XIII.2

Properties of $K(r, r')$. Here we only write down the results which are useful when $r \to \infty$. It turns out that $K(r, r')$ and its derivative with respect to r can be put in the following form:

$$\left(1 - i\frac{\partial}{\partial r}\right)K(r, r') = -ir'\,\mathcal{K}(r')\exp[ir] + K_N(r, r'), \quad \text{(XIII.4.4)}$$

where $\mathcal{K}(r)$ is absolutely bounded by $C(1 + r)^{-1/2}$, and $K_N(r, r')$ is a "negligible" function when r goes to ∞, "negligible" meaning precisely:

$$\left.\begin{array}{ll} K_N(r, r') \to 0 & (r \to \infty), \\[2mm] \int_0^r |K_N(r, r')| [r'(1 + r')]^{-1}\,dr' \to 0 & (r \to \infty). \end{array}\right\} \quad \text{(XIII.4.5)}$$

The interest of this result can be seen immediately in the formula (**XI.2.4**) (with $V_0 = 0$). Letting $r \to \infty$ in (**XI.2.4**), with the help of (**XIII.4.4**), (**XIII.4.5**) and (**XI.1.8**), we readily obtain the remarkable formula:

$$\int_0^\infty \mathcal{K}(\rho)u_l(\rho)\rho^{-1}\,d\rho = \exp\left[-il\frac{\pi}{2}\right](1 - A_l\exp[i\delta_l]). \quad \text{(XIII.4.6)}$$

Thus, at this point, we have related a function of one variable to a sequence $\{\delta_l\}$, and we can hope to reduce the inverse problem to solving a moment problem and so construct V from $\mathcal{K}$. But this approach, although possible "in principle," is very difficult and we shall not follow it through.

Properties of $f(r, r')$. The properties of $f(r, r')$ can be derived in a similar way from the partial differential equation (**XI.3.1**) instead of (**XIII.4.1**); one readily obtains

$$f(r, r') = -\tfrac{1}{2}(rr')^{1/2}\int_0^{(rr')^{1/2}} J_0[(r - r')(1 - \rho^2/rr')^{1/2}]\rho\omega(\rho)d\rho, \quad \text{(XIII.4.7)}$$

where

$$\omega(r) = -2r^{-1}\frac{d}{dr}\left[r^{-1}f(r, r)\right]. \quad \text{(XIII.4.8)}$$

Thus $f(r, r')$ is related to $\omega(r)$ in the same way $K_0(r, r')$ is related to $V(r)$ (and $K_0(r, r')$ satisfies the symmetric partial differential equation). If the Born approximation were exact $K_0(r, r')$ would yield $V(r)$ when it replaces $f(r, r')$ in the machinery. (For related results, see section XVI.2.) The properties of $f(r, r')$ can also be derived from those of $K(r, r')$ using (**XI.5.6**), which reads:

$$f(r, r') = K(r, r') + \int_0^r K(\tau, r)K(\tau, r')\tau^{-2}\,d\tau. \quad \text{(XIII.4.9)}$$

In particular, bounds on $K(r, r')$ when r' goes to ∞ readily show that $r^{-1}[f(r, r) - K(r, r)]$ goes to zero when r goes to ∞. Thus

$$\int_0^\infty \rho\omega(\rho)d\rho = \int_0^\infty \rho V(\rho)d\rho = V_0. \quad \text{(XIII.4.10)}$$

Other properties of $\omega(r)$ are also derived from (**XIII.4.9**):

(a) $r\omega(r)$ is continuous for any r;

(b) $r\omega(r) = O(r^\varepsilon)$ for $r \to 0$;

(c) there exist real numbers α, β and A, defining the function $\Omega(r) = 4r^{-2}(\alpha \cos 2r + \beta \sin 2r)\theta(r - A)$, such that $\int_0^\infty \rho\Omega(\rho)d\rho = 0$, and such that for any positive number ε', $\omega(r) + \Omega(r)$ is $O(r^{-3+\varepsilon'})$ as $r \to \infty$;

(d) $r\omega'(r)$ exists as a continuous function for any $r \geq 0$ and is $O(r^{-1+\varepsilon'})$ as $r \to \infty$.

Let Ω be the class of functions $\omega(r)$ satisfying these bounds. Now we see a way to study the inverse problem: we know that the matrix method of section XII.2 requires that $r\omega(r)$ belongs to a particular subclass of Ω (even entire functions, of exponential type 2), so that it is completely characterized by a sequence $\{c_l\}$; thus, we may try splitting $r\omega(r)$ into two parts— one part that will belong to this particular subclass, and the other one which can be chosen in its complement.

To do this it is convenient to introduce the sine Fourier transform of $r^{-1}f(r, r)$. Using the Parseval theorem, we easily transform (**XIII.4.7**) into

$$f(r, r') = 4rr' \lim \int_0^\infty w^{-1} \sin wuF(u)du, \qquad \text{(\textbf{XIII.4.11})}$$

where

$$w = [(r - r')^2 + 4rr'u^2]^{1/2} \qquad \text{(\textbf{XIII.4.12})}$$

so that

$$\tfrac{1}{2}r^{-1}f(r, r) = \text{l.i.m.} \int_0^\infty \sin[2ru]F(u)du. \qquad \text{(\textbf{XIII.4.13})}$$

Using (**XIII.4.10**), (**XIII.4.13**), and the properties (a)–(c) of $\omega(r)$, it is easy to show that $F(u)$ is of the form:

$$F(u) = -(2\pi)^{-1}V_0(u^{-1} - 1)\theta(1 - u) + \psi(u), \qquad \text{(\textbf{XIII.4.14})}$$

where θ is the Heaviside function, and $\psi(u)$ has the following properties:

(a') $\psi(u)$ and $u\psi(u)$ belongs to $L_2(0, \infty)$;

(b') $u\psi(u)$ is continuous except perhaps at $u = 1$ (possible logarithmic singularity), and goes to zero as u goes to ∞;

(c') there exists a number γ (which is related to α and β of $\Omega(r)$), such that $[u\psi'(u) - \gamma(1 - u)^{-1}]$ belongs to $L_{1/\varepsilon'}(0, \infty)$ for any number ε' $(0 < \varepsilon' < 1)$.

In addition, using property (d), it is easily proved that:

(d') for any number $\varepsilon'(0 < \varepsilon' < 1)$, $u\psi(u)$ is $O(u^{-1+\varepsilon'})$ as u goes to ∞.

Other properties of $f(r, r')$ can be derived from those of $K(r, r')$ by using (**XIII.4.9**). In particular, one obtains a formula similar to (**XIII.4.4**):

$$\left(1 - i\frac{\partial}{\partial r}\right)f(r, r') = -ir'\mathscr{F}(r')\exp[ir] + f_N(r, r'), \qquad \text{(\textbf{XIII.4.15})}$$

where $f_N(r, r')$ is "negligible" when r goes to ∞.

XIII.4.3   Direct Problem for $V \in \bar{\mathscr{E}}$

If V belongs to $\mathscr{E}$, the whole machinery of chapter XI works, and V can be related bijectively to $f(r, r)$, or, through (**XIII.4.11**) and its inverse, to $F(u)$. Thus, because of (**XIII.4.14**), V_0 and $\psi(u)$ completely define V in $\mathscr{E}$. In the following, we assume that V belongs to a certain class $\bar{\mathscr{E}}$ of potentials which is defined by the following conditions:

(a″) the machinery of chapter XI holds at almost every positive r, r';

(b″) the equations (**XIII.4.4**) to (**XIII.4.10**) are valid, with $\omega(r)$ satisfying conditions (a)–(d);

(c″) V_0 is finite, but not necessarily $\int_0^\infty \rho |V(\rho)| d\rho$.

Clearly, $\mathscr{E} \subset \bar{\mathscr{E}}$. Besides, equations (**XIII.4.10**)–(**XIII.4.15**) hold for a potential of $\bar{\mathscr{E}}$, and $\psi(u)$ satisfies conditions (a′)–(d′). Now, from (**XIII.4.4**), (**XIII.4.15**) and (**XI.3.4**), we obtain

$$\mathscr{K}(r') = \mathscr{F}(r') - \int_0^\infty \mathscr{K}(\rho)\phi(r, r')d\rho, \qquad \text{(XIII.4.16)}$$

where $\phi(r, r')$ has been defined by (**XI.3.6**).

This equation is the "asymptotic" form of the Regge–Newton equation. Thus it replaces (**XII.2.5a**) for $V \in \bar{\mathscr{E}}$. We now force (**XII.2.5a**) to appear as a part of (**XIII.4.16**). This is done by splitting $F(u)$ into two parts:

$$F(u) = F_I(u) + F_E(u), \qquad \text{(XIII.4.17)}$$

where

$$F_I(u) = F(u)\theta(1 - u). \qquad \text{(XIII.4.18)}$$

Because of (d′), the right-hand side of (**XIII.4.13**) converges uniformly. We can use the splitting of $F(u)$ to define two functions:

$$\bar{\phi}_I(r, r') = 4 \int_0^1 w^{-1} \sin(w)uF(u)du + \delta(rr')^{-1} \sin r \sin r', \qquad \text{(XIII.4.19)}$$

$$\bar{\phi}_E(r, r') = 4 \int_1^\infty w^{-1} \sin(w)uF(u)du - \delta(rr')^{-1} \sin r \sin r', \qquad \text{(XIII.4.20)}$$

where

$$F(u) = 2\pi^{-1} \int_0^\infty \sin(2ru)r^{-1}\phi(r, r)dr \qquad \text{(XIII.4.21)}$$

and the constant δ has been chosen in such a way that, for a potential of $\bar{\mathscr{E}}$, $\bar{\phi}_I(r, r)$ and $\bar{\phi}_E(r, r)$, like $\phi(r, r)$, are $O(r^\varepsilon)$ as $r \to 0$:

$$\delta = 2 \lim_{r \to 0} r^{-1} \int_1^\infty \sin(2ru)F(u)du = -4 \int_0^1 uF(u)du. \qquad \text{(XIII.4.22)}$$

From $\bar{\phi}_I(r, r')$ and $\bar{\phi}_E(r, r')$, we define $\mathscr{F}_I(r)$ and $\mathscr{F}_E(r)$, in the same way as we defined $\mathscr{F}(r)$ from $\phi(r, r')$ through (**XIII.4.15**).

Now, we *define* $\mathcal{K}_l(r)$ and $\mathcal{K}_E(r)$ by the formulas:

$$\mathcal{K}_l(r) = \mathcal{F}_l(r) - \int_0^\infty \mathcal{K}(\rho)\phi_l(\rho, r)d\rho, \tag{XIII.4.23}$$

$$\mathcal{K}_E(r) = \mathcal{F}_E(r) - \int_0^\infty \mathcal{K}(\rho)\phi_E(\rho, r)d\rho. \tag{XIII.4.24}$$

Clearly, (**XIII.4.16**) is satisfied if, and only if:

$$\mathcal{K}_E(r) + \mathcal{K}_l(r) = \mathcal{K}(r). \tag{XIII.4.25}$$

We now investigate properties of $\mathcal{K}_l(r)$. From (**XIII.4.19**) and a well-known addition formula, we readily obtain:

$$\bar{\phi}_l(r, r') = (rr')^{-1} \sum_0^\infty c_l u_l(r) u_l(r'), \tag{XIII.4.26}$$

where

$$c_l = 4(2l + 1) \int_0^1 F(u) P_l(1 - 2u^2) u \, du + \delta\delta_0^l, \tag{XIII.4.27}$$

where δ_0^l is the Kronecker symbol. With the choice (**XIII.4.22**) for δ, c_0 is zero. A similar formula is obtained for $\mathcal{F}_l(r)$. Inserting these formulas into (**XIII.4.23**), and taking (**XIII.4.6**) into account we obtain

$$\mathcal{K}_l(r) = \sum_{l=0}^\infty c_l A_l \exp\left[i\left(\delta_l - l\frac{\pi}{2}\right)\right] r^{-1} u_l(r). \tag{XIII.4.28}$$

Now, inserting (**XIII.4.28**) into (**XIII.4.6**) yields the infinite system which replaces (**XII.2.5a**) in this general case:

$$A_l \exp\left[i\left(\delta_l - l\frac{\pi}{2}\right)\right] = \exp\left[-il\frac{\pi}{2}\right] - \int_0^\infty \mathcal{K}_E(\rho) u_l(\rho)\rho^{-1} \, d\rho$$
$$- \sum_{l'} c_{l'} L_{ll'}(\infty) A_{l'} \exp\left[i\left(\delta_{l'} - l'\frac{\pi}{2}\right)\right]. \tag{XIII.4.29}$$

Using (**XIII.4.14**) and the properties (a′)–(c′), it is easy to show from (**XIII.4.27**) that, for a potential of $\bar{\mathscr{E}}$, there exists a positive number ε'' such that $[c_l - (2\pi)^{-1}V_0]$ is $O(l^{-1-\varepsilon''})$ as $l \to \infty$.

Suppose now that we are given a function $\bar{\phi}_E(r, r')$. Such a function can be constructed, for instance, by choosing *any* function $\omega(r)$ satisfying (a), (d) and constructing $\bar{\phi}_E$ through (**XIII.4.21**) and (**XIII.4.19**). Let $\mathcal{Y}_E(r, r')$ be the resolvent of $\bar{\phi}_E(r, r')$:

$$\mathcal{Y}_E(r, r') = \bar{\phi}_E(r, r') - \int_0^\infty \mathcal{Y}_E(r, \rho)\bar{\phi}_E(\rho, r')d\rho. \tag{XIII.4.30}$$

$\mathcal{Y}_E(r, r')$ can be used to obtain, from (**XIII.4.24**) and (**XIII.4.25**), a formula relating $\mathcal{K}_E(r)$ to $\bar{\phi}_E(r, r')$:

$$\mathcal{K}_E(r) = \mathcal{Y}_E(r) - \int_0^\infty \mathcal{K}_l(\rho)\mathcal{Y}_E(\rho, r)d\rho, \tag{XIII.4.31}$$

where

$$\mathscr{Y}_E(r) = \mathscr{F}_E(r) - \int_0^\infty \mathscr{F}_E(\rho)\mathscr{Y}_E(\rho, r)d\rho. \qquad \textbf{(XIII.4.32)}$$

Once $\bar{\phi}_E(r, r')$ has been chosen, $\mathscr{Y}_E(r, r')$ can be constructed in almost every case ((Sabatier 1972): **(XIII.4.30)** is a Fredholm equation with a Hilbert–Schmidt kernel) and $\mathscr{Y}_E(r)$ is thus obtained. Inserting **(XIII.4.31)** into **(XIII.4.29)** we obtain the infinite system of linear equations:

$$A_l \exp\left[i\left(\delta_l - l\frac{\pi}{2}\right)\right] = (1 - B_l \exp[i\sigma_l])\exp\left[\left(il\frac{\pi}{2}\right)\right]$$

$$- \sum_{l'} c_{l'}(L_{ll'}(\infty) - g_{ll'})A_{l'} \exp\left[i\left(\delta_{l'} - l'\frac{\pi}{2}\right)\right],$$

$$\textbf{(XIII.4.33)}$$

where B_l, σ_l, and $g_{ll'}$ are defined by

$$B_l \exp\left[i\left(\sigma_l - l\frac{\pi}{2}\right)\right] = \int_0^\infty \mathscr{Y}_E(\rho)u_l(\rho)\rho^{-1} \, d\rho, \qquad \textbf{(XIII.4.34)}$$

$$g_{ll'} = \int_0^\infty \int_0^\infty \mathscr{Y}_E(x, y)u_l(x)u_{l'}(y)x^{-1}y^{-1} \, dx \, dy. \qquad \textbf{(XIII.4.35)}$$

Actually, it is also possible to obtain $g_{ll'}$ from B_l and σ_l:

$$g_{ll'} = [l(l + 1) - l'(l' + 1)]^{-1}\{B_{l'} \sin[\sigma_{l'} + \tfrac{1}{2}\pi(l - l')]$$

$$- B_l \sin[\sigma_l + \tfrac{1}{2}\pi(l' - l)] + B_l B_{l'} \sin[\sigma_l - \sigma_{l'} - \tfrac{1}{2}\pi(l - l')]\}$$

$$\textbf{(XIII.4.36)}$$

and to show that there exists a positive number ε'', such that B_l is $O(l^{-1-\varepsilon''})$ as $l \to \infty$ (Sabatier 1972).

XIII.4.4   Inverse Problem in $\bar{\mathcal{O}}$

Let $\mathcal{O}_E$ be the class of functions $\bar{\phi}_E(r, r')$ which can be obtained, from a function $\omega(r)$ satisfying (a)–(d), by using **(XIII.4.21)** and **(XIII.4.19)**. We define the class $\bar{\mathcal{O}}$ of potentials as the class of potentials for which:

(a''') the c_l's are uniformly bounded;
(b''') $\bar{\phi}_E(r, r')$ belongs to $\mathcal{O}_E$;
(c''') the machinery works.

Clearly $\mathscr{E} \subset \bar{\mathscr{E}} \subset \bar{\mathcal{O}}$. Since c_0 can be $\neq 0$, $\bar{\mathcal{O}}$ contains potentials which are $O(r^{-1})$ when r goes to zero. Since we do not restrict the asymptotic behavior of the c_l's, $\bar{\mathcal{O}}$ contains potentials which are only $O(r^{-3/2})$ (or even worse) when $r \to \infty$. Thus $\bar{\mathcal{O}}$ is certainly larger than $\mathscr{E}$. (cf figure XIII.3).

$\mathscr{E}$ is defined by	$\bar{\mathscr{E}}$ is defined by	$\bar{\mathcal{O}}$ is defined by
(1) a set of simple properties of $V(r)$ (or $K(r, r)$) which are such that the following hold: (2) The machinery works (3) Properties of $\omega(r)$ (or $f(r, r)$) (a)–(d) (4) Properties of $\psi(u)$ (a′)–(d′)	(1′) $\int_0^r \rho V(\rho)d\rho$ finite for any r (2) the machinery works (3) Properties of $\omega(r)$ (a)–(d) and such that (5) $\bar{\phi}_E(r, r') \in \mathcal{O}_E$ (6) $c_0 = 0,\ c_l \to \lim(l \to \infty)$	(2) The machinery works (5) $\bar{\phi}_E(r, r') \in \mathcal{O}_E$ (6) c_l uniformly bounded

$$\boxed{\mathscr{E} \subset \bar{\mathscr{E}} \subset \bar{\mathcal{O}}}$$

Figure XIII.3 Definitions for the sets used in section XIII.4

Now suppose we are given a set of phase shifts $\{\delta_l\}$. Let us choose a function $\bar{\phi}_E(r, r')$ in $\mathcal{O}_E$, and let us construct $\mathscr{Y}_E(r, r')$ through (**XIII.4.30**) and $\mathscr{Y}_E(r)$ through (**XIII.4.31**). Thus B_l, σ_l, $g_{ll'}$, are obtained ((**XIII.4.34**)–(**XIII.4.36**)). Now for convenience, as in chapter XII, let us set $a_l = c_l A_l \cos \delta_l$, and let us introduce the compact notation:

$$\eta_l = B_l(\sin \sigma_l - \cos \sigma_l \tan \delta_l),$$

$$\bar{h}_{ll'} = \begin{cases} (-1)^{1/2(l-l')}g_{ll'} & \text{for } (l - l') \text{ even} \neq 0, \\ 0 & \text{for } l - l' \text{ odd or } 0, \end{cases} \quad \textbf{(XIII.4.37)}$$

$$h_{ll'} = \begin{cases} (-1)^{1/2(l-l'-1)}g_{ll'} & \text{for } (l - l') \text{ odd}, \\ 0 & \text{for } (l - l') \text{ even}. \end{cases}$$

Multiplying both sides of (**XIII.4.33**) by $\exp[-i(\delta_l - l(\pi/2))]$ yields the equivalent systems:

$$A_l[1 + c_l(L_{ll}(\infty) - g_{ll})] = \exp[-i\delta_l] - B_l \exp[i(\sigma_l - \delta_l)]$$

$$- \sum_{l' \neq l} c_{l'}(L_{ll'}(\infty) - g_{ll'})A_{l'} \exp\left[i(\delta_{l'} - \delta_l) + i(l - l')\frac{\pi}{2}\right]. \quad \textbf{(XIII.4.38)}$$

The imaginary part of (**XIII.4.38**) yields the fundamental system:

$$\tan \delta_l + \eta_l = \sum_{l'} M_{ll'} a_{l'}(1 + \tan \delta_l \tan \delta_{l'})$$

$$+ \sum_{l'} a_{l'}[h_{ll'}(1 + \tan \delta_l \tan \delta_{l'}) + \bar{h}_{ll'}(\tan \delta_{l'} - \tan \delta_l)]$$

$$\textbf{(XIII.4.39)}$$

or, in matrix notation

$$(\eta + \tan \Delta)\mathbf{e} = M(1 + S)\mathbf{a}, \quad \textbf{(XIII.4.40)}$$

where

$$S = M^{-1} \tan \Delta M \tan \Delta + M^{-1} \tan \Delta h \tan \Delta$$
$$+ M^{-1}h + M^{-1}\bar{h} \tan \Delta - M^{-1} \tan \Delta \bar{h}. \quad \textbf{(XIII.4.41)}$$

This obviously generalizes (**XII.2.9**).

Suppose now the phase shifts are $O(l^{-3-\varepsilon})$ as l goes to ∞. With the bounds which have been obtained for the B_l's, it is possible to prove (Sabatier 1972) that S is a Hilbert–Schmidt matrix, so that $1 + S$ is invertible for almost every set of phase shifts and $\mathbf{a}$ is obtained from the formula

$$\mathbf{a} = (1 + S)^{-1}\{M^{-1}(\eta + \tan \Delta)\mathbf{e} + \alpha \mathbf{v}\}, \qquad \text{(XIII.4.42)}$$

where α is chosen arbitrarily and $\mathbf{v}$ is given by (XII.3.3). Once a_l is obtained, A_l is readily derived from the real part of (XIII.4.38), and c_l from A_l and a_l.

The analysis of the results can be performed as in sections XII.4 and XII.5. It essentially reduces to this previous analysis if $\phi_E(r, r')$ has been chosen in such a way that B_l is $O(l^{-3-\varepsilon})$ as l goes to ∞. In the most general case, it has as yet not been proved that *all* the potentials which can be obtained are physically meaningful, although this result is very probable.

XIII.4.5   Larger Foundations of the Complete Method

Our study of the complete method was based on the existence of the equations (XIII.4.4)–(XIII.4.10) and the properties (a)–(d) of $\omega(r)$. Both are guaranteed for $V \in \mathscr{E}$. On the other hand, the properties (a′)–(d′) of $F(u)$ have been derived from (a)–(d), and are a little bit weaker. If one tries to prove (XIII.4.4)–(XIII.4.10) and the properties of $\omega(r)$ directly from the properties of $F(u)$ one only finds that a stronger asymptotic behavior ($F(u)$ and $F'(u)$ be $O(u^{-3-\varepsilon})$), would be sufficient to guarantee them. But, in the inverse problem we have worked in a class $\overline{\mathcal{O}}$ which is much larger than $\mathscr{E}$. Suppose we were able to prove, for a certain class of potentials, the existence of $F(u)$ in such a way that:

1. its properties for $u \in [0, 1]$ guarantee the existence of $\{c_l\}$;
2. it decreases as $u \to \infty$ strongly enough to give an appraisal of the deviation of equivalent potentials from each other.

Then, the inverse problem could be studied in this class by the complete method we have described above (except that we could not be sure, *a priori*, that it is consistent for *all* the potentials which could be constructed, but this difficulty could be overcome *a posteriori*).

It is of interest to prove from Loeffel's results that $\mathcal{O}$ is a class of potentials fulfilling the above properties. This program was carried through by Jean and Sabatier (1973). Starting from the Regge–Loeffel expansion (XIII.3.22), and proving that (XI.5.14) holds even with the weak assumptions of this theory, they were able to obtain the following expansion for $f_0^V(r, r')$ (which converges, as does (XIII.3.22), as an operator on L_2):

$$(rr')^{-1}f(r, r') \sim \sum_{k=1}^{\infty} d_k \phi_0(\lambda_k, r)\phi_0(\lambda_k, r')$$

$$- (2\pi)^{-1} \underset{\omega \to \infty}{\text{l.i.m.}} \int_{-\omega}^{\omega} \gamma(i\tau)\phi_0(i\tau; r)\phi_0(i\tau, r')d\tau, \qquad \text{(XIII.4.43)}$$

where $k \to \lambda_k$ is the unique strictly increasing mapping of the set N_+ of the positive integers into the set $\{\lambda_k^0, \lambda_k^V, k \in N_+\}$, d_k is defined by

$$
d_k = \begin{cases} d_h^V - d_h^0 & \text{if } \lambda_k \in \{\lambda_h^V\} \cap \{\lambda_h^0\}, \\ d_h^V & \text{if } \lambda_k \in \{\lambda_h^V\}, \\ -d_h^0 & \text{if } \lambda_k \in \{\lambda_h^0\}, \end{cases} \tag{XIII.4.44}
$$

and

$$
\gamma(\lambda) = \gamma_V(\lambda) - \gamma_0(\lambda). \tag{XIII.4.45}
$$

Notice that the selected values λ_k need not be integers. Now, in order to derive the fundamental representation (XIII.4.11), it suffices to substitute into (XIII.4.43) the following representation of the products of Bessel functions:

$$
J_\lambda(r)J_\lambda(r') = \pi^{-1}(rr')^{1/2} \int_0^1 w^{-1} \sin[w] P_{\lambda-1/2}(1 - 2u^2)u\, du
$$

$$
+ 2(rr')^{1/2} \cos[\pi\lambda] \int_1^\infty w^{-1} \sin[w] \mathscr{D}_{\lambda-1/2}(2u^2 - 1)u\, du,
$$

$$
\tag{XIII.4.46}
$$

where P and $\mathscr{D}$ are Legendre functions (usual and associated), and to prove the convergence of everything that needs to converge. This can be done, and (XIII.4.43) converges towards an absolutely continuous function, if V belongs to $\mathcal{O}' \subset \mathcal{O}$ defined by the conditions:

$$
\int_0^1 \rho^{1-2\varepsilon} |V(\rho)| d\rho + \int_1^\infty \rho^{1+2\varepsilon} |V(\rho)| d\rho < \infty,
$$

$$
\int_0^\infty \rho^{3+4\varepsilon} V^2(\rho) d\rho < \infty \qquad (\varepsilon > 0). \tag{XIII.4.47}
$$

Then, if $0 < \varepsilon < \frac{1}{4}$, $\psi(u) \in L_{1/(1-2\varepsilon)}(0, \infty)$. If $\varepsilon \geq \frac{1}{4}$, $\psi(u) \in L_2(0, \infty)$. If, in addition (subclass $\mathcal{O}'' \subset \mathcal{O}'$),

$$
\int_0^1 \rho^{-2\varepsilon'} |V(\rho)| d\rho| + \int_1^\infty \rho^{2+2\varepsilon'} (\text{Log } \rho)^2 |V(\rho)| d\rho < \infty, \qquad \varepsilon' > 0,
$$

$$
\tag{XIII.4.48}
$$

then $f(r, r')$ has absolutely continuous derivatives, and $u\psi(u) \in L_2(0, \infty)$. It may be of interest to notice that the first term on the right-hand side of (XIII.4.46) readily yields $\bar{\phi}_l(r, r')$, which has been used above.

Founding the complete solution on the Regge–Loeffel method, and thus on the Gel'fand–Levitan method, has not only given a generalization of the class of potentials (from $\mathscr{E}$ to $\mathcal{O}''$ and $\mathcal{O}'$ with some care). It has also furnished a link between all the methods of attack of inverse scattering problems, a result that had been wished for by several specialists (see, e.g, Cornille 1973).

XIII.4.6   Miscellaneous Results

The "complete" method gives answers to all the questions which have been formulated in section XI.1, for the large class of potentials satisfying (**XIII.4.47**). The answers to questions A, C, and D are positive, and the answer to B is negative. The construction is not easy, nor is the choice of a potential satisfying additional *a priori* physical requirements. It is of interest to study the effects of a strong constraint that restores uniqueness. The most simple one is obtained by truncating the potential, as has been done in Wigner's R-matrix theory. Assume $V(r) = 0$ for $x \geq a$. A glance at (**XI.2.5**) shows that for $r \geq a$, $K_0^V(r, r')$ is, like $f_0^V(r, r')$, a solution of the symmetric partial differential equation

$$\left[r^2 \left(\frac{\partial^2}{\partial r^2} + 1 \right) - r'^2 \left(\frac{\partial^2}{\partial r'^2} + 1 \right) \right] K_0^V(r, r') = 0. \qquad \textbf{(XIII.4.49)}$$

Hence, $K_0^V(r, r')$ can be constructed from its boundary values in the domains $r \geq a$, $rr' \leq a^2$ of the complex plane. It is possible to do it precisely after reducing to this case the integral equations similar to (**XIII.4.1**), obtained by Sabatier (1971) with the help of Riemann's representation. The "structure function" $\mathscr{K}(r)$, which has been defined by (**XIII.4.4**), can then be related (Loubatières, 1985) to an l-independent function $G(x)$ that generates the phase shifts and amplitudes through formulas like

$$A_1 \sin \delta_1 = \int_0^a G(x) \left(\frac{d}{dx} x^{-1} [u_1(x)]^2 \right) dx. \qquad \textbf{(XIII.4.50)}$$

In this formula, a is the range of the potential. The formula could also be obtained by means of the Wronskian formula and by using (**XII.4.7**) in the particular case $f(r, r') = u_1(r)u_1(r')$. A somewhat more complicated formula relates $G(x)$ and $a_1 \cos \delta_1$. The two formulas together show that the problem is overdetermined.

[*Hint*: Multiply both sides of (**XIII.4.50**) by $(l + \frac{1}{2})P_l(\cos \theta)$ and sum over l: we obtain, on the left-hand side, an entire function of exponential type which cannot fit the right-hand side unless the $A_l \sin \delta_l$ satisfy strong asymptotic conditions.]

If there is a solution, it is unique. But there is not a solution for an arbitrary set of phase shifts, and when there is one, the solution dependence on results is unstable and we easily get oscillations (we also get oscillations if a finite number of phase shifts is used, although the potential is then certainly not finite range, see Reignier (1979)). A lot of miscellaneous results have also been obtained in numerical studies of the inverse problem at fixed energy. They will be sketched in chapter XIX below.

XIII.5   Remarks on the Methods

$1°$ The methods of Regge–Loeffel and Burdet Giffon give good examples of the interest and difficulties of handling the connection described in section XI.7 between the problem at fixed E and the one at fixed l. In addition to the difficulties noticed in section XI.7, a new one arises from the quantization of λ. Since the only physical values of λ are $\lambda = l + \frac{1}{2}$, with integer λ, interpolation is a very important specific feature of the inverse problem at fixed E.

2° The complete solution is founded on the properties of $f(r, r)$. It is interesting to see what this would mean for methods like Marchenko's or Gel'fand–Levitan's. Take for example the method of Marchenko. The symmetric kernel $F(x + y)$ reduces, for $x = y$, $x \geq 0$, to

$$F(2x) = \frac{1}{2\pi} \int_{-\infty}^{+\infty} [S(k) - 1]e^{2ikx}\, dk + \sum_n S_n e^{-2\xi_n x}. \qquad \textbf{(XIII.5.1)}$$

Now, using the Fourier transform readily yields (if $e^{-2\xi_n x}$ is continued by 0 for $x < 0$, and this of course is an arbitrary choice!)

$$F(2x) = \frac{1}{2\pi} \int_{-\infty}^{+\infty} \exp[2ikx]\left\{ S(k) - 1 + \sum_n \frac{1}{\xi_n + ik} \right\}dk. \qquad \textbf{(XIII.5.2)}$$

We would be in a situation similar to the one described in section XIII.4 if the only information we had were that the Fourier transform of $F(x + x)$ is in the space $L_2(-\infty, +\infty)$. Thus, it is not unimaginable that we will be able one day to disentangle the physics in $\omega(r)$ or $\psi(u)$ as in Gel'fand–Levitan's or Marchenko's method. Maybe one will be able to show that the phase shifts and Regge poles appear very simply in $\psi(u)$. But this is still to be done.

CHAPTER XIV

The Three-Dimensional Inverse Problem

XIV.1 Introduction

In a time-independent formulation of nonrelativistic scattering, one starts with the Schrödinger equation

$$-\Delta\Psi + V(\mathbf{r})\Psi = E\Psi \qquad \text{(XIV.1.1)}$$

for the wave function Ψ, and one assumes that the properties of $V(\mathbf{r})$ are sufficient to guarantee the following asymptotic form of Ψ:

$$\Psi(\mathbf{k}, \mathbf{r}) = \exp[i\mathbf{k}\cdot\mathbf{r}] + r^{-1}\exp[ikr]A(\hat{r}, \mathbf{k}) + o(r^{-1}). \qquad \text{(XIV.1.2)}$$

(For any vector $\mathbf{v}$, we write v for its length, $\hat{v}$ for $\mathbf{v}/v$. In the scattering amplitude, we use $\mathbf{k}'$ for $k\hat{r}$, and, more generally, we use $\mathbf{k}'$ for a vector of length k, and which is not $\mathbf{k}$.)

It is assumed that $A(\hat{r}, \mathbf{k})$ has been obtained from the measured quantities (which actually give its modulus; see chapter X). The present inverse problem is to obtain $V(\mathbf{r})$ from $A(\hat{r}, \mathbf{k})$. Since $A(\hat{r}, \mathbf{k})$ depends on 5 variables, and $V(\mathbf{r})$ only depends on 3, we necessarily have an existence problem. In other words, the existence of an underlying local potential must lead to strong restrictions on $A(\hat{r}, \mathbf{k})$.

If we start with an amplitude $A(\hat{r}, \mathbf{k})$ that belongs to an underlying local potential, and we wish to reconstruct that potential, then we need not worry about the existence problem. On the other hand, the uniqueness problem can be easily solved since, for any "reasonable" potential, the scattering amplitude should approach its Born approximation at large energies

$$A \simeq -(4\pi)^{-1} \int d\mathbf{r}' V(\mathbf{r}') \exp[i\tau \cdot \mathbf{r}'], \qquad \text{(XIV.1.3)}$$

where $\tau = \mathbf{k} - \mathbf{k}'$. Since $\tau_{max} = 2k$, A approaches the whole Fourier transform of V more and more closely as E increases. Thus its high-energy limit determines the potential uniquely. The reason for which we do not content ourselves with this "solution" of the problem is that the high-energy limit of A is not known, and furthermore, we know that the Schrödinger equation (XIV.1.1) is not a physically valid description of particle behavior at high energies. One would hope to find a solution to the inverse problem that is not too sensitive to high-energy data.

If the scattering amplitude determines the potential uniquely, then it must *a fortiori* determine the bound states, that is, the point spectrum of the Hamiltonian H. For a potential satisfying $|V(\mathbf{r})| \le C|1 + r|^{-3-\varepsilon}$, the point spectrum can be obtained for example, from the forward scattering amplitude $A(k) \equiv A(\hat{k}, \mathbf{k})$. On the real axis $A(k)$ is the boundary value of an analytic function which is regular in the upper half plane except for poles on the positive imaginary axis at $k = iK_n$, where $-K_n^2$ are the eigenvalues of H and $A(k)$ goes to a constant as $|k| \to \infty$, (Im $k \ge 0$). These properties enable us to obtain the position of the poles from $A(k)$ on the real axis and hence, the point spectrum from the forward scattering amplitude. Thus the full scattering amplitude depends on the point spectrum, and we cannot expect to be able to prescribe them separately.

We first present, in section XIV.2, previously reviewed Gel'fand–Levitan or Marchenko procedures in a way that emphasizes their general organization. This helps us to sketch, in section XIV.3, the approach to the three-dimensional inverse problem due to Faddeev (1971) and Newton (1973, 1974). This study is interesting not because it is convenient for solving inverse problems (the results are much too complicated), but rather because it contains:

(a) a remarkable generalization of potential scattering studies to the case where $V(\mathbf{r})$ is not spherically symmetric;
(b) a remarkable generalization of the Gel'fand–Levitan procedure to this case.

In section XIV.4, this study is replaced in the framework of modern "$\bar{\partial}$-approaches." These enabled, in particular, Henkin and Novikov (1988a) and Lavine and Nachman (1987a) to simplify the study and to analyze more transparently the difficult existence, uniqueness, and consistency questions. Two other approaches in the frequency domain are sketched in section XIV.5. Section XIV.6 is devoted to time-domain approaches, whose range of application is that of problems without bound states. The chapter is closed with the usual historical comments.

XIV.2   New Outline of One-Dimensional Methods

As we already noticed, a very general machinery for solving inverse problems of the Schrödinger equation is shown in figure XIV.1.

This diagram does not explain how to obtain the centrally important "transformation kernel" and "symmetric kernel," but throughout this book, we have distinguished two ways to obtain them:

1. In the first, the kernels are studied as solutions of certain hyperbolic partial differential equations (not the Schrödinger equation).
2. In the second, the kernels are derived from the completeness relations of a set of solutions of the Schrödinger equation.

These two ways have been schematized in the figure. The former was studied in detail in chapter XI and applied to the inverse problem at fixed energy.

Let us now study the second way, which readily yields answers to the questions of existence, uniqueness, and constructability. We note that in the cases that were studied in detail (the Gel'fand–Levitan and Marchenko methods, the one-dimensional "line" problem), the details are essentially the same. They can be roughly divided into three sets, which correspond to three parts of the diagram given above, or to three steps in the method of solution, which we shall successively study for the example of the S-wave

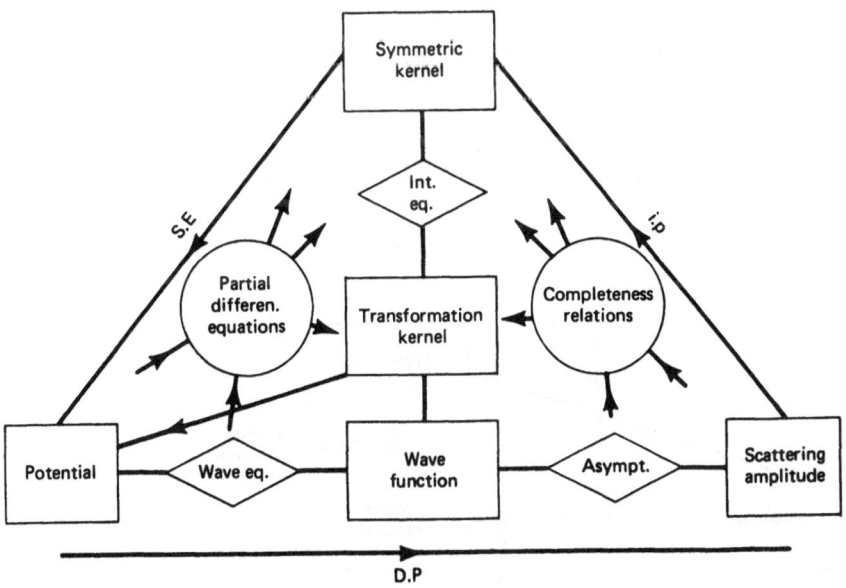

Figure XIV.1

Gel'fand–Levitan method. The words written in italics below are the key-words of the analogies we want to present, and are used as subtitles in section XIV.3.

XIV.2.1   The "Wave-Function" Step

In the following, we call "*wave function*" the solution of the wave equation which is the center of our study. It is not necessarily the "physical" wave function (see for example the Marchenko method or the one-dimensional inverse "line" problem). But in the Gel'fand–Levitan method, it is indeed proportional to the physical wave function, and vanishes at $r = 0$:

$$H\varphi \equiv \varphi''(k, r) - V\varphi(k, r) = -k^2 \varphi(k, r) \qquad \text{(XIV.2.1)}$$

and fixing the normalization at $r = 0$ amounts to *choosing a Green's function*.

$$\varphi(k, r) = \varphi_0(k, r) + \int_0^r G_0(k; r, \rho)V(\rho)\varphi_0(k, \rho)d\rho, \qquad \text{(XIV.2.2)}$$

where

$$\varphi_0(k, r) = \frac{\sin kr}{k}; \qquad G_0(k; r, \rho) = \frac{\sin k(r - \rho)}{k}. \qquad \text{(XIV.2.3)}$$

From (**XIV.2.2**) and (**XIV.2.3**) it can be shown that $\varphi(k, r)$ is an even entire function of k, and that, in the k complex plane,

$$\varphi(k, r) - \frac{\sin kr}{k} = o\left(\frac{\sin kr}{k}\right) \qquad (|k| \to \infty). \qquad \text{(XIV.2.4)}$$

From these *analytic properties*, it readily follows that a *triangular kernel* $K(r, r')$ exists, with the following properties (Paley–Wiener's theorem):

1. $K(r, r')$ is independent of k.
2. $K(r, r') = 0$ for $r' \geq r$.
3. $\varphi(k, r) = \varphi_0(k, r) + \int_0^r K(r, r')\varphi_0(k, r')dr'. \qquad \text{(XIV.2.5)}$

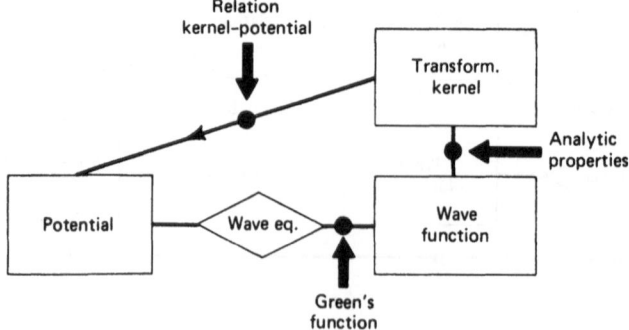

Figure XIV.2

By inserting (**XIV.2.5**) into (**XIV.2.2**) one obtains the *relation between* $V(r)$ *and the transformation kernel* K:

$$K(r, r) = \frac{1}{2} \int_0^r V(\rho)d\rho. \qquad \text{(XIV.2.6)}$$

XIV.2.2 The "Asymptotics and Completeness" Step

The *Jost function* $F(k)$ can be defined in two ways.

1. From the asymptotic behavior of $\varphi(k, r)$ as $r \to \infty$:

$$2ik\varphi(k, r) = \exp[ikr]F(-k) - \exp[-ikr]F(k) + o(1). \quad \text{(XIV.2.7)}$$

2. By introducing the *Lippmann–Schwinger integral equation* which defines the physical wave function

$$\psi(k, r) = \varphi_0(k, r) + \int_0^\infty g(k, r, \rho)V(\rho)\psi(k, \rho)d\rho, \quad \text{(XIV.2.8)}$$

where

$$g(k, r, \rho) = -k^{-1}\exp[ikr_>]\sin[kr_<]. \qquad \text{(XIV.2.9)}$$

This yields

$$\varphi(k, r) = F(k)\psi(k, r). \qquad \text{(XIV.2.10)}$$

Thus $\varphi(k, r)$ depends linearly on the solution of the Fredholm integral equation (**XIV.2.8**), with coefficient $F(k)$. It is furthermore possible to prove that $F(k)$ is *the Fredholm determinant* of equation (**XIV.2.8**). From this fact it follows that the zeros of $F(k)$ in the upper half plane occur exactly at those values $k = iK_n$ for which $-K_n^2$ is an eigenvalue of the Hamiltonian, and each zero has a multiplicity equal to the degeneracy of the corresponding eigenvalue.

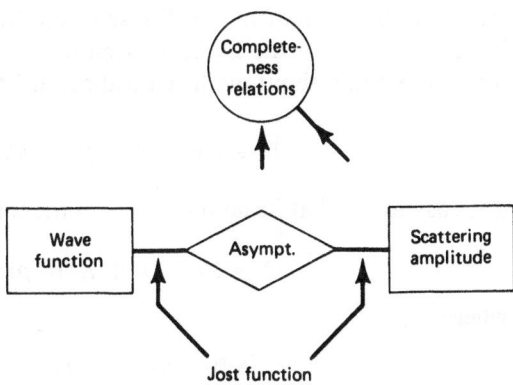

Figure XIV.3

One can prove the *completeness of the functions* $\psi(k, r)$ *together with the eigenfunctions of the Hamiltonian*:

$$\frac{2}{\pi} \int_0^\infty \psi(k, r)\psi^*(k, r')k^2 \, dk + \sum_n \psi^{(n)}(r)\psi^{(n)}(r') = \delta(r - r'), \quad \textbf{(XIV.2.11)}$$

where $\psi^{(n)}(r)$ is the nth normalized eigenfunction of H

$$\int_0^\infty |\psi^{(n)}(r)|^2 \, dr = 1. \quad\quad \textbf{(XIV.2.12)}$$

From these relations, and **(XIV.2.10)**, one can write the completeness in terms of φ as a Stieltjes integral:

$$\int \varphi(k, r)\varphi(k, r')d\rho(E) = \delta(r' - r'), \quad\quad \textbf{(XIV.2.13)}$$

where the *spectral function* $\rho(E)$ is defined by

$$\frac{d\rho}{dE} = \begin{cases} k\pi^{-1}|F(k)|^{-2}, & E \geq 0, \\ \sum_n \left[\int_0^\infty \varphi^2(iK_n, r)dr \right]^{-1} \delta(E - E_n), & E < 0. \end{cases} \quad \textbf{(XIV.2.14)}$$

There is, of course a similar completeness of the functions φ_0, with a spectral function ρ_0 such that $d\rho_0/dE$ equals 0 for $E < 0$, and k/π for $E \geq 0$.

XIV.2.3   The "Integral Equation" Step

Preliminary Remark. Let E be a linear space where the scalar product $\langle a, b \rangle = \int a(\rho)b(\rho)d\rho$ has been defined, and let T be a linear operator in E. We define its adjoint T^* by imposing that

$$\int \overline{Ta(\rho)}b(\rho)d\rho = \int \overline{a(\rho)}T^*b(\rho)d\rho \quad\quad \textbf{(XIV.2.15)}$$

for all couples a, b of $E \times E$. We shall use the "function notation" and the integral symbol for the Dirac distribution, and more generally we shall not worry too much about mathematical details. Now, if U is an operator of the form,

$$U = \delta(r - \rho) + K(r, \rho)\theta(r - \rho)\theta(\rho) \quad\quad \textbf{(XIV.2.16)}$$

it is easy to see that its adjoint U^* is of the form

$$U^* = \delta(r - \rho) + K^*(r, \rho)\theta(\rho - r)\theta(r), \quad\quad \textbf{(XIV.2.17)}$$

where

$$K^*(r, \rho) = K(\rho, r) \quad (r < \rho). \quad\quad \textbf{(XIV.2.18)}$$

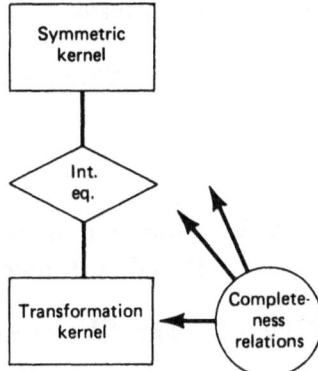

Figure XIV.4

In other words, the adjoint of a triangular kernel is a triangular kernel. In the following, we write **(XIV.2.15)** in the simple form

$$\overline{Ta(\rho)} = a(\rho)T^*. \tag{XIV.2.19}$$

Clearly $\delta(r - \rho)$ is the kernel of the unit operator hereafter denoted by **1**.

Formal Solution of the Problem. The representation **(XIV.2.5)** can be written

$$\varphi_k = U\varphi_{0k}. \tag{XIV.2.20}$$

Since any vector of L_2 can be expanded in terms of the φ_{0k}, **(XIV.2.20)** implies that U is *an intertwining operator*

$$HU = UH_0. \tag{XIV.2.21}$$

If we write the completeness relation of the φ_k's and the φ_{0k}'s symbolically:

$$\int d\rho \varphi_k \bar{\varphi}_k = 1, \quad \int d\rho \varphi_k^0 \bar{\varphi}_k^0 = 1, \tag{XIV.2.22}$$

and insert **(XII.4.20)**, we get

$$UgU^* = 1 - UU^*, \tag{XIV.2.23}$$

where

$$g = \int d(\rho - \rho_0)\varphi_{0k}\bar{\varphi}_{0k} \tag{XIV.2.24}$$

is the *symmetric kernel*. This could be done with any intertwining operator. But since K is triangular, we can construct an inverse

$$U^{-1} = 1 + K' \tag{XIV.2.25}$$

and K' is also triangular. Hence we have

$$Ug = K'^* - K \tag{XIV.2.26}$$

which for $r' < r$ reads

$$K(r, r') + g(r, r') + \int K(r, \rho)g(\rho, r')d\rho = 0 \qquad \textbf{(XIV.2.27)}$$

and this is the Gel'fand–Levitan *integral equation*.

XIV.3   Approach of the Three-Dimensional Problem Based on Faddeev's Green Function

XIV.3.1   The "Wave Function" Step (Figure XIV.2)

We have to choose a solution of **(XIV.1.1)** in such a way that the whole method works. It amounts to finding a *convenient* Green's function. This difficulty was overcome by Faddeev (1966).

The Green's Function. For any vector v, we use the notations already introduced in section XIV.1 ($\mathbf{v}$, length v, unit vector $\hat{v} = v^{-1}\mathbf{v}$). Let us now introduce a fixed real unit vector $\hat{y}$, and let q be a real nonnegative number. The Green's function of Faddeev is given by:

$$G_{q\hat{y}}(K^2; \mathbf{r} - \mathbf{r}') = \frac{1}{(2\pi)^3} \int d\mathbf{p} \, \frac{\exp[i(\mathbf{p} + iq\hat{y}) \cdot (\mathbf{r} - \mathbf{r}')]}{K^2 - (\mathbf{p} + iq\hat{y})^2}. \qquad \textbf{(XIV.3.1)}$$

It is easy to see that G satisfies, as expected, the equation

$$(\Delta + K^2)G(\mathbf{r}, \mathbf{r}') = \delta(\mathbf{r} - \mathbf{r}'). \qquad \textbf{(XIV.3.2)}$$

G is actually the resolvent of $-\Delta$ on the space $L^2_W(\mathbb{R}_3)$, with the weight $W = \exp[2q\hat{y} \cdot \mathbf{r}]$.

We now write, for any vector $\mathbf{v}$, $v_0 = \mathbf{v} \cdot \hat{y}$, $\mathbf{v}_\perp = \mathbf{v} - v_0\hat{y}$. After a shift in the variables of integration, we can then write

$$G_{q\hat{y}}[(\mathbf{k} + iq\hat{y})^2; \mathbf{r}] \equiv G_{\hat{y}}(\mu^2, s; \mathbf{r})$$

$$= \frac{1}{(2\pi)^3} \int d\mathbf{p} \, \frac{\exp[i\mathbf{p} \cdot \mathbf{r} + isr_0]}{\mu^2 - p^2 - 2sp_0}, \qquad \textbf{(XIV.3.3)}$$

where $\mu^2 = k_\perp^2, s = k_0 + iq$. Thus $\text{Im } s \geq 0$. One can easily see from **(XIV.3.3)** that $G_{\hat{y}}(\mu^2, s; \mathbf{r})$, for fixed μ, $\hat{y}$, $\mathbf{r}$, is an analytic function of s regular in the upper half-plane. It is also regular in the lower half-plane, but there is a cut from $-\infty$ to $+\infty$, and the values in the two half-planes are not one another's continuation. We are interested in $G_{\hat{y}}$ only in the upper half-plane and, particularly, its boundary value for real s. The asymptotic values of $G_{\hat{y}}$ for large r and real s can be obtained from **(XIV.3.3)**. The result is

$$G_{\hat{y}}(\mu^2, s; \mathbf{r}) = -(4\pi r)^{-1}[e^{ikr}\theta(sr - kr_0) + e^{-ikr}\theta(-sr - kr_0)] + o(r^{-1}),$$

$$\textbf{(XIV.3.4)}$$

where θ is the well-known Heaviside function.

The Wave Function. We use the Green's function (**XIV.3.1**) to define a solution of (**XIV.1.1**) through the integral equation.

$$\varphi_{\hat{y}}(\mathbf{k}_\perp, s, \mathbf{r}) = e^{i\mathbf{k}_\perp \cdot \mathbf{r}_\perp + isr_0} + \int G_{\hat{y}}(\mu^2, s; \mathbf{r} - \mathbf{r}')V(\mathbf{r}')\varphi_{\hat{y}}(\mathbf{k}_\perp, s, \mathbf{r}')d\mathbf{r}' \quad (\textbf{XIV.3.5})$$

or, setting

$$\boldsymbol{\varphi}_{\hat{y}} = |V|^{1/2}\exp[-isr_0]\varphi_{\hat{y}}, \quad\quad\quad (\textbf{XIV.3.6})$$

$$\Gamma_{\hat{y}}(\mu^2, s; \mathbf{r}, \mathbf{r}') = |V(\mathbf{r})|^{1/2}G_{\hat{y}}(\mu^2, s; \mathbf{r} - \mathbf{r}')e^{is(r_0' - r_0)}|V(\mathbf{r}')|^{-1/2}V(\mathbf{r}') \quad (\textbf{XIV.3.7})$$

through the integral equation:

$$\boldsymbol{\varphi}_{\hat{y}}(\mathbf{k}_\perp, s, \mathbf{r}) = |V(\mathbf{r})|^{1/2}e^{i\mathbf{k}_\perp \cdot \mathbf{r}} + \int \Gamma_{\hat{y}}(\mu^2, s; \mathbf{r}, \mathbf{r}')\boldsymbol{\varphi}_{\hat{y}}(\mathbf{k}_\perp, s; \mathbf{r}')d\mathbf{r}'. \quad (\textbf{XIV.3.8})$$

It is possible to prove that if $V(\mathbf{r})$ belongs to the Rollnick class

$$\int \frac{|V(\mathbf{r})||V(\mathbf{r}')|}{|\mathbf{r} - \mathbf{r}'|^2}\, d\mathbf{r}\, d\mathbf{r}' < \infty, \quad\quad\quad (\textbf{XIV.3.9})$$

then for all $\hat{y}$, μ, s, (with Im $s \geq 0$), the kernel $\Gamma_{\hat{y}}$ belongs to $L^2(\mathbb{R}_3 \times \mathbb{R}_3)$. Hence the integral equation (**XIV.3.8**) has a unique solution in $L^2(\mathbb{R}_3)$ for all s with Im $s \geq 0$, except at those values of s where the homogeneous form of (**XIV.3.8**) has a solution. These are the "*exceptional points*" where the modified Fredholm determinant of (**XIV.3.8**) vanishes:

$$D_{\hat{y}}(\mathbf{k}) = D_{\hat{y}}(\mu, s) = \det_{\text{mod}}(1 - G_{\hat{y}}(\mathbf{k})V) = 0. \quad\quad (\textbf{XIV.3.10})$$

Analytic Properties of the Wave Function. Like $G_{\hat{y}}$, $\Gamma_{\hat{y}}$ is holomorphic in the upper half s-plane. It follows that if V satisfies (**XIV.3.9**) and if

$$\int r|V(\mathbf{r})|d\mathbf{r} < \infty, \quad\quad\quad (\textbf{XIV.3.11})$$

then $\boldsymbol{\varphi}_{\hat{y}}$ is an analytic function of s, meromorphic in the upper half-plane, with poles at the exceptional points $s = s_m$ defined by (**XIV.3.10**).

From (**XIV.3.6**) it follows that, for each fixed μ^2, $\hat{y}$ and $\hat{r}$, $\varphi_{\hat{y}}$ is an analytic function of s meromorphic in the upper half-plane. Its properties for real s, and its properties as a function of $\mathbf{r}$, can be derived from (**XIV.3.6**), (**XIV.3.8**), and the estimates on $G_{\hat{y}}$. Thus it can be shown that $\varphi_{\hat{y}}$ is a twice differentiable function of $\mathbf{r}$ for all s with Im $s \geq 0$, a solution of the Schrödinger equation, and a continuous function of s for real s, *save for the exceptional points*. For large $|s|$, Im $s \geq 0$, and the assumptions (**XIV.3.9**) and (**XIV.3.11**), it is easy to see that $\|\Gamma_{\hat{y}}\|$ goes to zero as $|s| \to \infty$, and $D_{\hat{y}}$ goes to 1, so that

$$\boldsymbol{\varphi}_{\hat{y}}(\mathbf{k}, s, \mathbf{r}) = |V(\mathbf{r})|^{1/2}e^{i\mathbf{k} \cdot \mathbf{r}} + o(1). \quad\quad (\textbf{XIV.3.12})$$

Hence, the regularized solution

$$\phi_{\hat{y}}(\mathbf{k}_\perp, s; \mathbf{r}) = P_{\hat{y}}(\mu, s)\varphi_{\hat{y}}(\mathbf{k}_\perp, s; \mathbf{r}), \quad\quad (\textbf{XIV.3.13})$$

where

$$P_{\hat{y}}(\mu, s) = \prod_n \frac{s - s_n}{s - \bar{s}_n} \qquad \text{(XIV.3.14)}$$

is a continuous function of s on the real axis, holomorphic in the upper half-plane, and, for large $|s|$:

$$\phi_{\hat{y}}(\mathbf{k}_\perp, s; \mathbf{r}) = e^{i\mathbf{k}_\perp \cdot \mathbf{r} + isr_0} + o(e^{isr_0}) \qquad (|s| \to \infty). \qquad \text{(XIV.3.15)}$$

The Triangular Kernel. For the sake of convenience, we now use the notation k for $k_\perp + s\hat{y}$, and $\phi_{\hat{y}}(\mathbf{k}, \mathbf{r})$ for $\phi_{\hat{y}}(\mathbf{k}_\perp, s; \mathbf{r})$. It is clear from **(XIV.3.15)** that the support of the Fourier transform of $s \to (\phi_{\hat{y}}(\mathbf{k}, \mathbf{r}) - \exp[i\mathbf{k} \cdot \mathbf{r}])$ lies to the right of r_0. Thus, there exists a representation of the form

$$\phi_{\hat{y}}(\mathbf{k}, \mathbf{r}) = e^{i\mathbf{k} \cdot \mathbf{r}} - \int_{r_0}^{\infty} \tilde{K}_{\hat{y}}(\mathbf{r}; \mathbf{k}_\perp, r_0') e^{isr_0'} dr_0' \qquad \text{(XIV.3.16)}$$

which can also be written

$$\phi_{\hat{y}}(\mathbf{k}, \mathbf{r}) = \int_{-\infty}^{+\infty} \tilde{U}_{\hat{y}}(\mathbf{r}; \mathbf{k}_\perp, t) e^{ist} dt \qquad \text{(XIV.3.17)}$$

with

$$\tilde{U}_{\hat{y}}(r, \mathbf{k}_\perp, t) = \delta(t - r_0) e^{i\mathbf{k}_\perp \cdot \mathbf{r}_\perp} - \tilde{K}_{\hat{y}}(\mathbf{r}; \mathbf{k}_\perp, t) \theta(t - r_0) \qquad \text{(XIV.3.18)}$$

and

$$\tilde{U}_{\hat{y}}(r, \mathbf{k}_\perp, t) = (2\pi)^{-1} \int_{-\infty}^{+\infty} \phi_{\hat{y}}(\mathbf{k}, \mathbf{r}) e^{-ik_0 t} dk_0. \qquad \text{(XIV.3.19)}$$

Obviously $\tilde{K}$ and $\tilde{U}$ do not depend on s.

Fourier Transformed Representation. Things can be made more symmetric if we Fourier-transform $\mathbf{k}_\perp \to \phi_{\hat{y}}$ and $\mathbf{k}_\perp \to \tilde{U}_{\hat{y}}$, obtaining

$$\phi_{\hat{y}}(\mathbf{k}, \mathbf{r}) = e^{i\mathbf{k} \cdot \mathbf{r}} - \int K_{\hat{y}}(\mathbf{r}, \mathbf{r}') \theta[\hat{y} \cdot (\mathbf{r}' - \mathbf{r})] e^{i\mathbf{k} \cdot \mathbf{r}'} d\mathbf{r}'$$

$$= \int U_{\hat{y}}(\mathbf{r}, \mathbf{r}') e^{i\mathbf{k} \cdot \mathbf{r}'} d\mathbf{r}'. \qquad \text{(XIV.3.20)}$$

Clearly, the important fact that $\tilde{K}_{\hat{y}}$ (or $K_{\hat{y}}$) is triangular with respect to the direction $\hat{y}$ in $\mathbb{R}_3$ is a consequence of the definition of $\varphi_{\hat{y}}$ by means of Faddeev's Green's function, thanks to which $\phi_{\hat{y}}$ has the desirable analytic properties in the k_0-plane. The connection between $\tilde{K}$ or K and the potential will be studied below **(XIV.3.57)**, **(XIV.3.58)**.

On the Exceptional Points. Since $D_{\hat{y}}$ goes to 1 for $|s| \to \infty$, Im $s \geq 0$, there exists a semicircle in the upper half-plane around the origin outside of which there are no exceptional points. If the number of exceptional points on the real axis is finite, then the total number of exceptional points must be finite. It is easily proved that this is the case if $|V(\mathbf{r})|$ decreases exponentially as $r \to \infty$, because then $D_{\hat{y}}(\mu, s)$ has an analytic continuation to a neighborhood of the real axis.

For $\mu \neq 0$, the exceptional points occur either on the imaginary axis or in pairs symmetric with respect to it. As $\mu \to 0$ they approach the zeros of the modified Fredholm determinant of $1 - G(s)V$, where $G(s)$ is the resolvent of the Laplacean, viz., the zeros of the modified Fredholm determinant of the Lippmann–Schwinger equation. Hence, in the limit $\mu \to 0$, they approach values $s_n = iK_n$ on the imaginary axis, where $-K_n^2 = E_n$ are the eigenvalues of H. However, there can also be exceptional points on the real axis, as we shall see later. We assume in the following that their number is finite.

XIV.3.2 The "Asymptotics and Completeness" Step (Figure XIV.3)

The Usual Scattering Problem. For real s, we obtain from (**XIV.3.4**) and (**XIV.3.5**)

$$\varphi_{\hat{y}}(\mathbf{k}, \mathbf{r}) - \exp[i\mathbf{k} \cdot \mathbf{r}] = r^{-1}[e^{ikr}\theta(sr - kr_0)h_{\hat{y}}(\hat{r}, \mathbf{k})$$
$$+ e^{-ikr}\theta(-sr - kr_0)h_{\hat{y}}(-\hat{r}, \mathbf{k})] + o(r^{-1}),$$

(**XIV.3.21**)

where

$$h_{\hat{y}}(\hat{k}', \mathbf{k}) = -\frac{1}{4\pi}\int e^{-i\mathbf{k}' \cdot \mathbf{r}}V(\mathbf{r})\varphi_{\hat{y}}(\mathbf{k}, \mathbf{r})d\mathbf{r}. \qquad (\textbf{XIV.3.22})$$

We must now relate $\varphi_{\hat{y}}$ to the usual outgoing wave solution of the Schrödinger equation, both to derive a completeness relation and to connect $\varphi_{\hat{y}}$ with the scattering amplitude. This is done with the help of the incoming and outgoing functions $\varphi^{\pm}(\mathbf{k}, \mathbf{r})$, which are the well-known solutions of the Lippmann–Schwinger equation:

$$\varphi^{\pm}(\mathbf{k}, \mathbf{r}) = e^{i\mathbf{k} \cdot \mathbf{r}} + \int G^{\pm}(k; \mathbf{r} - \mathbf{r}')V(\mathbf{r}')\varphi^{\pm}(\mathbf{k}, \mathbf{r}')d\mathbf{r}', \quad (\textbf{XIV.3.23})$$

where

$$G^{\pm}(k, \mathbf{r} - \mathbf{r}') = -\frac{\exp[\pm ik|\mathbf{r} - \mathbf{r}'|]}{4\pi|\mathbf{r} - \mathbf{r}'|}. \qquad (\textbf{XIV.3.24})$$

The asymptotic form of φ^{+} is given by (**XIV.1.2**), where the scattering amplitude is given by

$$A(\hat{r}, \mathbf{k}) \equiv A_k(\hat{k}', \hat{k})$$

$$= -(4\pi)^{-1}\int \exp[-i\mathbf{k}' \cdot \mathbf{r}]V(\mathbf{r})\varphi^{+}(\mathbf{k}, \mathbf{r})d\mathbf{r}. \qquad (\textbf{XIV.3.25})$$

The S-matrix is defined from the scattering amplitude as an integral operator on the unit sphere:

$$S_k(\hat{k}', \hat{k}) = \delta(\hat{k}', \hat{k}) - \frac{k}{2\pi i} A_k(\hat{k}', \hat{k}) \tag{XIV.3.26}$$

and it connects φ^- with φ^+:

$$\varphi^+(\mathbf{k}, \mathbf{r}) = \int S_k(\hat{k}', \hat{k})\varphi^-(\mathbf{k}', \mathbf{r})d\hat{k}'. \tag{XIV.3.27}$$

"Jost Kernels". For a given value of k^2, the functions φ^+ and φ^-, separately, span the space of solutions of (**XIV.1.1**). Therefore $\varphi_{\hat{y}}$ must be representable in terms of either φ^+ or φ^-:

$$\varphi_{\hat{y}}(\mathbf{k}, \mathbf{r}) = \int F_{\hat{y}}^{\pm}(k, \hat{k}', \hat{k})\varphi^{\pm}(\mathbf{k}', \mathbf{r})d\hat{k}'. \tag{XIV.3.28}$$

The "kernels" $F_{\hat{y}}^{\pm}$ clearly correspond to the Jost functions in ordinary potential scattering. Since $\varphi_{\hat{y}}$ and $\varphi^{\pm}$ have the same plane wave $\exp[i\mathbf{k}\cdot\mathbf{r}]$ as their leading term for $r \to \infty$, we are led to define a probably more regular kernel, $q_{\hat{y}}$, by:

$$F_{\hat{y}}^{\pm}(k, \hat{k}', \hat{k}) = \delta(\hat{k}', \hat{k}) \pm \frac{k}{2\pi i} q_{\hat{y}}^{\pm}(k; \hat{k}', \hat{k}). \tag{XIV.3.29}$$

Inserting (**XIV.3.29**) in (**XIV.3.28**) and using the asymptotic behaviors (**XIV.3.21**) and (**XIV.1.2**), we obtain:

$$q_{\hat{y}}^{\pm}(k; \hat{k}', \hat{k}) = \theta[\hat{y}\cdot(\hat{k}' - \hat{k})]h_{\hat{y}}(\hat{k}', \mathbf{k}) \tag{XIV.3.30}$$

and

$$h_{\hat{y}}(\hat{k}', \mathbf{k}) = A_k(\hat{k}', \hat{k}) + \frac{k}{2\pi i}\int A_k(\hat{k}', \hat{k}'')h_{\hat{y}}(\hat{k}'', \mathbf{k})\theta[\hat{y}\cdot(\hat{k}'' - \hat{k})]d\hat{k}''. \tag{XIV.3.31}$$

This may be regarded as an integral equation for $h_{\hat{y}}$ or for $q_{\hat{y}}^+$, if the scattering amplitude A is given. Thus the Jost kernels can be derived from the scattering amplitude.

According to (**XIV.3.30**), the operator $F - 1$ is "triangular." Thus (**XIV.3.28**) is a Volterra's equation and can be solved. It is equivalent to say that F can be inverted, so that we obtain:

$$\varphi_{\hat{y}}^{\pm}(\mathbf{k}, \mathbf{r}) = \int F_{\hat{y}}^{\pm(-1)}(k; \hat{k}', \hat{k})\varphi_{\hat{y}}(\mathbf{k}', \mathbf{r})d\hat{k}'. \tag{XIV.3.32}$$

Substitution in (**XIV.3.27**) yields the decomposition

$$S_k(\hat{k}', \hat{k}) = \int F_{\hat{y}}^-(k, \hat{k}', \hat{k}'')F_{\hat{y}}^{+(-1)}(k; \hat{k}'', \hat{k})d\hat{k}''. \tag{XIV.3.33}$$

This representation of the S-matrix is the three-dimensional analog of the well-known Jost-function decomposition in the spherically symmetric case. The analogy appears also in the analytic properties since it follows from (XIV.3.32) and the analytic properties of $\varphi_{\hat{y}}$ that, as a function of s, $h_{\hat{y}}(\hat{k}', k_\perp, s)D_{\hat{y}}(\mu, s)$ is the continuous boundary value of an analytic function holomorphic in the upper half plane. The analogy appears also in symmetry properties with respect to $\hat{k}$ and $\hat{k}'$, and their connection with the unitarity and reciprocity theorem of the S-matrix (Newton 1974).

A symmetry relation with respect to $\hat{y}$ can also be obtained using the well-known orthogonality:

$$(2\pi)^{-3} \int \overline{\varphi^\pm}(\mathbf{k}, \mathbf{r})\varphi^\pm(\mathbf{k}', \mathbf{r})d\mathbf{r} = \delta(\mathbf{k} - \mathbf{k}') \qquad \text{(XIV.3.34)}$$

of the functions $\varphi^\pm$. With this and (XIV.3.27), it is not difficult to show that

$$(2\pi)^{-3} \int \overline{\varphi}_{-\hat{y}}(\mathbf{k}, \mathbf{r})\varphi_{\hat{y}}(\mathbf{k}', \mathbf{r})d\mathbf{r} = \delta(\mathbf{k} - \mathbf{k}') \qquad \text{(XIV.3.35)}$$

and

$$\int \overline{F}^\pm_{-\hat{y}}(k; \hat{k}'', \hat{k})F^\pm_{\hat{y}}(k; \hat{k}'', \hat{k}')d\hat{k}'' = \delta(\hat{k}', \hat{k}). \qquad \text{(XIV.3.36)}$$

In other words

$$F^{\pm\,-1}_{\hat{y}} = F^{\pm*}_{-\hat{y}}. \qquad \text{(XIV.3.37)}$$

Completeness. We have from (XIV.3.32)

$$\int \varphi^+(\mathbf{k}, \mathbf{r})\overline{\varphi}^+(\mathbf{k}, \mathbf{r}')d\mathbf{k}$$

$$= \int F^{-1}_{-\hat{y}}(k; \hat{k}', \hat{k})\overline{F}^{-1}_{\hat{y}}(k; \hat{k}'', \hat{k})\varphi_{-\hat{y}}(\mathbf{k}', \mathbf{r})\overline{\varphi}_{\hat{y}}(\mathbf{k}'', \mathbf{r}')d\mathbf{k}\,d\hat{k}'\,d\hat{k}''$$

$$= \int \varphi_{-\hat{y}}(\mathbf{k}, \mathbf{r})\overline{\varphi}_{\hat{y}}(\mathbf{k}, \mathbf{r}')d\mathbf{k} \qquad \text{(XIV.3.38)}$$

because of (XIV.3.37). But it follows from (XIV.3.28), (XIV.3.32) and (XIV.3.37) that

$$\varphi_{-\hat{y}}(\mathbf{k}, \mathbf{r}) = \int \varphi_{\hat{y}}(\mathbf{k}', \mathbf{r})M'_{\hat{y}}(k; \hat{k}', \hat{k})d\hat{k}', \qquad \text{(XIV.3.39)}$$

where

$$M'_{\hat{y}}(k; \hat{k}', \hat{k}) = \int \overline{F}_{-\hat{y}}(k; \hat{k}'', \hat{k}')F_{-\hat{y}}(k; \hat{k}'', \hat{k})d\hat{k}''. \qquad \text{(XIV.3.40)}$$

Therefore

$$(2\pi)^{-3} \int \varphi^+(\mathbf{k}, \mathbf{r})\bar{\varphi}^+(\mathbf{k}, \mathbf{r}')d\mathbf{k}$$

$$= (2\pi)^{-3} \int \varphi_{\hat{\gamma}}(\mathbf{k}', \mathbf{r})M'_{\hat{\gamma}}(k; \hat{k}', \hat{k})\bar{\varphi}_{\hat{\gamma}}(\mathbf{k}, \mathbf{r}')d\mathbf{k}\, d\hat{k}'. \quad \text{(XIV.3.41)}$$

But the well-known completeness of the functions φ^+ (Newton 1966) implies that the left-hand side is equal to $\delta(\mathbf{r} - \mathbf{r}') - \sum_n \varphi_n(\mathbf{r})\varphi_n(\mathbf{r}')$, where φ_n are normalized eigenfunctions of H. We therefore have the completeness relation

$$(2\pi)^{-3} \int \varphi_{\hat{\gamma}}(\mathbf{k}', \mathbf{r})M'_{\hat{\gamma}}(k; \hat{k}', \hat{k})\bar{\varphi}_{\hat{\gamma}}(\mathbf{k}, \mathbf{r}')d\mathbf{k}\, d\hat{k}' + \sum_n \varphi_n(\mathbf{r})\varphi_n(\mathbf{r}') = \delta(\mathbf{r} - \mathbf{r}').$$
$$\text{(XIV.3.42)}$$

Spectral Kernel. From (**XIV.3.3**) it is possible to show that $G_{\hat{\gamma}}(0, s; \mathbf{r})$ is equal for real positive s to the ordinary Green's function $G^+(s; \mathbf{r})$ given in (**XIV.3.24**). Thus it follows from (**XIV.3.5**) that for $\mathbf{k} = k\hat{\gamma}$ the function $\varphi_{\hat{\gamma}}$ goes over into φ^+. We must have for $k = k_n = iK_n$

$$\phi_{\hat{\gamma}}(i\hat{\gamma}K_n, \mathbf{r}) = C_n(\hat{\gamma})\varphi_n(\mathbf{r}). \quad \text{(XIV.3.43)}$$

Thus we may write (**XIV.3.42**)

$$(2\pi)^{-3} \int \phi_{\hat{\gamma}}(\mathbf{k}', \mathbf{r})M_{\hat{\gamma}}(k; \hat{k}', \hat{k})\bar{\phi}_{\hat{\gamma}}(\mathbf{k}, \mathbf{r}')d\mathbf{k}\, d\hat{k}'$$

$$+ \sum_n \phi_{\hat{\gamma}}(iK_n\hat{\gamma}, \mathbf{r}) \frac{\phi_{\hat{\gamma}}(iK_n\hat{\gamma}, \mathbf{r}')}{|C_n(\hat{\gamma})|^2} = \delta(\mathbf{r} - \mathbf{r}'), \quad \text{(XIV.3.44)}$$

where

$$M_{\hat{\gamma}}(k; \hat{k}', \hat{k}) = \frac{1}{P_{\hat{\gamma}}(\mathbf{k}')} M'_{\hat{\gamma}}(k; \hat{k}', \hat{k}) \frac{1}{\overline{P_{\hat{\gamma}}(\mathbf{k})}}. \quad \text{(XIV.3.45)}$$

We shall also use the notation $(2/k)M_{\hat{\gamma}}(\mathbf{k}, \mathbf{k}')$ instead of (**XIV.3.45**).

Fourier-Transformed Representation of Completeness Relations. Formula (**XIV.3.43**) and the representation (**XIV.3.20**) imply

$$\varphi_n(\mathbf{r}) = \int U_{\hat{\gamma}}(\mathbf{r}, \mathbf{r}')\alpha_{\hat{\gamma}}^{(n)}(\mathbf{r}')d\mathbf{r}', \quad \text{(XIV.3.46)}$$

where

$$\alpha_{\hat{\gamma}}^{(n)}(\mathbf{r}) = \frac{\exp[-K_n\hat{\gamma} \cdot \mathbf{r}]}{C_n(\hat{\gamma})}. \quad \text{(XIV.3.47)}$$

Insertion of (**XIV.3.20**) and (**XIV.3.46**) in (**XIV.3.44**) yields

$$\int U_{\hat{\gamma}}(\mathbf{r}, \boldsymbol{\rho})W_{\hat{\gamma}}(\boldsymbol{\rho}, \boldsymbol{\sigma})\bar{U}_{\hat{\gamma}}(\boldsymbol{\sigma}, \mathbf{r}')d\boldsymbol{\rho}\, d\boldsymbol{\sigma} = \delta(\mathbf{r} - \mathbf{r}'), \quad \text{(XIV.3.48)}$$

where

$$W_{\hat{y}}(\mathbf{\rho}, \mathbf{\sigma}) = (2\pi)^{-2} \int \delta(k^2 - k'^2) M_{\hat{y}}(\mathbf{k}, \mathbf{k}') e^{i(\mathbf{k}\cdot\mathbf{\rho} - \mathbf{k}'\cdot\mathbf{\sigma})} \, d\mathbf{k} \, d\mathbf{k}' + \sum_n \alpha_{\hat{y}}^{(n)}(\mathbf{\rho}) \alpha_{\hat{y}}^{(n)}(\mathbf{\sigma}).$$

$$(\textbf{XIV.3.49})$$

Now defining $m_{\hat{y}}(\mathbf{k}, \mathbf{k}')$ by

$$M_{\hat{y}}(\mathbf{k}, \mathbf{k}') = \frac{k}{2} \, \delta(\hat{k}, \hat{k}') + m_{\hat{y}}(\mathbf{k}, \mathbf{k}') \qquad (\textbf{XIV.3.50})$$

we obtain from (**XIV.3.49**)

$$W_{\hat{y}}(\mathbf{r}, \mathbf{r}') = \delta(\mathbf{r} - \mathbf{r}') + g_{\hat{y}}(\mathbf{r}, \mathbf{r}'), \qquad (\textbf{XIV.3.51})$$

where

$$g_{\hat{y}}(\mathbf{r}, \mathbf{r}') = \frac{(2\pi)^{-3}}{\delta(k^2 - k'^2)} \, m_{\hat{y}}(\mathbf{k}, \mathbf{k}') e^{i(\mathbf{k}\cdot\mathbf{r} - \mathbf{k}'\cdot\mathbf{r}')} \, d\mathbf{k} \, d\mathbf{k}' + \sum_n \alpha_{\hat{y}}^{(n)}(\mathbf{r}) \alpha_{\hat{y}}^{(n)}(\mathbf{r}').$$

$$(\textbf{XIV.3.52})$$

XIV.3.3   The "Integral Equation" Step (Figure XIV.4)

Looking at (**XIV.3.20**) and (**XIV.3.44**), we see that we are now in the position outlined in the section XIV.2.3. The equation (**XIV.3.48**) corresponds to (**XIV.2.22**) and, taking into account (**XIV.3.45**), (**XIV.3.50**), (**XIV.3.51**), we readily obtain the analogue of the Gel'fand–Levitan equation: for $\mathbf{r} \cdot \hat{y} < \mathbf{r}' \cdot \hat{y}$ we have

$$K_{\hat{y}}(\mathbf{r}, \mathbf{r}') = g_{\hat{y}}(\mathbf{r}, \mathbf{r}') + \int \theta[\hat{y} \cdot (\mathbf{\rho} - \mathbf{r})] K_{\hat{y}}(\mathbf{r}, \mathbf{\rho}) g_{\hat{y}}(\mathbf{\rho}, \mathbf{r}') d\mathbf{\rho}. \qquad (\textbf{XIV.3.53})$$

Using (**XIV.3.52**) in (**XIV.3.53**) yields the expansion

$$K_{\hat{y}}(\mathbf{r}, \mathbf{r}') = (2\pi)^{-3} \int \delta(k^2 - k'^2) m_{\hat{y}}(\mathbf{k}, \mathbf{k}') \phi_{\hat{y}}(\mathbf{k}, \mathbf{r}) e^{-i\mathbf{k}'\cdot\mathbf{r}'} \, d\mathbf{k} \, d\mathbf{k}' + \sum_n \varphi_n(\mathbf{r}) \overline{\alpha_{\hat{y}}^{(n)}(\mathbf{r}')}.$$

$$(\textbf{XIV.3.54})$$

Thus $K_{\hat{y}}$ and $g_{\hat{y}}$ satisfy the hyperbolic partial differential equations:

$$[\Delta - \Delta' - V(\mathbf{r})] K_{\hat{y}}(\mathbf{r}, \mathbf{r}') = 0, \qquad (\textbf{XIV.3.55})$$

$$(\Delta - \Delta') g_{\hat{y}}(\mathbf{r}, \mathbf{r}') = 0. \qquad (\textbf{XIV.3.56})$$

Using (**XIV.3.55**) and (**XIV.3.56**), together with integration by parts, leads to

$$2 \frac{\partial}{\partial r_0} K_{\hat{y}}(\mathbf{r}; r'_\perp, r_0) = V(\mathbf{r}) \delta(r_\perp - r'). \qquad (\textbf{XIV.3.57})$$

It must be noticed that the left-hand side of (**XIV.3.57**) has to come out independent of $\hat{y}$.

One may also work with the kernel $\tilde{K}_{\hat{y}}$ of (XIV.3.16). A Gel'fand–Levitan equation is obtained (Newton 1974) and the potential is related to $\tilde{K}_{\hat{y}}$ by

$$V(r) = 2\frac{\partial}{\partial r_0} \tilde{K}_{\hat{y}}(\mathbf{r}; \mathbf{k}_\perp, r_0)e^{-i\mathbf{k}_\perp \cdot \mathbf{r}_\perp}. \qquad \text{(XIV.3.58)}$$

XIV.3.4 Exceptional Points and Bound States

The following results are not specially linked with the present method but rather to the direct three-dimensional scattering problem.

Exceptional Points. Clearly one has to derive them from the scattering amplitude, so as to construct $P_{\hat{y}}$. We only give the results here, referring the reader to Newton (1974) for the derivation. Let $D^+(s)$ be the modified Fredholm determinant of the Lippmann–Schwinger equation. $D_{\hat{y}}(\mathbf{k})$ is related to $D^+(\mathbf{k})$ by the formula

$$D_{\hat{y}}(\mathbf{k}) = D^+(k)d_{\hat{y}}(\mathbf{k})\exp[\beta_{\hat{y}}(\mathbf{k})], \qquad \text{(XIV.3.59)}$$

where

$$\beta_{\hat{y}}(\mathbf{k}) = \frac{i}{4\pi} (k - \mathbf{k} \cdot \hat{y}) \int V(\mathbf{r})d\mathbf{r}, \qquad \text{(XIV.3.60)}$$

$$d_{\hat{y}}(\mathbf{k}) = \det(1 - A'_{\hat{y}}), \qquad \text{(XIV.3.61)}$$

$$A'_{\hat{y}}(\mathbf{k}; \hat{k}, \hat{k}') = \frac{k}{2\pi i} \theta[\hat{y} \cdot (\hat{k}' - \hat{k})]A_k(\hat{k}, \hat{k}').$$

Thus, $d_{\hat{y}}(\mathbf{k})$ is the Fredholm determinant of (XIV.3.31).

As for $D^+(k)$, it can itself be constructed by the formula

$$D^+(k) = \lim_{\varepsilon \to 0^+} \prod_n \left(1 - \frac{E_n}{E}\right)\exp\left[\frac{1}{\pi}\int_0^\infty \frac{\arg D^+(k')dE'}{E' - E - i\varepsilon}\right], \qquad \text{(XIV.3.62)}$$

where $\arg D^+(k)$ is related to the Fredholm determinant of the S operator (XIV.3.26) and to the potential $V(\mathbf{r})$ by the formula

$$\arg D^+(k) = \tfrac{1}{2}i \operatorname{Log} \det S - \frac{k}{4\pi}\int V(\mathbf{r})d\mathbf{r} \qquad \text{(XIV.3.63)}$$

and $E_n = -K_n^2$ are the eigenvalues of H.

Thus $D_{\hat{y}}(\mathbf{k})$ can be constructed from the scattering amplitude. Once $D_{\hat{y}}(\mathbf{k}) \equiv D_{\hat{y}}(\mu, s)$ is known on the real axis, we can find the exceptional points s'_n, its zeros in the upper half-plane. Because $D_{\hat{y}}$ goes to 1 as $|s| \to \infty$, this may be done, for example, by taking the Fourier transform of $(1/D_{\hat{y}} - 1)$:

$$\frac{1}{2\pi i}\int_{-\infty}^{+\infty} \left\{\frac{1}{D_{\hat{y}}(\mu, s)} - 1\right\}e^{ist} ds = \sum_n \frac{e^{its_n}}{D'_{\hat{y}}(\mu, s_n)} \qquad \text{(XIV.3.64)}$$

for $t > 0$, where $D'_{\hat{y}}$ is the derivative of $D_{\hat{y}}$ with respect to s.

Bound States and Normalization Constants. We recall that the bound states are zeros of the modified Fredholm determinant of the Lippmann–Schwinger equation (**XIV.3.23**) in $\mathbb{C}^+$. This determinant may also have zeros on the real axis, which are called *exceptional points* of the Lippmann–Schwinger equation, but it has been proved (Newton, 1982a) that they cannot exist except at $k = 0$ if the potential belongs to a wide class of potentials, including in particular those for which there exist $\varepsilon > 0$, $a > 0$ such that $(a + |r|)^{3+\varepsilon}|V(r)| < C < \infty$. In addition, for the sake of simplicity, we do not consider here the case of an exceptional point at $k = 0$. With these assumptions, it is possible to show that the K_n's are given from the forward scattering amplitude by the formula, valid for $t > 0$,

$$\int_{-\infty}^{+\infty} A(\mathbf{k}, \mathbf{k}) e^{ikt}\, dt + \int d\mathbf{x}\, V(\mathbf{x}) = -\sum_n K_n |c_n(\hat{k})|^2 \exp[-K_n t] \quad \textbf{(XIV.3.65)}$$

which also allows us to obtain a set of coefficients $|c_n(\hat{k})|$ that yields in turn $C_n(\hat{\gamma})$ according to

$$C_n(\hat{\gamma}) = c_n(\hat{\gamma}) \frac{i}{2K_n} \prod_m \frac{K_m - K_n}{K_m + K_n}. \quad \textbf{(XIV.3.66)}$$

Once the bound states and normalization constants have been obtained from (**XIV.3.65**) and (**XV.3.66**), the exceptional points from (**XIV.3.64**), we may be able to solve the linear integral equation (**XIV.3.31**) for $h_{\hat{\gamma}}$. That gives us $F_{\hat{\gamma}}$ and thus all the input in the Gel'fand–Levitan equation, from which we get K and the potential.

XIV.4 Consistency and $\bar{\partial}$-Approaches

We have to study the integral equations (**XIV.3.31**) and (**XIV.3.53**). If we assume that the scattering amplitude we are starting from is associated with a potential, then we know that solutions exist. However, if these solutions are not unique, then we do not know if the machinery will work.

XIV.4.1 Consistency of the Original Approach

If, at a given energy, $A_k(\hat{k}', \hat{k})$, as an integral kernel on the unit sphere, belongs to L_2, then (**XIV.3.31**) is a regular Fredholm equation. That will be the case if

$$\int \sigma_{\text{tot}}(\mathbf{k}) d\hat{k} = \int |A_k(\hat{k}', \hat{k})|^2\, d\hat{k}\, d\hat{k}' < \infty. \quad \textbf{(XIV.4.1)}$$

If furthermore $\int d\hat{k} A_k(\hat{k}, \hat{k})$ exists, then $d_{\hat{\gamma}}(\mathbf{k})$ exists. It may be zero: if that happens, $D_{\hat{\gamma}}(\mathbf{k})$ is also zero and there is an exceptional point on the real axis.

Then the kernels of the Gel'fand–Levitan equations are not locally square integrable. So exceptional points have to be ruled out by assumption. Clearly, a (very strong) sufficient condition is

$$\left(\frac{k}{2\pi}\right)^2 \int \sigma_{\text{tot}}(\hat{k})d\hat{k} < 1. \qquad \text{(XIV.4.2)}$$

The Gel'fand–Levitan equation introduced by Faddeev (1971) to solve the present problem and the two equations introduced by Newton (1974) were studied by Newton (1974). He was able to prove that the solution, *if it exists*, is unique, but in general was not able to prove the existence of a solution.

Faddeev (1971) has given a necessary and sufficient condition for the existence of an underlying *local* potential, but with only a rough outline of the proof (as regards the "sufficiency"). The condition is that the function $h_{\hat{\gamma}}(\mathbf{k'}, \mathbf{k})$ be such that $h_{\hat{\gamma}}(\mathbf{k'_\perp}, s; \mathbf{k_\perp}, s)D_{\hat{\gamma}}(k_\perp, s)$ is the boundary value of an analytic function of s which is regular in the upper half-plane.

XIV.4.2   The Problem of the Exceptional Points

The following three theorems give the main results on existence or nonexistence of exceptional points (real or complex) that have been proved in addition to the sufficient condition of section XIV.4.1 (for proofs, see Lavine and Nachman (1987a,b) and Henkin and Novikov (1988a)).

Theorem A. *There exists a constant C such that if $C \int |V(\mathbf{x})|^{3/2} \, d\mathbf{x} < 1$, there are no exceptional points. C can be given explicitly.*

Theorem B. *Suppose $|V(\mathbf{x})| \le C(1 + |\mathbf{x}|^2)^{-1/2-\varepsilon}$. There are no exceptional points k with $|\mathbf{k}| \ge Cd_n(\varepsilon)$. The proof gives an explicit constant $d_n(\varepsilon)$.*

Theorem C. *Assuming that $V \in L^{3/2}(\mathbb{R}^3)$, Lavine and Nachman show that if V has at least one bound state, then for any γ of unit length there is at least one pair of reals (s, l) with $l > 0$ such that every $k = s\hat{\gamma} + k_\perp$ with $|k_\perp| = l$ is an exceptional point. With the assumption that $V \in \mathscr{C}_\beta^\alpha$ ($\alpha \ge 4, \beta \ge 4$),*

$$\mathscr{C}_\beta^\alpha = \{V | \forall p', p' \in (0, \alpha), \exists C: |D_{\mathbf{x}}^{p'} V| \le C(1 + |\mathbf{x}|)^{-\beta}\}. \qquad \text{(XIV.4.3)}$$

Henkin and Novikov obtained similar results independently and emphasize that there exist real exceptional points *if* there exist complex exceptional points, e.g., bound states.

XIV.4.3   Characterization Problem and
Shortenings of the Approach

The original Faddeev–Newton method is so complicated that we can hardly expect to use it once for solving an inverse problem. Furthermore, although the method is a little reduced in the case of axial symmetry, it does not simplify up to the radial Gel'fand–Levitan method in the case of spherical symmetry. Finally, when there are bound states, we know now that there are also exceptional points on the real axis, and we do not know how the main tools in the method are to be constructed. There is a need for a better method, effective at least when there is no real exceptional point. Such a method has been obtained by means of $\bar{\partial}$-methods (presented on simpler problems in section XVII.2.4 below). In order to realize how this method is shorter, let us first remark that the only parts of section XIV.3 we need to know are those on the Green's function, the wave function and its analytic properties in section XIV.3.1, the part on the scattering problem, and the Jost kernel in section XIV.3.2. This last function is in fact the essential tool but we prefer now to introduce its "off-shell" continuation, which we shall call the $\hat{\gamma}$-scattering transform, and which is defined by

$$h_{\hat{\gamma}}(\mathbf{k}', \mathbf{k}) \equiv -2\pi^2 H_{\hat{\gamma}}(\mathbf{k}, \mathbf{k}' - \mathbf{k}) = -\frac{1}{4\pi}\int e^{-i\mathbf{k}'\cdot\mathbf{r}}V(\mathbf{r})\varphi_{\hat{\gamma}}(\mathbf{k}, \mathbf{r})d\mathbf{r}, \quad \text{(XIV.4.4)}$$

which is similar to (XIV.3.22) but now holds for any two vectors $\mathbf{k}'$, $\mathbf{k}$. $(H_{\hat{\gamma}})$ is the function used by Henkin and Novikov, whereas the Lavine and Nachman "scattering transform" $T(\mathbf{k}', \mathbf{k}, \hat{\gamma})$ is simply $-4\pi h_{\hat{\gamma}}(\mathbf{k}', \mathbf{k}))$. The off-shell scattering amplitude $A(\mathbf{k}', \mathbf{k})$ can also be defined from (XVII.3.25) (and could be seen as a $\hat{k}$-scattering transform: the Lavine–Nachman notation A differs from ours only by the factor -4π). The scattering amplitude $A(\mathbf{k}', \mathbf{k})$ can be read off "on-shell" (that is, for those vectors $\mathbf{k}'$ in $\mathbb{R}^3$ with $k' = k$) from the asymptotic form of the *physical wave function*, i.e., equation (XIV.1.2), which we rewrite for clarity, with the new notation

$$\Psi(\mathbf{k}, \mathbf{r}) = \exp[i\mathbf{k}\cdot\mathbf{r}] + r^{-1}\exp[ikr]A\left(k\frac{\mathbf{r}}{r}, \mathbf{k}\right) + o(r^{-1}). \quad \text{(XIV.4.5)}$$

It is clear that $h_{\hat{\gamma}}$ is related to the $\hat{\gamma}$ wave function $\varphi_{\hat{\gamma}}$ exactly as A is related to the physical wave function Ψ (compare (XIV.3.21) with (XIV.1.2) or (XIV.4.5) and (XIV.4.4) or (XIV.3.22) with (XIV.3.25)).

Now we would like to characterize amongst all function $A(\mathbf{k}', \mathbf{k})$ (given for all $\mathbf{k}'$, $\mathbf{k}$ in $\mathbb{R}^3$ *with* $k' = k$) those which correspond to local (real-valued, short-range) potentials V. Some necessary conditions are well known: unitarity (the generalized optical theorem), reciprocity, and forward analyticity; they are also known not to be sufficient. Unfortunately, there still exists *no direct characterization* of A. The $\bar{\partial}$-method will enable us to understand the overdetermined problem in $\mathbb{R}^3$ by relating it to the linear overdetermined system of the Cauchy–Riemann equations, and obtaining in this way *a characterization of the scattering transform* $h_{\hat{\gamma}}$. This is the beautiful body of results derived in the works of Henkin and Novikov (1987, 1988a), and Lavine and Nachman (1987a,b). We now try to sketch them.

Let $\mathbf{k} \in \mathbb{C}^3$, write $\mathbf{k} = \mathbf{k}_R + i\mathbf{k}_I$ with $\mathbf{k}_R, \mathbf{k}_I \in \mathbb{R}^3$, and assume, to begin with, that $\mathbf{k}_I \neq 0$. We construct a family $\psi(\mathbf{k}, \mathbf{x})$ of solutions of

$$[-\Delta + V(\mathbf{x})]\psi(\mathbf{k}, \mathbf{x}) = \mathbf{k}^2\psi(\mathbf{k}, \mathbf{x}), \quad \text{(XIV.4.6)}$$

or, setting $\psi(\mathbf{k}, \mathbf{x}) = \mu(\mathbf{k}, \mathbf{x})e^{i\mathbf{k}\cdot\mathbf{x}}$,

$$(\Delta + 2i\mathbf{k}\cdot\nabla)\mu = V\mu. \qquad\qquad \textbf{(XIV.4.7)}$$

Define a Green's function for **(XIV.4.7)** by

$$G(\mathbf{k}, \mathbf{x}) = -(2\pi)^{-3}\int e^{i\mathbf{x}\cdot\xi}(\xi^2 + 2\mathbf{k}\cdot\xi)^{-1}\, d\xi. \qquad\qquad \textbf{(XIV.4.8)}$$

It is not hard to check that, since $\mathbf{k}_I \neq 0$, the function $\xi \mapsto (\xi^2 + 2\mathbf{k}\cdot\xi)^{-1}$ is locally integrable so that G is well defined, at least as a tempered distribution, and satisfies

$$(\Delta + 2i\mathbf{k}\cdot\nabla)G(\mathbf{k}, \mathbf{x}) = \delta(\mathbf{x}). \qquad\qquad \textbf{(XIV.4.9)}$$

For $\hat{y} \in \mathbb{R}^3 \backslash 0$ and $\mathbf{k}_I = q\hat{y}$, comparing **(XIV.4.8)** and **(XIV.3.3)** shows a trivial relation between this Green's function and the Faddeev–Newton's one

$$[\exp[i\mathbf{k}\cdot\mathbf{x}]G(\mathbf{k}, \mathbf{x})]_{\mathbf{k}=\mathbf{k}_R + iq\hat{y}} = G_{q\hat{y}}[(\mathbf{k} + iq\hat{y})^2, \mathbf{x}]. \qquad \textbf{(XIV.4.10)}$$

We can use G to derive μ by solving the integral equation

$$\mu(\mathbf{k}, \mathbf{x}) = 1 + \int G(\mathbf{k}, \mathbf{x} - \mathbf{y})V(\mathbf{y})\mu(\mathbf{k}, \mathbf{y})d\mathbf{y}. \qquad\qquad \textbf{(XIV.4.11)}$$

We assume that $V \in L^{3/2} \cap L^1$. Equation **(XIV.4.11)** has a unique solution unless $\mathbf{k}$ is an exceptional point. We discard this case. Writing down briefly the right-hand side of **(XIV.4.11)** in the form $G * V\mu$, and using **(XIV.4.8)**, we successively obtain

$$\frac{\partial \mu}{\partial \bar{k}_j} = \frac{\partial G}{\partial \bar{k}_j} * V\mu + G * V\frac{\partial \mu}{\partial \bar{k}_j}, \qquad\qquad \textbf{(XIV.4.12j)}$$

$$\frac{\partial G}{\partial \bar{k}_j}(\mathbf{k}, \mathbf{x}) = -(2\pi)^{-2}\int e^{i\mathbf{x}\xi}\xi_j\delta(\xi^2 + 2\mathbf{k}\cdot\xi)d\xi, \qquad \textbf{(XIV.4.13j)}$$

$$\frac{\partial G}{\partial \bar{k}_j} * V\mu = -(2\pi)^{-2}\int e^{i\mathbf{x}\cdot\xi}\xi_j\delta(\xi^2 + 2\mathbf{k}\cdot\xi)t(\xi, \mathbf{k})d\xi, \qquad \textbf{(XIV.4.14j)}$$

where we have defined

$$t(\xi, \mathbf{k}) = \int e^{-i\xi\cdot\mathbf{y}}V(\mathbf{y})\mu(\mathbf{y}, \mathbf{k})d\mathbf{y}. \qquad\qquad \textbf{(XIV.4.15)}$$

Inserting **(XIV.4.14j)** into **(XIV.4.12j)** yields an integral equation in $\partial\mu/\partial\bar{k}_j$ where the inhomogeneous term is given by **(XIV.4.14j)**. Since it is a linear combination of $e^{i\mathbf{x}\cdot\xi}$ with $\xi^2 + 2\mathbf{k}\cdot\xi = 0$ (that is, $\xi^2 + 2\mathbf{k}_R\cdot\xi = 0$ and $k_I\cdot\xi = 0$), we calculate the solution by the superposition principle from the solutions of

$$v(\mathbf{k}, \xi, \mathbf{x}) = e^{i\mathbf{x}\xi} + G(\mathbf{k}, \cdot) * Vv, \qquad\qquad \textbf{(XIV.4.16)}$$

where ξ is such that $\xi^2 + 2\mathbf{k}\cdot\xi = 0$. For such ξ, $e^{-i\mathbf{x}\xi}G(\mathbf{k}, \mathbf{x}) = G(\mathbf{k} + \xi, \mathbf{x})$, so that, setting $\tilde{v}(\mathbf{k}, \xi, \mathbf{x}) = e^{-i\mathbf{x}\cdot\xi}v(\mathbf{k}, \xi, \mathbf{x})$, we get

$$\tilde{v}(\mathbf{k}, \xi, \mathbf{x}) = 1 + G(\mathbf{k} + \xi, \cdot) * V\tilde{v}. \qquad\qquad \textbf{(XIV.4.17)}$$

It is easy to see that if $\mathbf{k}$ is not exceptional, neither is $\mathbf{k} + \xi$, for any ξ with $\xi^2 + 2\mathbf{k}\cdot\xi = 0$. Thus **(XIV.4.11)** with $\mathbf{k} + \xi$ replacing $\mathbf{k}$ has a unique solution and $v(\mathbf{k}, \xi, \mathbf{x}) = e^{i\mathbf{x}\cdot\xi}(\mathbf{x}, \mathbf{k} + \xi)$. The solution of (**XIV.4.12j**) is therefore given by the superposition principle

$$\frac{\partial \mu}{\partial \bar{k}_j}(\mathbf{k}, \mathbf{x}) = -(2\pi)^{-2} \int e^{i\mathbf{x}\cdot\xi}\xi_j \delta(\xi^2 + 2\mathbf{k}\cdot\xi)t(\xi, \mathbf{k})\mu(\mathbf{x}, \mathbf{k} + \xi)d\xi. \quad \text{(XIV.4.16j)}$$

Differentiating formula (**XIV.4.15**) and using (**XIV.4.16j**) gives

$$\frac{\partial t}{\partial \bar{k}_j}(\mathbf{k}, \xi) = -(2\pi)^{-2} \int \eta_j \delta(\eta^2 + 2\mathbf{k}\cdot\eta)t(\eta, \mathbf{k})t(\xi - \eta, \mathbf{k} + \eta)d\eta \quad \text{(XIV.4.17j)}$$

for $\mathbf{k} \in C^3 \backslash \mathbb{R}^3$, $\mathbf{k}$ not exceptional.

The function t, defined by (**XIV.4.15**), is trivially related to the scattering transform: from (**XIV.4.10**), (**XIV.4.11**), and (**XIV.4.15**), compared with (**XIV.4.4**), we easily obtain, for $\mathbf{k}, \mathbf{k}' \in \mathbb{R}^3$,

$$h_{\hat{y}}(\mathbf{k}', \mathbf{k}) = -\frac{1}{4\pi} \lim_{q \to 0^+} t(\mathbf{k}' - \mathbf{k}, \mathbf{k} + iq\hat{y}). \quad \text{(XIV.4.18)}$$

The potential can itself be related directly to t and hence to $h_{\hat{y}}$.

Using the generalized Cauchy formula (**XVII.2.45**) and the asymptotic behavior of μ, we obtain from (**XIV.4.16j**)

$$\mu = 1 - \frac{1}{4\pi^3} \iiint \frac{\xi_j e^{i\mathbf{x}\cdot\xi}}{k_j - k_j'} \delta(\xi^2 + 2\mathbf{k}'\cdot\xi)t(\xi, \mathbf{k}')\mu(\mathbf{x}, \mathbf{k}' + \xi)d\xi\, dk'_{Ij}\, dk'_{Rj},$$
$$\text{(XIV.4.19)}$$

where it is well understood that the right-hand side does not depend on the choice of j. Comparing the large k_j behavior in (**XIV.4.7**) and (**XIV.4.19**) yields

$$V(\mathbf{x}) = \frac{2i}{\pi} \frac{\partial}{\partial x_j} \iint \frac{\partial \mu}{\partial \bar{k}_j}(\mathbf{k}, \mathbf{x})dk_{Ij}\, dk_{Rj}. \quad \text{(XIV.4.20)}$$

The fact that the right-hand side of (**XIV.4.20**) does not depend on j is analogous to the "miracle" in the procedure of Newton; however, in our case, it appears naturally and may be "guaranteed" by the characterization equations which follow. They are found from the compatibility conditions $\partial^2 \mu/\partial \bar{k}_i \partial \bar{k}_j = \partial^2 \mu/\partial \bar{k}_j \partial \bar{k}_i$. They lead to the following equation for t restricted to $\{(\xi, \mathbf{k}): \xi^2 + 2\mathbf{k}\cdot\xi = 0, k_I \neq 0\}$

$$\xi_j \frac{\partial t}{\partial \bar{k}_i} - \xi_i \frac{\partial t}{\partial \bar{k}_j} = -\frac{1}{4\pi^2} \int (\xi_j \eta_i - \xi_i \eta_j)\delta(\eta^2 + 2\mathbf{k}\cdot\eta)t(\eta, \mathbf{k})t(\xi - \eta, \mathbf{k} + \eta)d\eta.$$
$$\text{(XIV.4.21)}$$

After some algebra, condition (**XIV.4.21**) is expressed in terms of $h_{\hat{y}}$ by imposing that for all vectors $\mathbf{w}$ such that $\mathbf{w}(\xi - \mathbf{k}_R) = 0$

$$\mathbf{w} \cdot \nabla_{\hat{y}} h_{\hat{y}}(\xi, \mathbf{k}_R) = -i\pi^{-1} \int \mathbf{w} \cdot (\eta - \mathbf{k}_R)\delta(\eta^2 - k_R^2), \quad \text{(XIV.4.22a)}$$

$$\delta[\hat{y} \cdot (\eta - \mathbf{k}_R)]h_{\hat{y}}(\eta, \mathbf{k}_R)h_{\hat{y}}(\xi, \eta)d\eta = -i\pi^{-1}\mathbf{w} \cdot N[h](\hat{y}, \mathbf{k}_R, \xi). \quad \text{(XIV.4.22b)}$$

The results on the direct problem given above were derived by various authors for various classes of potentials, all containing at least the class $S = C_\infty^\infty$. It seems that they are all valid in the much larger class $L^{3/2}(\mathbb{R}^3)$. The first step of the inverse problem is always the construction of the scattering transform $h_{\hat{y}}$ on-shell from the scattering amplitude A on-shell (the supposed experimental result) by solving the integral equation (**XIV.3.31**). The scattering transform $h_{\hat{y}}$ can correspond to a local real-valued potential if $h_{\hat{y}}(s\hat{y} + \xi_\perp, s\hat{y} + k_\perp)$ is analytic in s, and if the function t, which is its continuation, see (**XIV.4.18**), satisfies conditions (**XIV.4.21**) or (**XIV.4.17j**). μ and V can then be reconstructed by means of (**XIV.4.19**) and (**XIV.4.20**) and the "miracle" is guaranteed. There is a more condensed formulation of the characterization and reconstruction formulas, which is derived from (**XIV.4.22**) by the generalized Cauchy integration (Lavine–Nachman); we only state it. $h_{\hat{y}}$ is acceptable if there exists a function $\hat{V}(\mathbf{p})$ in $L^3(\mathbb{R}^3)$ that eventually turns out to be the Fourier transform of the potential, and is such that for any ξ, $\mathbf{k}$ with $\xi^2 = k^2$, $\hat{y}\cdot\xi = \hat{y}\cdot\mathbf{k}$, and every $\mathbf{w}$ perpendicular to both ξ and $\mathbf{k}$, but not parallel to $\hat{y}$,

$$-\hat{V}(\xi - \mathbf{k}) = 4\pi h_{\hat{y}}(\xi, \mathbf{k}) + \lim_{\varepsilon \to 0^+} \int\!\!\int \mathbf{w}\cdot N[h](u\mathbf{w} + \hat{y}, s\mathbf{w} + \xi, s\mathbf{w} + \mathbf{k})\frac{du\,ds}{s + i\varepsilon u}.$$
$$\text{(XIV.4.23)}$$

The method extends to the case of exceptional points: h is then a meromorphic function and the characterization equation has additional terms corresponding to the poles of h at the exceptional points. We may continue to recover $\hat{V}$ directly (Lavine and Nachman, 1987b; Henkin and Novikov, 1988a). The possible characterization of $h_{\hat{y}}$ and the easy construction of V once $h_{\hat{y}}$ belongs to the proper class suggests that it is possible, starting from a reasonable "experimental" scattering amplitude A, to obtain a quasi-solution of the problem by first constructing an acceptable scattering transform which is closest to the function $h_{\hat{y}}$ derived from A by solving (**XIV.2.25**).

XIV.5   Other Approaches in the Frequency Domain

Even in its simplest form, the approach based on Faddeev's Green function shows difficulties due to the existence of exceptional points on the real k-axis outside the origin. Since these exceptional points may have no real physical meaning, it is interesting to avoid them by means of other approaches. However, up to now, the price to be paid is such that no method can be considered definitely convincing and completely superseding the previous approach in the form it has been now reduced to. We shall first sketch two examples, due to R. G. Newton, giving only an outline of the first one and describing a little more seriously the second one. Then we shall briefly review other approaches.

XIV.5.1   The Gel'fand–Levitan-Type Approach

Starting from the S-matrix (**XIV.3.26**), which is an operator on $\mathbf{h} = L^2(S^2)$, also depending on k and written briefly as $S(k)$, Newton (1980a,b,c, 1982b,c,

1983) defines another operator on **h**, called the "Jost function" $J(k)$, as *the solution* of the Riemann–Hilbert problem (for more details on this problem, see section XIV.5.2)

$$\left.\begin{aligned} & J(-k) = QJ(k)QS \qquad (k \in \mathbb{R}), \\ & J \text{ holomorphic in } \mathbb{C}^+, \\ & \lim_{|k| \to \infty} \|J(k) - I\| = 0, \end{aligned}\right\}$$ (XIV.5.1)

where Q is the operator such that for any $f \in \mathbf{h}$ and any $\theta \in S^2$, $(Qf)(\theta) = f(-\theta)$; $\|\cdot\|$ is a convenient norm for operators on **h**, and I is the unit operator on **h**. Clearly, θ can be equivalently considered as a point on the unit sphere S^2 or as a unit vector in $\mathbb{R}^3$. $J(k)$ is then used to define (from the *scattering function* $\varphi(k, \theta, \mathbf{x})$ ($\equiv \varphi^+(\mathbf{k}, \mathbf{x})$, with $\theta = \mathbf{k}/k$) which appeared in (XIV.3.23), the *wave function* $\phi(k, \theta, \mathbf{x})$)

$$\phi(k, \theta, \mathbf{x}) = J(k)\varphi(k, \theta, \mathbf{x}).$$ (XIV.5.2)

Under conditions that are stated by Newton (1982a), and are fulfilled if $\exists a,\ \varepsilon,\ C,\ \mu > 0$ such that for all $\mathbf{x} \in \mathbb{R}^3$, $|V(\mathbf{x})| \le C(a + |\mathbf{x}|)^{-3-\varepsilon}$ and $|\nabla V(\mathbf{x})| \le C(a + |\mathbf{x}|)^{-2-\varepsilon}$ and in addition the Fourier transform $\tilde{V}$ verifies $|\tilde{V}(\mathbf{k})| \le C(\mu + |\mathbf{k}|^2)^{-1}$, the bounds and analytic properties of $k \to \varphi(k, \cdot)$ are sufficient to yield the Povzner–Levitan representation

$$\phi(k, \theta, \mathbf{x}) = \exp[ik\theta \cdot \mathbf{x}] - \int_{-|\mathbf{x}|}^{|\mathbf{x}|} d\alpha \exp[ik\alpha] w(\alpha, \theta, \mathbf{x}).$$ (XIV.5.3)

In fact, the wave function ϕ is an entire function of k, of exponential type $|\mathbf{x}|$. It is the appropriate generalization of the solution which in one dimension is defined by boundary conditions at $x = 0$. Its existence, however, depends on the existence of a solution to the Riemann–Hilbert problem.

The definition (XIV.5.2) and the Lippmann–Schwinger equation (XIV.3.23) yield the integral equation for ϕ, which can be Fourier-transformed to obtain properties of $w(\alpha, \theta, \mathbf{x})$. As a result ∇w has the discontinuity

$$2\theta \cdot \nabla[w(\theta \cdot \mathbf{x}^+, \theta, \mathbf{x}) - w(\theta \cdot \mathbf{x}^-, \theta, \mathbf{x})] = V(\mathbf{x})$$ (XIV.5.4)

and the symmetry property

$$w(-\alpha, -\theta, \mathbf{x}) = w(\alpha, \theta, \mathbf{x}),$$ (XIV.5.5)

whereas, in a weak sense, we also obtain the partial differential equation

$$\left(\Delta - \frac{\partial^2}{\partial \alpha^2}\right) w = Vw \qquad (|\alpha| < |\mathbf{x}|).$$ (XIV.5.6)

Newton also introduced a distribution $h(\mathbf{x}, \mathbf{y})$ defined as the three-dimensional Fourier transform of $\phi(k, \theta, \mathbf{x})$ with regard to $\mathbf{k} = k\theta$

$$h(\mathbf{x}, \mathbf{y}) = (2\pi)^{-3} \int d\mathbf{k} e^{-i\mathbf{k}\cdot\mathbf{y}} [e^{i\mathbf{k}\cdot\mathbf{x}} - \phi(k, \theta, \mathbf{x})],$$ (XIV.5.7)

whose Radon transform is w

$$w(\alpha, \mathbf{\theta}, \mathbf{x}) = \int d\mathbf{y} h(\mathbf{x}, \mathbf{y}) \delta(\alpha - \mathbf{\theta} \cdot \mathbf{y}). \tag{XIV.5.8}$$

There are conditions sufficient to guarantee that the support of $h(x, y)$ is included inside $|y| \le x$, yielding for ϕ a three-dimensional Povzner–Levitan representation

$$\phi(\mathbf{k}, \mathbf{x}) = e^{i\mathbf{k} \cdot \mathbf{x}} - \int_{|\mathbf{y}| \le |\mathbf{x}|} d\mathbf{y} \, h(\mathbf{x}, \mathbf{y}) e^{i\mathbf{k}\mathbf{y}}. \tag{XIV.5.9}$$

Bound states and the Levinson theorem have been studied. A completeness relation has been proved, which for any function $f \in L^2(\mathbb{R}^3)$, reads

$$\left.\begin{aligned} f(x) &= \int (\hat{f}(\mathbf{k}), d\rho(\mathbf{k}^2)\phi(\mathbf{k}, \mathbf{x})), \\[2mm] \hat{f}(\mathbf{k}) &= \int d\mathbf{x} \, \overline{f(\mathbf{x})} \phi(\mathbf{k}, \mathbf{x}), \end{aligned}\right\} \tag{XIV.5.10}$$

where $(\cdot, \cdot)$ is the inner product in $L^2(S^2)$, i.e., means an integration on $\mathbf{\theta}$ over S_2, and the *spectral function* ρ is defined by

$$\frac{d\rho(k^2)}{dk^2} = \begin{cases} \frac{1}{16}\pi^{-3}k[J(k)J^*(k)]^{-1}, & k^2 > 0, \\[2mm] \sum_n M_n \delta(E - E_n), & k^2 < 0, \end{cases} \tag{XIV.5.11}$$

where J^* is the adjoint of J on $L^2(S^2)$, $-k_n^2$ are the eigenvalues, and M_n is an operator on $L^2(S^2)$ such that if $u_n^{(i)}(x)$ are the normalized eigenfunctions with eigenvalues $-|k_n|^2 = E_n$, then

$$\sum_i u_n^{(i)}(x) u_n^{(i)}(y) = (\phi(i\mathbf{k}_n, \mathbf{y}), M_n \phi(i\mathbf{k}_n, \mathbf{y})). \tag{XIV.5.12}$$

In the inverse problem, the position of bound states, i.e., eigenvalues $-k_n^2$ and the asymptotic values of the eigenvectors, the so-called "characters" (see in (XIV.5.2) below), are obtained from the S-matrix by using forward analyticity, and this information enables us in principle to construct the Jost function, which in turn achieves the determination of the M_n's. Once the spectral measure $d\rho$ is constructed we can subtract $d\rho_0$, which is given by (XIV.5.11) with $M_n = 0$ and $J = I$, and derive the symmetric kernel

$$g(\mathbf{x}, \mathbf{y}) = \int (e^{i\mathbf{k} \cdot \mathbf{x}}, d(\rho - \rho_0)e^{i\mathbf{k} \cdot \mathbf{y}}), \tag{XIV.5.13}$$

where, as above, the inner product is the $\mathbf{\theta}$ integral over S^2, and the integral sign is for the k integral. The Gel'fand–Levitan equation is obtained in the standard way, and reads ($|\mathbf{x}| > |\mathbf{y}|$)

$$h(\mathbf{x}, \mathbf{y}) = g(\mathbf{x}, \mathbf{y}) - \int_{|\mathbf{z}| < |\mathbf{x}|} d\mathbf{z} \, h(\mathbf{x}, \mathbf{z}) g(\mathbf{z}, \mathbf{y}), \tag{XIV.5.14}$$

w can then be determined by means of (**XIV.5.8**) and $V(\mathbf{x})$ by (**XIV.5.4**) or the equivalent formula

$$V(\mathbf{x}) = 2\theta \cdot \nabla[w(\theta \cdot \mathbf{x}^+, \theta, \mathbf{x}) - w(\theta \cdot \mathbf{x}^-, \theta, \mathbf{x})]. \qquad \text{(XIV.5.15)}$$

The reader will notice the similarity of (**XIV.5.15**) with (**XIV.2.6**) and (**XIV.3.57**) and, as in the latter equation, the obvious effect of the problem ill-posedness in the fact that the left-hand side of (**XIV.5.15**) must not depend on θ, whereas the right-hand side involves this vector condition, called by R. G. Newton the (necessary) "miracle." Hence the reconstruction of an underlying potential from its scattering amplitude is possible. *With central potentials it is very interesting to see that the method reduces to the usual Gel'fand–Levitan procedure for the radial fixed l problem.* The lth component of the triangular transformation kernel being $K_l(r, r)$, the "miracle" condition then reduces to impose the left-hand side of

$$-2\frac{d}{dr}K_l(r, r) = V(r) \qquad \text{(XIV.5.16)}$$

to be independent of l. By the way, notice that the radial problem illustrates how bound states are not a source of underdetermination in the three-dimensional problem, whereas they are in the one-dimensional problem: Indeed, whatever the size of V, there exists a value l_0 of l above which the centrifugal barrier is large enough to prevent any bound state, so that V is completely determined by any lth component of the scattering amplitude with $l > l_0$.

For further details on the method, the reader is referred to Newton (1983a, 1984a). The method obviously is complicated and the conditions for which it works do not seem as yet to be closed in a wide class of potentials comparable to the Rollnick one.

XIV.5.2   The Marchenko-Type Approach

This is certainly the simplest approach, particularly in the case without bound states. It relies on properties of the scattering function $\varphi^+(\mathbf{k}, \mathbf{x})$ (or $\varphi(k, \theta, \mathbf{x})$) for a large class of potentials, $\mathscr{C}$, to be described below. The function $\varphi(-k, \theta, \mathbf{x})$ is related to $\varphi(k, \theta, \mathbf{x})$ by formula (**XIV.3.27**), which can be written as

$$\varphi(-k, \theta, \mathbf{x}) = \int d\theta' \, S^{-1}(k, \theta, -\theta')\varphi(k, \theta', \mathbf{x}) \qquad \text{(XIV.5.17a)}$$

$$= \int d\theta' \, S(-k, -\theta, \theta')\varphi(k, \theta', \mathbf{x}), \qquad \text{(XIV.5.17b)}$$

the equality between the operators in these two formulas being a consequence of the well-known (Newton 1982a) unitarity of the S-matrix. In operator notation, (**XIV.5.17a**) also reads as

$$\varphi(-k) = S^{-1}Q\varphi(k). \qquad\qquad \text{(XIV.5.17c)}$$

As in the previous example, it should be kept in mind that φ is an element of a convenient Banach space, $\mathbf{h}$, of functions of $\boldsymbol\theta$, which itself can also be considered as a unit vector in $\mathbb{R}^3$ or a point of the unit sphere S^2. S is an operator on $\mathbf{h}$, and so is the flipping operator Q which transforms any function $f(\boldsymbol\theta)$ into $f(-\boldsymbol\theta)$. In addition, these functions or operators may depend on x, which is fixed, and on k. Let us define the function

$$\gamma(k, \boldsymbol\theta, \mathbf{x}) = \varphi(k, \boldsymbol\theta, \mathbf{x})\exp[-ik\boldsymbol\theta\cdot\mathbf{x}]. \qquad\qquad \text{(XIV.5.18)}$$

For any $V \in \mathscr{C}, x \in \mathbb{R}^3, \boldsymbol\theta \in S^2$, and providing $k = 0$ is not an exceptional point, which we shall assume, γ has the following properties:

(a) for $k \in \mathbb{R}$, γ is a continuous uniformly bounded function of k and $\boldsymbol\theta$;

(b) $$\int_{-\infty}^{+\infty} dk\, |\gamma(k, \boldsymbol\theta, \mathbf{x}) - 1|^2 < \infty; \qquad\qquad \text{(XIV.5.19)}$$

(c) $k \mapsto \gamma$ has an analytic continuation into $\mathbb{C}^+$ that is meromorphic there, with simple poles at those points $h = i\kappa$ where $-\kappa^2$ is a bound state eigenvalue, and, furthermore

$$\lim_{|k|\to\infty} \gamma(k, \boldsymbol\theta, \mathbf{x}) = 1. \qquad\qquad \text{(XIV.5.20)}$$

We can write **(XIV.5.17)** as an equation for γ

$$\gamma(-k) = S_{\mathbf{x}}^{-1}\, Q\gamma(k), \qquad\qquad \text{(XIV.5.21)}$$

where $S_{\mathbf{x}}$ is the operator whose kernel is

$$S_{\mathbf{x}}(k, \boldsymbol\theta, \boldsymbol\theta') = S(k, \boldsymbol\theta, \boldsymbol\theta')\exp[ik(\boldsymbol\theta - \boldsymbol\theta')\cdot\mathbf{x}], \qquad\qquad \text{(XIV.5.22)}$$

and, as such, satisfies the condition $S_{\mathbf{x}}^{-1}(-k) = QS_{\mathbf{x}}(k)Q$. We easily see that $S_{\mathbf{x}}$ is the S-matrix that corresponds to a shifted potential $V_{\mathbf{x}}(\mathbf{r}) = V(\mathbf{x} + \mathbf{r})$. Now, suppose we are given S for real k, with the evidence on S that there is no bound state, and assume we know that an underlying potential exists belonging to the class $\mathscr{C}$ which guarantees the properties studied above. Then equation **(XIV.5.21)**, together with the requirement that $\gamma(k)$ be holomorphic in $\mathbb{C}^+$ and have the asymptotics **(XIV.5.20)**, constitutes a Riemann–Hilbert problem, and since similar problems are encountered several times in inverse scattering (Newton 1983a, 1982c), let us briefly explain it.

The Riemann–Hilbert Problem in Scattering Studies. A very general problem of this kind can be studied on the following scheme, where we use the notations above (but they may now represent different spaces). $\mathbf{h}$ is a Banach space of functions, say, $f(\boldsymbol\theta)$. $\mathscr{B}$ is a space of bounded linear operators on $\mathbf{h}$, so that, if they are written in integral form, their kernels are functions of $\boldsymbol\theta$ and $\boldsymbol\theta'$. Q is the flipping operator and I is the identity operator. We are interested in elements of $\mathscr{B}$ that depend on a variable k, i.e., functions of k with values in $\mathscr{B}$. Being given, for $k \in \mathbb{R}$, one of these functions, say $\Omega(k)$, inversible but satisfying the *strong compatibility* condition $\Omega(-k) = Q[\Omega(k)]^{-1}Q$, there are two forms of this problem which are related to a classical Riemann–Hilbert problem (and also to a Wiener–Hopf problem).

A. Inhomogeneous Operator Problem Without Poles. We seek in $\mathcal{B}$ the family of operators $\Gamma(k)$ such that $\Gamma(k) - I$ is a continuous function of $k \in \mathbb{R}$ which can be continued into a holomorphic function in $\mathbb{C}^+$, goes to zero as $|k| \to \infty$ in $\mathbb{C}^+$ and $\mathbb{R}$, and satisfies the equation

$$\Gamma(-k) = \Omega(k)Q\Gamma(k)Q. \tag{XIV.5.23}$$

B. Homogeneous Operator Problem Without Poles. The properties required in the preceding problem for $\Gamma(k) - I$ are required for $\Gamma(k)$ itself.

These problems can be handled if $\Omega(k) - I$ and the assumed properties of $\Gamma(k) - I$ for $k \in \mathbb{R}$ are such that the Fourier transformation can be used (the most simple situation being, of course, if they belong to $L^2(\mathbb{R})$). If this is so, let us set

$$\omega(\alpha) = (2\pi)^{-1} \int_{-\infty}^{+\infty} dk \, \exp[ik\alpha][\Omega(k) - I]Q, \tag{XIV.5.24a}$$

$$\gamma(\alpha) = (2\pi)^{-1} \int_{-\infty}^{+\infty} dk \, \exp[-ik\alpha][\Gamma(k) - I]. \tag{XIV.5.24b}$$

We assume that the rate of convergence to zero of $(\Gamma(k) - I)$ as $|k| \to \infty$ is sufficient to enable contour integration, so that $\gamma(\alpha)$ vanishes for $\alpha < 0$

$$\Gamma(k) - I = \int_0^\infty d\alpha \, \exp[ik\alpha]\gamma(\alpha). \tag{XIV.5.24c}$$

Now if we reset (**XIV.5.13**) in the form

$$\Gamma(-k) - I = (\Omega(k) - I)Q(\Gamma(k) - I)Q + Q(\Gamma(k) - I)Q + \Omega - I, \tag{XIV.5.25}$$

and use inverse Fourier transformation, we obtain for ω and γ a convolution equation which reads

$$\gamma(\alpha) = \omega(\alpha)Q + \int_0^\infty d\beta \, \omega(\alpha + \beta)\gamma(\beta)Q + Q\gamma(-\alpha)Q. \tag{XIV.5.26}$$

The last term disappears for positive α, whereas the homogeneous problem yields (if $\Gamma(k) - I$ is replaced by $\Gamma(k)$, γ by γ_0 in (**XIV.5.24b**), and α is positive)

$$\gamma_0(\alpha) = \int_0^\infty d\beta \, \omega(\alpha + \beta)\gamma_0(\beta)Q, \tag{XIV.5.27}$$

which is the homogeneous form of the Fredholm equation (**XIV.5.26**). If the operator ω in these equations is compact, the Fredholm alternative holds and we can deduce, in particular, that if the inhomogeneous operator Riemann–Hilbert problem has a solution, it is unique if and only if the homogeneous problem has only the trivial solution $\Gamma = 0$.

C.   Function Problem. The solution of the operator Riemann–Hilbert problem can be used to construct the solution of the function Riemann–Hilbert problem, where we seek a function $F \in \mathbf{h}$ that depends on k in such a way that $k \to F$ is continuous on $\mathbb{R}$, can be continued into $\mathbb{C}^+$ into a holomorphic function, satisfies conditions of the form **(XIV.5.19)** and **(XIV.5.20)**, and satisfies the equation

$$F(-k) = \Omega(k)QF(k). \tag{XIV.5.28}$$

It suffices to note that if 1 is a function of θ equal to the number 1, and $\Gamma(k)$ is a solution of **(XIV.5.23)**, then $\Gamma(k)1$ is a solution of **(XIV.5.28)** (notice that $Q1 = 1$). The convolution equation now reads

$$f(\alpha) = \omega(\alpha)1 + \int_0^\infty d\beta\, \omega(\alpha + \beta)f(\beta) + Qf(\alpha). \tag{XIV.5.29}$$

Again the last term disappears for $\alpha > 0$, which is the only interesting range in the case without pole. The Riemann–Hilbert problem can also be studied with the assumption that Γ or F are meromorphic with simple poles satisfying certain conditions (Newton 1982b, 1983a, 1985a). These poles can then either be treated and removed step after step or be dealt with in a generalized convolution equation (see below).

The Newton–Marchenko Equation. The problem which was described by **(XIV.5.19)**–**(XIV.5.21)**, and where S_x^{-1} satisfies the strong compatibility condition, is a function Riemann–Hilbert problem. The Fourier transformations can be justified and the Fredholm alternative holds if the potential $V(x)$, $\mathbb{R}^3 \to \mathbb{R}$, in the Schrödinger equation obeyed by φ satisfies a number of conditions stated in Newton (1981a, 1983a, 1985a) (all of which are satisfied if $\exists C, a > 0, \mu > 3, \nu > 7/2$, such that $|V(x)| \le C(a + |x|)^{-\mu}$, $|\mathbf{grad}\, V(x)| \le C(a + |x|)^{-\nu}$, $\Delta V \in L^1(\mathbb{R}^3)$. We can take these conditions as a definition of $\mathscr{C}$. With them, equation **(XIV.5.29)** holds, for $\alpha > 0$, with $f \in L^2(\mathbb{R} \times S^2)$, $\omega_x 1 \in L^2(\mathbb{R} \times S^2)$

$$f(\alpha, \boldsymbol{\theta}, \mathbf{x}) = \frac{1}{2\pi} \int_{-\infty}^{+\infty} \exp[-ik\alpha][\gamma(k, \boldsymbol{\theta}, \mathbf{x}) - 1], \tag{XIV.5.30}$$

$$\omega_x(\alpha, \boldsymbol{\theta}, \boldsymbol{\theta}') = [i/(2\pi)^2] \int_{-\infty}^{+\infty} dk\, kA_x(k, -\boldsymbol{\theta}, \boldsymbol{\theta}')\exp[-ik\alpha], \tag{XIV.5.31a}$$

$$(\omega_x 1)(\alpha, \theta) = \int_{S^2} d\theta'\, \omega_x(\alpha, \boldsymbol{\theta}, \boldsymbol{\theta}'), \tag{XIV.5.31b}$$

where the shifted potential scattering amplitude is defined by

$$(k/2\pi i)A_x(k, \boldsymbol{\theta}, \boldsymbol{\theta}') = \delta(\boldsymbol{\theta}, \boldsymbol{\theta}') - S_x(k, \boldsymbol{\theta}, \boldsymbol{\theta}'). \tag{XIV.5.32}$$

With these operators, equation **(XIV.5.29)** is nothing but a Marchenko equation relating the scattering amplitude A and a Fourier transform of the wave function. We shall call it the Newton–Marchenko equation

$$f(\alpha, \boldsymbol{\theta}, \mathbf{x}) = \int_{S^2} d\boldsymbol{\theta}'\, \omega_{\mathbf{x}}(\alpha, \boldsymbol{\theta}, \boldsymbol{\theta}')$$

$$+ \int_0^\infty d\beta \int_{S^2} d\boldsymbol{\theta}' \omega_{\mathbf{x}}(\alpha + \beta, \boldsymbol{\theta}, \boldsymbol{\theta}')f(\beta, \boldsymbol{\theta}', \mathbf{x}) \qquad (\alpha > 0). \quad \textbf{(XIV.5.33)}$$

Equation (**XIV.5.24c**) can be rewritten, by using (**XIV.5.18**), as

$$\varphi(k, \boldsymbol{\theta}, \mathbf{x}) = \exp[ik\boldsymbol{\theta} \cdot \mathbf{x}] + \int_{\boldsymbol{\theta} \cdot \mathbf{x}}^\infty d\alpha \exp[ik\alpha] f(\alpha - \boldsymbol{\theta} \cdot \mathbf{x}, \mathbf{x}, \boldsymbol{\theta}) \quad \textbf{(XIV.5.34)}$$

in which we recognize once again a transformation formula. Inserting this representation of φ into the Schrödinger equation we find that f must satisfy the partial differential equation

$$\left(\Delta - \frac{\partial^2}{\partial \alpha^2}\right) f(\alpha - \boldsymbol{\theta} \cdot \mathbf{x}, \mathbf{x}, \boldsymbol{\theta}) = V(\mathbf{x})f(\alpha - \boldsymbol{\theta} \cdot \mathbf{x}, \mathbf{x}, \boldsymbol{\theta}) \quad \textbf{(XIV.5.35)}$$

and the boundary condition

$$-2\boldsymbol{\theta} \cdot \mathbf{grad}\, f(0^+, \mathbf{x}, \boldsymbol{\theta}) = V(\mathbf{x}). \quad \textbf{(XIV.5.36)}$$

Equation (**XIV.5.35**), however, must be considered in a weak sense; f is not necessarily twice differentiable. Nevertheless, equation (**XIV.5.36**) can be justified. The *reconstruction* of a potential of $\mathscr{C}$ that underlies a given scattering amplitude therefore proceeds by calculating ω by means of (**XIV.5.31**), then f as the solution of the Newton–Marchenko equation (**XIV.5.33**), then V from f by means of (**XIV.5.35**), all this being possible through convergent algorithms because the operator norm of $\omega_{\mathbf{x}}$ is smaller than 1. Again the ill-posedness of the problem shows up in the fact that the left-hand side of (**XIV.5.36**) "depends" on $\boldsymbol{\theta}$ whereas the right-hand side should not.

Now suppose the modified Fredholm determinant of the Lippmann–Schwinger equation (**XIV.3.23**) has zeros in $\mathbb{C}^+$. With the assumption on V, the only zero on the real axis is the possible exceptional point at $k = 0$, which we discard here for the sake of simplicity. Thus, zeros corresponding to the bound states are located at $k = i\kappa_n$, $\kappa_n > 0$, $n = 1, 2, \ldots, N$. Let ρ_n be the degeneracy of the jth eigenvalue, $u_n^h(\mathbf{r})$, $h = 1, 2, \ldots, \rho_n$, the corresponding orthonormalized eigenvectors. The asymptotic form of $u_n^h(\mathbf{r})$ is given by (Newton, 1982a)

$$4\pi r e^{\kappa_n r} u_n^h(r\boldsymbol{\theta}) = -Y_n^h(\boldsymbol{\theta}) + o(1), \quad \textbf{(XIV.5.37)}$$

and it can be seen that the set of angle functions $Y_n^h(\boldsymbol{\theta})$, called the *characters* of the bound state, completely determine the eigenspace. It can also be shown (Newton, 1982a, 1983a) that the forward analyticity enables us to determine not only the positions of the bound states, but also their characters from the right-hand side of (**XIV.3.65**). We can use this information by means of a generalized equation (**XIV.5.29**) which still has the same form, but is now understood for any $\alpha \in \mathbb{R}$, the part $\alpha < 0$ being submitted to the additional constraint (coming from the poles in the contour used to calculate the Fourier transform)

$$(\alpha < 0) \qquad f(\alpha, \boldsymbol{\theta}) = \sum_{n,b} P_{nb}(\mathbf{x}) \exp[\kappa_n(\alpha + \boldsymbol{\theta} \cdot \mathbf{x})] Y_n^h(-\boldsymbol{\theta}). \qquad \textbf{(XIV.5.38)}$$

Formula (**XIV.5.36**) also has to be modified to include bound states (Newton, 1983a). To conclude, it is known how to reconstruct an underlying potential in $\mathscr{C}$ from the corresponding scattering amplitude in all cases.

The *construction* problem of a potential in $\mathscr{C}$ corresponding to a given scattering amplitude is a much more difficult question, which of course cannot be separated from the *characterization* problem of a scattering amplitude $A(k, \boldsymbol{\theta}, \boldsymbol{\theta}')$ without bound states, satisfying the various symmetry relations that characterize the S-matrix (given from (**XIV.3.26**) or from (**XIV.5.32**) and (**XIV.5.22**)), and guaranteeing the "strong compatibility condition" in the Riemann–Hilbert problem. Suppose, in addition, that $\omega_{\mathbf{x}}(\alpha\boldsymbol{\theta}, \boldsymbol{\theta}')$ can be defined through (**XIV.5.31a**), preferably as the kernel of an L^2 operator in (**XIV.5.23**), in any case as the kernel of a compact operator, and suppose that the homogeneous form of the Newton–Marchenko equation (**XIV.5.33**) has no nontrivial solution. Then the Newton–Marchenko equation has a unique solution $f(\alpha)$ which gives a function $\gamma(k, \mathbf{x}, \boldsymbol{\theta})$ by the Fourier transformation

$$\gamma(k, \mathbf{x}, \boldsymbol{\theta}) = 1 + \int_0^\infty d\alpha \, e^{ik\alpha} f(\alpha, \mathbf{x}, \boldsymbol{\theta}). \qquad \textbf{(XIV.5.39)}$$

Comparing the representation of f, obtained by inverting this Fourier transformation, and that obtained from (**XIV.5.29**), we can see that $\gamma - 1$ is the boundary value of an analytic function that is holomorphic in the upper half-plane and tends to 0 for large $|k|$, and using the properties of S, we can also prove that $\gamma(-k) - S_x^{-1} Q \gamma(k)$ vanishes. The Schrödinger equation (**XIV.1.1**) is then obtained if and only if $\boldsymbol{\theta} \cdot \operatorname{grad} f(0^+, \mathbf{x}, \boldsymbol{\theta})$ is independent of $\boldsymbol{\theta}$. Then it is possible to show (Newton, 1983a) that the Schrödinger equation with $V(\mathbf{x})$ defined by (**XIV.5.36**) is solved by the function φ related to γ by (**XIV.5.18**). Hence, in this case with no bound state, the consistency condition on the right-hand side of (**XIV.5.36**) is an (indirect) characterization of $A(k, \mathbf{x}, \boldsymbol{\theta})$, to be added of course to the symmetry and L^2 conditions. It is obvious that if we stick to deriving an *exact solution* of the inverse problem it must be very unstable for reasonable topologies. Things are probably better for a quasi-solution, which could be derived by first solving the Newton–Marchenko equation from a real scattering amplitude (interpolating data and satisfying symmetry and L^2 conditions), then calculating a $V(\mathbf{r})$ which determines, in the left-hand side of (**XIV.5.36**), the closest point to $f(\alpha, \mathbf{x}, \boldsymbol{\theta})$ in $L^2(\mathbb{R}, S^2)$. Although it is not yet of practical interest, the fact that a cost functional reaches a *global* maximum (Newton, 1985c) at the solution of the Newton–Marchenko equation should be noted.

XIV.5.3   Other Approaches

The aim of all the previous approaches was to give a *general method* starting
from data which *do not rely only* on "measurements" to be made in extreme
conditions (e.g., very high frequency experiments), because these conditions
may not be consistent with the model or may correspond to a very low
signal-to-noise ratio. We may accept this situation or look for methods that
are not general in their purpose but can be enlarged if necessary. We see now
examples of either point of view.

Saito (1982a,b, 1984, 1986) further studied the relations between $V(\mathbf{r})$ and
the large k behavior of the scattering amplitude $A(k, \theta, \theta)$. He was able to add
a new formula to the Born formula (**XIV.1.3**). For $\mathbf{x} \in \mathbb{R}^3$, let us introduce the
function

$$f(\mathbf{x}, k) = k^2 \int_{S^2} d\theta\, e^{-ik\theta \cdot \mathbf{x}} \int_{S^2} d\theta'\, e^{-ik\theta' \cdot \mathbf{x}} A(k, \theta, \theta'), \quad \textbf{(XIV.5.40)}$$

and assume that for any $\mathbf{r} \in \mathbb{R}^3$, $V(\mathbf{r}) \le C_0 (1 + |\mathbf{r}|)^{-\beta}$, with $\beta > 1$. Then the
limit

$$f(x, \infty) = \lim_{k \to \infty} f(x, k) = -2\pi \int_{\mathbb{R}^3} V(\mathbf{y})|\mathbf{y} - \mathbf{x}|^{-2}\, d\mathbf{y} \quad \textbf{(XIV.5.41)}$$

exists. The potential $V(\mathbf{x})$ can be recovered by the double Fourier transform,
written here in integral form,

$$V(\mathbf{x}) = -(8\pi^6)^{-1} \int d\xi\, |\xi| e^{i\mathbf{x} \cdot \xi} \int dy\, e^{-iy \cdot \xi} f(y, \infty). \quad \textbf{(XIV.5.42)}$$

The left-hand side is, in general, well defined only as a tempered distribution.
However, if $\beta > 7/4$ in the bound for V, and if the two first derivatives of V
also satisfy this bound, but with $\beta > 5/2$, it is possible to derive bounds for
$f(\mathbf{x}, k) \quad f(\mathbf{x}, \infty)$ and its Fourier transform such that, for a given value of C_0,
we can approximate the potential V with any desired accuracy by choosing a
sufficiently high energy number.

Prosser (1980, 1982) made a further study of the formal methods he had
previously given (1969, 1976) to inverse the Born series of the scattering
amplitude in powers of the potential. He was able to get convergence proofs
and errors estimates for reasonable norms (at least in the three-dimensional
case, where the Green's function is bounded), provided the scattering data are
sufficiently restricted.

XIV.6   Time-Domain Inverse Scattering Theory

An elementary approach to time-domain inverse scattering in one dimension
is given in section XVII.3.5. Here we sketch only the main ideas on the example
of a scattering problem which in certain conditions is equivalent to the
Schrödinger problem.

Any scattering process involves time evolution. The pulse is emitted from the source and propagates to the scatterer. There it interacts with the scattering potential and then it propagates to the detector, see, e.g., Lax and Phillips (1967). The life of a given pulse begins at the source and ends as the counter records its scattering pattern, whether in angles and energy or in angles and time. Conceptually, scattering is clearer in the time domain. In fact, it is frequently described as an evolution process, even when a frequency domain equation is being discussed.

The fundamental physical requirement in the time domain is that of causality. *Primitive causality* is the condition that the output of linear systems cannot precede the input in time. This concept plays an important role in special relativity and in quantum field theory, but those ideas are not directly relevant to the nonrelativistic quantum mechanics which is under discussion here. For a classical wave, primitive causality requires that "signals carrying information cannot travel with velocities greater than c, the speed of light." Unfortunately, this is usually, but not quite always, too simple for quantum mechanical scattering. In nonrelativistic quantum mechanics no limiting velocity exists, it is impossible to have a sharp wave packet for any finite time interval, the waves are complex rather than real, and when there are bound states additional complications arise.

The first correct statement of causality for quantum mechanics seems to have been made by Van Kampen (1953) and Wigner and Von Neumann (1954). Wigner (1964) has called this form of causality "phenomenological causality" although "nonrelativistic quantum causality" would have been more descriptive.

Van Kampen Causality. If an incoming Schrödinger wave is normalized to one at $t \to -\infty$, then the total probability of the wave outside a potential cannot be greater than one at any later time. Van Kampen proved that as a direct consequence of his form of causality, the scattering operator $S(k, \hat{e}', \hat{e})$ at fixed $\hat{e}', \hat{e} \in S^2$ is analytic in the first quadrant of the complex k-plane, $k = +(2\hbar E)^{1/2}$. In contrast to classical waves, quantum causality says nothing about continuing $S(k, \cdot, \cdot)$ beyond the first quadrant of the k-plane. Since bound states give poles in S on the positive imaginary k-axis, it is well known that the spectral singularities of S live on that axis. Bound states have L^2 eigenfunctions but do not satisfy the reality conditions of a classical wave, $S(-k) = S^*(k)$, for k real. If the existence of a single-valued analytic continuation of $S(k)$ is assumed and the system Hamiltonian is self-adjoint, then $S(k) = S^*(-k^*)$ for all k in the upper half-plane (UHP) in the quantum case. The meromorphy of $S(k)$ in the upper half-plane of k will depend crucially upon the number and nature of exceptional points.

Wigner has given an alternative form of Van Kampen causality using his R-matrix. In the R-matrix theory, the solutions of a system of Schrödinger equations, representing a multichannel process governed by short-range interactions, and their first-order derivatives, are related with each other on generalizes what would be the logarithmic derivative of the wave function at a fixed point $b \geq a$ in a one-dimensional radial problem governed by a potential of range a. In this most simple case, the meromorphic properties of R are obvious and easily related to causality. Wigner's causality is, however, equivalent to Van Kampen's, see, e.g., Nussenzweig (1972), but since Wigner has added significantly to this concept, it will be called Van Kampen–Wigner causality hereafter. The Wigner R-matrix approach (1955) gives an elementary proof of the Wigner causal inequality

$$\frac{d\delta}{dk} \geq -\left[a + \frac{1}{2k}\right], \qquad \text{(XIV.6.1)}$$

where $\delta(k)$ is the phase shift, a is the channel radius at which $R(E)$ is defined, and k is the wave number. The physical interpretation of equation (XIV.6.1) is as a time-delay or advance. Although there is no maximum time-delay since a particle could be captured, the minimum time-delay of the center of a wave packet is $\Delta t \cong 2(d\delta/dk)$. From this discussion, it is clear that although the causality condition has not been used explicitly as yet, the analyticity in k of S and $\psi^{\pm}$ which it provides have been used throughout this monograph. But recently, several authors have worked directly in the time domain. The quantum mechanical results which they obtained will be very briefly discussed in this subsection.

The plasma wave equation (PWE) is a model for the electron density in the Earth's atmosphere. It is assumed that a scalar electromagnetic wave (no polarization changes, i.e., azimuthal symmetry) propagates in a region where magnetic fields are negligible ($\vec{v} \times \vec{B} \approx \vec{0}$), and where the electron density is sufficiently low that electron–electron collisions can be neglected. Taking the excess electron density as a potential ($V(\vec{x})$), the plasma wave equation for the time-domain wave $u(t, \vec{x})$ becomes

$$[\Delta - \partial_t^2 - V(\vec{x})]u(t, \vec{x}) = 0, \qquad \text{(XIV.6.2)}$$

where Δ is the Laplacian and $\partial_t^2 = \partial^2/\partial t^2$. This is a linear, nondispersive, hyperbolic wave equation with phase speed $c = 1$. Suppose that the potential $V(\vec{x})$ is repulsive (V can even be attractive so long as $-\Delta + V(\vec{x})$ supports no bound states) and goes to zero as $|\vec{x}| \to \infty$ fast enough for a scattering operator to exist. Then direct and inverse scattering theories can be studied following the pioneering paper by Morawetz (1981). For the present purposes three of Morawetz's four models will not be discussed. They are interesting but correspond to classical wave models rather than quantum systems. Rose et al. (1984) considered the relation between time-domain and frequency-domain inverse scattering. This gave the first connection of Morawetz's work to quantum mechanics. In order for Fourier transforms between time and frequency to exist there can be no bound states. This is already a basic property of the plasma wave equation, but in order to start the isometry with Schrödinger's equation all bound and half-bound states must have been ruled out.

It is worth noticing that the plasma wave equation, as a linear, nondispersive, wave equation with straight-line characteristics will satisfy primitive causality, whereas Schrödinger's equation satisfies Van Kampen–Wigner causality instead.

The incoming–outgoing free Green's functions for (**XIV.6.2**) are

$$G_0^{\pm}(t, \hat{e} \cdot \overline{x}) = -\frac{\delta(t \mp \hat{e} \cdot \overline{x})}{4\pi |\overline{x}|} \qquad \text{(XIV.6.3)}$$

in agreement with primitive causality. In equation (**XIV.6.3**), G_0^+ propagates forward in time whereas G_0^- propagates backward in time. The time-domain analogue of the monochromatic plane waves $\exp(\pm i\overline{k} \cdot \overline{x})$ is the sharp pulse in time incident from the direction $\hat{e} \in S^2$

$$u_0^{\pm} = \delta(t \mp \hat{e} \cdot \overline{x}). \qquad \text{(XIV.6.4)}$$

If a scattering process starts at large negative times given by the sharp pulse, the time-domain scattering equation becomes

$$u^{\pm}(t, \hat{e}, \overline{x}) = \delta(t - \hat{e} \cdot \overline{x}) + \int G_0^{\pm}(t - t', |\overline{x} - \overline{y}|) V(\overline{y}) u^{\pm}(t', \hat{e}, \overline{y}) dt' \, d^3\overline{y}. \qquad \text{(XIV.6.5)}$$

The far-field impulse response R plays the role of the time-domain scattering amplitude and is given by

$$R(\hat{e}', \hat{e}, \tau) = \lim_{\substack{t, |\overline{x}| \to \infty \\ \tau = t - |\overline{x}|}} \{|\overline{x}| [u(t, \hat{e}, \overline{x}) - \delta(t - \hat{e} \cdot \overline{x})]\}, \qquad \text{(XIV.6.6)}$$

then

$$u^+(t, \hat{e}, \overline{x}) =: \delta(t - \hat{e} \cdot \overline{x}) + u^{sc}(t, \hat{e}, \overline{x})$$

$$\underset{x \to \infty}{\sim} \delta(t - \hat{e} \cdot \overline{x}) + \frac{R(\hat{e}', \hat{e}, t - \hat{e} \cdot \overline{x})}{x} + \cdots, \qquad \text{(XIV.6.7)}$$

where u^{sc} is the scattered wave, $x = |\overline{x}|$ and $\sim$ is the far-field limit of equation (**XIV.6.6**). The fields u, u^{sc} are necessarily real because they represent classical waves, but following Rose, Cheney, and De Facio the *quantum* causal extension to negative frequencies

$$u^+(t, \hat{e}, \overline{x}) = u^-(-t, -\hat{e}, \overline{x})$$

will be taken. This mixed system will be useful.

Taking $u^{+,sc}(t, \hat{e}, \overline{x}) - u^{-,sc}(t, \hat{e}, \overline{x})$ and using the above extension yields

$$u^{+,sc}(t, \hat{e}, \overline{x}) = u^+(-t, -\hat{e}, \overline{x}) - \frac{1}{2\pi} \int_{S^2} d^2\hat{e}' \, \partial_t R(\hat{e}', \hat{e}, t - \hat{e} \cdot \overline{x})$$

$$-\frac{1}{2\pi} \int_{S^2} d^2\hat{e}' \int_{-\infty}^{\infty} d\tau \, \partial_t R(\hat{e}', \hat{e}, t + \tau) u^{+,sc}(\tau, -\hat{e}', \overline{x}),$$

where

$$u^-(t, \hat{e}, \vec{x}) \sim \delta(t + \hat{e} \cdot \vec{x}) + \frac{R(\hat{e}, \hat{e}, t + \tau)}{x} \qquad (\tau = \hat{e} \cdot \vec{x}).$$

Now the causality which gave the propagators of, e.g., (**XIV.6.3**), requires that $u(t, \hat{e}, \vec{x}) \equiv 0$, for $t < \hat{e} \cdot \vec{x}$, reducing the previous equation to

$$u^{+,sc}(t, \hat{e}, \vec{x}) = -\frac{1}{2\pi} \int_{S^2} d^2\hat{e}'\, \partial_t R(\hat{e}', \hat{e}, t - \hat{e} \cdot \vec{x})$$

$$= \frac{1}{2\pi} \int_{S^2} d^2\hat{e}' \int_{-\infty}^{\infty} d\tau\, \partial_t R(\hat{e}', \hat{e}, t + \tau) u^{+,sc}(\tau, -\hat{e}', \vec{x})$$

$$\text{(XIV.6.8)}$$

for all $t > \hat{e} \cdot \vec{x}$. The reader is reminded that $\dot{R} = \partial_t R$ is to be expected in the time-domain integral equation since the (formal) Fourier transform of $\dot{R}(\hat{e}, \hat{e}, t - \hat{e}, \vec{x})$ is $kA(\hat{e}', \hat{e}, k - k\hat{e} \cdot \hat{e}')$. Let $\alpha := t - \hat{e} \cdot \vec{x}$ and define the quantities

$$\eta(\alpha, \hat{e}, \vec{x}) := u^{+,sc}(t, \hat{e}, \vec{x})$$

$$= u^+(t, \hat{e}, \vec{x}) - \delta(\alpha), \qquad \text{(XIV.6.9)}$$

and

$$M(\alpha, \hat{e}', \hat{e}, \vec{x}) := -\frac{1}{2\pi} \frac{\partial}{\partial\alpha} R(\hat{e}', \hat{e}, \alpha + (\hat{e} - \hat{e}') \cdot \vec{x})$$

$$= -\frac{1}{2\pi} \frac{\partial}{\partial t} R(\hat{e}', \hat{e}, t - \hat{e}' \cdot \vec{x}). \qquad \text{(XIV.6.10)}$$

substituting equation (**XIV.6.4**) and (**XIV.6.10**) into equation (**XIV.6.8**) yields the time-domain equation for the scattered wave

$$\eta(\alpha, \hat{e}, \vec{x}) = -\frac{1}{2\pi} \int d^2\hat{e}'\, M(\alpha, \hat{e}', \hat{e}, \vec{x})$$

$$= \frac{1}{2\pi} \int d^2\hat{e}' \int_{-\infty}^{\infty} d\tau\, M(\alpha - \tau, \hat{e}', \hat{e}, \vec{x}) \eta(-\tau, -\hat{e}, \vec{x}). \qquad \text{(XIV.6.11)}$$

As mentioned earlier the L^2 solutions to the plasma wave equation automatically have Fourier transforms almost everywhere in L^2. Denote the inverse Fourier transform by "hut", e.g., $\check{\eta}$, which will constrast them from the hatted unit vectors $\hat{e}'$ and $\hat{e}$, and take the inverse Fourier transform of equation (**XIV.6.11**) to obtain

$$\check{\eta}(\check{\alpha}, \hat{e}, \vec{x}) = -\int_{S^2} d^2\hat{e}'\, \check{M}(\check{\alpha}, \hat{e}', \hat{e}, \vec{x})$$

$$- \int_{S^2} d^2\hat{e}' \int_{-\infty}^{\infty} d\delta\, \check{M}(\check{\alpha} - \delta, \hat{e}', \hat{e}, \vec{x}) \check{\eta}(-\delta, \hat{e}', \vec{x}). \qquad \text{(XIV.6.12)}$$

Equation (**XIV.6.12**) is exactly Newton's five variable "Marchenko" inverse scattering equation. The primitive causality of (**XIV.6.2**) guarantees that $\check{\eta}$ is analytic in the upper half $\check{\alpha}$-plane. Rose et al. and De Facio and Rose (1985) followed Morawetz's idea of using the singular decomposition of the solution near the forward boundary of the wave front set. The wave front set $WF(u)$ is the set of all spatial points in the causal support of the solution $u(t, \hat{e}, \vec{x})$ at the fixed finite time t. They used the fact that causality, combined with the sharp incident pulse, gives the expansion of u near $t \approx \hat{e} \cdot \vec{x}$

$$u(t, \hat{e}, \vec{x}) = \delta(t - \hat{e} \cdot \vec{x}) + A_1(\hat{e}, \vec{x})\theta(t - \hat{e} \cdot \vec{x}) + A_2(\hat{e}, \vec{x})R_A(t - \hat{e} \cdot \vec{x}) + \cdots,$$
$$(\textbf{XIV.6.13})$$

where θ is the Heaviside function ($\theta' = \delta$), R_A is the ramp function $R'_A = \theta$, and ... indicates less singular terms. A linear hyperbolic partial differential equation propagates its discontinuities without smoothing or worsening them. This works as follows. Substituting equation (**XIV.6.13**) into (**XIV.6.2**) it is found that the coefficients of δ' and δ'' are identically zero. The coefficient of δ is

$$V(\vec{x}) = 2\hat{e} \cdot \vec{\nabla}[A, (\hat{e}, \vec{x})], \qquad (\textbf{XIV.6.14})$$

and the coefficient of θ is

$$2\hat{e} \cdot \vec{\nabla}[A_2(\hat{e}, \vec{x})] = [-\Delta + V(\vec{x})]A_1(\hat{e}, \vec{x}). \qquad (\textbf{XIV.6.15})$$

These equations are called the wave front condition or Morawetz's identity and the first transport equation, respectively.

The first of these two equations can be further simplified by using equation (**XIV.6.13**) near the wave front in the form

$$A_1(\hat{e}, \vec{x}) = \left\{ \lim_{t \uparrow \hat{e} \cdot \vec{x}} [u(t, \hat{e}, \vec{x}) - \delta(t - \hat{e} \cdot \vec{x})] \right\}.$$

Substituting this last equation into equation (**XIV.6.15**) yields *Morawetz's identity*

$$V(\vec{x}) = 2\hat{e} \cdot \vec{\nabla} \left\{ \lim_{t \uparrow \hat{e} \cdot \vec{x}} [u(t, \hat{e}, \vec{x}) - \delta(t - \hat{e} \cdot \vec{x})] \right\}, \qquad (\textbf{XIV.6.16})$$

which was called the "key identity" and "fundamental identity." This is a short-time or high-frequency inversion. The Morawetz identity, equation (**XIV.6.16**), is a direct consequence of the propagation of singularities of a linear, hyperbolic system. If the proper high frequency data is available it furnishes a three-variables inversion. In addition, the Fourier transform of this fundamental identity using quantities in equation (**XIV.5.12**) is just Newton's so-called miracle. The first transport equation, (**XIV.6.15**), relates the discontinuity of A_2 at the wave front set to a zero-energy Schrödinger operator acting upon A_1.

Thus the *causality* of the plasma wave equation gives the analyticity (triangularity) of η and the miracle. It gives the leading corrections near the wave front set and, by iterating this procedure, all corrections.

Layer-Stripping Methods. These methods appeared in one-dimensional problems and are based on the idea of reconstructing the scatterer layer-by-layer (see chapter XVII below). This can be done either by downward continuation (reconstruction of the wave field at successively deeper layers) or by invariant imbedding (imbedding of the scatterers in a family of scatterers). Both of these methods use the idea of decomposing the wave into components traveling in different directions. Yagle and Levy (1986) and Yagle (1986) have proposed inverse scattering algorithms based on splitting the Schrödinger equation into upward- and downward-going wave components. Similar wave decompositions in the case of a three-dimensional wave, propagating in a medium that depends on only one dimension, have been found by Weston (1987) and Yagle and Levy (1985). Cheney and Kristensson (1988) considered the fully three-dimensional scattering problems for the Schrödinger equation and the reduced wave equation with variable wave velocity. They derive a simple layer-stripping formula that tells how the wave field changes when the scatterer is truncated and they use it to show that various inverse scattering problems are ill-posed.

Remark. If we look at the quantum scattering theory from the physical point of view, independently of the mathematical approach, we see that it differs from the acoustical or other classical scattering theories, not only because the causality concept somewhat differs but also because the strong discontinuities, which are essential in most classical problems, are lacking in the quantum theory (they would correspond, in the Schrödinger equation, to unphysical distributions more singular than Dirac ones). The physical problem which would correspond in the quantum theory to classical scattering is that of chains of Schrödinger equations, which are separated in $\mathbb{R}^3$ by "singular" surfaces through which ψ and its derivative are discontinuous but special currents are conserved (Sabatier, 1988). It has been shown only very recently (Sabatier, 1989) that the three-dimensional scattering amplitude can then be separated into two parts: the first one is due to the discontinuities through singular surfaces—and physically corresponds to "classical scattering," or scattering from "reflectors." The second one is due to the potential scattering in the presence of discontinuities. This "diffuse scattering" is not essentially different from quantum scattering, to which it reduces when discontinuities are missing. From the point of view of the inverse problem there are two essential points. First, the leading terms of the scattering pattern at large distances may be related to fine properties of the surface discontinuities and not only to their approximate positions. This property explains the well-known difficulties in radar patterns, which are essentially due to specular points, i.e., lack of analytic properties in the target surface. The second point is that for band limited and not precise data, random or complicated distributions of discontinuities and diffuse scattering may be almost equivalent—but never quite equivalent. This property explains why many studies can be done either by means of stratified models (i.e., piecewise constant, discontinuous media) or by means of continuous models, with equal success—or lack of success: in all cases a more refined analysis, or more precision, requires the mixed—realistic—model. Finally, the recent study of chains of Schrödinger equations explains why studies of Helmholtz scattering problems (see, e.g., Colton and Kress, 1986) and those of quantum scattering problems have grown quite independently, in spite of their strong relation.

XIV.7   Comments and References

A first study of the Gel'fand–Levitan method for the three-dimensional scattering problem has been given by Kay and Moses (1961a,b). Their method, if it works, does not necessarily lead to a local potential, even if one exists.

In 1971, Faddeev gave a method which is essentially the one which is given here, except that several functions are Fourier transformed (examples of these equivalent ways to study the problem are given in section XIV.3). The method was rediscovered independently by R. G. Newton, who wrote his paper, and communicated it to one of the authors (PCS), on February 2, 1972, then received Faddeev's preprint at the end of the same month. In 1973 and 1974 R. G. Newton gave lectures mixing Faddeev's and his own work.

After 1975, it was obvious that this method was so complicated that it had almost no chance of being used for practical applications. The compatibility condition was so unlikely to be satisfied that Newton called it "miracle," and the problem of exceptional points on the real axis was open. Newton (1977) first showed that there exists a class of potentials for which the modified Fredholm determinant det $D(k)$ cannot vanish on the real k-axis for any $k \neq 0$. And at $k = 0$ either:

(1) there exists an exceptional point "of first kind" with n-fold degenerate bound state (near $k = 0^+$, $D(k) \simeq k^{2n}$, $d\sigma/d\Omega(k) \simeq M < \infty$, det $S(k) \simeq 1$);
(2) there exists an exceptional point "of second kind" with a half-bound state (near $k = 0^+$, $D(k) \simeq k$, $d\sigma/d\Omega \to \infty$, det $S(k) \simeq -1$);
(3) the other case is a combination of the first two with $D(k) \simeq k^{2n+1}$, $d\sigma/d\Omega \to \infty$, det $S(k) \simeq -1$.

A study of the Levinson theorem has also been proposed by Osborn and Bollé (1977) in a class ($L^2 \cap L^1$) of potentials larger than Newton's one. Besides, in connection with these problems of resolvent analycity, several results also exist in the mathematical literature (see, e.g., Simon (1971), Kato (1986), Agmon (1986)). Newton's attempt (1980c) to write down a Cauchy formula for the Jost solution that could be used as an input for a method of solution in the three-dimensional problem as in the one-dimensional problem, and his attempt (1979) to formulate inverse scattering problems as Riemann–Hilbert problems, led him to introduce new methods, either of the Gel'fand–Levitan or of the Marchenko form (Newton 1980a,b, 1981, 1982b,c). Meanwhile, Prosser (1980, 1982) defined the domain where his inversing Born series is justified, and Moses (1979, 1980, 1981) further studied his method of solution of the three-dimensional inverse problem. In particular, in the case without bound state, Moses introduced new Jost solutions, proved their completeness, and constructed a Gel'fand–Levitan transformation kernel. Saito (1982a,b,

1984, 1986) also showed that the high frequency analysis of scattering may be much more meaningful than had been previously felt. After 1983, studies of multidimensional inverse problems connected with the inverse method showed how it is interesting, in the frequency domain, to reformulate the problem as a "D-bar problem," where the holomorphic parts of the function of interest are related to the nonholomorphic part. This seems to have begun with Beals and Coifman (1981, 1984), who also treated (1985) first-order systems which would correspond in our Schrödinger problem to the case $k = 0$. However, the case of first-order systems which would correspond to the general Schrödinger problem was first studied by Nachman and Ablowitz (1984a, 1984b) who showed, by the way, how the Faddeev's Green function arises very naturally in this framework. Lavine and Nachman (1987a,b) on the one hand, and Henkin and Novikov (1986, 1987, 1988) on the other, were then able to solve the problem of exceptional points, to characterize admissible scattering amplitudes, and to give some uniqueness theorems (for instance, the uniqueness of a short-range potential yielding a given amplitude at fixed energy). Other results on admissible amplitudes and uniqueness results have appeared in Ramm (1987a,b, 1988b), Ramm and Weaver (1987), and Novikov (1987), whereas Nachman et al. have given an n-dimensional Borg–Levinson theorem (1988). Newton (1985b) related his Marchenko–type approach (to the n-dimensional problem) to Nachman–Ablowitz results and gave a variational formulation (1985c) of the problem. The $\bar{\partial}$-method also enabled a new formulation of the problem in terms of standing wave solutions and the reactance matrix (Somersalo, 1987). Other problems, related to the direct and inverse multidimensional scattering of the Schrödinger equation (for instance, the analysis of bound states, backward scattering, and various characterization problems), are treated by methods of functional analysis in Melin (1987) and Saito (1987). It should be noticed at this point that the problem of bound states was often treated as an "unessential" complication until their connection with the "essential" complication of real exceptional points was known. Notice by the way that inequalities on V, or on σ, that are sufficient to guarantee the nonexistence of bound states have been known for a long time (Bargmann, 1949; Glaser et al., 1976; Martin and Sabatier, 1977; Newton, 1982a). An approach to multidimensional inverse problems using path integrals and Maslov asymptotics (1972) has also been proposed (De Facio and Brander, 1988).

It is tempting to try applications of frequency-domain techniques in cases where acoustic or electromagnetic equations, for example, are reducible to a Schrödinger-like equation. However, the main problems connected with these equations, even in the transmission case, are different (see, e.g., Angell et al., 1987) from those of the Schrödinger equation, except in the Born approximation (see, e.g., De Facio and Rose, 1986), and so are out of the scope of this book. In contrast, time-domain approaches to the acoustic wave equation, the plasma wave equation, and the Schrödinger equation are much more similar. They all rely on causality concepts which were covered in this chapter. Their history is in large part that of the one-dimensional case, treated in chapter XVII below. General mathematical studies of multidimensional problems of this kind already appear in Simon (1971) and Ens (1978). As for three-dimensional scattering, C. Morawetz, who had already studied related inverse problems in her lectures (e.g., Morawetz, 1975), formulated several classical wave problems, including the plasma wave equation in the time domain, in a fundamental paper (1981). This paper was the foundation of several studies on the connection between time- and frequency-domain three-dimensional inverse scattering methods and applications to plasma and variable velocity wave equations (Rose et al., 1984, 1985), and theoretical physics of the time domain for the plasma wave equation with the introduction of the Radon transform (Rose et al., 1985; De Facio and Rose, 1985), whereas more mathematical studies appeared in Morawetz and Kriegsmann (1983). The inverse scattering problem of the Schrödinger equation in dimension two has also been solved by time-domain techniques (Cheney, 1984). Layer-stripping methods, like most other techniques, first appear in the one-dimensional problems (see chapter XVII below), then arrive in multidimensional problems (Bruckstein et al., 1985; Yagle and Levy, 1985, 1986; Yagle, 1986, 1987; Cheney and Kristenson, 1988; Weston, 1987).

As a last remark, notice that most results obtained in the three-dimensional case are easily extended to the $(n \geq 3)$-dimensional case. In the time domain, there is clearly a qualitative difference between a medium $(n = 1)$ where a hard obstacle, or a hole, cuts off the information, and a medium $(n \geq 2)$ where it is possible to get around. In the frequency domain, the zero-frequency Green's function is singular in the $n = 1$ *and* $n = 2$ cases, and this has strong consequences on the existence of bound states (Simon, 1976) and the convergence of "simple" Born series (Prosser, 1980). In both domains the mathematical and "physical" situations do not modify beyond $n = 3$. We also understand why the mathematical analysis of multidimensional problems is sometimes simpler than that of the one-dimensional corresponding problems.

CHAPTER XV

Miscellaneous Approaches to Inverse Problems at Fixed Energy

As we have seen throughout this book, the key to the solution of inverse problems usually is some machinery centered on a "transformation kernel." But how can this machinery be obtained? We give here two general approaches which have proved useful. The first one makes use of the interpolation properties of the wave functions. It has been applied either directly (Miodek's approach to the inverse plasma problem) or as a tool to obtain the machinery (Sabatier–Hooshyar construction of spin-orbit potentials, and generalizations), which in turn enabled the development, by analogy, of the Jaulent–Jean method for inverse problems at fixed l with potentials depending linearly on $\sqrt{E}$.

In the second approach, following Levitan, Coudray and Coz applied a method extracted from the theory of generalized translation operators to conveniently chosen inverse problems.

XV.1   Methods Using Interpolation Properties

We use the notations (**XI.4.10**), except for $\varphi_\lambda^1(r)$, which we replace by $s_\lambda(r)$: according to (**XII.8.6**), $s_\lambda(r)$ is proportional to $r^{\lambda+1/2}$. Now for $V_0 = 1$ the interpolation formula (**XII.8.11**) reads

$$\frac{\pi\lambda}{\sin[\pi\lambda]}\,\varphi_\lambda^V(r)s_{-\lambda}(r) = \varphi_0^V(r)s_0(r) + \sum_{\mu\,\in\,S}\frac{\lambda}{\lambda+\mu}\,\frac{\pi\gamma_\mu(V,1)}{2\mu}\,\varphi_\mu^V(r)s_\mu(r). \quad \text{(XV.1.1)}$$

If we let $|\lambda| \to \infty$ in both sides of (XV.1.1), because of (XII.8.3) we obtain

$$\tfrac{1}{2}\pi r = \varphi_0^V(r)s_0(r) + \sum_{\mu \in S} \frac{\pi \gamma_\mu(V, 1)}{2\mu} \varphi_\mu^V(r)s_\mu(r). \qquad \text{(XV.1.2)}$$

By subtracting (XV.1.2) from (XV.1.1), and multiplying the result by $(\sin[\pi\lambda]/\pi\lambda)[s_{-\lambda}(r)]^{-1}$, we obtain

$$\varphi_\lambda^V(r) = s_\lambda(r) - \frac{\pi}{2} \sum_{\mu \in S} \frac{(\tfrac{1}{2}r)^{\lambda+\mu}}{\Gamma(1 + \lambda)\Gamma(1 + \mu)(\lambda + \mu)} \gamma_\mu(V, 1)\varphi_\mu^V(r), \qquad \text{(XV.1.3)}$$

where (XII.8.6) has been taken into account. But, according to (XI.4.14), this formula is equivalent to the transformation formula

$$\varphi_\lambda^V(r) = \varphi_\lambda^1(r) - \int_0^r K_1^V(r, \rho)\varphi_\lambda^1(\rho)\rho^{-2} \, d\rho, \qquad \text{(XV.1.4)}$$

where

$$K_1^V(r, \rho) = \sum_{\mu \in S} \gamma_\mu(V, 1)\varphi_\mu^V(r)\varphi_\mu^1(\rho) \qquad \text{(XV.1.5)}$$

and from our analysis of chapter XI, it is clear from (XV.1.4), that not only the whole machinery starting from the "potential" 1 can be derived, but also all the machinery starting from any potential.

   Thus, interpolation formulae (XV.1.1) and (XV.1.2) for a particular class of potentials (here, analytic potentials), provide a key to inverse problems. We shall first recall the Frobenius method, which enables one to study the analytic properties of the solutions of a differential equation with analytic coefficients. Then we shall show how it can be used to obtain interpolation formulae and we successively study two possible uses of them.

XV.1.1   The Frobenius Method and Interpolation Formulas

It is easy to show that the regular solution $\varphi_\lambda^V(r)$ of the partial wave Schrödinger equation:

$$\left[r^2 T_r - r^{-1}\frac{d}{dr} + \frac{1}{4} \right]\varphi_\lambda^V(r) = \lambda^2\varphi_\lambda^V(r), \qquad \text{(XV.1.6)}$$

where

$$T_r = \frac{1}{r}\frac{d}{dr} r \frac{d}{dr} + 1 - V(r) \qquad \text{(XV.1.7)}$$

also satisfies the two following equations:

$$\left[T_r - 2\lambda r^{-1}\frac{d}{dr} \right][r^{-1}\varphi_\lambda^V(r)s_\lambda(r)] = 0 \qquad \text{(XV.1.8)}$$

$$\left[T_r + 2\lambda r^{-1}\frac{d}{dr} \right][r^{-1}\varphi_\lambda^V(r)s_{-\lambda}(r)] = 0. \qquad \text{(XV.1.9)}$$

Assume now that $V(r)$ is an analytic function in the disc $D(0, R_0)$. The function $[s_\lambda(r)]^{-1}\varphi_\lambda^V(r)$ is a solution of **(XV.1.9)**. We can derive its Taylor expansion by applying the Frobenius's method: we introduce the coefficients d_λ^p and g_λ^n such that:

$$[s_\lambda(r)]^{-1}\varphi_\lambda^V(r) = \sum_p d_\lambda^p \lambda^p \qquad (d_\lambda^0 = 1), \qquad \textbf{(XV.1.10)}$$

$$r^{-p}\left[T_r + 2\lambda r\frac{d}{dr} \right]r^p = \sum_{n=0}^{\infty} g_\lambda^n(p)r^n = p(2\lambda + p) + r^2(1 - V(r)). \quad \textbf{(XV.1.11)}$$

The g_λ^n's are known from the Taylor expansion of $V(r)$. Using Cauchy's inequality, we readily derive for any $R < R_0$

$$|g_\lambda^{n+1}| \le R^{-n}M(R), \qquad \textbf{(XV.1.12)}$$

where

$$M(R) = \sup_{|z|=R} |V'(z)|. \qquad \textbf{(XV.1.13)}$$

Inserting **(XV.1.11)** in **(XV.1.9)** we obtain a triangular system which yields the d_λ^p's from the g_λ^n's:

$$d_\lambda^0 = 1,$$
$$d_\lambda^1 g_\lambda^0(1) + d_\lambda^0 g_\lambda^1 = 0, \qquad \textbf{(XV.1.14)}$$
$$\vdots$$
$$d_\lambda^p g_\lambda^0(p) + d_\lambda^{p-1} g_\lambda^1 + \cdots + d_\lambda^0 g_\lambda^p = 0,$$

using a condensed notation for $g_\lambda^n(p)$ with $n \ge 1$, since they do not depend on p. Using **(XV.1.12)** we obtain the inequality[1]

$$|d_\lambda^{p+1}| \le \frac{M(R)}{|2\lambda + 1|} \prod_{n=1}^{n=p} R^{-1} \frac{RM(R) + n|2\lambda + n|}{(n + 1)|2\lambda + n + 1|}, \qquad \textbf{(XV.1.15)}$$

which is conveniently replaced by the stronger one:

$$|d_\lambda^{p+1}| \le \frac{M(R)}{|2\lambda + 1|} \prod_{n=1}^{n=p} \frac{R^{-1}}{n + 1} \sup[n, RM(R)]. \qquad \textbf{(XV.1.16)}$$

The right-hand side of **(XIV.1.16)** is readily calculated

$$|d_\lambda^{p+1}| \le \frac{C(R)R^{-p-1}}{|2\lambda + 1|(p + 1)} \le C(R)R^{-p-1}, \qquad \textbf{(XV.1.17)}$$

where $C(R)$ does not depend on λ or p for Re $\lambda \ge 0$. Now, let $V_1(r)$ be a potential which is also analytic in $D(0, R_0)$. From the second inequality **(XV.1.17)**, it is easy to see that the series

$$\bar{d}_{2\lambda}(r) \equiv \sum_n \bar{d}_{2\lambda}^n r_n = \left(\sum_p d_{\lambda, V_0}^p r^p \right)\left(\sum_q d_{\lambda, V_1}^q r^q \right) \qquad \textbf{(XV.1.18)}$$

---

[1] In the following derivations, for the sake of simplicity, we implicitly discard the case where 2λ is a negative integer. C is used as a general positive constant, $C(R)$ a general positive function of R.

is analytic in the disc $D(0, R')$ and that

$$|\bar{d}_{2\lambda}^n| \le C(R')R'^{-n}, \tag{XV.1.19}$$

where $C(R')$ does not depend on λ for Re $\lambda \ge 0$, and $\bar{d}_{2\lambda}(r) \le C(R')/(1 - r/R')$. If $f(r)$ is any function $\sum_0^\infty f_n r^n$ which is analytic in $D(0, R_0)$, it can be expanded in terms of the function $r^n \bar{d}_n(r)$:

$$f(z) = \sum_0^\infty b_n r^n \bar{d}_n(r). \tag{XV.1.20}$$

Proof. Formal identification of coefficients yields

$$b_k = f_k - \sum_{n=0}^{k-1} b_n \bar{d}_n^{k-n}. \tag{XV.1.21}$$

It is possible to define a function $C(R')$ such that $\bar{d}_{2\lambda}^n$ and f_n satisfy (**XV.1.19**). Using it in (**XV.1.21**) readily yields

$$|b_k| \le C(1 + C)^k R^{-k} \tag{XV.1.22}$$

and this bound is sufficient to show that there is a disc in which both sides of (**XV.1.20**) are equal analytic functions.

From (**XV.1.20**), it follows that we can construct a sequence a_n (with $a_0 = 1$), for which the equality

$$(\tfrac{1}{2}\pi r)^{-1} \sum_{\mu \in S^*} a_\mu \varphi_n^V(r) s_\mu(r) = 1 \qquad (S^* = \{SUO\}) \tag{XV.1.23}$$

holds in a disc centered at the origin. This formula is identical with (**XV.1.2**) if we set

$$a_\mu = \frac{\pi \gamma_\mu(V, 1)}{2\mu}. \tag{XV.1.24}$$

The interpolation formula (**XV.1.1**) is easily derived from (**XV.1.23**) since, applying $\{T_r + 2\lambda r^{-1}\, d/dr)r^{-1}$ to both sides of (**XV.1.1**), and using (**XV.1.9**) (**XV.1.8**), and (**XV.1.23**), we obtain 0, and the conditions at $r = 0$ are consistent with each other.   Q.E.D.

Thus, we see that the machinery of the inverse problem at fixed energy can be obtained by a derivation essentially using the classical Frobenius method.

XV.1.2   Indirect Uses of Interpolation Formulas

When considering the scattering of spin half particles by central and spin-orbit potentials. the Schrödinger equation can be written in the following form:

$$\Delta \Psi(\mathbf{r}) + [1 - U_0(r) - 2L \cdot S U_s(r)] \Psi(\mathbf{r}) = 0. \tag{XV.1.25}$$

The differential cross section is then given by

$$I(\theta) = |f(\theta)|^2 + |g(\theta)|^2 \tag{XV.1.26}$$

with

$$f(\theta) = \frac{1}{2i} \sum_0^\infty \{(l + 1)[\exp(2i\delta_l^+) - 1] + l[\exp(2i\delta_l^-) - 1]\}P_l(\cos\theta),$$

$$\text{(XV.1.27)}$$

$$g(\theta) = \frac{1}{2i} \sum_1^\infty [\exp(2i\delta_l^+) - \exp(2i\delta_l^-)]P_l^{\,1}(\cos\theta). \qquad \text{(XV.1.28)}$$

The inverse problem is to find the interaction U_c and U_s given the phase shifts. In other words, find the potentials in such a way that the asymptotic behavior of the regular solutions $\varphi_l^\pm(r)$, which satisfy the differential equations given below, will have the desired form:

$$\left[r^2 T_r^\pm - r^{-1}\frac{d}{dr} + \frac{1}{4}\right]\varphi_\lambda^\pm(r) = \lambda^2 \varphi_\lambda^\pm(r), \qquad \text{(XV.1.29)}$$

where

$$T_r^\pm = \frac{1}{r}\frac{d}{dr}r\frac{d}{dr} + 1 - (V(r) \pm 2\lambda Q(r)), \qquad \text{(XV.1.30)}$$

$$V(r) = U_c - \tfrac{1}{2}U_s,$$

$$Q(r) = \tfrac{1}{2}U_s(r), \qquad \text{(XV.1.31)}$$

and the indices $\pm$ correspond to each other according to their position. This convention will be used in the following.

The formula (XV.1.30) reduces to (XV.1.7) if $V(r)$ is replaced by $V(r) \pm 2\lambda Q(r)$. Thus, if $V(r)$ and $Q(r)$ are analytic functions in $D(0, R_0)$, the inequality (XV.1.15) holds with $M(R)$ replaced by $M_0(r)(2\lambda + 1)$ (we take $\lambda \geq 0$). It is easy to see that a bound which is independent of λ is obtained by replacing (XV.1.16) by

$$|d_\lambda^{p+1}| \leq M_0(R)\prod_{n=1}^{n=p} \frac{2R^{-1}}{(n + 1)} \sup[n, RM_0(r)]. \qquad \text{(XV.1.32)}$$

Thus, the second inequality (XV.1.17) holds with R replaced by $R/2$, and expansions like (XV.1.20) again are valid in a disc centered at 0. We introduce hereafter the expansions of two particular functions

$$F^\pm(r) = \exp\left[\pm \int_0^r \rho V(\rho)d\rho\right] = (\tfrac{1}{2}\pi r)^{-1}\sum_{\mu\in S^*} a_\mu^\pm \varphi_\mu^\mp(r)s_\mu(r) \qquad \text{(XV.1.33)}$$

which are generalizations of (XV.1.23). From (XV.1.33) we can derive the interpolation formulae (Sabatier 1967)

$$\frac{\pi\lambda}{\sin\pi\lambda}\,\varphi_\lambda^\pm(r)s_{-\lambda}(r) = \varphi_0(r)s_0(r) + \sum_{\mu\in S}\frac{\pi}{2\mu}\gamma_\mu(V^\pm, 1)\frac{\lambda}{\lambda + \mu}\,\varphi_\mu^\mp(r)s_\mu(r),$$

$$\text{(XV.1.34)}$$

where

$$\gamma_\mu(V^\pm, 1) = 2\mu\pi^{-1}a_\mu^\pm \qquad \text{(XV.1.35)}$$

and which are proved from (**XV.1.33**) in a way quite similar to the one used following (**XV.1.24**). These formulae are obviously generalizations of (**XV.1.1**). Subtracting (**XV.1.34**) from (**XV.1.33**), multiplying both sides by $s_\lambda(r)$, and noticing that $r \int_0^r s_\mu(\rho)s_\lambda(\rho)\rho^{-2} d\rho$ is nothing but $(\lambda + \mu)^{-1}s_\mu(r)s_\lambda(r)$, yields the following formula which can obviously be continued for Re $\lambda > -\frac{1}{2}$.

$$\varphi_\lambda^\pm(r) = F^\pm(r)s_\lambda(r) - \int_0^r K^\pm(r, \rho)s_\lambda(\rho)\rho^{-2} d\rho, \qquad \text{(XV.1.36)}$$

where

$$K^\pm(r, r') = \sum_{\mu \in S} \gamma_\mu(V^\pm, 1)\varphi_\mu^\mp(r)s_\mu(r'). \qquad \text{(XV.1.37)}$$

These formulae show that in the present problem, for analytic potentials, there exist transformation kernels independent of λ and they yield $\varphi_\lambda^\pm(r)$ from the known functions $s_\lambda(r)$. Let us now successively give λ all the values v belonging to the set S; for every v, multiply both sides of (**XV.1.36**) by $\gamma_v(V^\pm, 1) s_v(r')$, and sum all the terms. We obtain the system of integral equations

$$K^\pm(r, r') = F^\mp(r)f^\pm(r, r') - \int_0^r K^\mp(r, \rho)f^\pm(\rho, r')\rho^{-2} d\rho, \quad \text{(XV.1.38)}$$

where the "symmetric" kernels are

$$f^\pm(r, r') = \sum_{\mu \in S} \gamma_\mu(V^\pm, 1)s_\mu(r)s_\mu(r'). \qquad \text{(XV.1.39)}$$

It is easy to see, from (**XV.1.33**) and (**XV.1.36**), that the system (**XV.1.38**) can be completed by the following conditions on the potentials:

$$s_0(r)[F^+(r) - F^-(r)] = \int_0^r [K^+(r, \rho) - K^-(r, \rho)]s_0(\rho)\rho^{-2} d\rho, \quad \text{(XV.1.40)}$$

$$F^+(r)F^-(r) = 1, \qquad \text{(XV.1.41)}$$

and a formula which yields $V(r)$ from $F^\pm(r)$ and $K^\pm(r, r)$ (see formula (**XV.1.47**) below).

The system (**XV.1.38, XV.1.40**, and **XV.1.41**) replaces the Regge–Newton equation for the scattering of a spinning particle. It provides a way to derive the potentials V and Q from the input functions $f^\pm(r, r')$ (themselves defined from (**XV.1.39**) and the data of the γ_μ's), and to construct all the solutions of the Schrödinger equation. However, such a study would start from the potential $+1$ as a "comparison potential" (one can also talk of "zero energy base"), which is not always convenient. Fortunately, as has been done in section 12.8, it is possible to use the very special analytic properties in the λ plane which are implied by a formula like (**XV.1.36**) to derive interpolation formulae and integral equations starting from any comparison potential. The result is as follows

$$\varphi_\lambda^\pm(r) = F_\lambda^\pm(r)\varphi_\lambda^{V_0}(r) - \int_0^r K^\pm(r, \rho)\varphi_\lambda^{V_0}(\rho)\rho^{-2}\, d\rho, \qquad \text{(XV.1.42)}$$

where

$$K^\pm(r, r') = I^\pm(r, r') - J^\pm(r, r'), \qquad\qquad\qquad \text{(XV.1.43)}$$

$$J^\pm(r, r') = F^\pm(r)e(r, r') - \int_0^r K^\pm(r, \rho)e(\rho, r')\rho^{-2}\, d\rho, \qquad \text{(XV.1.44)}$$

$$I^\pm(r, r') = F^\pm(r)g^\pm(r, r') - \int_0^r K^\mp(r, \rho)g^\pm(\rho, r')\rho^{-2}\, d\rho, \quad \text{(XV.1.45)}$$

$$e(r, r') = \sum_{\mu \in S} \gamma_\mu(V_0, 1)\varphi_\mu^{V_0}(r)\varphi_\mu^{V_0}(r'), \qquad\qquad \text{(XV.1.46)}$$

$$g^\pm(r, r') = \sum_{\mu \in S} \gamma_\mu(V^\pm, 1)\varphi_\mu^{V_0}(r)\varphi_\mu^{V_0}(r'), \qquad\qquad \text{(XV.1.47)}$$

and the condition (XV.1.40 and XV.1.41) holds with the replacement $s_0 \to \varphi_0^{V_0}$. To calculate $V(r)$ they can be completed by the formula

$$K^\pm(r, r) = \tfrac{1}{2}rF^\pm(r)\left\{\pm r^2 Q(r) + \int_0^r [\rho^3 Q^2(\rho) - \rho V(\rho)]d\rho\right\}. \quad \text{(XV.1.48)}$$

Furthermore, $I^\pm$ and $J^\pm$ can be expanded in terms of the wave functions, as was done in the case of the central potential:

$$I^\pm(r, r') = \sum_{\mu \in S} \gamma_\mu(V^\pm, 1)\varphi_\mu^\mp(r)\varphi_\mu^{V_0}(r') \qquad\qquad \text{(XV.1.49)}$$

$$J^+(r, r') = \sum_{\mu \in S} \gamma_\mu(V_0, 1)\varphi_\mu^\pm(r)\varphi_\mu^{V_0}(r'). \qquad\qquad \text{(XV.1.50)}$$

Substituting these expansions in (XV.1.42) yields the infinite system

$$\varphi_\lambda^\pm(r) = F^\pm(r)\varphi_\lambda^{V_0}(r) - \sum_{\mu \in S}[\varphi_\mu^\mp(r)\gamma_\mu(V^\pm, 1) - \varphi_\mu^\pm(r)\gamma_\mu(V_0, 1)]\Lambda_{\lambda\mu}^{V_0}(r),$$

$$\text{(XV.1.51)}$$

where Λ was defined by (XI.4.14).

Thus, it has been proved (Sabatier 1967) that:

1. Given analytic functions V and Q, one can find a set of coefficients $\gamma_\mu(V^\pm, 1)$ in such a way that the functions $K^\pm(r, r')$ defined by equations (XV.1.43–47) will satisfy (XV.1.48) and the regular solutions of (XV.1.29) can be represented by them through (XV.1.42).
2. Given a set of coefficient $\gamma_\mu(V^\pm, 1)$ in such a way that $g^\pm(r, r')$ is an analytic function of r and r', and a comparison potential $V_0(r)$ that is itself analytic, the solution of the system (XV.1.44–XV.1.47) exists and is unique unless "exceptional" conditions are satisfied; exceptional conditions cannot hold for r small enough, and correspond to isolated values in the complex r plane.

Thus Sabatier was able to generate exactly solvable examples, give a generating function for the $\psi_\lambda^\pm(r)$, and to show that the inverse problem for central + spin orbit potentials can be solved by machinery following the diagram (14.1). But there remained the important step of constructing the γ_μ's, and hence potentials, from the phase shifts. This was done by Hooshyar (1975). This author starts from a zero reference potential, so that $\psi_\lambda^{V_0}(r)$ reduces to $u_\lambda(r)$, and he assumes that

$$
\left.
\begin{aligned}
\left| \int_0^r sQ(s)ds \right| &< \infty \qquad \text{for all } r, \\
Q(r) &= O(r^{-3}) \qquad \text{for large } r.
\end{aligned}
\right\}
\tag{XV.1.52}
$$

Because $V_0 = 0$, $\gamma_\mu(V_0, 1)$ is simply $2\lambda/\pi$ for positive integer λ and 0 otherwise. For convenience let us introduce coefficients $c_\lambda^\pm$ and $d_\lambda^\pm$ by the relations:

$$
\gamma_\lambda(V^\pm, 1) = h_\pm^2(\gamma_\lambda(V_0, 1) + c_\lambda^\pm) = h_\pm^2 \, d_\lambda^\pm,
\tag{XV.1.53}
$$

where

$$
h_\pm = \exp\left[\pm \int_0^\infty \rho Q(\rho)d\rho \right].
\tag{XV.1.54}
$$

The study is restricted to potentials generated by coefficients $c_\lambda^\pm$ satisfying the condition

$$
c_\lambda^\pm = O(\lambda^{1/3}) \quad \text{for large } \lambda.
\tag{XV.1.55}
$$

With this assumption, it is possible to take the limit $r \to \infty$ in (XV.1.51), and equate the coefficients of e^{ir} and e^{-ir} separately. Doing so, we get

$$
\begin{aligned}
A_\lambda^\pm \exp\left[i\left(\delta_\lambda^\pm - \pi \frac{\lambda}{2} \right) \right] = {} & h_\pm \exp\left(-i\pi \frac{\lambda}{2} \right) \\
& - \sum_{\mu \in S} [A_\mu^\mp h_\pm^2 \, d_\mu^\pm \exp(i\delta_\mu^\mp) - A_\mu^\pm \gamma_\mu(V_0, 1)\exp(i\delta_\mu^\pm)] \\
& \times e^{-i\mu\pi/2} \Lambda_{\lambda\mu}^{V_0}(\infty).
\end{aligned}
\tag{XV.1.56}
$$

Let us call

$$
B_\lambda^\pm = A_\lambda^\pm h_\mp.
\tag{XV.1.57}
$$

Next multiply (XV.1.56) by $h_\mp$. Now since $h_- h_+ = 1$, Eq. (XV.1.56) reduces to

$$
\begin{aligned}
B_\lambda^\pm \exp\left[i\left(\delta_\lambda^\pm - \lambda \frac{\pi}{2} \right) \right] = {} & \exp\left(-i\lambda \frac{\pi}{2} \right) \\
& - \sum_{\mu \in S} [B_\mu^\mp \, d_\mu^\pm \exp(i\delta_\mu^\mp) - B_\mu^\pm \gamma_\mu(V_0, 1)\exp(i\delta_\mu^\pm)] \\
& \times e^{-i\mu\pi/2} \Lambda_{\lambda\mu}^{V_0}(\infty).
\end{aligned}
\tag{XV.1.58}
$$

Now, assume that the values of $\delta_\lambda^\pm$ are given for physical values of λ, $\lambda \in S_p = \{\frac{1}{2}, \frac{3}{2}, \ldots\}$, and that the values of $d_\lambda^\pm$ are chosen for nonphysical values of $\lambda \in S_0 = \{1, 2, 3, \ldots\}$. In this case, Hooshyar proved that it is possible to reduce the system (**XV.1.58**) to the inversion of infinite matrices and this can be done in general because they are either very similar to the matrix M studied in the chapter XII, or related to a Hilbert–Schmidt kernel that depends on the phase shifts. Hooshyar also studied the connection of the $d_\mu^\pm$ with the potentials, and made an interesting remark: Suppose we take $\delta_\lambda^+ = \delta_\lambda^-$ for physical values of λ, and we do not choose $d_\lambda^+ = d_\lambda^-$ for nonphysical values of λ. In this case, in general, the method for central + spin orbit potential does not reduce to the method for central potentials. In other words, even when $\delta_\lambda^+ = \delta_\lambda^-$ for all physical values of λ, it is still possible to have spin-orbit interaction. These types of potentials can be called "spin-orbit-transparent" potentials because their spin-orbit part cannot be detected from a change of polarization or any other information on spin, at this fixed energy.

XV.1.3   Direct Use of Interpolation Formulas

Setting

$$h(\lambda, r) = [s_{(1/2)\lambda}(r)]^{-1}\varphi_{(1/2)\lambda}^V(r) \qquad \text{(XV.1.59)}$$

in (**XV.1.3**) and using (**XII.8.6**), we readily obtain

$$h(\lambda, r) = 1 + \sum_{n>0} c_n r^n \frac{h(n; r)}{\lambda + n}, \qquad \text{(XV.1.60)}$$

where

$$c_n = -\pi \gamma_{n/2}(V, 1)(\tfrac{1}{2})^n [\Gamma(1 + \tfrac{1}{2}n)]^{-2} \qquad \text{(XV.1.61)}$$

Miodek (1976) tried to use the formula (**XV.1.60**) (and some of its generalizations) directly to construct the potential from the logarithmic derivatives of the physical wave functions at a fixed value of r, $r = R_1$. Clearly, if the potential is supposed to be zero outside the ball $(B(0, r_1)$, these logarithmic derivatives can be readily obtained from the scattering amplitude, so that Miodek's problem makes sense in the ordinary inverse problem at fixed energy (if one assumes *a priori* that the potential has a cut-off at R_1 and is analytic for r less than R_1). It also makes sense in other problems, and, in particular, in a special inverse plasma problem that Miodek considered in detail, and in which the sequence $\{c_n\}$ and the logarithmic derivatives are complex numbers.

Thus, in the Miodek's method, which is not restricted to real sequences $\{c_n\}$, the "experimental data" give

$$\beta(n) = \left[r \frac{\partial}{\partial r} \text{Log } h(n; r)\right]_{n = R_1}. \qquad \text{(XV.1.62)}$$

Differentiate **(XV.1.60)** and evaluate both **(XV.1.60)** and its derivative at $r = R_1$

$$h(\lambda; R_1) = 1 + \sum_{n>0} \frac{b_n}{\lambda + n}, \qquad \text{(XV.1.63)}$$

$$\beta(\lambda)h(\lambda; R_1) = \sum_{n>0} \frac{b_n}{\lambda + n} [\beta(n) + n], \qquad \text{(XV.1.64)}$$

where

$$b_n = c_n R_1^n h(n; R) \in C \qquad \text{(XV.1.65)}$$

If we can find $\{b_n\}$ from $\{\beta(n)\}$, we will be done because **(XV.1.63)** will then give us $h(n; R_1)$, and $\{c_n\}$ can then be obtained from the definition of b_n. Clearly **(XV.1.10 and XV.1.11)** give us

$$\beta(\lambda) = \frac{\sum_n (b_n'/\lambda + n)}{1 + \sum_m (b_m/\lambda + m)}, \qquad \text{(XV.1.66)}$$

where

$$\frac{b_n'}{b_n} = \beta(-n) = \beta(n) + n. \qquad \text{(XV.1.67)}$$

It remains to construct this interpolation of $\beta(n)$. Equations **(XV.1.66)** and **(XV.1.67)** are the necessary and sufficient conditions for the existence of a potential $u(r) = \sum_{n>0} u_n r^n$ which will generate $\beta(n)$ ($n > 0$) via the regular solutions of the radial equations. Miodek (1976) proved that such an interpolation can be constructed for any finite set: let us consider

$$\beta^N(\lambda) = \frac{\sum_{0<n<N} (b_n'^N/\lambda + n)}{1 + \sum_{0<m<N} (b_m^N/\lambda + m)}. \qquad \text{(XV.1.68)}$$

This corresponds to the case $c_n \equiv 0$ for $n > N$. There are a finite number of poles for $h(\lambda, r)$ at $\lambda = n$ ($0 < n \le N$). $\beta^N(\lambda)$ is a rational function of λ. It is the ratio of two polynomials $Q^{N-1}(\lambda)/P^N(\lambda)$, of order $N - 1$ and N, respectively, which are in one to one correspondence with the $2N$ constants $\{b_n'^N, b_n^N\}$

$$P^N(\lambda) \equiv p^N(\lambda)\left(1 + \frac{\sum_n b_n^N}{(\lambda + n)}\right), \qquad \text{(XV.1.69)}$$

$$Q^{N-1}(\lambda) = p^N(\lambda)\frac{\sum_n b_n'^N}{(\lambda + n)}, \qquad \text{(XV.1.70)}$$

where $p^N(\lambda) = \prod_{0<n<N} (\lambda + n)$. This rational function has $2N$ free parameters and can be constructed uniquely from its values at $\lambda = \pm n(n > 0)$.

We need only interpolate $\{\beta^N(n)$ and $\beta^N(-n) = \beta^N(n) + n\}$ by $Q^{N-1}(\lambda)/P^N(\lambda) = \beta^N(\lambda)$ to satisfy **(XV.2.66)** and **(XV.2.67)**. Then

$$b_m^N = \lim_{\lambda \to -n} \frac{(\lambda + n)P^N(\lambda)}{p^N(\lambda)} \qquad \text{(XV.1.71)}$$

gives us a solution to the inverse problem for the finite data set $\{\beta^N(n)\}$ $(0 < n \le N)$.

Unfortunately, the stability of these interpolations as $N \to \infty$ is still an open question, and it cannot be guaranteed that the data generated by (XV.1.66), $\beta^N(n)$ $(n > N)$, will be small. Preliminary studies do not lead Miodek to expect them to be small in general. On the other hand, Padé approximants may give a stable way to numerically construct these rational approximations and, in any case, the method is already interesting because of its simplicity.

XV.1.4   Further Generalizations

The structure of integral equations and the machinery obtained by using interpolation formulas can be used to suggest machinery applying to more general problems. This remark has been used with great success by Hooshyar, who studied the case of L^2-dependent potentials (1972), obtaining relatively simple results, and also (1971) the inverse scattering problem for potentials containing central, spin orbit, and tensor terms altogether. Clearly, this problem is of great interest in nuclear physics. Hooshyar (1971) was able to construct machinery which follows the general diagram (14.1) but which is extremely complicated in detail. Because of this, some points, in particular the construction of potentials, are not completely achieved as yet.

Using the standard analogy between the inverse problem at fixed E and the one at fixed l (see remark at the end of chapter XI), it is possible to derive machinery applying to the second problem from a machinery constructed for the first one. This remark has been used by Jaulent and Jean (1972), 1974, 1975) to solve the inverse problem at fixed l for potentials that depend linearly on $E^{1/2}$, through machinery which is similar to the one introduced by Sabatier to study potentials that depend linearly on λ at fixed energy, and which was described above.

XV.2   Methods Using Generalized Translation Operators

The theory of generalized translation operators was reviewed and extended by Levitan (1951, 1964), who applied it to differential operators, in particular second-order differential operators. Levitan also showed how generalized translation operators can be used to study inverse problems. Actually, we already met an example of application in the intertwining operator U used in chapter XIV. The method has been applied to some inverse problems at fixed energy by Coudray and Coz (1970, 1972), who were able to generalize Newton's method to the Klein–Gordon and Dirac equations. Here we rapidly show how generalized translation operators (more particularly the so-called "transformation operators") can be used to study inverse problems, then we give the Coudray and Coz method for the Klein–Gordon equation.

XV.2.1   Formal Use of Transformation Operators

Let $\mathscr{E} \subset \mathscr{E}'$, $\mathscr{F} \subset \mathscr{F}'$ be linear function spaces. Let A be a linear (differential) operator mapping $\mathscr{E}$ into $\mathscr{E}'$, B be a linear (differential) operator mapping $\mathscr{F}$ into $\mathscr{F}'$. We are interested in any linear operator, X mapping $\mathscr{F}$ onto $\mathscr{E}$ bijectively, and $\mathscr{F}'$ into $\mathscr{E}'$, and such that

$$AX = XB,$$
$$X^{-1} \quad \text{exists as a mapping of } \mathscr{E} \text{ onto } \mathscr{F}. \tag{XV.2.1}$$

Clearly, (XV.2.1) means that for any f ($f \in \mathscr{F}$), we have in $\mathscr{F}'$ the equality

$$A(Xf) = X(Bf). \tag{XV.2.2}$$

In particular, if f is a solution of $Bf = 0$, $g = Xf$ is a solution of $Ag = 0$. Thus X is a transformation operator generating g from f, whereas X^{-1} generates f from g.

From (XV.2.1), it is obvious that if X is a transformation operator for A and B, it is so for $A + \gamma$ and $B + \gamma$, where γ is an arbitrary constant. Thus if f_γ is a solution of

$$(B + \gamma)f_\gamma = 0, \tag{XV.2.3}$$

$g_\gamma = Xf_\gamma$ is a solution of

$$(A + \gamma)g_\gamma = 0. \tag{XV.2.4}$$

To guarantee that X is invertible, one usually requires that X be equal to $\mathbf{1} - \mathbf{K}$, where $\mathbf{K}$ is an integral operator with triangular kernel. Thus

$$(Xf)(r) = f(r) - \int_0^r K(r, \rho)\omega(\rho)f(\rho)d\rho. \tag{XV.2.5}$$

EXAMPLE. Let us introduce the differential operators

$$A = [h(r)]^{-1}\left[\frac{d^2}{dr^2} + W(r)\right],$$
$$B = [h(r)]^{-1}\left[\frac{d^2}{dr^2} + W_0(r)\right]. \tag{XV.2.6}$$

Let $\mathscr{E} = \mathscr{F}$ be the set of functions that are twice differentiable on $\mathbb{R}^+$ and equal to zero at $r = 0$, and let $\mathscr{E}' = A(\mathscr{E})$, $\mathscr{F}' = B(\mathscr{F})$. Setting $\omega(r) = h(r)$, (and assuming that $h(r)$, $W_0(r)$ and $W(r)$ have properties sufficient to allow the following derivation), we obtain (after a double integration by part, and some algebra):

$$(AX - XB)f = -K(r, 0)f'(0)$$

$$+ \frac{1}{h(r)}\left[W(r) - W_0(r) - h'(r)K(r, r) - 2h(r)\frac{d}{dr}K(r, r)\right]f(r)$$

$$+ \int_0^r \left\{\left[\frac{1}{h(\rho)}\left(\frac{\partial^2}{\partial \rho^2} + W_0(\rho)\right) - \frac{1}{h(r)}\left(\frac{\partial^2}{\partial r^2} + W(r)\right)\right]K(r, \rho)\right\}f(\rho)d\rho.$$

$$\tag{XV.2.7}$$

Thus, AX is equal to XB if and only if

$$K(r, 0) = 0, \qquad \text{(XV.2.8)}$$

$$W(r) - W_0(r) = h'(r)K(r, r) + 2h(r)\frac{d}{dr}K(r, r) \qquad \text{(XV.2.9)}$$

$$\times \left[\frac{1}{h(r)}\left(\frac{\partial^2}{\partial r^2} + W(r)\right) - \frac{1}{h(\rho)}\left(\frac{\partial^2}{\partial \rho^2} + W_0(\rho)\right) \right]K(r, \rho) = 0.$$
$$\text{(XV.2.10)}$$

If $h(r) = 1$, and the centrifugal term is included in W or W_0, one obtains the Gel'fand–Levitan transformation kernel for fixed l inverse problem, and with $\gamma = k^2$ in (XV.2.3), (XV.2.4).

If $h(r) = -r^{-2}$, $W_0(r) = k^2$, $W(r) = k^2 - V(r)$, one obtains the Regge–Newton transformation kernel for fixed energy inverse problems, with $\gamma = l(l + 1)$.

XV.2.2   Application to the Klein–Gordon Equation

The energy-independent Klein–Gordon equation reads:

$$[E - V(r)]^2\phi(\mathbf{r}) = [\mu^2 - \Delta]\phi(\mathbf{r}) \qquad \text{(XV.2.11)}$$

and the partial wave expansion, in the spherically symmetric case,

$$\phi(\mathbf{r}) = \sum_l r^{-1}\phi_l(r)P_l(\cos\theta) \qquad \text{(XV.2.12)}$$

yields the radial equations

$$[E - V(r)]^2\phi_l(r) = \left[\mu^2 + \frac{l(l + 1)}{r^2} - \frac{\partial^2}{\partial r^2}\right]\phi_l(r). \qquad \text{(XV.2.13)}$$

At fixed energy E, a set of differential equations is obtained in this way for each value of the angular momentum l.

Now, defining A and B as in (XV.2.6), with

$$h(r) = -r^{-2}, \qquad W_0(r) = [E - V_0(r)]^2 - \mu^2,$$
$$W(r) = [E - V(r)]^2 - \mu^2, \qquad \text{(XV.2.14)}$$

and setting $\gamma = l(l + 1)$ in (XV.2.3–XV.2.4), we see that the results just derived in the example readily apply. Thus, the kernel $K(r, r')$ of the transformation operator satisfies the equations

$$K(r, 0) = 0, \qquad \text{(XV.2.15)}$$

$$r^2\{[E - V(r)]^2 - [E - V_0(r)]^2\} = 2r\frac{d}{dr}[r^{-1}K(r, r)], \quad \text{(XV.2.16)}$$

$$\left\{ r^2\left[(E - V(r))^2 + \mu^2 - \frac{\partial^2}{\partial r^2}\right] - r'^2\left[(E - V_0(r'))^2 + \mu^2 - \frac{\partial^2}{\partial r'^2}\right] \right\}K(r, r') = 0.$$
$$\text{(XV.2.17)}$$

Setting

$$F(r, r') = \sum_l c_l \phi_l^0(r) \phi_l^0(r'), \qquad\qquad \text{(XV.2.18)}$$

$$K(r, r') = \sum_l c_l \phi_l(r) \phi_l^0(r'), \qquad\qquad \text{(XV.2.19)}$$

and using the representation of $\phi_l(r)$ involving $\phi_l^0(r)$ and the transformation kernel ($\phi_l = X \phi_l^0$):

$$\phi_l(r) = \phi_l^0(r) - \int_0^r K(r, s) \phi_l^0(s) s^{-2}\, ds \qquad\qquad \text{(XV.2.20)}$$

we readily obtain the Regge–Newton equation

$$K(r, r') = F(r, r') - \int_0^r K(r, \rho) F(\rho, r') \rho^{-2}\, d\rho. \qquad\qquad \text{(XV.2.21)}$$

The coefficients c_l give $F(r, r')$ via **(XV.2.18)**, and $K(r, r')$ can be obtained from $F(r, r')$ by applying **(XV.2.21)**. Hence, the inverse problem reduces to determining the coefficients c_l from the experimental results. Clearly, by inserting **(XV.2.18)** in **(XV.2.20)**, we again obtain the results of section XII.2:

$$\phi_l(r) = \phi_l^0(r) - \sum_{l'} c_{l'} L_{ll'}(r) \phi_{l'}(r), \qquad\qquad \text{(XV.2.22)}$$

where

$$L_{ll'}(r) = \frac{\phi_{l'}^0(r)(d/dr)\phi_l^0(r) - \phi_l^0(r)(d/dr)\phi_{l'}^0(r)}{l(l+1) - l'(l'+1)}. \qquad\qquad \text{(XV.2.23)}$$

Let us now allow $r \to \infty$ in **(XV.2.22)**. Assuming for simplicity that $V_0(r) = 0$, we know that the wave functions have the following asymptotic forms:

$$\phi_l^0(r) \sim \sin[\sqrt{E^2 - \mu^2}\, r - l\pi/2], \qquad r \to \infty, \qquad \text{(XV.2.24)}$$

$$\phi_l(r) \sim A_l \sin[\sqrt{E - \mu^2}\, r - l\pi/2 + \delta_l], \qquad r \to \infty, \qquad \text{(XV.2.25)}$$

so that the whole method of section XII.2 applies.

For the Dirac equation, the procedure given in the example **(XV.2.6)** has to be slightly modified ($d^2/dr^2 \to d/dr$), but the principle is the same. Again $K(r, r')$ is expanded in terms of products of physical wave functions, with coefficients c_l, and more complicated results.

XV.2.3   Miscellaneous

Thanks to the abstract presentation of transformation operators, (also called transmutations), which goes back to Delsarte and Lions (Delsarte, 1938; Delsarte and Lions, 1957; Lions, 1958), through Levitan (1964b), the mathematician Carroll (1985) was able to construct these operators for large classes of differential equations or systems (Carroll, 1979a,b, 1981a, 1982a,b, 1984a, 1986). He used the results to analyze scattering techniques and to connect them with special functions, and to afford new mathematical points of view on the Gel'fand–Levitan and Marchenko equations (for instance, he considered the latter as a minimizing procedure (Carroll, 1982c,d, 1983, 1984a,b; Carroll and Gilbert, 1981)). The method being of a purely geometric nature, it can easily be transposed from one inverse problem to another one and enabled Carroll to study the Newton–Sabatier method (Carroll and Delic, 1986) as well as some inverse problems connected with geophysics (Carroll and Santosa, 1981, 1982, 1984; Carroll and Raphael, 1986). It is fair to say that we give here only a few selected references, interesting for quantum inverse theory—there are many other applications to special functions, filtering theory, integral equations, etc.

XV.3 Remark on the Results Given in This Chapter

None of the inverse problems studied in the present chapter is completely solved. Formal procedures are given for all of them, and a method to construct one, or several, potentials from the experimental information, is given for most of them, but the general questions concerning the inverse problem have in no case been answered completely. Answering these questions is certainly the most difficult part of an inverse problem. But as long as they are not answered, the difference between a "solution" of the problem and a special fitting procedure is not very clear.

Besides, it is unlikely that very complicated machinery will ever be used in practical applications. This probably explains why the methods which have been constructed to generate inversion techniques have in fact been applied only to cases in which $V(r)$ has a very simple dependence on E or l. The method of Cornille (1976), who starts from a relation between the Fredholm determinants of two kernels, does not escape this limitation, although it could in principle be used in more general cases.

Several approaches to quantum inverse problems at fixed energy were recently developed for numerical computations. They will be reviewed in chapter XIX below. Other approaches were developed either to generalize the class of potentials to be obtained (e.g., the construction of L^2-dependent potentials by Hooshyar (1980)), or to apply to inverse problems which are not of quantum inverse scattering but use similar techniques (e.g., Hooshyar and Razavy, 1982, 1984).

CHAPTER XVI

Approximate Methods

XVI.1 Introduction

We are again interested in scattering experiments where a beam of particles A is scattered by a target made up of particles B, at a "non-relativistic" given energy E (except in sections which are devoted to particular problems). For the quantum mechanical description of the system A–B, we use the Schrödinger equation, and one or several functions depending on the relative distance between the particles. Our "inverse problem" is the construction of these "potentials" from the experimental measurements.

Throughout this chapter, we use the Schrödinger equation with a spherically symmetric potential:

$$\Delta^2 \psi + [k^2 - U(r)]\psi = 0, \tag{XVI.1.1}$$

where $k^2 = 2mE/h^2$, $U(r) = 2mV(r)/h^2$. The well-known partial wave decomposition leads one to the partial wave equations:

$$\frac{d^2 \phi_l}{dr^2} + [k^2 - U(r) - l(l+1)r^{-2}]\phi_l = 0, \tag{XVI.1.2}$$

where $\phi_l(r)$ behaves at the origin like:

$$u_l(r) = \left(\frac{\pi}{2} kr\right)^{1/2} J_{l+1/2}(kr). \tag{XVI.1.3}$$

It is assumed that $U(r)$ is taken in the largest class of potentials for which

ϕ_l and its derivative behave asymptotically as follows, (if there is no Coulomb potential):

$$\left.\begin{array}{l} \phi_l(r) = A_l \sin(kr - l\pi/2 + \delta_l) + o(1), \\[2mm] \dfrac{d}{dr}\, \phi_l(r) = kA_l \cos(kr - l\pi/2 + \delta_l) + o(1), \end{array}\right\} \qquad \text{as } r \to \infty. \quad \textbf{(XVI.1.4)}$$

In the presence of a Coulomb potential, these equations have to be modified. **(XVI.1.2)** and **(XVI.1.4)** are replaced by:

$$\frac{d^2\phi_l}{dr^2} + [k^2 - U(r) - 2\eta r^{-1} - l(l+1)r^{-2}]\varphi_l(r) = 0, \quad \textbf{(XVI.1.5)}$$

$$\left\{\begin{array}{l} \phi_l = A_l \sin[kr - l\pi/2 - \eta \, \mathrm{Log}(2\,kr) + \gamma_l + \delta_l] + o(1), \\[3mm] \dfrac{d}{dr}\, \phi_l = kA_l \cos[kr - l\pi/2 - \eta \, \mathrm{Log}(2\,kr) + \gamma_l + \delta_l] + o(1), \end{array}\right. \qquad \textbf{(XVI.1.6)}$$

where

$$\gamma_l = \arg \Gamma(l + 1 + i\eta). \qquad \textbf{(XVI.1.7)}$$

The cross section is related to the phase shifts by the formulas

$$f(\theta) = (k)^{-1} \sum_{l=0}^{\infty} (2l + 1)e^{i\delta_l} \sin \delta_l P_l(\cos \theta), \qquad \textbf{(XVI.1.8)}$$

$$\sigma(\theta) = |f(\theta)|^2. \qquad \textbf{(XVI.1.9)}$$

The physicist who wants to fit a scattering cross section by means of a complex potential usually chooses a family $\mathscr{F}$ of potentials depending on a few parameters, and manages to get an optimal fit of the cross section by a potential in this family. If the number of adjustable parameters is small enough, this can usually be done without ambiguity. Unfortunately, restricting the class of permissible potentials to $\mathscr{F}$ is a drastic working assumption, whose justification depends much more on previous successes than on physical grounds. It is generally felt that the potentials which are physically acceptable are probably not very different from the one which is constructed in $\mathscr{F}$. The aim of the present discussion is to investigate the basis for this feeling using the *simplest approaches*. We shall therefore avoid the general theories of inverse scattering problems, in favor of very simple and well-known approximations. We cannot pretend to exhibit *all the possible ambiguities* but we do give classes of ambiguities which are particularly interesting and physically meaningful.

At any step of the construction of potentials from the cross sections, there may be ambiguities due to the uncertainties on the results. From their very definition, these ambiguities can be reduced to zero in the limit of perfect measurements. On the other hand, and since no measurement is perfect, it is important to study how and where they can introduce uncertainties in the potential to be constructed. Take for example the scattering of a particle by a spherical complex potential with an imaginary part located in the surface. Because of the strong absorption, the coefficients $\exp[2i\delta_l]$ have a very small modulus for the first few values of l for which the impact parameter is inside the range of the potential. Therefore small errors in the cross section can lead to larger errors on the first δ_l's, which may lead in turn to much larger uncertainties on the inner part of the potential than on its surface. One may speak of a certain "screening" effect due to the imaginary part of the potential. This effect however, should not be overestimated, since it is compensated for by the precision of the experimental measurements.

A general study of the uncertainties—leading eventually to ambiguities— due to experimental uncertainties is too difficult at present and each experimental case needs to be studied separately. We shall therefore not study them in this present chapter.

We assume in the following that V belongs to a class of potentials such that the scattering amplitude $f(\theta)$ exists and is given with good precision by a certain method of approximation. The case of the Born approximation is studied in section XVI.2. The case of the semiclassical approximation is studied in sections XVI.3, XVI.4, XVI.5. Ambiguities obtained by these two methods are compared with exact solutions in section XVI.6. Some points of comparison with other inverse problems are given in section XVI.7.

The three types of semiclassical approximations correspond to different interests. One can see a decrease in mathematical justifications and a modification of the nature of physical applications.

It should be borne in mind that the use of approximations in the direct problem makes it difficult to formulate the inverse problem correctly. If these approximations are simply taken for granted, one may include solutions irrelevant to the correct problem, and one may forget relevant ones. Take for example the semiclassical analysis. In the inverse problems of quantum mechanics, the quantum effects are never negligible. Besides, using in heavy-ion scattering and in molecular scattering a single real potential is itself but an approximation. It follows that physicists have not tried to derive the set of *all the solutions* of quantum semiclassical scattering, but rather to exhibit typical solutions or typical ambiguities. As we see in section 7, the situation may be different in other fields.

XVI.2   Born Approximation

We first assume that the potentials to be found are small enough to use the simple Born approximation, and we describe the set of equivalent potentials thus obtained. For the sake of simplicity, we do not study the Coulomb case here, because it introduces several uninteresting complications.

In the non-Coulomb case, the phase shift and the scattering amplitude are given by (see, e.g., Mott and Massey 1965)

$$\delta_l = -k^{-1} \int_0^\infty U(\rho)[u_l(\rho)]^2 \, d\rho, \qquad (\textbf{XVI.2.1})$$

$$f(\theta) = -\left(2k \sin \frac{\theta}{2}\right)^{-1} \int_0^\infty \sin\left[2kr \sin \frac{\theta}{2}\right] U(r) r \, dr, \qquad (\textbf{XVI.2.2})$$

$f(\theta)$ is defined by (**XVI.2.2**) as a function of $2k \sin \theta/2$. Setting $t = 2k \sin \theta/2$, and $g(t) = -(4/\pi)k \sin(\theta/2) f(\theta)$, we see that $g(t)$ is the sine Fourier transform of $rU(r)$:

$$g(t) = \frac{2}{\pi} \int_0^\infty \sin(rt) r U(r) dr. \qquad (\textbf{XVI.2.3})$$

Now since $g(t)$ can, at best, be known only for $t \in (0, 2k)$, $rU(r)$ can be obtained from $g(t)$ only if a convenient continuation of $g(t)$ is chosen for $t > 2k$. If $rU(r)$ is sought in $L_2(0, \infty)$, this continuation can be any arbitrarily chosen function in $L_2(2k, \infty)$. When $g(t)$ is thus continued, $rU(r)$ is given by:

$$rU(r) = \text{l.i.m.} \int_0^\infty \sin(rt) g(t) dt. \qquad (\textbf{XVI.2.4})$$

But experimentally the Fourier transform of $rU(r)$ is known only on $(0, 2k)$. This situation is similar to the one in optics when one tries to reconstruct the shape of an object from its image. It is also similar to the numerous experiments on electronic devices (lines, wave guides, amplifiers, etc.) through which a signal is transmitted. In all these cases, the physical experiment is equivalent to a linear filter, whose width is $2k$ in our case.

The simplest way of continuing $g(t)$ beyond $t = 2k$ is to set $g(t) = 0$ for $t > 2k$. This yields $rU_a(r)$ equal to:

$$rU_a(r) = \int_0^{2k} \sin(rt) g(t) dt \qquad (\textbf{XVI.2.5})$$

and $rU_a(r)$ is an entire function of r, of order 1, type $2k$, in $L_2(0, \infty)$. It follows from Paley–Wiener's theorem that (**XVI.2.5**) is the unique solution of (**XVI.2.3**) in the space E_{2k} containing all the entire functions of exponential type $2k$ belonging to $L_2(0, \infty)$. Therefore we can construct all the potentials equivalent to $rU_a(r)$ (viz, giving the same phase shifts in the Born approximation) and belonging to $L_2(0, \infty)$, simply by choosing arbitrary functions of $L_2(2k, \infty)$.

Not only is the function $rU_a(r)$ bounded on the real axis but also its derivatives are as well. Conversely it follows from well-known theorems that bounds on the derivatives of $rU(r)$ can be related to the behavior of $g(t)$ for large t, and in particular, for t larger than $2k$. Now suppose that instead of looking for $rU(r)$ in $L_2(0, \infty)$, we impose *smoothness* conditions on $rU(r)$, requiring for instance that for any $n < N$ and any $k' < \frac{1}{2}N$:

$$\int_0^\infty \left| \frac{d^n}{d\rho^n} \rho U(\rho) \right|^2 d\rho \leq v_n^2 < \infty, \tag{XVI.2.6}$$

$$\left[\frac{d^{2k'}}{d\rho^{2k'}} \rho U(\rho) \right]_{\rho = 0} = 0, \tag{XVI.2.7}$$

then it is easy to see from integration by parts and Parseval's theorem that:

$$\frac{\pi}{2} \int_{2k}^\infty |t^n g(t)|^2 \, dt < v_n^2 \qquad (n < N). \tag{XVI.2.8}$$

If the coefficients v_n are given as *a priori* "smoothness" conditions, it is possible to prove from **(XVI.2.8)** that all the equivalent potentials are approximations of one of them, and their mean square deviation from each other goes to zero either when $\sup_{n<N}[v_n^2]$ goes to zero, E being fixed, or when the energy goes to infinity, for fixed $\sup[v_n^2]$. The first condition means that the physicist has increased the smoothness he requires for the potential. The second means that for larger energy, the width $2k$ of the window increases. The proof follows readily from **(XVI.2.4)**, **(XVI.2.8)**, and the Parseval's theorem, which yields:

$$\int_0^\infty r^2 [U(r) - U_a(r)]^2 \, dr = \frac{\pi}{2} \int_{2k}^\infty g^2(t) dt \leq [v_n^2(2k)^{-2n}] \qquad (n \leq N). \tag{XVI.2.9}$$

The most simple and commonly used condition is a single bound on the derivative and $U(0)$ finite. Then the conditions **(XVI.2.6)** and **(XVI.2.7)** are fulfilled with $k = 0$, $N = 1$, and **(XVI.2.9)** holds with $N = 1$, so that the deviation of the equivalent solutions goes to zero like E^{-1} as E goes to infinity.

In order to analyse the lack of uniqueness of the problem, we could also describe the set of transparent potentials, viz., the set of potentials yielding *all* phase shifts equal to zero. This way of proceeding is equivalent to the previous one: transparent potentials actually correspond to functions $g(t)$ equal to zero on $(0, 2k)$ and belonging to $L_2(2k, \infty)$. That the corresponding δ_l's equal zero can be checked readily from **(XVI.2.1)** using the Parseval identity and noticing that the sine Fourier transform of $\rho^{-1}[u_l(\rho)]^2$ is equal to zero for $t > 2k$ (and to the Legendre polynomial $\frac{1}{2}P_l(1 - \frac{1}{2}k^{-2}t^2)$ for $0 \leq t \leq 2k$).

It should be noticed that because of the linearity of the Born approximation, an imaginary part of the potential would not give rise to a special problem.

Generalization. In the Born approximation, it is assumed that, starting from a zero potential, one introduces the potential $U(r)$ as a small perturbation. It is also possible to start from a potential U_0, that yields a set of phase shifts

$\delta_l^{(0)}$, and to add a perturbation ΔU that is small enough to justify the generalized Born approximation. The perturbations to the phase shifts are then obtained as a linear transform of ΔU. The ambiguities that are involved in the inversion of this linear transform have also been studied (Sabatier 1973) and are somewhat similar to the ones obtained in the simple case.

It may be worth noticing that if the scattering is completely elastic (no absorption), and if the Born approximation applies, the only ambiguity in the step $\sigma \rightarrow \{\delta_l\}$ is the trivial one (see chapter X), since $f(\theta)$ is real.

For results concerning the three-dimensional Born approximation see section XIV.1.

XVI.3   The Semiclassical Approximation I

In the present section, we explore the space of all potentials that give a set $\{\delta_l\}$ by means of the JWKB approximation.

We first recall results of the direct problem in section XVI.3.1.

In section XVI.3.2, we derive some inversion formulas and some necessary and sufficient constraints on the functions of interest. We are then in position to study the ambiguities. The case in which several turning points are present is studied in section XVI.3.3. In this case, the contribution to the phase shifts of certain parts of the potential that are located between two turning points, and the contribution of the remaining part of the potential, behave differently, and it is possible to balance a modification of the former by a modification of the later, so as to shift certain phase shifts by $K\pi$, where K is an integer. Hence several equivalent potentials are outlined. Other ambiguities arise both in the case of several turning points or only one, and are also related to the shifting of certain phase shifts by $K\pi$ or to the lack of uniqueness of their interpolation.

XVI.3.1   Direct Problem

(a)   Non-Coulomb Case. Our working assumption is that the phase shift Δ_l is related to the potential by the JWKB formula:

$$\Delta_l = \delta(l + \tfrac{1}{2}) = \lim_{r \to \infty} \left\{ \int_{r_0}^r [\bar{\lambda}^2(k\rho) - (l + \tfrac{1}{2})^2]^{1/2} \rho^{-1} \, d\rho \right.$$

$$\left. - \int_{(l+1/2)/k}^r [k^2\rho^2 - (l + \tfrac{1}{2})^2]^{1/2} \rho^{-1} \, d\rho \right\}, \quad \textbf{(XVI.3.1)}$$

where

$$\bar{\lambda}^2(x) = x^2 \left[1 - k^{-2} U\left(\frac{x}{k}\right) \right] \qquad \textbf{(XVI.3.2)}$$

and r_0 is the largest "turning point," viz., the largest zero, for $\lambda = l + \frac{1}{2}$, of the function $\rho \to [\bar{\lambda}^2(k\rho) - \lambda^2]$. Notice that $\bar{\lambda}(x)$ also depends on k. In the following k is assumed to be fixed, U is assumed to be continuous on $[0, \infty]$ and going to zero more rapidly than $x^{-1-\varepsilon}$ as $x \to \infty$.

Let us now put (XVI.3.1) in the form of an Abel transform, using a trick introduced several years ago (Sabatier 1965) which yields, as we know, a very simple inversion procedure (see the notes at the end of the chapter for historical details on the JWKB approach to inverse problems).

Let $x_{\min}$ be the largest zero of $\bar{\lambda}(x)$. In agreement with the JWKB approximation and the well-known properties of a potential, it is reasonable to suppose that, for $x \geq x_{\min}$ $\bar{\lambda}(x)$ is a continuous and piecewise differentiable function, with only a finite number of maxima and minima, and that it behaves asymptotically like x. Let x_3 be the smallest value such that $\bar{\lambda}(x)$ is monotonically increasing for $x \geq x_3$, and let λ_0 be equal to $\bar{\lambda}(x_3)$. For $(x > x_3, \lambda > \lambda_0)$, the mapping $x \to \lambda = \bar{\lambda}(x)$ has an inverse, which we denote as

$$\lambda \to x = \psi(\lambda) \qquad (\lambda > \lambda_0). \qquad \text{(XVI.3.3)}$$

$\psi(\lambda)$ is a continuous, piecewise differentiable, and monotonically increasing function. Using λ as a new variable in (XVI.3.1), we obtain for $p > \lambda_0$

$$\delta(p) = \lim_{r \to \infty} \left\{ \int_p^{\bar{\lambda}(kr)} [\lambda^2 - p^2]^{1/2} \left[\frac{\psi'(\lambda)}{\psi(\lambda)} \right] d\lambda - \int_p^{kr} [\lambda^2 - p^2]^{1/2} \lambda^{-1} \, d\lambda \right\}.$$

$$\text{(XVI.3.4)}$$

Using the fact that $U(r)$ goes to zero more rapidly than $r^{-1-\varepsilon}$ as r goes to infinity, from (XVI.3.4) we derive straightforwardly the following formulas, valid for $p > \lambda_0$:

$$\delta(p) = -\int_p^{\infty} [\lambda^2 - p^2]^{1/2} H'(\lambda) d\lambda, \qquad \text{(XVI.3.5)}$$

$$\delta(p) = \int_p^{\infty} [\lambda^2 - p^2]^{-1/2} \lambda H(\lambda) d\lambda, \qquad \text{(XVI.3.6)}$$

where

$$H(\lambda) = \text{Log} \left[\frac{\lambda}{\psi(\lambda)} \right]. \qquad \text{(XVI.3.7)}$$

Assume now that $\bar{\lambda}(x)$ has a single maximum (figure XVI.1) for x between $x_{\min}$ and x_3, say, at x_2, with $\bar{\lambda}(x_2) = \lambda_1$, and let x_1 be the value of $(x_{\min}, x_2)$ for which $\bar{\lambda}(x) = \lambda_0$. For $p < \lambda_0$, the JWKB phase shift is equal to:

$$\delta(p) = \delta(p) + \delta_{\lambda_0}(p)\theta(\lambda_0 - p), \qquad \text{(XVI.3.8)}$$

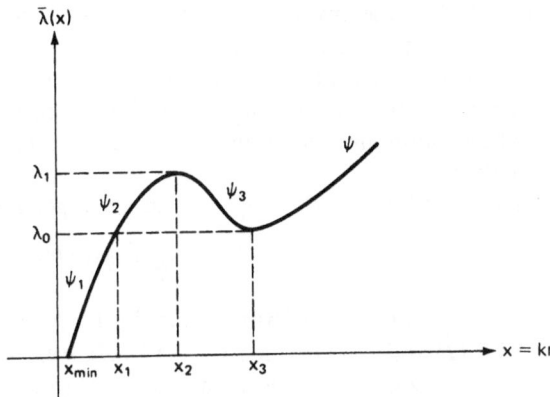

Figure XVI.1

where θ is the Heaviside step function, and

$$\delta(p) = \int_{x_0}^{x_1} [\bar{\lambda}^2(x) - p^2]^{1/2} x^{-1} \, dx + \int_{x_1}^{x_3} (\lambda_0^2 - p^2)^{1/2} x^{-1} \, dx$$

$$+ \lim_{X \to \infty} \left\{ \int_{x_3}^{X} [\bar{\lambda}^2(x) - p^2]^{1/2} x^{-1} \, dx - \int_{p}^{X} [x^2 - p^2]^{1/2} x^{-1} \, dx \right\},$$

$$\text{(XVI.3.9)}$$

x_0 being equal to $kr_0(p)$, and

$$\delta_{\lambda_0}(p) = \int_{x_1}^{x_3} [\bar{\lambda}^2(x) - p^2]^{1/2} x^{-1} \, dx - \int_{x_1}^{x_3} (\lambda_0^2 - p^2)^{1/2} x^{-1} \, dx. \quad \text{(XVI.3.10)}$$

Let us denote the inverse mappings of $x \to \lambda = \bar{\lambda}(x)$ by

$$\left. \begin{array}{lll} \lambda \to x - \psi_1(\lambda) & \text{for} \quad (x \in (x_{\min}, x_1), & \lambda \in (0, \lambda_0)), \\ \lambda \to x = \psi_2(\lambda) & \text{for} \quad (x \in (x_1, x_2), & \lambda \in (\lambda_0, \lambda_1)), \\ \lambda \to x = \psi_3(\lambda) & \text{for} \quad (x \in (x_2, x_3), & \lambda \in (\lambda_0, \lambda_1)), \end{array} \right\} \quad \text{(XVI.3.11)}$$

and let us define $\psi(\lambda)$ as equal to $\psi_1(\lambda)$ for $\lambda < \lambda_0$. The function $\psi(\lambda)$ defined by this and (XVI.3.3) is piecewise continuous and piecewise differentiable. From (XVI.3.9) and (XVI.3.10) we easily derive the formulas:

$$\delta(p) = \int_{p}^{\infty} [\lambda^2 - p^2]^{-1/2} \lambda H(\lambda) d\lambda, \quad \text{(XVI.3.12)}$$

where $H(\lambda)$ is still given by the formula (XVI.3.7), and

$$\delta_{\lambda_0}(p) = \int_{\lambda_0}^{\lambda_1} [\lambda^2 - p^2]^{-1/2} \lambda h(\lambda) d\lambda, \quad \text{(XVI.3.13)}$$

where

$$h(\lambda) = \text{Log} \left[\frac{\psi_3(\lambda)}{\psi_2(\lambda)} \right]. \quad \text{(XVI.3.14)}$$

Thus $\psi(\lambda)$ is the inverse function of $x \to \lambda_s(x)$, where $\lambda_s(x)$ is the largest nondecreasing function such that $\lambda_s(x) \le \bar\lambda(x)$ for any $x \ge x_{\min}$.

These formulae can be generalized for an arbitrary number of maximums and minimums between the two points of ordinate λ_0, provided that the smallest minimum is above λ_0, and that $h(\lambda)$ is conveniently defined (Sabatier 1973).

The JWKB phase shift $\delta(p)$ is then given generally by

$$\delta(p) = \bar\delta(p) + \sum_m \delta_{\lambda_0^{(m)}}(p)\theta(\lambda_0^{(m)} - p), \qquad \textbf{(XVI.3.15)}$$

where $\delta_{\lambda_0^{(m)}}$ is given by formulas similar to **(XVI.3.13)**, $\bar\delta(p)$ is again given by **(XVI.3.12)** and $H(\lambda)$ defined by **(XVI.3.7)**, were $\psi(\lambda)$ is the inverse mapping of $x \to \lambda = \lambda_s(x)$. It should be noticed that $\psi(\lambda)$ is therefore a positive piecewise continuous and piecewise differentiable function, and that $\psi'(\lambda)$ satisfies the inequality

$$\psi'(\lambda) \ge 0. \qquad \textbf{(XVI.3.16)}$$

It is interesting to notice that all the results obtained above hold even when $\bar\lambda(x)$ has constant parts and/or first kind discontinuities.

(b)   Coulomb Case. In the JWKB approximation, the difference between the phase of the wave function and the phase of the free wave function (viz. $[kr - l\pi/2]$), is still given for large r by the formula

$$\Delta_l(r) = \int_{r_0}^r [\bar\lambda^2(k\rho) - (l + \tfrac{1}{2})^2]^{1/2}\rho^{-1}\,d\rho - \int_{(l+1/2)/k}^r [k^2\rho^2 - (l + \tfrac{1}{2})^2]^{1/2}\rho^{-1}\,d\rho,$$

$$\textbf{(XVI.3.17)}$$

where

$$\bar\lambda^2(x) = x^2\left[1 - 2\eta x^{-1} - k^{-2}U\left(\frac{x}{k}\right)\right] \qquad \textbf{(VXI.3.18)}$$

and r_0 is the largest turning point, and η is the usual Coulomb parameter $(Z_1 Z_2 e^2/\hbar v)$. When $U(x)$ is equal to zero, $\bar\lambda(x)$ is monotonic from $x_{\min} = 2\eta$, and the inverse mapping of $x \to \lambda = \bar\lambda(x)$ is

$$\lambda \to x = \psi_c(\lambda) = \eta + (\eta^2 + \lambda^2)^{1/2}. \qquad \textbf{(XVI.3.19)}$$

Hence, for $U(r) \equiv 0$, $\Delta_l(r)$ reduces to

$$\Delta_l^c(r) = \int_p^{[k^2r^2 - 2\eta kr]^{1/2}} [\lambda^2 - p^2]^{1/2}\left[\frac{\psi_c'(\lambda)}{\psi_c(\lambda)} - \lambda^{-1}\right]d\lambda$$

$$- \int_{[k^2r^2 - 2\eta kr]^{1/2}}^{kr} [\lambda^2 - p^2]^{1/2}\lambda^{-1}\,d\lambda, \qquad \textbf{(XVI.3.20)}$$

where p stands for $(l + \tfrac{1}{2})$. For large r, and $(l^2 + \eta^2)$ not small it is easy to see that the second term in the right-hand side of **(XVI.3.20)** goes to η, and the first one is asymptotic to $-\eta \, \text{Log}[2kr] + \text{Arg}[\Gamma(l + 1 + i\eta)]$, which gives the exact phase of the purely Coulomb wave function. Now, if a potential $U(r)$ is present, the "modified" JWKB phase shift is defined by the difference:

$$\delta_c(l + \tfrac{1}{2}) = \lim_{r \to \alpha} [\Delta_l(r) - \Delta_l^c(r)]. \qquad \text{(XVI.3.21)}$$

Using (XVI.3.17) in (XVI.3.21), we easily see that we are led to study the functions $\bar{\lambda}(x)$ just as we were in the non-Coulomb case. In particular, we can again define $\lambda_s(x)$ as the largest monotonically increasing function such that $\lambda_s(x) \leq \bar{\lambda}(x)$, and $\psi(\lambda)$ the inverse function of $\lambda_s(x)$. Hence $\delta(p)$ is still given by the formula (XVI.3.15), and it follows from (XVI.3.21) and (XVI.3.20) that $\delta(p)$ is again given by (XVI.3.12), but with

$$H(\lambda) = \text{Log}\left[\frac{\psi_c(\lambda)}{\psi(\lambda)}\right]. \qquad \text{(XVI.3.22)}$$

XVI.3.2   Inversion Formulas and Consistency Inequalities

Suppose that $H(\lambda)$ is piecewise differentiable. Then it follows from (XVI.3.5) that, except may be at certain isolated values of p, $\delta(p)$ is differentiable, and its derivative is given by the formula:

$$\delta'(p) = p \int_p^\infty (\lambda^2 - p^2)^{-1/2} H'(\lambda) d\lambda. \qquad \text{(XVI.3.23)}$$

It is easy to see in (XVI.3.23) that $\delta'(p)$ is continuous for any p if $H'(\lambda)$ has no singularity λ_s more divergent that $|\lambda - \lambda_s|^{-1/2+\varepsilon}$. Such singularities do not exist if $\lambda'(x)$ is continuous and everywhere positive.

Now, suppose *we know* that $\delta(p)$ is produced by a monotonic function $\bar{\lambda}(x)$, and suppose *we know* $\delta'(p)$ from $p = 0$ to $p = \infty$. Then $\delta(p)$ must be related to $\bar{\lambda}(x)$ by (XVI.3.6), (XVI.3.7) and (XVI.3.3). It is possible to inverse the Abel transform (XVI.3.6), obtaining:

$$H(\lambda) = -2\pi^{-1} \int_\lambda^\infty (p^2 - \lambda^2)^{-1/2} \delta'(p) dp. \qquad \text{(XVI.3.24)}$$

If $\delta'(p)$ has itself a piecewise continuous derivative, $H(\lambda)$ and $H'(\lambda)$ can be obtained from $(d/dp)p^{-1}\delta'(p)$:

$$H(\lambda) = 2\pi^{-1} \int_\lambda^\infty (p^2 - \lambda^2)^{1/2} \frac{d}{dp} [p^{-1}\delta'(p)], \qquad \text{(XVI.3.25)}$$

$$H'(\lambda) = -2\pi^{-1}\lambda \int_\lambda^\infty (p^2 - \lambda^2)^{-1/2} \frac{d}{dp} [p^{-1}\delta'(p)] dp. \qquad \text{(XVI.3.26)}$$

It is essential to notice at this point that if $\delta'(p)$ is taken at random for instance in $C_1(0, \infty)$, the function $H(\lambda)$ may correspond to no potential. To obtain a potential, in the JWKB approximation, $H(\lambda)$ must be differentiable, except at a finite number of points, and $\psi(\lambda)$ must be a nondecreasing positive function for $\lambda > 0$. Now, from a real $H(\lambda)$, we can always derive a positive $\psi(\lambda)$ through the formulas:

$$\psi(\lambda) = \begin{cases} \lambda \exp[-H(\lambda)] & \text{(n.c.c.)}, \\ \psi_c(\lambda)\exp[-H(\lambda)] & \text{(c.c.)}, \end{cases} \quad \textbf{(XVI.3.27)}$$

which hold respectively in the non-Coulomb and in the Coulomb case. A *necessary and sufficient condition* to obtain $\lambda(r)$ is therefore **(XVI.3.16)**, or

$$H'(\lambda) < \begin{cases} \lambda^{-1} & \text{(n.c.c.)}, \\ \lambda[\lambda^2 + \eta^2 + \eta(\lambda^2 + \eta^2)^{1/2}]^{-1} & \text{(c.c.)}. \end{cases} \quad \textbf{(XVI.3.28)}$$

When inserted in **(XVI.3.23)**, these inequalities in turn give *necessary inequalities* for $\delta'(p)$. In the non-Coulomb case, we get for instance

$$\delta'(p) < \tfrac{1}{2}\pi. \quad \textbf{(XVI.3.29)}$$

However this condition is not sufficient to yield a nondecreasing $\psi(\lambda)$. Take for instance $\delta'(p) = (\pi/2)p$ for $p \le p_0 < 1$ and 0 for $p > p_0$. Then $H(\lambda)$ is equal to $(p_0^2 - \lambda^2)^{1/2}\theta(p_0 - \lambda)$ (where θ is the Heaviside step function), and $H'(\lambda)$ is equal to $\lambda(p_0^2 - \lambda^2)^{-1/2}\theta(p_0 - \lambda)$, which is obviously not consistent with **(XVI.3.28)** for $\lambda \to p_0$.

On the other hand, if $d/dp[p^{-1}\delta'(p)]$ is piecewise continuous, or nowhere more singular than $|\lambda - \lambda_s|^{-(1/2)+\varepsilon}$, a *sufficient* condition to get a nondecreasing $\psi(\lambda)$ (still in the non-Coulomb case), is

$$\frac{d}{dp}[p^{-1}\delta'(p)] > -\tfrac{1}{2}\pi p^{-2}. \quad \textbf{(XVI.3.30)}$$

When inserted in **(XVI.3.26)**, this inequality yields **(XVI.3.28)** (n.c.c.). Since $\delta'(p)$ is finite, obviously **(XVI.3.29)** follows from **(XVI.3.30)**. Similar, but more complicated, results, can be obtained in the Coulomb case.

The constraint that a function $h(\lambda)$ must satisfy readily follows from the formula **(XVI.3.14)**, since the $\psi_{2i+1}(\lambda)$ are decreasing and the $\psi_{2i}(\lambda)$ are increasing. One must have:

$$h'(\lambda) \le 0. \quad \textbf{(XVI.3.31)}$$

Now, if $h(\lambda)$ fulfill **(XVI.3.31)**, one may wonder whether $H(\lambda)$ obtained from $\delta(p)$ will lead to a function $\bar{\lambda}(x)$ which is continuous with the part giving $h(\lambda)$. The condition for this is that (figure XVI.1):

$$h(\lambda_0) = \text{Log}\,\frac{\psi_3(\lambda_0)}{\psi_1(\lambda_0)} = \lim_{\lambda \to \lambda_0^-} H(\lambda) - \lim_{\lambda \to \lambda_0^+} H(\lambda). \quad \textbf{(XVI.3.32)}$$

Actually this condition is fulfilled provided that $\delta'(p)$ has no divergence stronger than $|p - p_s|^{-1/2+\varepsilon}$.

XVI.3.3   Balanced Ambiguities

Let us assume, for the sake of simplicity, that $\bar{\lambda}(x)$ has a single maximum (figure XVI.1) so that $\delta(p)$ is given by (**XVI.3.8**), with $\bar{\delta}(p)$ and $\delta_{\lambda_0}(p)$ given by (**XVI.3.12**) and (**XVI.3.13**). Suppose that $h(\lambda)$ is modified:

$$h(\lambda) \to h^+(\lambda) = h(\lambda) + \Delta h(\lambda), \qquad (\mathbf{XVI.3.33})$$

where $\Delta h(\lambda) = 0$ for $\lambda < \lambda_0$, and $= 0$ for λ large enough, so that we define λ_0^M as the smallest value of λ above which $h(\lambda)$ and $\Delta h(\lambda)$ are identically zero. Besides, $h^+(\lambda)$ must satisfy (**XVI.3.31**). Suppose also that $H(\lambda)$ is modified

$$H(\lambda) \to H^+(\lambda) = H(\lambda) + \Delta H(\lambda), \qquad (\mathbf{XVI.3.34})$$

where $\Delta H(\lambda) = 0$ for $\lambda > \lambda_0$, and the continuity condition (**XVI.3.37**) is fulfilled, viz.

$$\Delta H(\lambda_0) = \Delta h(\lambda_0). \qquad (\mathbf{XVI.3.35})$$

Owing to these perturbations, the phase shift is modified by

$$\delta(p) \to \delta^+(p) = \delta(p) + \Delta\delta(p) \qquad (\mathbf{XVI.3.36})$$

and according to (**XVI.3.8**), (**XVI.3.12**) and (**XVI.3.13**), $\Delta\delta(p)$ is equal to zero for $p > \lambda_0$, and

$$\Delta\delta(p) = \Delta\bar{\delta}(p) + \Delta\delta_{\lambda_0}(p) = \int_p^{\lambda_0} (\lambda^2 - p^2)^{-1/2} \lambda \Delta H(\lambda) d\lambda$$

$$+ \int_{\lambda_0}^{\lambda_0^M} (\lambda^2 - p^2)^{-1/2} \lambda \Delta h(\lambda) d\lambda \quad \text{for} \quad p < \lambda_0. \quad (\mathbf{XVI.3.37})$$

Now, for a given scattering function $\exp[2i\delta(p)]$, it is clear that there may be several solutions $(H(\lambda), h(\lambda))$, that correspond to the perturbations:

$$\Delta\delta(p) = K\pi \qquad (p < \lambda_0), \qquad (\mathbf{XVI.3.38})$$

where K is an integer, positive, or negative, or zero. The first term in (**XVI.3.37**) is zero for $p = 0$. Hence, to obtain a solution, we have to choose $\Delta h(\lambda)$ in such a way that

$$\Delta\delta_{\lambda_0}(\lambda_0) = \int_{\lambda_0}^{\lambda_0^M} (\lambda^2 - \lambda_0^2)^{-1/2} \lambda \Delta h(\lambda) d\lambda = K\pi \qquad (\mathbf{XVI.3.39})$$

and then calculate $\Delta H(\lambda)$ so as to obtain $(\Delta\delta(p) - \Delta\delta(\lambda_0))$ equal to zero for any $p < \lambda_0$. According to (**XVI.3.39**), this is achieved if

$$\Delta H(\lambda) = -2\pi^{-1} \int_\lambda^{\lambda_0} (p^2 - \lambda^2)^{-1/2} \frac{d}{dp} [\Delta\delta_{\lambda_0}(\lambda_0) - \Delta\delta_{\lambda_0}(p)] \qquad (\mathbf{XVI.3.40})$$

or, after some algebra

$$\Delta H(\lambda) = 2\pi^{-1}(\lambda_0^2 - \lambda^2)^{1/2} \int_{\lambda_0}^{\lambda_0^M} \Delta h(\mu)(\mu^2 - \lambda_0^2)^{-1/2}(\mu^2 - \lambda^2)^{-1}\mu \, d\mu.$$

$$(\textbf{XVI.3.41})$$

From (**XVI.3.41**), we can derive a rough approximation of $\Delta H(\lambda)$ for $\lambda \ll \lambda_0$ by neglecting the variations of $\mu^2 - \lambda^2$ on (λ_0, λ_0^M).

This is valid if $\lambda_0^M - \lambda_0 \ll \lambda_0$. The result is:

$$\Delta H(\lambda \ll \lambda_0) \sim 2\pi^{-1}(\lambda_0^2 - \lambda^2)^{-1/2} \int_{\lambda_0}^{\lambda_0^M} \Delta h(\mu)(\mu^2 - \lambda_0^2)^{-1/2}\mu \, d\mu. \quad (\textbf{XVI.3.42})$$

Now, taking into account (**XVI.3.39**), we obtain the very remarkable result

$$\Delta H(\lambda \ll \lambda_0) \sim 2K(\lambda_0^2 - \lambda^2)^{-1}. \qquad (\textbf{XVI.3.43})$$

Hence, the equivalent solutions corresponding to the perturbation (**XVI.3.38**) of the phase shifts can be divided into classes labelled by K. All the functions $\Delta H(\lambda)$ of a given class have roughly the same behavior as λ goes to zero. This behavior is independent of the function $h(\lambda)$ which has been used in (**XVI.3.41**), and also of the function $H(\lambda)$. It merely depends on K! Hence, one can talk of layers of equivalent solutions, and of a layered ambiguity. Besides, suppose that $H(\lambda)$ has been obtained from an attractive potential going to a constant—say, U_0—as r goes to zero. Then the potential corresponding to $H^+(\lambda)$ goes to a constant U_K:

$$H^+(0) = \tfrac{1}{2} \text{Log}(1 - k^2 U_K). \qquad (\textbf{XVI.3.44})$$

Hence the layering of $H(0)$ corresponds to a layering of $U(0)$ through the formula:

$$-U_K = -U_0 \exp\left[\frac{4K}{\lambda_0}\right] + k^2\left(\exp\left[\frac{4K}{\lambda_0} - 1\right]\right). \qquad (\textbf{XVI.3.45})$$

It is interesting to notice that if this layering is considered at each energy, one may see that it varies smoothly with the energy.

We have pointed out that the layering is a layering of *classes* of equivalent potentials, each class being indexed by K. Inside a given class, the various potentials correspond to the various functions $h^+(\lambda)$ consistent with (**XVI.3.32**) for the given value of K. That this continuous ambiguity does exist can easily be checked with simple forms of $\Delta H(\lambda)$ provided that the variation is small enough. It may be noticed that a further ambiguity comes from the fact that $h(\lambda)$ itself is related to $\bar{\lambda}(x)$ ambiguously, since giving $h(\lambda)$ only determines $\psi_3(\lambda)/\psi_2(\lambda)$. Within certain bounds, one can choose $\psi_2(\lambda)$ and then calculate $\psi_3(\lambda)$. Besides, since $\Delta h(\lambda)$ is not completely determined by (**XVI.3.39**), $\Delta H(\lambda)$ is not completely determined by (**XVI.3.41**) and (**XVI.3.39**). The corresponding ambiguity, for a given K, essentially results in variation of the surface of the potential ("shape ambiguity").

The physical meaning of the layering is not yet very difficult to understand. The formula (**XVI.3.9**) when the upper bound X is a fixed number, as well as the formula (**XVI.3.17**), gives the phase of the regular wave function as a function of r. Hence one sees easily that $\delta_{\lambda_0}(p)$ is nothing but the phase shift due to the part of the effective potential (viz. $V(r)$ minus the centrifugal term) located between the two turning points at λ_0. Thus, when K increases by 1, the number of nodes of the wave function increases by 1. The matching condition with the wave function in the internal part remains roughly the same provided that the local value of the effective wave number be modified to fit the new frequency (since this fit is imposed by JWKB approximation). Hence $-V(r)$ has to increase by a fixed quantity. Let us notice also that the number of nodes in the barrier is obviously related to the number of states in the inverted barrier.

Other examples of ambiguities have been exhibited. They correspond either to the invariance of scattering amplitudes under translations of the phase-shifts modulo π (layered ambiguities), or to the existence of a continuum of interpolations between values at integer l's. The smoothness conditions that are required by the JWKB analysis are unfortunately consistent with such a continuum, within certain limits. The layered ambiguities that appear in the cases where one turning point only is allowed correspond to enormous modifications of the potential, and they are probably easily discarded using physical arguments.

XVI.3.4   Other Results

It is interesting to say a few words on the problem of ambiguities in optical-model analyses. This problem is more complicated than the one studied above because absorption comes in, with two consequences:

1. The scattering coefficients for small values of l have a small amplitude, so that relative errors are increased.
2. The imaginary part of the potential is an additional continuous parameter, which adds to the real part to cause ambiguities. However, results show that the ambiguities due to the real part are still essential.

Ambiguities have been studied in most cases by computer analyses, and the relevance of JWKB analysis in giving a qualitative explanation of certain ambiguities, and its validity when absorption is present, have been noticed (Austern 1961, Singh et al. 1969, 1972, Drisko et al. 1963, Cuer 1976).

To finish this section, it may be worth noticing that the following order of the JWKB approximation again leads to a formula similar to (**XVI.3.6**), so that the inversion procedure again applies (Sabatier 1965b).

XVI.4   Semiclassical Analysis II

In the previous section, we only assumed that the phase shifts can be calculated through the JWKB approximation. A further step is taken if the number of nonnegligible phase shifts is so large, and their dependence on l so smooth that the following additional approximations can be made (Ford and Wheeler 1959a,b,c).

1. The use of an asymptotic approximation to the Legendre polynomials (for angles not too close to 0 or π)

$$P_l(\cos \theta) = [\tfrac{1}{2}(l + \tfrac{1}{2})\pi \sin \theta]^{-1/2} \sin\left[(l + \tfrac{1}{2})\theta + \frac{\pi}{4}\right]. \quad \textbf{(XVI.4.1)}$$

2. The replacement of the sum in the scattering amplitude (**XVI.1.8**) by an integral:

$$f(\theta) = -(2\pi k^2 \sin \theta)^{-1/2} \int_0^\infty (l + \tfrac{1}{2})(e^{i\varphi^+} - e^{i\varphi^-})dl, \quad \textbf{(XVI.4.2)}$$

$$\varphi^\pm = 2\delta_l \pm (l + \tfrac{1}{2})\theta \pm \frac{\pi}{4}. \quad \textbf{(XVI.4.3)}$$

3. and the use of the stationary phase approximation to evaluate the integral in equation (**XVI.4.2**).

If the stationary phase points are isolated, we then obtain the following expression for the semiclassical approximation to the differential cross section:

$$\sigma(\theta) = \sum_{i=1}^{n} \sigma_i(\theta) + \sum_{i>j} 2(\sigma_i\sigma_j)^{1/2} \cos(\beta_i - \beta_j), \quad \textbf{(XVI.4.4)}$$

where

$$\sigma_i = \frac{(l_i + \tfrac{1}{2})}{2k^2 \sin \theta \delta''_{l_i}}, \quad \textbf{(XVI.4.5)}$$

$$\beta_k = \left[2\delta_l - 2(l + \tfrac{1}{2})\delta'_l - \left(2 - \frac{\delta''_l}{|\delta''_l|} - \frac{\delta'_l}{|\delta'_l|}\right)\frac{\pi}{4}\right]_{l=l_k}. \quad \textbf{(XVI.4.6)}$$

The primes are differentiation symbols. The l_i are determined for each angle by the values of l that satisfy

$$\Theta_l = 2\delta'_l \equiv \frac{2d\delta_l}{dl} = \pm\theta. \quad \textbf{(XVI.4.7)}$$

Θ_l is called the deflection function. As an example, in figure XVI.2a, if θ is small enough, there are three branches. It may be noticed that for identical particles, $\sigma(\theta) = |f(\theta)|^2$ must be replaced by $\sigma(\theta) = |f(\theta) + f(\pi - \theta)|^2$. This, in the case above, adds a fourth branch.

When the deflection function Θ_l is nonmonotonic, contributing two or more branches to the same deflection angle, it follows that $\Theta(l) = \Theta_l$ must possess either singularities or maxima or minima.

Let us first study the case in which $\Theta(l)$ has an extremum. Since the classical cross section (**XVI.4.5**) contains a factor $(d\Theta_l/dl)^{-1}$, it will have a singularity

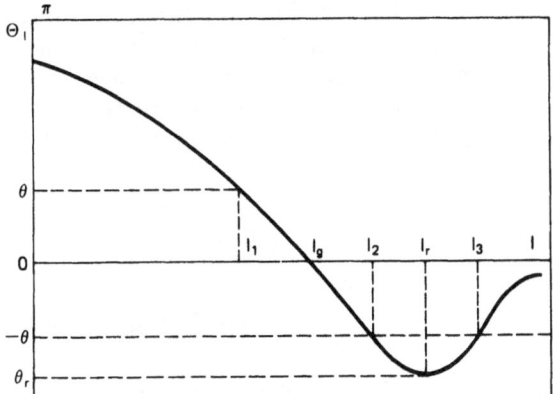

Figure XVI.2a
Deflection function

when $d\Theta/dl$ vanishes. In optics this phenomenon is responsible for the rainbows and the name rainbow angle is appropriate for such an extremal of the deflection function. We refer to the scattering in the neighborhood of a rainbow angle as rainbow scattering.

The classical intensity at a rainbow would become infinite on the bright side, and zero on the dark side (assuming no additional contributing branches), just as in geometrical optics the scattering of light from water droplets near a rainbow angle.

In the semiclassical analysis, just as in the Airy–Mie theory of an optical rainbow, a careful evaluation of the integral (**XVI.4.2**) shows that on the dark side there is a fall-off of intensity more rapid than exponential, and on the bright side an oscillatory behavior that corresponds under low resolution to the geometrical–optical prediction. The oscillation, of course, originates in the interference between the two branches of the classical deflection curve. The fall-off, of course, can be seen only as long as the contribution of an additional branch, if any, is smaller.

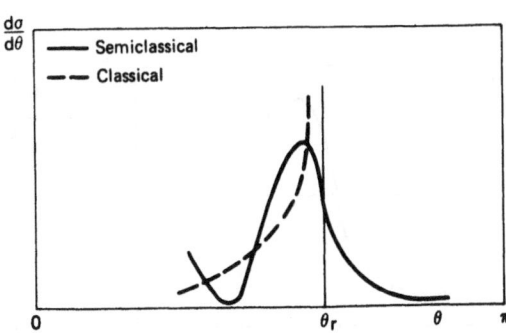

Figure XVI.2b
$d\sigma/d\theta$ (or $I(\theta)$) vs θ near $\theta = \theta_r$

The deflection function can also possess singularities say at some critical value l_1 corresponding to an infinite spiralling of the classical orbit towards a limiting circular orbit. Such a phenomenon is related to the existence of a double turning point like in the situation studied in section XVI.3. Under these circumstances, the classical deflection function varies logarithmically near l_1:

$$\Theta(l) = \begin{cases} \theta_1 + b \, \mathrm{Log}\!\left(\dfrac{l-l_1}{l_1}\right), & l > l_1, \\[2ex] \theta_2 + 2b \, \mathrm{Log}\!\left(\dfrac{l_1-l}{l_1}\right), & l < l_1, \end{cases} \tag{XVI.4.8}$$

where θ_1 and θ_2 depend on the nature of the potential within the barrier. For l somewhat greater than l_1 the particle will spiral in and out, not quite reaching the barrier top. For l somewhat less than l_1, the centrifugal barrier will be lower, and the particle will spiral over the barrier, into an inner region, and out over the barrier again. There is no simplification of the integral. We give here some illustrations of rainbow scattering. In (XVI.3a), an attractive potential $V(r)$ containing a barrier and the corresponding deflection function are shown. In (XVI.3b), the ratio of the elastic cross section of 22 Mev α particles incident on Ag to their Coulomb cross section is compared to a "rainbow" calculation. In (XVI.3c), we show the quantum and the classical angular distribution of scattering by a Lennard–Jones (**XII.6**) potential near the rainbow angle ($\theta_r = 87°$).

If the deflection function passes smoothly through 0, or through $\pm\pi$, etc., a new effect arises since, classically, this would lead to a singularity in the cross section for forward or backward scattering. In the language of optics, this is a "glory." It may sometimes be observed in powerful backscattering of solar radiation from the mist beneath one's airplane, outlining the airplane's shadow.

All these typical behaviors can be used for practical inversion problems. There are two fields in physics in which good conditions for such an analysis are met: ion-nucleon scattering (with $E \gtrsim 10$ Mev/nucleon), and atom scattering (with $T \gtrsim 300°K$). In the first case, theoretical arguments for a

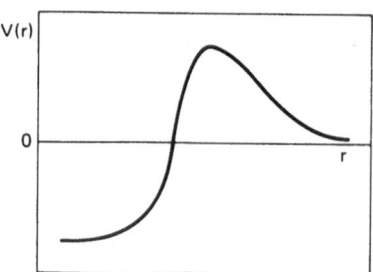
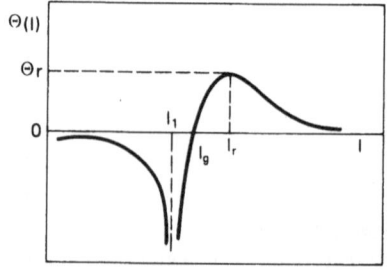

Figure XVI.3a (from Eisberg and Porter 1961)

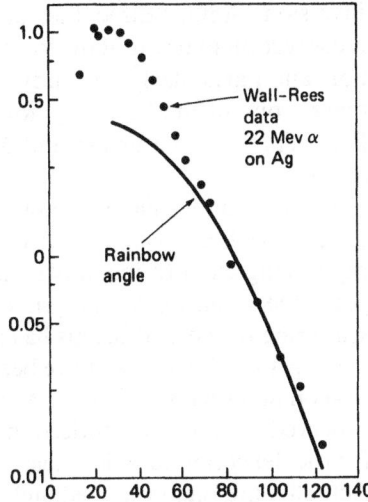

Figure XVI.3b (from Eisberg and Porter 1961)

single potential are poor, unless a very strong absorption is assumed. One can see typical rainbow behaviors (see, e.g., Eisberg and Porter 1961) in certain scattering curves, but the physical meaning of the corresponding potentials can be questioned.

In the second case, theoretical chemists are faced with the following situation: Values of δ_l are important for several hundred values of l. Chemists sometimes try to obtain the differential cross section in terms of parameters which may be points in a spline-fit curve of the phase shifts δ_l. But even in the case of six-parameter potentials, one is usually not able to fit fully resolved differential cross-section data. At the same time, some angular regions of σ are very insensitive to the parameters. This tends to create a very unstable

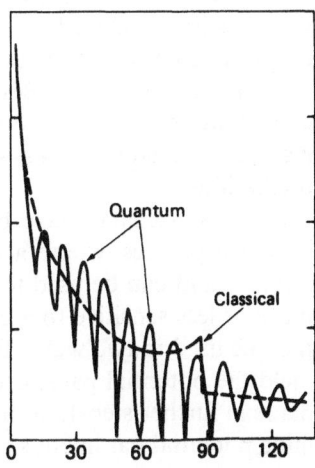

Figure XVI.3c

inversion. Similar difficulties are experienced with a representation of the cross section in terms of the Regge poles. Hence the semiclassical approximation, and particularly the study of oscillations and of the region of rainbows remain one of the best ways for the practical inversion problem. (See, e.g., Bernstein 1966, Bernstein and Muckerman (1968), Mason and Monchick, (1968).

The first molecular rainbow was probably seen, but not recognized, by Knauer (1933). This apparently went unrecognized for over 20 years, until a ray-tracing exploration was made (Mason 1957), preceding by a couple of years the remarkable analysis of Ford and Wheeler (1959). Recently, the scattering of thermal beams has again been undertaken experimentally, and many primary rainbows have been seen, and have been used to give information on potential well depths. More complicated interferences have also been observed. It should be noticed that the Airy theory (parabola approximation for the deflection function near a maximum) is of very limited applicability. Actually, the region immediately around the rainbow is extremely sensitive to the potential details, and an exact quantum mechanical treatment often seems to be required.

Since the rainbows are usually related to the behavior of the potential in the so-called "surface" region, they can give good information on intermediate-range forces, but not on long-range forces. Such information, however, can be supplied by the behavior of the cross section near the zero angle, viz., in the "shadow." Glories interfere in some case with those phenomena.

Some other remarks can be made on the various contributions to (**XVI.4.4**), (in the direct problem):

To begin with, since for a given θ, equation (**XVI.4.7**) defines an l_i for each particular branch then we can look at the l_i as functions of θ. Thus it is easily seen that for a given angular region of the differential cross section, only certain regions of he phase-shift curve are important.

Besides, since the magnitudes and shapes of the various σ_i's are usually very different, the range of l which corresponds to a given range of θ can usually still be reduced for a first approach. Hence one can be confident that a given angular region gives us specific information about a particular region of the phase-shift curve. Besides, in most regions, only one interference term contributes significantly to $\sigma_{sc}(\theta)$, so that it may be easy to explain, why, for instance, a differential cross section past θ_r eventually maintains evenly spaced oscillations.

All these remarks have been applied to practical inversion problems in chemical physics, where a rough shape of the potential is usually known *a priori*, and can be used to determine the parts of σ which are likely to be more or less sensitive to a given part of the potential. This information can then be used in two ways: Either it is used to define local "inversions" that yield the potential parameters approximately part by part (refinements are made in further steps); or, it is used to design the best experiments to determine potential parameters.

In practice three methods have been used by chemists to obtain $\Theta(l)$ or $\{\delta_l\}$ from experimental data (Buck 1974):

(a)   Deriving the Unknown Coefficients of a Parametrized Phase Shift or S Matrix through an Optimization Procedure on Measured and Calculated Data. One usually assumes the asymptotic behavior $V(r) \sim -Cr^{-n}$ as $r \to \infty$, so that $\eta \sim Cl^{-n-1}$ as $l \to \infty$ (Pauly and Toennies 1965, Bernstein and Muckerman 1967, Pauly 1973), and one introduces convenient parametrizations for the phase shifts or the scattering functions. (Barckett et al. 1963, Vollmer 1969, Remler 1971, Rich et al. 1971, Klingbeil 1972a), The cross section is then calculated by using the exact formulas given by quantum mechanics. Thus, this method is not relevant to the present section (XVI.4) but is either a fitting procedure or a step in an analysis relevant to our section (XVI.3).

(b)   Comparing the Semiclassical Features of the Cross Section with Calculated Values Obtained via the Parametrization of the Deflection Function. Here the main problem arises most of the expressions for the semiclassical cross section contain $l(\theta)$ rather than $\Theta(l)$, $l(\theta)$ being in general a multivalued function. In order to unfold this multivalued character, the deflection function is separated into monotonic functions $\Theta_i(l)$, which are approximated by simple curves. Procedures have been proposed which try to use all measurable quantities and/or try to treat the most remarkable features of the semiclassical cross sections (Buck and Pauly 1971, Boyle 1971, Miller 1971).

(c)   Directly Determining $\Theta(l)$ or a Part of $\Theta(l)$ by Using Semiclassical Cross Sections. This direct determination is possible if the cross section is monotonic. From **(XVI.4.5)** and **(XVI.4.7)** we see that $\sigma \sin \theta$ is then proportional to $(l + \frac{1}{2})dl/d\theta$, so that the deflection function is obtained by direct integration over the measured cross section.

$$(l(\theta) + \tfrac{1}{2})^2 = 2k^2 \int_\theta^\pi \sigma(\theta')\sin \theta' \, d\theta'. \tag{XVI.4.9}$$

The potential is determined by applying one of the procedures given in section XVI.3. It is also possible to obtain direct information on the deflection function through the interference data (F. T. Smith et al. 1965, 1966, Pritchard 1972).

For a good review of these methods, their generalizations to identical particles, exchange cross sections, potential crossings, and a review of results obtained in practical problems of finding molecular potentials from actual data sets, the reader is referred to Buck (1974).

Mathematical Justification of the Semiclassical Analysis. The overall relevance of the semiclassical analysis has been studied (see, e.g., Sabatier 1965, Berry 1969) from the mathematical point of view by deriving next orders in all the steps of the approximation, including the replacement of a discrete sum by an integral. It has been shown that this approximation differs from exact theory by errors of the order of $(ka)^{-1}$ at most, where a is a measure of the range of the potential (ka is approximately the number of non-negligible δ_l's). Actually, since the two first orders in the JWKB approximation have the same form, the error is probably still smaller.

XVI.5   Semiclassical Analysis III

The JWKB and semiclassical methods are not restricted to the inverse scattering problem at fixed energy. They can just as well be applied to all inverse scattering problems of quantum theory. We successively study two of them: the inverse spectrum problem and the inverse scattering problem at fixed l. In both cases, the approximate theory was the first to be introduced, and yet its relevance in problems of current interest is still very great. Besides, to some extent the structure of the method makes possible interesting comparisons with the exact theory.

XVI.5.1   The Inverse Spectrum Problem

Dealing with the Schrödinger equation (**XVI.1.1**), we are interested in the potential energy $V(r)$ for r such that $V(r)$ is negative ("attractive"). For negative E, we assume that there are only two turning points $r = TP_1$ and $r = TP_2$, where $E = V(r)$. This means that, for the sake of convenience, we defer the case where the potential has more than one minimum. Now, following Wheeler (1976), we introduce the three following quantities (see figure XVI.4): the "*inclusion*" $I(E)$, which is defined as the area

$$I(E) = \int_{TP_1}^{TP_2} [E - V(r)]dr \qquad \textbf{(XVI.5.1)}$$

included between the potential curve and the available energy E, and terminated at the turning points; the *phase change* of the wave function between the first point and the second one

$$[n(E) + \tfrac{1}{2}]\pi = \left(\frac{2m}{h^2}\right)^{1/2} \int_{TP_1}^{TP_2} [E - V(r)]^{1/2}\, dr \qquad \textbf{(XVI.5.2)}$$

the *excursion* of the r-coordinate

$$X(E) = \int_{TP_1}^{TP_2} dr = (r_2 - r_1). \qquad \textbf{(XVI.5.3)}$$

The S-wave energy levels, $E_0, E_1, \ldots, E_n$, correspond to the non-negative integral values of $n(E)$ in (**XVI.5.2**). In the inverse spectrum problem, one tries to derive $V(r)$ from the ladder of energy levels.

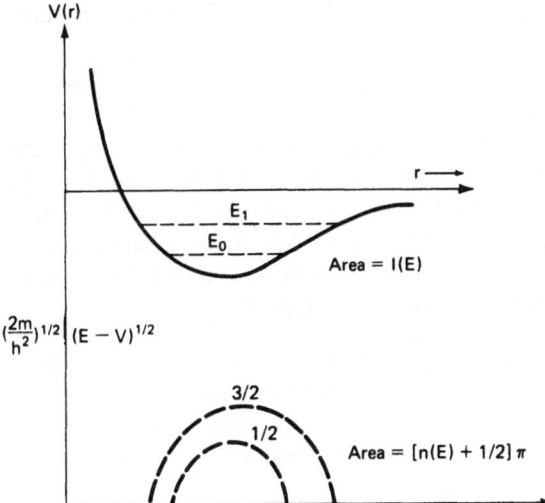

Figure XVI.4

The JWKB solution to this problem is not difficult:

1. Since we know E_n, we "know" $E(n)$ within a reasonable approximation and thus $n(E) + \frac{1}{2}$. By extrapolation this defines the energy $E_{\min}$ which corresponds to the minimum of the potential energy curve:

$$\min(V(r)) = E_{\min},$$
$$[n(E_{\min}) + \tfrac{1}{2}]\pi = 0. \tag{XVI.5.4}$$

2. We now manage to invert the Abel transform appearing on the right-hand side of (**XVI.5.2**), as we did in section (**XVI.3**). Since this is a $\frac{1}{2}$-order integral, it is natural to apply to both sides of (**XVI.5.2**) the semi-integration operator $D^{-1/2}$ defined by

$$[D^{-1/2}f]_{(E)} = \pi^{-1/2} \int_{E_{\min}}^{E} (E - E')^{-1/2} f(E') dE'. \tag{XVI.5.5}$$

On the right-hand side, we obtain the integral

$$\pi^{-1/2} \left(\frac{2m}{h^2}\right)^{1/2} \int_{E_{\min}}^{E} (E - E')^{-1/2} \left\{ \int_{TP_1(E')}^{TP_2(E')} (E' - V(r))^{1/2} \, dr \right\} dE'. \tag{XVI.5.6}$$

Inverting the order of integrations (an elementary elliptic integral appears) readily yields the "inclusion":

$$I(E) = \int_{TP_1(E)}^{TP_2(E)} [E - V(r)] dr = 2\pi^{-1/2} \left(\frac{h^2}{2m}\right)^{1/2} [D^{-1/2}(n(E) + \tfrac{1}{2})\pi]. \tag{XVI.5.7}$$

3. From $I(E)$, $X(E)$ follows after one differentiation

$$X(E) = \frac{\partial I(E)}{\partial E} = (r_2 - r_1). \qquad \textbf{(XVI.5.8)}$$

Hence, the energy levels $E_0, E_1, \ldots, E_n$, determine neither the left-hand branch of the potential energy curve nor the right-hand branch but only the separation between them. Arbitrarily we specify the left-hand branch by giving $r_1(E)$ as some smooth decreasing function of energy. Let this function decrease with energy no faster than the "excursion" $X(E)$ increases. Then the right-hand turning point, $r_2(E)$,

1. is fully determined: $r_2(E) = r_1(E) + X(E)$;
2. is a nondecreasing function of energy;
3. provides the unique acceptable continuation of the left-hand branch of the potential energy curve from E_{min} to the right.

A similar derivation can be given for an $l \neq 0$ spectrum kernel, provided that $V(r)$ is replaced by the effective potential

$$V_{eff}(r) = V(r) + \frac{h^2}{2m} \frac{(l + \frac{1}{2})^2}{r^2}. \qquad \textbf{(XVI.5.9)}$$

Suppose now that the inclusion I is determined not only as a function of E, as heretofore, but also as a function of angular momentum or as a function of the parameter

$$L^2 = (l + \frac{1}{2})^2. \qquad \textbf{(XVI.5.10)}$$

Then, following from (**XVI.5.1**) and (**XVI.5.9**), and according to the classical Rydberg–Klein–Rees method of analyzing data on atom–atom interactions, equation (**XVI.5.8**) can be completed by the following:

$$-\frac{2m}{h^2} \frac{\partial I(E, L^2)}{\partial L^2} = \int_{r_1}^{r_2} \frac{dr}{r^2} = \frac{1}{r_1} - \frac{1}{r_2}. \qquad \textbf{(XVI.5.11)}$$

Equations (**XVI.5.8**) and (**XVI.5.11**) together determine both turning points, allowing one to construct the potential energy $V(r)$ completely throughout the region of interaction.

The information which is required to derive $I(E, L^2)$ is given in atomic physics by vibration-rotation spectra, and thus the method is used for practical determinations. Although higher order JWKB approximations have been studied, it has been shown that this first order result is remarkably accurate. This first accuracy can be rationalized: "Near the minimum, it is known that the first-order JWKB energy levels are exact. Near the dissociation limit, the motion becomes more clearly classical and first order JWKB results should also be accurate." (Mason and Monchick 1967: these authors give a very good review of the Rydberg–Klein–Rees method and its applications).

When $I(E)$ is known for a given value of l only, it determines only a family of spectrum equivalent potentials. The parametrization of this family is most conveniently expressed by giving up the idea of writing V as a function of r and instead writing r as a function of V. Thus, if, for a parabolic barrier one has

$$r = r_0 \pm \left(\frac{2V}{m\omega^2}\right)^{1/2} \qquad \text{(XVI.5.12)}$$

all equivalent solutions are given by

$$r = r_0 - \left(\frac{2V}{m\omega^2}\right)^{1/2} + f(V) \qquad \text{(left branch)},$$

$$r = r_0 + \left(\frac{2V}{m\omega^2}\right)^{1/2} + f(V) \qquad \text{(right branch)}. \qquad \text{(XVI.5.13)}$$

Here $f(V)$ is understood to be a function of V that is continuous, and has a well-defined derivative, limited by the inequality:

$$\left|\frac{df(V)}{dV}\right| < \frac{d}{dV}\left(\frac{2V}{m\omega^2}\right)^{1/2} \qquad \text{(XVI.5.14)}$$

("no overhanging cliffs" in the potential energy curve $V(r)$.)

XVI.5.2   Inverse Scattering Problem

We use "$V(r)$" for the effective potential (potential energy, including Coulomb barrier, plus centrifugal JWKB terms $(l + \frac{1}{2})^2 r^{-2}$). The actual potential contains these three terms. The "reference" potential contains only the Coulomb term and the centrifugal term. Now, according to (XVI.3.1) and (XVI.3.17), the JWKB phase shift is given by the integral:

$$\delta(E) = \left(\frac{2m}{\hbar^2}\right)^{1/2} \lim_{R \to \infty} \left\{ \int_{TP}^{R} (E - V_0(r))^{1/2}\, dr - \int_{TP_0}^{R} (E - V_0(r))^{1/2}\, dr \right\},$$

$$\text{(XVI.5.15)}$$

where TP and TP_0 are the turning points of the actual and the reference potential (see figure XVI.5, which shows a position of TP and TP_0 that we assume in the following, for the sake of simplicity). The integrals in (XVI.5.15) are similar to the integral (XVI.5.2), which for negative E, still gives the phase change. Thus, there corresponds to the phase shift and the "phase change" an "inclusion" which is the area enclosed, under the line $V = E = $ constant, between the actual potential curve and the reference potential curve:

$$I(E) = \lim_{R \to \infty} \left\{ \int_{TP}^{R} (E - V(r))dr - \int_{TP_0}^{R} (E - V_0(r))dr \right\}. \quad \text{(XVI.5.16)}$$

Clearly, the first integral on the right-hand side of (XVI.5.16) is made of two parts. The part below the line $E = 0$ is the inclusion as calculated by one semi-integration from the phase difference $[n(E) + \frac{1}{2}]\pi$ between turning point and turning point. The new part, above $E = 0$, together with the integral

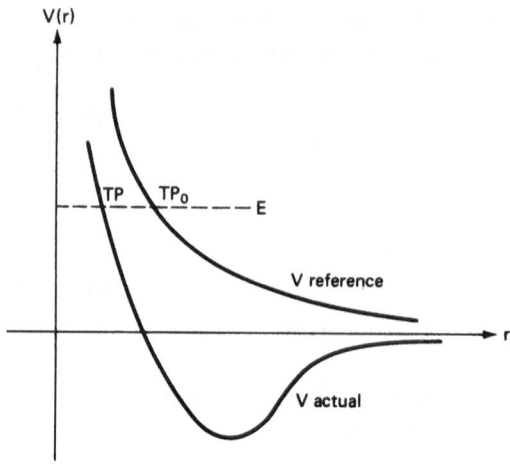

Figure XVI.5

containing V_0, arise in the same way from the phase shift (the two terms in
(**XVI.5.16**) are the JWKB eigenvalue relations for the potential V and the
potential V_0, respectively, with an inpenetrable vertical wall at $r = T$). Thus
the total "inclusion" $I(E)$ is given as before by a single semi-integration:

$$I(E) = \frac{2}{\pi}\left(\frac{h^2}{2m}\right)^{1/2}\left\{\int_{E_{\min}}^{0} \frac{[n(E') + \frac{1}{2}]\pi}{(E - E')^{1/2}} + \int_{0}^{E} \frac{\delta(E')}{(E - E')^{1/2}}\,dE'\right\} \quad (\textbf{XVI.5.17})$$

and from this, the position of the turning point TP follows after one differentiation:

$$TP_0 - TP = \frac{\partial I(E)}{\partial E}. \quad (\textbf{XVI.5.18})$$

Since TP_0 is known, this is the solution of the "inverse scattering problem"
in the semiclassical approximation.

Thus we see that, for negative energies, two unknown turning points are
involved, with only one condition to fix them, whereas, for positive energies,
only one unknown real turning point comes in, so that there is a perfectly
well-defined result. In other words, in the JWKB approximation, phase
equivalent potentials differ "below the line," but not "above it" (Wheeler
1976). On the other hand, the differentiation of E_n with respect to l (RKR
method) suppresses all ambiguities.

This compares with the exact theory, where the position of bound states,
together with the phase shift as a function of E, are not sufficient to completely
determine the potential, but additional parameters enable one to do it.

XVI.5.3   Other Problems and Remarks

Inverse problems starting from other experimental sets are tractable by
approximate methods using the JWKB approximation in the high-energy
limit.

A Method Starting from Differential Cross Sections at High Energy. From (**XVI.3.2**), (**XVI.3.3**) and (**XVI.3.7**) we see that, in this limit

$$H(\lambda) \sim -\tfrac{1}{2}E^{-1}V\left(\frac{\lambda}{k}\right) + O(E^{-2}). \tag{XVI.5.19}$$

Thus, using the impact parameter $b = (l + \tfrac{1}{2})k^{-1}$, we readily derive from (**XVI.3.23**) the high-energy limit of the deflection function:

$$\tau(E, b) \equiv E\Theta(E, b) = 2Ek^{-1}\frac{d\eta}{db} = \tau_0(b) + O(E^{-1}), \tag{XVI.5.20}$$

$$\tau_0(b) = -b\int_b^\infty (r^2 - b^2)^{-1/2}\frac{dV}{dr}\,dr. \tag{XVI.5.21}$$

Actually, this asymptotic form is valid either for high energies at a fixed angle of scattering or at small angles for any energy (Smith et al. 1966, Smith 1969). It can be used to obtain the potential from the differential cross section in the following way: from the classical cross section

$$\rho(\tau, E) = \theta \sin \theta\sigma(\theta, E) = 0.5\tau\left|\frac{db^2}{d\tau}\right| = \rho_0(\tau) + O(E^{-1}), \tag{XVI.5.22}$$

we see that $b(\tau_0)$ is given by

$$[b(\tau_0)]^2 = 2\int_{\tau_0}^\infty \rho_0(\tau)\frac{d\tau}{\tau}. \tag{XVI.5.23}$$

The quantity $\rho_0(\tau)$ can be obtained from a plot of $\rho(E, \tau)$ versus $\Theta(E)$ for several energies E. All curves should be asymptotic to $\rho_0(\tau)$ at small τ. Figure XVI.6 shows this behavior. Clearly, this procedure combines all data from a wide range of energies into a single curve, a fact which is especially important if the data at one energy are incomplete. A similar expansion at small b for backward scattering processes is also available (Smith et al. 1966).

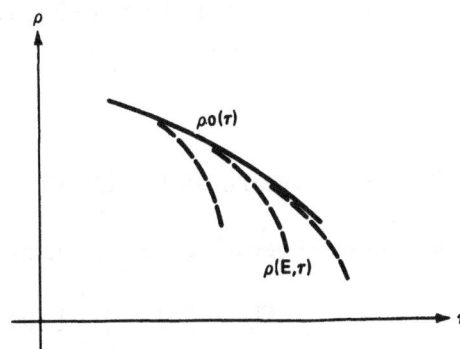

Figure XVI.6

A Method Starting from the Total Cross Section. From **(XVI.5.19)** and **(XVI.3.6)**, we obtain the high-energy limit of the phase shift:

$$\delta(b) = -(2k)^{-1}[\omega(b) + O(E^{-1})], \qquad \text{(XVI.5.24)}$$

where

$$\omega(b) = \frac{2m}{h^2} \int_b^\infty (r^2 - b^2)^{-1/2} r V(r) dr. \qquad \text{(XVI.5.25)}$$

This is Abel's integral and can be inverted (Miller 1969) to give

$$V(r) = -\frac{h^2}{\pi m} \int_r^\infty (b^2 - r^2)^{-1/2} \omega'(b) db. \qquad \text{(XVI.5.26)}$$

On the other hand, the total cross section is given, in integral form, by

$$\sigma(k) = 4\pi \int_0^\infty \sin^2 \frac{\omega(b)}{2k} (2b) db. \qquad \text{(XVI.5.27)}$$

If $\omega(b)$ is a monotonic function (e.g., monotonic repulsive potential), ω can be taken as a new variable, and the Fourier sine transform thus obtained in **(XVI.5.27)** can be inverted to give $b(\omega)$ from the total cross section:

$$\bar{\sigma}(\omega) = [b(\omega)]^2 = \pi^{-2} \int \sigma(k) \sin \frac{\omega}{k} k^{-1} dk \qquad \text{(XVI.5.28)}$$

and changing the integration variables in **(XVI.5.26)** leads to

$$V(r) = \frac{h^2}{\pi m} \int_0^{TP} [\bar{\sigma}(\omega) - r^2]^{-1/2} d\omega,$$

where the upper limit TP of the integral is the zero of the radicand. This procedure is due to Miller (1969) who extended it to cases in which $\omega(b)$ is not monotonic but the glory angle is known. Similar procedures are due to Henry et al. (1973) and to Sigmund and Lillemark (1974), who studied the inverse problem for the incomplete cross section

$$\sigma_{\theta_c}(E) = \int_{\theta > \theta_c} \frac{d\sigma(E, \theta)}{d\theta} d\theta. \qquad \text{(XVI.5.29)}$$

Besides, it is clear that the procedure can be applied to cases other than the high-energy limit if V is replaced by $H(\lambda)$, and **(XVI.5.26)** by **(XVI.3.6)** (author's remark).

XVI.5.4   Remark on the Derivations in Section XVI.3

The right-hand side of **(XVI.3.1)** contains integrals of the form

$$F[(l + \tfrac{1}{2})^2] = \int_{TP_1}^{TP_2} [f(r) - (l + \tfrac{1}{2})]^{1/2} \frac{dr}{r}. \qquad \text{(XVI.5.30)}$$

These integrals are quite similar to the ones which are treated in section XVI.5a (with $E \to (l + \frac{1}{2})^2$ and $r \to \text{Log } r$). One can treat them by applying the appropriate semi-integration:

$$\frac{2}{\pi} \int_{(l+1/2)^2}^{f(r)} \frac{[f(r) - (l' + \frac{1}{2})^2]^{1/2}}{[(l' + \frac{1}{2})^2 - (l + \frac{1}{2})^2]^{1/2}} \, d(l' + \frac{1}{2})^2 = [f(r) - (l + \frac{1}{2})^2].$$

$$(\textbf{XVI.5.31})$$

Integrating this with respect to Log r from turning point to turning point gives the new "inclusion," which yields by differentiation the excursion in Log r (Wheeler 1976). This proves at hand that from $F(l + \frac{1}{2})^2$, one will never be able to find out more about the function f than the amount of excursion in Log r between the two turning points at which $f(r) = (l + \frac{1}{2})^2$.

Clearly, this remark is part of the final remark of chapter XI.

XVI.6   From Approximate to Exact Methods

XVI.6.1   Uniqueness Problems

It has been shown by Loeffel (1968) that there can be only one potential yielding a given interpolation of $l \to \delta_l$ for any Re $l > 0$. The results in section 2 and 3 give good examples to illustrate this theorem. One may easily check, for instance, that the potentials that are "transparent" in Born approximation correspond to $\delta_l \neq 0$ for noninteger l. In the Born approximation, the step $\sigma \to \{\delta_l\}$ is trivially ambiguous. In the semiclassical method of section XVI.3, the step $\sigma \to \{\delta_l\}$ is like that in the exact theory. Besides, $p \to \delta(p)$ has a branch point at $p = \lambda_0$, its values in the positive half plane are not unique for given values on the positive real axis, and the step $h(\lambda) \to V(r)$ is itself ambiguous. The corresponding ambiguities remain in the method used in section XVI.4, where the step $\sigma \to \Theta_l$ itself has never

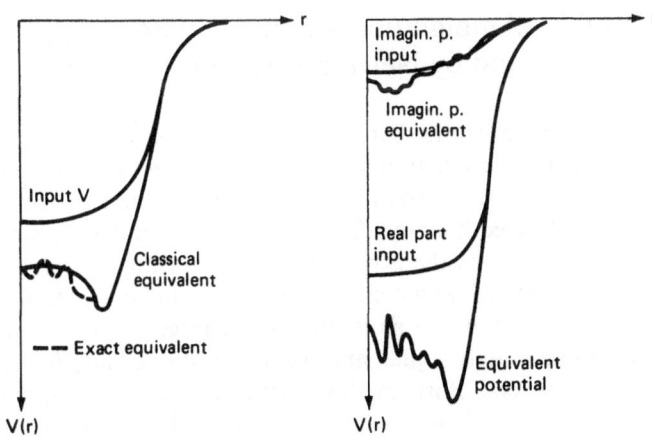

Figure XVI.7 $V(r)$ $V(r)$

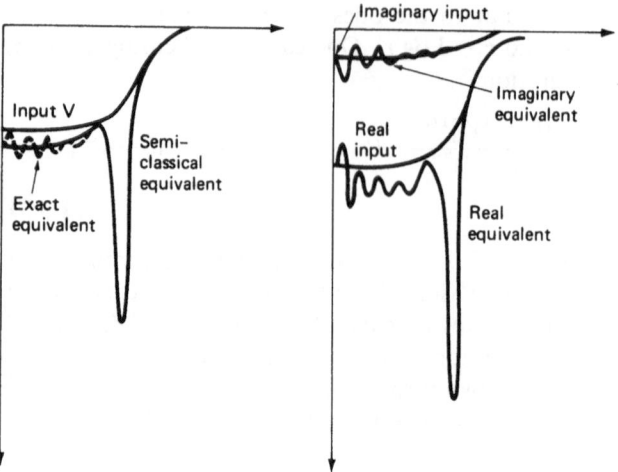

Figure XVI.8

been systematically studied. Since the approximate method of section XVI.3 yields a convenient way to study the ambiguities, it is interesting to use it as a basis to get exact ambiguities in inversion procedures. Such a program has recently been accomplished (Cuer 1976). It involves:

1. A systematic analysis of all the possible semiclassical ambiguities.
2. An iterative program going from the semiclassical results to exact ones.
3. A detailed classification of ambiguities, and a comparison with the results obtained from computer analyses. In particular, it is shown that the "layered ambiguities" of section XVI.3 are the "potential depth ambiguities" of these analyses, the "shape ambiguity" is one of their potential surface" ambiguities, and that certain "rules of thumb." e.g., the VR^n rule, can be obtained from a semiclassical analysis. Transparent potentials are given, up to the classical limit (Cuer 1977).

XVI.6.2   Ambiguities Obtained by Born Approximation and by Newton's Method

It has been shown (Sabatier 1972) that Newton's method (1962) amounts to replacing the function $f(r, r)$ associated with the potential (see chapter XIII) by its best approximation in $L_2(0, 2k)$, and to construct a potential from this filtered function. Such a process is completely similar to that done on the potential in the inversion of Born formula. Now, suppose one starts from a potential $V(r)$, which gives a set $\{\delta_l\}$, and one obtains $V_a(r)$ through the Newton's method. If this is done for a large value of k, $f(r, r)$ and $K(r, r)$ are approximately equal and $V_a(r)$ is nothing but the solution obtained from the inverse Born method, with the function $f(\theta)$ continued by zero. Hence, for large k, the Newton's method should exhibit the same features

as the inverse Born method does (the smoother $V(r)$, or the larger E for a given $V(r)$, the better is the approximation $V_a(r)$). A systematic computer investigation has been done (Sabatier and Quyen Van Phu 1971), which proves that these features are present over a large range, where $f(r, r)$ and $K(r, r)$ can be very different (see figures XVI.4).

XVI.6.3   From Exact Inverse Methods to Approximate Methods

The semiclassical approaches, in particular those of section XVI.6, suggest that certain features of exact methods are present in approximate methods. Besides, in approximate methods the physics is more intuitive. Thus is would be extremely interesting to derive the JWKB inverse analysis directly from exact methods (Gel'fand–Levitan or Marchenko methods in the inverse problem at fixed l complete solutions of chapter XIII in the inverse problem at fixed E). Unfortunately, this program, which would throw light on the physics contained in exact methods, has never been accomplished. Let us quote J. A. Wheeler (1976): "The integral equation that Marchenko got by the Gel'fand Levitan method is a great and beautiful achievement. It is a measure of its greatness and its beauty that it has telescoped away and hidden inside so much physics that has never been unfolded and brought to light."

XVI.7   Semiclassical Studies in Other Fields

Linear approximations and inversions of the linear transforms that they give have been obtained in so many fields of interest that we do not try to quote them here. The semiclassical studies are also present in other fields, and usually yield formulas somewhat similar to those which have been given in sections XVI.3–XVI.5. However, it should be pointed out that the centers of interest are not always identical to ours.

In classical scattering, optics, atmospheric and ionospheric soundings, and plasma diagnosis, the Abel equation appears in a form similar to the one we have studied (see, e.g., Keller, Kay and Schmoys (1956), Cremers and Birkebak (1966), Herlitz 1963, Maldonado et al. (1965, 1966), Freeman and Katz (1960), Phinney 1970, Paul 1967, 1968 and see in the Workshop on Mathematics of Profile immersion, (L. Colin 1971), the lectures by Kliore, Gross and Pirraglia, Wallio and Grossi, Stewart and Hogan, St Germain, Jackson, Culley, Paul, Nielson and Watt, Kay, Shmoys and Pirraglia); but the problems of interest are much more stability problems than others (see, e.g., Minerbo and Levy (1969), Gorenflo and Kovetz (1966) and the lectures quoted above).

In seismology, the situation is very interesting, because the semi-classical approximation is so good (except of course in the crust, where no approximation would be good), that we really can take it for granted and try to

obtain all the solutions. Besides, there are several possible positions of the "source" (whereas in our problem, of course, incident particles come from ∞). The study has been done by Gerver and Markushevich (1966, 1967). They reduced the problem of determination of the velocity-depth distribution $v(r)$ from a travel-time curve in a spherically symmetric Earth to the problem of determining a velocity $u(y)$ in a half plane $y > 0$. They did this by introducing the variables (R = Earth radius, θ = source-observer angle)

$$x = \frac{R\theta}{v(R)}; \qquad y = \frac{R \operatorname{Log} R/r}{v(R)}; \qquad u(y) = \frac{v(Re^{-yv(R)/R})}{v(R)e^{-yv(R)/R}}; \quad \textbf{(XVI.7.1)}$$

and considering a travel time curve Γ in the x, t plane. The functions $X(p)$, $T(p)$ are determined from a given Γ in the following way: let $(2X, 2T)$ be the point on Γ and l be the tangent to Γ at this point. They put

$$p = \tan(\widehat{l, x}). \qquad \textbf{(XVI.7.2)}$$

The function $\tau(p) = T(p) - pX(p)$ is introduced. If $\tau(p)$ has no points of discontinuity, we are in the situation that has been studied originally by Herglotz and Wiechert (1907), $u(y)$ is monotone, $X(p)$ and $T(p)$ are Abel transforms of $[u(y)]^{-1}$, and the inversion of any of them (just as in section 3 in the case of one turning point only), yields $u(y)$ unambiguously. If $\tau(p)$ has points of discontinuity $p_1, \ldots, p_n$, one introduces

$$\sigma_k = \tau(p_k^-) - \tau(p_k^+). \qquad \textbf{(XVI.7.3)}$$

The function σ is somewhat analogous with $\delta_{\lambda_0}(p)$ that we have introduced in section 3. It enables the authors to obtain the various "inverses" $u_v(y)$ of the function $X(p)$ (or $T(p)$) when $u(y)$ is not monotone. A whole set $\{u_v(y)\}$ of solutions exists. Gerver and Markushevich were able to show that the plots of all solutions, in the (u, y) plane, can be bounded. An example is given in the figure XVI.9a. One can indicate the depth precisely y_1, where the first waveguide starts, and one can determine the velocity $u(y)$ for $y \in (0, y_1)$ by the Herglotz–Wiechert method. For $y > y_1$, let us introduce

$$N(p) = \frac{2}{\pi} \int_p^1 \frac{X(q)dq}{\sqrt{q^2 - p^2}}, \qquad \textbf{(XVI.7.4)}$$

$$\tilde{H}(k) = \inf\left\{\frac{T(p)}{p}, p < p_k\right\} - N(\bar{p}_k), \qquad 1 \le k \le n, \qquad \textbf{(XVI.7.5)}$$

and let u be any number from the interval (p_k^{-1}, p_{k+1}^{-1}). It has been proved that:

1. The velocity may take the value u on the interval

$$N(u^{-1}) \le y \le N(u^{-1}) + \tilde{H}_k.$$

2. With $y < N(u^{-1})$, the velocity is strictly less than u.

Finally the position of the wave guides (numbers $y_2, \ldots, y_n, \bar{y}_2, \ldots, \bar{y}_n$) is also determined only approximately. Travel-time curves from deep sources give a way to reduce the ambiguity in the determination of the velocity $u(y)$.

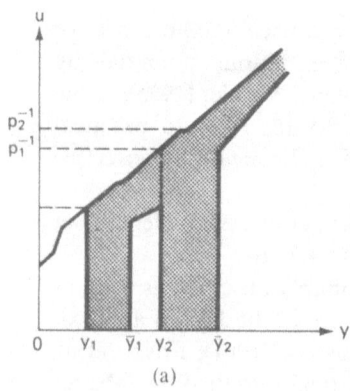

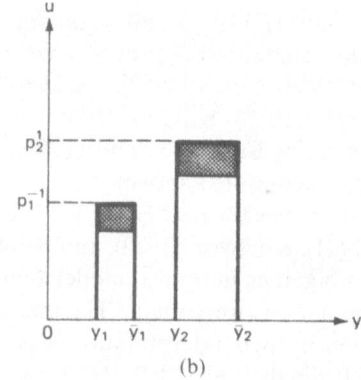

Figure XVI.9

An example of the results is given in figure XVI.9b. The same authors also gave restrictions on Γ that are sufficient to guarantee existence of a solution.

Some simple modifications of this theory would give the corresponding results in the quantum mechanical scattering problem (Demkov et al. 1971). However, in our opinion, such a study needs to be accompanied by:

1. a study of quantum corrections;
2. at least a semiquantitative study of the deviations from the single potential model.

To finish this chapter let us recall that the Abel transform was introduced in the oldest inverse problem which was ever solved in the literature (Abel 1826).

Historical Notes. Asymptotic approximations for linear second-order differential equations go back to Green (1837) Liouville (1837), Stokes (1864) (for general textbooks see Erdelyi 1956, Heading 1962, Olver 1974). Extensions and applications in physics are generally attached to the names of Jeffreys (1923), Wentzel (1926), Kramers (1926), Brilloin (1926). The first extensive study of the semiclassical approximation as a whole is due to Ford and Wheeler (1959b,c). Mathematical justifications and improvements have been afforded to the theory and its application to differential equations (Langer 1932, 1934, 1937, 1949, Olver 1961, 1965, Pike 1964), to phase shifts (Hecht and Mayer (1957), Erdelyi 1960, Rosen and Yennie (1964), Fröman and Fröman (1965), Yennie Boos and Ravenhall (1965)), to barrier phenomena (Goldman and Migdal (1954), Ford et al. 1959, Fröman and Fröman (1965)), to the semiclassical approximation as a whole (Sabatier 1965, Berry 1969), as well as to more practical scattering problems (see the reviews by Pauly and Toenies (1965), Bernstein 1966, Mason and Monchick (1967), Berstein and Muckerman (1967), Berry and Mount (1972), Child 1974, Buck 1974).

Use of the JWKB method in the inverse spectrum problem goes back to Rydberg (1931, 1933), Klein (1932), Rees (1947), and has been extended by

Miller (1971), Flugge and Vollmer (1971). The semiclassical method to solve the inverse scattering problem has been applied to monotonic potentials by Hoyt (1939), Firsov (1953), (see also Keller et al., 1956), Wheeler (1955), to the general case by Sabatier (1966), Vollmer (1969), Miller 1969, to practical problems by Buck and Pauly (1969) and several other chemical-physicists (see the reviews quoted above).

Other relevant results are, for instance, Kalinskin et al. (1963) Demkov et al. (1971), Kujawski (1973), and Knoll and Schaeffer (1976).

Ambiguities in optical model analyses were probably observed as soon as computations were made. The interest and relevance of the JWKB approximation in their interpretation was noticed by Austern (1961), Frisko et al. (1963), Singh et al. (1969, 1972), and more systematically studied by Sabatier (1973a,b) and Cuer (1976, 1977).

In section XVI.7 we gave references to several applications of the semiclassical method to inverse problems in other fields. For a comparison of approximate (and exact) methods applied to inverse problems in various fields, one can refer to a lecture by Sabatier (1971).

New Comments and References. The development of "approximate methods" leading to "profile inversion" became slower and slower with the increasing development of "numerical methods," themselves favored by the expansion of computers, and which will be reviewed in chapter XIX. This is a pity, because approximate methods often go deeper into the physics of the problem. Besides, in complicated problems, they are still often the only realistic way of understanding its structure, by producing *global relations* between parameters and results. Among the few relevant results we can cite a numerical verification by Cuer (1979) of the ambiguities previously described by Sabatier (1973a,b) in the JWKB approximation, and reviewed here in section XVI.3. We also find an interesting study of the classical limit of methods used in the quantum inverse scattering problem (Bogdanov, 1985), and a refined application (Dammert, 1983) of the "phase integral method" of Fröman and Fröman (which is an extension of the JWKB method) to transmission problems through a system of potential barriers. This however is more relevant to the direct than to the inverse problem.

Path integrals have been introduced by Feynman to give a Lagrangian form of quantum theory. Recently, De Facio and Brander used the mathematical structure of stochastic quantum mechanics to unify the inverse problems in a coherent-state path integral formulation (De Facio and Brander 1988a,b). The view taken there is that noise and filtering are the sources of stochasticity. A path integral is used for asymptotics "à la Maslov." Although there are only preliminary results up to now, it is permissible to think that an approximate method of the semiclassical form can grow out of this point of view: in the direct problem, for instance, there is already a path-integral representation for the S-matrix (Hammer et al., 1978), and a path-integral calculation of glory scattering from a black hole (Matzner et al., 1985).

CHAPTER XVII

Inverse Problems in One Dimension

XVII.1   Introduction

The scattering problem in one dimension on the entire real line $-\infty < x < \infty$ is known to have some fundamental differences from scattering in three dimensions, and its radial one-dimensional counterpart on the half-line $0 \leq r < \infty$. It has given rise to numerous studies because of its applications in many electrical engineering problem, among which we may mention:

1. Nonuniform transmission line in which either the inductance or the capacitance is constant, and in which we would like to reconstruct the line from the knowledge of the reflection coefficient $r(\omega)$ supposed to be known for all frequencies.
2. Under certain conditions, the reconstruction of a waveguide of variable cross section from some part of the scattering matrix, usually the reflection coefficient.
3. The problem in which a variable stratified dielectric is to be constructed so that the reflection coefficient of an electromagnetic wave will vary in a prescribed way with frequency.
4. Determination of the ionization density of a stratified ionosphere from the time delay of a pulse radio wave which has been transmitted from the earth and reflected back to the earth by the ionosphere.

In this section, we shall study the direct problem, and give all the properties of the S-matrix needed for solving the inverse problem. The problem at hand is again the second-order differential equation

$$\varphi''(x) + k^2\varphi(x) = V(x)\varphi(x), \qquad -\infty < x < \infty. \qquad \textbf{(XVII.1.1)}$$

Many of the technical aspects of the problems being very similar to those encountered previously with the radial Jost solutions, we shall content ourselves here with enumerating the results. We shall assume that the potential is real, locally integrable, and satisfies

$$\int_{-\infty}^{\infty} (1 + |x|) |V(x)| dx < \infty. \qquad \text{(XVII.1.2)}$$

This class of potentials is called the "Faddeev class," or L_1^1, whereas, more generally, L_p^1 is the set $\{V|\int_{-\infty}^{+\infty} dx (1 + |x|^p)|V(x)| < \infty\}$. As long as we do not set a new assumption, we shall work in L_1^1. However, the reader should keep in mind that a complete study has also been made, with somewhat simpler results, in $L_2^1 \subset L_1^1$; and that many weaker results have been obtained either in wider classes, or in connection with problems that are "almost" equivalent to the one-dimensional problem. Noting these approaches is our first insight of the most remarkable character of this problem: it is an atelier where methods of solution that are planned for more complicated problems have been studied. The object of the present chapter is not only to present the direct and inverse scattering problem for equation (XVII.1.1), but also to guide the reader through these various developments. For $V \in L_1^1$, it should be kept in mind that a "solution" of (XVII.1.2) must have, for any $x \in \mathbb{R}$, a continuous derivative.

We seek scattering solutions ψ_1 and ψ_2 of (XVII.1.1) having the following asymptotic forms for $x \to \pm\infty$:

$$\psi_1(k, x) \sim \begin{cases} e^{ikx} + s_{12}(k)e^{-ikx}, & x \to -\infty, \\ s_{11}(k)e^{ikx}, & x \to +\infty, \end{cases} \qquad \text{(XVII.1.3)}$$

$$\psi_2(k, x) \sim \begin{cases} s_{22}(k)e^{-ikx}, & x \to -\infty, \\ e^{-ikx} + s_{21}(k)e^{ikx}, & x \to +\infty. \end{cases} \qquad \text{(XVII.1.4)}$$

In the time-dependent description, the solution ψ_1 is the one which corresponds to a wave e^{ikx} incoming at $t = -\infty$ from the left (i.e., at $x = -\infty$ for $t = -\infty$), and propagating to the right. At $t = +\infty$, part of it is reflected to the left, and gives rise to the term $s_{12}(k)e^{-ikx}$, where $s_{12}(k)$ is the reflection coefficient to the left, whereas another part is transmitted to the right, and corresponds to $s_{11}(k)e^{ikx}$, $s_{11}(k)$ being the transmission coefficient to the right. The alternative notations $s_{11}(k) = T(k)$, $s_{12}(k) = R_-(k)$ ... etc. are very frequently used in the literature (T for transmission coefficients, and R for reflection coefficients).

Similarly for $\psi_2(k, x)$, where now $x = +\infty$ and $x = -\infty$ have been exchanged. $s_{22}(k)$ is the transmission coefficient to the left, and $s_{21}(k)$ the reflection coefficient to the right.

We call the matrix

$$S(k) = \begin{pmatrix} s_{11}(k) & s_{21}(k) \\ s_{12}(k) & s_{22}(k) \end{pmatrix} = \begin{pmatrix} T(k) & R_+(k) \\ R_-(k) & T(k) \end{pmatrix} \qquad \text{(XVII.1.5)}$$

the S-matrix of the equation (XVII.1.1). The question is: to what extent does the S-matrix determine the potential V? For this purpose, we must first study the properties of the S-matrix. It then turns out, in analogy with the Marchenko method of chapter V, that analogous equations exist which would permit us to solve the inverse problem.

Before entering into a detailed study of the properties of the S-matrix, let us make the following heuristic remark. As a consequence of the general scattering theory, it follows from the reality of the potential that the S-matrix is unitary (conservation of probabilities, or of the number of particles), and must satisfy the symmetry condition

$$s_{11}(k) = s_{22}(k) \tag{XVII.1.6}$$

which is due to time-reversal invariance. It follows that the S-matrix depends only on three independent real functions of k, $0 \le k < \infty$. On the other hand, the potential $V(x)$, $-\infty < x < \infty$, may be considered as two real functions of x in the interval $[0, \infty]$. Since the potential determines completely the S-matrix, it seems plausible that there should exist one relation among the three independent real elements of the S-matrix in terms of which it can be expressed. This additional property will turn out to be the analyticity of the transmission coefficients $s_{11}(k) = s_{22}(k)$ in the upper half-plane Im $k > 0$. We shall see that this entails that the whole S-matrix is essentially determined by the specification of one of the reflection coefficients $s_{12}(k)$ or $s_{21}(k)$.

XVII.1.1   Jost Solutions and Transformation Kernels

As is well known, all the solutions of the Schrödinger equation (**XVII.1.1**) can be obtained by linear combinations of some fundamental solutions satisfying precise boundary conditions. For these, we take

$$f_+(k, x): \quad \lim_{x \to \infty} e^{-ikx} f_+(k, x) = 1, \tag{XVII.1.7}$$

$$f_-(k, x): \quad \lim_{x \to -\infty} e^{ikx} f_-(k, x) = 1. \tag{XVII.1.8}$$

These solutions are somewhat the analogue of the Jost solutions of the radial Schrödinger equation. They satisfy the integral equations

$$f_+(k, x) = e^{ikx} - \int_x^\infty \frac{\sin k(x - t)}{k} V(t) f_+(k, t) dt, \tag{XVII.1.9}$$

$$f_-(k, x) = e^{-ikx} + \int_{-\infty}^x \frac{\sin k(x - t)}{k} V(t) f_-(k, t) dt. \tag{XVII.1.10}$$

Setting, with the signs at a given level corresponding to each other,

$$F_\pm(k, x) = \exp[\mp ikx] f_\pm(k, x), \tag{XVII.1.11}$$

we can also study the integral equations

$$F_\pm(k, x) = 1 - \int_{-\infty}^{+\infty} dy \, \theta[\pm(y - x)] L_\pm(x, y) F_\pm(k, y) V(y), \tag{XVII.1.12}$$

where θ is the Heaviside function and

$$L_+(x, y) = L_-(y, x) = k^{-1} \sin k(x - y)\exp[ik(y - x)]. \quad \textbf{(XVII.1.13)}$$

The iterated series $\sum_0^\infty F_\pm^{(n)}$ solving **(XVII.1.2)** is constructed through the algorithm

$$F_\pm^{(n)}(k, x) = -\int_{-\infty}^{+\infty} dy\, \theta[\pm(y - x)]L_+(x, y)F_\pm^{(n-1)}(k, y)V(y) \quad \textbf{(XVII.1.14)}$$

starting at $F_\pm^{(0)} = 1$.

Now setting

$$I(x) = J(-x) = 1 + |x|\theta(-x), \quad \textbf{(XVII.1.15)}$$

we notice that for any $y \geq x$, $y - x$ is smaller than or equal to $I(x)J(y)$. Using this result, and equation **(I.25)**, we obtain for $y \geq x$ the inequalities

$$|L_+(x, y)V(y)| \leq C|V(y)|\exp[(y - x)\sigma(k)] \begin{cases} |k|^{-1}, & \textbf{(XVII.1.16a)} \\ I(x)J(y), & \textbf{(XVII.1.16b)} \end{cases}$$

where C is a number and $\sigma(k) = |\mathrm{Im}\, k| - \mathrm{Im}\, k$. Similar inequalities are obtained for L_- and $x \geq y$. All these bounds are of the form $\alpha(x)\beta(y)$. If this is used in **(XVII.1.14)**, we obtain, as a majorant $\bar{F}_n$ of $|F_\pm^{(n)}|$, the corresponding term of a series expansion to the solution of

$$\bar{F}(x) = 1 + \int_x^\infty dy\, \alpha(x)\beta(y)\bar{F}(y) \quad \textbf{(XVII.1.17)}$$

which can be written down exactly, yielding for $n \geq 0$

$$n!\, \bar{F}_{n+1} = \alpha(x) \int_x^\infty du\, \beta(u)\left[\int_x^u dt\, \alpha(t)\beta(t)\right]^n. \quad \textbf{(XVII.1.18)}$$

Notice that

$$\alpha(x)\beta(x) = C|V(x)| \begin{cases} |k|^{-1}, \\ (1 + |x|). \end{cases} \quad \textbf{(XVII.1.19)}$$

Thus, for $V \in L_1^1$, it is clear that for $\mathrm{Im}\, k \geq 0$ the integral on the right-hand side of **(XVII.1.18)** is uniformly bounded, and each iterated term $F_\pm^{(n)}$ is absolutely bounded by the nth term of a numerical convergent series. Since it also follows from **(XVII.1.14)** that the $F^{(n)}$'s are entire functions of k, we conclude that $F_+(k, x)$ is a holomorphic function of k in $\mathrm{Im}\, k \geq 0$, continuous as $\mathrm{Im}\, k \to 0$. A quite similar analysis shows the same result for $F_-(k, x)$. In addition, the bounds **(XVII.1.18)**, **(XVII.1.19)** and the ones for $F_-(k, x)$ show that, for $\mathrm{Im}\, k \geq 0$, uniformly in x and k,

$$|F_+(k, x) - 1| \leq C \int_x^\infty dy\, |V(y)| \begin{cases} |k|^{-1}, \\ I(x)I(y), \end{cases} \quad \textbf{(XVII.1.20)}$$

$$|F_-(k, x) - 1| \leq C \int_{-\infty}^x dy\, |V(y)| \begin{cases} |k|^{-1}, \\ I(x)I(y). \end{cases} \quad \textbf{(XVII.1.21)}$$

Thus we can define, at least in $L^2(\mathbb{R})$, the so-called transformation kernels $K_\pm$

by the Fourier transforms

$$K_\pm(x, y) = (2\pi)^{-1} \int_{-\infty}^{+\infty} dk \, \exp[\mp ik(y - x)](F_\pm(k, x) - 1). \quad \text{(XVII.1.22)}$$

If we close the contour of the inegrals (**XVII.1.22**) by a half-circle in the upper half-plane, we can show that K_+(resp. K_-) vanishes for $y < x$ (resp. $y > x$), so that the inverse Fourier transforms yield for $F_\pm(k, x)$ the Levin representations

$$F_+(k, x) = 1 + \int_x^\infty dy \, \exp[ik(y - x)]K_+(x, y), \quad \text{(XVII.1.23)}$$

$$F_-(k, x) = 1 + \int_{-\infty}^x dy \, \exp[ik(x - y)]K_-(x, y). \quad \text{(XVII.1.24)}$$

Let us now recall the well-known formula

$$\beta(x, y) =: (2\pi)^{-1} \int_{-\infty}^{+\infty} \cos kx \sin kyk^{-1} \, dk = \tfrac{1}{2}[\theta(x + y) - \theta(x - y)].$$
$$\text{(XVII.1.25)}$$

Substituting (**XVII.1.12**) into (**XVII.1.1**), then using (**XVII.1.23**), (**XVII.1.24**) to eliminate $F_\pm$ in favour of $K_\pm$, and performing the Fourier integrals by means of (**XVII.1.25**), we obtain the integral equations

$$K_+(x, y) = \frac{1}{2} \int_{1/2(x+y)}^\infty ds \, \{V(s) + 2 \int_0^{1/2(y-x)} dr \, V(s - r)K_+(s - r, s + r)\},$$

$$(y > x), \quad \text{(XVII.1.26)}$$

$$K_-(x, y) = \frac{1}{2} \int_{-\infty}^{1/2(x+y)} ds \, \{V(s) + 2 \int_{1/2(y-x)}^0 dr \, V(s - r)K_-(s - r, s + r)\}.$$
$$\text{(XVII.1.27)}$$

The transformation kernel can be constructed from V by solving these Volterra equations and yields the Jost solutions through (**XVII.1.23**), (**XVII.1.24**) giving thus all the solutions of the direct problem at least for $k \neq 0$, since $f_+(k, x)$ and $f_+(-k, x)$ are linearly independent. In turn, V is trivially obtained from K_+ or K_- since it follows from (**XVII.1.26**), (**XVII.1.27**) that

$$K_+(x, x^+) = \frac{1}{2} \int_x^\infty ds \, V(s), \quad \text{(XVII.1.28)}$$

$$K_-(x, x^-) = \frac{1}{2} \int_{-\infty}^x ds \, V(s). \quad \text{(XVII.1.29)}$$

For possible generalizations, it is interesting to notice that (**XVII.1.28**), (**XVII.1.29**) can be obtained more rapidly: for instance, the expansion obtained through (**XVII.1.14**) gives,

$$F_+(k, x) = 1 - \int_x^\infty \frac{\sin k(x - y)}{k} e^{ik(y-x)} V(y)[1 + O(k^{-1})] dy, \quad \text{(XVII.1.30)}$$

which can be inserted into **(XVII.1.22)** and, by using **(XVII.1.25)**, can be proved to give **(XVII.1.28)** at the limit $y \to x^+$. In a *more heuristic* way, we can also compare the asymptotic behavior of $F_+(k, x)$, as it can be *directly* obtained from **(XVII.1.1)** and as it can be derived by integrating **(XVII.1.23)** once by parts. It should also be noticed that the statement of L^2 convergence of Fourier integrals like **(XVII.1.22)** (resp. of their derivatives) can be replaced by a strong convergence statement if we expand $F_\pm(k, x)$ up to the order k^{-2} (resp. k^{-3}), and we calculate exactly the contributing first terms (see **(XVII.1.30)** for F^+). This, as well as a direct study of the integral equations **(XVII.1.26)**, **(XVII.1.27)**, shows that $K_+(x, y)$ and $K_-(x, y)$ are at least absolutely continuous. At least, in a weak sense, we can derive from equations **(XVII.1.26)**, **(XVII.1.27)** the partial differential equation

$$\left[\frac{\partial^2}{\partial x^2} - \frac{\partial^2}{\partial y^2} - V(x) \right] K_\pm(x, y) = 0 \qquad \text{(XVII.1.31)}$$

which must be completed by the boundary conditions **(XVII.1.28)**, **(XVII.1.29)** and the asymptotic condition

$$K_\pm(x, \pm\infty) = 0. \qquad \text{(XVII.1.32)}$$

XVII.1.2   Scattering Coefficients

For real values of $k \neq 0$, $f_+(\pm k, x)$ and $f_-(\pm k, x)$ are two fundamental systems of solutions of the differential equation **(XVII.1.1)** because

$$W[f_\pm(k, x), f_\pm(-k, x)] = \mp 2ik. \qquad \text{(XVII.1.33)}$$

Hence we can write

$$f_-(k, x) = \frac{1}{T(k)} f_+(-k, x) + \frac{R_+(k)}{T(k)} f_+(k, x), \qquad \text{(XVII.1.34)}$$

$$f_+(k, x) = \frac{1}{T(k)} f_-(-k, x) + \frac{R_-(k)}{T(k)} f_-(k, x), \qquad \text{(XVII.1.35)}$$

where

$$\frac{2ik}{T(k)} = W[f_-(k, x), f_+(k, x)], \qquad \text{(XVII.1.36)}$$

$$2ik \frac{R_+(k)}{T(k)} = -W[f_-(k, x), f_+(-k, x)], \qquad \text{(XVII.1.37a)}$$

$$2ik \frac{R_-(k)}{T(k)} = W[f_+(k, x), f_-(-k, x)]. \qquad \text{(XVII.1.37b)}$$

A comparison of **(XVII.1.34)** and **(XVII.1.35)** with **(XVII.1.4)** and **(XVII.1.3)** shows that $T(k) = s_{11}(k) = s_{22}(k)$, $R^+(k) = s_{21}(k)$, $R-(k) = s_{12}(k)$, and explains the name "transmission" or "reflection" coefficients. From the formulas

(**XVII.1.37**), simple calculations yield

$$R_+(k)T(-k) + R_-(-k)T(k) = 0, \quad (\textbf{XVII.1.38a})$$

$$R_+(k)R_+(-k) + T(k)T(-k) = R_-(k)R_-(-k) + T(k)T(-k) = 1. \quad (\textbf{XVII.1.38b})$$

Because of the reality of $V(x)$, we have in addition, for real k

$$R_\pm(-k) = [R_\pm(k)]^*; \qquad T(-k) = [T(k)]^*, \qquad (\textbf{XVII.1.39})$$

which can be combined with (**XVII.1.38b**) to produce the following relations between the scattering coefficients on the k real axis

$$|T(k)|^{-2} = 1 + |R_+(k)/T(k)|^2 = 1 + |R_-(k)/T(k)|^2, \qquad (\textbf{XVII.1.40})$$

on the other hand, it is clear in (**XVII.1.34**) and (**XVII.1.35**) that these coefficients play a role analogous to the role of the Jost function $F(k)$ of previous chapters. They also admit, as expected, integral representations similar to that of the Jost function. Let us obtain them now in the more general framework of a perturbation theory. Let $V \in L_1^1$. Because of the bounds derived in the formulas (**XVII.1.12**)–(**XVII.1.21**), formulas like (**XVII.1.30**) and the derived formulas show that for k real and nonvanishing, $e^{-ikx} f_+(k, x)$ and $(ik)^{-1} e^{-ikx} f_+'(k, x)$ (resp. $e^{ikx} f_-(k, x)$ and $(ik)^{-1} e^{ikx} f_-'(k, x)$) go to $1 + O(|x|^{-1})$ as $x \to +\infty$ (resp. $x \to -\infty$). Now let V and $\tilde{V}$ be two potentials in L_1^1. It follows from (**XVII.1.34**) and the limits we have just given that

$$\lim_{x \to \infty} W[f_-(k, x), \tilde{f}_-(k, x)] = 2ik \frac{\tilde{R}_+(k) - R_+(k)}{T(k)\tilde{T}(k)}. \qquad (\textbf{XVII.1.41})$$

We can then evaluate the Wronskian from equation (**XVII.1.1**) written for f_-, V, and $\tilde{f}_-$, $\tilde{V}$, so that we obtain

$$2ik \frac{R^+(k) - \tilde{R}^+(k)}{T(k)\tilde{T}(k)} = \int_{-\infty}^{+\infty} [V(x) - \tilde{V}(x)]\tilde{f}_-(k, x)f_-(k, x)dx. \qquad (\textbf{XVII.1.42a})$$

Similar studies of $\lim_{x \to \infty} W[(\tilde{f}^-(k, x) - f_-(k, x), f_+(k, x)]$ and of Wronskians of functions where the plus and minus signs have been flipped yield the formulas

$$2ik\left(\frac{1}{\tilde{T}(k)} - \frac{1}{T(k)}\right) = \int_{-\infty}^{+\infty} [V(x) - \tilde{V}(x)]\tilde{f}_-(k, x)f_+(k, x)dx, \quad (\textbf{XVII.1.42b})$$

$$2ik \frac{R_-(k) - \tilde{R}_-(k)}{T(k)\tilde{T}(k)} = \int_{-\infty}^{+\infty} [V(x) - \tilde{V}(x)]\tilde{f}_+(k, x)f_+(k, x)dx, \quad (\textbf{XVII.1.42c})$$

$$2ik\left(\frac{1}{\tilde{T}(k)} - \frac{1}{T(k)}\right) = \int_{-\infty}^{+\infty} [V(x) - \tilde{V}(x)]\tilde{f}_+(k, x)f_-(k, x)dx, \quad (\textbf{XVII.1.42d})$$

which for $\tilde{V} = 0$, so that $\tilde{T} = 1$ and $\tilde{R}_\pm = 0$, yield the integral representations of the Jost coefficients

$$2ikR^+(k)/T(k) = \int_{-\infty}^{+\infty} V(x)e^{-ikx} f_-(k, x)dx, \qquad (\textbf{XVII.1.43})$$

$$2ik[1 - 1/T(k)] = \begin{cases} \displaystyle\int_{-\infty}^{+\infty} V(x)e^{-ikx}f_{+}(k, x)dx, & \textbf{(XVII.1.44a)} \\ \displaystyle\int_{-\infty}^{+\infty} V(x)e^{ikx}f_{-}(k, x)dx, & \textbf{(XVII.1.44b)} \end{cases}$$

$$2ikR^{-}(k)/T(k) = \int_{-\infty}^{+\infty} V(x)e^{ikx}f_{+}(k, x)dx. \qquad \textbf{(XVII.1.45)}$$

The functions of k which appear in (**XVII.1.44**) are in fact $F_{+}(k, x)$ and $F_{-}(k, x)$, which are holomorphic and uniformly bounded for Im $k \geq 0$. It follows that $k/T(k)$ is holomorphic in Im $k > 0$, and continuous down to the real axis. Using the estimates given earlier for $F_{\pm}(k, x)$, it is also seen that we have the following asymptotic properties in the upper half-plane

$$T(k) = 1 + \frac{1}{2ik}\int_{-\infty}^{+\infty} V(x)dx + O(k^{-2}), \qquad \textbf{(XVII.1.46)}$$

whereas, *on the real axis*, it follows from (**XVII.1.43**) and the Riemann–Lebesgue theorem that

$$R_{\pm}(k) = o(k^{-N-1}), \qquad k \to \pm\infty, \qquad \textbf{(XVII.1.47)}$$

where $N = 0$ corresponds to $V \in L_{1}^{1}$, and $N > 0$ to a potential having N derivatives in $L^{1}(\mathbb{R})$

XVII.1.3   Bound States

It follows from (**XVII.1.40**) that $[T(k)]^{-1}$ cannot vanish on the real axis. On the other hand, the right-hand side of (**XVII.1.36**) is the Wronskian of two functions of k that are holomorphic in the upper half-plane. It is therefore itself holomorphic in the upper half-plane and can have isolated zeros. Their number is necessarily finite because for large $|k|$, the Wronskian goes to $2ik$. At each zero $k = k_{i}$ of $W(f_{-}, f_{+})$, the function $f_{-}(k, x)$, which exponentially vanishes as $x \to -\infty$, and the function $f_{+}(k, x)$, which exponentially vanishes as $x \to +\infty$, are proportional, say, $f_{-} = C^{-}\psi, f_{+} = C^{+}\psi$. ψ belongs to $L^{2}(\mathbb{R})$ and will be standardized by imposing ψ real and $||\psi|| = 1$, Hence, the points k_{i} are eigenvalues of the Schrödinger operator appearing in (**XVII.1.1**), and the corresponding functions ψ are the corresponding normalized eigenfunctions, or bound-state functions. These eigenvalues are simple (the Wronskian of two different eigenfunctions corresponding to the same k_{i} would be zero). They are purely imaginary because from (**XVII.1.1**) and its complex conjugate we readily show that for $k = k_{i}$,

$$0 = W[\psi^{*}(k, x), \psi(k, x)]\Big|_{-\infty}^{+\infty} = (k^{*2} - k^{2})\int_{-\infty}^{+\infty} |\psi(k, x)|^{2}\, dx, \qquad \textbf{(XVII.1.48)}$$

which impose Im $k = 0$ or Re $k = 0$. We shall denote the eigenvalues different from zero by $k = i\varkappa_{1}, i\varkappa_{2}, \ldots, i\varkappa_{n}$. $T(k)$ has a simple pole at each point

$k_p = i\varkappa_p$. According to equation **(XVII.1.36)**, the residue of $T(k)$ is readily related to the k-derivative of $W(f_-, f_+)$ and hence to

$$\lim_{k \to k_p} \left\{ W\left[\frac{f_-(k, x) - f_-(k_p, x)}{k - k_p}, f_+(k_p, x) \right] + W\left[f_-(k_p, x), \frac{f_+(k, x) - f_+(k_p, x)}{k - k_p} \right] \right\}.$$
$$\text{(XVII.1.49)}$$

From **(XVII.1.1)** we can write the equations

$$f_-''(k, x) - f_-''(k_p, x) + (k_p^2 - V)[f_-(k, x) - f_-(k_p, x)] + (k^2 - k_p^2)f_-(k, x) = 0,$$
$$f_+''(k_p, x) + (k_p^2 - V)f_+(k_p, x) = 0,$$
$$\text{(XVII.1.50)}$$

and use them to evaluate the first Wronskian (which is equal to its value between $-\infty$ and x). The second Wronskian is evaluated in the same way, after permuting $+$ and $-$, $+\infty$ and $-\infty$. The result is

$$\lim_{k \to k_p} (k + k_p) \left[\int_{-\infty}^{x} f_-(k, y)f_+(k_p, y)dy + \int_{x}^{\infty} f_-(k_p, y)f_+(k, y)dy \right],$$
$$\text{(XVII.1.51)}$$

which gives

$$\text{Res } T(k) = -i \int_{-\infty}^{+\infty} dy\, f_-(k_p, y)f_+(k_p, y) = -i[C_p^+ C_p^-]. \quad \text{(XVII.1.52)}$$

Notice that $f_\pm(k_p, x)$ and $C_p^\pm$ are real.

XVII.1.4   Zero-Energy Behavior of Scattering Coefficients

The zero-energy behavior of scattering coefficients is easy to study if we assume that the potential belongs to $L_2^1(\mathbb{R})$. Instead of the Jost solutions, let us introduce the solutions $\sigma^\pm(k, x)$ and $\gamma^\pm(k, x)$ of the Schrödinger equation **(XVII.1.1)** which are defined at least on the semiaxes $x \geq 0$ (case $+$) or $x \leq 0$ (case $-$) by the integral equations

$$\sigma^+(k, x) = k^{-1} \sin kx + \int_{x}^{\infty} dy\, k^{-1} \sin[k(y - x)] V(y)\sigma^+(k, y), \quad \text{(XVII.1.53a)}$$

$$\sigma^-(k, x) = k^{-1} \sin kx + \int_{-\infty}^{x} dy\, k^{-1} \sin[k(x - y)] V(y)\sigma^-(k, y), \quad \text{(XVII.1.53b)}$$

$$\gamma^+(k, x) = \cos kx + \int_{x}^{\infty} dy\, k^{-1} \sin[k(y - x)] V(y)\gamma^+(k, y), \quad \text{(XVII.1.53c)}$$

$$\gamma^-(k, x) = \cos kx + \int_{-\infty}^{x} dy\, k^{-1} \sin[k(x - y)] V(y)\gamma^-(k, y), \quad \text{(XVII.1.53d)}$$

and can be defined at any other finite point at least by integrating **(XVII.1.1)** All these are even functions of k and the iterative solutions of **(XVII.1.53)** converge as soon as $V \in L_1^1(\mathbb{R})$. But it is only if $V \in L_2^1(\mathbb{R})$ that the limit forms of (**XVII.1.53**) as $k \to 0$, and of their iterative solutions, are obtained in a continuous limit process and yield zero-energy solutions of (**XVII.1.1**)

$$\sigma^+(0, x) \equiv \tilde{\sigma}^+(x) = x + \int_x^\infty dy \, (y - x) V(y) \tilde{\sigma}^+(y), \qquad \text{(XVII.1.54a)}$$

$$\sigma^-(0, x) \equiv \tilde{\sigma}^-(x) = x + \int_{-\infty}^x dy \, (x - y) V(y) \tilde{\sigma}^-(y), \qquad \text{(XVII.1.54b)}$$

$$\gamma^+(0, x) \equiv \tilde{\gamma}^+(x) = 1 + \int_x^\infty dy \, (y - x) V(y) \tilde{\gamma}^+(y), \qquad \text{(XVII.1.54c)}$$

$$\gamma^-(0, x) \equiv \tilde{\gamma}^-(x) = 1 + \int_{-\infty}^x dy \, (x - y) V(y) \tilde{\gamma}^-(y), \qquad \text{(XVII.1.54d)}$$

For $k \neq 0$, the functions $\sigma^\pm(k, x)$ and $\gamma^\pm(k, x)$ are combinations of the Jost solutions. In addition, $\sigma^+(k, x)$ and $\gamma^+(k, x)$, as well as $\sigma^-(k, x)$ and $\gamma^-(k, x)$, are fundamental systems of solutions, so that we can write

$$\begin{pmatrix} \sigma^+(k, x) \\ \gamma^+(k, x) \end{pmatrix} = \begin{pmatrix} Q^{11}(k) & Q^{12}(k) \\ Q^{21}(k) & Q^{22}(k) \end{pmatrix} \begin{pmatrix} \sigma^-(k, x) \\ \gamma^-(k, x) \end{pmatrix}, \qquad \text{(XVII.1.55)}$$

and check that $\det Q = 1$ and that

$$T(k) = 2ik[Q^{21}(k) + ikQ^{11}(k) + ikQ^{22}(k) - k^2 Q^{12}(k)], \qquad \text{(XVII.1.56)}$$

$$R_\pm(k) = -\frac{Q^{21}(k) \mp ikQ^{11}(k) \pm ikQ^{22}(k) + k^2 Q^{12}(k)}{Q^{21}(k) + ikQ^{11}(k) + ikQ^{22}(k) - k^2 Q^{12}(k)}. \qquad \text{(XVII.1.57)}$$

As $k \to 0$, Q^{ij} goes to q^{ij}, and the unitarity reads

$$\det q = 1 = q^{11} q^{22} - q^{12} q^{21}. \qquad \text{(XVII.1.58)}$$

Now we shall define as a "*zero-energy bound state*" a solution of the Schrödinger equation at $k = 0$ which is everywhere bounded on the whole x-axis and we shall say that $k = 0$ is an "*exceptional point*." A glance at the asymptotic behaviors of $\tilde{\sigma}^\pm$ and $\tilde{\gamma}^\pm$ shows that such a solution exists if and only if

$$W[\tilde{\gamma}^+(x), \tilde{\gamma}^-(x)] = -q^{21} = 0. \qquad \text{(XVII.1.59)}$$

Then (**XVII.1.58**) implies that q^{11} cannot vanish and $q^{22} = (q^{11})^{-1}$. It is now possible to determine the behavior of $T(k)$ and $R_\pm(k)$ as $k \to 0$ for $V \in L_2^1$.

(a) *No zero-energy bound state*

Since $q^{21} \neq 0$, the formulas (**XVII.1.56**) and (**XVII.1.57**) readily yield

$$T(k) = 2(q^{21})^{-1} ik[1 + O(k)], \qquad \text{(XVII.1.60a)}$$

$$R^\pm(k) = -1 + O(k). \qquad \text{(XVII.1.60b)}$$

(b) *Zero-energy bound state*

Setting $q^{22} = \tan(\alpha/2)$, we obtain

$$T(0) = \sin \alpha; \qquad R_\pm(0) = \pm \cos \alpha. \qquad \text{(XVII.1.61)}$$

XVII.1.5   Reconstruction of the Scattering Coefficients from One Reflection Coefficient

The properties of the S-matrix for $V \in L_1^1$ make it possible to reconstruct all its elements from the knowledge of only one of the reflection coefficients $R_\pm(k)$. Indeed, the modulus of $T(k)$ on the real axis is readily obtained from, say, $R^+(k)$, by means of (**XVII.1.38**) and (**XVII.1.39**)

$$|T(k)| = [1 - |R_+(k)|^2]^{1/2}. \qquad (\textbf{XVII.1.62})$$

Since the transmission coefficient is meromorphic in the upper half-plane, with simple poles at the bound states, it is possible to establish a Levinson theorem. In addition, a Hilbert relation relates $T(k)$ to $|T(k)|$, which can be obtained in Im $k > 0$ from the Cauchy formula (see section II.2). We write it in the case with no zero-energy bound state

$$T(k) = \lim_{\varepsilon \to 0^+} \exp \left[\frac{1}{2i\pi} \int_{-\infty}^{+\infty} \frac{\log(1 - |R_+(k')|^2)^{1/2}}{k' - k - i\varepsilon} \, dk' \right] \prod_{j=1}^{n} \frac{k + i\varkappa_j}{k - i\varkappa_j}, \qquad (\textbf{XVII.1.63})$$

$R_-(k)$ can then be derived from $T(k)$ and $R_+(k)$ by using (**XVII.1.38a**).

XVII.1.6   Spectral Theorem and Completeness Formula

The spectral theorem for the self-adjoint extension of $-d^2/dx^2 + V(x)$ in $L^2(\mathbb{R})$ ($V \in L_1^1$, the Friedrichs extension) can be obtained in a standard way (Deift and Trubowitz, 1979). The corresponding completeness formula is

$$\delta(x - y) = \frac{1}{4\pi} \int_{-\infty}^{+\infty} dk \, |T(k)|^2 [\bar{f}_+(k, x)f_+(k, y) + f_-(k, x)f_-(k, y)]$$

$$+ \sum_{p=1}^{n} \psi(i\varkappa_p, x)\psi(i\varkappa_p, y). \qquad (\textbf{XVII.1.64})$$

Completeness formulas for products of two Jost solutions are also available (see section XVII.3.3).

XVII.2   The Inverse Problem: Approaches Related to a Marchenko Equation

Several approaches to this problem are based on analytic properties of $F^\pm(k, x)$ and eventually yield what is now called the Faddeev–Marchenko equation, or an equivalent result. However, these approaches have *different* generalizations in more complicated problems (see, in particular, chapter XIV). Hence, in view of the "atelier" character of the present problem, we shall sketch them separately, but we study the reconstruction problem only in the framework of the Faddeev–Marchenko theory.

XVII.2.1 Integral Equations Based on the Cauchy Formula

According to formulas **(XVII.1.20)** and the surrounding analysis, $F_\pm(k, x)$ are holomorphic functions of k for $\operatorname{Im} k \geq 0$, going to $1 + O(|k|^{-1})$ as $|k| \to \infty$. It follows that the Cauchy formula holds

$$F_\pm(k, x) - 1 = \lim_{\varepsilon \to 0^+} \frac{1}{2\pi i} \int_{-\infty}^{+\infty} \frac{F_\pm(k', x) - 1}{k' - k - i\varepsilon} dk'. \qquad \textbf{(XVII.2.1)}$$

The right-hand side of **(XVII.2.1)** can be transformed by making $k' \to -k'$ and by inserting **(XVII.1.34)** or **(XVII.1.35)** into the result, obtaining

$$F_\pm(k, x) - 1 = \frac{1}{2\pi i} \lim_{\varepsilon \to 0^+} \int_{-\infty}^{+\infty} \frac{1 - T(k')F_\mp(k', x)}{k' + k + i\varepsilon} dk'$$

$$+ \frac{1}{2\pi i} \lim_{\varepsilon \to 0^+} \int_{-\infty}^{+\infty} \frac{R_\pm(k')F_\pm(k', x)\exp[\pm 2ik'x]}{k' + k + i\varepsilon} dk'.$$

$$\textbf{(XVII.2.2)}$$

If we know that there is *no bound state*, $1 - T(k')F_\mp(k, \cdot)$ is holomorphic in the upper half-plane and $O(|k|^{-1})$ as $|k| \to \infty$, so that the first integral of **(XVII.2.2)** is equal to that on a contour made of the real axis and an infinite half-circle in the upper half-plane, which is itself zero. Hence the formula **(XVII.2.2)**, if there is no bound state, reduces to

$$F_\pm(k, x) - 1 = \frac{1}{2\pi i} \lim_{\varepsilon \to 0^+} \int_{-\infty}^{+\infty} \frac{R_\pm(k')F_\pm(k', x)\exp[\pm 2ik'x]}{k' + k + i\varepsilon} dk',$$

$$\textbf{(XVII.2.3)}$$

which is a singular integral equation relating $R_+(k)$ (resp. $R_-(k)$) to $F_+(k, x)$ (resp. $F_-(k, x)$) and hence yields a way of reconstructing V from R_+ or R_-. If there are bound states, the first integral on the right-hand side of **(XVII.2.2)** is still equal to the contour integral, which is now equal to $2\pi i$ times the sum of residues at the poles on the imaginary k-axis.

For the sake of simplicity, here and in the following, we leave the case of a zero-energy bound state as an exercise for the reader and we exclude it by assumption in our results. Taking into account **(XVII.1.52)**, we readily obtain

$$F_\pm(k, x) - 1 = -i \sum_{p=1}^{n} \frac{(C_p^\pm)^{-2} e^{\mp 2\varkappa_p x} F_\pm(i\varkappa_p, x)}{k + i\varkappa_p}$$

$$+ \frac{1}{2\pi i} \lim_{\varepsilon \to 0^+} \int_{-\infty}^{+\infty} \frac{R_\pm(k')F_\pm(k', x) \exp[\pm 2ik'x]}{k' + k + i\varepsilon} dk'.$$

$$\textbf{(XVII.2.4)}$$

The inverse problem could be studied by means of the singular integral equations **(XVII.2.3)** and **(XVII.2.4)**. But it is more convenient to transform them, as we shall do now.

XVII.2.2   The Faddeev–Marchenko Theory

Let us multiply both sides of **(XVII.2.4)** by $(2\pi)^{-1}\exp[\mp ik(y-x)]$ and integrate over $k \in \mathbb{R}$. The left-hand side is readily obtained from **(XVII.1.26)** and the k-integral of the right-hand side can be evaluated by a contour integration in the lower half-plane. Replacing also $F_\pm(k,x) - 1$ by its value **(XVII.1.23)**, **(XVII.1.24)** we readily obtain

$$K_+(x,y) + M_+(x+y) + \int_x^\infty dz \, K_+(x,z)M_+(z+y) = 0 \qquad (y > x),$$

$$\text{(XVII.2.4a)}$$

$$K_-(x,y) + M_-(x+y) + \int_{-\infty}^x dz \, K_-(x,z)M_-(z+y) = 0 \qquad (y < x),$$

$$\text{(XVII.2.4b)}$$

where

$$M_\pm(x) = (2\pi)^{-1}\int_{-\infty}^{+\infty} dk \, R_\pm(k)\exp[\pm ikx] + \sum_{p=1}^N \rho_p^\pm \exp[\mp \varkappa_p x],$$

$$\text{(XVII.2.5)}$$

where

$$\rho_p^\pm = (C_p^\pm)^{-2} = \left[\int_{-\infty}^{+\infty} dx \, f_\pm(i\varkappa_p, x)^2\right]^{-1}. \qquad \text{(XVII.2.6)}$$

Clearly the term containing the bound states disappears if there is no bound state.

Equation **(XVII.2.4)** relates a function M_+ which is known from the so-called spectral data, i.e., respectively, $R_+(k)$, $\{\varkappa_p\}$, $\{\rho_p'\}$, or $R_-(k)$, $\{\varkappa_p\}$, $\{\rho_p^-\}$, to the function $K_+(x,y)$ or $K_-(x,y)$ which is the transformation kernel of the direct problem and yields all the quantities of interest in the direct problem.

At this point, and if our assumptions are correct, the Faddeev–Marchenko equations **(XVII.2.4)** should give us a way to answer the following questions: reconstruction problem, characterization problem, construction problem. Let us sketch the answers when there is no bound state, so that $R_+(k)$, or $R_-(k)$, is the only spectral data.

Reconstruction Problem. For a potential in L_1^1, it follows from the bounds of $R^\pm(k)$ (by 1 everywhere, by **(XVII.1.47)** at ∞) that $M^\pm(x)$ is absolutely bounded and that in particular, the operator

$$\mathscr{M}_x^+ : \mathscr{M}_x^+ g = \int_x^\infty M^+(x+y)\bar{g}(y)dy \qquad \text{(XVII.2.7)}$$

is defined on $L^1(x,\infty)$ and, thanks to the Fréchet–Kolmogorov theorem (Yosida, 1978), is compact. The Fredholm alternative holds and equation

(**XVII.2.4a**) has a solution if and only if

$$g(x) = \int_x^\infty M^+(x+y)\bar{g}(y)dy \qquad \text{(XVII.2.8)}$$

has only the zero solution. Now if g is a solution of (**XVII.2.8**), we can continue it on $\mathbb{R}$ by 0, and take its Fourier transform, which is a bounded continuous function. We can also notice that u takes L^1 functions into bounded functions, so that $g \in L^2(x, \infty)$ also. From (**XVII.2.8**) we obtain

$$\int_x^\infty |g(t)|^2 \, dt = \int_x^\infty \bar{g}(t) \int_x^\infty M^+(t+u)\bar{g}(u)du, \qquad \text{(XVII.2.9)}$$

and from (**XVII.2.5**) (with no bound state) we get

$$\begin{aligned}
\int_{-\infty}^{+\infty} dk \, |\tilde{g}(k)|^2 &= \int_{-\infty}^{+\infty} dk \, R^+(k)[\overline{\tilde{g}(k)}]^2 \\
&= \int_{-\infty}^{+\infty} dk \, \bar{R}^+(k)[\tilde{g}(k)]^2,
\end{aligned} \qquad \text{(XVII.2.10)}$$

where

$$\tilde{g}(k) = \frac{1}{2\pi} \int_{-\infty}^{+\infty} dt \, \bar{g}(t)\theta(t-x)\exp[-ikt]. \qquad \text{(XVII.2.11)}$$

From (**XVII.2.10**), using the unitarity relation (**XVII.1.40**), we readily derive

$$\int_{-\infty}^{+\infty} dk \, [|\tilde{g}(k) - R^+(k)\bar{\tilde{g}}(k)|^2 + |T(k)\tilde{g}(k)|^2] = 0, \qquad \text{(XVII.2.12)}$$

and since $T(k)$ does not vanish on $\mathbb{R}$, this implies that $\tilde{g}$ is zero.   Q.E.D.

Hence the Marchenko equation (**XVII.2.4a**) has one solution, and one only, $K_+(x, y)$. Since the transformation kernel associated with the potential V to be reconstructed should satisfy the same equation, it is nothing but $K_+(x, y)$. The method is easily transposed to equation (**XVII.2.4b**) and $K_-(x, y)$.

The case with bound states whose values $\varkappa_p$ and the (positive) parameters ρ_p^+ (resp. ρ_p^-) are known, can be reduced to the case with no bound state by a method using n successive Darboux transformations (see section XVII.3.2 and (Deift, 1978)). We can also solve the Marchenko equation and it has been proved that the potential which is obtained yields these bound states, with the corresponding parameters.

Characterization Problem
Characterization in L_2^1. Deift and Trubowitz (1979) have given conditions on R_+, R_-, and T sufficient to guarantee that a potential V yielding no discrete spectrum exist and is unique in L_2^1, where it can be constructed by solving the Marchenko equation, and yields R_+, R_-, and T in the scattering problem. These conditions are:

A1. $T(k)$, $R_+(k)$, $R_-(k)$, satisfy on the real axis equations **(XVII.1.38)**–**(XVII.1.40)**, and are continuous functions of k.

A2. $T(k)$ can be continued as a holomorphic function in $\operatorname{Im} k > 0$, continuous down to the real axis

A3.
$$T(k) - 1 = O(|k|^{-1}) \qquad (\operatorname{Im} k \geq 0),$$
$$R_\pm(k) = O(|k|^{-1}) \qquad (k \in \mathbb{R}).$$

A4.
$$|T(k)| > 0 \qquad (\operatorname{Im} k \geq 0, k \neq 0),$$

either $|T(k)| > 0$ for $k = 0$ or there exists $\dot{T}(0)$ and $\rho^\pm$ such that
$$T(k) = k\dot{T}(0) + o(k) \quad (\operatorname{Im} k \geq 0) \qquad \text{and} \qquad 1 + R_\pm(k)$$
$$= k\rho_\pm + o(k) \quad (k \in \mathbb{R}).$$

A5. $M_\pm(x)$ is absolutely continuous and there exists $c(a)$ finite such that

$$\int_{-\infty}^{+\infty} dx\, \theta[\pm(x-a)]M'_\pm(x)(1+x^2) \leq c(a) \qquad \text{for any } a \in \mathbb{R}.$$

Characterization in L_1^1. Proofs given above for the direct problem are easily extended for $V \in L_1^1(\mathbb{R})$ except for the behavior at $k = 0$, where we assumed $V \in L_2^1$, and further studies should be done in the case of a zero-energy bound state (see references below). The inverse problem has also been studied for $V \in L_1^1(\mathbb{R})$. It has been proved that any $V \in L_1^1(\mathbb{R})$ is determined by its scattering matrix, eigenvalues and norming constants, and a complete characterization of these scattering data has been given (Melin, 1985). The Faddeev–Marchenko method then enables the reconstruction of V. As we shall see later, the asymptotic behavior of V is the determining obstacle to extending uniqueness results beyond h^1. As soon as an x^{-2} behavior at ∞ is allowed, which takes V to the border of L_1^1, uniqueness statements break down, and counter examples are available (Degasperis and Sabatier, 1986). In addition to the very important study of Melin cited above, the reader will find useful information in Bollé et al. (1983, 1985, 1987), Klaus (1977, 1979, 1981, 1987), and Faddeev and Takhtadjian (1987). In particular, it has been proved that $S(k)$ is continuous at $k \to 0$ for any $q \in L_p^1$, with $p \geq 1$.

Construction Problem. Suppose we are given on the real axis a reflection coefficient satisfying the part of assumptions A which contain explicitly the reflection coefficient, including assumption A5, and which is absolutely bounded by 1, and we know there is no bound state. We do not need to construct $T(k)$ by means of **(XVII.1.60)**. The derivations of **(XVII.2.8)**–**(XVII.2.12)** are valid and prove that the Marchenko equation has a unique solution. Furthermore, since they also prove that the L^2 norm of $\mathcal{M}$ is smaller than or equal to 1 and that ± 1 cannot be eigenvalues, the Neumann series of the Marchenko equation converges in L^2. Admittedly, this is not enough to guarantee the needed properties of of $K_+(x, y)$, but it is enough to guarantee a weak stability of $K_+(x, x)$. Actually, in all practical cases, we do not need so much: $R^+(k)$ or $R^-(k)$ is usually given as a function that goes rapidly to zero as $k \to \infty$, and the reader can easily prove that the solution $K_+(x, y)$ of the Marchenko equation (**XVII.2.4**) is then a solution of the partial differential equation (**XVII.1.32**) with $V(x)$ defined by (**XVII.1.30**), and satisfying (**XVII.1.33**). By means of similar straightforward calculations, the functions $f_\pm(k, x)$ defined from (**XVII.1.27**) are proved to be a solution of the Schrödinger equation (**XVII.1.1**). The same remarks hold if we study the Marchenko equation with bound states. As a straightforward but interesting example, we can study the case of $R_+(k) = 0$, one bound state, characterized by its position $\varkappa$ and parameter ρ. Hence $\mathscr{M}_+(x)$ reduces to $\rho \exp[-\varkappa x]$. The Marchenko equation has a separable kernel and it can be solved exactly

$$K_+(x, y) = -\frac{\varkappa \exp[-\varkappa(y - x_0)]}{\cosh[\varkappa(x - x_0)]}, \tag{XVII.2.13}$$

$$V(x) = -\frac{2\varkappa^2}{[\cosh[\varkappa(x - x_0)]]^2}, \tag{XVII.2.14}$$

where

$$x_0 = (2\varkappa)^{-1} \log[\rho/2\varkappa]. \tag{XVII.2.15}$$

For such a potential, which is obviously well behaved, the reflection coefficients are zero, and the transmission coefficient is 1. The potential can be called a "transparent potential." The same remark holds true for the potential obtained by solving the Marchenko equation with $M^+(x)$ or $M^-(x)$ equal to the sum $\sum_1^n \rho_n \exp[-\varkappa_n x]$, with $\rho_n, \varkappa_n \in \mathbb{R}^+$, which corresponds to bound states, and to $R_\pm(k) = 0$. Here again, the Marchenko equation is a Fredholm equation with a degenerate kernel, whose solution is readily obtained as a quotient of two determinants.

Further Reading on the Faddeev–Marchenko Approach. The Faddeev study (1964) for $V \in L_1^1$ was essentially complete and rigorous but some upper bounds (in particular, (**XVII.1.20**)) had to be reassessed, together with some characterization conditions. This has been done in L_2^1 by Deift and Trubowitz (1979) and in L_1^1 by Melin (1985) and by Faddeev and Takhtadjian (1987). The reader will find several complements to the characterization and reconstruction problems in these references. Other complements to the reconstruction problem appear in Jaulent and Jean (1976) and to the stability problem in Marchenko (1968), Prosser (1983), and Atkosun (1987b).

However, we did not follow closely any of these references in our exposé (section XVII.1) of the direct problem or in our derivation of the Marchenko equation. The general organization of the lecture has been inspired by Atkinson (1979) who followed the ideas of Karlsson (1978) of using directly the Cauchy formulas to derive equations of the Gel'fand–Levitan or Marchenko type. The proofs are due to successive efforts at simplification, in particular, the trick (**XVII.1.15**) (Sabatier, 1983) which greatly shortens the derivation of absolute bounds for Volterra's equations that appear in this kind of problem. The notations are those where physics is most transparent, and most commonly used in current papers on inverse methods (the main users of this problem).

XVII.2.3   Approaches as a Riemann–Hilbert Problem

We have seen in section XVII.5.2 a definition of the Riemann–Hilbert problem for operators on a Banach space **h** of functions defined on S^2. A simpler form appears in connection with the one-dimensional inverse problem. Let us sketch the "inhomogeneous-without-poles" case: **h** is now the space $\mathbb{R}^2$ where $e = \left(\begin{smallmatrix}1\\1\end{smallmatrix}\right)$, operators on **h** are 2 × 2 matrices (set $\mathscr{B}$) the identity being $I = \left(\begin{smallmatrix}1 & 0\\0 & 1\end{smallmatrix}\right)$, the flipping operator $Q = \left(\begin{smallmatrix}0 & 1\\1 & 0\end{smallmatrix}\right)$, and the Pauli operator $\sigma = \left(\begin{smallmatrix}1 & 0\\0 & -1\end{smallmatrix}\right)$. Being given a 2 × 2 matrix-valued function $\Omega(k)$, $k \in \mathbb{R}$, inversible and satisfying the strong compatibility condition

$$\Omega(-k) = Q[\Omega(k)]^{-1}Q, \qquad \text{(XVII.2.16)}$$

the Riemann–Hilbert problem is one of seeking in $\mathscr{B}$ a matrix-valued function $\Gamma(k)$ such that $\Gamma(k) - I$ is continuous on $\mathbb{R}$ and can be continued into a holomorphic function in $\mathbb{C}^+$, going to zero as $|k| \to \infty$, Im $k \geq 0$, and satisfying the equation

$$\Gamma(-k) = \Omega(k)Q\Gamma(k)Q. \qquad \text{(XVII.2.17)}$$

We impose the additional assumptions that $\Omega(k) - I$ and its derivative belong to $L^2(\mathbb{R})$ and the additional constraint that $\Gamma(k) - I \in L^2(\mathbb{R})$. It can then be proved that if this problem has a solution, then the solution is unique and can be obtained by means of a variant of the Marchenko equation. So to do it, we write (**XVII.2.18**) in the form

$$\Gamma(-k) - I = \Omega(k)Q\Gamma(k)Q + Q\Gamma(k)Q + \Omega(k). \qquad \text{(XVII.2.18)}$$

Setting

$$\left.\begin{aligned}
\Gamma^T(\alpha) &= (2\pi)^{-1}\int_{-\infty}^{+\infty} dk\,\exp[-ik\alpha]\,[\Gamma(k) - 1],\\[4pt]
\Omega^T(-\alpha) &= (2\pi)^{-1}\int_{-\infty}^{+\infty} dk\,\exp[-ik\alpha]\,[\Omega(k) - 1]Q,
\end{aligned}\right\} \qquad \text{(XVII.2.19)}$$

both of which are well defined in the L^2 sense, we notice that thanks to the properties required for $\Gamma - I$, $\Gamma^T(\alpha)$ vanishes for negative α. Applying $(1/2\pi)\int_{-\infty}^{+\infty} dk\,e^{ik\alpha}$ to both members of (**XVII.2.18**) we obtain for positive α

$$\Gamma^T(\alpha) = \Omega^T(\alpha)Q + \int_0^\infty d\beta\,\Omega^T(\alpha - \beta)\Gamma^T(\beta)Q. \qquad \text{(XVII.2.20)}$$

With the assumptions on $\Omega(k) - I$ and its derivative, it is easy to check that $\int_0^\infty\int_0^\infty d\alpha\,d\beta[\Omega^T(\alpha + \beta)]^2 < \infty$, so that the Fredholm alternative holds for equation (**XVII.2.20**), which proves our point.

Let us now show how the Riemann–Hilbert problem may be used in our study, sticking to cases with *no bound state* (or zero-energy exceptional point). We write sets of two solutions of the Schrödinger equation (**XVII.1.1**) in vectorial form. For instance, the vector 'scattering solution"

$$\psi(k, x) = \begin{pmatrix} T(k)f_+(k, x) \\ T(k)f_-(k, x) \end{pmatrix} \tag{XVII.2.21}$$

is the solution of the Lippmann–Schwinger equation

$$\psi(k, x) = \psi_0(k, x) - \frac{i}{2k} \int_{-\infty}^{+\infty} dy\, e^{ik|x-y|}\, V(y)\psi(k, y), \tag{XVII.2.22}$$

where

$$\psi_0(k, x) = \begin{pmatrix} \exp[ikx] \\ \exp[-ikx] \end{pmatrix}. \tag{XVII.2.23}$$

From (**XVII.1.34**), (**XVII.1.35**) it is easy to show that

$$\varphi(-k, x) = [S(k)]^{-1} q\psi(k, x), \tag{XVII.2.24}$$

where $S(k)$ is the S-matrix defined by (**XVII.1.5**). There are several approaches to the problem where a Riemann–Hilbert equation appears. If a solution is defined from fixed conditions at a given point, for instance,

$$\varphi(k, x) = \psi_0(k, x) + k^{-1} \int_0^x dy \sin k(x - y) V(y)\varphi(k, y). \tag{XVII.2.25}$$

We can define a "Jost function" $J(k)$, which has its values in $\mathscr{B}$, by

$$J(k)\psi(k, x) = \varphi(k, x). \tag{XVII.2.26}$$

$J(k)$ is trivially related with the Wronskian of $\tilde{\psi}$ and φ, and since it follows from (**XVII.2.25**) that φ is an entire function of k, $J(k)$ is holomorphic in Im $k \geq 0$. Furthermore, as $k \to \pm\infty$ or $|k| \to \infty$ in $\mathbb{C}^+$, $J(k) - I = O(|k|^{-1})$. From (**XVII.2.24**) and (**XVII.2.26**) we obtain the equation

$$J^{-1}(-k) = S^{-1}(k)Q\, J^{-1}(k)Q, \tag{XVII.2.27}$$

which is a Riemann–Hilbert problem for $J(k)$, knowing $S(k)$. Its solution is the cornerstone of a Gel'fand–Levitan method which is here rather artificial (the point $x = 0$ corresponds to an arbitrary choice, unless the potential vanishes for all $x > 0$ or for all $x < 0$), but which can be generalized to the three-dimensional case (see section XIV.5.1). Much more physical is a construction of the Jost solution from the S-matrix. The key to it is equation (**XVII.2.24**). Let us define the function

$$g(k, x) = \exp[-ikx\sigma]\psi(k, x) = T(k)\begin{pmatrix} F_+(k, x) \\ F_-(k, x) \end{pmatrix},$$

then g satisfies the relation

$$g(-k, \cdot) = S_x^{-1}(k)Qg(k, \cdot), \tag{XVII.2.28}$$

where $S_x(k) = \exp[-ikx\sigma]S(k)\exp[ikx\sigma]$, i.e.,

$$S_x = \begin{pmatrix} T & R_+e^{2ikx} \\ R_-e^{-2ikx} & T \end{pmatrix}. \qquad \text{(XVII.2.29)}$$

Notice that S_x is the S-matrix that would correspond to a potential $V_x(z) = V(x + z)$ and that $S_x^{-1}(k) = \Omega(k)$ satisfies the compatibility condition (XVII.2.16) (proof: use formulas (XVII.1.38). Furthermore, the results obtained in section XVII.1 show that $g(k, x)$ has a holomorphic continuation into $\mathbb{C}^+$ and that $(g - e) \in L^2(\mathbb{R})$ as a function of k, and goes to zero as $|k| \to \infty$ in $\mathbb{C}^+$. Hence, if the Riemann–Hilbert problem (XVII.2.17) has a solution $\Gamma(k)$ for $\Omega(k) = S_x^{-1}$, $\Gamma(k)e$ is a solution of (XVII.2.28). It is equivalent to saying that the solution $g(k, x)$ of (XVII.2.28) is given by a Fourier transform from

$$g^T(\alpha) = (2\pi)^{-1} \int_{-\infty}^{+\infty} dk \, \exp[-ik\alpha][g(k, x) - e], \qquad \text{(XVII.2.30)}$$

which obeys the Marchenko equation ($\alpha > 0$)

$$g^T(\alpha) = \Omega^T(\alpha)e + \int_0^\infty d\beta \, \Omega^T(\alpha + \beta)g^T(\beta), \qquad \text{(XVII.2.31)}$$

where

$$\Omega^T(\alpha) = (2\pi)^{-1} \int_{-\infty}^{+\infty} dk \, \exp[ik\alpha][S_x^{-1}(k) - I]Q. \qquad \text{(XVII.2.32)}$$

As in the usual Marchenko equation, it is possible to interpret $g(\alpha)$ in terms of a transformation kernel and to write partial differential equations. But (XVII.2.31) remains a vectorial integral equation, i.e., a set of coupled integral equations and is not very convenient. If, instead of $g(k, x)$, we use the vectorial Jost solution $f(k, x) = \begin{pmatrix} F_+(k, x) \\ F_-(k, x) \end{pmatrix}$, relation (XVII.2.28) is replaced by

$$f(-k, x) = \dot{S}_x(k)Qf(k, x). \qquad \text{(XVII.2.33)}$$

The properties of g and those of S_x^{-1} defining the Riemann–Hilbert problem are readily transposed to f and $\dot{S}_x$. The Marchenko equation now reads ($\alpha > 0$)

$$f^T(\alpha) = \Omega^T(\alpha)e + \int_0^\infty d\beta \, \Omega^T(\alpha + \beta)f^T(\beta), \qquad \text{(XVII.2.34)}$$

where

$$f^T(\alpha) = (2\pi)^{-1} \int_{-\infty}^{+\infty} dk \, \exp[-ik\alpha]\begin{pmatrix} F_+(k, x) - 1 \\ F_-(k, x) - 1 \end{pmatrix}, \qquad \text{(XVII.2.35)}$$

$$\Omega^T(\alpha) = (2\pi)^{-1} \int_{-\infty}^{+\infty} dk \, \exp[ik\alpha]\begin{pmatrix} -R_+(k) & T(k) - 1 \\ T(k) - 1 & -R_-(k) \end{pmatrix}, \qquad \text{(XVII.2.36)}$$

In fact

$$\Omega^T(\alpha) = -(2\pi)^{-1} \int_{-\infty}^{+\infty} dk \, \exp[ik\alpha]\begin{pmatrix} R_+(k) & 0 \\ 0 & R_-(k) \end{pmatrix}, \qquad \text{(XVII.2.37)}$$

because, for $\alpha > 0$, and no bound state, the off-diagonal term vanishes. Hence, equation (XVII.2.34) separates into two uncoupled equations, which are nothing but (XVII.2.4a) and (XVII.2.4b).

XVII.2.4   D-Bar Approach

Let $k_1 = \mathrm{Re}\, k$, let $k_2 = \mathrm{Im}\, k$, and let $F(k)$ be a function of $k_1 + ik_2$, we define the D-bar ($\bar{\partial}$) derivative of $F(k)$ by the formula

$$\frac{\partial F}{\partial \bar{k}} = \frac{1}{2}\left(\frac{\partial F}{\partial k_1} + i\frac{\partial F}{\partial k_2}\right). \tag{XVII.2.38}$$

It is clear that if F is a holomorphic function of k (i.e., $F'(k)$ exists as a continuous function), $\partial F/\partial\bar{k}$ vanishes. Hence the D-bar derivative is a "measure" of the defect of holomorphy of F. The reader can easily check that if $F(k)$ is discontinuous across the real axis, its $\bar{\partial}$-derivative is proportional to the Dirac measure $\delta(k_2)$

$$\frac{\partial}{\partial \bar{k}}[F^+(k)\theta(k_2) + F^-(k)\theta(-k_2)] = \tfrac{1}{2}i[F^+(k) - F^-(k)]\delta(k_2). \tag{XVII.2.39}$$

Another useful formula is

$$\frac{\partial}{\partial \bar{k}}(k - k_0)^{-1} = \pi\delta(k - k_0). \tag{XVII.2.40}$$

Notice also that the Leibniz formula for products applies to $\bar{\partial}$.

The "D-bar approach" to an inverse problem is the most compact representation of any approach to this problem that is based on analytic properties. Hence, the structure of the inverse problem is expressed in the simplest way and can be disentangled from the effects of extra parameters or operators which may come, for instance, in constructing inverse methods for solving nonlinear partial differential equations. To be more specific, let us consider the function

$$\mathscr{F}(k, x) = \begin{cases} T(k)F_-(k, x), & \mathrm{Im}\, k > 0, \\ F_+(-k, x), & \mathrm{Im}\, k \leq 0, \end{cases} \tag{XVII.2.41}$$

(notations of (XVII.1.11)). It is clear that $\mathscr{F}(k, x)$ is holomorphic in the lower half-plane. It is also holomorphic in the upper half-plane except at the poles of $T(k)$, and it is discontinuous through the real axis. Using (XVII.2.39) and (XVII.2.40), then (XVII.1.35) and (XVII.1.52), we obtain the $\bar{\partial}$-equation

$$\frac{\partial}{\partial \bar{k}}\mathscr{F}(k, x) = \mathscr{R}(k)e^{2ikx}\mathscr{F}(-k, x), \tag{XVII.2.42}$$

where

$$R(k) = \tfrac{1}{2}iR_+(k)\delta(k_2) + \sum_{n=1}^{N}\pi\rho_n^+\delta(k - k_n), \tag{XVII.2.43}$$

the $\bar{\partial}$-equation should be completed by the "normalization" condition at ∞

$$\mathscr{F}(k, x) \to 1 \qquad (|k| \to \infty). \tag{XVII.2.44}$$

A generalized Cauchy formula enables us to construct a function $F(k)$ from its defect of holomorphy. If C is a closed contour (with the kind of regularity used in the theory of analytic functions) around a domain D, and k an interior point of D, we can write

$$F(k) = (2i\pi)^{-1}\left[\int_C \frac{F(l)dl}{l-k} + \int\int_D \frac{\partial F(l)/\partial \bar{l}}{l-k} \, dl \wedge d\bar{l}\right]. \qquad \textbf{(XVII.2.45)}$$

Applying formula **(XVII.2.45)** to both sides of **(XVII.2.42)** and taking into account **(XVII.2.44)** readily yields equation **(XVII.2.4)** $(dl \wedge d\bar{l}$ simply reduces to $-2i \, dl_1 \, dl_2)$.

The generality of $\bar{\partial}$-approaches is such that they can also be adapted to general first-order systems (section XVII.4.4 below) and to the three-dimensional problem (section XIV.3.2 above). It is also possible to use it to connect algebraic or geometric structures in the spaces $\mathscr{C}$ and $\mathscr{E}$ by the direct and inverse problem. The connection can be used in turn to derive hierarchies of equations that can be solved by the inverse method (section XVII.4.4). For further details on the $\bar{\partial}$-method, see Henkin and Novikov (1987).

XVII.3   The Inverse Problem: Other Approaches

If the class of potentials is reduced by an *a priori* assumption the problem becomes ill-posed, but if we are interested only in a reconstruction process or in a quasi-solution, this reconstruction—or the construction of this quasi-solution—may be simpler than in the general case.

XVII.3.1   Reduced Support Potentials

A strong physical *a priori* assumption which appears quite naturally in sounding problems is that of a potential vanishing on half of the axis (the half where the "source" of progressive waves which sound the medium is located). Assume, for instance, $V = 0$ for $x > 0$, and no bound state. It follows from **(XVII.1.17)** that the integral containing $\beta(y)$, which is bounded up by $|V(y)|$ $\exp[y\sigma(k)]J(y)$, converges even for $\mathrm{Im}\, k < 0$ $(\sigma(k) = 2|\mathrm{Im}\, k|)$ since the upper bound of the interval is 0 instead of $+\infty$. Hence $f_+(k, x)$ is an entire function, and, according to **(XVII.1.37)**, $R^+(k)$ can be continued in the upper half-plane as a holomorphic function, going to zero like $O(|k|^{-1})$ as $|k| \to \infty$. Hence, $M^+(x)$, defined by **(XVII.2.5)**, vanishes for $x > 0$, and **(XVII.2.4a)** reduces to

$$K_+(x, y) = 0 \qquad \text{for} \quad y > x \geq 0 \quad \text{and} \quad y > -x > 0, \qquad \textbf{(XVII.3.1)}$$

whereas, for $x \leq 0$ and $x < y < -x$,

$$K_+(x, y) + M_+(x + y) + \int_x^{-y} dz \, K_+(x, z)M_+(z + y) = 0 \qquad \textbf{(XVII.3.2)}$$

is the Marchenko equation. It is in this case that it has been obtained for the first time in the one-dimensional problem (Kay, 1955; Kay and Moses, 1982), and misnamed a Gel'fand–Levitan equation (such an equation should contain a spectral measure or a Jost function). This misnomer is still common among many research scientists of electromagnetic theory and nonlinear partial differential equations. If the potential is cut on both sides, both Jost solutions are entire function—of exponential type related to the width a of the support—and this is in turn a characterization of a potential of finite support a. Both Marchenko equations (**XVII.2.4**) take special forms similar to (**XVII.3.1**), (**XVII.3.2**) and the additional properties of $R_{\pm}(k)$ and $M_{\pm}(x)$ can be used to characterize the support a. We leave this study as an exercise for the reader who can check his results in the papers by Portinari (1967) and Ramm (1965).

XVII.3.2   Darboux Transformations and Rational Reflection Coefficients

Scattering coefficients which are rational fractions of k can be produced by a special class of potentials. If we impose them to satisfy consistently the relations (**XVII.1.38**) and to satisfy $R_{+}(0) = R_{-}(0) = -1$ and $T(\pm\infty) = 1$, and if there is no bound state or exceptional point, we can easily show that they must be of the form

$$R_{+}(k) = \frac{P(-k)}{\prod\limits_{j=1}^{q} (\lambda_j - k)} \prod_{\mu_i \in \mathbf{M}^+} \frac{\mu_i + k}{\mu_i - k} \prod_{\lambda_l \in \mathbf{L}^+} \frac{\lambda_l + k}{\lambda_l - k}, \qquad \text{(XVII.3.3a)}$$

$$T(k) = (-1)^q \frac{\prod\limits_{i=1}^{q} (\mu_i + k)}{\prod\limits_{j=1}^{q} (\lambda_j - k)}, \qquad \text{(XVII.3.4a)}$$

$$R_{-}(k) = \frac{P(k)}{\prod\limits_{j=1}^{q} (\lambda_j - k)} \prod_{\mu_i \in \mathbf{M}^-} \frac{\mu_i - k}{\mu_i + k} \prod_{\lambda_l \in \mathbf{L}^-} \frac{\lambda_l - k}{\lambda_l + k}, \qquad \text{(XVII.3.3b)}$$

where the polynomial P degree is smaller than q, $\operatorname{Im} \mu_i > 0$ except $\mu_i = 0$, $\operatorname{Im} \lambda_l < 0$, $T(k)$ is an irreducible fraction, and $\mathbf{M}^+$, $\mathbf{M}^-$, $\mathbf{L}^+$, $\mathbf{L}^-$ are sets of numbers $\neq 0$ which may or may not contain the zeros or the poles of $T(k)$ and $T(-k)$, once or several times, and which are coupled in such a way that (**XVII.1.38a**) is satisfied (see examples in Appendix XVII.A). Furthermore, if the potential is real, either μ_k, λ_k are pure imaginary or for each μ_k, each λ_k which is not imaginary, there exists $-\mu_k^*$, $-\lambda_k^*$, whereas $[P(-k)]^* = \pm P(k)$, the plus or minus sign being determined by the numbers of μ_k, λ_k of each kind. If there are N (nonvanishing) bound states, $T(k)$ is *multiplied* by the phase factor (a phase factor $\Phi(k)$ is by definition equal to $[\Phi(-k)]^{-1}$)

$$T_B(k) = \prod_{p=1}^{N} \frac{i\varkappa_p + k}{i\varkappa_p - k}(-1)^N, \qquad \text{(XVII.3.4b)}$$

whereas $R_+(k)$ (resp. $R_-(k)$) is *multiplied* by the phase factor $R_+^B(k)$ (resp. $R_-^B(k)$) such that

$$R_+^B(k)/R_-^B(-k) = T_B(k)/T_B(-k) = [T_B(k)]^2. \qquad \text{(XVII.3.3c)}$$

At this point we must realize that the rational character of coefficients is not an intrinsic property, but depends on the choice of the origin in the sense that a translation of the origin from 0 to a does modify the reflection coefficients: according to (XVII.1.34), (XVII.1.35) they are multiplied by a phase factor which is not a rational fraction. Hence it is not surprising that simplifications occur on both sides of $x = 0$ in a different way. For the sake of simplicity, assume that all the poles v_j^+ (resp. v_j^-) of $R_+(k)$ (resp. $R_-(k)$) in the upper half-plane are simple, with residues R_+^j (resp. R_-^j). The integrals on the right-hand side of (XVII.2.2) can be evaluated by contour integration in the upper half-plane for $x > 0$ in the case $+$ and $x < 0$ in the case $-$. The result is

$$F_+(k, x) - 1 = \sum_{v_+^j \in \mathcal{P}^+} \frac{R_+^j F_+(v_+^j, x)\exp[2iv_+^j x]}{k + v_+^j}$$

$$-i \sum_{p=1}^{N} (C_p^+)^{-2} \frac{F_+(i\varkappa_p, x)\exp[-2\varkappa_p x]}{k + i\varkappa_p} \quad (x > 0), \quad \text{(XVII.3.5)}$$

$$F_-(k, x) - 1 = \sum_{v_j^- \in \mathcal{P}^-} \frac{R_-^j F_-(v_-^j, x)\exp[-2iv_-^j x]}{k + v_-^j}$$

$$-i \sum_{p=1}^{N} (C_p^-)^{-2} \frac{F_-(i\varkappa_p, x)\exp[+2\varkappa_p x]}{k + i\varkappa_p} \quad (x < 0), \quad \text{(XVII.3.6)}$$

where $\mathcal{P}^+$ (resp. $\mathcal{P}^-$) is the set of poles v_j^+ (resp. v_j^-) of R_+ (resp. R_-) in the upper half-plane. The terms corresponding to the bound states and those corresponding to the simple poles are similar and can be formally identified if we enumerate the "resonances" $v_j^\pm$ from $j = N + 1$ on, and set $i\varkappa_p = v_p^\pm$, $-i(C_p^\pm)^{-2} = R_\pm^p$. Suppose we do it and let k run through the v_i^+ or through the v_i^-. we obtain two systems of linear algebraic equations that determine $F_+(v_+^j, x)$ (resp. $F_-(v_-^j, x)$) for all values of j and $x > 0$ (resp. $x < 0$). The solutions of these systems can be written in the form of quotients of determinants. $K_+(x, y)$ and $K_-(x, y)$ can then be identified by comparing (XVII.3.5), (XVII.3.6) with (XVII.1.23), (XVII.1.26).

The final result is

$$\begin{aligned} K_+(x, x) &= \mathcal{D}_+'(x)/\mathcal{D}_+(x) \qquad (x > 0), \\ K_-(x, x) &= -\mathcal{D}_-'(x)/\mathcal{D}_-(x) \qquad (x < 0), \end{aligned} \qquad \text{(XVII.3.7)}$$

where

$$\mathscr{D}_+(x) = \det\{\delta_{mj} - (v_+^m + v_+^j)^{-1} R_+^j \exp[2iv_+^j x]\},$$
$$\mathscr{D}_-(x) = \det\{\delta_{mj} - (v_-^m + v_-^j)^{-1} R_-^j \exp[2iv_-^j x]\}. \qquad \textbf{(XVII.3.8)}$$

Hence $V(x)$ can be given in a closed form. If the result on both sides of $x = 0$ belongs to a class where the inverse problem has a unique solution, this closed form is the unique solution in this class. This is the case, for instance, if the reflection coefficients $R_\pm(k)$ vanish on the real axis. Then it follows from **(XVII.3.8)** that $\mathscr{D}(x)$ are combinations of decreasing exponentials whose derivative exponentially decrease (respectively, at $+\infty$ and $-\infty$. The corresponding potential, which is often called a "multisolitonic" potential, therefore belongs to L_2^1. It is a transparent potential, and the N-terms generalization of that given by **(XVII.2.14)**.

Such a potential is holomorphic in a strip containing the real axis. In the case of nontransparent potentials, the poles of $R_+(k)$ and those of $R_-(k)$ in the upper half-plane, together with their residues, can be chosen almost independently with large degrees of freedom and, according to **(XVII.1.47)**, if R_+ or R_- is $O(k^{-n-1})$ at $\pm\infty$, we cannot expect that V has more than n derivatives in $L_1(-\infty, +\infty)$. If they are in L_2^1, the two closed forms derived on both sides of $x = 0$ cannot have more than n continuous derivatives across $x = 0$. A typical case is that where R_+ (resp. R_-) has no pole in Im $k \geq 0$. Then $V(r)$ vanishes identically for $x > 0$ (resp. $x < 0$) if there is no bound state.

Transformations. Since rational scattering coefficients and the corresponding potentials depend on a finite number of parameters, it is expected that we can construct them by a convenient algorithm after a finite number of steps. This is achieved by means of Darboux transformations.

Let μ be a complex number, with Im $\mu > 0$, let u be any given solution of the equation

$$u''(\mu, x) + (\mu^2 - V)u(\mu, x) = 0, \qquad \textbf{(XVII.3.9)}$$

and let f be a solution of the equation

$$f''(k, x) + (k^2 - V)f(k, x) = 0. \qquad \textbf{(XVII.3.10)}$$

It is easy to check that if

$$f^T(k, x) = \frac{d}{dx} f(k, x) - \frac{u'(\mu, x)}{u(\mu, x)} f(k, x), \qquad \textbf{(XVII.3.11)}$$

then

$$\frac{d^2 f^T(k, x)}{dx^2} + (k^2 - \tilde{V}(x))f^T(k, x) = 0, \qquad \textbf{(XVII.3.12)}$$

where

$$\tilde{V}(x) = V(x) - 2\frac{d}{dx}\left(\frac{u'(\mu, x)}{u(\mu, x)}\right). \qquad \textbf{(XVII.3.13)}$$

Hence, if a given one-dimensional Schrödinger equation is exactly solvable, so is the transformed equation and its solutions are readily derived from the previous equation by means of **(XVII.3.10)**. In particular, a Jost solution, $f_\pm(k, x)$ of **(XVII.3.10)** yields a Jost solution $\tilde{f}_\pm(k, x)$ of **(XVII.3.12)** by means of **(XVII.3.11)**, except that a convenient factor A_+ must be introduced to $f_\pm^T(k, x)$ to restore the asymptotic normalization **(XVII.1.7)**, **(XVII.1.8)**. We can then identify reflection and transmission coefficients of $\tilde{V}$. The result depends on the value of $\alpha_-, \beta_-,$ or α_+, β_+, in the following relation:

$$u(\mu, x) = \begin{cases} \alpha_- f_-(\mu, x) + \beta_- f_-(-\mu, x), \\ \alpha_+ f_+(\mu, x) + \beta_+ f_+(-\mu, x). \end{cases} \quad \text{(XVII.3.14)}$$

There are four possible cases:

(a) If $\beta_- \neq 0$, any α_- except $\alpha_- = \beta_- R_-(\mu)$ [i.e., $\beta_+ \neq 0$] then

$$\tilde{T}(k) = -\frac{\mu + k}{\mu - k} T(k); \qquad \tilde{R}_+(k) = \frac{\mu + k}{\mu - k} R_-(k). \quad \text{(XVII.3.15)}$$

(b) If $\beta_- \neq 0, \alpha_- = \beta_- R_-(\mu)$ [i.e., $\beta_+ = 0$], then

$$\tilde{T}(k) = T(k); \qquad \tilde{R}_+(k) = \frac{\mu - k}{\mu + k} R_+(k). \quad \text{(XVII.3.16)}$$

(c) If $\beta_- = 0$ and μ is not a bound state [i.e., $\beta_+ \neq 0$], then

$$\tilde{T}(k) = T(k); \qquad \tilde{R}_+(k) = \frac{\mu + k}{\mu - k} R_+(k). \quad \text{(XVII.3.17)}$$

(d) If $\beta_- = 0$, and μ is a bound state [i.e., $\beta_+ = 0$]

$$\tilde{T}(k) = -\frac{\mu - k}{\mu + k} T(k); \qquad \tilde{R}_+(k) = \frac{\mu - k}{\mu + k} R_+(k). \quad \text{(XVII.3.18)}$$

The transformations of R_- can be readily derived from those of T, R_+, and the formulas **(XVII.1.38)**. It is clear that these transformations can only introduce—or destroy—rational "phase factors" to the scattering coefficients. The transformations of the form (a) introduce a bound state. There is obviously a free parameter α_-/β_- which will determine the normalizing constant of the new bound state. The transformation of the form (d) suppresses a bound state. Using transformations, successively, makes it possible to suppress the bound states, to construct the potential by a method which assumes no bound state, and then to reintroduce the bound states with their normalizing constants.

The transformations of form (b) or (c) are "isospectral," i.e., they leave bound states (if any) and $T(k)$ invariant. They multiply $R_+(k)$ by a phase factor or its reverse (and this is very useful since we assumed Im $\mu > 0$). Thus, they enable us to introduce poles in R_+ and $R_-(k)$. Hence if we start from a reference reflection coefficient $R_+^0(k)$ that has no pole in Im $k \geq 0$, we can introduce successively N poles in this half-plane by means of N phase factors. In the "no bound state" case, the potential corresponding to $R_+^0(k)$ vanishes on $x > 0$. The effect of the N successive Darboux transformations, i.e., of N steps of the algorithm, constructs the part of $V(x)$ for $x > 0$ in a closed form. If we are given $R_+(k)$ rational and arbitrary, we can always write it as the product of R_+^0 by a convenient phase factor and complete this task. Doing it again for $R_-(k)$ yields the part of $V(x)$ on the negative x side. So as to obtain a real potential at each step it is convenient to combine in one step the effect of two symmetric resonances (i.e., two poles $a + ib$ and $-a + ib$)—not a difficult trick. The closed forms we obtain are, of course, equivalent to **(XVII.3.7)**, but the cited algorithms are the shortest way to calculate them, since the number of elementary operations is $O(P^2)$, where P is the number of poles. Hence these algorithms are short and easy to implement, *but* only if R^+, R^-, and T have been exactly and consistently given as rational fractions. For more details, see Sabatier (1983b, 1985b). There is also in the literature a very large number of studies where the direct or the inverse problem is worked out in closed form for a few poles, simple or multiple (see section XVII.5).

XVII.3.3 Generalized Born Approximations

From equations **(XVII.1.42)**, we see that if $\delta V = \tilde{V} - V$ is an infinitesimal of first order, $\tilde{T} - T$ is $O(\delta V)$, $\tilde{f}_\pm - f_\pm$ is $O(\delta V)$, and the distorted wave Born approximation for δR_+ is

$$\tilde{R}_+(k) - R_+(k) = \frac{T^2(k)}{2ik} \int_{-\infty}^{+\infty} \delta V(x) [f_-(k, x)]^2 \, dx + O(\delta V)^2, \quad \textbf{(XVII.3.19a)}$$

$$\tilde{R}_-(k) - R_-(k) = \frac{T^2(k)}{2ik} \int_{-\infty}^{+\infty} \delta V(x) [f_+(k, x)]^2 \, dx + O(\delta V)^2. \quad \textbf{(XVII.3.19b)}$$

Admittedly, these formulas are reliable only for $\delta V(x)$ "small enough," and k "large" and yield "small" perturbations of $R_\pm$. In other words, they give a solution of the direct problem in a neighborhood of $(V, R_\pm)$. Conversely, in such a neighborhood, we may use inversion formulas applying to **(XVII.3.19)**. Inversion formulas can be obtained from completeness formulas which involve products of two Jost solutions. Let

$$\Phi_+(k, x) = f_+(k, x)\tilde{f}_+(k, x), \quad \textbf{(XVII.3.20a)}$$

$$\Phi_-(k, x) = f_-(k, x)\tilde{f}_-(k, x), \quad \textbf{(XVII.3.20b)}$$

where f_+ and f_- (resp. $\tilde{f}_+$ and $\tilde{f}_-$), respectively, correspond to V and $\tilde{V}$. It has been shown that if no bound state comes in and $f \in L^1(\mathbb{R})$

$$\int_{-\infty}^{x} f(y)dy = -\lim_{R \to \infty} \frac{1}{2\pi i} \int_{-R}^{R} \Phi_-(k, x)(f, \Phi_+(k, \cdot)) \frac{k \, dk}{T(k)\tilde{T}(k)}, \quad \textbf{(XVII.3.21a)}$$

$$\int_{x}^{\infty} f(y)dy = -\lim_{R \to \infty} \frac{1}{2\pi i} \int_{-R}^{R} \Phi_+(k, x)(f, \Phi_-(k, \cdot)) \frac{k \, dk}{T(k)\tilde{T}(k)}, \quad \textbf{(XVII.3.21b)}$$

where

$$(f, \Phi_+(k, \cdot)) = \int_{-\infty}^{+\infty} f(x)\Phi_+(k, x)dx, \quad \textbf{(XVII.3.22a)}$$

$$(f, \Phi_-(k, \cdot)) = \int_{-\infty}^{+\infty} f(x)\Phi_-(k, x)dx. \qquad \textbf{(XVII.3.22b)}$$

These formulas, which are also known with bound states (Christov, 1981), are exact formulas which can give the exact inversion formulas corresponding to **(XVII.1.42)**. If we write them down and keep only the first δV (or δR) order we obtain

$$\int_x^{\infty} \delta V(t)dt = -\pi^{-1} \int_{-\infty}^{+\infty} [f_+(k, x)]^2 \delta R_+(k)dk. \qquad \textbf{(XVII.3.23a)}$$

$$\int_{-\infty}^{\infty} \delta V(t)dt = -\pi^{-1} \int_{-\infty}^{+\infty} [f_-(k, x)]^2 \delta R_-(k)dk. \qquad \textbf{(XVII.3.23b)}$$

The formulas **(XVII.3.23)** enable us to solve the linearized inverse problem on the line. Hence they give a local inversion which may be combined with a global way of exploring the set of parameters, in particular, the algorithms based on rational reflection coefficients. This is particularly appropriate because the Jost solutions are produced by these algorithms (Sabatier, 1985b).

XVII.3.4   Trace Method

The starting point of the trace method is the trace formula, which we write for $R_+(k)$ and no bound state. It follows from the asymptotic properties proved in section XVII.1 that

$$F_+(k, x)F_-(k, x) = 1 + \tfrac{1}{2}k^{-2}V(x) + O(|k|^{-2}). \qquad \textbf{(XVII.3.24)}$$

A contour integration in the upper half-plane therefore yields, if there are no bound states,

$$V(x) = \frac{2i}{\pi} \int_{-\infty}^{+\infty} k[T(k)F_+(k, x)F_-(k, x) - 1]dk. \qquad \textbf{(XVII.3.25)}$$

Then using **(XVII.1.34)** we obtain the trace formula

$$V(x) = \frac{2i}{\pi} \int_{-\infty}^{+\infty} kR_+(k)e^{2ikx}[F_+(k, x)]^2 \, dk, \qquad \textbf{(XVII.3.26)}$$

or

$$V(x) = Q_{R_+}(x, F_+). \qquad \textbf{(XVII.3.27)}$$

Q_{R_+} is a nonlinear functional on a space of functions F_+. We may also consider a space Y of two-component vectorial functions $y = \begin{pmatrix} y_1 \\ y_2 \end{pmatrix}$. Setting $y_1 = F_+(k, x)$, $y_2 = e^{2ikx}F_+'(k, x)$, and using **(XVII.3.27)**, we can write the Schödinger equation **(XVII.1.1)** as

$$y' = V(x, y), \qquad \textbf{(XVII.3.28)}$$

where

$$V(x, y) = \begin{pmatrix} e^{-2ikx}y_2 \\ e^{2ikx}y_1 Q_R(x, y_1) \end{pmatrix}. \qquad \textbf{(XVII.3.29)}$$

Equation (**XVII.3.28**) is a nonlinear differential equation which has to be solved with an initial condition at $+\infty$. Hence the problem is to determine its solution or its solutions. It is easy to prove that $\mathbf{V}(x, \mathbf{y})$ is Lipschitz in any ball $\subset Y$. Hence the local existence and construction problem is solved (Deift and Trubowitz, 1979).

The global problem has also been solved.

XVII.3.5   Time-Domain Approach

The one-dimensional problem of quantum scattering theory is equivalent to one-dimensional problems of wave propagation which can be formulated either in the frequency domain or in the time domain. In the latter case, the "energy parameter" k^2 is replaced by $-\partial^2/\partial t^2$, so that the resulting equation is hyperbolic. In addition, the physical problems that appear in this framework usually involve potentials whose support is either finite or extended to a half-axis only. Hence *time* and *causality* are ingredients that appear in these problems. They show in the existence of solutions which satisfy initial conditions given on a characteristic surface, i.e., here, a ray $x \pm t = \text{constant}$, and vanishing identically outside of a characteristic cone, i.e., here, the oriented angle between two rays $(t - \tau)^2 - (x - \xi)^2 = 0$. The time-domain approach has been extended to all problems discussed in this book and several others, provided bound states are missing or eliminated in a first step by Crum–Darboux transformations. We give here only a simple example.

EXAMPLE. Instead of the Schrödinger equation (**XVII.1.1**), we consider the equation

$$\frac{\partial^2 u}{\partial t^2} - \frac{\partial^2 u}{\partial x^2} + Vu = 0 \tag{XVII.3.30}$$

with

$$V \in L_1^1(\mathbb{R}), \qquad V(x) = 0 \qquad (x < 0). \tag{XVII.3.31}$$

For $x < 0$, any solution of (**XVII.3.30**) is the sum of a pure right-going solution $f(t - x)$ and a pure left-going solution $g(t + x)$, whose determination is achieved by a boundary condition at $x = 0$. Let us introduce a (noncausal) right-going impulse solution

$$\left. \begin{array}{l} G_-(x, t) = \delta(t - x) + K_-(x, t), \\[4pt] \text{suppt } K_-(x, t) = \{|t| \le x\}, \end{array} \right\} \tag{XVII.3.32}$$

where K_- is bounded if $V \in L_1^1$. Clearly, $G_-(x, -t)$ is left-going, $G_-(0, t) = \delta(t)$, and

$$\left[\left(\frac{\partial}{\partial t} - \frac{\partial}{\partial t} \right) G_-(x, t) \right]_{x=0} = 0. \tag{XVII.3.33}$$

Any solution $u(x, t)$ of (**XVII.3.30**) that satisfies (**XVII.3.33**) is a superposition of these functions

$$u(x, t) = \int G_-(x, -\tau)u(0, t - \tau)d\tau$$

$$= u(0, t + x) + \int_{-x}^{x} K(x, \tau)u(0, t + \tau)d\tau. \tag{XVII.3.34}$$

Now let us introduce a causal impulse solution

$$G(x, t) = \delta(t - x) + K(x, t), \Big\}$$
$$\text{suppt } K(x, t) = \{|x| \le t\}, \Big\} \qquad \textbf{(XVII.3.35)}$$

where K is bounded if $q \in L_1^1$. Setting

$$v(x, t) = G(x, t) - G_-(x, t) = K(x, t) - K_-(x, t) \qquad \textbf{(XVII.3.36a)}$$
$$= -K_-(x, t) \qquad \text{for} \quad |t| < x. \qquad \textbf{(XVII.3.36b)}$$

We see that v is a solution of **(XVII.3.30)** and **(XVII.3.33)**, so that we can apply **(XVII.3.34)**

$$v(x, t) = v(0, t + x) + \int_{-x}^{x} K_-(x, \tau)v(0, t + \tau)d\tau. \qquad \textbf{(XVII.3.37)}$$

Substituting **(XVII.3.36b)** and setting

$$v(0, t) = K(0, t) \equiv R(t) \qquad \textbf{(XVII.3.38)}$$

we obtain for $t < x$

$$K_-(x, t) + R(t + x) + \int_{-t}^{x} K_-(x, \tau)R(t + \tau)d\tau, \qquad \textbf{(XVII.3.39)}$$

which is nothing but the Marchenko equation when V is supported on $x > 0$ (compare the symmetric case, which produced **(XVII.3.2.)**. We also notice that $\exp[-ikt]f_-(k, x)$ is a solution of **(XVII.3.30)** and **(XVII.3.33)**, so that **(XVII.3.34)** yields

$$f_-(k, x) = \exp[-ikx] + \int_{-x}^{x} K_-(x, t)\exp[-ikx]dt. \qquad \textbf{(XVII.3.40)}$$

As an exercise the reader may also derive the relation between $K_-(x, x)$ and $V(x)$.

Time-Domain Approach and Layer Stripping. To understand how the time-domain approach enables us to analyze a layered medium by stripping layers in a sequential way, let us analyze a simple problem. We are given the response, of a pile of slabs of thickness T_l and index v_l, to a "classical" plane wave incident normally upon the first slab at a distance L from the observer. The world "classical" is used to mean that the signal can always be identified and is delayed between two points by its propagation time. Now let $f(t)$ be the variation with time of the incident wave at the observation point $z = 0$ and let η be the admittance of the slab relative to free space. Let us first consider the case of one slab only (lower indices omitted). Then the field in the neighborhood of the observer is of the form

$$F(t, z) = f\left(t - \frac{z}{c}\right) + \frac{1 - \eta}{1 + \eta}f\left(t - \frac{2L}{c} + \frac{z}{c}\right)$$

$$- 4\eta \sum_{k=1}^{\infty} \frac{(1 - \eta)^{2n-1}}{(1 + \eta)^{2n+1}}f\left(t - \frac{2L}{c} - \frac{2kvT}{c} + \frac{z}{c}\right). \qquad \textbf{(XVII.3.41)}$$

The first term in **(XVII.3.41)** is the incident wave. The second term is the wave initially reflected from the outer face of the slab. The nth term in the summation is the wave that has been reflected n times from the inner face of the slab before being transmitted back to the observer. All these "multiple" terms can (in principle) be separated—this is obvious if the "signal" f is a δ-function—but with increasing errors in real cases. However, this is not so important since the first two reflections provide all the information that can be gathered by the observer, namely L, η, and the optical thickness vT of the slab. The information provided by multiples is redundant, and this explains why they are omitted in usual seismic analyses made by ray methods and time-domain approaches.

Suppose now we have a pile of slabs. The signal will contain arrivals (quite separate if the incident wave is a δ-function) corresponding to the various faces and the multiples. The arrivals corresponding to the first two reflections on the first slab in (**XVII.3.41**) are still the first arrivals, and thus can be used to identify the parameters of the first slab. It is then easy to subtract the synthetic response made out of this slab from the observed response and to pursue the analysis on the second slab. Hence, we see that causality and the time-domain approach give rise in a straightforward way to a sequential approach, called a "layer-stripping approach," to reconstruct a pile of slabs—or a layered medium.

In physical problems, this method cannot be used without modifications involving filtering and cleaning techniques at each step, and strong stabilizing levers. Obviously, if this is not done, the sequential algorithms which are introduced will have a poor numerical stability, because errors made in the analysis of first-arrived signals will be reported on the following signals, for which the signal is progressively smaller. Furthermore, the source is never a δ-function, so that the signals are never completely identified and the signal-to-noise ratio is progressively smaller. In theoretical problems, the method can be transformed into a sequence of differential equations enabling us to analyze a layered medium by layer stripping. Thus, this kind of layer-stripping method, which may also be called a downward continuation, has been developed in several fields (see, for example, Bates and Millane (1981), Bruckstein et al. (1985), and Symes (1986)). Other references will be given in section XVII below.

A different time-domain approach to inverse scattering has been proposed by Corones et al. (1983). The starting point is the definition of the "reflection kernel" due to a part of a medium characterized by the continuous velocity $c(z)$—say the part between $z = a$ and $z = b > a$. This is achieved by means of the following "thought experiment". We imbed this portion of the medium in a new medium obtained by continuing the velocity $c(z)$ below $z = a$ and above $z = b$, respectively, by its values at a and b. In this fictitious medium, because of causality and invariance under time translation, the left-going wave reflected wave at $z = a$, $g[(z - a)/c(a) + t]$ $(z < a, t > 0)$, and the right-going incident wave, $f[-(z - a)/c(a) + t]$, are related by a convolution equation

$$g(t) = \int_{-\infty}^{t} f(t - s)R^{+}(a, b, s)ds, \qquad \text{(XVII.3.42)}$$

and the "reflection kernel" (for waves incident from the left, for the subinterval a, b), $R^+(a, b, s)$ represents reflection processes which occur inside the interval $[a, b]$. Hence the "invariant imbedding" method, like the "layer-stripping methods," relies on causality and proceeds through decomposing the waves into components traveling in different directions. It leads to an integro-differential equation which is satisfied by the reflection kernel. Discretizing it yields stable numerical schemes for solving the direct and inverse problems.

XVII.4   More General One-Dimensional Problems

XVII.4.1   Dirac Measures and Impedance Equations

Up to now we have enforced that a solution of (XVII.1.1) should be continuous with a continuous derivative. This is no longer possible if we allow the potential $V(x)$ to contain Dirac measures. We assume that there is a finite number of them, located at separated points

$$V(x) = q(x) + \sum_{n=0}^{N} q_n \, \delta(x - x_n), \qquad q \in L_1^1, \qquad \text{(XVII.4.1)}$$

and we now define a solution of (XVII.1.1), say, $\varphi(x)$, as a continuous function everywhere, with continuous derivatives everywhere except at the points x_n, solving (XVII.1.1), and such that

$$\varphi'(x_n^+) - \varphi'(x_n^-) = q_n \, \varphi(x_n), \qquad \text{(XVII.4.2)}$$

where $\varphi'(x_n^+)$ (resp. $\varphi'(x_n^-)$) is the limit of $\varphi'(x)$ as x goes to x_n by upper (resp. lower) values. Thanks to this definition, we can go through the theory of the direct problem given in section XVII.1 with only minor, and trivial, modifications. So as to give the reader a feeling of them, we sketch a few results obtained in the particular case $q(x) = 0$: this case gives exactly solvable examples because the integral equations for $F_{\pm}(k, x)$ reduce to triangular linear algebraic systems

$$F_+(k, x) = 1 + \sum_{n=0}^{N} q_n \, F_+(k, x_n)\theta(x_n - x)\frac{e^{2ik(x_n-x)} - 1}{2ik}, \qquad \text{(XVII.4.3a)}$$

$$F_-(k, x) = 1 + \sum_{n=0}^{N} q_n \, F_-(k, x_n)\theta(x - x_n)\frac{e^{2ik(x-x_n)} - 1}{2ik}. \qquad \text{(XVII.4.3b)}$$

From equations (XVII.4.3) it is easy to see that the generic case is still $T(0) = 0$, $R_\pm(0) = -1$, and that for $|k| \to \infty$

$$T(k) = 1 + (2ik)^{-1} \sum_{n=0}^{N} q_n + O(k^{-2}), \qquad \text{(XVII.4.4)}$$

$$R_+(k) = (2ik)^{-1} \sum_{n=0}^{N} q_n e^{-2ikx_n} + O(k^{-2}). \qquad \text{(XVII.4.5)}$$

The asymptotic behavior (XVII.4.4) is of course a flat generalization of

(XVII.1.4), (XVII.1.6). But (XVII.4.5) (which remains the leading term even if $0 \neq q(x) \in L_1^1$: compare (XVII.1.4) or see section (XVII.4 below) shows that the weights and positions of δ-measures are identified from the asymptotic behavior of $R_+(k)$. The most simple solvable example is that of one δ-measure localized at $x_0 = 0$, which yields

$$T(k) = \frac{2ik}{2ik - q_0}, \qquad R_+(k) = \frac{q_0}{2ik - q_0}. \qquad \textbf{(XVII.4.6)}$$

A bound state is obtained for negative q_0 at $k = -\frac{1}{2}iq_0$. Notice the non-analyticity as a function of q_0 if $h \to 0$, related with the divergence of the Green's function. This example is also the most simple example of a potential with rational scattering coefficients. A slightly more complicated example is defined by means of two δ-measures

$$T(k) = \left[1 - \frac{q_1 + q_0}{2ik} - \frac{q_0 q_1}{4k^2} (1 - e^{2ik(x_1 - x_0)}) \right]^{-1}, \qquad \textbf{(XVII.4.7a)}$$

$$R_+(k)/T(k) = \left[\frac{q_1}{2ik} e^{-2ikx_1} + \frac{q_0}{2ik} e^{-2ikx_0} - \frac{q_0 q_1}{4k^2} (e^{-2ikx_0} - e^{-2ikx_1}) \right].$$
$$\textbf{(XVII.4.7b)}$$

It is clear that the effect of δ-measures is to introduce special conservation laws between intervals where the Schrödinger equation is integrated in the usual way. Another method of doing it is to integrate the mixed impedance-potential equation

$$\left(\alpha^{-2} \frac{d}{dx} \alpha^2 \frac{d}{dx} + k^2 - V \right) p(k, x) = 0, \qquad \textbf{(XVII.4.8)}$$

whose solution p must satisfy the conditions

$$p, \ \alpha^2 \frac{dp}{dx} \quad \text{continuous everywhere} \qquad \textbf{(XVII.4.9)}$$

and whose parameters α and V satisfy the assumptions:

A1. The "impedance factor" $\alpha(x)$ is bounded away from zero at any finite point. Since only α^2 is used, there is no loss of generality in setting $\alpha > 0$.

A2. $\alpha(x)$ is continuous and has a continuous derivative everywhere on $\mathbb{R}$, except at a finite number of points, $x_0, x_1, \ldots, x_N$, where α and α' show jumps

$$t_n^{-1} = \frac{1}{2} \left(\frac{\alpha_n^+}{\alpha_n^-} + \frac{\alpha_n^-}{\alpha_n^+} \right), \qquad \textbf{(XVII.4.10a)}$$

$$t_n^{-1} r_n = \frac{1}{2} \left(\frac{\alpha_n^+}{\alpha_n^-} - \frac{\alpha_n^-}{\alpha_n^+} \right), \qquad \textbf{(XVII.4.10b)}$$

$$t_n^{-1} s_n = \frac{1}{2} \left(\frac{\alpha_n'^-}{\alpha_n^+} - \frac{\alpha_n'^+}{\alpha_n^-} \right), \qquad \textbf{(XVII.4.10c)}$$

where α_n^+ stands for $\alpha(x_n^+)$, $\alpha_n'^+$ for $\alpha'(x_n^+)$, etc. Note that $r_n^2 + t_n^2 = 1$, so that $t_n = \sqrt{1 - r_n^2}$. r_n, s_n, t_n are called "singular factors".

A3. $V \in L_1^1(\mathbb{R})$.

We must keep in mind the following:

Equivalence Theorem. *Let* $V \in L_1^1(\mathbb{R})$, *and let* α, β *be differentiable positive functions everywhere except for possible jumps at the points* $\{x_n\}$, *such that* β/α *is continuous everywhere and* $\alpha\beta' - \beta\alpha' = W(x)$ *is absolutely continuous. Then if* p *is a solution* $(p, \alpha^2\, dp/dx$ *continuous) of* **(XVII.4.8)**, $q = \alpha p/\beta$ *is a solution* $(q, \beta^2\, dq/dx$ *continuous) of*

$$\left[\beta^{-2} \frac{d}{dx} \beta^2 \frac{d}{dx} + k^2 - V + (\alpha\beta)^{-1} \frac{dW}{dx}\right] q(k, x) = 0. \quad \textbf{(XVII.4.11)}$$

Besides, β *and* α *have the same "singular factors."*

Proof. Check that $\beta^2\, dq/dx = \alpha\beta\, dp/dx - pW$ is continuous. The equation follows. Checking the singular factors is easy. $\qquad\square$

Consequences

C1. Setting $\beta = 1$ at a point where α is twice differentiable, we see that equation **(XVII.4.8)** is equivalent to the Schrödinger equation (continuous f, f')

$$\left(\frac{d^2}{dx} + k^2 - V - \alpha^{-1} \frac{d^2\alpha}{dx^2}\right) f(k, x) = 0, \quad \textbf{(XVII.4.12)}$$

where $f(k, x) = \alpha(x) p(k, x)$.

This transformation can be done inside each regular interval. Across any singular point, $\alpha^{-1} f$ and $W(\alpha, f)$ should be continuous. Hence we see that this equation is equivalent to a chain of Schrödinger equations with appropriate continuity conditions across singular points. We call "background potential"

$$V_B = V + \alpha^{-1} \frac{d^2\alpha}{dx^2} = V + V_\alpha \quad \textbf{(XVII.4.13)}$$

the potential which is defined almost everywhere in this chain of Schrödinger equations. We call "impedance equation" the equation **(XVII.4.8)** with $V = 0$. We see that an impedance equation is always equivalent to a chain of Schrödinger equations but that the reverse is true, with $V_B = \alpha^{-1}(d^2\alpha/dx^2)$, only if this last equation can be solved by a function free of zeros at finite distance, i.e., provided there is no bound state to V_B.

C2. Let α and β be two piecewise-differentiable positive functions such that β/α is continuous and $W = \alpha\beta' - \beta\alpha'$ is a constant number. Then, if p is a solution of **(XVII.4.8)**, $q = \alpha p/\beta$ is a solution of the same equation with β^2 replacing α^2. Hence there is a "standard" equivalence in the impedances family. If α is one of them, others are given by the formula

$$\beta = W\alpha \int^x \frac{dt}{\alpha^2(t)} + C\alpha, \qquad \text{(XVII.4.14)}$$

where C and W are only constrained by the sign condition.

C3. We introduce the "singular data function" (SD function) $\sigma(x)$ which is equal to 1 for $x > x_N$, is constant on each regular interval, and shows the reflection and transmission factors r_n and t_n at each singular point. This function can be written explicitly

$$\sigma(x) = \exp\left[-\sum_{n=0}^{N} \rho_j \, \theta(x_j - x)\right], \qquad \text{(XVII.4.15)}$$

where

$$\rho_j = \log[(1 + r_j)/t_j] = \tfrac{1}{2}\log[(1 + r_j)/(1 - r_j)]. \quad \text{(XVII.4.16)}$$

A simple modification of the proof of the equivalence theorem shows that the function $r(k, x)$, equal to $\alpha p/\sigma$, is the continuous solution of

$$\left[\frac{d}{dx}\sigma^2\frac{d}{dx} + \sigma^2(k^2 - V_\alpha - V) + 2\sum_0^N \sigma_i^+\sigma_i^- t_i^{-1}s_i\,\delta(x - x_i)\right]r(k, x) = 0,$$

$$\text{(XVII.4.17)}$$

whose derivative satisfies the requirement (XVII.4.2). In particular, if α is continuous, $\sigma = 1$, and we obtain an equation similar to (XVII.1.1) with the potential (XVII.4.1). We should not conclude, of course, that arbitrary q_i's at arbitrary points x_i can be identified with the coefficients s_i/t_i corresponding to a strictly positive impedance factor.

The scattering problem for equation (XVII.4.8) can be defined like that of the Schrödinger equation, by means of Jost solutions $p_\pm(k, x)$,

$$p_\pm(k, x) \sim \alpha^{-1} \exp[\pm ikx] \qquad (x \to \pm\infty), \qquad \text{(XVII.4.18)}$$

and the relations

$$\left.\begin{array}{l} T(k)p_-(k, x) = p_+(-k, x) + R_+(k)p_+(k, x), \\ p_-(-k, x) + R_-(k)p_-(k, x) = T(k)p_+(k, x), \end{array}\right\} \qquad \text{(XVII.4.19)}$$

which define the reflection and transmission coefficients. They can be expressed in terms of the Wronskian of p_+ and p_- (note that $\alpha^2 W$ is constant). Formulas (XVII.1.38), (XVII.1.40) still hold. We can also write the scattering problem for the Schrödinger equation (XVII.4.12) with the background potential, and define the corresponding Jost solutions $f_\pm$ and scattering coefficients $R_\pm^B$, $T_\pm^B$. It is possible (Sabatier, 1983; Degasperis and Sabatier, 1986) to calculate $R_\pm$, $T_\pm$ from $R_\pm^B$, $T_\pm^B$ and to calculate the values of $f_\pm$ at the points x_n, and to study the impedance scattering problem (or the mixed impedance-potential scattering problem) by this method. It is also possible to proceed as in section XVII.1, i.e., to write integral equations for the Jost solutions (Degasperis and Sabatier, 1987; Sabatier and Dolveck-Guilpart, 1988). Let us sketch it. We start from (XVII.4.17), where we set for convenience

$V + V_a = U$ and $2\sigma(x_n^+)\sigma(x_n^-)s_n/t_n = u_n$, and we write it for $r_+ = \alpha p_+/\sigma$ and $r_- = \alpha p_-/\sigma$

$$\left[\frac{d}{dx}\sigma^2\frac{d}{dx} + \sigma^2(k^2 - U) + \sum_0^N u_i\delta(x - x_i)\right]r_\pm(k, x) = 0. \quad \textbf{(XVII.4.20)}$$

Then setting

$$\tau_\pm(k, x) = \exp[\mp iks]r_\pm(k, x) \quad \textbf{(XVII.4.21)}$$

we obtain the differential equations

$$\left[e^{\pm 2ikx}\frac{d}{dx}e^{\mp 2ikx}\sigma^2\frac{d}{dx} - \sigma^2 U + \sum_0^N W_n^\mp \delta(x - x_n)\right]\tau_\mp(k, x) = 0,$$

$$\textbf{(XVII.4.22)}$$

where

$$W_n^\pm \equiv u_n \pm ikv_n = 2\sigma(x_n^+)\sigma(x_n^-)(s_n \pm ikr_n)/t_n, \quad \textbf{(XVII.4.23)}$$

and hence the integral equations

$$\tau_+(k, x) = 1 + \int_x^\infty dt\, \tau_+(k, t)W_+(k, t)\int_x^t du\,[\sigma(u)]^{-2}\exp[2ik(t - u)],$$

$$\textbf{(XVII.4.24a)}$$

$$\tau_-(k, x) = \gamma + \int_{-\infty}^x dt\, \tau_-(k, t)W_-(k, t)\int_t^x du\,[\sigma(u)]^{-2}\exp[2ik(u - t)],$$

$$\textbf{(XVII.4.24b)}$$

where

$$W_\pm(k, x) = \sigma^2 U - \sum_0^N W_n^\pm\delta(x - x_n), \quad \textbf{(XVII.4.25)}$$

$$\gamma - 1/\sigma(-\infty). \quad \textbf{(XVII.4.26)}$$

Equations **(XVII.4.24)** can be studied as were equations **(XVII.1.14)**, since for Im $k \geq 0$, finite k, the kernels can be bounded in a similar way, for instance,

$$\left|W_+(k, t)\int_x^t du\,[\sigma(u)]^{-2}\exp[2ik(t - u)]\right|$$

$$\leq C\left\{[1 + |x|\theta(-x)][1 + |t|\theta(t)]\left[|U(t)| + (1 + |k|)\sum_{n=0}^N \delta(x_n - t)\right]\right\},$$

$$\textbf{(XVII.4.27)}$$

where C is an appropriate number. This yields an exactly solvable majorant equation and absolute bounds for the iterative solutions of **(XVII.4.24)** by a convergent series of numbers. Hence $\tau_+(k, x)$ is holomorphic and uniformly bounded in the upper half-plane, and $T(k)$ is meromorphic there. Let us look for "bound states," i.e., solutions of **(XVII.4.8)** belonging to $L_{a^2}^2(\mathbb{R})$. If p is such a solution, then

$$\left[p^*\alpha^2\frac{dp}{dx} - p\alpha^2\frac{dp^*}{dx}\right]_{-\infty}^{+\infty} = (k^{*2} - k^2)\int_{-\infty}^{+\infty}\alpha^2 pp^*\,dx = 0, \quad \textbf{(XVII.4.28)}$$

so that k is pure imaginary. Positions of bound states are zeros of $W(p^+, p^-)$. They are simple and isolated (proofs are like those in section XVII.1. Furthermore, the minimum of the energy functional

$$\mathscr{E}(f) = \int_{-\infty}^{+\infty} [\alpha^2(f'(x))^2 + V(x)f^2(x)]dx \qquad \textbf{(XVII.4.29)}$$

is obtained at the first bound state of **(XVII.4.8)**, f then being equal to the solution p of **(XVII.4.8)** (normalized to 1 in $L_{\alpha^2}^2(\mathbb{R})$), and $\mathscr{E}$ being equal to $E = k^2 = -\varkappa_1^2$. In the case of a pure impedance equation (i.e., $V = 0$ in **(XVII.4.8)** or $U = V_\alpha$ in **(XVII.4.20)**, or a mixed equation with a repulsive potential ($V > 0$)), this minimum cannot be negative and the equation cannot have a true bound state.

The asymptotic behavior of $\tau_\pm$, unfortunately, is not so simple. For $\mathrm{Im}\, k \to +\infty$, we still get finite limits for $\tau_\pm$, depending on x. But on the real k-axis, the asymptotic behavior is oscillating. The leading term can be obtained by solving the triangular linear algebraic systems obtained by cancelling U and all the s_n'''s, and making successively $x = x_0, x_1, \ldots, x_N$, in

$$\tau_+(k, x) = 1 - ik \sum_{n=0}^{N} v_n \tau_+(k, x_n)\theta(x_n - x) \int_x^{x_n} dy\, \sigma^{-2}(y)e^{2ik(x_n-y)},$$

$$\textbf{(XVII.4.30a)}$$

$$\tau_-(k, x) = \gamma + ik \sum_{n=0}^{N} v_n \tau_-(k, x_n)\theta(x - x_n) \int_{x_n}^{x} dy\, \sigma^{-2}(y)e^{2ik(y-x_n)}.$$

$$\textbf{(XVII.4.30b)}$$

Equations **(XVII.4.30)** also yield the exact solution of the impedance problem of N layers with constant parameters. To show the reader how the scattering coefficients behave, the simplest way is to give him their exact value in the case of two layers with constant parameters

$$T(k) = \frac{t_0 t_1}{1 + r_0 r_1 \exp[2ik(x_1 - x_0)]}, \qquad \textbf{(XVII.4.31)}$$

$$R_+(k)/T(k) = (t_0 t_1)^{-1}[r_1 e^{-2ikx_1} + r_0 e^{-2ikx_0}]. \qquad \textbf{(XVII.4.32)}$$

Notice that $T(k)$ and $R^+(k)$ are almost-periodic functions and that the leading asymptotic behavior of $R_+(k)/T(k)$ on the real axis yields the positions of the singular points, together with the reflection and transmission factors. This remains true in more complicated cases and with general impedances. For any positive impedance, the singular factors and singular points can be extracted from the leading asymptotic behavior of $R_+(k)$. It is then possible to construct $R_+^B(k)$ and since there is no bound state, α''/α can be reconstructed. Hence the inverse problem for impedances that generalize bound-state-free potentials has a unique solution (Sabatier, 1988). These problems are of first importance in geophysics but, of course, in practical cases, the asymptotic behavior cannot be obtained.

The impedance equation (and the mixed equation) are structurally invariant by the following Darboux transformation: starting from (**XVII.4.8**) with $V = 0$, i.e.,

$$\left(\alpha^{-2}\frac{d}{dx}\alpha^2\frac{d}{dx} + k^2\right)p(k, x) = 0 \qquad \text{(XVII.4.33)}$$

with $p(\mu, x)$ being a particular solution of this equation, and with μ instead of k, Im $\mu > 0$, we can readily prove that the transformed function

$$p^T(k, x) = p(k, x) - \frac{p'(k, x)p(\mu, x)}{p'(\mu, x)} \qquad \text{(XVII.4.34)}$$

is a solution of (**XVII.4.33**), with $\tilde{\alpha}$ instead of α, and

$$\tilde{\alpha}^2 = \left[\frac{\alpha p'(\mu, x)}{p(\mu, x)}\right]^2. \qquad \text{(XVII.4.35)}$$

There are four cases, as in the potential case, and if, in particular, $p_-(\mu, x)$ is used, the transformation keeps the transmission coefficient, whereas the new reflection coefficient is

$$\tilde{R}_+(k) = \frac{\mu + k}{\mu - k} R_+(k). \qquad \text{(XVII.4.36)}$$

Nothing prevents $p'(\mu, x)$ or $p(\mu, x)$ having zeros, so that the transformation may lead to unallowed impedances.

XVII.4.2   Wider Classes of Parameters and Ambiguities

The classes of potentials explicitly or implicitly introduced in section XVII.4.1 are more general than L_1^1 as far as local properties are concerned: for instance, a blind use of formula (**XVII.4.3**) suggests that discontinuities of α and α' correspond to δ' and δ distributions. We have seen that in the "no bound state case," this does not modify the inverse problem uniqueness property. Yet, special cases where this property was violated have been given since the first paper of Abraham et al. (1981). These exactly solvable examples show potentials which have no true bound state and yield the same reflection coefficient $R_+(k)$ as a function defined on the real axis. We call them "ambiguities." A way to construct ambiguous potentials, starting from any potential of L_1^1 or L_1^1 + Dirac measures, has been given (Sabatier, 1984; Degasperis and Sabatier, 1986). We sketch it here. The basic idea is: For potentials in L_1^1, the only ambiguities we know of are due to (true) bound states and are described for each bound state by one parameter. Suppose we go to a limit where the bound state disappears, might it be possible to keep the ambiguity? Now, the Darboux transformation (**XVII.3.15**), or the similar transformation in impedance theory, is a way of introducing one bound state at $\mu = i\gamma(\gamma > 0)$, with one arbitrary parameter. Letting $\gamma \to 0^+$ yields a limit transformation where everything can be explicitly written down, and which is called the "reflection-flip" transformation. The simplest way to show it starts from the family of standard-equivalent impedances defined by (XVII.4.14), and to consider $\tilde{\alpha} = \beta^{-1}$. Let p be a solution of (XVII.4.33), it is easily checked that

$$p^T(k, x) = p(k, x) - \alpha\beta W^{-1} p'(k, x) \qquad \text{(XVII.4.37)}$$

is a solution, with p^T and $\tilde{\alpha}^2 (dp^T/dx)$ continuous, of

$$\left[\tilde{\alpha}^{-2} \frac{d}{dx} \tilde{\alpha}^2 \frac{d}{dx} + k^2 \right] p^T(k, x) = 0. \qquad \text{(XVII.4.38)}$$

The singular points are unchanged. The Jost solutions are transformed like any other, and, after normalization, give the new scattering coefficients

$$\tilde{T}(k) = T(k), \qquad \text{(XVII.4.39a)}$$

$$\tilde{R}_+(k) = -R_+(k). \qquad \text{(XVII.4.39b)}$$

The singular data are also transformed according to the rule

$$\tilde{t}_n = t_n, \qquad \text{(XVII.4.40a)}$$

$$\tilde{r}_n = -r_n, \qquad \text{(XVII.4.40b)}$$

$$\tilde{s}_n = -t_n^{-2}(1 + r_n^2 s_n) - 2t_n^{-2} r_n q_n, \qquad \text{(XVII.4.40c)}$$

$$\tilde{q}_n = -t_n^{-2}(1 + r_n^2 q_n) - 2t_n^{-2} r_n s_n, \qquad \text{(XVII.4.40d)}$$

where

$$q_n = \tfrac{1}{2} t_n \left(\frac{\alpha_n'^-}{\alpha_n^+} + \frac{\alpha_n'^+}{\alpha_n^-} \right). \qquad \text{(XVII.4.41)}$$

The transformation can be studied without loss of generality by first considering $W = 1$, so that

$$\tilde{\alpha}(x) = [\alpha(x)]^{-1} \left[\int_0^x \alpha^{-2}(t)dt + c \right]^{-1}, \qquad \text{(XVII.4.42)}$$

then $W = 0$, $c = 1$, so that $\tilde{\alpha}(x)$ is simply $[\alpha(x)]^{-1}$, and the limit form of (XVII.4.37) is

$$p^T(k, x) = \alpha^2 p'(k, x). \qquad \text{(XVII.4.43)}$$

This is called the "special transformation." From the impedance factor $\tilde{\alpha}$, and if there is no singular point (resp. if there are only "weak" singularities, i.e., $\tilde{r}_n = 0$, $\tilde{t}_n = 1$), we can construct a potential, which is a function of x equal to $\alpha''/\tilde{\alpha}$ (resp. this function plus Dirac measures at the singular points). The potential is a regular measure at any finite point of $\mathbb{R}$, unless $\tilde{\alpha}$ happens to vanish, usually then yielding a double pole to the potential. Although this case could be used to introduce a class of singular potentials, it is generally discarded because the physical scattering problem has no meaning.

Ambiguities. The general transformation (**XVII.4.42**) depends on the arbitrary parameter c. Hence all the derived impedance factors $\alpha(c, x)$, together with α^{-1} (which can be associated with $c = \infty$), correspond to the same reflection coefficient $-R^+(k)$. $T(k)$ being unchanged, no true bound state may have been created by the transformation. Thus, provided they belong to the scattering class $\mathscr{A}$, the new $\tilde{\alpha}(c, x)$ form a one-parameter class of ambiguities. So too, do the corresponding potentials $\tilde{\alpha}''/\tilde{\alpha}$, provided they belong to the 'scattering class" $\mathscr{V}$. The class L_1^1 (resp. the corresponding class of impedances) is transformed, and the new classes contain potentials (resp. impedances), with asymptotic behaviors that are not allowed in L_2^1 (resp. in the corresponding class of impedances). Let $\mathscr{P}(l_-, l_+)$ be the class of potentials u such that

$$\left. \begin{array}{l} u(x) - l_-(l_- + 1)x^{-2} \in L_1^1(-a, -\infty) \\ u(x) - l_+(l_+ + 1)x^{-2} \in L_1^1(a, +\infty) \end{array} \quad (a > 0) \right\} \mathscr{P}(l_-, l_+). \quad \textbf{(XVII.4.44)}$$

Without loss of generality, we can limit out study to nonnegative l_-, l_+. The class $L_1^1(\mathbb{R})$ corresponds to $l_- = l_+ = 0$. Now let $[p, q]$ be the class of impedance factors asymptotic to Cx^p as $x \to -\infty$, and to $C'x^q$ as $x \to +\infty$, where C, C' are unspecified positive constants, p, q are nonnegative. For $u \in L_1^1(\mathbb{R})$, it is not difficult to prove that the impedance factor is generally a linear combination of one in $[1, 0]$ and one in $[0, 1]$. There are, however, special potentials such that α goes to a limit at $x \to \pm\infty$. It has been proved (Degasperis and Sabatier, 1987) that α is then the Jost solution corresponding to a zero-energy bound state of this potential: (Hint: Notice that $\int_{-\infty}^{+\infty} (\alpha'^2 + V\alpha^2)dx = 0$). Indeed, there also exists in this case a zero-energy bound state of the impedance equation, since $p = 1$ obviously satisfies

$$\left\{ \begin{array}{l} \alpha^{-2} \dfrac{d}{dx} \alpha^2 \dfrac{dp}{dx} = 0, \\[3mm] \displaystyle\int_{-\infty}^{+\infty} dx \, \alpha^2 \left(\dfrac{dp}{dx}\right)^2 = 0. \end{array} \right. \qquad \textbf{(XVII.4.45)}$$

If a potential u is asymptotic to $p(p - 1)x^{-2}$ as $x \to +\infty$, there exists an impedance factor α such that $\alpha''/\alpha = u$ and going to infinity like Cx^p, and another impedance factor α obtained from the first one by the formula $\alpha(x)\int_x^\infty \alpha^{-2}(t)dt$ and going to zero like $C'x^{-p-1}$. The same results hold at $-\infty$. "Transforming" the impedance factors α and α yields impedance factors $\tilde{\alpha}$ and $\tilde{\alpha}$ which have almost everywhere an absolutely continuous derivative. The asymptotic behavior of the corresponding potentials is readily derived from that of the impedance factors in this sense that $q(q - 1)x^{-2}$ corresponds to x^q. Hence, we can derive the table of transformations of potential classes from that of transformations of impedance classes and we shall give only the latter one.

$$[0, 0] \overset{*}{\underset{}{\diagdown}} \begin{array}{l} \longrightarrow [0, 0] \\ \longrightarrow [-1, -1] \quad \text{with a pole on } \mathbb{R}, \end{array} \qquad \textbf{(XVII.4.46a)}$$

$$[1, 0] \left\{ \begin{array}{l} \nearrow [-1, 0] \\ * \\ \rightarrow [-1, -1] \quad \text{with a pole on } \mathbb{R}, \text{ if } c < \int_{-\infty}^{0} dt/\alpha^2(t), \\ \rightarrow [0, -1] \qquad \text{if } c = \int_{-\infty}^{0} dt/\alpha^2(t), \\ \searrow [-1, -1] \quad \text{if } c > \int_{-\infty}^{0} dt/\alpha^2(t), \end{array} \right.$$

(XVII.4.46b)

$$[0, 1] \left\{ \begin{array}{l} \nearrow [0, -1] \\ * \\ \rightarrow [-1, -1] \quad \text{if } c < \int_{0}^{\infty} dt/\alpha^2(t), \\ \rightarrow [-1, 0] \qquad \text{if } c = -\int_{0}^{\infty} dt/\alpha^2(t). \\ \searrow [-1, -1] \quad \text{with a pole on } \mathbb{R}, \text{ if } c > -\int_{0}^{\infty} dt/\alpha^2(t), \end{array} \right.$$

(XVII.4.46c)

if p, q are nonnegative integers

$$[-p, -q] \underset{*}{\overset{\nearrow [p, q]}{\searrow}} \quad [-(p+1), -(q+1)] \quad \text{with a pole on } \mathbb{R}.$$

(XVII.4.47)

If p, q, are integers ≥ 1

$$[p, q] \left\{ \begin{array}{l} \nearrow [-p, -q] \\ \nearrow [-p, -q] \qquad \text{if } c < -\int_{0}^{\infty} dt/\alpha^2(t), \\ * \\ \nearrow [-p, q-1] \quad \text{if } c = -\int_{0}^{\infty} dt/\alpha^2(t), \\ \rightarrow [-p, -q] \qquad \text{with a pole on } \mathbb{R}, \\ \qquad\qquad\qquad \text{if } -\int_{0}^{\infty} dt/\alpha^2(t) < c < \int_{-\infty}^{0} dt/\alpha^2(t), \\ \searrow [p-1, -q] \quad \text{if } c = \int_{-\infty}^{0} dt/\alpha^2(t), \\ \searrow [-p, -q] \quad \text{if } c > \int_{-\infty}^{0} dt/\alpha^2(t). \end{array} \right.$$

(XVII.4.8)

We should notice the values which break down the classes of impedances (or potentials) into unphysical impedances and "ghosts."

"Zero-energy bound states" corresponding to these cases and their characterization have been thoroughly discussed (Degasperis and Sabatier, 1986). The method which is discussed here is an example of a "geometrical" method, since it proceeds through geometrical transformations inside the "set of results" (here the set of reflection coefficients). Interest in this kind of method is in their generality (Sabatier, 1986). Particular methods using algebraic manipulations of the Schrödinger equation were often used to derive examples of ambiguities; most of them, if not all, can be generated by the geometrical method just discussed. To conclude this section we give an example without comment, and which is left to the reader as an exercise.

For the reflection coefficient

$$R_+(k) = -ik(1 - ik)^{-1}, \qquad \text{(XVII.4.49)}$$

there exist at least three equivalent potentials

$$V_1(x) = -2\,\delta(x) + 8(2x + 1)^{-2}\theta(x), \qquad \text{(XVII.4.50a)}$$

$$V_2(x) = -2\,\delta(x) + 8(1 - 2x)^{-2}\theta(-x), \qquad \text{(XVII.4.50b)}$$

$$V_3(x) = -2\,\delta(x) + 2(1 + |x|)^{-2}. \qquad \text{(XVII.4.50c)}$$

Equations (**XVII.4.50a and b**) are due to Abraham et al. (1981) and (**XVII.4.5c**) is due to Brownstein (1983).

XVII.4.3   Weakly Coupled Systems

Coupled or matrix problems can all be reduced to first-order systems but it is not always convenient to do so. Hence the simplest system is that of two first-order equations equivalent to the Schrödinger equation (we saw it in section XVII.2.3), or to the impedance equation. First, we shall see the impedance system, then a generalization of the Schrödinger equation to k-dependent potentials.

First-Order Impedance System. This system has been studied by Howard (1981) (who studied sound propagation) in the case of α continuous-positive, $\alpha' \in L^2(\mathbb{R}) \cap L'(\mathbb{R})$, $\alpha' = 0$ for $x \le 0$. The system is

$$\frac{d}{dx}\mathbf{u} = k\mathbf{Au}, \qquad \text{(XVII.4.51)}$$

where

$$\mathbf{u} = \begin{pmatrix} u_1 \\ u_2 \end{pmatrix}, \qquad \mathbf{A} = \begin{pmatrix} 0 & -\alpha^2 \\ \alpha^{-2} & 0 \end{pmatrix}. \qquad \text{(XVII.4.52)}$$

Setting $u_1 = \alpha^2(dp/dx)$, $u_2 = kp$, we easily get back to the impedance equation (**XVII.4.33**), and thus to the Schrödinger equation with a potential $V = \alpha''/\alpha$. The system (**XVII.4.51**) can also be written

$$\frac{d}{dx}\boldsymbol{\varphi} = ik\sigma\boldsymbol{\varphi} - \gamma\mathbf{Q}\boldsymbol{\varphi}, \qquad \text{(XVII.4.53)}$$

where

$$\sigma = \begin{pmatrix} 1 & 0 \\ 0 & -1 \end{pmatrix}, \qquad Q = \begin{pmatrix} 0 & 1 \\ 1 & 0 \end{pmatrix}, \qquad \gamma = \frac{\alpha'}{\alpha}. \qquad \text{(XVII.4.54)}$$

Notice that

$$e^{ik\sigma x} = \begin{pmatrix} e^{ikx} & 0 \\ 0 & e^{-ik\sigma x} \end{pmatrix}. \qquad \text{(XVII.4.55)}$$

The system (**XVII.4.53**) can be studied in a standard way by introducing the Jost solutions

$$\mathbf{f}_+(k, x) = e^{ik\sigma x}\begin{pmatrix} 1 \\ 0 \end{pmatrix} + \int_x^\infty dt\, \gamma(t)e^{ik\sigma(x-t)}Q\mathbf{f}_+(k, t), \qquad \text{(XVII.4.56a)}$$

$$\mathbf{f}_-(k, x) = e^{ik\sigma x}\begin{pmatrix} 0 \\ 1 \end{pmatrix} - \int_{-\infty}^x dt\, \gamma(t)e^{ik\sigma(x-t)}Q\mathbf{f}_-(k, t), \qquad \text{(XVII.4.56b)}$$

and $\bar{\mathbf{f}}_+$, $\bar{\mathbf{f}}_-$ which are given by the same equations with $\begin{pmatrix} 1 \\ 0 \end{pmatrix}$ and $\begin{pmatrix} 0 \\ 1 \end{pmatrix}$ interchanged. The scattering coefficients are introduced by the equality

$$\mathbf{f}_+ = T^{-1}(\bar{\mathbf{f}}_- + R_-\mathbf{f}_-) \qquad \text{(XVII.4.57)}$$

and the symmetric one ($R_-(k)$ is the only reflection coefficient of interest in this special half-axis problem). The usual analytic and asymptotic properties are derived. They justify the existence of a vectorial transformation kernel $H(x, y)$ (which in this case also vanishes for negative x because γ does)

$$\mathbf{f}_-(k, x) - e^{-ikx}\begin{pmatrix} 0 \\ 1 \end{pmatrix} = \int_{-\infty}^x dy\, \mathbf{H}(x, y)e^{-iky}. \qquad \text{(XVII.4.58)}$$

Clearly, $H(x, y)$ has been defined as a Fourier transform of the left-hand side of (**XVII.4.58**). Taking into account (**XVII.4.57**), a few calculations yield the vectorial Marchenko equation

$$\mathbf{H}(x, y) = \tilde{R}_-(x + y)\begin{pmatrix} 1 \\ 0 \end{pmatrix} - \int_{-y}^x Q\mathbf{H}(x, t)\tilde{R}_-(t - y)dt, \qquad \text{(XVII.4.59)}$$

where

$$\tilde{R}_-(x) = (2\pi)^{-1}\int_{-\infty}^{+\infty} R_-(k)e^{-ikx}\,dk \qquad \text{(XVII.4.60)}$$

then

$$\gamma(x) = 2H_1(x, x). \qquad \text{(XVII.4.61)}$$

With Howard assumptions, $|R_-(k)|$ is bounded by 1, which is reached only at $k = 0$, and the Neumann series converges to the solution of the Marchenko equation. Therefore $\alpha(x)$ can be reconstructed from $R_-(k)$ by solving equation (**XVII.4.59**) and using the first component of **H** to derive $\gamma(x)$.

Schrödinger Equation with k-Dependent Potential. The basic equations, and their generalizations, were mainly studied by Jaulent and Jean, who shows that there exists an inversion equation of the coupled form already introduced in section XV.1.2. We are led to study two equations simultaneously, namely,

$$\frac{d^2}{dx^2}y^+(k, x) + [k^2 - V^+(k, x)]y^+(k, x) = 0, \qquad \text{(XVII.4.62a)}$$

$$\frac{d^2}{dx^2} y^-(k, x) + [k^2 - V^-(k, x)]y^-(k, x) = 0, \qquad \text{(XVII.4.62b)}$$

where

$$V^\pm(k, x) = U(x) \pm 2kQ(x). \qquad \text{(XVII.4.63)}$$

We define,[1] for real k, the Jost solutions $f_+^\pm(k, x)$ which are asymptotic to e^{ikx} as $x \to \infty$, and those $f_-^\pm(k, x)$ which are asymptotic to e^{-ikx} as $x \to -\infty$. Clearly $f_+^+(k, x)$ and $f_+^-(-k, x)$ are linearly independent solutions of (XVII.4.62a), as well as $f_-^+(k, x)$ and $f_-^-(-k, x)$. The same remarks holds for the functions f^- and (XVII.4.62b). If we keep in mind the general scattering schemes expressed by (XVII.1.3) and (XVII.1.4), we are led to define the scattering coefficients of (XVII.4.62a,b) by

$$R_-^\pm(k)f_-^\pm(k, x) + f_-^\mp(-k, x) = T^\pm(k)f_+^\pm(k, x), \qquad \text{(XVII.4.64a)}$$

$$T^\pm(k)f_-^\pm(k, x) = R_+^\pm(k)f_+^\pm(k, x) + f_+^\mp(-k, x), \qquad \text{(XVII.4.64b)}$$

where the upper indices $\pm$ correspond to the cases a and b in (XVII.4.62), and the lower indices are the ones we have used in other parts of chapter XVII. These formulas enable us to express the scattering coefficients in the form of Wronskians of the Jost solutions and, to show that, if U and Q are real,

$$T^\pm(k) = \overline{T^\mp(-k)}, \qquad \text{(XVII.4.65a)}$$

$$R_\pm^\mp(k) = \overline{R_\pm^\mp(-k)} \qquad (k \in \mathbb{R}), \qquad \text{(XVII.4.65b)}$$

$$[S^\mp(k)^t]S^\mp(-k) = I, \qquad \text{(XVII.4.65c)}$$

where "t" means transposed, I is the identity matrix, and

$$S^\pm(k) = \begin{pmatrix} T^\pm(k) & R_+^\pm(k) \\ R_-^\pm(k) & T^\pm(k) \end{pmatrix} \qquad \text{(XVII.4.65d)}$$

The Jost solutions have a "coupled representation" (similar to (XV.1.36))

$$f_+^\pm(k, x) = F_+^\pm(x)e^{ikx} + \int_x^\infty K_+^\pm(x, t)e^{ikt}\, dt, \qquad \text{(XVII.4.66b)}$$

$$f_-^\pm(k, x) = F_-^\pm(x)e^{-ikx} + \int_{-\infty}^x K_-^\pm(x, t)e^{-ikt}\, dt. \qquad \text{(XVII.4.66b)}$$

Inserting (XVII.4.66) into the usual Volterra integral equations for the Jost solutions defined as in section XVII.1 yields

$$F_+^\pm(x) = \exp\left[\pm i \int_x^\infty Q(t)dt \right],$$

$$\qquad \qquad \qquad \qquad \text{(XVII.4.67)}$$

$$F_-^\pm(x) = \exp\left[\pm i \int_{-\infty}^x Q(t)dt \right],$$

---

[1] Jaulent and Jean's original definitions for f_1 and f_2 are the complex conjugate of ours for f_+ and f_-, so that the analytic properties of their functions are in Im $k \le 0$.

and the integral equation (written here for K_+, F_+, only)

$$K_{\mp}^{\pm}(x, t) = \frac{1}{2} \int_{(x+t)/2}^{\infty} F_{\mp}^{\pm}(y) U(y) dy,$$

$$+ \frac{1}{2} \int_{x}^{(x+t)/2} U(y) dy \int_{t+x-y}^{t+y-x} K_{\mp}^{\pm}(y, u) du,$$

$$+ \frac{1}{2} \int_{(x+t)/2}^{\infty} U(y) dy \int_{y}^{t+y-x} K_{\mp}^{\pm}(y, u) du,$$

$$\mp \frac{1}{2} i F_{\mp}^{\pm} \left(\frac{x+t}{2} \right) Q\left(\frac{x+t}{2} \right) \pm i \int_{x}^{\infty} Q(y) K_{\mp}^{\pm}(y, t + y - x) dy,$$

$$\mp i \int_{x}^{(x+t)/2} Q(y) K_{\mp}^{\pm}(y, t + x - y) dy, \qquad t \geq x, \quad x \in \mathbb{R},$$

$$\text{(XVII.4.68)}$$

whose Neumann series converges, proving the existence and uniqueness of solutions, and hence the representation (**XVII.4.66a**). From (**XVII.4.68**), differentiations yield the partial differential equation

$$\left[\frac{\partial^2}{\partial x^2} - \frac{\partial^2}{\partial t^2} - U(x) \mp 2iQ(x) \frac{\partial}{\partial t} \right] K_{\mp}^{\pm}(x, t) = 0, \qquad \text{(XVII.4.69)}$$

and the equation

$$U(x) = f_+^{\pm}(x)[F_+^{\pm}(x)]^{-1}, \qquad \text{(XVII.4.70)}$$

where

$$f_+^{\pm}(x) = \frac{d^2}{dx^2} F_+^{\pm}(x) - 2 \frac{d}{dx} K_+^{\pm}(x, x) + 2K_+^{\pm}(x, x)(F_+^{\pm}(x))^{-1} \frac{d}{dx} F_+^{\pm}(x).$$

$$\text{(XVII.4.71)}$$

Similar results are true for K_- and F_-. Let us now state the principal results on the direct and inverse problems for (**XVII.4.62a**). Assume that U and Q satisfy the following conditions, which define a class $\mathbb{C}$:

A1. $U(x)$ $(x \in \mathbb{R})$ is real, continuously differentiable, and $x^2 U(x)$ and $x U'(x)$ are integrable in $\mathbb{R}$.

A2. $Q(x)$ $(x \in \mathbb{R})$ is real, continuously differentiable, goes to zero as $|x| \to \infty$, and $x^2 Q'(x)$ and $x Q''(x)$ are integrable in $\mathbb{R}$.

A3. The function $[T^+(k)]^{-1} = (2ik)^{-1} W[f_-^+(k, x), f_+^+(k, x)]$ has no zero for Im $k > 0$, i.e., there is no bound state for equation (**XVII.4.62a**)

It is possible to prove that the matrix $S^+(k)$ (see (**XVII.4.65d**))

(1) is unitary for every $k \in \mathbb{R}$;

(2) is a continuous function of k, $k \in \mathbb{R}$;

(3) $T^+(k)$ can be continued into a holomorphic function in Im $k \geq 0$, and does not vanish for any $k \neq 0$;

(4) if $T^+(k) = 0$, $k^{-1} T^+(k)$, $k^{-1}[1 + R_+^+(k)]$, $k^{-1}[1 - R_-^+(k)]$, go to finite limits as $k \to 0$ $(k \in \mathbb{R})$;

(5) there exists a complex number of modulus one $F^+(k)$ such that $T^+(k) - F^+(k) = O(|k|^{-1})$, (Im $k \geq 0$), $R_+^+(k) = O(|k|^{-1})$, $R_-^\pm(k) = O(|k|^{-1})$, ($k \in \mathbb{R}$);

(6) the functions $r_+^+(t)$ and $r_-^\pm(t)$ defined (at least in L^2) by

$$r_+^+(t) = -\frac{1}{2\pi} \int_{-\infty}^{+\infty} \overline{R_+^+(k)} e^{ikt}\, dk, \qquad \text{(XVII.4.72a)}$$

$$r_-^\pm(t) = -\frac{1}{2\pi} \int_{-\infty}^{+\infty} R_-^\pm(k) e^{ikt}\, dk, \qquad \text{(XVII.4.72b)}$$

are twice differentiable for $t \in \mathbb{R}$. Furthermore, x_0 being any real number, the functions $t^2\, r_+^{+\prime}(t)$ and $tr_+^{+\prime\prime}(t)$ are integrable in $[x_0, \infty[$, and the functions $t^2\, r_-^{\pm\prime}(t)$ and $tr_-^{\pm\prime\prime}(t)$ are integrable in $]-\infty, x_0]$.

These properties define a class $\mathscr{S}$ of scattering matrices, and a class $\mathscr{R}$ of reflection coefficients (which must be in the correct position in the scattering matrix). In addition, the representation (**XVII.4.66**) holds (together with (**XVII.4.67**), (**XVII.4.68**) and similar ones), and the transformation kernels are related to r_+^+ and $r_-^\pm$ by the "coupled" Marchenko equation (written here only for r_+^+, K_+^+)

$$K_+^+(x, t) = F_+^+(x) r_+^+(x + t) + \int_x^\infty r_+^+(t + u)\overline{K_+^+(x, u)}\,du, \quad \text{(XVII.4.73a)}$$

$$f_+^+(x)\overline{F_+^+(x)} = \overline{f_+^+(x)} F_+^+(x) \qquad (x \in \mathbb{R}), \qquad \text{(XVII.4.73b)}$$

where $f_+^+(x)$ has been defined by (**XVII.4.71**).

Conversely, suppose $R_+^+(k)$ is given on the whole real axis (see (**XVII.4.65**) to realize what this information means), and suppose that it belongs to the class $\mathscr{R}$, it is possible to show that U and Q are uniquely determined, and can be reconstructed through the following steps:

(1) Calculate $r_+^+(x)$ from $R_+^+(k)$ by means of (**XVII.4.72a**).

(2) Solve the Fredholm equations

$$a_+^+(x, t) = r_+^+(x + t) + \int_x^\infty r_+^+(t + u) a_+^+(x, u)\,du,$$

$$\text{(XVII.4.74)}$$

$$b_+^+(x, t) = -ir_+^+(x + t) + \int_x^\infty r_+^+(t + u) b_+^+(x, u)\,du.$$

Then, setting

$$\alpha_+(x, t) = \frac{a_+^+(x, t) - ib_+^+(x, t)}{2}, \qquad \text{(XVII.4.75a)}$$

$$\beta_+(x, t) = \frac{a_+^+(x, t) + ib_+^+(x, t)}{2}, \qquad \text{(XVII.4.75b)}$$

it is readily seen that

$$K_+^+(x, t) = \overline{F_+^+(x)}\alpha_+(x, t) + F_+^+(x)\beta_+(x, t) \qquad \text{(XVII.4.76)}$$

solves (**XVII.4.73a**). Then, setting $F_+^+(x) = \exp[\tfrac{1}{2}iz(x)]$ and inserting it, with

(XVII.4.76), into **(XVII.4.73b)** we get a second-order differential equation for z, which can readily be integrated once, thanks to the boundary condition $z'(\infty) = 0$, and which yields the equation

$$z' + 2i[\alpha_+ e^{-iz} - \overline{\alpha_+}\, e^{iz}] + 2i[\overline{\beta}_+ - \beta_+] = 0,$$
$$z(\infty) = 0.$$

$$\textbf{(XVII.4.77)}$$

The Cauchy–Lipschitz theorem enables us to prove that this equation has a solution and that it is unique and can be constructed by an iterative algorithm.

More generally, the whole consistency of the method has been proved (Jaulent and Jean, 1976). Generalizations to cases where bound states are present, to cases where Q is pure imaginary, and to k-polynomial dependances of the potential, are also available, and applications to inverse problems in absorbing media (Jaulent, 1976).

XVII.4.4   General Matrix Systems and the Inverse Method

The Inverse Method. The matrix systems presented here have been studied almost only in connection with the inverse method and they cannot be considered "inverse problems of quantum scattering theory." Hence we shall only sketch the main results, doing it because they are natural generalizations of those obtained in the remainder of the book. So as to understand what the inverse method is, the simplest way is to consider a "scattering problem" described by the equation

$$L\psi = k^2\psi \qquad (\psi \in \mathcal{B}),$$

$$\textbf{(XVII.4.78)}$$

where the linear operator L depends on $V (V \in \mathcal{V})$, and is related to a "spectral measure" $d\rho(k)$, whose support in the complex plane is Σ. We assume that the "parameter" V determines ρ (direct problem, say, $\rho = \mathcal{M}(V)$), and that it can itself be reconstructed from ρ (inverse problem, say, $V = \mathcal{M}^{-1}(\rho)$) in a well-defined way, which is related to the structure of L. Hence, to a one-parameter family $V(t)$, there corresponds bijectively a one-parameter family $\rho(t)$. If $\rho(t)$ is known, $V(t)$ can be constructed by the inversion formula

$$V(t) = \mathcal{M}^{-1}[\rho(t)].$$

$$\textbf{(XVII.4.79)}$$

A particularly simple case is met when the family $V(t)$ is isospectral, i.e., it keeps Σ and the nature of the measure $d\rho(k)$ on Σ, modifying only the weights of $d\rho(k)$—for example, in the problems of this book—a "scattering coefficient" $r(k)$ on a the "continuous spectrum," and "normalizing parameters" r_i at the eigenvalues. We shall call $d\rho(k)$, or, in the isospectral case, the weights $r(k)$ and r_i defined on Σ, the "spectral data." Their dependence on t can be called their "evolution" on a "trajectory" and $V(t)$ is the corresponding evolution of V. Since this evolution preserves the support of the point spectrum, we expect that the trajectories $V(t)$, that correspond to a pure point spectrum and especially those corresponding to one eigenvalue only, will be particularly interesting. They are often (improperly) called solitons.

To be more specific, we can consider the case $L(t) = -\partial^2/\partial x^2 + V(x, t)$, which is the one-dimensional Schrödinger operator studied in section XVII.1. It is easy to see that $L(t)$ is isospectral if there exists a unitary operator N acting in B such that $\psi(x, t) = N(t)\psi(x, 0)$ and $L(t) = N(t)L(0)N^{-1}(t)$. Differentiating these equalities yields the Lax equations

$$\frac{\partial V}{\partial t} = \left[B, -\frac{\partial^2}{\partial x^2} + V(x, t) \right], \qquad \text{(XVII.4.80a)}$$

$$\frac{\partial \psi(x, t)}{\partial t} = B\psi(x, t), \qquad \text{(XVII.4.80b)}$$

where $[\cdot, \cdot]$ is the "commutator" and $B = (\partial N/\partial t)N^{-1}$ should be an antisymmetric operator such that the left-hand side of (XVII.4.80) is a multiplicative operator. Setting $V(x, t) = -\frac{1}{6}u(x, t)$, the simplest nontrivial example is

$$B = -4\frac{\partial^3}{\partial x^3} - u\frac{\partial}{\partial x} - \frac{1}{2}\frac{\partial u}{\partial x}. \qquad \text{(XVII.4.81)}$$

It yields the trajectory of u by means of (XVII.4.80a) with the result

$$\frac{\partial u}{\partial t} + u\frac{\partial u}{\partial x} + \frac{\partial^3 u}{\partial x^3} = 0. \qquad \text{(XVII.4.82)}$$

The trajectory of R_+, say, can be derived from (XVII.4.80b). Applying this equation to the asymptotic behavior of $\psi(x, t, k)$

$$\psi(x, t, k) \sim \begin{cases} A^-(t)e^{ikx} + B^-(t)e^{-ikx}, & x \to -\infty, \\ A^+(t)e^{ikx} + B^+(t)e^{-ikx}, & x \to +\infty, \end{cases} \qquad \text{(XVII.4.83)}$$

since $B \sim -4\,\partial^3/\partial x^3$ as $x \to \pm\infty$, we find that

$$A^\pm(t) = A^\pm(0)e^{4ik^3t}, \qquad B^\pm(t) = B^\pm(0)e^{-4ik^3t}. \qquad \text{(XVII.4.84)}$$

Starting from $A^-(0) = 0$, $B^-(0) = T(k)$, $A^+(0) = R_+(k)$, $B^+(0) = 1$, and dividing the result by $B^+(t)$, to write again the scattering function, we obtain

$$R_+(k, t) = R_+(k)e^{8ik^3t}. \qquad \text{(XVII.4.85a)}$$

The trajectory of $(C_p^+)^2$ is easy to obtain since the unitary transformation N takes from a normalized eigenfunction $\psi(iK_p, x, 0)$ to a normalized eigenfunction $\psi(iK_p, x, t)$. Since this function is asymptotic to $[C^\pm(t)]^{-1}\exp[\pm|K_p|x]$ at $x \to \pm\infty$, applying (XVII.4.80b) readily yields

$$\rho_p^\pm(t) = [C_p^\pm(t)]^{-2} = [C_p^\pm(0)]^{-2}\exp[\pm 8|K|^3 t]. \qquad \text{(XVII.4.85b)}$$

The spectral data trajectory, given by (XVII.4.85b), is particularly simple and our goal is reached: using the inversion formulas of section XVII.2 gives a solution of the nonlinear equation (XVII.4.82) as the image of the spectral data trajectory. We can even solve the so-called "Cauchy problem," i.e., we can start from any $u(x, 0)$ in L_1^1, then calculate $\{R_+(k), \rho_p^+(0)\}$, which readily yield $\{R_+(k, t), \rho_p^+(t)\}$, and calculate $u(x, t)$ by means of the Marchenko equation. We can also generate exact solutions. The simplest one corresponds to $R_+(k) = 0$ (transparent potential) and one eigenvalue only (K, ρ). According to (**XVII.2.14**) we have

$$u(x, t) = \frac{12K^2}{\{\cosh[K(x - x_0(t))]\}^2}, \qquad \text{(XVII.4.86a)}$$

where

$$x_0(t) = (2K)^{-1}[\log(\rho/2K) + 4K^2 t]. \qquad \text{(XVII.4.86b)}$$

Hence $u(x, t)$ propagates with a constant velocity and without deformation. It is a solitary wave. The waves that correspond to several eigenvalues have a different velocity, cross each other, and at large distances are separated and propagate like solitary waves. Solitary waves that can cross each other (collide), and at large distances recover their shape and propagate without deformation are called "solitons." But this has also become a generic name for solutions associated with the point spectrum.

Equation (**XVII.4.82**) is the Korteweg–de Vries equation, the simplest nonlinear equation that can be solved by the inverse method of the Schrödinger one-dimensional problem. An infinity of other equations (belonging to the so-called KdV hierarchy) can be obtained by choosing other operators B. In the next part of this section, the inverse method exists but will not be described. It enables us to solve a large number of interesting nonlinear equations of mathematical physics (the nonlinear Schrödinger equation, the Sine–Gordon equation, the Boussinesq equation, the Burgers equation, the Kadomtsev–Petiashvili equation, etc.) which appear in modeling important physical problems, and an infinity of other equations which belong to their "hierarchies." In order to solve the Cauchy problem for these equations, the inverse problems should have been studied with the same care as has been done in the Schrödinger one-dimensional case. Unfortunately, this has not always been true, because many authors were mostly interested in special solutions (solitons or their analogues) of the nonlinear equations.

A few years ago, the exact solvability of interesting nonlinear partial differential equations, by a method which was nothing more than a wide generalization of the Fourier method, was a great surprise. It has now been understood that these special nonlinear equations are limit forms of very general models of wave propagation, in one or several dimensions, so that it is sufficient that one of these models is solvable to warrant that the limit form is solvable too (Calogero and Eckans, 1987; Calogero and Sabatier, 1987).

The Matrix Schrödinger Problem. The basic equation is

$$\frac{d^2\psi}{dx^2} = [Q - k^2]\psi, \qquad \text{(XVII.4.87)}$$

where the $N \times N$ matrix Q depends on the real variable x, and the N-vector ψ depends on x and k. The best-known case is that where Q is Hermitian

$$Q^+ = Q \qquad \text{(XVII.4.88)}$$

which is the only one where:

(a) eigenvalues $k^{(j)}$ are purely imaginary, so that their square, i.e., the actual eigenvalue, is real and negative;

(b) the direct and inverse spectral problems are fully in hand.

We give results only with this assumption (although many results hold in other cases). We also assume, for the sake of simplicity, that the matrix $\mathbf{Q}(x)$ is finite valued (for real x) and that it vanishes asymptotically exponentially or faster, i.e., we assume that, for some positive ε,

$$\lim_{x \to \pm \infty} [\exp[\varepsilon|x|]\mathbf{Q}(x)] = 0. \tag{XVII.4.89}$$

The continuous part of the spectrum associated with the matrix Schrödinger problem (**XVII.4.87**) is characterized by the asymptotic conditions

$$\psi(k, x) = \begin{cases} \mathbf{I} \exp[-ikx] + \mathbf{R}_+(k)\exp[ikx] + O(1) & (x \to \infty) \\ \mathbf{T}(k)\exp[-ikx] + O(1) & (x \to -\infty) \end{cases} \quad (k \in \mathbb{R}^+), \tag{XVII.4.90}$$

where $\psi(k, x)$ is a matrix solution of (**XVII.4.87**), i.e., a matrix built out of N-vector solutions, used as columns of the $N \times N$ matrix ψ. The "reflection" coefficient $\mathbf{R}$ and the "transmission coefficient" $\mathbf{T}$ are matrices of rank N.

The discrete part of the spectrum consists of a finite number of purely imaginary eigenvalues

$$k^{(j)} = iK^{(j)} \tag{XVII.4.91}$$

whose corresponding eigenfunctions satisfy the Schrödinger equation

$$\frac{d^2 \psi^{(j)}}{dx^2} = [\mathbf{Q} + (K^{(j)})^2]\psi^{(j)} \tag{XVII.4.92}$$

and the normalization condition

$$\int_{-\infty}^{+\infty} dx \langle \psi^{(j)}, \psi^{(j)} \rangle = 1. \tag{XVII.4.93}$$

The asymptotic behavior of any discrete spectrum eigenfunction is clearly

$$\psi(x) \xrightarrow[x \to +\infty]{} \mathbf{c}_+ \exp[-px], \tag{XVII.4.94a}$$

$$\psi(x) \xrightarrow[x \to -\infty]{} \mathbf{c}_- \exp[px], \tag{XVII.4.94b}$$

and it introduces the two N-vectors $\mathbf{c}^+$ and $\mathbf{c}^-$. The discrete eigenvalues $k^{(j)}$ correspond to the poles of $\mathbf{T}(k)$ in the upper half-plane, where this function of k is meromorphic. We assume they are all simple.

If $\mathbf{Q}(x)$ is a given matrix of rank N, the matrices $\mathbf{R}_+(k)$ and $\mathbf{T}(k)$ and the vectors $K^{(j)}$, $\mathbf{c}_+^{(j)}$, $\mathbf{c}_-^{(j)}$, characterizing the discrete part of the spectrum (if any), are uniquely determined. Such a determination constitutes the direct one-dimensional matrix Schrödinger problem.

Conversely, the input data that are sufficient to determine $\mathbf{Q}(x)$ uniquely are the reflection coefficient $\mathbf{R}_+(k)$, the discrete eigenvalues $K^{(j)}$ (if any), and the corresponding matrices $\mathbf{c}^{(j)}\mathbf{c}^{(j)}$ (dyadic notation). We report here the

Marchenko formula that solves the reconstruction problem, referring to Wadati and Kamijo (1974) for all details

$$\mathbf{M}(x) = \sum_j \mathbf{c}^{(j)} \mathbf{c}^{(j)+} \exp[-K^{(j)}x] + (2\pi)^{-1} \int_{-\infty}^{+\infty} dk\, \mathbf{R}_+(k) e^{ikx}, \quad \text{(XVII.4.95a)}$$

$$\mathbf{K}(x, y) + \mathbf{M}(x + y) + \int_x^\infty dt\, \mathbf{K}(x, t) \mathbf{M}(t + y) = 0, \quad \text{(XVII.4.95b)}$$

$$\mathbf{Q}(x) = -2 \frac{d}{dx} \mathbf{K}(x, x). \quad \text{(XVII.4.96)}$$

In our case of Hermitian $\mathbf{Q}(x)$, $\mathbf{R}_+(-k) = \mathbf{R}_+(k)$, so that only the values of $\mathbf{R}_+(k)$ for positive k are needed.

The nonlinear partial differential equations that can be solved by the inverse method associated with the matrix Schrödinger equation have been studied mainly by Calogero and Degasperis (1976, 1977) who showed the curious properties of some of the corresponding "solitons" (boomerons, zoomerons, etc.).

The Zakharov–Shabat Problem. The basic equation is

$$\frac{\partial}{\partial x} \mathbf{F}(k, x) = (-ik\boldsymbol{\sigma} + \mathbf{Q}) \mathbf{F}(k, x), \quad \text{(XVII.4.97)}$$

where

$$\boldsymbol{\sigma} = \begin{pmatrix} 1 & 0 \\ 0 & -1 \end{pmatrix}, \quad \mathbf{Q}(x) = \begin{pmatrix} 0 & q^{(-)}(x) \\ q^{(+)}(x) & 0 \end{pmatrix}, \quad \text{and} \quad \mathbf{F} = \begin{pmatrix} F_{11} & F_{12} \\ F_{21} & F_{22} \end{pmatrix}$$

is a "matrix solution" whose columns are vector solutions of equation (XVII.4.97). k is the spectral parameter. The "potential" $\mathbf{Q}$ satisfies the condition

$$\int_{-\infty}^{+\infty} dx\, |q^\pm(x)| < \infty. \quad \text{(XVII.4.98)}$$

The Jost solutions $\mathbf{F}^{(\pm)}(k, x)$ are defined by their asymptotic behavior as $x \to \pm\infty$, for real k,

$$\mathbf{F}^{(\pm)}(k, x) \xrightarrow[x \to \pm\infty]{} \exp[-ik\boldsymbol{\sigma}x], \quad \text{(XVII.4.99)}$$

where $\exp[-ik\boldsymbol{\sigma}x] = \begin{pmatrix} e^{-ikx} & 0 \\ 0 & e^{ikx} \end{pmatrix}$. It is clear that $\det \mathbf{F}^{(\pm)}(k, x) = 1$ is constant. The S-matrix is defined from the Jost solutions by their linear dependence on each other (each one defines a complete system of solutions of (XVII.4.97)

$$\mathbf{F}^{(-)}(k, x) = \mathbf{F}^{(+)}(k, x) \mathbf{S}(k). \quad \text{(XVII.4.100)}$$

It is clear that $\det \mathbf{S}(k) = 1$ ("unitarity"). The scattering solution, defined in the usual way (see (XVII.1.3), (XVII.1.4)), is related to the Jost solutions

$$\Psi(k, x) = \mathbf{F}^{(-)}(k, x) \mathbf{t}(k) = \mathbf{F}^{(+)}(k, x)[1 + \mathbf{r}(k)], \quad \text{(XVII.4.101)}$$

where

$$\mathbf{r}(k) = \begin{pmatrix} 0 & r^{(-)}(k) \\ r^{(+)}(k) & 0 \end{pmatrix}$$

is the reflection matrix, and

$$\mathbf{t}(k) = \begin{pmatrix} t^{(+)}(k) & 0 \\ 0 & t^{(-)}(k) \end{pmatrix}$$

is the transmission matrix. Comparing (XVII.4.100) and (XVII.4.101) yields the expression of $\mathbf{S}(k)$

$$\mathbf{S}(k) = [\mathbf{I} + \mathbf{r}(k)]^{-1}\mathbf{t}(k) = \begin{pmatrix} 1/t^{(+)} & r^{(-)}/t^{(-)} \\ r^{(+)}/t^{(+)} & 1/t^{(-)} \end{pmatrix}. \quad \text{(XVII.4.102)}$$

The "unitarity" property reads, in terms of scattering coefficients,

$$r^{(+)}(k)r^{(-)}(k) + t^{(+)}(k)t^{(-)}(k) = 1. \quad \text{(XVII.4.103)}$$

The Jost solutions can also be defined by the Volterra integral equation

$$\mathbf{F}^{(\pm)}(k, x) = \exp[-ikx\sigma] + \int_{\pm\infty}^{x} dy \exp[ik(y-x)\sigma]\mathbf{Q}(y)\mathbf{F}^{(\pm)}(k, y), \quad \text{(XVII.4.104)}$$

which easily follows from (XVII.4.97) and (XVII.4.99) (noticing that $\mathbf{Q}(y) \times \exp[-iky\sigma] = \exp[iky\sigma]\mathbf{Q}(y)$). Combining (XVII.4.104) with (XVII.4.101) we obtain two alternative Volterra equations for the scattering solution Ψ and, by letting x go to $\pm\infty$, the integral representation (for $k \in \mathbb{R}$)

$$\mathbf{t}(k) - \mathbf{r}(k) = \mathbf{I} - \int_{-\infty}^{+\infty} dx \exp[ikx\sigma]\mathbf{Q}(x)\Psi(k, x). \quad \text{(XVII.4.105)}$$

Since $\mathbf{t}$ is diagonal and $\mathbf{r}$ is antidiagonal, relation (XVII.4.105) yields two separate integral expressions for $\mathbf{t}(k)$ and $\mathbf{r}(k)$, which do not need to be written here.

In addition to the continuous spectrum, the Zakharov–Shabat spectral problem may (or may not) possess a discrete spectrum, which is the set of $N^{(+)} + N^{(-)}$ eigenvalues $k = k_n^{(+)}$, $n = 1, 2, \ldots, N^{(+)}$, and $k = k_n^{(-)}$, $n = 1, 2, \ldots, N^{(-)}$, with

$$\pm \text{Im } k_n^{(\pm)} > 0, \qquad n = 1, 2, \ldots, N^{(\pm)}. \quad \text{(XVII.4.106)}$$

For each eigenvalue $k_n^{(\pm)}$ there exists a vector eigenfunction $\varphi_n^{(\pm)}(x)$ such that

$$\frac{\partial}{\partial x} \varphi_n^{(\pm)}(x) = [\mathbf{Q}(x) - ik_n^{(\pm)}\sigma]\varphi_n^{(\pm)}(x), \quad \text{(XVII.4.107a)}$$

$$\varphi_n^{(\pm)}(x) \to \gamma_n^{(\pm)} \exp[\pm ik_n^{(\pm)}x]\chi_{\mp} \qquad (x \to +\infty), \quad \text{(XVII.4.107b)}$$

$$\varphi_n^{(\pm)}(x) \to \delta_n^{(\pm)} \exp[\pm ik_n^{(\pm)}x]\chi_{\pm} \qquad (x \to -\infty), \quad \text{(XVII.4.107c)}$$

$$\int_{-\infty}^{+\infty} dx \, [\varphi_n^{(\pm)}(x)]^T \mathbf{q}\varphi_n^{(\pm)}(x) = 1, \quad \text{(XVII.4.107d)}$$

where

$$\chi_+ = \begin{pmatrix} 1 \\ 0 \end{pmatrix}, \qquad \chi_- = \begin{pmatrix} 0 \\ 1 \end{pmatrix}, \qquad q = \begin{pmatrix} 0 & 1 \\ 1 & 0 \end{pmatrix}. \qquad \text{(XVII.4.108)}$$

We shall exclude from our study "exceptional cases" where the integral on the left-hand side of (XVII.4.107d) is zero and hence the eigenvector is isotropic.

From the integral equation (XVII.4.104), it is possible to obtain the analytic properties of $F^{(\pm)}$ as functions of k. The simplest way is to consider separately the first column $F^{(\pm)}\chi_+$ and then the second one $F^{(\pm)}\chi_-$, and then to use the standard argument (solution obtained by a series of iterative terms, all analytic, and absolutely bounded by a numerical convergent series). The result is that the vector solutions $F^{(+)}(k, x)\chi_+$ and $F^{(-)}(k, x)\chi_-$ are analytic for Im $k < 0$, while the vector solutions $F^{(+)}(k, x)\chi_-$ and $F^{(-)}(k, x)\chi_+$ are analytic for Im $k > 0$. The condition (XVII.4.98) also implies continuity on the real axis, but analyticity in a strip including the real axis is obtained only if the functions $q^{(\pm)}(x)$ vanish (at least) exponentially as $|x| \to \infty$

$$\lim_{|x| \to \infty} [\exp(+\varepsilon|x|)q^{(\pm)}(x)] = 0 \qquad (\varepsilon > 0). \qquad \text{(XVII.4.109)}$$

From these results and from formula (XVII.4.100) we easily infer the analyticity of $[(t^{(+)}(k)]^{-1}$ in the upper half k-plane, where its zeros coincide with the discrete eigenvalues $k_n^{(+)}$, and that of $[t^{(-)}(k)]^{-1}$ in the lower half k-plane, where its zeros coincide with the discrete eigenvalues $k_n^{(-)}$. In the following, we shall assume these zeros are simple. If they are, we can easily calculate the residues of the corresponding poles of $t^{\pm}(k)$

$$\lim_{k \to k_n^{(\pm)}} [(k - k_n^{(\pm)})t^{(\pm)}(k)] = \pm i\gamma_n^{(\pm)}\delta_n^{(\pm)}. \qquad \text{(XVII.4.110)}$$

The reflection coefficients are continuous on the real axis. They can be continued as meromorphic functions in a strip if condition (XVII.4.109) is fulfilled. More important, if this condition is fulfilled (no matter how small ε is), the numbers $N^{(+)}$ and $N^{(-)}$ of discrete eigenvalues are finite because they cannot accumulate to a point on the real axis.

Thus, the scattering coefficients $r^{(\pm)}(k)$, $t^{(\pm)}(k)$, the eigenvalues $k_n^{(\pm)}$, and the normalizing constants $\gamma_n^{(\pm)}$, $\delta_n^{(\pm)}$ are "spectral data" which can be determined from $Q(x)$ by solving the "direct problem." They are not independent (unitarity, symmetries) and we can expect that giving the reflection coefficients, the eigenvalues, and the set of numbers $\rho_n^{(\pm)} = [\gamma_n^{(\pm)}]^2$ is sufficient to reconstruct $Q(x)$. This is true, and the reconstruction can again be done by means of the Marchenko equation

$$K(x, y) + M(x + y) + \int_x^\infty dz\, K(x, z)M(z + y) = 0 \qquad (x \leq y),$$

$$\text{(XVII.4.111)}$$

where

$$M(z) = \sum_{n=1}^{N^{(+)}} \rho_n^{(+)} \exp[ik_n^{(+)}z] \begin{bmatrix} 0 & 0 \\ 1 & 0 \end{bmatrix},$$

$$+ \sum_{n=1}^{N^{(-)}} \rho_n^{(-)} \exp[-ik_n^{(-)}z] \begin{bmatrix} 0 & 1 \\ 0 & 0 \end{bmatrix},$$

$$+ (2\pi)^{-1} \int_{-\infty}^{+\infty} dk \, \mathbf{r}(k) \exp[ikz\sigma]. \qquad \text{(XVII.4.112)}$$

The solution $\mathbf{K}(x, y)$ of this (matrix) Fredholm equation enables us to construct $\mathbf{Q}(x)$. Let $\mathbf{K}^{(a)}(x, y)$ be the off-diagonal part of $\mathbf{K}(x, y)$. It is not difficult to prove that

$$\mathbf{Q}(x) = -2 \lim_{\varepsilon \to 0^+} \mathbf{K}^{(a)}(x, x + \varepsilon) \equiv -2\mathbf{K}^{(a)}(x, x^+). \qquad \text{(XVII.4.113)}$$

Moreover, the diagonal part $\mathbf{K}^{(c)}(x, y)$ is also related to $\mathbf{Q}(x)$, via the formula

$$\mathbf{K}^{(c)}(x, x^+) = \frac{1}{2} \int_x^\infty dy \, [\mathbf{Q}(y)]^2 = \frac{1}{2} \int_x^\infty dy \, q^{(+)}(y) q^{(-)}(y)\mathbf{I}. \qquad \text{(XVII.4.114)}$$

$\mathbf{K}(x, y)$ is also a transformation kernel which enables us to construct the Jost solution $\mathbf{F}^{(+)}$

$$\mathbf{F}^{(+)}(k, x) = \exp[ikx]\left[\mathbf{I} + \int_0^\infty dy \, \mathbf{K}(x, x + y)\exp[iky]\right]\chi_-. \qquad \text{(XVII.4.115)}$$

Technical details and further results on the Zacharov–Shabat problem are given in Calogero and Degasperis (1989) and in the references of section XVII.5 below. In particular, a Levinson theorem for the eigenvalues, several Backlund and Darboux transformations, are available. The main reason which decided us to give the main results here is to show their similarity with those of the Schrödinger one-dimensional problems, although the operator is not self-adjoint! It is of interest to notice that the Schrödinger problem can be considered a special case of the Zakharov–Shabat problem. This is true even for matricial cases and Jaulent–Jean generalizations. More simply, if $q = q^+ = q^-$, setting

$$\psi = \tfrac{1}{2}(F_{11}^+ + F_{21}^+), \qquad \phi = \tfrac{1}{2}(F_{11}^+ - F_{21}^+) \qquad \text{(XVII.4.116)}$$

shows that ψ and ϕ should be solutions of Schrödinger equations

$$\frac{d^2\psi}{dx^2} = (-k^2 + q^2 + q')\psi, \qquad \text{(XVII.4.117a)}$$

$$\frac{d^2\phi}{dx^2} = (-k^2 + q^2 + q')\phi, \qquad \text{(XVII.4.117b)}$$

and that $0 = ik^{-1}(\psi' - q\psi)$ (this formula and (XVII.4.17) show that ψ and ρ are related by the limit transformation of section XVII 4.3.

Then, playing with the asymptotic behaviors enables us to write the Jost matrix solutions of the Zacharov–Shabat problem in terms of the Jost solutions of problem (**XVII.4.117a**)

$$2F_{11}^+ = f_+(-k, x) + ik^{-1}\left(\frac{\partial}{\partial x} - q\right)f_+(-k, x),$$

$$2F_{21}^+ = f_+(-k, x) - ik^{-1}\left(\frac{\partial}{\partial x} - q\right)f_+(-k, x),$$

$$2F_{12}^+ = f_+(k, x) + ik^{-1}\left(\frac{\partial}{\partial x} - q\right)f_+(k, x),$$

$$2F_{22}^+ = f_+(k, x) - ik^{-1}\left(\frac{\partial}{\partial x} - q\right)f_+(k, x),$$

$$\qquad\qquad (\textbf{XVII.4.118a})$$

$$2F_{11}^- = f_-(k, x) + ik^{-1}\left(\frac{\partial}{\partial x} - q\right)f_-(k, x),$$

$$2F_{21}^- = f_-(k, x) - ik^{-1}\left(\frac{\partial}{\partial x} - q\right)f_-(k, x),$$

$$2F_{12}^- = f_-(-k, x) + ik^{-1}\left(\frac{\partial}{\partial x} - q\right)f_-(-k, x),$$

$$2F_{22}^- = f_-(-k, x) - ik^{-1}\left(\frac{\partial}{\partial x} - q\right)f_-(-k, x),$$

$$\qquad\qquad (\textbf{XVII.4.118b})$$

There is another (more trivial) case which reduces the Zakharov–Shabat problem to a Schrödinger problem. (*Hint*: $q^\pm = 1$, $q^- = q$, (**XVII.4.98**) is not fulfilled). We leave it as an exercise for the reader, who will also be amused in calculating exactly solvable examples by inserting "transparent" $\mu(z)$, with only one or two eigenvalues, into the Marchenko equation (**XVII.4.111**).

The inverse method associated with the Zakharov–Shabat problem has given exact solutions of the nonlinear Schrödinger equation

$$i\frac{\partial u}{\partial t} + \frac{\partial^2 u}{\partial x^2} + 2|u|^2 = 0, \qquad u \equiv u(x, t), \qquad (\textbf{XVII.4.119})$$

and the sine–Gordon equation

$$\frac{\partial^2 u}{\partial x\, \partial t} = \sin u, \qquad u \equiv u(x, t), \qquad (\textbf{XVII.4.120})$$

and, of course, large hierarchies of other equations.

General First-Order Systems. The basic equation is

$$\frac{d\psi}{dx} = \xi \mathbf{J}\Psi + \mathbf{Q}\Psi \qquad (x \in \mathbb{R} \text{ a.e.}), \qquad (\textbf{XVII.4.121})$$

where the "spectral parameter" ξ is a complex number, $\mathbf{J}$, $\mathbf{Q}$ and the unknown

Ψ are $N \times N$ matrices; $\mathbf{J}$ is diagonal, $J_{ij} = \lambda_i \delta_{ij}$ with distinct complex eigenvalues, $\mathbf{Q}$ is a zero-diagonal matrix, the "potential," and the solution Ψ should have a nonvanishing determinant. We assume that $\mathbf{Q}$ satisfies at least

$$\mathbf{Q} \in \mathscr{B} = \left\{ \mathbf{Q} \left| \int_{-\infty}^{+\infty} \sum_{i,j} |Q_{ij}(x)| dx = \|\mathbf{Q}\| < \infty \right. \right\}. \qquad \textbf{(XVII.4.122)}$$

If $\mathbf{Q}$ vanished, the solution of **(XVII.4.12)** would be of the form $e^{x\xi \mathbf{J}} \mathbf{A}(\xi)$ (and is generally an unbounded function). We are led to determine a fundamental matrix ψ of the form

$$\psi(x, \xi) = \mathbf{m}(x, \xi) e^{\xi \times \mathbf{J}}, \qquad \textbf{(XVII.4.123)}$$

where $\mathbf{m}(\cdot, \xi)$ is bounded and absolutely continuous, and goes to $\mathbf{I}$ as $x \to -\infty$. Equation **(XVII.4.121)** is equivalent to

$$\frac{d}{dx}\mathbf{m} = [\mathbf{I}, \mathbf{m}] + \mathbf{Q}\mathbf{m} \qquad (x \in \mathbb{R} \text{ a.e.}). \qquad \textbf{(XVII.4.124)}$$

Let Σ be the union of lines through the origin in $\mathbb{C}$ and such that, for pairs $j \neq k$, $\mathrm{Re}[\xi(\lambda_j - \lambda_k)] = 0$. A number of theorems concerning the spectral problem have been given by Beals and Coifman (1987).

We reproduce only a few results, valid for "generic" potentials (for instance, those of the Schwartz class):

(1°) (a) Except on Σ and a finite set Z of points, the problem **(XVII.4.121)**, **(XVII.4.123)**, **(XVII.4.124)** has a unique solution $\mathbf{m}(\cdot, \xi)$; and for every x, $\mathbf{m}(x, \cdot)$ is meromorphic in $\mathbb{C} \setminus \Sigma$ and goes to $\mathbf{I}$ as $\xi \to \infty$.

(b) The poles are simple, and distinct columns of $\mathbf{m}$ have distinct poles. Hence $\mathbf{m}$ is an *eigenfunction associated with* $\mathbf{Q}$. In addition, $\mathbf{m}(x, \cdot)$ has continuous extensions to the borders of $\mathbb{C} \setminus \Sigma$.

(2°) Let Ω_ν be a sector component of $\mathbb{C}/\Sigma$: we order the Ω_ν's in the positive sense about the origin. Let Σ_ν be the closed ray border of Ω_ν and $\Omega_{\nu+1}$. We call $\mathbf{m}_\nu^-(x, \cdot)$ the limit of $\mathbf{m}(x, \cdot)$ on Σ_ν from Ω_ν, and we call $\mathbf{m}_\nu^+(x, \cdot)$ the limit from Ω_ν. Now suppose $\mathbf{Q}$ is a generic potential with associated eigenfunction $\mathbf{m}$:

(a) For $\xi \in \Sigma_\nu$, there is a unique matrix $\mathbf{v}_\nu(\xi)$ such that for all x

$$\mathbf{m}_\nu^+(x, \xi) = \mathbf{m}_\nu^-(x, \xi) e^{x\xi \mathbf{J}} \mathbf{v}_\nu(\xi) e^{-x\xi \mathbf{J}}. \qquad \textbf{(XVII.4.125)}$$

(b) If $\mathbf{m}(x, \cdot)$ has poles at $\{\xi_1, \ldots, \xi_N\}$, then for each ξ, there is a matrix $\mathbf{v}(\xi_j)$ such that the residue satisfies

$$\mathrm{Res}(\mathbf{m}(x, \cdot); \xi_j) = \lim_{\xi \to \xi_j} \mathbf{m}(x, \xi) e^{\xi_j x \mathbf{J}} \mathbf{v}(\xi_j) e^{-\xi_j x \mathbf{J}}. \qquad \textbf{(XVII.4.126)}$$

(c) The potential $\mathbf{Q}$ is uniquely determined by the functions $\{\mathbf{v}_\nu\}$, the singularities by $\{\xi_j\}$, and the matrices by $\{v(\xi_j)\}$.

Given $\mathbf{Q}$, we denote

$$\mathbf{v} = (v_1, \ldots, v_r, \xi_1, \ldots, \xi_N, v(\xi_1), \ldots, v(\xi_N))$$

and call $\mathbf{v}$ the *scattering data* associated with Q. We note that $v_\nu \in C(\Sigma_\nu)$,

$\mathbf{v}_v(\xi) \to I$ as $\xi \to \infty$. We understand now that the direct problem is the determination of the scattering data from the potential, and the inverse problem is the reverse one. Theorem $(2°)$ shows that $\mathbf{Q}$ in the generic class is uniquely determined from $\mathbf{v}$. Part of the scattering data may itself be recovered from asymptotic information on the singular set.

We stop our introduction to the problem here. There exists a number of results available in the literature, which generalize in a weak sense the result obtained for the Zakharov–Shabat problem. However, a scattering problem of the physical kind, with an S-matrix, does not exist in general and the generalization is never straightforward. The D-bar approach is a very useful guide in most cases. Results on the general theory are given in Beals and Coifman (1984, 1985, 1987).

XVII.5   Comments and References

The history of the one-dimensional Schrödinger problem is very long and very rich. The first papers appeared in the early fifties. What we have called the Faddeev–Marchenko equation (because its form is identical to the equation obtained by Marchenko in the radial problem, and Faddeev (1964) wrote it in the general one-dimensional problem) was apparently first written by Kay to solve the truncated one-dimensional inverse problem (1955). He called it a "Gel'fand–Levitan equation," and this misleading name unfortunately became fairly popular. Related papers, up to 1964, are due to Bloch (1953), Krein (1953a), Kay and Moses (in particular, 1956b, 1957), Faddeev (1958), and several other papers by Kay and Moses now collected in a book (Kay and Moses, 1982). Faddeev's study (1964) of the problem was complete, but some upper bounds and conditions had to be reassessed for $V \in L_1^1$. This had been noticed and partially corrected in this class by Jaulent and Jean (1975, 1976). Characterization conditions and a complete reconstruction were then given for $V \in L_2^1$ by Deift and Trubowitz (1979) and for $V \in L_1^1$ by Melin (1985). Related studies appear in Bollé et al. (1983, 1985, 1987), Faddeev and Takhtadjian (1986), Klaus (1977, 1979, 1981, 1987), Simon (1976), and Klaus and Simon (1980). Extensions to classes where V is only $O(x^{-2})$ at ∞ appear in Degasperis and Sabatier (1987) and Sabatier (1986). Extensions to classes where V goes to a nonzero constant at $+\infty$ or $-\infty$, and is L_2^1 "on the other side," appear in Buslaev and Fomin (1962), Hruslov (1976), Legendre (1982), and Cohen and Kappeler (1985). As is mentioned in the text, the presentation of Faddeev's method, which is given here, has been corrected, simplified, and made more transparent after the lectures of Karlsson (1978), Atkinson (1979), and Sabatier (1983a,b,c).

Concerning the convergence of the Neumann series of the Marchenko equation we must consider separately the "reconstruction" theorem, "the Neumann series of the Marchenko equation converges if the input has been obtained by solving the direct problem for a potential of a given class," and the "construction" theorem: "the Neumann series converges if the input satisfies the following constraints ...". The two ones are equivalent only if we state the characterization properties. But there are conditions which are hardly stronger, simpler, and which avoid stating all the characterization properties. If we take into account the multiple forms of the Marchenko equation, we may understand why there are many different related statements of this property. In the one-dimensional problem which is studied here, and in the construction form, with the assumption that $|R^+(k)| = 1$ only at $k = 0$ (corresponding to condition A4), the result was known by Marchenko for a long time (private communication to P.C.S.). It appears to have been stated in a paper for the first time in Western countries by Segur (1973). See also Jaulent and Jean (1975, 1976), Newton (1982a), Xu Bang Qing and Shao Chang Qui (1985), and Sabatier (1985b). The stability of the Marchenko equation in the radial case had received a complete study by Marchenko (1968). In the one-dimensional case, the stability problem was studied by Deift and Trubowitz (1979), Prosser (1983, 1984), and, in connection with the elastic wave equation, by Symes (1981), and Bamberger et al. (1979). See also Atkosun (1987b). The fact that bound states can be introduced by Crum–Krein transformations (which are of the "Darboux type") has been known in radial problems for a very long time (see chapter IV), and holds true in "one-dimensional" problems on the line. It has been used particularly by Deift and Trubowitz (1979), and has been shown to be a special application of a very general relation between the spectra of two operators (Deift, 1978).

Approaches to one-dimensional problems as Riemann–Hilbert problems were much studied by Newton (1979, 1980a, 1982a,c, 1983), who used them to prepare his studies of the three-dimensional problem. Although the $\bar{\partial}$-approaches eventually became the key to three-dimensional problems, they were primarily introduced to study general functions of two variables, and functions on complex manifolds (Henkin and Leiterer, 1984), then used to study the direct and inverse scattering problems for systems of N first-order differential equations (with one-space dimension) generalizing the 2×2 spectral problem of Zakharov and Shabat (1972). Studies of these systems by the $\bar{\partial}$-method are due to Beals and Coifman (1981, 1984, 1985, 1987). Related studies, which were the basis of "three-dimensional" studies, are due to Nachman and Ablowitz (1983, 1984). Other studies are Henkin and Novikov (1984, 1986), Jaulent and Manna (1987), and Jaulent et al. (1988).

As was already recalled, the first study of the one-dimensional problem containing what we have called the Faddeev–Marchenko equation, and which can also be considered the first form of a "time-domain" approach, studied a potential vanishing on the half-line (Kay, 1955). In the case of truncated potentials, Portinari (1967) obtained the special form of the scattering coefficients $[T(k)]^{-1}$ and $R^+(k)/T(k)$, which are integral transforms of a truncated function, as in the usual radial case (Ramm, 1965) and in the fixed-energy radial case (Loubatieres, 1985).

The solution of the one-dimensional inverse problem for rational reflection coefficients was given by Kay (1960) (in the case $V(x) = 0$ for $x < 0$). General expressions of the solution (equivalent to a quotient of determinants) also appear in Calogero and Degasperis (1982) and Prosser (1983, 1984) who gave an approximation theory, whereas Sabatier (1983b,c) began on this example a general study of inverse problems (Sabatier, 1984) by means of transformations (in this case, Darboux transformations), and showed (Sabatier, 1985b; Guilpart, 1988) that the corresponding algorithms (once the reflection coefficient is known exactly as a rational fraction) for N poles are at least N-times shorter than calculating the general solution in its determinant form. Cases where only a few poles are present were treated by several authors (see, for instance, Jordan and Ahn (1979), Szu et al. (1976), Pechenik and Cohen (1981, 1983a,b), and Sabatier (1985c)). Rational coefficients have also been studied in the Zakharov–Shabat inverse problem (Calogero and Degasperis, 1989; Léon, 1988). We have followed Sabatier (1983b) for the presentation of these problems and his introduction to the one-dimensional inverse problem of Darboux transformation techniques. Transformations which are related to those introduced by Darboux (1984) and to those introduced by Backlund (1980) have a long history in inverse problems and have already been studied in chapter IV. Those which are used in the one-dimensional problem are really of the Darboux form, whose interest had been revived by Matveev (1980) for the study of inverse problems connected with inverse methods. It is interesting to remark at this point that we can use the technique of transformations for solving inverse problems (Sabatier, 1984, 1987b) as soon as transformations of this kind are available and they are available in many inverse problems (see, for example, Calogero and Degasperis (1982) and their references on Backlund–Darboux transformations, Laddomada and Tu Gui Zhang (1982), Boiti et al. (1983), Zheng (1984), Levi et al. (1982, 1984), and Andrianov et al. (1985).

The inversion of expansion formulas on products of Jost functions are due to Christov (1981) and have been used to complete transformation algorithms by Sabatier (1984, 1985a,b). The trace method is due to Deift and Trubowitz (1979), and some three-dimensional studies in the time domain (see section XIV.6) are related to it (see Morawetz, 1981).

Our simple presentation of the time-domain approach to the one-dimensional scattering inverse problems follows Bates and Millane (1981), and introduces layer-stripping methods (Bruckstein et al., 1985; Yagle and Levy, 1984a,b, 1985a,b,c; Symes, 1986), whereas we used the original paper of Corones et al. (1983) to present the invariant imbedding method. Admittedly, these methods are "designed" for wave equations and not for the Schrödinger equation, but in the one-dimensional case, and since bound states must have been ruled out, it is not difficult to go from one of these equations to the others; the reader may (also) like to see references like Moses (1984), Kristensson and Krueger (1986a,b), and Corones et al. (1984).

The general study of one-dimensional problems where the potential is made of Dirac measures and one-dimensional impedance problems is due to Sabatier (1983b, 1984), Degasperis and Sabatier (1987), Sabatier (1987d, 1988), and Sabatier and Dolveck-Guilpart (1988). A large part of this study is made by means of Darboux transformations.

Ambiguities in the potentials corresponding to a reflection coefficient $R(k)$, known as an exact function on the real k-axis, appeared in the literature with a paper by Abraham et al. (1981) (see also Abraham et al., 1983; De Facio and Moses, 1980). This paper was a trigger for a series of papers giving other examples of ambiguities (Brownstein, 1982; Moses, 1983, 1984; Atkosun and Newton, 1985; Atkosun, 1987). The discovery of ambiguities can be put into the general frame of attempts to extend the classes where the equations solving the inverse problem are valid (see, e.g., Moses, 1978, 1979; Newton, 1983). It has been shown (Sabatier, 1984) that all previously known ambiguities can be obtained by means of a limit Darboux transformation, which in fact can yield a one-parameter class of potentials corresponding to any reflection coefficient $R^+(k)$, but presenting either a singularity at finite real x or an x^{-2} asymptotic behavior. A complete study of the method is given in Degasperis and Sabatier (1987).

The first-order impedance systems were studied by Howard (1981, 1983). They belong to the general frame of one-dimensional studies of acoustic or elastic problems: many other related papers are cited below in section XIX.2.4, because the main object of their authors was computational aspects rather than theory.

The Schrödinger problem on the line with k-dependent potential was deeply studied by Jaulent and Jean (1975, 1976) who extended to this case the method devised for radial problems (which is itself related in the fixed E problem to the spin-orbit cases previously solved). Jean (1983) gave a generalization of the one-dimensional completeness theorem, whereas Jaulent and co-workers produced several applications and extensions of the method (Jaulent, 1976, 1982; Jaulent and Jean, 1981a,b; Guillaume and Jaulent, 1982) and the method was used for solving other problems (Léon, 1980): in fact, the Schrödinger potential with k-linear or k-polynomial dependence can be proved to be equivalent to the Zakharov–Shabat problem, but not in a trivial way (Jaulent and Miodek, 1977; Léon, 1981).

The studies of inverse problems, specially designed to apply inverse methods to solving nonlinear partial differential equations, belong more to this domain than to the scope of the present book. Hence we think it is preferable to refer the reader to the very good treatises or extended reviews already published on the subject, in particular, Ablowitz et al. (1974), Miura (1976), Ablowitz (1978), Longren and Scott (1978), Bullough and Caudrey (1980), Ablowitz and Segur (1981), Calogero and Degasperis (1982), Eckaus and Van Harten (1981), Newell (1985), Faddeev and Takhtadjian (1986), Konopelchenko (1987), and to give only a few references for the two studies of general systems that we present:

(a) The general matrix Schrödinger problems which were studied for the first time by Wadati and Kamijo (1974). A complete study appears in Calogero and Degasperis (1976, 1977) who also gave their Backlund transformations.

(b) The general first-order systems generalizing the Zakharov–Shabat problem. We have seen above their studies by means of the $\bar{\partial}$-method. Another method is due to Caudrey (1982). The Darboux transformations were given by Sabatier (1986) and Beals and Coifman (1987). A very complete study of the Zakharov–Shabat problem is to be published in Calogero and Degasperis (1989).

Among the theoretical papers connected with the one-dimensional problem, and which have not been cited in this chapter, we should not forget the inverse problems connected with the randomness of the medium (Kohler, 1986; Burridge et al., 1988), the inverse problem for periodic potentials (Trubowitz, 1977) or for impurities in a periodic potential (Newton, 1983b, 1985d), and the discrete forms of one-dimensional problems (Case, 1974; Geronimo, 1982). Particular problems connected with the one-dimensional problems have also been treated, e.g., asymptotic behaviors of the scattering coefficients for the scattering problem corresponding to the string $\mu'' - \lambda^2 qy = 0$ (Fedorjuk, 1965), asymptotic or JWKB approximation methods (Hecht, 1971; Pike, 1964), tunneling through one-dimensional barriers (Kowalski and Fry, 1987), a variational principle (Moses, 1978) and, curiously, a "plausibility argument for the Marchenko equation" (Scott, 1978).

XVII.A   Appendix (Exercises for Readers)

We noticed that the one-dimensional problem is a workshop in the study of inverse problems. The rational coefficients case is a workshop in the study of one-dimensional inverse problems. Hence it would be interesting to have a complete study of this case, including in a self-contained way all the proofs necessary for constructing a potential from scattering data and proving the uniqueness of this reconstruction. The papers of Kay (1960) and Sabatier (1983) are parts of this program. The results, which are scattered in the present chapter XVII together with some demonstrations to be done, can be used as a kit for such a theory. We propose that the reader do it himself and if he needs some help, we provide a guide here.

(1) First read sections XVII.1.1, XVII.1.2, XVII.1.3, XVII.1.5, and XVII.2.1.

(2) Introduce a rational reflection coefficient $R_+(k)$ and N (regular, i.e., $\varkappa \neq 0$) bound states with their normalizing constants. If $N = 0$, check that $T(k)$ can be constructed in the form (**XVII.3.4a**) by formula (**XVII.1.60**), and with the extra factor (**XVII.3.4b**) if there are N regular bound states. $R_-(k)$ is then obtained by means of (**XVII.3.3b**).

Also construct $T(k)$ from $R_+(k)$ and (**XVII.1.38b**) by algebraic methods, i.e., by writing $\prod_{j=1}^q (\lambda_j^2 - k^2) - \prod_{i=1}^q (\mu_i^2 - k^2)$ as the product of factors $P(k)P(-k)$.

EXAMPLE

$$R_+(k) = -\frac{i\varkappa + k}{i\varkappa - k}\frac{i\mu - k}{i\mu + k}\frac{a^2 + b^2}{a^2 + (b - ik)^2},$$

$$R_-(k) = -\frac{i\mu + k}{i\mu - k}\frac{a^2 + b^2}{a^2 + (b - ik)^2}, \qquad \text{(XVII.A.1)}$$

$$T(k) = -\frac{k(k + i\varkappa)}{a^2 + (b - ik)^2},$$

where Im $\mu > 0$, $\varkappa = [2(b^2 - a^2)]^{1/2}$, $b \geq a \geq 0$.

(3) The integral equations (**XVII.2.2**) reduce to equations (**XVII.3.5**), (**XVII.3.6**) for rational coefficients. In the chapter, we have used the general theory in L_2^1 and claimed that these equations solve the inverse problem. The result can also be proved by independent, self-contained ways which we sketch here. First, rewrite (**XVII.3.5**) and (**XVII.3.6**) after using the notation ρ_+^j or ρ_-^j for any pole in the upper half-plane, and ic_+^j or ic_-^j for the corresponding coefficient

$$f_+(k, x) = \exp[ikx] + i \sum c_+^j \frac{f_+(\rho_+^j, x)\exp[i(\rho_+^j + k)x]}{\rho_+^j + k}, \qquad \text{(XVII.A.2)}$$

$$f_-(k, x) = \exp[-ikx] + i \sum c_-^j \frac{f_-(\rho_-^j, x)\exp[-i(\rho_-^j + k)x]}{\rho_-^j + k}. \qquad \text{(XVII.A.3)}$$

We study, for instance, (**XVII.A.2**). The number of terms in the sum, which extends to poles of $R_+(k)$ in the upper half-plane and bound states, is finite. Assume for a while that $f_+(\rho_+^j, x)$ is twice differentiable. We can differentiate twice and introducing any function $W(x)$ of x we obtain, by straightforward calculations,

$$f_+''(k, x) + [k^2 - W(x)]f_+(k, x)$$

$$= i \sum c_j^+ \frac{[f_+''(\rho_+^j, x) + (\rho_+^j)^2 f_+(\rho_+^j, x) - W(x)f_+(\rho_+^j, x)]\exp[i(\rho_+^j + k)x]}{\rho_+^j + k}$$

$$-\exp[ikx]\left\{W(x) + 2\sum c_+^j \frac{d}{dx}[\exp(i\rho_+^j x)f_+(\rho_+^j, x)]\right\}. \qquad \text{(XVII.A.4)}$$

On the other hand, it is clear that (**XVII.A.2**) is solved if we can solve the algebraic system

$$f_+(\rho_+^l, x) = \exp[i\rho_+^l x] + i \sum c_+^j \frac{f_+(\rho_+^j, x)\exp[i(\rho_+^l + \rho_+^j)x]}{\rho_+^j + \rho_+^l}, \qquad \text{(XVII.A.5)}$$

where l runs through the values of j, the ρ_+^j's and the c_+^j's are given, and the vector $\mathbf{f}_+(x) = \{f_+(\rho_+^l, x)\}$ must be determined for each x. Equation (**XVII.A.5**) is solved if we can inverse the matrix

$$\mathbf{M}^+ = \{M_{lj}^+\} = \left\{\delta_{\rho_j} - ic_+^j \frac{\exp[i(\rho_+^j + \rho_+^l)x]}{\rho_+^j + \rho_+^l}\right\}.$$

Let $\mathscr{D}_+(x)$ be the determinant of $\mathbf{M}^+(x)$. The solution of the system (**XVII.A.5**), also written as

$$\sum_j M_{lj}^+ f_+^j(x) = \exp[i\rho_+^l x], \qquad \text{(XVII.A.6)}$$

is

$$f_+^j = (\mathscr{D}_+)^{-1} \sum m_{lj} \exp[i\rho_+^l x], \qquad \text{(XVII.A.7)}$$

where m_{lj} is the cofactor of M_{lj}^+ in $\mathbf{M}^+$. Comparing with

$$\frac{d}{dx}\mathscr{D}_+(x) = \sum_{l=1}^{n} \sum_{j=1}^{n} \left(\frac{d}{dx}M_{lj}^+(x)\right)m_{lj}$$

$$= \sum_{l=1}^{n} \sum_{j=1}^{n} c_+^j m_{lj} \exp[i(\rho_+^l + \rho_+^j)x], \qquad \text{(XVII.A.8)}$$

we see that

$$\sum c_+^j \exp[i\rho_+^j x] f_+(\rho_+^j, x) = \frac{\mathscr{D}_+'(x)}{\mathscr{D}_+(x)}. \qquad \text{(XVII.A.9)}$$

Now, it is clear that $\mathscr{D}_+(x)$ is an entire function of x. If we consider a narrow strip containing the positive axis in the x complex plane, $\mathscr{D}(x)$ can have there only a finite number of zeros because $\mathbf{M}^+$ goes to $\mathbf{I}$ as x goes to $+\infty$ (remember that $\mathrm{Im}\,\rho_+^j > 0$). Between these points $f_+(\rho_+^j, x)$ is twice differentiable, and if $W(x)$ is defined so as to cancel the last term in (XVII.A.4)

$$W(x) = -2\sum c_+^j \frac{d}{dx}[\exp(i\rho_+^j x) f_+(\rho_+^j, x)], \qquad \text{(XVII.A.10)}$$

$f_+(k, x)$ obeys, for $x > 0$, the Schrödinger equation with the potential $W(x)$. Besides, (XVII.A.2) is the transformation formula with

$$K_+(x, y) = \sum c_+^j f_+(\rho_+^j, x)\exp[i\rho_+^j y]. \qquad \text{(XVII.A.11)}$$

Hence, if $R_+(k)$ and the information given on bound states are such that $\mathscr{D}(x)$ has no zero on $[0, \infty)$, there exists a unique solution of (XVII.A.2), and it satisfies the Schrödinger equation with the potential $W(x)$ on $[0, \infty)$.

This potential goes to zero exponentially as $x \to +\infty$. The corresponding equation (XVII.2.4) (label $+$) therefore exists and should be identified with (XVII.A.2). This is possible only if the c_+^j's and ρ_+^j's come from the poles and residues of $R_+(k)$ and from the bound states. Hence the construction problem is solved for $x \in (0, \infty)$. The same analysis applies for $x \in (-\infty, 0)$, except that $W(x) = 2(d/dx)K_-(x, x)$ and $K_-(x, x) = -\mathscr{D}_-'(x)/\mathscr{D}_-(x)$. We may question the continuity of $f_+(k, x)$ and $f_+'(k, x)$ at $x = 0$, it being understood that we assume $\mathscr{D}_\pm(x) \neq 0$ on $\mathbb{R}$. From (XVII.A.2) we can then construct $f_+(k, x)$ and $f_-(-k, x)$ for any $x \in \mathbb{R}^+$, and clearly it is sufficient to prove that $f_-(k, x)$ and its derivative are respectively equal to $[T(k)]^{-1}f_+(-k, x) + [T(k)]^{-1} R_+(k)f_+(k, x)$ and its derivative at $x = 0$, and that the same equality holds if the index $+$ and $-$ are reversed. We shall now see this point.

(4) The reader will need prove that $f_+(k, x)$ (defined for $x > 0$ by equation (XVII.A.2)) and $[T(k)]^{-1}[f_-(-k, x) + R_-(k)f_-(k, x)]$ (defined for $x < 0$ by equation (XVII.A.3)), together form a continuously differentiable function at $x = 0$. Here we give the method for the case $R_+(k)$ which has no pole in the upper half-plane with no bound state (the other cases follow by means of the algorithm defined below in (5) or are treated similarly). If $R_+(k)$ has no pole in the upper half-plane, $f_+(k, 0)$ and its derivative are simply 1 and ik, so that we must first prove that

$$\varphi(k) = \frac{1}{T(k)}f_-(-k, 0) + \frac{R_-(k)}{T(k)}f_-(k, 0) \qquad \text{(XVII.A.12)}$$

is equal to 1. From (XVII.A.3) we get

$$f_-(\pm k, 0) = 1 + \sum R_-^j \frac{f_-(v_-^j, 0)}{v_-^j \pm k}. \qquad \text{(XVII.A.13)}$$

Hence $\varphi(k)$ is a rational fraction, with no poles on the real axis but at $k = 0$. For Im $k > 0$, its poles are at v^i_- and it readily follows from **(XVII.A.13)** and **(XVII.A.12)** that the residues vanish. At $k = 0$, $T(k)$ is O(k), but we can write

$$f_-(-k, 0) + R_-(k)f_-(k, 0) = [1 + R_-(k)]\left[1 + \sum \frac{R^j_-}{v^j_-} f_-(v^j_-, 0)\right]$$

$$+ \sum \frac{R^j_-}{(v^j_-)^2} f_-(v^j_-, 0)\left[\frac{k}{v^j_- - k} - \frac{kR_-(k)}{v^j_- + k}\right],$$
$$\text{(XVII.A.14)}$$

and since $1 + R_-(k)$ is O(k), it follows that the residue of $\varphi(k)$ is still zero. For Im $k < 0$, we first notice that

$$[T(k)]^{-1} = -\frac{R_-(-k)}{T(-k)R_+(k)}. \qquad \text{(XVII.A.15)}$$

Since $T(k)$ has no zero in the upper half-plane but $k = 0$, the poles of $[T(k)]^{-1}$ in the lower half-plane belong to the set of numbers $-v^j_-$ of $R_-(k)$ or the zeros of $R_+(k)$. But if $R_+(k)$ and $T(k)$ vanish, $R_+(-k) \sim [R_+(k)]^{-1}$, and this cannot be in the lower half-plane since $R_+(k)$ has no pole in the upper half-plane. Hence, the poles of $\varphi(k)$ in the lower half-plane belong to the set $\{-v^j_-\}$. Using **(XVII.A.15)** in **(XVII.A.12)** we easily obtain

$$\varphi(k \to -v^i_-) \sim f_-(v^i_-, 0)\left[\frac{R^i_-}{T(v^i_-)R_+(-v^i_-)(k + v^i_-)} - \frac{R^i_- R_+(v^i_-)}{T(v^i_-)(v^i_- + k)}\right],$$
$$\text{(XVII.A.16)}$$

and since $T(-v^i_-) = 0$ makes $R_+(v^i_-) = [R_+(-v^i_-)]^{-1}$, it follows that the residue of φ is zero. Hence φ is an entire function, which goes to 1 as $|k| \to \infty$ and is therefore 1. Q.E.D.

A similar study is easily done for

$$\psi(k) = [T(k)]^{-1}[f'_-(-k, 0) + R_-(k)f'_-(k, 0)] \qquad \text{(XVII.A.17)}$$

by using

$$f'_-(k, 0) = -ik + \sum \frac{R^j_- f'_-(v^j_-, 0)}{v^j_- + k} - i\alpha,$$

$$\alpha = \sum R^j_- f_-(v^j_-, 0), \qquad \text{(XVII.A.18)}$$

the only new point is the asymptotic behavior of $\psi(k)$: setting $T(k) \sim 1 + ak^{-1} + O(k^{-2})$, we can check from **(XVII.1.46)** and **(XVII.A.10)** (remembering that in our case $V = 0$ for $x > 0$), that

$$\alpha = -a = \tfrac{1}{2}i \int_{-\infty}^{0} V(y)dy. \qquad \text{(XVII.A.19)}$$

The asymptotic behavior of $\psi(k)$ being

$$\psi(k) \sim ik[1 - (a + \alpha)k^{-1} - R_-(k) + O(k^{-2})], \qquad \text{(XVII.A.20)}$$

we see with our assumption that $R_-(k) = O(k^{-2})$, and we can ascertain that $\psi(k)$ is equal to ik.   Q.E.D.

Remark. Thus it is shown, without any reference to the Deift–Trubowitz theorem, that rational fractions that satisfy the algebraic requirements *and* the assumption $R_\pm(k) = O(k^{-2})$ enable us to construct f_+ and f_-, respectively, on $(0, \infty)$ and $(-\infty, 0)$ by means of, respectively, (**XVII.A.2**) and (**XVII.A.3**), and that they are continued through $x = 0$ (in the sense given above) consistently with the Schrödinger equation. Hence the inverse problem is completely solved in this case by means of (**XVII.A.2**) for $x > 0$ and by (**XVII.A.3**) for $x < 0$, and this is one of the justifications of Sabatier's algorithm. We must notice that the extension to $R_\pm(k) = O(k^{-1})$ makes it necessary to allow a Dirac measure in the potential, i.e., a discontinuity of f'_+ or f'_-, at the origin, and this is also obvious in the results above, since it follows from (**XVII.A.20**) that $\psi(k) - ik$ is then still a constant, but not zero.

(5) The main justification of algorithms based on Darboux transformations lies in the following result, which is now proposed to the reader with a sketch of its proof. The reader will generalize to the case with bound states, to the case of limit Darboux transformations, and to the case where $R_\pm(k) = O(k^{-1})$.

Theorem. *Let* $R_+(k)$, $R_-(k)$, $T(k)$ *be consistent with each other, with no bound state and with $R_\pm(k)$ being $O(k^{-2})$. Suppose that there exists a solution $f_+(k, x)$ $(x \geq 0)$ of the $(+)$ equation (**XVII.A.2**) and a solution $f_-(k, x)$ $(x \leq 0)$ of the $(-)$ equation (**XVII.A.3**), with the following properties:*

(a) $T(k) f_-(k, x) (x \leq 0)$ *and* $f_+(-k, x) + R_+(k)f_+(k, x)$ $(x \geq 0)$ *together form a continuously differentiable function for any $x \in \mathbb{R}$. Because of assumption* (b) *below, it is equivalent to assume that $T(k)f_+(k, x)$ $(x \geq 0)$ and $f_-(-k, x) + R_-(k)f_-(k, x)$ $(x \leq 0)$ together form a continuously differentiable function for any $x \in \mathbb{R}$.*

(b) $f_+(k, x)$ $(x \geq 0)$ *and* $f_-(k, x)$ $(x \leq 0)$ *are solutions of the equation $f'' + (k^2 - V)f = 0$, with $V(x)$ real and regular for any $x \in \mathbb{R}$.*

(c) $f_+(k, x)$ $(x \geq 0)$, *and its "continuation" at $x < 0$, $f_-(k, x)$ $(x \leq 0)$, and its "continuation" at $x > 0$, are holomorphic in $\mathrm{Im}\, k \geq 0$, and, as $|k| \to \infty$ in the upper half-plane $(0 < \mathrm{Arg}\, k < \pi)$:*

$$e^{-ikx}f_+(k, x) - 1 \to 0, \qquad e^{-ikx}f'_+(k, x) - ik \to 0,$$

$$e^{ikx}f_-(k, x) - 1 \to 0, \qquad e^{ikx}f'_-(k, x) + ik \to 0.$$

Then if $\mathrm{Im}\, \mu > 0$ and $u(x) = f_-(\mu, x)$, we can claim that the coefficients

$$\tilde{R}_+(k) = \frac{\mu + k}{\mu - k} R_+(k), \qquad \tilde{T}(k) = \tilde{T}(k),$$

$$\tilde{R}_-(\mu) = \frac{\mu - k}{\mu + k} R_-(k), \qquad \qquad \text{(XVII.A.21)}$$

satisfy all the properties of the assumption, and correspond to the new solu-

tions of equations (**XVII.A.2**), (**XVII.A.3**)

$$\tilde{f}_+(k, x) = \frac{f'_+(k, x) - f_+(k, x)u'/u}{i(\mu + k)},$$ (**XVII.A.22**)

$$\tilde{f}_-(k, x) = \frac{f'_-(k, x) - f_-(k, x)u'/u}{i(\mu - k)}.$$ (**XVII.A.23**)

Hints for the Proof. There is no difficulty in proving the consistency of the new coefficients with each other, and the conservation of the continuity properties of the functions and their derivatives at $x = 0$. Nor is there any difficulty in proving the holomorphy of $\tilde{f}_+$ and $\tilde{f}_-$ (notice that $\mu - k$ vanishes in the upper half-plane but the numerator of (**XVII.A.23**) vanishes simultaneously. It remains to prove, for instance, that $\tilde{f}_+(k, x)$ satisfies, for $x \geq 0$, $k \in \mathbb{R}$, the equation

$$\tilde{f}_+(k, x) - e^{ikx} = \frac{1}{2\pi i} \int_{-\infty}^{+\infty} dk' \, \tilde{R}_+(k') \frac{\tilde{f}_+(k', x)e^{i(k'+k)x}}{k' + k + i\varepsilon}.$$ (**XVII.A.24**)

Setting $u'/u = \eta$, we can see after some algebra that the right-hand side of (**XVII.A.24**) can be written as

$$-\frac{e^{ikx}}{2\pi} \int_{-\infty}^{+\infty} \frac{dk'}{(\mu - k')(k' + k + i\varepsilon)} [-(e^{ik'x}f'_+(-k', x) + ik')$$

$$+ \eta f_+(-k', x)e^{ik'x} + T(k')(e^{ik'x}f'_-(k', x) + ik')$$

$$- \eta T(k')e^{ik'x}f(k', x) + ik'(1 - T(k'))].$$ (**XVII.A.25**)

Each term can be calculated by means of a half-circle either in the upper half-plane or in the lower half-plane, and the result proves (**XVII.A.24**).

(6) As we said, Sabatier's algorithm solves equation (**XVII.A.2**) on $x \geq 0$ and equation (**XVII.A.3**) on $x \leq 0$, by successively applying Darboux transformations respectively to a case where $R_+(k)$ has no pole in the upper half-plane, and a case where $R_-(k)$ has no pole in the upper half-plane. The interest is that derivations are always done on products of exponentials by polynomials so that the algorithm reduces to a finite number of additions and trivial multiplications by k (Sabatier, 1983, 1984; Dolveck and Guilpart, 1988). To make all this clear, it is suggested that the reader look at Example (**XVII.A.1**). In this example, $R_+(k)$ has only the pole $k = i\varkappa$ in the upper half-plane, and $R_-(k)$ has only the pole $k = i\mu$. For the same $T(k)$ we can write

$$R_+^0(k) = -\frac{i\mu - k}{i\mu + k} \frac{a^2 + b^2}{a^2 + (b - ik)^2},$$ (**XVII.A.26**)

$$R_-^2(k) = -\frac{a^2 + b^2}{a^2 - (b - ik)^2},$$ (**XVII.A.27**)

and we straightforwardly derive (remember $R_+(k)/R_-(-k)$ is invariant)

$$R_-^0(k) = -\frac{i\varkappa + k}{i\varkappa - k} \frac{i\mu + k}{i\mu - k} \frac{a^2 + b^2}{a^2 - (b - ik)^2},$$ (**XVII.A.28**)

$$R_+^2(k) = -\frac{i\varkappa + k}{i\varkappa - k} \frac{a^2 + b^2}{a^2 + (b - ik)^2}.$$ (XVII.A.29)

The problem which corresponds to the reflection coefficient $R_+^0(k)$ is the starting point for determining V for *positive* x. Then $f_+^0(k, x) = e^{ikx}$ $(x \geq 0)$, $f_-^0(\varkappa, x)$ is given $(x \geq 0)$ as $[T(\varkappa)]^{-1}[f_+^0(-\varkappa), +R_+^0(\varkappa)f_+^0(\varkappa, x)]$ and $V(x)$ on $x \geq 0$ follows by the formula $V(x) = -2(d/dx)(f_-^{\prime 0}/f_-^0)$. The same process gives $V(x)$ for $x \leq 0$, starting from $R_-^2(k)$.

Overall Remark. In the "no bound state" case, for instance, writing $T(k) = kQ(k)/S(k)$ where $Q(k)$ and $S(k)$ have no zero in Im $k > 0$, we see that if $P(k)$ is constrained to have no zero in Im $k > 0$, it is uniquely determined by $P(k)P(-k) = S(k)S(-k) + k^2Q(k)Q(-k)$ and so is $r(k) = P(k)/S(k)$. Then $R_+(k)$ is necessarily of the form $r(k)F_I(k)F_E(k)$ where $F_I(k)$ is a phase factor whose zeros are zeros or poles of $S(k)$, $P(k)$, or $Q(k)$, and $F_E(k)$ involves other zeros and poles. $R_-(k)$ is readily calculated as $-R_+(-k)T(k)/T(-k)$ and we easily see that multiple poles may appear. For multiple poles, equations (XVII.A.2) and (XVII.A.3) are obviously modified, but a limiting process with simple poles could deal with this case, as well as repeated Darboux transformations of the Marchenko equation. However, the reader may see from this remark, and from all our hints to him introducing other Darboux transformations (e.g., to study bound states) or limit transformations (e.g., to study ambiguities), that an extensive study of the inverse problem with rational reflection coefficients would be tedious and not very useful: the Marchenko method, the set of equations (XVII.A.2), (XVII.A.3), or the algorithms based on Darboux transformations, give all the methods of study, and detailed applications should go through *many* cases. We think it is better to consider such a study as a stock of exercises.

CHAPTER XVIII

Problems Connected with Discrete Spectra

XVIII.1   Introduction

When a vibration is created in a finite structure, its energy may be divided into a set of well-defined vibrations that are called modes of the system. The mathematical model is usually the spectral problem of a linear operator. The modes are then associated with eigenvalues and eigenfunctions (e.g., vibrating strings, etc.). A "continuous extension" of these problems is observed in scattering studies (see chapter II). There are linear operators showing a discrete spectrum, a continuous spectrum, and an exceptional spectrum; and there are also nonlinear systems exhibiting a generalized discrete spectrum. However, we shall stay with the "generic" (most simple) cases and their relations to quantum inverse scattering.

XVIII.1.1   Modes and Spectrum of Operators

From the physical point of view, an important characteristic property of a mode is that the "vibration" identically vanishes on a set of fixed surfaces, called nodal surfaces, which bound domains, called nodal domains. This "nodal" property is somewhat related to the orthogonality of eigenfunctions when it holds (e.g., for self-adjoint operators). From the mathematical point of view, the most important property of modes is that (together with the continuous spectrum functions, if there is a continuous spectrum) they enable us to expand along them an arbitrary function belonging to the operator image (completeness relation).

To be more precise, let us recall the spectral properties of self-adjoint operators.

Definitions. We are given a Hilbert space H (inner product $\langle \cdot, \cdot \rangle$. We call "operators" on H the linear mappings of H into H. Such an operator T is "continuous" if, for any $x, y \in H$, $\| Tx - Ty \| \to 0$ as $\| x - y \| \to 0$, where $\| \cdot \|$ is the canonical norm $\| x \| = [\langle x, x \rangle]^{1/2}$. Any linear continuous operator is bounded, i.e., there exists a number $\| T \|$ such that, $\forall x \in H$, $\| Tx \| \le \| T \| \| x \|$. The spectrum of a linear operator $T: H \to H$ is the set $\sigma(T)$ of a complex number λ for which $T - \lambda T$ has no bounded inverse. The spectral radius $|\sigma(T)|$ of T is the upper bound of $|\lambda|$ in $\sigma(T)$. There usually exists in $\sigma(T)$ numbers λ, for which $Tx = \lambda x$ for a nonvanishing vector x, called the "eigenvector" corresponding to the "eigenvalue" λ. It is possible to show that if T is a bounded linear operator, $\sigma(T)$ is bounded by $\| T \|$ and closed. A self-adjoint operator is an operator T defined in $\mathscr{D} \in H$ such that

$$(\forall x, \forall y, x \in \mathscr{D}, y \in \mathscr{D}) \qquad (\langle Tx, y \rangle = \langle x, Ty \rangle). \qquad \text{(XVIII.1.1)}$$

It is possible to show that if T is bounded and self-adjoint, $\sigma(T)$ is a nonempty set of real numbers and $|\sigma(T)| = \| T \|$.

Now we shall state two spectral theorems corresponding to more or less particular self-adjoint operators.

Compact Self-Adjoint Operator. A "compact" (or "completely continuous") operator K is an operator mapping any bounded set of its domain $\mathscr{D}$ onto a relatively compact set. In other words, for each bounded sequence $\{x_n\}$ $(x_n \in \mathscr{D})$, the sequence $\{Kx_n\}$ is either convergent or contains a convergent "extracted subsequence." Now let K be a self-adjoint compact operator, then each element of the spectrum is an eigenvalue of K. The nonvanishing eigenvalues are all isolated points, the only possible clustering point in $\sigma(K)$ being zero. For each nonvanishing eigenvalue λ the null space of $(\lambda I - K)$, called the eigenspace of λ, is a finite-dimensional space whose orthonormal eigenvectors form a basis. Ranking the eigenvalue by decreasing absolute value and sign, and repeating them according to the dimension of the eigenspace, we associate with the sequence of eigenvalues a sequence of orthonormal eigenvectors $w_1, w_2, \ldots, w_n$. *This sequence is a basis of* $R(K)$ ("range," or "image" of K, i.e., $K(\mathscr{D})$)

$$(\forall x)(x \in \mathscr{D}), \qquad Kx = \sum_n \lambda_n \langle x, w_n \rangle w_n. \qquad \text{(XVIII.1.2)}$$

This decomposition is nothing other than what we have called, in other places, a completeness relation for the w_n's. This theorem is obviously dependent on the "discrete" property of the spectrum. To get a generalization, we need the concepts of *spectral measure* and *spectral function*.

We call a *"family of projections"* any function E_λ of $\lambda \in \mathbb{R}$, with values in the orthogonal projectors in H, which satisfies the following properties:

(a) $E_{-\infty} = 0$; $E_\infty = 1$, which means that for any x in H,

$$\lim_{\lambda \to -\infty} \| E_\lambda x \| = 0 \quad \text{and} \quad \lim_{\lambda \to +\infty} \| E_\lambda x - x \| = 0.$$

(b) $(\lambda \leq \mu) \Rightarrow E_\lambda \leq E_\mu$, which means that for any x in H,

$$\langle (E_\mu - E_\lambda)x, x \rangle \geq 0.$$

(c) $x \to E_\lambda x$ is continuous on the right, which means that for any x in H,

$$\lim_{\varepsilon \to 0} \| E_{\lambda+\varepsilon}x - E_\lambda x \| = 0.$$

Let f be any continuous complex-valued function defined on $[a, b]$. Using convenient partitions of $[a, b]$ and a limiting process as in any other integral definition, it is possible to define integrals of the form

$$\int_a^b f(\lambda)dE_\lambda = \lim \sum_{i=1}^n f(\xi_i)(E_{\lambda_i} - E_{\lambda_{i-1}}), \qquad \text{(XVIII.1.3a)}$$

$$\int_a^b f(\lambda)dE_\lambda x = \lim \sum_{i=1}^n f(\xi_i)(E_{\lambda_i}x - E_{\lambda_{i-1}}x), \qquad \text{(XVIII.1.3b)}$$

and to show, for instance, that

$$\left\| \int_{-\infty}^{+\infty} f(\lambda)E_\lambda x \right\|^2 = \int_{-\infty}^{+\infty} |f(\lambda)|^2 d\langle E_\lambda x, x \rangle. \qquad \text{(XVIII.1.4)}$$

Details and proofs can be found in several references and, in particular, in Helmberg (1969) and Dautray and Lions (1985). In particular, the following theorem is proved.

Spectral Theorem. *For each self-adjoint operator A on H, domain $\mathcal{D}_A$, there exists a spectral family E_λ with the following properties:*

(a) *x belongs to $\mathcal{D}_A$ if and only if $\int_{-\infty}^{+\infty} \lambda^2 d\langle E_\lambda x, x \rangle < \infty$;*
(b) *for any $x \in \mathcal{D}_A$, the following two formulas hold*

$$Ax = \int_{-\infty}^{+\infty} \lambda \, dE_\lambda x, \qquad \text{(XVIII.1.5a)}$$

$$\| Ax \|^2 = \int_{-\infty}^{+\infty} \lambda^2 d\langle E_\lambda x, x \rangle. \qquad \text{(XVIII.1.5b)}$$

Conversely, each operator which is defined by (a) and (b), and by a spectral family E_λ, is a self-adjoint operator in H. In the cases we shall study, the projectors may be written in terms of the eigenfunctions $\varphi(x, \lambda)$, for instance, in the Schrödinger problem on $(0, \infty)$

$$(E_\lambda f)(x) = \int_{-\infty}^{\lambda} F(\lambda')\varphi(x, \lambda')d\rho(\lambda'), \qquad \text{(XVIII.1.6a)}$$

$$F(\lambda) = \int_0^{\infty} f(t)\varphi(t, \lambda)dt, \qquad \text{(XVIII.1.6b)}$$

where f is any element of $L^2(0, \infty)$, $\rho(\lambda)$ is nondecreasing, almost everywhere, differentiable function, and the support of $d\rho(\lambda)$ is the spectrum $\sigma(A)$ of A. $\rho(\lambda)$ is called the *spectral function* of A and $d\rho(\lambda)$ its *spectral measure*.

Since $E_\infty f = f$, the mapping $f \leftrightarrow F$ is a bijection between $L^2_{d\rho}(-\infty, +\infty)$ and $L^2(0, \infty)$. It is also an isometry, and Parseval's equality

$$\int_{-\infty}^{\infty} |F(\lambda)|^2 \, d\rho(\lambda) = \int_0^{\infty} |f(t)|^2 \, dt \qquad \text{(XVIII.1.7)}$$

holds, with its consequences (conservation of the scalar product, so that this isometry is also an isomorphism). Thus (XVIII.1.6) is a generalization of the Fourier transformation. For "normal modes," when the spectrum is made up of simple eigenvalues, the spectral measure is a sum of Dirac measures, and formula (XVIII.1.6) gives the expansion of an arbitrary function f along with the normal modes. We see that the coefficient of the Dirac measure at λ_n in $d\rho(\lambda)$ is then simply the inverse of the nth mode normalization coefficient, i.e.,

$$\|\varphi_n\|^{-2} = \left[\int_0^{+\infty} |\varphi(x, \lambda_n)|^2 \, dx \right]^{-1}, \qquad \text{(XVIII.1.8)}$$

and we see that the projector has the expected form $\sum \varphi_n(x) \varphi_n(t) \|\varphi_n\|^{-2}$. Clearly, (XVIII.1.5) and (XVIII.1.6) also yield, in a formal way, the response to a given excitation, i.e., the solution of

$$Ay - \lambda_0 y = f(y) \qquad \text{(XVIII.1.9)}$$

as

$$y(x) = \text{l.i.m.} \int_{-\infty}^{+\infty} \frac{F(\lambda)}{\lambda - \lambda_0} \varphi(x, \lambda) d\rho(\lambda), \qquad \text{(XVIII.1.10)}$$

where l.i.m. means that the limit of the integral over infinite domains is understood in the sense of the quadratic norm. Hence, if a formally self-adjoint operator D depends on some unknown parameters, which are to be determined from some set of spectral data, a particularly appropriate intermediate step is the spectral function ρ_A of any given self-adjoint extension A of D Constructing A from ρ_A is therefore a very important inverse problem. For second-order differential operators D, this problem is solved completely by the Gel'fand–Levitan method, and it is no surprise if so many complete studies of inverse problems begin, in fact, with constructing the spectral function.

XVIII.1.2   The Classical String Model

The simplest problem governed by a formally self-adjoint second-order differential operator of a general form is

$$-\frac{d}{dx}\left[p(x) \frac{dY}{dX} \right] + Q(X)Y = \mu R(X)Y, \qquad \text{(XVIII.1.11)}$$

$$Y(0)\cos A + Y(0)\sin A = 0, \qquad \text{(XVIII.1.12a)}$$

$$Y(L)\cos B + Y'(L)\sin B = 0, \qquad \text{(XVIII.1.12b)}$$

where Y and pY' are absolutely continuous functions on $(0, L)$. The assumptions are $p > 0$, $R > 0$, p is piecewise continuous on $(0, L)$; in particular, p is continuous near 0 and near L, p, Q, R integrable on $[0, L]$. With these assumptions it is possible to construct an equation equivalent to (**XVIII.1.11**) by setting $x = \int_0^{\bar{x}} [p(t)]^{-1} \, dt$, the so-called *Sturm–Liouville equation*

$$\left[-\frac{d^2}{dx^2} + q(x) \right] y(x) - \lambda r(x) y(x) = 0, \qquad \textbf{(XVIII.1.13)}$$

$$y(0)\cos \alpha + y'(0)\sin \alpha = 0, \qquad \textbf{(XVIII.1.14a)}$$

$$y(l)\cos \beta + y'(l)\sin \beta = 0. \qquad \textbf{(XVIII.1.14b)}$$

If some additional assumptions are valid, the stronger one being that $[R(X)p(X)]''$ exists at least in L^1, the Liouville transform $x = \int_0^X dt \, [R(t)]^{1/2} [p(t)]^{-1/2}$ yields the *Schrödinger equation*

$$\left[-\frac{d^2}{dx^2} + q(x) \right] y(x) - \lambda y(x) = 0, \qquad \textbf{(XVIII.1.15)}$$

which looks like (**XVIII.1.13**) with $r = 1$. On the contrary, we can meet (**XVIII.1.13**) with $q = 0$. It is then called the classical string equation and it is convenient to write it in the form

$$-[M'(x)]^{-1} y'' = \mu y, \qquad \textbf{(XVIII.1.16)}$$

where $M(x)$ is any nondecreasing finite function on $(0, l)$, strictly increasing in the neighborhood of 0 and l. This amounts to saying that $dM(x)$ is a positive measure. The problem is completed by two homogeneous boundary conditions at 0 and l. It is possible to show that, in the classical string problem, if l is finite, the spectrum is that of a self-adjoint compact operator obtained by setting the differential equation *and* the boundary conditions, and therefore this spectrum depends on the boundary conditions. The corresponding physical problem would be that of a string of variable mass: the boundary conditions are obtained by fixing the string at 0 and l and, in physical cases, they are obviously not arbitrary. The inverse problem is: "being given the frequencies of a string of variable mass $\mu(x)$, i.e., the eigenvalues, is it possible to construct $\mu(x)$?" The correct answer is *no* (Ambarzumian, 1929), although heuristic remarks suggest *yes* (Lord Rayleigh, 1877). We need either two independent sets of eigenvalues (different boundary conditions) or one set and additional information (e.g., symmetry).

The existence and uniqueness theorem of Kac and Krein (1974), which concluded several years of study by Krein, completely fixes this inverse problem, even in the case when $l \to \infty$, but it does not give a general construction method of the string. Methods are known in simpler cases, the two extreme ones being:

(a) The case where $dM(x)$ is a finite sum of Dirac measures. The number of eigenfrequencies is then finite and the problem can be treated by linear algebra.
(b) The case where the string problem can be reduced to the Schrödinger

form studied below, and for which the Gel'fand–Levitan construction applies.

The string problem has been generalized to drums, cavities, and more general vibrating systems, with or without holes, in multidimensional spaces. The famous lecture by Mark Kac (1966) on "Can one hear the shape of a drum?" shows the interest in this subject, but which is beyond the scope of the present book.

XVIII.1.3   The Regular Sturm–Liouville Problem

We shall study the regular Sturm–Liouville problem in the "Schrödinger form" (XVIII.1.15) only. Setting $\lambda = k^2$ ($k \in \mathbb{C}$), and writing the boundary conditions, we study the problem

$$\left[-\frac{d^2}{dx^2} + q(x) \right] y(k, x) = k^2 y(k, x), \qquad \text{(XVIII.1.17)}$$

$$\sin \alpha y'(k, 0) + \cos \alpha y(k, 0) = 0, \qquad \text{(XVIII.1.18a)}$$

$$\sin \beta y'(k, l) + \cos \beta y(k, l) = 0. \qquad \text{(XVIII.1.18b)}$$

Let us now briefly review the results on the direct problem.

Properties of Solutions of (XVIII.17). If it is given, at a point $c \in (0, l)$, the values $y(k, c)$ and $y'(k, c)$, transforming (XVIII.1.17) into a Volterra integral equation shows that the solution $y(k, x)$ exists and is unique in $C(0, l)$. Let $y_1(k, x)$ and $y_2(k, x)$ be two solutions on $[a, b]$ of two equations (XVIII.1.17) depending on the sets of parameters $q_1, k_1^2; q_2, k_2^2$. Elementary calculations yield the Lagrange identity

$$W(y_2, y_1)]_a^b = \int_a^b (q_1 - q_2 + k_2^2 - k_1^2) y_1 y_2 \, dx. \qquad \text{(XVIII.1.19)}$$

Eigenfunctions of (XVIII.1.17), (XVIII.1.18). This system has solutions only for particular values of k^2, the eigenvalues—the solutions are then the corresponding eigenfunctions. By means of the Lagrange identity it is a simple exercise to show that the eigenfunctions corresponding to two different eigenvalues are orthogonal, that the eigenvalues are all real (*hint*: study the complex conjugate of an eigenfunction) and that each eigenspace is one-dimensional (*hint*: calculate the Wronskian of two eigenfunctions which would correspond to the same eigenvalue).

The Green Function. The Green function of the problem (XVIII.1.17), (XVIII.1.8) is the resolvent kernel $G(x, y, k)$ of the inhomogeneous problem obtained by adding $-f(x)$ on the right-hand side of (XVIII.1.17): the solution of such a problem should be $\int_0^l G(x, y, k) f(y) dy$. It is well known (Courant and

Hilbert, 1932) that G can be constructed from the Jost solutions u and v which are defined by

$$u(k, 0) = \sin \alpha; \qquad u'(k, 0) = -\cos \alpha, \qquad \text{(XVIII.1.19a)}$$

$$v(k, l) = \sin \beta; \qquad v'(k, l) = -\cos \beta. \qquad \text{(XVIII.1.19b)}$$

The Green function is for any k such that $W(u, v) \neq 0$

$$G(x, y, k) = \begin{cases} [W(u, v)]^{-1}u(k, x)v(k, y), & (x \leq y), \\ [W(u, v)]^{-1}u(k, y)v(k, x), & (x \geq y). \end{cases} \qquad \text{(XVIII.1.20)}$$

G also enables us to transform the problem (**XVIII.1.17**), (**XVIII.1.18**) into the integral equation

$$y(k, x) + k^2 \int_0^l G(x, t, k)y(k, t)dt = 0 \qquad \text{(XVIII.1.21)}$$

(it has been implicitly assumed that 0 is not an eigenvalue). Thanks to (**XVIII.1.20**), the integral operator in (**XVIII.1.21**) maps $C(0, l)$ into a set of equicontinuous functions, and it follows that the operator is compact. Hence the spectrum is a countable set of eigenvalues. It is not difficult to see that this set contains a lowest element k_0^2 if $q(x)$ is bounded below, and that the eigenvalues are the zeros of $W[u, v]$.

Other Properties of Spectrum. We shall order the eigenvalues by increasing values $k_0^2 < k_1^2 < \cdots$, it being understood that the first eigenvalues may be negative. If it is possible to show (one-dimensional form of the *nodal* property) that the eigenfunction corresponding to λ_n has exactly n zeros in $[0, l]$, it is also possible to give the asymptotic distribution of eigenvalues by means of the asymptotic behavior of the problem solutions. We consider two cases.

(a) $\sin \alpha \neq 0$

Setting $h = -\cot \alpha$, $H = \cot \beta$, we define the "regular solution" of (**XVIII.1.12**) (with $r = 1$), $\varphi(x, \lambda)$ by the initial conditions $\varphi(0, \lambda) = 1$; $\varphi'(0, \lambda) = h$. Thus this is the solution in $C(0, l)$ of

$$\varphi(x, \lambda) = \cos kx + h\frac{\sin kx}{k} + \int_0^x \frac{\sin[k(x - t)]}{k} q(t)\varphi(t, \lambda)dt. \qquad \text{(XVIII.1.22)}$$

This Volterra equation is solved by its Neumann series $\sum_n \varphi_n(x, \lambda)$, and the majorant series shows that $\varphi(x, k^2)$ is entire and

$$\left| \varphi(x, k^2) - \cos kx - h\frac{\sin kx}{k} - \int_0^x \frac{\sin k(x - t)\cos kt}{k} q(t)dt \right|$$

$$\leq \frac{C \exp[|\text{Im } k|x]}{k^2} \qquad (k \in \mathbb{C}), \qquad \text{(XVIII.1.23)}$$

and similar bounds for $(\partial/\partial x)\varphi$, $(\partial/\partial k)\varphi$, $\partial^2 \varphi/\partial x\, \partial k$. Setting, for finite H,

$$\mathscr{D}(k) = H\varphi(l, k^2) + \varphi'(l, k^2) = (\sin \alpha \sin \beta)^{-1}\omega(k^2), \qquad \text{(XVIII.1.24)}$$

we can obtain the asymptotic form of this entire function of k, whose zeros k_n

396   Problems Connected with Discrete Spectra

are eigenvalue root squares, and similarly for $H = \infty$, obtaining

$$(H \neq \infty) \quad k_n = \frac{n\pi}{l} + (n\pi)^{-1}\left[h + H + \frac{1}{2}\int_0^l q(t)dt\right] + o(n^{-1}),$$

$$(H = \infty) \quad k_n = (n + \tfrac{1}{2})\frac{\pi}{l} + (n\pi)^{-1}\left[h + \frac{1}{2}\int_0^l q(t)dt\right] + o(n^{-1}),$$

$$(\text{any } H) \quad \|\varphi(\cdot, \lambda_n)\| = \left[\int_0^l [\varphi(x, \lambda_n)]^2\, dx\right]^{1/2} = \left(\frac{l}{2}\right)^{1/2}[1 + o(n^{-1})].$$

$$(\text{XVIII.1.25})$$

(b) $\sin \alpha = 0$

The regular solution in this case (labeled $h = \infty$) is defined by (XVIII.1.14) and $\varphi_\infty(0, \lambda) = 0$, $\varphi'_\infty(0, \lambda) = 1$. This is the solution of the integral equation

$$\varphi_\infty(x, \lambda) = k^{-1} \sin kx + k^{-1}\int_0^x \sin k(x - t)q(t)\varphi_\infty(t, \lambda)dt, \quad (\text{XVIII.1.26})$$

which can be studied as φ, has the same analytic properties, and the bound

$$\left|k\varphi_\infty(x, \lambda) - \sin kx - k^{-1}\int_0^x \sin k(x-t)\sin kt\, q(t)dt\right| < C|k|^{-2}\exp[|\text{Im } k|x],$$

$$(\text{XVIII.1.27})$$

with similar results for the derivatives. Hence

$$(H \neq \infty) \quad k_n = (n + \tfrac{1}{2})\frac{\pi}{l} + (n\pi)^{-1}\left[H + \frac{1}{2}\int_0^l q(t)dt\right] + o(n^{-1}),$$

$$(H = \infty) \quad k_n = \frac{n\pi}{l} - \tfrac{1}{2}(n\pi)^{-1}\int_0^l q(t)dt + o(n^{-1}),$$

$$(\text{any } H) \quad \|\varphi(\cdot, \lambda_n)\| = \left(\frac{l}{2}\right)^{1/2}[1 + o(n^{-1})].$$

$$(\text{XVIII.1.28})$$

It is important to notice that in all these formulas, the remainder $o(n^{-1})$ is in fact the sum of a term $O(n^{-2})$ and an integral, which may be of either form

$$\int_0^a \cos ptq(t)dt \quad \text{or} \quad \int_0^a \sin ptq(t)dt, \quad (\text{XVIII.1.29})$$

where $0 \leq a \leq l$ and p is proportional to n or $(n + \tfrac{1}{2})$. The order $o(n^{-1})$ follows from the integrability of q and the Riemann–Lebesgue theorem. Convergence checks with this single assumption require careful appraisals. But if q is more regular, the remainder is smaller and convergence checks in the studies below are very easy, as soon as the remainder is $O(n^{-1-\varepsilon})$ with $\varepsilon > 0$. If q has an integrable derivative, the remainder is $O(n^{-2})$. The reader who wants further results for more regular q will have to continue the expansions in (XVIII.1.23) and (XVIII.1.27)—not a difficult task.

Expansion Theorem. Thanks to the estimates (**XVIII.1.23**)–(**XVIII.1.28**), it is easy to see that the series of normalized eigenfunctions products

$$\sum_{n=0}^{\infty} (\lambda_n)^{-1} \bar{\varphi}(x) \bar{\varphi}_n(y) := p(x, y)$$

is absolutely and uniformly convergent towards a continuous function, so that $p(x, y) + G(x, y, 0) := Q(x, y)$ is a continuous symmetric function and we can calculate $\int_0^l Q(x, y) \bar{\varphi}_n(y) dy$ term after term. Because of (**XVII.1.7**), the result is zero. It follows that any eigenfunction $u(x)$ of Q is orthogonal to every $\bar{\varphi}_n$, whereas it is possible to check that $u(x)$ is also an eigenfunction of $G(x, y, 0)$. Hence u is one of the φ_n's and is orthogonal to all of them. Therefore it vanishes, so that Q has no spectrum and it vanishes, yielding the expansion theorem for the Green's function

$$G(x, y, 0) = - \sum_{n=0}^{\infty} (\lambda_n)^{-1} \bar{\varphi}_n(x) \bar{\varphi}_n(y). \qquad \textbf{(XVIII.1.30)}$$

Let f be a real-valued function in $C_2(0, l)$, satisfying the boundary conditions (**XVIII.1.18**). Set $g(x) = f''(x) - q(x)f(x)$. Using the resolvent properties of $G(x, y, 0)$ together with (**XVIII.1.30**), we obtain a series expansion for $g(x)$ and hence for $f(x)$

$$f(x) = \sum_{n=0}^{\infty} f_n \varphi(x, k_n^2)/\|\varphi(\cdot, k_n^2)\|, \qquad \textbf{(XVIII.1.31a)}$$

$$f_n = \int_0^l dx\, f(x)\varphi(x, k_n^2)/\|\varphi(\cdot, k_n^2)\|, \qquad \textbf{(XVIII.1.31b)}$$

$$\int_0^l |f(x)|^2\, dx = \sum_{n=0}^{\infty} |f_n|^2. \qquad \textbf{(XVIII.1.31c)}$$

Using the density in $L^2(0, l)$ of the class we chose for f, we can extend these results to any element of $L^2(0, l)$, and write them down as in the spectral theorem (**XVIII.1.5**), (**XVIII.1.7**)

$$\int_0^l |f(x)|^2 = \int_{-\infty}^{+\infty} |F(\lambda)|^2\, d\rho(\lambda), \qquad \textbf{(XVIII.1.32)}$$

$$F(\lambda) = \int_0^l f(x)\varphi(x, \lambda)dx, \qquad \textbf{(XVIII.1.33)}$$

$$\rho(\lambda) = \text{sgn}(\lambda) \sum_{\lambda_n < \lambda} \|\varphi(\cdot, k^2)\|^{-2}, \qquad \textbf{(XVIII.1.34)}$$

where sgn λ is the sign of λ, and $\lambda_n = k_n^2$ may be positive or negative. In addition, from (**XVIII.1.32**), we readily show the canonical isomorphism between $L^2(0, l)$ and $L_{d\rho}^2(-\infty, +\infty)$ (which is written here for real-valued functions)

$$\int_0^l f(x)g(x)dx = \int_{-\infty}^{+\infty} F(\lambda)G(\lambda)d\rho(\lambda). \qquad \textbf{(XVIII.1.35)}$$

Hence the spectral function $\rho(\lambda)$ has a very transparent interpretation in the

Sturm–Liouville problems, where it yields a natural generalization of Fourier series through formulas (**XVIII.1.31**) and (**XV.III.1.33**). Gel'fand and Levitan used this property to reconstruct q from the spectral function. For that they needed:

The Transformation Kernels. We already know that a "transformation kernel" is the kernel of a "transformation operator," which is able to transform a regular solution of the problem (**XVIII.1.7**), k being arbitrary and $q = 0$, into a regular solution of the same problem with the same k and $q \neq 0$.

Even if it is imposed, in addition, that the transformation kernel be triangular, several transformation kernels can be constructed if the Sturm–Liouville problem is only defined by (**XVIII.1.17**). Not only is it possible to choose the boundary conditions (**XVIII.1.18**), including the interval, but it is even possible to define transformation operators that can change them. The reader can reconstruct these results (or see them, for instance, in Sabatier (1987c))—from formulas (**XVIII.1.22**)–(**XVIII.1.25**)—by following the line of study we shall give here in the case when $\sin \alpha = 0$. This case is of course closer to those of quantum scattering theory, since the "regular wave function" is defined as in quantum mechanics. For the sake of simplicity we shall take off, in the following, the subscript ∞ used in (**XVIII.1.26**), (**XVIII.1.27**).

From (**XVIII.1.26**) we see that

$$\Phi(k) =: k\varphi(x, k^2) - \sin kx - \int_0^t dt\, q(t)\sin k(x - t)k^{-1} \sin kt \quad \textbf{(XVIII.1.36)}$$

is an odd entire function of k of exponential type x, whose restriction to the real axis is real and belongs to $L^2(\mathbb{R}) \cap L^1(\mathbb{R})$. The last term on the right-hand side of (**XVIII.1.36**) belongs to $L^2(\mathbb{R})$, so that

$$K(x, y) = (i\pi)^{-1} \int_{-\infty}^{+\infty} dk\, [k\varphi(x, k^2) - \sin kx]\exp[-iky] \quad \textbf{(XVIII.1.37)}$$

exists, at least in $L^2(\mathbb{R})$, as an odd function of y. The Fourier transform of $\Phi(k)$ is a continuous function of y; it can be evaluated for $|y| < |x|$ by a contour integration and shown to vanish. The Fourier transform of the last term in (**XVIII.1.36**) can be calculated exactly and also vanishes for $|y| < |x|$. Hence, inserting (**XVIII.1.36**) into (**XVIII.1.37**), we show that $K(x, y)$ vanishes for any $|y| < |x|$ and that

$$K(x, x^-) = \frac{1}{2} \int_0^x q(t)dt, \quad \textbf{(XVIII.1.38)}$$

where $|x^-| = \lim_{\varepsilon \to 0^+} |x| - \varepsilon$. Inverting (**XVIII.1.37**), and taking into account the parity, we get

$$\varphi(x, k^2) = k^{-1} \sin kx + k^{-1} \int_0^x K(x, t)\sin kt\, dt. \quad \textbf{(XVIII.1.39)}$$

Thus, the operator whose kernel is $\delta(x - t) + K(x, t)$ transforms $k^{-1} \sin kx$ into $\varphi(x, k^2)$. Inverting it (i.e., solving equation (**XVIII.1.39**) for $k^{-1} \sin kx$) yields a new kernel $\tilde{K}$, which transforms φ into $k^{-1} \sin kx$, is triangular (i.e., $= 0$ for $|t| > |x|$), and is given for $x \geq t \geq 0$ by the equation

$$K(x, t) + \tilde{K}(x, t) = -\int_t^x \tilde{K}(x, u)K(u, t)du = -\int_t^x K(x, u)\tilde{K}(u, t)du.$$

$$(\text{XVIII.1.40})$$

It is not difficult to derive from (**XVIII.1.26**), (**XVIII.1.27**), and (**XVIII.1.17**) the following properties of the transformation kernel:

(1) For any $x > t \geq 0$, $K(x, t)$ and $\tilde{K}(x, t)$ are continuous functions of x and t, and have $(m + 1)$ locally integrable derivatives if $q(x)$ has m locally integrable derivatives.

(2) For $x > y \geq 0$, $K(x, y)$ is the solution of the integral equation

$$K(x, y) = \frac{1}{2}\int_{(x-y)/2}^{(x+y)/2} q(t)dt + \int_{(x-y)/2}^{(x+y)/2} dr \int_0^{(x+y)/2} ds\, q(r+s)K(r+s, r-s)$$

$$(\text{XVIII.1.41})$$

and the partial differential equation

$$\left(\frac{\partial^2}{\partial x^2} - q(x) - \frac{\partial^2}{\partial y^2}\right)K(x, y) = 0. \qquad (\text{XVIII.1.42})$$

$\tilde{K}(x, y)$ is the solution of the transposed equation. The boundary condition on the diagonal border x, x^- is given by (**XVIII.1.38**). The boundary condition at $y = 0$ is

$$K(x, 0) = \left[\frac{\partial}{\partial y}K(x, y)\right]_{y=0} = 0. \qquad (\text{XVIII.1.43})$$

The property (**XVIII.1.42**) and the boundary conditions can be derived from the integral equation (**XVIII.1.41**). Conversely, using the Riemann method enables us to show that the solutions of (**XVIII.1.42**), completed by the diagonal boundary condition and (**XVII.1.43**), satisfy the integral equation (**XVIII.1.41**).

(3) Let $K(x, y)$ have locally integrable second derivatives which satisfy (**XVIII.1.41**). Then the function φ defined by (**XVIII.1.39**) is a regular solution of (**XVIII.1.17**), i.e., it satisfies this equation and is $O(x)$ as $x \to 0$.

Similar properties are obtained for the transformation kernels that correspond to other conditions at $x = 0$. They can be seen, for instance, with proofs of the stated properties, in Sabatier (1987c).

XVIII.1.4   The Inverse Sturm–Liouville Problem

As we have already stated, physical inverse problems arise for Sturm–Liouville equations when information is given on spectra of eigenvalues or on the response to a given source. All these problems can be reduced to or compared with "spectral function" inverse problems, where the input is information on the spectral function. It has been shown (Gel'fand–Levitan, 1951) that this function makes possible the reconstruction of $q(x)$. We shall first see how this can is done.

Since we are only interested in $q(x)$ on $(0, l)$, it is possible to use various wave functions, differing by the boundary conditions at $x = 0$, and to deal with various problems defined on intervals larger than $(0, l)$. It is even possible to use (for the unperturbed problem and for the perturbed one) wave functions that fit different boundary conditions at $x = 0$. However, for the sake of simplicity, we shall use the same wave function for the $q = 0$ and $q \neq 0$ problems, namely the "regular wave function," which is $O(x)$ as $x \to 0$. We assume that one gives the spectral function $\rho(\lambda)$ corresponding to $q(x)$ and to the Sturm–Liouville problem with the $y = 0$ boundary condition at $x = 0$, and another (homogeneous) one at $x = l$. The way we proceed is quite similar to the formal way described in section XIV.2.

Let $\rho^0(\lambda)$ be a spectral function corresponding to $(0, l)$, $q = 0$, and the boundary condition $y = 0$ at $x = 0$. Let F be the symmetrical function

$$F(x, y) = \int d[\rho(\mu) - \rho^0(\mu)]\varphi_0(x, \mu)\varphi_0(y, \mu), \qquad \textbf{(XVIII.1.44)}$$

where $\varphi_0(x, \mu) = k^{-1} \sin kx$, $\mu = k^2$. Taking into account **(XVIII.1.34)**, we get

$$F(x, y) = \sum_0^\infty \left[\frac{\varphi_0(x, \mu_n)\varphi_0(y, \mu_n)}{\|\varphi(\cdot, \mu_n)\|^2} - \frac{\varphi_0(x, \mu_n^0)\varphi_0(y, \mu_n^0)}{\|\varphi(\cdot, \mu_n^0)\|^2} \right]. \qquad \textbf{(XVIII.1.45)}$$

For integrable q, $F(x, y)$ is continuous, but the two terms in **(XVIII.1.45)** cannot be summed separately since their convergence is guaranteed only in the sense of operators, i.e., if we calculate $\langle g, F, f \rangle$ for any two elements f, g of $C_2(0, l)$. (*Hint:* For $f \in C_2(0, l)$, $\int f\varphi = \mu^{-1} \int fD\varphi = \mu^{-1} \int \varphi Df$ is $O(n^{-2})$ for $\mu = \mu_n$.)

The Gel'fand–Levitan Equation. Let T be an operator, say, $\delta(x - y) + K(x, y)\theta(y)\theta(x - y)$, which transforms $\varphi_0(x, y)$ into $\varphi(x, y)$ $(x > 0)$. Let T^* be the adjoint of T, say, $\delta(x - y) + K^*(x, y)\theta(x)\theta(y - x)$ and $(T^*)^{-1} = \delta(x - y) + \tilde{K}^*(x, y)\theta(x)\theta(y - x)$ the inverse of T^*. For differentiable q these operators exist, they map C_2 into C_2 and $K(x, y)$ is twice differentiable for $0 < y < x$. Now, for any f and g in C_2, it is easy to show that

$$\langle f, TFT^*g \rangle = \langle f, g \rangle - \langle f, TT^*g \rangle \qquad \textbf{(XVIII.1.46)}$$

by using, on the left-hand side, formula **(XVIII.1.45)** and the expansion theorem. Hence $TF - (T^*)^{-1} + T$ vanishes in the sense of the operator. The kernel of this operator vanishes almost everywhere. If we write it for $0 < y < x$, we see that it is a continuous function and its disappearance yields the Gel'fand–Levitan equation

$$F(x, y) + \int_0^x dt\, K(x, t)F(t, y) + K(x, y) = 0 \qquad (0 < y < x). \quad \textbf{(XVIII.1.47)}$$

Reconstruction of $q(x)$. The "potential" $q(x)$ can be reconstructed from its spectral function $\rho(\mu)$ ($\rho(\mu)$ is determined by q and also by l and the boundary conditions—here we have fixed the condition at $x = 0$ by imposing the wave function as $O(x)$). The reconstruction is achieved by constructing $F(x, y)$ by means of (**XVIII.1.44**), solving the Gel'fand–Levitan equation (**XVIII.1.46**), and calculating $q(x)$ from $K(x, y)$ by means of (**XVIII.1.38**). The unperturbed spectral function $\rho^0(\mu)$ can be calculated from its definition (**XVIII.1.34**). For $l = \infty$, any boundary condition ρ^0 can be written in a simple form (see below). For finite l, with the condition $O(x)$ at $x = 0$, ρ^0 can be written in a simple form if the condition is also zero at $x = l$, so that $k_n^{-1} \sin k_n l = 0$ and $k_n = n\pi/l$. Then we can write

$$\frac{d\rho_0(\mu)}{d\mu} = 2 \sum_{n=1}^{\infty} \mu_n \delta(\mu - \mu_n). \qquad \text{(XVIII.1.48)}$$

Let us now go through the reconstruction problem. For the sake of simplicity, assume $q(x)$ is continuous, and assume we know $d\rho(\mu)/d\mu$. If we construct $F(x, y)$, we know from our previous study that it is continuous. Then the Gel'fand–Levitan equation is a Fredholm equation, with compact kernel, and the Fredholm alternative holds. It is possible to rule out trivial solutions of the homogeneous equation by using the positivity of the spectral measure. The inversion is possible and is stable. The result can be generalized to summable q, and the consistency of q with the spectral function can be proved. For more details, we refer to Sabatier (1987c).

It is of interest to note that from the bilinear expansion (**XVII.1.44**) we can easily derive the following results, which hold at least in the sense of operators,

$$K(x, y) = \int_{-\infty}^{+\infty} d[\rho^0(\mu) - \rho(\mu)]\varphi(x, \mu)\varphi_0(y, \mu), \quad 0 < y < x, \quad \text{(XVIII.1.49)}$$

$$\cdot \left(\frac{\partial^2}{\partial x^2} - \frac{\partial^2}{\partial t^2}\right) F(x, t) = 0,$$

$$F(x, 0) = 0, \qquad \text{(XVIII.1.50)}$$

and it may be of interest for the reader to prove the following result: If F has integrable second-order derivatives satisfying (**XVIII.1.50**), the solution of the Gel'fand–Levitan equation is the transformation operator K corresponding to the potential $q(x) = 2(d/dx)K(x, x)$. Such a result shows that, as in the previous chapters, the Gel'fand–Levitan machinery can be used with F as an input to construct examples of potentials.

Inverse Problem Starting from Two Spectra. From the physical point of view, reconstructing $q(x)$ from its spectral function is determining a distributed parameter from the medium response to an ideally localized source (a Dirac measure, in fact). In some way, this amounts to determining two sequences, the sequence of eigenvalues and that of the response amplitudes for each of these modes. We can also ask whether two "different" complete sets of eigenvalues determine the spectral function of a Sturm–Liouville problem; for instance, those corresponding to the same (homogeneous) boundary condition at $x = l$ and two different boundary conditions at $x = 0$ say, $\varphi'/\varphi = h_1$ and $\varphi'/\varphi = h_2$. The answer is yes. These two spectra (which have interlacing values) enable us to reconstruct a spectral function of $q(x)$ and hence determine the potential. Again we refer to our other book or to original references (see section XVIII.4) for more details.

XVIII.2   Relations with Other Problems and Extensions

XVIII.2.1   Relations with Radial Problems
The Half-Axis Sturm–Liouville Problem

Let us first study the Sturm–Liouville equation (**XVIII.1.17**) with the regular condition at $x = 0$, say,

$$D_0 \varphi = \left[-\frac{d^2}{dx^2} + q(x) \right] \varphi(x, k^2) = k^2 \varphi(x, k^2), \qquad \textbf{(XVIII.2.1)}$$

$$x^{-1} \varphi(x, k^2) \to 1 \qquad (x \to 0). \quad \textbf{(XVIII.2.2)}$$

We have already obtained a number of results concerning the operator in (**XVIII.1.51**) where $q \in L^1(0, \infty)$. To study its spectrum, we are led to introduce the Jost solution $f(x, k)$ which is by definition asymptotic to e^{ikx} as $x \to \infty$, real k. This has been studied in chapter I, in the framework of the S-wave problem. Recalling the results of chapters I and II also, we introduce the Green's function

$$G(x, y, \mu) = \begin{cases} -[F(k)]^{-1} \varphi(x, k^2) f(t, k) & (x \le t), \\ -[F(k)]^{-1} \varphi(t, k^2) f(x, k) & (x \ge k), \end{cases} \quad \textbf{(XVIII.2.3)}$$

where $k = \mu^{1/2}$ (Im $k \ge 0$). We shall never need to go into the second Riemann sheet. In the μ-plane, G is not defined on the half-axis $\mu \ge 0$, which is a "cut" across which G is not continuous. $F(k)$ is the Jost function $W\{f(x, k), \varphi(x, k^2)\}$ and, as we already know, is holomorphic for Im $k > 0$, continuous as Im $k \to 0^+$, and goes to $2ik$ as $|k| \to \infty$. Thus $F(k)$ has only a finite number of isolated zeros $\{k_n\}$ for Im $k \ge 0$. They can be identified with bound states and are necessarily simple and located on Re $k = 0$, Im $k \ge 0$. The resolvent of our problem does not exist at these points and on the real k-axis: the $\mu \ge 0$ half-axis is called the *continuous spectrum*, the zeros $\mu_n = -|k_n|^2$ define the discrete spectrum of the Schrödinger operator $-d^2/dx^2 + q(x)$ defined on the half-axis.

So as to identify completely the half-axis extension of a Sturm–Liouville problem and the radial problem studied in chapters I and II, let us show how the expansion theorem goes over to the completeness relations of section II.3. Let g be a finite support C_2 function on $\mathbb{R}^+$, and let $h = g'' - qg$. For any

μ which is not in the spectrum, the solution of

$$\left[-\frac{d^2}{dx^2} + q(x) - \mu \right] g(x) = -[h(x) + \mu g(x)] \qquad \textbf{(XVIII.2.4)}$$

is

$$\mu^{-1} g(x) = \int_0^\infty G(x, t, \mu) g(t) dt + \mu^{-1} \int_0^\infty G(x, t, \mu) h(t) dt. \qquad \textbf{(XVIII.2.5)}$$

Integrating both sides of (**XVIII.2.5**) in the μ-plane on a path made of a large circle indented along the real positive axis, we obtain the expansion formula

$$g(x) = \int_{-\infty}^{+\infty} d\rho(\mu) \int_0^{+\infty} dt \, \varphi(x, \mu) \varphi(t, \mu) g(t), \qquad \textbf{(XVIII.2.6)}$$

where the spectral measure $d\rho(\mu)$

$$d\rho(\mu) = \begin{cases} \pi^{-1} \mu^{1/2} |F(\mu^{1/2})|^{-2} \, d\mu & (\mu \geq 0), \\ \sum \|\varphi(\cdot, \mu_n)\|^{-2} \delta(\mu - \mu_n) & (\mu < 0), \end{cases} \qquad \textbf{(XVIII.2.7)}$$

is clearly made of two parts—one that corresponds to the continuous spectrum and one that corresponds to the discrete spectrum. We should notice that the regular (finite l) Sturm–Liouville case is not a trivial restriction on the half-axis case, since $F(k)$ is not defined by a trivial limit of the boundary condition at l. Yet, it is not difficult to see that the finite l spectral functions go to the half-axis spectral function as $l \to \infty$.

It is easy to make a similar study for any homogeneous boundary condition at $x = 0$, i.e., for a wave function proportional to $\cos kx + h \sin kx/k$, $h \in \mathbb{R}$. It is also easy to extend the Gel'fand–Levitan theory to the half-axis case. As a matter of fact, the whole theory of section XVIII.1.4 can be reproduced without any modification provided we work "in the sense of operators." In the "finite h" case, it is even possible to use $\cos kx$ instead of $\cos kx + (h/k) \sin kx$ in the symmetrical kernel F, since this amounts to modifying the theory by a zero operator. For wave functions $\cos kx$ and for regular wave functions $k^{-1} \sin kx$, it is easy to calculate the spectral measure (that corresponds to $q = 0$) in a closed form. There is no discrete spectrum and

$$\frac{d\rho^0}{d\mu} = \begin{cases} \pi^{-1} \mu^{-1/2} & (\varphi_0 = \cos kx), & \textbf{(XVIII.2.8a)} \\ \pi^{-1} \mu^{1/2} & (\varphi_0 = k^{-1} \sin kx). & \textbf{(XVIII.2.8b)} \end{cases}$$

Using (**XVIII.2.8b**) in (**XVIII.1.43**), we again obtain the symmetrical kernel used in chapter III, and the Gel'fand–Levitan theory constructs a transformation kernel similar to that of chapter III, in agreement with the transformation formula (**XVIII.1.39**).

Normal Modes of a Ball. Spectral problems with spherical symmetry are shown in the theory of Earth normal modes. To see how they are involved, we consider the simplest (and academic) example of a (liquid) ball where (acoustical) waves are propagated by the equation

$$\rho \frac{\partial^2 u}{\partial t^2} = \text{grad}(\lambda \text{ div } u), \qquad \text{(XVIII.2.9)}$$

u is a vector, and the density ρ and the Lamé parameter λ are functions that depend only on the radius r. We are interested by motions u that derive $(u = \rho^{-1} \text{ grad } P)$ from an harmonic potential $P = p(x, \omega)e^{i\omega t}$

$$-\omega^2 p = \lambda \text{ div}[\rho^{-1} \text{ grad } p]. \qquad \text{(XVIII.2.10)}$$

Now, in a spherical coordinates system $(0, r, \theta, \varphi)$, we look for motions that do not depend on φ, expanding p on Legendre polynomials. The "partial wave" $p_l(\omega, r)$ should satisfy the equation

$$\left[-\rho r^{-2} \frac{d}{dr} r^2 \rho^{-1} \frac{d}{dr} + l(l+1)r^{-2} \right] p_l(\omega, r) = \omega^2 \rho \lambda^{-1} p_l(\omega, r) \quad \text{(XVIII.2.11)}$$

in such a way that p_l and $\rho^{-1} d/dr \, p_l$ are differentiable and that the boundary conditions

$$[rp_l(\omega, r)]_{r=0} = p_l(\omega, R) = 0, \qquad \text{(XVIII.2.12)}$$

where R is the ball radius. Clearly this is a spectral problem which has a solution only if ω belongs to a set $\{\omega_{ln}\}$ of eigenvalues (modes) and p_l is the corresponding eigenfunction. The problem equations look like the radial Schrödinger equations. If λ and ρ are strictly positive and C^2 on $(0, R)$, introducing the local velocity $c = (\lambda/\rho)^{1/2}$, arrival time $\tau = \int_0^R dx/c(x)$, and impedance $Z = r^{-1}(\rho c)^{1/2}$, we indeed obtain (Sabatier, 1978a) Schrödinger equations

$$\left[\frac{d^2}{d\tau^2} + \omega^2 - V(\tau) + l(l+1)U(\rho) - l(l+1)\tau^{-2} \right] \varphi_l(\omega, \tau) = 0 \quad \text{(XVIII.2.13)}$$

where $V(\tau) = Z \, d^2/d\tau^2 \, Z^{-1}$ and $U(\tau) = \tau^{-2} - r^{-2}c^2$ are continuous on $[0, \tau(R)]$. The boundary condition on φ_l is that it should vanish at both ends. Thus we have a set of Sturm–Liouville spectral problems, with the Schrödinger formulation, but they are not regular, except in the case $l = 0$. The idea of Sabatier (1978a,b) is to continue these problems into scattering problems. λ and ρ being continued by constants beyond $r = R$, equations (XVIII.2.13) define (for a plane wave $\exp[i\omega\tau \cos \theta]$) a scattering problem by an l-dependent potential $W(\tau) = V(\tau) - l(l+1)U(\tau)$. This potential being $O(\tau^{-3})$ as $\tau \to \infty$, an S-matrix can be defined, and it turns out that $S_l(\omega)$ is completely determined by the sequence $\{(\omega_{ln})\}$ *and* by a related sequence of number, for instance, the sequence of derivatives of $d/d\tau \, \varphi_l(\omega, \tau)$ at $\tau = \tau(R)$. The two sequences of course correspond to the sequence of eigenvalues and amplitudes, or to the two independent sequences of eigenvalues, or to the spectral function, in the Sturm–Liouville spectral problems. From the mathematical point of view, the two sequences yield the poles and residues of the wave function logarithmic derivative at $\tau = \tau(R)$. As is well known, since Wigner's R-matrix theory, this logarithmic derivative, given on the border $\tau = \tau(R)$ of the interaction, determines the S-matrix and, conversely, an S-matrix which has consistent meromorphic properties determines this logarithmic derivative. Hence, we can consider that in these problems, S_l gives complete information on the spectral measure for the lth spectral problem on $[0, R]$. The generality of these remarks suggests that the method can be extended to cases that are more physical than the case treated above. In the Earth normal mode problems, we may consider only elastic waves, a spherically symmetric medium characterized by its density $\rho(r)$, and Lamé parameters $\lambda(r)$ and $\mu(r)$. We then have several kinds of modes, and hence several kinds of spectral problems. Those which are governed by second-order differential equations have been shown by Sabatier (1978a,b) to be equivalent to scattering problems of the Schrödinger form, and the direct and inverse problems are now completely understood. For those which are governed by fourth-order differential equations, studies of the inverse problem are certainly not in a satisfactory state. If we try to study better models of Earth vibrations, we must take into account gravity effects, the liquid nucleus, and transitions between it and the solid mantle, together with various shifts from sphericity. A few existence and uniqueness results only are available for these (nonlinear) spectral problems. Detailed references are given in section XVIII.4 below.

XVIII.2.2   Further Extensions and Other Related Problems

Scattering problems are related to spectral problems which are "more general," in the sense that for wave motions governed by partial differential equations, spectral problems may be defined in several cases where scattering problems can not. Hence it is easy to find spectral problems which are physically interesting, and whose inverse problem is treated by methods related to those used in the present book. However, these problems or methods are far from inverse problems of quantum scattering theory: we shall only sketch them, with a few references; other problems will be given in section XVIII.4 below.

The beam problem is a generalization of the string problem and is also related to the fourth-order mode problems of the elastic Earth. Both beam problems and string problems are closely related to spectral problems of oscillatory matrices A, i.e., regular matrices whose minors are nonnegative, and the quasi-principal elements (i.e., $(a_{i,i+1})$, $(a_{i+1,i})$) are all positive. These matrices have positive distinct eigenvalues, and the inverse problem is the art of reconstructing the matrix from information related to eigenvalues. We refer the reader to the book by Gladwell (1986) for a review of these questions. Many mathematical studies have been inspired or encouraged by general spectral problems dealing with the determination of cavities, or other "n-dimensional drums," with or without holes, or by the Weyl problem of the eigenvalue distribution of a wave operator for finite domains (Kac, 1966). Others stay with Sturm–Liouville problems, but in more general cases, e.g.,

$q(x)$ is periodic (Trubowitz, 1977), or is complex-valued (Marchenko, 1987), or if the Laplacian operator replaces d^2/dx^2 (Poschel and Trubowitz, 1987).

Higher order wave equations, and the corresponding direct and inverse problems, have also been studied with unequal success. These inverse spectral problems were renewed by studies of integrable nonlinear systems. In the inverse spectral transform, we usually need to determine a parameter from the defect of holomorphy of a function in the spectral domain. The reader can refer to section XVIII.4 for examples and references. The generalized translation operators, already cited in section XV.2, were developed and used to construct formal solutions of inverse problems following the lines of Gel'fand–Levitan or Marchenko methods. Their natural framework is spectral problems rather than scattering problems.

Finally, we notice that problems of physics which are far from being spectral problems, sometimes led to analyses of inverse problems, which are related to those of inverse spectral problems because the equations are formally related. An example is given in electromagnetic soundings of the Earth where the equation is

$$
\left.
\begin{aligned}
\frac{d^2}{dz^2} E(\omega, z) &= i\mu\lambda_0 \sigma E(\omega, z), \\
E(0, \omega) &= 1, \qquad E(-\infty, \omega) = 0,
\end{aligned}
\right\}
\qquad \textbf{(XVIII.2.14)}
$$

where E is the field, ω is the frequency, and μ_0 and σ are real. When the inverse problem $dE/dz\,(0, \omega) \to \sigma$ is discretized, it leads to analyses which recall those of the string problem. When the parameters are C^2, generalized translation operators can be used. Thus, in both cases, the "algebraic" properties of the derivations do not depend heavily on the real or complex nature of the "spectral parameter," contrary to what could have been expected by considering the usual derivations. This encourages us to think that many inverse problems in physics can still be reduced to known inverse spectral problems by appropriate transformations. We shall now see another interesting example.

XVIII.3   Inverse Problem in the Coupling Constant

As an application of the results of previous sections, we consider the inverse problem in the coupling constant in its simplest form, which is as follows. We consider the Schrödinger equation on a finite interval with Dirichlet boundary conditions at both ends

$$
\varphi''(E_0, g; x) + E_0 \varphi(E_0, g; x) = g V(x)\varphi(E_0, g; x), \qquad x \in [0, R], \quad \textbf{(XVIII.3.1)}
$$

$$
\varphi(E_0, g; 0) = 0, \qquad \dot\varphi'(E_0, g; 0) = 1; \qquad \varphi(E_0, g; R) = 0. \quad \textbf{(XVIII.3.2)}
$$

The potential $V(x)$ is assumed to be real, purely attractive $(V < 0)$, twice differentiable, and regular at $x = 0$: $xV(x) \in L^1$. As we saw in previous sections, we must assume in general $V \in L^1$ with h finite for $x = 0$. However, since we are dealing here with Dirichlet conditions at $x = 0$, we can assume $xV \in L^1$ without difficulty, as is shown in general in chapters I and II. We also assume that

$$I = \int_0^R |V(x)|^{1/2}\, dx < \infty. \qquad \text{(XVIII.3.3)}$$

E_0 (real) is considered to be fixed, and what we vary here is g (the coupling constant). As we know, the conditions at $x = 0$ determine φ in a unique way for all $x \in [0, R]$, all E_0 (real or complex), and all g (real or complex). In fact, we know that φ is a real entire function of E_0 and g for every fixed x. By writing $\varphi(E_0, g; R) = 0$, we determine the "spectrum" in g: these are the values $g_n(E_0)$ of g for which the operator $-(d^2/dx^2) + gV(x)$, together with the boundary conditions, has E_0 as one of its eigenvalues. That the "spectrum" $\{g_n\}$ is real and discrete is of course clear from the self-adjointness of the operator, the finiteness of the interval, and the continuity properties of φ. It is also a consequence of this fact that the roots of the transcendental entire functions (here $\varphi(E_0, g; R)$) are isolated, infinite in number, and have infinity at a limit point only. The inverse problem in g is then to see whether V can be determined from the knowledge of $\{g_n\}$.

Henceforth, to simplify the algebra, we shall set $E_0 = 0$. Since the potential is attractive, and the operator $-(d^2/dx^2)$ with Dirichlet boundary conditions is positive, it is clear that $g_n > 0$ for all n, and $g_n \to \infty$ as $n \to \infty$. The key to the solution of the problem is now the Liouville transformation

$$x \to Z = Z(x) = \int_0^x |V(t)|^{1/2}\, dt, \qquad \varphi(x) \to \psi(Z) = \lambda[-V(x)]^{1/4}\varphi(x),$$

$$\text{(XVIII.3.4)}$$

where λ is an arbitrary (positive) parameter. This transformation, which is one-to-one and C^2, leads to

$$\ddot{\psi}(G; Z) + G^2\psi(G; Z) = \tilde{V}(Z)\psi(G; Z), \qquad \dot{} = \frac{d}{dZ},$$

$$g = G^2, \qquad Z \in [0, I], \qquad \text{(XVIII.3.5)}$$

$$\hat{V}(Z) = \frac{\ddot{F}(Z)}{F(Z)}, \qquad F(Z) = [-V(x)]^{1/4}\Big|_{x=x(Z)} \qquad \text{(XVIII.3.6)}$$

This is exactly what we need since here g plays the role of an eigenvalue. We can therefore apply the results of previous sections to (XVIII.3.5) without further ado.

From our assumptions on $V(x)$, it is clear that ψ also satisfies the Dirichlet boundary condition at both ends of the interval $Z \in [0, I]$. Indeed, at $x = 0$, φ behaves like x, and since $V(x)$ must be less singular than x^{-2}, we have $\psi(0) = 0$. At $x = R$, V being integrable and $\varphi(R) = 0$, we again have $\psi(Z = I) = 0$. We should notice here that the Liouville transformation, being a simple change of variable and function, does not modify the self-adjoint character of our problem, even though it may introduce singularities in $\tilde{V}(Z)$. The most important of these is the singularity at the origin. Indeed, if $V(x) \simeq x^{-\alpha}$, $\alpha < 2$ as $x \to 0$, we immediately obtain $\tilde{V}(Z) \simeq \Lambda/Z^2$ as $Z \to 0$ with $\Lambda > -\frac{1}{4}$. If we write $\Lambda = l(l+1)$, then $l > -\frac{1}{2}$. However, since $\psi(0) = 0$, we have a unique self-adjoint extension of $-(d^2/Z^2) + \tilde{V}(Z)$ in $L^2(0, I)$, even when $l < 0$ (the limit circle case, Titchmarsh, 1962, pp. 81–84, with $c = \infty$). In order to use the results of previous sections without generalizing them to include the centrifugal term $l(l+1)Z^{-2}$, we assume that $V(0) = V_0 \neq 0$ (finite). We shall return later to the study of the general case. In short, we now assume that $Z\tilde{V} \in L^1(0, I)$. Since we have to deal with the "S-wave," we normalize ψ, quite independently of G and λ, by

$$\dot{\psi}(0) = 1. \tag{XVIII.3.7}$$

It then follows that

$$\lambda = \lim_{x \to 0} \frac{\psi(Z(x))}{[-V(x)]^{1/4} \varphi(x)} = \lim_{x \to 0} \frac{Z(x)}{x[-V(x)]^{1/4}}. \tag{XVIII.3.8}$$

This shows that λ is independent of G. At any rate, the spectrum $\{G_n\}$ has the asymptotic property $(g = G^2)$

$$G_n = \frac{n\pi}{I} + O(1), \qquad n \to \infty. \tag{XVIII.3.9}$$

As is now clear, there are two steps for solving the inverse problem for $V(x)$. The first is to use the Gel'fand–Levitan theory of the previous section to recover ψ and the potential $\tilde{V}$ from the spectral data $\{G_n, \tilde{C}_n\}$ where $\tilde{C}_n$, the so-called normalization constants, are defined by

$$(\tilde{C}_n)^{-1} = \int_0^I \psi_n^2(Z)dZ. \tag{XVIII.3.10}$$

They have the asymptotic behavior

$$\tilde{C}_n = \frac{2\pi^2 n^2}{I^3} + nO(1), \qquad n \to \infty. \tag{XVIII.3.11}$$

In the x-variable, we have

$$\tilde{C}_n = -\lambda^2 \int_0^R V(x)\varphi_n^2(x)dx = -\frac{\lambda^2}{G_n^2} \int_0^R \varphi_n'^2(x)dx. \tag{XVIII.3.12}$$

Having solved the Gel'fand–Levitan integral equation to find the kernel $K(Z, Z')$, we obtain $\tilde{V} = 2(d/dZ)K(Z, Z)$ and

$$\psi(G; Z) = \frac{\sin GZ}{G} + \int_0^Z K(Z, Z')\frac{\sin GZ'}{G} dZ'. \tag{XVIII.3.13}$$

The second step is now to go from ψ to V. Making $G = 0$ in (XVIII.3.13), we find

$$\omega(Z) \equiv \psi(0; Z) = Z + \int_0^Z K(Z', Z')Z' \, dZ'. \tag{XVIII.3.14}$$

On the other hand, from (XVIII.3.1) and (XVIII.3.2), with $g = 0$, it follows that

$\omega(Z) = \lambda x[-V(x)]^{1/4}$. Therefore,

$$dZ = [-V(x)]^{1/2}\, dx = \frac{\omega^2(Z)dx}{\lambda^2 x^2}. \qquad \textbf{(XVIII.3.15)}$$

Integrating this, and taking into account the fact that $Z(r = R) = I$, we find

$$x(Z) = \frac{1}{\dfrac{1}{R} + \lambda^2 \displaystyle\int_Z^I \frac{dZ'}{\omega^2(Z')}}. \qquad \textbf{(XVIII.3.16)}$$

Inverting this relation, we find $Z = Z(x)$, from which we get, finally,

$$V(x) = -\left(\frac{dZ}{dx}\right)^2 = \frac{-1}{\left(\dfrac{dx}{dZ}\right)^2}. \qquad \textbf{(XVIII.3.17)}$$

We have to make several remarks here. The first is that $\omega(Z) \simeq Z$ as $Z \to 0$. It then follows from (**XVIII.3.16**) that when $Z \to 0$, $x \simeq Z/\lambda^2 \to 0$, as it should. Second, inverting (**XVIII.3.16**) is possible, and leads to a unique solution $Z(x)$ since $x(Z)$ is an increasing function of Z. Finally, both $x(Z)$ and $Z(x)$ are differentiable, and (**XVIII.3.17**) has therefore a meaning.

So far, λ is arbitrary, and we must give it also. Or else, we may give the other normalization constants

$$\tilde{C}'_n = \lambda^{-2}\tilde{C}_n = -g_n^{-1}\int_0^R \varphi_n'^2\, dx \underset{n\to\infty}{=} \frac{2\pi^2 n^2}{\lambda^2 I^3} + n\,o(1) \quad \textbf{(XVIII.3.18)}$$

from the last part of which we obtain λ.

As is now clear, the strategy for solving our inverse problem is as follows. From the asymptotic form of spectral data, (**XVIII.3.9**) and (**XVIII.3.18**), we obtain I, the length of the interval in Z and λ. We can then solve the Gel'fand–Levitan integral equation, and proceed as shown above.

Up to now, we have assumed that R is finite. However, it can easily be seen that our results are valid provided $xV(x) \in L^1(0, \infty)$, and

$$I = \int_0^\infty |V(x)|^{1/2}\, dx < \infty. \qquad \textbf{(XVIII.3.19)}$$

What is important is the finiteness of the interval in Z in order to have a discrete spectrum in g. At $Z = 0$, we always have $\psi(0) = 0$. However, now, at $x = \infty$, we have, according to (**II.1.5c**) and for $g = g_n$,

$$\varphi_n(x) \underset{x\to\infty}{=} A_n + o(1), \qquad \textbf{(XVIII.3.20)}$$

where A_n ($\neq 0$) is finite. In fact, at $x = \infty$, the boundary condition for the S-wave is the Neumann condition $\varphi_n'(\infty) = 0$. In the notations of chapter II, $\varphi'(\infty) = F(0, g)$ where $F(k, g)$ is the Jost function. Nevertheless, because of $|V|^{1/4}$ in the definition of ψ, we again have $\psi(I) = 0$. So, as for the case of finite R, we are always dealing with Dirichlet boundary conditions for ψ at both ends of the interval in Z. All the previous results are therefore valid. Making

$R = \infty$ in (**XVIII.3.16**), we find now

$$x(Z) = \frac{1}{\lambda^2 \displaystyle\int_Z^I \frac{dZ'}{\omega^2(Z')}} .$$ (**XVIII.3.21**)

Another generalization would be when the term $l(l + 1)Z^{-2}$ is present in $\tilde{V}$. Here we have to use Gel'fand–Levitan theory for higher waves ($l > 0$) developed in chapters II and III. We leave the reader to work out the details. The main point is that now sine and cosine have to be replaced by appropriate Bessel and Neumann functions. The main change in our previous formulas concerns (**XVIII.3.14**), which now becomes

$$\omega(Z) = Z^{l+1} + \int_0^Z K(Z, Z')Z'^{l+1} \, dZ'.$$ (**XVIII.3.22**)

Again, the results are the same for R finite or infinite, as long as (**XVIII.3.19**) holds.

We end our remarks by mentioning the integral representation

$$\varphi(g; x) = x + \int_0^{I(x)} \tilde{K}(x, y)\cos \sqrt{g}y \, dy,$$ (**XVIII.3.23**)

where the kernel $\tilde{K}$ does not depend on g, is at least once continuously differentiable, and satisfies the hyperbolic equation

$$\frac{\partial^2 \tilde{K}}{\partial x^2} + V(x)\frac{\partial^2 \tilde{K}}{\partial y^2} = 0, \qquad x \in [0, R], \qquad 0 \le y \le I(x), \quad (\textbf{XVIII.3.24})$$

and the boundary conditions

$$\tilde{K}(x, I(x)) = 0, \qquad \frac{\partial \tilde{K}}{\partial y}\bigg|_{y=+0} = x.$$ (**XVIII.3.25**)

Like the Gel'fand–Levitan differential equation (**III.1.18**), the above equation can also be used for the study of the direct problem. If V satisfies the usual conditions $xV \in L^1$ and $V \in L^{1/2}$, its solution is unique. As for the inverse problem, it is simpler to use the Gel'fand–Levitan integral equation. The integral equation for $\tilde{K}$ is less practical, although equivalent to the Gel'fand–Levitan equation. In fact, making the appropriate changes of variables and functions, and after three integrations by parts, we end up with the Povzner–Levitan representation (**XVIII.3.13**), from which we can infer the equivalence of the two methods.

As examples, we consider three cases. The first is the potential

$$V(x) = -\frac{a^2}{(1 + ax)^4}, \qquad x \in [0, R],$$ (**XVIII.3.26**)

for which we immediately find

$$Z(x) = \frac{ax}{1 + ax}, \qquad I = \frac{aR}{1 + aR}, \qquad \tilde{V}(Z) = 0. \quad (\textbf{XVIII.3.27})$$

In the Z-variable, therefore, we have the free case

$$\psi = \frac{(\sin GZ)}{G}, \quad G_n = G_{n0} = \frac{n\pi}{I}, \quad \tilde{C}_n = \tilde{C}_{n0} = \frac{2\pi^2 n^2}{I^3}, \quad n = 1, 2, \ldots,$$

and also $\lambda = a^{1/2}$. Note that $I < 1$ and $a = I/R(1 - I)$.

The inverse problem in Z: $\{G_n, \tilde{C}_n\} \Rightarrow V(x)$ is, of course, very simple here. We get $\tilde{K}(Z, Z') = 0$, $\psi = (\sin GZ)/G$, and $\omega(Z) = Z$. This, when substituted into (XVIII.3.16), with $\lambda = a^{1/2} = I^{1/2}/\sqrt{(R(1 - I))}$, leads to

$$Z = \frac{xI}{R + I(x - R)} \tag{XVIII.3.28}$$

from which we deduce $V(x) = -(dZ/dx)^2$, (XVIII.3.26), by putting $I = aR/(1 + aR)$. Finally,

$$\varphi = \frac{1 + ax}{aG} \sin\left(G \frac{ax}{1 + ax}\right). \tag{XVIII.3.29}$$

For $R = \infty$, we have $I = 1$, and it can be verified that all the limits exist.

The second example is the exponential potential

$$V(x) = -e^{-x}, \quad x \in [0, \infty). \tag{XVIII.3.30}$$

Here, to avoid complicated algebra, we take $R = \infty$. Then

$$Z(x) = 2(1 - e^{-x/2}), \quad I = Z(\infty) = 2, \quad \tilde{V}(Z) = \frac{-1}{4(2 - Z)^2}. \tag{XVIII.3.31}$$

Again, we are dealing with the "S-wave" in the Z-variable, and at $Z = 2$ we have the limit circle case of Weyl. However, since here again $\psi(2) = 0$, we have a unique self-adjoint extension of our differential operator, as we saw previously. The wave function is given by the first two parts of (XVIII.3.2)

$$\varphi(g; x) = \pi[J_0(2GX)N_0(2G) - N_0(2GX)J_0(2G)], \quad X = e^{-x/2}. \tag{XVIII.3.32}$$

where J and N are Bessel functions.

It is easily seen that the spectral data are given by

$$J_0(2G_n) = 0, \quad \tilde{C}_n = G_n^2. \tag{XVIII.3.33}$$

They lead, in turn, to free spectral data

$$G_{n0} = \frac{n\pi}{2}, \quad \tilde{C}_{n0} = \frac{n^2\pi^2}{4} = G_{n0}^2. \tag{XVIII.3.34}$$

Here we have $\lambda = 1$.

For the inverse problem, a tedious but straightforward calculation shows that we have

$$K(Z, Z') = \frac{-Z'}{16(2 - Z)} {}_2F_1\left(\tfrac{3}{2}, \tfrac{3}{2}; 2, \frac{Z'^2 - Z^2}{8(2 - Z)}\right), \tag{XVIII.3.35}$$

${}_2F_1$ being the hypergeometric function. Obviously, $K(Z, -Z') = -K(Z, Z')$, $0 \le Z' \le Z \le I$. Using the differential equation satisfied by this function, it can be verified that, indeed,

$$\frac{\partial^2 K}{\partial Z^2} - \frac{\partial^2 K}{\partial Z'^2} = \tilde{V}(Z)K(Z, Z'), \qquad \text{(XVIII.3.36)}$$

and

$$K(Z, Z) = -\frac{Z}{16(2 - Z)} = \frac{1}{2}\int_0^z \tilde{V}(t)dt, \qquad \text{(XVIII.3.37)}$$

in total conformity with Gel'fand–Levitan theory. Now, using K in (XVIII.3.14), we find $\omega(Z) = -\sqrt{(2(2 - Z))}\log(1 - Z/2)$. From this we deduce, via (XVIII.3.16) with $\lambda = 1$,

$$x = -2\log\left(1 - \frac{Z}{2}\right), \qquad Z = 2(1 - e^{-x/2}). \qquad \text{(XVIII.3.38)}$$

We can now calculate V by differentiating this expression. Once again, we find that $V = -e^{-x}$.

The last example is the potential

$$V(x) = -\frac{1}{(1 + x^2)^2}. \qquad \text{(XVIII.3.39)}$$

For this potential, we have first

$$Z(x) = \arctan x, \quad I = Z(\infty) = \frac{\pi}{2}, \quad \tilde{V}(Z) = -1, \quad \text{(XVIII.3.40)}$$

and $(l = 0)$

$$\psi(g, Z) = \frac{\sin\sqrt{g + 1}Z}{\sqrt{g + 1}}. \qquad \text{(XVIII.3.41)}$$

The spectral data are given by

$$g_n = 4(n + \tfrac{1}{2})(n + \tfrac{3}{2}), \qquad \tilde{C}_n = \frac{16(n + 1)^2}{\pi}. \qquad \text{(XVIII.3.42)}$$

It follows that $\lambda = 1$, and

$$K(Z, Z') = -Z'\frac{J_1(\sqrt{Z^2 - Z'^2})}{(Z^2 - Z'^2)^{1/2}}. \qquad \text{(XVIII.3.43)}$$

It is easily seen that we again have (XVIII.3.13), with $\tilde{V}(Z) = -1$,

$$K(Z, Z) = -\frac{Z}{2} = \frac{1}{2}\int_0^z \tilde{V}(Z)dZ. \qquad \text{(XVIII.3.44)}$$

We now obtain $\omega(Z) = \psi(0, Z) = \sin Z$, $x(Z) = \text{tg } Z$. Therefore $(dZ/dx) = (1 + x^2)^{-1}$ and $V(x) = -(1 + x^2)^{-2}$. We can solve exactly the inverse problem for many more potentials, for instance,

$$V(x) = -\theta(1 - x)\frac{1}{x} \qquad \text{(XVIII.3.45)}$$

for which we refer the reader to the literature.

XVIII.4   Comments and References

We have sketched here only the parts of the subject which are connected with inverse problems of quantum theory. Even so, these problems have a very rich history, that we cannot pretend to exhaust here, and which is not completely exhausted either by the excellent treatises published on the subject: that of Levitan (1984) on inverse Sturm–Liouville problems which completed "on the inverse side" the classical book by Levitan and Sargsjan (1975), that of Marchenko (1986) which studies these problems even in the (non-self-adjoint) case of a complex potential, those of Gladwell (1986a), more physical, and easy to read, and of Poschel and Trubowitz (1987), more abstract. Thus, we shall only try to give some guidelines in this history. Lord Rayleigh guessed (1877) that one Sturm–Liouville spectrum (i.e., the set of all eigenvalues corresponding to fixed homogeneous boundary conditions at the end points) was sufficient to determine the (variable) density of a string. That his heuristic argument was not correct and two spectra are necessary (unless we know that the string is symmetrical) was proved by Ambarzumian (1929). Meanwhile the asymptotics of vibrating domain eigenvalues had been studied by mathematicians and led them to a famous formula, giving the leading terms of the number $N(k)$ of modes corresponding to eigenvalues k_n not exceeding a certain upper limit k. For instance, the leading term of $N(k)$ as $k \to \infty$ has been proved by Weyl in the years 1911–1915 to be proportional to the area, or the volume, of the vibrating domain. Next orders were then evaluated by Titchmarsh and Kodaira. An easy-to-read survey of these results is available in the famous lecture by Mark Kac (1966), and for detailed results see Baltes and Hilf (1976). After Ambarzumian it became clearer and clearer, in particular with the results of Borg (1946, 1949a,b), that two "independent" spectra are sufficient for reconstructing a string. In 1951, the famous paper by Gel'fand and Levitan showed how the potential $q(x)$ appearing in a Sturm–Liouville operator of the Schrödinger form (i.e., the operator $-d^2/dx^2 + q(x)$ completed by homogeneous boundary conditions at two end points) can be constructed from its spectral function (i.e., the eigenvalues forming one spectrum *and* the corresponding eigenfunctions normalizing constants—clearly two independent sets of numbers). Since 1951, the problem in its Schrödinger form has been the object of many publications, already cited in this book in most cases, and whose mathematical elements are well described in the two Faddeev review papers (1963, 1976) and, for more recent results, in Marchenko (1986), and in papers by Trubowitz and coworkers (Isaasson et al. 1983, 1984; Dahlberg et al. 1984). We shall add here some papers by Levitan (1956, 1963, 1964a), which deal with the determination of q from two spectra, Marchenko (1958), Hochstadt (1977), and Levitan (1973a,b), which deal with the well-posedness of the inverse Sturm–Liouville problem and completeness theorems, and also Symes (1979), Li Yi Shen (1965, 1981), Benedek and Panzone (1980), and Panzone (1984), who treated related problems.

The string form of the inverse Sturm–Liouville problem, i.e., the determination of a positive function $\rho(x)$ from the eigenvalues of the operator $-\rho^{-1}(d^2/dx^2)$, or more general problems, have been developed almost independently of the Schrödinger form after 1951. After many short papers, some of them available only in Russian, published in the fifties and the sixties, M. G. Krein produced, with I. S. Kac, a very complete study (1974a) of the existence and uniqueness problem, proving the bijection between densities ρ and R-functions (a special class of meromorphic functions which appear in the problem as the weight of the inner product in a Hilbert space which is isometrically carried through the operators into a Hilbert space whose inner product is weighted by ρ). Kac and Krein (1974a,b) and Krein (1974) also studied the connection of this problem with continued fractions, the problem of moments, and Chebychef–Markov inequalities. Studies, more oriented towards the construction problems of ρ from its spectra in discrete or continuous cases, were then produced by Hochstadt (1976), Barcilon (1974, 1980, 1983), and Hald (1977a, 1978a,b) who considered symmetric cases (where one spectrum may be sufficient). Authors then concentrated their efforts on inverse eigenvalue problems that are more complicated, either because the equation is more complicated (vibrating-beam problem, Earth mode problems, etc.) or because discontinuities are admitted. In the first category we find papers by Barcilon (1976a,b,c, 1977, 1978a,b), Hald (1977b, 1970, 1983), Gladwell (1984, 1985, 1986b), Willis (1986), and Symes (1981). In the second category, discontinuities were introduced very slowly—one or two in the first papers (Hald, 1984; Willis, 1985), then the general case (Andersson, 1987a,b). Authors also tried to introduce a singularity on the interval, but no exact solution of the inverse problem was produced (Boyd, 1981). The interest in inverse Sturm–Liouville problems was then shifted to slightly different forms where, for instance, the input is the value of a particular eigenvalue for an infinite set of different boundary conditions, or other unusual, but physical, input (McLaughlin, 1986, 1987; McLaughlin and Rundell, 1987). It is not surprising to find applications in some of these approaches of Darboux–Backlund or equivalent transformations, or of Deift's remarks (1978) on the spectra of operators related by a convenient commutation formula. It is more surprising to see that the strong relation between spectral and scattering problems for a given parameter was not fully exploited before Sabatier (1974b,c, 1978a,b) who used it to prove that the full response only (spectra and amplitudes) corresponding to the lowest radial mode and two toroidal modes determine the elastic parameters λ, μ, ρ as functions of R in a "spherically symmetric" elastic ball (a similar study might be used to relate impedance spectral problems and the impedance scattering problems which are now reasonably well understood). The numerical problems connected with inverse eigenvalue problems (see section XIX.2.5) were also the origin of theoretical papers on the densities made of Dirac positive measures (Sabatier, 1979; Barcilon, 1980; Barcilon and Turchetti, 1979). Other approaches to numerical problems led to theoretical studies of Jacobi matrices or related matrices and their inverse problems (see, for example, Boley and Golub (1978, 1987), Hald (1976), Antoun and Grünbaum (1980), and Grünbaum (1978, 1985, 1986). Curiously, a by-product of the inverse methods in partial differential equations was a new way of computing Sturm–Liouville eigenvalues (Calogero, 1983). The main difficulty of solving inverse problems of higher order lies in the difficulty of defining a Povzner–Levitan representation. This has been done by Zachary (1981, 1984) for classes of equations with analytic coefficients, and progress in this direction has also been carried out in transmutation theory (Carroll, 1985).

Let us finally cite the applications of spectral inverse theory to special physical problems (confining potentials) by Grosse and Martin (1979), Baumgartner et al. (1985), and Common et al. (1987). See also Kristensson (1986). For the inverse problems in the coupling constant, see Chadan and Grosse (1984), Chadan and Kobayashi (1986), and Chadan and Musette (1987, 1988). See also Chadan (1984).

CHAPTER XIX

Numerical Problem

XIX.1   Introduction

XIX.1.1   Foundations

We are interested in the quantum scattering of particles. The experimental results are cross sections corresponding to "long" wave packets that justify a time-independent formulation. We assume that we know a good model to describe the interaction, for instance, the Schrödinger equation, where the parameter describing the interaction is called "the potential." Hence, the central rule of our work is a mathematical model enabling us to define a mapping $\mathcal{M}$ from a "*set of parameters*," $\mathscr{C}$, into a "*set of results*," $\mathscr{E}$.

If the model was not fixed, we would assume that its indeterminacy can be represented by a set of additional parameters, the "*model parameters*," which would also have to be included in $\mathscr{C}$. In fact, it is always convenient to choose a large set $\mathscr{C}$, defined in a simple way, and to characterize the elements of $\mathscr{C}$ that are physically meaningful by further constraints, called the constraints due to "*a priori information*." Other constraints may be necessary to make the mathematical model self-consistent (for instance, if the inverse problem is linearized in the *neighborhood* of a reference parameter), they are called "*consistency constraints*." The "set of results" $\mathscr{E}$ should contain all the "*computed results*," i.e., the image of $\mathscr{C}$ by the mapping $\mathcal{M}$, augmented in order to include all "*physical results*," i.e., all the possible results of measurements, involving *possible errors*. Like $\mathscr{C}$, $\mathscr{E}$ is still enlarged to be defined simply.

It is also necessary to have convenient *topologies* in $\mathscr{C}$ and $\mathscr{E}$. In most cases, we can make them *metric spaces*—often *normed spaces*. It is extremely important that the distances in $\mathscr{C}$ and $\mathscr{E}$ are "*reasonable*," i.e., they correspond to sound physical concepts. For instance, using in $\mathscr{E}$ a topology which would involve the properties of an infinite number of derivatives would not be consistent with our knowledge of results—necessarily at discrete points and with errors! It is surprising that this trivial remark is so often overlooked.

Deriving the "*computed results*" from the parameters is called "*solving the direct problem.*" *Describing the reciprocal images*, in $\mathscr{E}$, of all physical results, and giving algorithms to construct them, is called "*solving the inverse problem.*" In the remainder of this book, our main concern was to answer *existence*, *uniqueness*, and *construction* questions from ($\mathscr{E}$) back to $\mathscr{C}$, i.e., to study the existence and extent of reciprocal images of elements of $\mathcal{M}(\mathscr{C})$ and to exhibit *exact inverse mappings* $\mathcal{M}^{-1}$ (what we have called the *reconstruction problem*). However, this is not sufficient for a numerical analysis of real inverse problems. Again, in the real world, measurements are finitely many and with errors. Hence, we are led to define a set $\mathscr{E}$ of results that does not coincide with $\mathcal{M}(\mathscr{C})$, and even if it does, does not with $\mathcal{M}(\mathscr{C}_0)$ where ($\mathscr{C}_0$) is the subset of $\mathscr{C}$ constrained by *a priori* information and consistency constraints. In addition, we must keep only mappings from $\mathscr{E}$ to $\mathscr{C}_0$ that are *continuous* with respect to the distance introduced in $\mathscr{E}$: if this "*stability*" condition was not fullfilled, a consequence of data errors, no matter how small, would be that parameter values would be uncontrolled. Numerical errors, due to the finiteness of computers, would have the same effect.

There is no inverse problem which is "*perfectly well-posed,*" i.e., where $\mathcal{M}$ is a bicontinuous bijection between $\mathcal{M}_0$ and $\mathscr{E} = \mathcal{M}(\mathscr{C}_0)$. In real inverse problems, we must redefine the concept of solution, and this is tantamount to defining stable mappings $\bar{\mathcal{M}}$ from $\mathscr{E}$ to $\mathscr{C}$ which are, in some sense, the inverse of $\mathcal{M}$. The two most usual ways to make this "*regularization*" is by introducing either "*quasi solutions*" or "*approximate solutions.*"

Quasi Solutions. For any element e of $\mathscr{E}$, suppose we can construct a point $e' = PP(e)$ in $\mathcal{M}(\mathscr{C}_0)$ which achieves the distance of e to $\mathcal{M}(\mathscr{C}_0)$—then let us determine the reciprocal image $\mathcal{M}^{-1}(e')$. If this defines one element only in $\mathscr{C}_0$ that corresponds to e, and if this element $\bar{\mathcal{M}}(e)$ depends continuously on e, it is called the quasi solution of the inverse problem corresponding to e

$$PP_{\mathcal{M}(\mathscr{C}_0)}(e) = \{e' | e' \in \mathcal{M}(\mathscr{C}_0); \|e - e'\|_{\text{minimum}}\},$$

$$x = \bar{\mathcal{M}}(e) = \mathcal{M}^{-1}[PP_{\mathcal{M}(\mathscr{C}_0)}(e)]. \tag{XIX.1.1}$$

As a common example, the quasi solution (XIX.1.1) is well defined if $\mathscr{C}$ and $\mathscr{E}$ are Hilbert spaces, $\mathcal{M}(\mathscr{C}_0)$ is a closed convex subset of $\mathscr{E}$, and $\mathcal{M}^{-1}$ is a continuous mapping of $\mathcal{M}(\mathscr{C}_0)$ into ($\mathscr{C}_0$): $PP_{\mathcal{M}(\mathscr{C}_0)}(e)$ is then uniquely defined as a continuous mapping of $\mathscr{E}$ into $\mathcal{M}(\mathscr{C}_0)$ (the "projection").

Approximate Solutions. In the quasi solution concept, one "physical" criterion for a generalized solution was met, namely the quality of the "*fit*" in $\mathcal{E}$, since this quality can be measured by the distance of e to the "*computed result*" e', and e' was defined so as to minimize this distance. But the remaining *a priori* information on the solution x was used only by imposing that $x(e)$ must be kept if it belongs to $\mathcal{C}_0$ and rejected if it does not. This requirement obviously is too harsh, and the likelihood of a parameter should be appraised more smoothly. Hence, physics suggests putting in the space of parameters one or several *likelihood criteria* which compete with the *fit criterion* so as to define a generalized solution. An "*approximate solution*" is a parameter that achieves the *minimum* of a well-defined convex combination of these criteria. A common example, the so-called "generalized least squares solution" is well defined if $\mathcal{C}$ and $\mathcal{E}$ are Hilbert spaces, by minimizing over $\mathcal{C}_0$ a linear convex combination of two positive definite hermitian forms, q and f, respectively, defined on $\mathcal{C}$ and $\mathcal{E}$ and called the "*a priori quality*" of the solution (or its *likelihood*) and the "*fit*"

$$F(x) = \lambda q(x) + (1 - \lambda)f(\mathcal{M}x - e), \qquad \textbf{(XIX.1.2)}$$

$$0 < \lambda < 1, \qquad \textbf{(XIX.1.3)}$$

$$x \equiv \bar{\mathcal{M}}(e): F(x) \text{ minimum } (x \in \mathcal{C}_0). \qquad \textbf{(XIX.1.4)}$$

The "*trade-off parameter*" λ may be chosen to maximize stability. When these equations and requirements define, for each physical result e, *one* solution $x(e)$ that depends continuously on e it is called an approximate solution $\bar{\mathcal{M}}(e)$ of the inverse problem. This is true, in particular, if $\mathcal{M}$ is linear so that $F(x)$ is a quadratic functional, and $\mathcal{C}_0$ is compact.

Both quasi solutions and approximate solutions, when they are well defined, circumvent the various kinds of problems of ill-posedness: "*overdetermination*," i.e., results that cannot be reproduced exactly by the model because they do not belong to $\mathcal{M}(\mathcal{C})$—"*underdetermination*," i.e., results which are known only by a finite (usually small) number of estimates. Before proceeding we shall now recall the ill-posedness of the main inverse problems we met in this book.

XIX.1.2   The Status of Quantum Inverse Scattering Problems

We briefly recall results concerning the main problems, and see how they could be predicted without too much mathematics.

Three-Dimensional Inverse Problem. A rule of the thumb says that a function defined on $\mathbb{R}^n$ cannot be bijectively related to a function defined on $\mathbb{R}^{n'}$ if n' and n are different. This argument correctly predicts that the inverse problem for a local potential $U(x)$ (three variables) is overdetermined if the full scattering amplitude (five variables) is given. In the case of spherical symmetry, $U(x)$ depends on one variable, x and $f(k, \theta)$ on two variables, and the conclusion is similar. Hence, for a local potential, the inverse problem is "overdetermined." For a "nonlocal potential," the term $U(x)\psi(x)$ in the Schrödinger equation describing the interaction becomes $\int V(x, x')\psi(x')dx''$, so that the full inverse problem is no longer overdetermined and, if some additional constraint is imposed, may be well-posed.

One-Dimensional Radial Problems. The potential is a function of one variable on the half-axis and so is the S-matrix as a function of k (because of unitarity, etc.), but it has an infinity of elements, depending on l, say, $\exp[ri\delta_l(k)]$.

Now, first suppose that we are interested by the inverse problem from a phase shift $\delta_l(k)$ ($k \in \mathbb{R}$, l fixed) to $V(r)$ and assume first there is only one constraint, $\int_0^\infty x|V(x)|dx$ finite. We easily see that the problem is generally underdetermined by using the following argument. Apply a Crum transformation of the form adding a new bound state (formulas **(IV.1.22)** ↔ **(IV.1.25)**). This transformation depends on two parameters—the energy $-\gamma^2$ of the new bound state and its normalization constant C. It leaves the phase shift unchanged for $k \in \mathbb{R}$. Hence, for $l = 0$, the inverse problem $\delta_0(E) \to U$ is *underdetermined*. The same result can be obtained in a similar way for $l \neq 0$. More precisely, it has been proved that $k \to \delta_0(k)$, the energies $E_1, E_2, \ldots, E_n$ of the bound states and their normalizing constants, are *sufficient* to determine the potential in the class of functions whose first moment is finite. Marchenko's theorem (see section IV.4) enables us to extend this result to $l \neq 0$.

Stronger *A Priori* Constraint. Suppose now we have the additional *a priori* information that $V(x)$ satisfies the inequality

$$|x^2 V(x)| < C^2. \qquad \textbf{(XIX.1.5)}$$

If $l \geq C$, $V - l(l + 1)r^{-2}$ is everywhere negative, there is no bound state and the corresponding Jost function $F_l(k)$ has no zero. Thanks to Marchenko's theorem, F also corresponds for $l = 0$ to a potential V_0, which is completely determined and, going back to V through l transformations, V is completely determined. Hence V is completely determined by giving any function $k \to \delta_l$ with $l \geq C$. Thus, if we seek *a bounded potential*, the overdetermination of the spherically symmetric three-dimensional problem is such that for all values of l, except a finite number, the functions $\delta_l(k)$ are determined from each other! The same conclusion holds if the integral of the potential, or any of its moments, is *a priori* bounded. Notice, finally, that apart from the weak extensions given in section 1.6 there exist several results concerning the inverse theory for strongly singular potentials in which the singular part is repulsive (see sections II.6.5 and III.5). Needless to say, these problems are also ill-posed.

Radial Inverse Problem at Fixed Energy. This problem has been thoroughly studied and its ill-posedness was made clear in chapter XIII. Let us summarize the results. We know a sequence $\delta_l(k)$ for fixed k, all l's. First suppose that U belongs to the class of twice continuously differentiable potentials such that $xU(x)$ is locally integrable and $O(x^{-2})$ as $x \to \infty$, and that $\delta_l = O(l^{-3})$ as $l \to \infty$. The main ill-posedness appears in the function $f(r, r)$, given, for instance, in **(XIII.4.13)**. A complete knowledge of $f(r, r)$ yields $f(r, r')$ by means of **(XIII.4.11)**, and thus yields $V(r)$ by means of the Regge–Newton equation. From $V(r)$ we readily get $U(x) = k^{-2} V(r/k)$ (remember that the radial equation **(XI.1.6)** was obtained after using k^{-1} as the unit of length). Conversely, $U(x)$ determines $f(r, r)$ which is therefore a piece of information equivalent to $U(x)$. Now, it follows from **(XIII.4.14)** that $\{\delta_l(k)\}$, fixed k, gives only a partial statement on

$$\Omega(x) = -2k^{-1} \frac{d}{dx} [x^{-1} f(kx, kx)] \qquad \text{(XIX.1.6a)}$$

$$= -4 \frac{d}{dx} \int_0^\infty \sin 2vx G(v)dv, \qquad \text{(XIX.1.6b)}$$

where

$$G(v) = kF(kv) = \theta(k - v)G(v) + \theta(v - k)G(v). \qquad \text{(XIX.1.7)}$$

Comparing **(XIII.4.14)** and **(XIX.1.7)**, we see that this partial statement amounts to giving $G(v)$ for $v \in (0, k)$ and that $G(v)$ $(v > k)$ is arbitrary in a class of functions contained in $L^2(k, \infty)$, and depending more precisely on the general constraints we may put on U and its derivatives. The problem is underdetermined for $\Omega(x)$ like the construction of a signal from its image through a low-pass filter, the bandwidth of this filter being k. In the Born approximation $V(x) = \Omega(x)$, and the ill-posedness for Ω is that for V.

It is of interest to know that "transformation" methods enable us to understand this ill-posedness by reducing it to the one which appears in linearized inversion formulas, and has been studied in section XVI.2. The method, which enables us to replace a global study of ill-posedness by a local study (Sabatier, 1985), can be widely used and hence deserves some attention. The idea is: Suppose we know transformations T_i in the space $\mathscr{E}$ of results associated with transformations t_i in the space $\mathscr{C}$ of parameters, in such a way that if e is the ($\mathscr{M}$) image of x, $T_i e$ is that of $t_i x$. Suppose each T_i and t_i depend on a parameter $a_i \in \mathbb{R}$. For a given (reference) couple of points (x_0, e_0) (with $x_0 \in \mathscr{C}$, $e_0 \in \mathscr{E}$, $e_0 = \mathscr{M} x_0$), there exists a set of points of $\mathscr{E}$ that can be obtained by applying finitely many successive transformations $T_i(a_i)$ to e_0, and which are the images of points in $\mathscr{C}$ obtained by applying the same sequence of transformations t_i, for the same values of parameters a_i, to x_0

$$(e_0 = \mathscr{M} x_0) \quad \Rightarrow \quad (T_i(a_i) \dots T_1(a_1)e_0 = t_i(a_i) \dots t_1(a_1)x_0. \qquad \text{(XIX.1.8)}$$

Hence for each parameter x "that has finitely many coordinates $a_1, \dots, a_i$, relative to transformations $t_1 \dots t_i$ and to the reference parameter x_0," the direct problem is readily solved by applying $T_i(a_i) \dots T_1(a_1)$ to the image e_0 of x_0. Conversely, if a point e can be written in the form $T_i(a_i) \dots T_1(a_1)e_0$, a class of solutions of the inverse problem for e is obtained from the class of solutions of the problem for e_0. If a set X_0 is such that for each point $x_0 \in X_0$, $\mathscr{M} x_0 = e_0$, a whole set of solutions is generated from x_0 at each point e with finitely many coordinates. In the problems we meet, this method of study may become easier to implement if the transformations commute with each other and if the points reached from a given reference point e_0 fill a dense subset of $\mathscr{E}$.

Now, in order to apply this idea we need transformations that preserve the Schrödinger equation but which are different from the Crum transformations studied in chapter IV or from the Darboux transformations studied in section XVII. The differential equation

$$f'' + (k^2 - W)f = 0 \qquad \text{(XIX.1.9a)}$$

is equivalent to the Riccati equation for $F = f'/f$

$$F' + F^2 + k^2 - W = 0. \qquad \text{(XIX.1.9b)}$$

Let us introduce

$$\tilde{F} = F + \frac{F' + \alpha'}{F + \alpha} - \frac{\beta'}{\beta}, \qquad \text{(XIX.1.10)}$$

where α and β are such that there exists a constant C for which

$$k^2 - W(x) + \alpha^2 - \alpha' = C\beta^2 \qquad \text{(XIX.1.11)}$$

then $\tilde{F}$ is a solution of (**XIX.1.9**) with $\tilde{W}$ instead of W

$$\tilde{W}(x) = W(x) + 2\alpha' - \frac{2\alpha\beta'}{\beta} + \frac{2\beta'^2}{\beta^2} - \frac{\beta''}{\beta}. \qquad \text{(XIX.1.12)}$$

Hence, the transformation defined by going from F to $\tilde{F}$ preserves the structure of the Schrödinger equation and this remark can be used to construct the transformations we are looking for. We leave to the reader the exercise of checking that convenient choices of α and β enable him to generate Crum and Darboux transformations. Those of interest for the inverse problem at fixed energy were managed by Lipperheide and Fiedeldey (1978). They correspond to $W = V(x) + (\lambda^2 - \frac{1}{4})x^{-2}$, F_λ and $\bar{F}_\lambda$ are solutions of (**XIX.1.9**) that correspond to a given value of λ, but different asymptotic behaviors as $x \to \infty$, α and β are

$$\alpha(x) = \frac{\lambda^2 - v^2}{v^2 - \mu^2} F_\mu - \frac{\lambda^2 - \mu^2}{v^2 - \mu^2} \bar{F}_v; \qquad \beta(x) = F_\mu - \bar{F}_v, \quad \text{(XIX.1.13)}$$

where μ and v are complex numbers. To derive a real potential from a real potential, $\mu = v^*$. According to (**XIX.1.12**), the transformed potential $\tilde{W}(x)$ is

$$\tilde{W}(x) = W(x) - 2\frac{v^{*2} - v^2}{x} \frac{d}{dx} [x(F_{v^*} - \bar{F}_v)]^{-1}, \qquad \text{(XIX.1.14)}$$

and the new phase shift $\tilde{\delta}(\lambda)$ is such that

$$\exp[2i\tilde{\delta}(\lambda)] = \frac{\lambda^2 - v^{*2}}{\lambda^2 - v^2} \exp[2i\delta(\lambda)]. \qquad \text{(XIX.1.15)}$$

For $\lambda = l + \frac{1}{2}$, we obtain the physical phase shift $\delta(l + \frac{1}{2}) = \delta_l$. If they are transformed according to (**XIX.1.15**), it is easy to see that the scattering amplitude $F(k, \cos \theta)$, which is given by **(XI.1.5)** in terms of the phase shifts, is transformed according to the linear transformation

$$\tilde{F}(k, x) - F(k, x) = \frac{v^2 - v^{*2}}{2ik} \left\{ \sum_{l=0}^{\infty} (2l + 1) \frac{P_l(x)}{(l + \frac{1}{2})^2 - v^2} \right.$$

$$+ 2ik \int_{-1}^{+1} \int_{-1}^{+1} dy \, dz \, F(k, y) S(x, y, z) \sum_{l=0}^{\infty} (2l + 1) \frac{P_l(z)}{(l + \frac{1}{2})^2 - v^2} \right\},$$

$$\text{(XIX.1.16)}$$

where $S(x, y, z)$ is given by **(X.1.7)**. Notice that

$$\sum_{l=0}^{\infty} (2l + 1) \frac{P_l(x)}{(l + \frac{1}{2})^2 - v^2} = \frac{-\pi}{\sin \pi(v - \frac{1}{2})} P_{v-1/2}(-x). \quad \text{(XIX.1.17)}$$

Now assume we are given $\{\exp[2i\delta_l]\}$. It is generally possible (using, for instance, Padé approximants) to write

$$\exp[2i\delta_l] = \exp[2i\delta_l^0] \prod_{n=1}^{N} \frac{(l + \frac{1}{2})^2 - v_n^*}{(l + \frac{1}{2})^2 - v_n}, \quad \text{(XIX.1.18)}$$

where the v_n are all different, N is finite, and all these numbers are such that δ_l^0 is small enough to correspond to a potential in the Born approximation (see section XVII.2). Using now the algebraic expansion

$$\prod_{n=1}^{N} \frac{(l + \frac{1}{2})^2 - (v_n^*)^2}{(l + \frac{1}{2})^2 - v_n^2} - 1 = \sum_{n=1}^{N} \frac{A_n}{(l + \frac{1}{2})^2 - v_n^2}, \quad \text{(XIX.1.19)}$$

elementary calculations yield

$$2ikF(k, x) = \sum_{l=0}^{\infty} (2l + 1)(1 - \exp[2i\delta_l^0]) P_l(x)$$

$$+ \sum_{n=1}^{N} A_n \sum_{l=0}^{\infty} (2l + 1) P_l(x)(1 - \exp[2i\delta_l^0])[(l + \frac{1}{2})^2 - v_n^2]^{-1}$$

$$+ \sum_{n=1}^{N} A_n \sum_{l=0}^{\infty} (2l + 1) P_l(x)[(l + \frac{1}{2})^2 - v_n^2]^{-1}. \quad \text{(XIX.1.20)}$$

Up to this point, all the formulas are exact. The first term on the right-hand side of **(XIX.1.20)** is a scattering amplitude which, in the Born approximation (according to the discussion of section XVI.2), corresponds to an infinity of potentials, each of them reproducing the same sequence $\{\delta_l^0\}$. By applying the N transformations parametrized by $v_1, v_2, \ldots, v_n$, we can therefore construct an infinity of potentials that reproduce the same sequence $\{\delta_l\}$, and the nature of this "ambiguity" is similar to that obtained by the Born approximation, and hence to that of a signal reconstructed from its low-pass filtered image. It is of interest to note that if Im $v_n > 0$, the last term on the right-hand side of **(XIX.1.20)** can be interpreted as a sum of Regge poles, with the middle term as a composition of the first and third poles. F is reconstructed from F^0 by applying a sequence of transformations XIX.1.16.

Strong *A Priori* Constraint. Since the linearized inverse problem is justified for small V and is ill-posed, it would not be of any effect to impose an upper bound on the potential. The only effective *a priori* constraint is that of assuming a very short range to the potential; for instance, assuming that it vanishes completely beyond $x = a$. From the physical point of view, this assumption is not so strong and was generally assumed, for instance, in R-matrix theory. Elements of the S-matrix are of the form $(\mathscr{I} - R\mathscr{I}')/(\mathscr{E} - R\mathscr{E}')$, where $\mathscr{I}$ (resp. $\mathscr{E}$) is the value of the ingoing (resp. outgoing) unperturbed wave function at $x = a$ ($\mathscr{I} = e^{-ika}$, $\mathscr{E} = e^{ika}$ for the S-wave), and R^{-1} is the logarithmic derivative of the regular wave function $\varphi(k, x)$ at $x = a$. It follows from our analysis of chapter I that $\varphi(k, a)$ and $\varphi'(k, a)$ are entire functions of exponential type a, so that the elements of S satisfy very hard mathematical constraints, which are strongly related to the short-range property of V (and to causality). This has two consequences:

(a) If the scattering amplitude (and hence $\{\delta_l\}$) is given at fixed k, in a way which is consistent with a "short-range" potential, this potential is completely determined.

(b) The scattering amplitude must satisfy very strong constraints (difficult to "characterize") to be consistent with a "short-range" potential.

Hence uniqueness is restored, but *at the price of strong restrictions* on the image of $\mathscr{C}_0$ in $\mathscr{E}$. Because of them, stability for reasonable norms in $\mathscr{E}$ has disappeared. Hence the problem is still ill-posed, and in a way that is much more difficult to disentangle than the previous one. Strong regularizations, again usually defined after putting constraints, are necessary to obtain sound numerical results. These results hold true in the three-dimensional problem with no symmetry. If $V(\mathbf{x})$ is short range, the corresponding scattering amplitude at a fixed energy is given, and it completely determines $V(\mathbf{x})$ (see section XIX.1.2). Again, if we have experimental information on the scattering amplitude and assume *a priori* that the potential vanishes outside a finite hall, the problem of constructing this potential is not stable for any reasonable distance in $\mathscr{E}$. Other studies of ill-posedness can be found in this book with the corresponding inverse problems. Here we must now give some tutorial remarks because they are (unfortunately) often overlooked.

XIX.1.3   Ill-Posedness and Numerical Calculations

Underdetermined Problem. If a problem has an infinity of exact solutions, and if no *a priori* information enables you to choose among these solutions, the choice is made *by the algorithm you use*. This algorithm cannot be called a *reconstruction*, but the construction of a solution in a given class.

As an example, the Newton–Sabatier procedure *is not the reconstruction* of a potential V_1 but the construction of a potential V which gives exactly all the phase shifts that V_1 gives *at this energy*, but usually does not belong to the same class of functions. V_1 and V are different, as you and the father of your

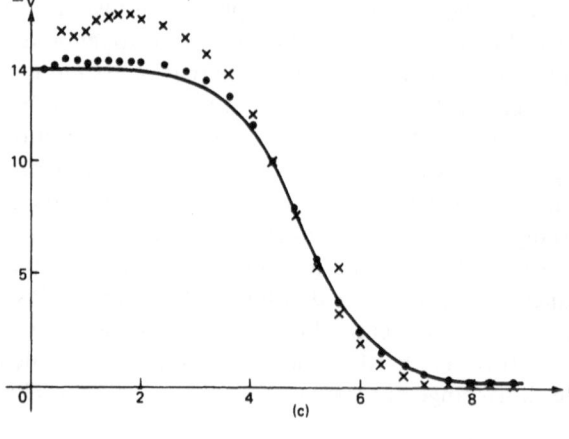

Figure XIX.1a Parameter shift (V, V_1) as a function of potential shape: $V_1 = -14[1 + e^{(r-5)/0.65}]^{-1}$; $E = 50$ MeV (crosses), 400 MeV (dots).

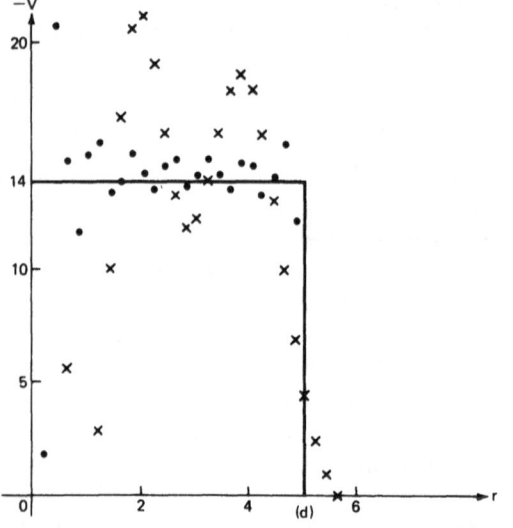

Figure XIX.1b Parameter shift (V, V_1) as a function of potential shape: square potential of depth 14 MeV, range 5 F; $E = 50$ MeV (crosses), 400 MeV (dots).

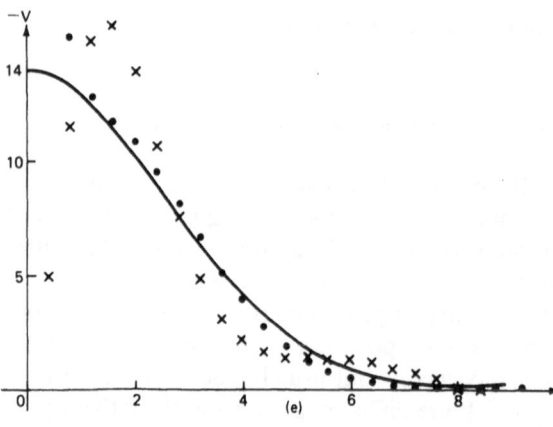

Figure XIX.1c Parameter shift (V, V_1) as a function of energy. $V_1 = -14e^{-(r/3.5)^2}$ (solid line). $E = 7.5$ MeV (crosses), 10 MeV (dots),

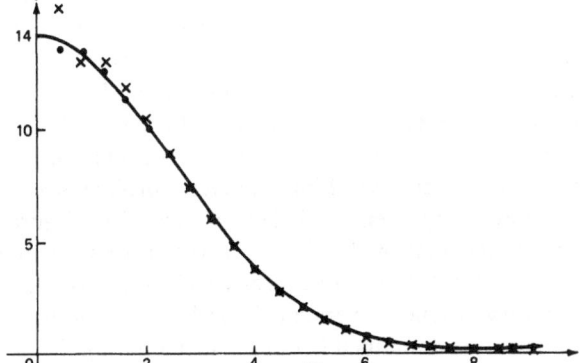

Figure XIX.1d Parameter shift (V, V_I) as a function of energy. $V_I = -14e^{-(r/3.5)^2}$ (solid line). $E = 50$ MeV (crosses), 200 MeV (dots).

son-in-law are different, even if together you are going to have the same infinity of heirs.

The Newton–Sabatier procedure cannot be called a *method* for solving the inverse problem at fixed energy either. It is a method for obtaining *a solution* in *a particular class* $\mathscr{C}$ of functions. Sabatier's uniqueness theorem for potentials, decreasing faster than $x^{-2-\varepsilon}$, holds *in this class* $\mathscr{C}$. The only remarks which can be made for practical applications is that it may be, in certain cases, a reliable class of approximation for any potential V_1 satisfying reasonable constraints. In other words, if V_1 satisfies certain constraints and V is determined by the phase shifts at a sufficiently large energy, it may be true that V_1 and V are not far from each other.

Figures XIX.1(a –d) are meant to show the usual shift $V - V_1$ and not to illustrate a "method" for solving a problem practically. In these figures we show V_1 (solid line) and V (crosses or points) which *exactly reproduce all the phase shifts* produced by V_1 at a given energy. Each figure shows the two potentials V, constructed from the scattering amplitudes, calculated at two different energies from a static potential V_1, and the various figures show the effect of the shape of V_1. They illustrate remarks on the effects of the second term of the right-hand side of **(XIX.1.7)** on the underdetermination: Sabatier and Quyen Van Phu (1971) noticed that this term is smaller for a smoother potential V_1 and that, for any given potential V_1, the term is less and less important when E, and thus the "filter bandwidth" k, increases.[1] For the very unsmooth square well, there are huge oscillations obviously related to the cut-off of the Fourier transform of $\Omega(x)$ at $v = k$ (only those inside the well are reproduced, since the potential tail is not shown here); as noticed by Reignier (1980), tail oscillations may also be related to using finite sets of phase shifts, for whatever class of potentials).

In fact, if a problem is underdetermined, either deeply or because the data are finitely many, and if we do not want to describe the set of equivalent solutions,

---

[1] The original results published by these authors quite correctly illustrated these trends but they were not exact, because of inadequate computers. Here we give modified results.

it is preposterous to look for one solution only without putting *clearly* additional constraints that restore uniqueness and preserve stability. If we do not want to do it, there is no alternative: we must deal with several solutions. (In all cases, we must try to collect new data.) In no case is there is a miraculous method, founded on statistics or information theory, that gives *the physical solution* in a problem where data are insufficient to determine it. For mathematical physicists working on inverse problems, it is a permanent surprise to see how many scientists believe in this kind of miraculous method, which is sometimes published in books as the "inverse problem theory," whereas it is only a set of fitting procedures which conceals a lack of information. On the contrary, a true "inversion theory" is a guide to appraising the amount of information contained in data, understanding the nature of ill-posedness, and defining and describing the sets of generalized solutions (Sabatier, 1987c).

Problems That Are Ill-Posed Both by Underdetermination and Overdetermination. In this kind of problem it is *first necessary* to define the concept of (generalized) solutions—as is suggested in section XIV.1.1. Then, as above, we must *either describe* the set of equivalent (generalized) solutions or put *clearly* constraints that establish the uniqueness of the (generalized) solution which is constructed. It is clear that representing a set by only one "average" point is meaningful only if the set of "equivalent solutions" is not too wide. If it is, we call the problem "strongly ill-posed" and there is no alternative but to describe the set either by its geometrical features (Sabatier, 1987b) or by using the technique of "well-posed questions for ill-posed problems" (Sabatier, 1983, 1984, 1987c).

XIX.2 Numerical Methods for Local Potentials

XIX.2.1 Introduction

Numerical methods were developed to solved the inverse problems of nuclear scattering theory. We shall sketch these methods. Reading them, we should keep several points in mind (Sabatier, 1985):

(1) All the inverse problems in nuclear scattering theory are ill-posed. If the available information is narrow (fixed energy, fixed l), *even with complete exact data*, they would be underdetermined. If the available information is large (all k, all l), *they are overdetermined*. If realistic information is to be used (range of values of k and of l, finite number of inexact, sometimes redundant or contradictory, experimental results), the inverse problem is always both overdetermined and underdetermined.

(2) Approaches which are "local" in the space of parameters, i.e., remain in some neighborhood of a reference potential, ignore most ill-posed aspects: the algorithm automatically selects *one* solution or quasi solution *only*, which is therefore "unique" but in a class of potentials which is not well defined; at best, the "neighborhood" of the reference parame-

ter is bounded by constraints due to the algorithm and not by physical "*a priori*" constraints. Hence these approaches *cannot* be called reconstructions of V in a well-defined class, the prefix *re* meaning that we would go *back* to this class.

(3) Global approaches, e.g., methods using transformations or matrix methods, enable us to make a good exploration of a well-defined subset of the set of parameters $\mathscr{C}$. They must not be confused with a complete method: uniqueness in a subset does not mean uniqueness in the set of all physical potentials.

(4) The degrees of freedom, introduced by allowing an imaginary part or a noncentral part in the potential, generally do not improve the problem's "posedness."

(5) Unless there are very few data, it is often convenient to introduce *a priori* constraints that restore uniqueness, even if the problem becomes overdetermined and unstable: then we restore stability by additional constraints and existence by defining a stable quasi solution. Clearly such a process comes close to the most refined fitting procedures. The difference is that constraints and ill-posedness are carefully introduced or recognized in the inverse method and their consequences are clearly seen, whereas they are not in a fitting procedure. Rather than new fitting techniques, this deep understanding of *methods* is what inversion theory has brought to nuclear science.

XIX.2.2   Fitting Procedures (Radial Problems)

The most common practical quantum inverse problem is that of constructing a nuclear "optical potential" from cross sections. Such a potential has a small imaginary part which takes into account (in the scattering channel) the effect of other open channels. The optical potential has been determined by trial and error in the first instance (early fifties), and since then has been mainly determined by systematic "fitting procedures": to get a good fit to a given set of data, we make calculations for a series of trial potentials until a satisfactory fit is obtained. It is, of course, important to make this fitting procedure as reliable and efficient as possible. In the early days, the experimental and calculated angular distributions were compared visually. It is often possible to find semiempirical rules giving the behavior of the calculated distribution, as one or other of the parameters vary, e.g.,

(a) the positions of the maxima and minima are determined mainly by the value of UR^2, R being the potential radius;

(b) the results are not sensitive to the shape of the absorbing part W but their amplitudes decrease as the size of W increases, etc.).

In more systematic fitting procedures, we begin with a physically reasonable potential, as determined by theoretical considerations or from previous analyses. An iteration is then carried out by calculation at each stage of what we shall call the "quadratic misfit," Δ, which is a measure of the discrepancy between the experimental data and the computed values at this stage

$$\Delta = N^{-1} \sum \left(\frac{\sigma_{exp} - \sigma_{th}}{\delta\sigma_{exp}} \right)^2, \qquad \textbf{(XIX.2.1)}$$

where $\sigma_{exp} \pm \delta\sigma_{exp}$ is the experimental cross section, σ_{th} is the theoretical cross section at a particular angle θ, and the summation runs over all N experimental points. The value of the parameters is systematically varied until the minimum value of Δ is found. The range of Δ for which a fit is acceptable is best judged by eye, but in general a value $\Delta \gtrsim 10$ is satisfactory. We first carry out a fitting on differential cross sections, which usually has greater statistical weight than the polarizations or reaction cross section: thus the central potential is first adjusted to minimize the misfit to the differential cross section. Then the spin-orbit potential is adjusted to fit the polarizations. Finally, the central potential is readjusted.

The process of minimization is usually begun by finding the direction of greatest descent from the point corresponding to the first parameter set. We then proceed until a minimum is found, and then repeat the process until the absolute minimum is located. The efficiency of this process clearly depends on the character of the terrain. The location of the lowest minimum is an additional difficulty. Together with the nonuniqueness due to insufficient information, it is a source of ambiguities. These ambiguities have been (partially) classified: the most important ones are due to lack of information and can be described by means of approximate methods, as we did in chapter XVI. For details concerning the numerical fitting procedures of optical model analyses, we refer to the textbooks of Hodgson (1963, 1971), From the point of view of inverse problems, we can describe the fitting procedures as the research of an approximate solution, by minimizing iteratively a cost function that involves only a misfit in the neighborhood of a reference parameter which represents the *a priori* information. This process is usually stable and only needs a good knowledge of the direct problem (mapping $\mathcal{M}$ of parameters into results) so that it can be very widely used. It fails if $\mathcal{M}$ is strongly nonsmooth or is such that the image of a parameter is too wide a set ("large nonuniqueness"); in other words, it fails in all cases in what we have called, in our geometrical analysis of inverse problems, "strongly ill-posed problems" (Sabatier, 1987b).

Another weak point is that the "*a priori* information" is greater than it should be, and, in fact, it is often guesswork as well as previous scientific analyses! Finally, a very weak point is that we cannot easily decide, from the observed ambiguities or lack of stability, whether they are due to poor numerical treatment (method implementation) or due to a poor way of analyzing the information contained in the data (method principles). Nevertheless, because of their generality as well as their comparatively easy implementation, fitting procedures will probably be used more than inverse methods in the future.

XIX.2.3   Inverse Methods (Radial Problems)

To solve the inverse problem we shall call (numerical) inverse methods:

(a) Methods that are based on exact solutions of the inverse problem in a given class of potentials, defined *a priori* (e.g., in the Gel'fand–Levitan method) or *a posteriori* (e.g., in the "Newton–Sabatier method").

(b) Methods that are based on exact or approximate inversion formulas.

We shall call (inverse) numerical methods the general numerical methods where the *a priori* information and the nonuniqueness are *clearly* described and controlled.

We shall sketch very briefly the methods of either kind which were developed (most of them in the last ten years) for radial potential inverse problems, and give references for all details.

(Numerical) Inverse Methods. The most physical applications were concerned with the inverse problem at fixed energy. In almost all of them, the input is not the cross section but the scattering amplitude or the equivalent phase shifts. In other words, it is assumed that the phase problem at fixed energy was solved first. We shall go through these methods and we shall cite the corresponding methods for the other radial problem:

(a) The matrix methods explained in chapter XII were used either directly (the "Newton–Sabatier" method) or after a modification leading to a more practical scheme (the "modified Newton–Sabatier method"). In particular, the first scheme was applied to synthetic data by Coudray (1977, 1979) and Pelosi (1978). It has also been extended to the spheroidal case (Funke and Zakhariev, 1987). The "modified Newton–Sabatier" method was initiated by Munchow and Scheid (1980) who introduced the "*a priori* assumption" that the potential has a given asymptotic behavior. If this behavior is imposed up to an exponentially decreasing error term, the uniqueness is restored as for finite range potentials (however, in spite of what is claimed in one reference, Sabatier's uniqueness result does not apply, since it proves uniqueness only inside the class obtained by Newton's method). The method is implemented by means of a reference potential $U_0(\rho)$, which has the desired asymptotic behavior, and the calculation of $U - U_0 = \Delta U$ through the Newton perturbed method with the additional requirement that ΔU must go to zero rapidly for $r > r_0$. Because of this requirement, the problem is overdetermined and a regularized quasi solution of the algebraic systems must be calculated (the authors use the least squares method). Results were compared to synthetic scattering amplitudes and, after extending them to cases containing a Coulomb potential, to real data (Munchow and Scheid, 1980; May et al., 1984).

(b) The problem at fixed energy can also be discretized by finite difference methods. The Schrödinger equation in the interaction region is replaced by a particular form of a finite difference equation. It is obtained after dividing the interaction range a, which is supposed finite and *a priori* known, into N equal parts Δ, and *choosing* a difference operator to replace the differential one. We notice at this stage that the finite range assumption is sufficient (in the central case) to restore uniqueness, but that the construction method should be handled with care to avoid instabilities. Here the R-matrix (i.e., in the simplest case, the inverse logarithmic derivative at $x = a$ of the regular solution) is obtained from the solution of the difference equation and compared with the R-matrix found empirically from the results. In the direct problem, a finite continued fraction expansion for the R-matrix is obtained from the potential. In the inverse problem, a Padé approximant of this fraction can be inverted by standard techniques to recover the potential. The technique has been applied to central + spin-orbit potential and to central + spin-orbit + tensor force. There is an optimum choice of Δ (or N, which varies like Δ^{-1}) below which error propagation blurs off the results. Nonuniqueness can be intrinsic (in particular, if noncentral terms are present) or can come from the choice of the finite difference equation: all of them would go to the same Schrödinger equation as $\Delta \to 0$, but they are necessarily used at finite values of Δ^{-1}. For details, see Hooshyar et al. (1982, 1983).

The finite difference methods were also applied to solving the other direct and inverse radial problems. They produce algebraic analogues of the Gel'fand–Levitan and Marchenko methods. Details can be found in a treatise on potential quantum scattering by Zachariev and Suzko (1985).

(c) Algebraic forms of the Gel'fand–Levitan and Marchenko methods are also obtained if the kernel of the integral equation is degenerate—this being the case, for instance (see sections IV.1, IV.2, VI.1), for Bargmann potentials and for rational scattering functions. If a large number of parameters is involved, this result can be developed in a way which obtains a potential as an approximate generalized Bargmann potential. However, it is more convenient to consider that the process implements a transformation method as described in section XIX.1: Crum—Krein transformations in the fixed l cases and others in the fixed-energy case. In the former case, the method has been developed by Rudyak, Suzko and Zachariev (Rudyak et al., 1984; Zachariev and Suzko, 1985; Pivagarchik et al., 1986).

In the latter case, Lipperheide and co-workers (Lipperheide et al., 1978, 1981, 1982; Bürger et al., 1983; Fiedeldey et al., 1982) first used the simple transformation shown in section XIX.1, and then introduced modified transformations which were slightly more complicated but which prevents the intrusion of nonphysical Regge poles. The results were applied to synthetic data (i.e., data computed from a potential) and to real data (Lipperheide et al., 1979; Fröbrich et al., 1979; Krappe and Lipperheide, 1985; Fiedeldey et al., 1985; May and Scheid, 1987).

It has been noticed that when the transformation methods or the variants of the Newton–Sabatier methods are used to analyze a finite number of phase shifts, they generate scattering amplitudes that are closed to each other but not identical; in the sense that they can be reduced to similar rational fractions of λ or closely related functions. This may be understood on the basis of the complete method: each method corresponds to a finite spectrum approximation of $f(r, r)$—or on the basis of interpolation properties—in the sense that a given interpolation in the complex λ-plane (for instance, a given interpolation by a rational fraction or a closely related function), completely defines the class of potentials which is reached.

(d) The semiclassical analysis was also used as a practical method of solving the inverse problem at a fixed energy for a complex potential. Kujawski (1972) used the JWKB approximation for phase shifts, in the form (**XVI.3.6**) and the corresponding inversion formula, and applied it successfully to real data in the noncritical range. The JWKB phase shift corresponds to a well-defined interpolation of phase shifts (Sabatier, 1965) which is not easy to derive from data, but the conditions of validity for the JWKB approximation are such that any smooth-enough interpolation may be used. Various interpolations have been used, for instance, that of the Lipperheide–Fiedeldey method (Allen and Bürger, 1984; Fiedeldey et al., 1984). Ambiguities predicted by semiclassical analysis (Sabatier, 1972; and see section XVI.2) have also be checked numerically (Cuer, 1976).

(e) The other forms of "exact" inversion methods at fixed energy do not seem to have produced numerical methods except for the Burdet–Giffon method (Malyarov et al., 1975). Another original method, connected with a "peeling" method of the potential, has been used by Gerber et al. (1980).

(f) For specific problems of phase reconstruction, see the lectures by Arndt, Rijken, Marty, Krappe and Rossner, De Swart et al. in Krappe and Lipperheide (1985).

(Inverse) Numerical Methods. Since the cross section $\sigma(\Omega)$, is related to the potential $V(x)$ by a functional relation $\sigma = M(V)$, we meet two standard ways to obtain in $\mathscr{C}$ an approximate solution V corresponding to a cross section σ:

(a) We can minimize a cost functional of the form (**XIX.1.2**)

$$F(V) = \lambda q(V) + (1 - \lambda) f[M(V) - \sigma], \qquad \text{(XIX.2.2)}$$

where $\lambda \in (0, 1)$ and q and f are positive definite hermitian forms. Let V_0 be a "first guess" for V, a simple choice for q and f is

$$F(V) = \lambda \int_0^\infty \omega(r) [V(r) - V_0(r)]^2 \, dr$$

$$+ (1 - \lambda) \sum w_i \{[M(V)]_{\Omega_i} - \sigma(\Omega_i)]\}^2, \qquad \text{(XIX.2.3)}$$

where $\omega(r)$ is a weight function and the w_i are weights where the index i runs through measurements. We can also use hermitian forms that involve covariance matrices. They can be justified by bayesian argument. Other convex cost functions appear in connection with other ways of appraising the information, e.g., in the so-called maximum entropy method. Variations of the trade-off parameter λ give a feeling of the problem nonuniqueness if the problem is not too *strongly ill-posed* (i.e., if the sets of equivalent solutions are connected sets and not too large). In most cases, modifying λ also modifies stability and there is an optimal choice. As for the choice of the cost functional, we can say that all are justified if the problem is not too strongly ill-posed, because all give an "average" solution inside the set of equivalent ones. But showing that the problem is so needs a preliminary and difficult study. On the other hand, the cost functions are not equivalent from the computational point of view. It is simpler to write the algorithms for positive definite quadratic cost functions which are differentiable if M is so, and convex if M is linear. The iteration from V_0 usually involves a descent algorithm along the gradient of the cost functional $F(V)$. For instance, in the simplest case, let λ be a (continuous) parameter labeling the algorithm steps, so that V runs through values C_λ from V_0 to the desired minimum V_∞, with a "velocity" $df/d\lambda$ which is given by the formula

$$\frac{df}{d\lambda} = \left\langle \operatorname{grad} F, \frac{\partial V_\lambda}{\partial \lambda} \right\rangle (V_\lambda), \qquad \text{(XIX.2.4)}$$

and is obviously negative if $\partial V_\lambda/\partial \lambda = -\operatorname{grad} F$. This simple gradient method (which is superseded by faster ones) clearly shows how a descent towards a minimum can be managed for a differentiable functional (in the special case (XIX.2.4), algorithms are obtained by setting $\lambda = n\varepsilon_n$, $V_{n\varepsilon} =: v_n$, and $v_{n+1} = v_n - (\varepsilon_n(\operatorname{grad} F)_n)$, where ε_n satisfies appropriate constraints which guarantee a good iteration).

If the mapping M is linear, F is a positive-definite quadratic functional of x so that convexity and uniqueness of the minimum make the calculation easy. If M is not linear, F may not be convex, and may show local minima that stop algorithms on a poor "approximate solution" or need much more computer time. There are also choices of a priori constraints that suggest in F the use of L or L^∞ norms. The algorithms can be handled by linear programming and are usually more robust and less sensitive to "high frequency details" of experimental results than those using quadratic norms (high frequency details of data σ are details which would correspond to small eigenvalues of MM^*: if we tried to invert M directly, these details would produce instabilities of $M^{-1}\sigma$). Some other norms in F are suggested by particular management of information (e.g., the "maximum entropy" method). They usually lead to more complicated algorithms. Again, let us say that all the methods give reasonable average solutions *if the problem is not too strongly ill-posed*. If the nonlinearity is too large, if the sets of solutions are not connected, and if the data errors are too large, we cannot escape a full analysis of the inverse problem!

(b) If we have strong arguments to support the first guess V_0, we may "linearize" the mapping μ before proceeding. Hence, a priori "theoretical constraints" are imposed in the set of potential: we look for V in a neighborhood of V_0, e.g., the ball $\|V - V_0\| \le \varepsilon$, where ε guarantees that the linear approximation is valid. Linear methods are then available. For instance, we can discretize and use algebraic methods, or we can regularize as above, but with the improvements due to convexity. In addition, we can use standard iteration methods. Of course, the remark given a few lines above still must be emphasized.

We can find many studies of radial problems using the numerical methods described above. We shall only cite some of them which in addition:

(a) compare them to other inversion methods, e.g., Mackintosh and Joannidès (1985), Krappe and Rossner (1985), and Krappe et al. (1987);
(b) discuss the regularizations, e.g., Korolyova and Zhigunov (1983), Turchin (1985) and other lectures in Krappe and Lipperheide (1985).

It should be noticed that inverse methods and nonlinear figures like "solitons" have also been applied to a direct understanding of nuclear matter and optical potential (Hefter et al. 1984).

XIX.2.4   Numerical Methods for Other Scattering Inverse Problems

Nonradial Problems of Nuclear Scattering. The methods we called "inverse numerical methods," i.e., those where an approximate solution is obtained by minimizing a cost functional, have been extended to the construction of a three-dimensional optical potential (without special symmetry)—see, e.g., Funke and Zachariev (1987).

One-Dimensional Problem on the Line. There have been several "sources" of numerical studies for this problem—electromagnetic wave scattering, acoustic wave scattering, and elastic wave scattering in geophysical problems are the main ones (problems of plane waves impinging a layered medium at normal incidence). There are other cases. In most cases the medium is either smooth (twice differentiable) or discretized into layers of equal width or equal time-shift (Goupillaud model). Now, we find in the literature:

(a) "Numerical Inverse Methods," based on the Marchenko method or other exact methods (Burridge, 1980; Newton, 1981; Sabatier, 1983; Ware and Aki, 1968; Gopinath and Sondhi, 1971; Krueger, 1981; Reilly and Jordan, 1981; Sondhi, 1982; Corones et al. 1985; and Kristensson and Krueger, 1986a,b). These equations are often solved numerically (particularly in electromagnetic problems) by means of a superposition of Bargmann–type potentials, which correspond to a degenerate kernel or a few poles in the reflection coefficient (Kay, 1956, 1960; Jordan and Kritikos, 1973; Ahn and Jordan, 1976; Reilly and Jordan, 1981; Pechenik and, Cohen, 1981, 1982, 1983; Kritikos et al. 1982; Ladouceur and Jordan, 1985; Jordan and Lakshamany, 1987; and Ge, 1987) or to a few Darboux transformations (Sabatier, 1983, 1984). Iterative methods have also been used either on the equation or directly as in the Jost–Kohn method (Razavy, 1975; Lefebvre, 1978; Kak et al. 1978; Riska, 1981, 1983), as well as the trace method (Stickler and Deift, 1981; Stickler, 1983).

A more original method uses a new spectral representation of the wave function, which enables constructing the solution by solving a second kind of

Fredholm equation and was used on synthetic data plus noise (Schaubert and Mittra, 1977). Modifications of the Gel'fand–Levitan or Marchenko methods are also used. In particular, it is possible to show that a solution of the zero-energy Schrödinger equation $\eta''(x) = V(x)\eta(x)$ is

$$\eta(x) = 1 + \int_0^x K(x, t)dt \qquad \textbf{(XIX.2.5)}$$

(this result follows from **(XVII.1.30)**, **(XVII.1.32)**, **(XVII.2.5)**). The function η can be reconstructed from the reflection coefficient in a more stable way than the potential by variants of the Marchenko method (Kay, 1972; Berryman and Greene, 1980; Coen, 1981a).

A similar representation has been used by Jaggard and Olson (1985) who used the so-called Balanis (1982) reflection coefficient and derived a sequential algorithm with iterative correction (see also Kritikos, 1982). Other approaches of this type were developed by Coen (1981b,c) on geophysical problems, by Corones, Kristensson, and Krueger on electromagnetic problems (Corones et al., 1985; Kristensson and Krueger, 1986a,b), and by Ammicht et al. in acoustical problems (1987). It should be noticed that the papers we cite here are either review papers or papers giving methods and numerical applications on "synthetic data" (theoretical papers are reviewed in chapter XVII); but most numerical inverse methods have not been applied, up to now, to experimental data.

(b) Time-Domain Techniques and Related Sequential Algorithms. As we have seen in section XVII.4, time-domain techniques are deeply founded on the physical characteristics of the sounding problem. The signals travel step after step and the information is extracted at each step, thus enabling us to strip each layer one after the other. From the mathematical point of view, a sequential algorithm propagates errors so that corrections, or more elaborate algorithms, must be implemented. A simplification is obtained by assuming a Goupillaud medium (equal time-shift layers). The problem is then simplified because functions of interest become entire functions of finite order of the time shift. Typically, we get algorithms of the "Levinson" type to solve Toeplitz systems.

This method has been pioneered by Goupillaud, Kunetz, and co-workers (Baranov and Kunetz, 1960; Goupillaud, 1961; Kunetz, 1961, 1963; Kunetz and d'Erceville, 1962; Wuenshel, 1960) who obtained synthetic seismograms (direct problem) and tried to recognize them in the experimental results. Although some of the results are already useful for inversion we cannot call this a method for solving the inverse problems before Claerbout (1968), Kunetz and Rocroy (1970), and Berryman and Greene (1980); and, more recently, Ursin and Bertenssend (1986) who compare the various forms of this method and give the state-of-the-art (noise, various source forms, non-Goupillaud models, etc).

In addition to the Levinson algorithm used by Claerbout (Claerbout, 1968; Mendel, 1986), there are implementations where the effect of the k first reflection factors is subtracted from the result to be analyzed by subtracting the (synthetic) result they would generate, and then other implementations where waves going downward and upward are separated and continued (Mendel and Hahihi Astragi, 1980, 1982). Still more popular methods are those where multiple reflections are neglected (see, e.g., Gazanhes et al., 1981) and which are related to multidimensional migration methods, widely used in geophysics, and beyond the scope of the present book. Similar techniques were used in other fields (see Bates and Millane (1981) and section XVII.4.5, and for a very large survey, see Tijhuis (1987)).

A class of parallel methods which either amounts to stripping the layers of a medium, but are obtained by an analysis of differential equations, or using directly the characteristic directions of the hyperbolic system, were developed by theoreticians and sometimes applied to noisy data (Bruckstein et al., 1986; Symes, 1981; Gray and Symes, 1985; Santosa and Schwetlick, 1982; Levy, 1985; Hahashy and Mittra, 1987; and Yagle, 1987). After discretizing they lead, in certain cases, to the Cholesky algorithm. Stability problems have been studied (Fawcett, 1985; Carrion, 1986).

(c) "Inverse Numerical Methods". It being understood that our choice is arbitrary, we can say that recurrent methods which involve strong controls and corrections are already inverse numerical methods, although they are related to the preceding ones—essentially because the aim of the authors is the realistic calculation of a solution instead of a study whose interest is half-way between the understanding of the problem's mathematical structure and the construction of approximate solutions.

In this framework, we shall refer to the works of Lesselier (1978, 1981), Bolomey et al. (1978, 1979), Bojarsky (1980), Tijhuis (1981, 1987) and to those of Hahashy and Mittra (1986), Corones et al. (1983, 1985), already cited, and Levy (1985).

However, the easiest to use, because they are the less "specific" inverse numerical methods, are the optimization methods, where the three essential and coupled problems are, in almost all cases, the choice of the cost functional to be minimized—the ways of going around secondary minima if any, and the stability problem. In the one-dimensional impedance problem, these problems have been deeply studied in the works of Bamberger et al. (1979, 1982), Macé and Lailly (1986), Kolb et al. (1986), Richard (1987), and in electromagnetic problems by Lesselier (1981, 1982a,b), Roger (1978), Tijhuis (1987, chap. VI), and Coen et al. (1981). These authors generally use gradient methods, with the so-called "adjoint state" method. Connected studies, sometimes including an analysis of real data, appear essentially in studies of electromagnetic wave problems, and treat either the original equations or the perturbed ones (Mostafavi and Mittra, 1972; Mittra and Itoh, 1975; Lesselier, 1981, 1982a,b,c; Tabbara, 1979), and we can find a large survey of them in Tijhuis (1987).

To finish this brief survey of methods introduced in geophysical, acoustical, or electromagnetic problems for the numerical treatment of data, we should mention the so-called resonance methods, where a characteristic part of the observed response, "a signature," is compared with synthetic ones that correspond to essential parameters of the medium (Braum, 1976; Block and Tijhuis, 1985; Fiorito et al., 1986; Jackins and Gaunaud, 1986). We should also notice that the simplicity of well-known approximations (the Born approximation, the distorted wave Born approximation, the WKB or Rytov approximation) is sometimes used for numerical computations.

Other Numerical Studies of the One-Dimensional Problem. An unexpected source of numerical studies of the one-dimensional Schrödinger problem has been the recent development of a class of nonlinear signal analyses. They use the inverse problem like the Fourier transform is used in linear (and usual) signal processing. Hence, they are theoretically founded on the inverse method described in section XVII.5.

From the numerical point of view, they need a good and (?) specific treatment of the Marchenko–Faddeev method or others methods, with a specific analysis of the discrete and the continuous spectrum. This treatment was implemented (to study one-dimensional water waves spectra) by A. R. Osborne and co-workers (Osborne et al., 1982a,b,c; Osborne, 1983; Osborne and Bergamasco, 1985, 1986). It is interesting to note that two-dimensional studies recently began with works by Segur and co-workers (1987).

Another unexpected source for theoretical and numerical methods which apply to the one-dimensional inverse problem (and the Schrödinger problem is again, in some way, a particular case) has been the "vocal tract" inverse problem (reconstructing the vocal tract from speech analysis). The relevant papers are Gopinath and Sonthi (1970), Sondhi and Gopinath (1970, 1971), Sondhi (1981, 1982). In addition to the problems of electromagnetic waves in plasmas on elastic waves in the Earth, it is also interesting to cite solutions of the electrodynamical inversion problem for vertically layered Earth (Riska 1981, 1982), the inversion problems of magnetic ore prospecting (Rajala and Riska 1981), and the problem of electromagnetic dipole radiation reflected from Earth (Vidberg and Riska, 1983; Vogelzang et al., 1983); not forgetting the very applied, very up-to-date, and almost all classified methods for minimizing the reflection of electromagnetic waves by surface impedance.

Mathematical Studies for Numerical Analyses of the One-Dimensional Problem. It is clear that any separation between mathematical papers, theoretical papers, papers of numerical analysis, and papers dealing with algorithms for computing a solution is arbitrary. However, we shall try to cite in this paragraph the mathematical papers dealing with mathematical aspects of "signal processing" and "numerical analyses" of the one-dimensional Schrödinger problems and the closely related papers in acoustics, electromagnetic theory, etc. These problems are:

(a) Uniqueness of a solution. We already know it in the "quantum" case. In the electromagnetic case, it has been studied by Païvärinta and Somersalo (1987).

(b) Existence and construction of the solution of the one-dimensional inverse problem corresponding to elastic wave propagation in a stratified medium (Berryman and Greene, 1980; Carroll and Santosa, 1981; Santosa, 1982), or to electromagnetic waves through an arbitrary number of scatterers (Ström, 1974a,b; Gopinath, 1976), or through a dispersive medium (Beezley and Krueger, 1985), or to the inverse problem of electrical conductivity (Hooshyar and Razavy, 1982).

(c) Stability problems (Carrion, 1986; Fawcett, 1985; Symes, 1981, 1985).

(d) Inversion from band-limited data (Santosa and Symes, 1983).

(e) Special mathematical problems related to certain numerical methods (Symes, 1983a,b).

XIX.3   Numerical Methods for Inverse Spectral Problems

There are many applied problems connected with "inverse spectral problems," since determining a mechanical structure from its spectra is also of interest in fundamental research (e.g., the inverse Earth mode problem) and in industrial research (e.g., stabilization of structures). The exact methods which were given in chapter XVIII are rarely useful for practical cases, where only a finite number of frequencies is known, with errors. The general numerical methods which have been sketched in section XIX.1.1 (least square solutions or other approximate solutions that minimize a given norm) are useful, but there are also two "branches" of methods specially devised for this kind of problem. The first one is, in some way, the modern and numerical development of an idea by Lord Rayleigh (1894–1896), who insisted in his lectures that the eigenvalues of a system should depend continuously on parameters like the resonator shape or the string density, so that a continuous deformation of this shape may reproduce a given spectrum.

Staying with the resonator shape, we can consider the domain Ω depending on a parameter t, and find a "steepest descent evolution" of Ω to minimize a cost functional of the difference $|\sigma_{mes} - \sigma_{cal}(\Omega)|$, where σ_{mes} is obtained from the spectrum measurements or requirements. This is the basic idea of the method developed, in particular, by Rousselet (1978, 1981, 1982) on numerical schemes proposed by Céa (1978).

Lord Rayleigh had also noticed that certain eigenvalues get their maximum value for certain shapes. On the other hand, let us consider inverse problems with "positivity constraints," for instance, those which are ruled by the equations

$$Ax = e; \qquad x_n \geq 0, \qquad n = 1, 2, \ldots, N, \qquad \text{(XIX.3.1)}$$

where $x \in \mathbb{R}^N$, $e \in \mathbb{R}^M$, $M < N$, and A is an $M \times N$ matrix.

In physical problems, M is the number of measurements, and N is the number of "finite elements" that define the "discretization." In this problem, it is easy to see that the set of solutions is convex, and can be represented by its extreme points (also called "extremals"), which can be calculated by linear programming, but there are too many for convenient practical calculations. There are also special solutions, called "ideal solutions," such that for one of them, say $x^{(i)}$, the upper bound $\mathrm{Sup}_n\, x_n(i)$ is *minimum over the set of solutions.* Clearly, this solution is interesting because the maximum value of any other solution should be larger. Furthermore, it is easy to see, for an ideal solution, that x_n may take only two values, 0 and $\mathrm{Sup}\, x_n$, except at M sides, so that, in the "continuous limit" $n \to \infty$, the solution has two values only. Interest in ideal solutions in gravity inverse problems was shown by Parker (1974, 1975), and their general properties and construction were demonstrated by Sabatier (1977). Now, the question was whether there exist, or not, ideal solutions of inverse spectral problems. For the string problem, Barcilon (1979b) proved that a string, whose density ρ is a linear combination of M Dirac measures, minimizes the total density over the set of strings whose M lowest eigenvalues are given, and Sabatier (1979) showed that only special moments of the density could lead to such an ideal body. Let us reproduce here a very simple argument to understand these results, and their possible extensions or limitations (Sabatier, 1980).

Suppose a spectral problem is ruled by a Fredholm integral equation, for instance,

$$w(x) = v \int_\Omega K(x, y)w(y)d\sigma_y, \tag{XIX.3.2}$$

where Ω is an interval of $\mathbb{R}$, K is a positive symmetric kernel, and $d\sigma_y$ is a positive measure. The N lowest eigenvalues v_i are given and the inverse problem is $\{v_i\} \to \sigma$. Since the v_i's are implicit functionals of σ, the problem is very complicated. Since $d\sigma$ can be any positive measure, and the v_i's are finitely many, the problem is (at least) underdetermined.

Now let us write the trace formulas

$$T_m = \sum_{i=1}^\infty (v_i)^{-m} = \int_\Omega K_m(y, y; \sigma)d\sigma_y, \tag{XIX.3.3}$$

where the kernel K_m is defined by the following recurrence

$$K_{m+1}(x, y; \sigma) = \int_\Omega K(x, t)K_m(t, y; \sigma)d\sigma_t \tag{XIX.3.4}$$

beginning $K_1(x, y; \sigma) = K(x, y)$. Since the v_i's are positive, the unknown exact value of T_m is at least equal to $\sum_{i=1}^M (v_i)^{-m}$, M being the number of measurements. Writing $T_m = \sum_{i=1}^M (v_i)^{-m}$ exactly for each m obviously corresponds on the one hand to a solution $d\sigma$ which, for all m, minimizes $\int_\Omega K_m(y, y; \sigma)d\sigma_y$ over the set of solutions—and on the other hand, to a system with only a finite number of eigenvalues. This solution is therefore a "Krein's density" of the form $\sum_1^M a_p\delta(x - x_p)$, and the a_p's can be calculated by solving algebraic systems. Another interest in this density (also called a "degenerate string") is that it helps to show physically how matrix inverse problems are related to physical spectral problems, the mathematical way being of course to replace the differential operator by a finite difference operator.

Miscellaneous Problems. Numerical studies of the Gel'fand–Levitan method were also developed in connection with problems of discrete spectra (Hald, 1979). However, these problems most often need numerical approaches that are related to special properties of matrix spectral problems (Hald, 1972; 1978, 1984; Morel, 1978a,b; Rousselet, 1976; Boley and Golub, 1977, 1978, 1987; Friedland, 1977, 1979; Friedland et al.. 1984).

The two- and three-dimensional problems of acoustic, electromagnetic, and elastic waves received much attention, and the time-domain approaches to the Schrödinger problem, as well as some frequency-domain approaches, are closely related to them. However, in the case of acoustic or electromagnetic waves, almost all the numerical methods were rather concerned with the determination of the scatterer shape—a problem which has no real equivalent in quantum scattering, where the surface scatterer is always "diffuse." Others were very specific about the nature of these waves. In geophysics, the discontinuous interfaces between media do not have their counterpart in quantum problems. Hence, we shall not go through the references of these works, citing, however, three papers because, very curiously, they give somewhat intermediate approches between one- and two-dimensional problems (in problems with no special symmetry): Millane (1983), Moses (1979), and Hooshyar and Razavy (1982).

XIX.4 Nonlocal Potentials

We shall not go through numerical methods which are systematic trial and error fits. There are many of them in nuclear physics, because extensive studies of the theory of nuclear matter have shown that the optical potential, when derived as the summation of two-body potentials, is nonlocal in character. A part of these studies is more closely connected with the point of view of inverse problems: that which proves the "equivalence," with respect to observable results, between a slightly nonlocal potential and a potential depending slowly on energy (for a review and references, see Hodgson (1965, 1972)).

In connection with the Faddeev equations for the three-body problems, and the more general equations for more particle problems, for which we need the off-the-energy-shell sub-t-matrices, considerable amounts of calculations have also been performed by nuclear physicists. In most, if not all, these calculations, extensive use is made of nonlocal separable potentials. We refer the reader to the papers by Oryu (1983) and Plessas (1986), already mentioned at the end of chapter VIII, and to current issues of the journal *Few Body Systems*.

Reference List

This reference list can be used as a personal authors index since each reference is followed by the number of the section in which it is quoted. The titles are given so as to make easy the recollection of subjects. "H" refers to comments or historical details at the end of chapters even when a special number has been given to these end sections.

Abel N. H. (1826): Résolution d'un problème de mécanique, *J. Reine Angew. Math.* 1, 97–101 (section XVI.7)

Ablowitz M. J. (1978): Lectures on the inverse scattering transform, *Stud. Appl. Math.* 58, 17–94 (sections XVII.5, XVII.7)

Ablowitz M. J., and Segur H. (1981): *Solitons and the Inverse Scattering Transform*, SIAM, Philadelphia (section XVII.5)

Ablowitz M. J., Kaup D. J., Newell A. C., and Segur H. (1974): The inverse scattering transform—Fourier analysis for nonlinear problems, *Stud. Appl. Math.* 53, 4, 249–315 (section XVII.5)

Abraham P. B., De Facio B., and Moses H. E. (1981): Two distinct local potentials with no bound states can have the same scattering operator: A nonuniqueness in inverse spectral transformations. *Phys. Rev. Lett.* 46, 1657–1659 (section XVII.5)

Abraham P. B., De Facio B., and Moses H. E. (1982): An explicit example of a local and a nonlocal potential whose hamiltonians are unitarily equivalent and whose scattering operators are identical, *Stud. Appl. Math.* 66, 45–51 (section XVII.5)

Abraham P. B., De Facio B., and Moses H. E. (1983): Parity-dependent potentials for the one-dimensional Schrödinger equation obtained from inverse spectral theory, *J. Phys. A* 16, 303–316 (section XVII.5)

Agmon S. (1986): Spectral theory of Schrödinger operators on euclidean and non-euclidean spaces, *Comm. Pure Appl. Math.* XXXIX, S3–S16 (section XIV.7)

Agranovich Z. S., and Marchenko V. A. (1957a): Reestablishment of the potential from the scattering matrix for a system of differential equations, *Dokl. Akad. Nauk. SSSR* 113, 951–954, [*Math. Rev.* 19, 7, 746 (1958)] (sections V.H, IX.2, Foreword)

Agranovich Z. S., and Marchenko V. A. (1957b): Reconstruction of the potential energy from the dispersion matrix, *Usp. Mat. Nauk.* 12, 143–145: translated in *Amer. Math. Soc. Transl.* 16, 2, 355–357 (1960) (sections III.H, V.H, IX.2, Foreword)

Agranovich Z. S., and Marchenko V. A. (1958): Construction of tensor forces from scattering data, *Dokl. Akad. Nauk. SSSR* 118, 1055–1058, [*Math. Rev.* 21, 6, 741–742 (1960)] (sections V.H, IX.2, Foreword)

Agranovich Z. S., and Marchenko V. A. (1963): *The Inverse Problem of Scattering Theory*: English translation Gordon and Breach, New York (sections III.H, V.H, IX.1, Foreword)

Aktosun T. (1987a): Marchenko inversion for perturbations, *Inverse Problems* 3, 523–553 and 565–575 (section XVII.5)

Aktosun T. (1987b): Stability of the Marchenko inversion, *Inverse Problems* 3, 554–564 (section XVII.5)

Aktosun T. (1987c): Examples of non-uniqueness in one-dimensional inverse scattering for which $T(k) = O(k^3)$ and $O(k^4)$ as $k \to 0$, *Inverse Problems* 3, L1–L3 (section XVII.5)

Aktosun T., and Newton R. G. (1985): Non-uniqueness in the one-dimensional inverse scattering problem, *Inverse Problems* 1, 291–300 (section XVII.5)

Allen L. J., and Burger H. (1984): Local potentials equivalent to matrix effective potentials for e–He scattering at 200 and 400 eV, *Phys. Rev. A* 30, 1237–1240 (section XIX.2)

Alvarez-Estrada R. F. (1971): On the construction of scattering amplitudes from unitary, analyticity and experimental data, *Ann. Phys.* 68, 222 (section X.1)

Alvarez-Estrada R. F., and Carreras B. (1973): On the rigorous determination of the spin-0-spin $\frac{1}{2}$ elastic scattering amplitudes from experimental data and unitarity, *Nucl. Phys.* B54, 237–262 (section X.6)

Ambarzumian V. (1929): Uber eine frage der eigenwertheorie, *Z. Phys.* 53, 690–695 (section 5, III.H, XVIII.4, Foreword)

Andersson L. A. (1988a): Inverse eigenvalue problems with discontinuous coefficients, *Inverse Problems* 4, 353–398 (section XVIII.4)

Andersson L. A. (1988b): Inverse eigenvalue problem for a Sturm–Liouville equation in impedance form, submitted to *Inverse Problems* (section XVIII.4)

Andrianov A. A., Borisov N. V., and Ioffe M. V. (1984): Factorization method and Darboux transformation for multidimensional hamiltonians, *Theoret. i Mat. Fiz.* 61, 183–198 (section XVII.5)

Angell T. S., Kleinman R. E. and Roach G. F. (1987): An inverse transmission problem for the Helmholtz equation, *Inverse Problems* 3, 149–181 (section XIV.7)

Antoun J. T., and Grünbaum F. A. (1980): The lowest eigenvalue of a random matrix, *SIAM J. Appl. Math.* 38, 168–174 (section XVIII.4)

Arutyunyan T. N. (1987): An inverse problem for the Dirac system on the semi-axis, with discrete spectrum, *Amer. Math. Soc. Transl.* (2) 136, 43–47 (section XVIII.4)

Atkinson D. (1979): Marchenko in one dimension, Internal report, Groningen University (section XVII.5)

Atkinson D., and Stefanescu I. S. (1985): Truncation of continuum ambiguities in phase-shift analysis, *Comm. Math. Phys.* 101, 291–304 (section X.7)

Atkinson D., Johnson P. W., Kok L. P., and De Roo M. (1974): Construction of unitary, analytic scattering amplitudes III. Practical application to scattering, *Nucl. Phys.* B77, 109–138 (sections X.4, X.6, Foreword)

Atkinson D., Johnson P. W., Mehta N., and De Roo M. (1973): Crichton's phase-shift ambiguity, *Nucl. Phys.* B55, 125–131 (section X.1, Foreword)

Atkinson D., Johnson P. W., and Warnock R. L. (1972): Determination of the scattering amplitude from the differential cross-section and unitarity, *Commun. Math. Phys.* 28, 133–158 (section X.1, Foreword)

Atkinson D., Johnson P. W., and Warnock R. L. (1973): Construction of analytic, unitary scattering amplitudes from a given differential cross-section: a refined analysis, *Commun. Math. Phys.* 33, 221–242 (section X.6, Foreword)

Atkinson D., Kaekebeke M., and De Roo M. (1975a): Phase-shifts as functions of the cross-section, *J. Math. Phys.* 16, 3, 685–693 (section X.1, X.4, Foreword)

Atkinson D., Kok L. P., and De Roo M. (1977): Crichton ambiguities with infinitely many partial waves. Phys. Rev. *D* 17, 2492–2498 (section X.1)

Atkinson D., Mahoux G., and Yndurain F. J. (1973a): Construction of a unitary analytic scattering amplitude I. Scalar particles, *Nucl. Phys.* B54, 263–284 (sections X.1, X.6, Foreword)

Atkinson D., Mahoux G., and Yndurain F. J. (1973b): Construction of a unitary analytic scattering amplitude II. Introduction of spin and isospin (spin 0–spin $\frac{1}{2}$), *Nucl. Phys.* B66, 429–463 (section X.6)

Atkinson D., Mahoux G., and Yndurain F. J. (1975): Construction of a unitary analytic scattering amplitude IV. Many two-body channels, *Nucl. Phys.* B98, 521–532 (section X.6)

Auberson G. (1988): On the reconstruction of a unitary matrix from its moduli. Existence of continuous ambiguities, to be publ. in *Phys. Lett. B*

Austern N. (1961): Optical model wave functions for strongly absorbing nuclei, *Ann. Phys.* 15, 299–313 (sections XVI.3, XVI.4)

Backlund A. V. (1880): Zür Theorie der Partiellen Differentialgleichungen erster Ordnung, *Math. Ann.* XVII, 285–328 (section XVII.4)

Balanis G. N. (1972): The plasma inverse problem, *J. Math. Phys.* 13, 1001–1005 (section XVII.4)

Baltes H. P., and Hilf E. R. (1976): Spectra of finite systems. A review of Weyl's problem: the eigenvalue distribution of the wave equation for finite domains and its applications on the physics of small systems, Bibliographisches Institut, Mannheim/Wien/Zurich, B. I. Wissenschafts Verlag (section XVIII.4)

Baeteman M. L., and Chadan K. (1975): The inverse scattering problem for singular oscillating potentials, *Nucl. Phys.* A255, 35–49 (sections I.H, II.H, III.H, V.H, VI.H)

Baeteman M. L., and Chadan K. (1976): Scattering theory with highly singular oscillating potentials, *Ann. Inst. Henri Poincaré* XXIV A, 1–16 (sections I.H, II.H, III.H, V.H, VI.H)

Bamberger A., Chavent G., Hemon Ch., and Lailly P. (1982): Inversion of normal incidence seismograms, *Geophys.* 47, 757–770 (section XIX.2)

Bamberger A., Chavent G., and Lailly P. (1979): About the stability of the inverse problem in a 1-D wave equation. Application to the interpretation of seismic profiles, *J. Appl. Math. Optim.* 5, 1–47 (section XIX.2)

Barakat R., and Newsam G. (1984): Necessary conditions for a unique solution to two-dimensional phase recovery, *J. Math. Phys.* 25, 3190–3193 (section X.7)

Baranov V., and Kunetz G. (1960): Film synthétique avec reflexions multiples. Théorie et calculs pratiques, *Geophys. Prospect.* 8, 315–325 (section XIX.2)

Barcilon V. (1974): Iterative solution of the inverse Sturm–Liouville problem, *J. Math. Phys.* 15, 429–436 (section XVIII.4)

Barcilon V. (1976a): A discrete model of the inverse Love wave problem, *Geophys. J. R. Astron. Soc.* 44, 61–76 (section XVIII.4)

Barcilon V. (1976b): Inverse problem for a vibrating beam, *J. Appl. Math. Phys.* 27, 347–358 (section XVIII.4)

Barcilon V. (1977): On impulse response data and the uniqueness of the inverse problem *J. Geophys.* 43, 139–152 (section XVIII.4)

Barcilon V. (1978a): On the information content of natural frequency spectra associated with different angular numbers, *Geophys. J. R. Astron. Soc.* 53, 623–641 (section XVIII.4)

Barcilon V. (1978b): Inverse problems for vibrating elastic structures, Invited Lecture to the Eighth U.S. National Congress of Applied Mechanics, UCLA, 1978 (section XVIII.4)

Barcilon V. (1979a): On the multiplicity of solutions of the inverse problem for a vibrating beam, *SIAM J. Appl. Math.* 37, 605–613 (section XVIII.4)

Barcilon V. (1979b): Ideal solution of an inverse normal mode problem with finite spectral data, *Geophys. J. R. Astron. Soc.* 56, 399–408 (section XVIII.4)

Barcilon V. (1980): Explicit solution of the inverse problem for a vibrating string, *J. Math. Anal. Appl.* 93, 222–234 (section XVIII.4)

Barcillon V. (1982): Inverse problem for the vibrating beam in the free-clamped configuration, *Philos. Trans. Roy. Soc. London Ser. A* 304, 211–251 (section XVIII.4)

Barcilon V. (1987a): Sufficient conditions for the solution of the inverse problem for a vibrating beam, *Inverse Problems* 3, 181–194 (section XVIII.4)

Barcilon V. (1987b): Inverse eigenvalue problems, in *Inverse Problems* (G. Talenti, ed.), Springer-Verlag, New York, pp. 1–51 (section XVIII.4)

Barcilon V., and Turchetti G. (1979): On an inverse eigenvalue problem with truncated spectra data, in *Inverse and Improperly Posed Problems in Differential Equations* (G. Anger, ed.), Akademie-Verlag, Berlin, pp. 25–34. See also idem. *Wave Motion* 2, 139–148 (1980) (section XVIII.4)

Bargmann V. (1949a): Remarks on the determination of a central field of force from the elastic scattering phase shifts, *Phys. Rev.* 75, 2, 301–303 (sections IV.H, VI.H, Foreword)

Bargmann V. (1949b): On the connection between phase shifts and scattering potential, *Rev. Mod. Phys.* 21, 3, 488–493 (sections IV.H, VI.H, Foreword)

Bargmann V. (1952): On the number of bound states in a central field of force. *Proc. Nat. Acad. Sci. U.S.A.* 38, 961–966 (sections II.1, IX.9)

Bart G. R., Johnson P. W., and Warnock R. L. (1973): Continuum ambiguity in the construction of unitary analytic amplitudes from fixed energy scattering data, *J. Math. Phys.* 14, 11, 1558–1565 (section X.6)

Bart G. R., Johnson P. W., and Warnock R. L. (1974): Unitary amplitudes with cut-plane analyticity from given scattering data, *Nucl. Phys.* B72, 329–359 (section X.6)

Bates H. T., and Millane R. P. (1981a): Time-domain approach to inverse scattering, *IEEE Trans. Antennas and Propagation* AP-29, 359–363 (section XIX.2)

Baumgartner H., Grosse H., and Martin A. (1985): The laplacian of the potential and the order of energy levels, *Nucl. Phys.* B 254, 528–542 (sections IV.5, XVIII.4)

Beals R. (1985): The inverse problem for ordinary differential operators, *Amer. J. Math.* 107, 281–366 (section III.7)

Beals R., and Coifman R. R. (1981): Scattering, transformations spectrales et equations non-linéaires I, II, Séminaire Goulaouic–Meyer–Schwartz, 1980–1981, Exposés 21, 22 (section XIV.7)

Beals R., and Coifman R. R. (1984a): Scattering and inverse scattering for first-order systems, *Comm. Pure Appl. Math.* 87, 39–90 (sections XIV.5, XVII.5)

Beals R., and Coifman R. R. (1985a): "Multidimensional inverse scatterings and non-linear partial differential equations", *Proc. Sympos. Pure Math.* 43, 45–70 (section XIV.7)

Beals R., and Coifman R. R. (1985b): Inverse scattering and evolution equation, *Comm. Pure Appl. Math.* 88, 29–42 (section XVII.5)

Beals R., and Coifman R. R. (1986): The *d*-bar approach to inverse scattering and nonlinear evolutions, *Physica D* 18, 242–249.

Beals R., and Coifman R. R. (1987): Scattering and inverse scattering for first-order systems II, *Inverse Problems* 3, 577–593 (section XVII.5)

Becher W. D., and Sharpe C. B. (1969): A synthesis approach to magnetotelluric exploration, *Radio Sci*, 4, 1089–1094 (section XVII.4)

Bednar B., Redner R., Robinson E., and Weglein A. (1983): *Conference on Inverse Scattering: Theory and Application*, SIAM, Philadelphia (section XIX.2)

Beezley R. S., and Krueger R. J. (1985a): An electromagnetic inverse problem for dispersive media, *J. Math. Phys.* 26, 317–325 (section XVII.5)

Benedek I., and Panzone R. (1980): On inverse eigenvalue problems for a second-order differential equation with parameter contained in the boundary conditions, *Notas Algebra Anal.* 9, 1–13 (section XVIII.4)

Benn J., and Scharf G. (1972): Determination of nucleon–nucleon potentials from coupled partial waves using the Marchenko theory, *Nucl. Phys.* A 183, 319–336 (section VI.H)

Berends F. A., and Ruijsenaars S. N. M. (1973a): Examples of phase-shift ambiguities for spinless elastic scattering, *Nucl. Phys.* B56, 507–524 (section X.1)

Berends F. A., and Ruijsenaars S. N. M. (1973b): Different sets of phase shifts for the same differential cross-section and polarization in spin 0–spin $\frac{1}{2}$ elastic scattering, *Nucl. Phys.* B56, 525–535 (section X.6)

Berezanskii J. M. (1955): On the inverse problem of spectral analysis for the Schrödinger equation, *Dokl. Akad. Nauk. SSSR* 105, 433–436, [*Math. Rev.* 17, 8, 1210 (1956)] (section III.H, Foreword)

Berezanskii J. M. (1958): The uniqueness theorem in the inverse problem of spectral analysis for the Schrödinger equation. *Trudy Moscow Mat. Obsc.* 7, 3–62, translated in *Amer. Math. Soc. Transl.* 35, 2, 167–235 (1964) (Foreword)

Bernstein R. B. (1966): Quantum effects in elastic molecular scattering, *Adv. Chem. Phys.* 11, 75–134 (sections XVI.4, XVI.H)

Bernstein R. B., and Muckerman J. T. (1967): Determination of intermolecular force via low-energy molecular beam scattering, *Adv. Chem. Phys.* 12, 389 (sections XVI.4, XVI.H)

Berry M. W. (1969): Uniform approximation for glory scattering and diffraction peaks, *J. Phys. B (Atom. Molec. Phys.)* Série 2, 2, 381–392 (sections XVI.4, XVI.H)

Berry M. V., and Mount K. E. (1972): Semiclassical approximations in wave mechanics, *Rep. Prog. Phys.* 35, 315–397 (section XVI.H)

Berryman J. G. (1979): Inverse methods for elastic waves in stratified media, *J. Appl. Phys.* 50, 6742–6744 (section XIX.2)

Berryman J. G., and Greene R. R. (1978): Discrete inverse scattering theory and the continuum limit, *Phys. Lett. A* 65, 13–15 (section XIX.2)

Berryman J. G., and Greene R. R. (1980): Discrete inverse method for elastic waves, *Geophys.* 45, 213–233 (section XIX.2)

Bessis J. D., and Martin A. (1967): A theorem of uniqueness, *Nuovo Cimento* 52A, 719–726 (section X.1)

Blagoveshchenskii A. S. (1967): The inverse problem in the theory of seismic wave propagation, Topics in Math. Phys. Vol. 1: Spectral theory and wave processes, ed. by M. S. Birman Consultants Bureau N.Y. (section XVII.4)

Blagoveshchenskii A. S. (1969): The inverse problem in the theory of seismic wave propagation, Topics in Math. Phys. Vol. 3: Spectral theory, ed. by M. S. Birman Consultants Bureau N.Y. (section XVII.4)

Blagoveshchenskii A. S. (1971a): The inverse problem for the wave equation with an unknown source, Topics in Math. Phys. Vol. 4: Spectral theory and wave processes, ed. by M. S. Birman Consultants Bureau N.Y. (section XVII.4)

Blagoveshchenskii A. S. (1971b): On the various formulations of the one-dimensional inverse problem for the telegraph equation, Topics in Math. Phys. Vol. 4: Spectral theory and wave processes, ed. by M. S. Birman Consultants Bureau (section XVII.4)

Blazek M. (1962a): The Gel'fand–Levitan equation and its application to the scattering of neutron on protons, *Czech. J. Phys.* B12, 249–257 (sections IV.H, VI.H, IX.1)

Blazek M. (1962b): Explicit determination of a potential with n bound states by means of the solution of the inverse problem, *Czech. J. Phys.* B12, 258–263 (sections IV.H, IX.9, Foreword)

Blazek M. (1963): Determination of the potential by means of the scattering amplitude analytic properties, *Mat. Fyz. Cas.* 13, 147 (section IX.1)

Blazek M. (1966): On a method for solving the inverse problem in potential scattering, *Commun. Math. Phys.* 3, 282–291 (section IX.1)

Bloch A. Sh. (1953): On the determination of a differential equation by its spectral matrix functions, *Dokl. Akad. Nauk. SSSR* 92, 209–212 (section III.H)

Boas R. P. (1954): *Entire functions*, Academic Press, New York (sections III.1, XII.8)

Bogdanov I. V. (1985): Classical limit transition in the quantum inverse scattering problem, *Teoret. i Mat. Fiz.* 65, 35–44 (section XVI.H)

Boiti M., Laddomada C., and Pempinelli F. (1986): Bäcklund transformations for the Schrödinger equation with a spectral dependence in the potential *J. Phys. A* 17, 3151–3167 (section XVII.5)

Boiti M., Laddomada C., Pempinelli F., and Tu G. Z. (1983): Bäcklund transformations related to the Kaup–Newell spectral problem, *Physica D* 9, 425–432 (section XVII.5)

Boiti M., Leon J. P. Manna M., and Pempinelli F. (1985): On the spectral transform Korteweg–de Vries equation in two space dimensions, *Inverse Problems* 2, 271–279 (section XIV.7)

Boiti M., Leon J. P., Manna M., and Pempinelli F. (1986): On a spectral transform of a KdV-like equation related to the Schrödinger operator in the plane, *Inverse Problems* 3, 25–36 (section XIV.7)

Boiti M., and Pempinelli F. (1980a): Similarity solutions and Bäcklund transformations of the Boussinesq equation, II, *Nuovo Cimento B* 56, 148–156 (section XVII.5)

Boiti M., and Pempinelli F. (1980b): Nonlinear Schrödinger equation, Bäcklund transformations and Painlevé transcendents, II, *Nuovo Cimento B* 59, 40–58 (section XVII.5)

Bojarsky N. N. (1980): One-dimensional direct and inverse scattering in causal space, *Wave Motion* 2, 115–124 (section XIX.2)

Boley D. L., and Golub G. H. (1977): *Inverse Eigenvalue Problem for Band Matrices*, Lecture Notes in Mathematics, Numerical Analysis Springer-Verlag, New York (section XIX.2)

Boley D. L., and Golub G. H. (1978): The matrix inverse eigenvalue problem for periodic Jacobi matrices, Presented at the Fourth Conference on Basic Problems of Numerical Analysis, Pilsen, Czechoslovakia (section XVIII.4)

Boley D. L., and Golub G. H. (1987): A survey of matrix inverse eigenvalue problems, *Inverse Problems* 3, 595–622 (sections XIX.2, XVIII.4)

Bollé D. (1986): Higher-order Levinson theorems and the Planck–Larkin partition function for reacting plasmas, in *Strongly Coupled Plasma Physico* (F. J. Rogers and H. E. de Witt, eds.), Plenum, New York (section XVII.5)

Bollé D., Gesztesy F., and Klaus M. (1987): Scattering theory for one-dimensional systems with $\int_s dx\, V(x) = 0$, *J. Math. Anal. Appl.* 122, 496–518 (sections XVIII.2, XVII.5)

Bollé D., Gesztesy F., and Wilk S. F. J. (1983): New results for scattering on the line, *Phys. Lett. A* 97, 30–34 (sections XVII.2, XVII.5)

Bollé D., Gesztesy F., and Wilk S. F. J. (1985): A complete treatment of low-energy scattering in one dimension, *J. Operator Theory* 13, 3–31 (sections XVII.2, XVII.5)

Bolomey J. Ch., Durix Ch., and Lesselier D. (1978): Time-domain integral equation approach for inhomogeneous and dispersive slab problems, *IEEE Trans. Antennas and Propagation* AP-26, 658–667 (section XIX.2)

Bolomey J. Ch., Durix Ch., and Lesselier D. (1978b): Determination of conductivity profiles by time-domain reflectometry, *IEEE Trans. Antennas and Propagation,* AP-27, 244–248 (section XIX.2)

Bolsterli M., and McKenzie J. (1965): Determination of separable potential from phase shift, *Physics* 2, 3, 141–149 (section VIII.H, Foreword)

Borg G. (1946): Eine Umkehrung der Sturm–Liouvilleschen Eigenwertaufgabe, *Acta Math.* 78, 1, 1–96 (sections III.H, XVIII.4)

Borg G. (1949): Uniqueness theorems in the spectral theory of $\{-(d^2/dx^2) + V(x)\}u = \lambda u$, Eleventh Congress of Scandinavian Mathematics, Trondheim, Aug. 22–25, 276–287 (section III.H, Foreword)

Borg G. (1949b): On the completeness of some sets of functions, *Acta Math.* 81, 265–283 (section XVIII.4)

Boyd J. P. (1981): Sturm–Liouville eigenproblems with an interior pole, *J. Math. Phys.* 22, 1575–1590 (section XVIII.4)

Boyle J. F. (1971): Semiclassical inversion of rainbow scattering data, *Mol. Phys.* 22, 993–1011 (section XVI.6)

Brackett J. W., Mueller C. R., and Sanders W. A. (1963): Direct determination of scattering phase shifts from differential cross sections, *J. Chem. Phys.* 39, 1, 2564–2571 (section XVI.4)

Brander O. (1981): High energy behaviour of phase shifts for scattering from singular potentials, *J. Math. Phys.* 22, 1229–1235 (section II.17)

Brander O., and Chadan K. (1980): A tauberian theorem in quantum mechanical inverse scattering theory, *Ann. Inst. H. Poincaré,* XXXIII, 283–299 (section III.H)

Brillouin L. (1926): La mécanique ondulatoire de Schrödinger; une méthode de résolution par approximations successives, *C. R. Acad. Sc. Paris* 183, 24 (section XVI.H)

Browstein K. R. (1982): Nonuniqueness of the inverse-scattering problem and the presence of $E = 0$ bound states, *Phys. Rev. D,* 25, 2704–2705 (section XVII.5)

Bridges T. J., Chen G., and Crosta G. (1987): Minimizing the reflection of electromagnetic waves by surface impedance, *Wave Motion* 9, 1–18 (section XIX.2)

Bruckstein A. M., Levy B. C., and Kailath T. (1985): Differential methods in inverse scattering, *SIAM J. Appl. Math.* 45, 312–335 (sections XIV.7, XVII.5, XIX.2)

Bube K. P. (1986): Numerical methods for reflection inverse problems: convergence and non-impulsive sources, *SIAM J. Numer. Anal.* 23, 227–258 (see also Bube K. P. (1984): Convergence of difference methods for one-dimensional inverse problems, *IEEE Trans. Geosci. Remote Sensing* GE-22, 674–682) (section XIX.2)

Buck U. (1971): Determination of intermolecular potentials by the inversion of molecular beam scattering data I. The inversion procedure, *J. Chem. Phys.* 54, 1923–1928 (section XVI.4, Foreword)

Buck U. (1974): Inversion of molecular scattering data, *Rev. Mod. Phys.* 46, 369–389 (section XVI.4, XVI.H, Foreword)

Buck U., and Pauly H. (1969): Determination of intermolecular potentials by inversion of molecular beam scattering data, *J. Chem. Phys.* 51, 1662–1664 (sections XVI.4, XVI.H, Foreword)

Buck U., and Pauly H. (1971): Determination of intermolecular potentials by the inversion of molecular beam scattering data II. High resolution measurements of differential scattering cross sections and the inversion of the data for Na–Hg, *J. Chem. Phys.* 54, 5, 1928–1936 (section XVI.4, Foreword)

Bullough R., and Caudrey P. (1980): *Solitons*, (Topics Current in Physics, Vol. 17, Springer-Verlag, New York (section XVII.5)

Burdet G., Dufour J., and Giffon M. (1964): Etude du caractère complet des solutions de l'équation de Schrödinger dans le plan du moment angulaire complexe, *C. R. Acad. Sc. Paris* 259, 2369–2371 (section XIII.3)

Burdet G., and Giffon M. (1964): Sur le problème de la construction du potentiel à partir des données dans le plan du moment angulaire complexe, *C. R. Acad. Sci. Paris* 259, 4, 3190–3192 (section XIII.3)

Burdet G., and Giffon M. (1965): A proof of the completeness of the solutions of the Schrödinger equation in the λ-plane, *Acta Phys. Hung.* 19, 263–267 (section XIII.3)

Burdet G., Giffon M., and Goldberg J. (1966): On the inversion problem in the λ-plane II, *Nuovo Cimento* 44 A, 138–146 (section XIII.3)

Burdet G., Giffon M., and Predazzi E. (1965): On the inversion problem in the λ-plane, I, *Nuovo Cimento* 36, 1337–1347 (section XIII.3)

Burger H., Allen L. J., and Sofianos S. A. (1983): Potentials obtained by inversion of e–He atomic scattering data, *Phys. Lett. A* 97, 39–41 (section XIX.2)

Burkhardt H. (1972): Removal of ambiguities from inelastic phase-shift analysis, *Nuovo Cimento* 10 A, 2, 379–388 (section X.6, Foreword)

Burkhardt H. (1974): Serious ambiguities in inelastic phase-shift analysis, Preprint CERN-TH. 1784 to be published by the Slovak Acad. of Sc. (section X.6, Foreword)

Burkhardt H., and Martin A. (1975): Unique inelastic phase-shift from general structure in two variables, *Nuovo Cimento* 29 A, 141 (section X.1, X.6, Foreword)

Burridge R. (1980): The Gel'fand–Levitan, the Marchenko, and the Gopinath–Sondhi integral equations of inverse scattering theory, regarded in the context of inverse impulse-response problems, *Wave Motion* 2, 305 323 (sections IX.H, XVII.5)

Burridge R., Papanicolou G., Sheng P., and White B. (1988): Probing a random medium with a pulse, Preprint, Courant Institute (section XVII.5)

Burridge R., Papanicolou G., and White B. S. (1988a): One-dimensional wave propagation in a highly discontinuous medium, *Wave Motion* 10, 19–44

Buslaev V. S. (1962): Trace formulas for Schrödinger's operator in three-space. *Sov. Phys. Dokl.* 7, 4, 295–297 (section VII.H, Foreword)

Buslaev V. S. (1967): The trace formulas and some asymptotic estimates of the resolvent kernel of the three-dimensional equation, Topics in *Math. Phys.* Vol. 1: ed. by M. S. Birman Consultants Bureau N. Y. (section VII.H)

Buslaev V. S. (1971): Spectral identities and the trace formula in the Friedrichs Model, Topics in *Math. Phys.* Vol. 4: ed. by M. S. Birman Consultants Bureau N.Y. (section VII.H)

Buslaev V. S., and Faddeev L. D. (1960): Formulas for traces for a singular Sturm–Liouville differential operator, *Sov. Math. Dokl.* 1, 451–454 (section VII.H, Foreword)

Calogero F. (1971): Explicit expressions of the potential and its derivatives at the origin in terms of the scattering data: in *Mathematics of profile inversion*, L. Colin ed. (section VII.H, Foreword)

Calogero F. (1976): Generalized Wronskian relations, one-dimensional Schrödinger equation and nonlinear partial differential equations solvable by the inverse-scattering method, *Nuovo Cimento B* 31, 229–249 (sections XVII.4, XVII.5)

Calogero F. (1983): Computation of Sturm–Liouville eigenvalues via lagrangian interpolation, *Lett. Nuovo Cimento* 37, 9–16 (section XVIII.4)

Calogero F., Corbella O. D., Degasperis A., and De Stefano M. B. (1968): Variational bounds for the potential in terms of the S-wave phase shift, *J. Math. Phys.* 9, 7, 1002–1006 (section VII.H, Foreword)

Calogero F., and Cox J. R. (1968): Evaluation of the potential and the S-wave scattering amplitude from the discontinuity of the latter across its left-hand cut, *Nuovo Cimento* 55, A 786–808 (section VII.H)

Calogero F., and Degasperis A. (1968): Values of the potential and its derivatives at the origin in terms of the S-wave phase shift and bound-state parameters. *J. Math. Phys.* 9, 1, 90–116 (section VII.H, Foreword)

Calogero F., and Degasperis A. (1976): Nonlinear equations solvable by the inverse spectral transform—I, *Nuovo Cimento B* 32, 201–242 (sections XVII.4, XVII.5)

Calogero F., and Degasperis A. (1977): Nonlinear evolution equations solvable by the inverse spectral transform—II, *Nuovo Cimento B* 39, 1–54 (section XVII.4, XVII.5)

Calogero F., and Degasperis A. (1982): *Solitons and the Spectral Transform I*, North-Holland, Amsterdam (section XVII.5)

Calogero F., and Degasperis A. (1989): *Solitons and the Spectral Transform II*, North-Holland, Amsterdam (section XVII.5, Preface)

Calogero F., and Sabatier P. C. (1987): Nonlinear modulation of a transversally trapped mode, in *Topics in Soliton Theory and Exactly Solvable Nonlinear Equations* (M. Ablowitz, B. Fuchssteiner, and M. Kruskal, eds.), World Scientific, Singapore, pp. 307–319 (section XVII.4)

Cannon J. R., and Hornung U. (1986): *Inverse Problems.* Birkhäuser, Basel–Boston–Stuttgart (Preface)

Carion P. (1986): On stability of 1D exact inverse methods, *Inverse Problems* 1–22 (section XIX.2)

Carroll R. W. (1979a): *Transmutation and Operator-Differential Equations*, Mathematical Studies, Vol. 67, North-Holland, Amsterdam (section XV.2)

Carroll R. (1979b): Transmutation and separation of variables, *Applicable Anal.* 8, 253–263 (section XV.2)

Carroll R. (1981): Remarks on the Gel'fand–Levitan and Marcenko equations, *Appl. Anal.* 12, 153–157 (section XV.2)

Carroll R. (1982a): Transmutation, generalized translation, and transform theory, I and II, *Osaka J. Math.* 19, 815–831, 833–861 (section XV.2)

Carroll R. (1982b): On the characterization of transmutations, *An. Acad. Brasil. Ciênc.* 54, 271–280 (section XV.2)

Carroll R. (1982c): The Gel'fand–Levitan and Marcenko equations via transmutation, *Rocky Mountain J. Math.* 12, 393–427 (section XV.2)

Carroll R. (1983): Some remarks on the generalized Gel'fand–Levitan equation, *J. Math. Anal. Appl.* 92, 410–426 (section XV.2)

Carroll R. (1984a): Transmutation, filtering, and scattering, *Proc. Japan Acad. Ser A*, 60, (section XV.2)

Carroll R. (1984b): Some inversion theorems of fourier type, *Rev. Roumaine Math. Pures Appl.* XXIX, 113–138 (section XV.2)

Carroll R. (1984c): Fourier type analysis and the Marchenko equation, *Applicable Anal.* 18, 39–54 (section XV.2)

Carroll R. (1984d): Least squares, transmutation, and the Marchenko equation, *C. R. Roy. Soc. Canada* 6, 85–88 (section XV.2)

Carroll R. (1985): *Transmutation Theory and Applications*, North-Holland, Amsterdam (sections XV.2, XVIII.4)

Carroll R. (1986): Patterns and structure in systems governed by linear second-order differential equations, *Acta Appl. Math.* 6, 109–184 (section XV.2)

Carroll R. (1988a): Some remarks on half-line scattering and forced integrable systems, Proceedings of the Howard University Symposium on Nonlinear Semigroups, Partial Differential Equations and Attractors (section XV.2)

Carroll R. (1988b): The Marcenko equation as a minimizing criterion, Preprint (section XV.2)

Carroll R., and Delic G. (1988): Some remarks on Newton–Sabatier methods, *Math. Methods Appl. Sci.* (section XV.2)

Carroll R., and Gilbert J. (1981a): Some remarks on transmutation, scattering theory and special functions, *Math. Ann.* 258, 39–54 (section XV.2)

Carroll R., and Gilbert J. E. (1981b): Scattering techniques in transmutation and some connection formulas for special functions, *Proc. Japan Acad.* 57, 34–37 (section XV.2)

Carroll R., and Raphael L. (1988): Some inverse problems with spectral limitations on data, *SIAM J. Appl. Math.*

Carroll R., and Santosa F. (1981a): Scattering techniques for a one-dimensional inverse problem in geophysics, *Math. Methods Appl. Sci.* 3, 145–171 (section XIX.2)

Carroll R., and Santosa F. (1981b): Résolution d'un problème qui détermine complètement les données géophysique, *C. R. Acad. Sci. Paris* 292, 23–26 (section XV.2)

Carroll R., and Santosa F. (1982): Stability for the one-dimensional inverse problem via the Gel'fand–Levitan equation, *Appl. Anal.* 13, 271–277 (section XV.2)

Carroll R., and Santosa F. (1984): Impedance profile recovery from transmission data, *J. Acoust. Soc. Amer.* 76, 935–941 (section XV.2)

Case K. M. (1973): On discrete inverse scattering problems II. *J. Math. Phys.* 14, 7, 916–920 (section IX.6, Foreword)

Case K. M. (1974a): The discrete inverse scattering problem in one dimension, *J. Math. Phys.* 15, 143–147 (section XVII.5, Foreword)

Case K. M. (1974b): Scattering theory, orthogonal polynomials and the transport equation, *J. Math. Phys.* 15, 974–983 (section IX.6, Foreword)

Case K. M. (1974c): Orthogonal polynomials from the viewpoint of scattering theory, *J. Math. Phys.* 15, 2166–2174 (section IX.6, Foreword)

Case K. M., and Kac M. (1973): A discrete version of the inverse scattering problem, *J. Math. Phys.* 14, 594–603 (section IX.6, Foreword)

Caudrey P. J. (1982): The inverse problem for a general $n \times n$ spectral equation, *Physica D* 6, 51–66 (section XVII.5)

Chadan K. (1955): Potentiel neutron–proton et photodésintégration du deutéron, *Journal de Physique et le radium* 16, 843–848 (section VI.H)

Chadan K. (1956): Potentiel neutron–proton et capture des neutrons thermiques par les protons, *C. R. Acad. Sc. Paris* 242, 1964–1966 (section VI.H)

Chadan K. (1958a): On the connection between the S-matrix and a class of non-local interaction I. *Nuovo Cimento* 10, 5, 892–908 (section VIII.H, Foreword)

Chadan K. (1958b): Thèse de Doctorat d'Etat—Paris (section VIII.H, Foreword)

Chadan K. (1962): The left-hand cut discontinuity and equivalent potentials, *Nuovo Cimento* 24, 3, 379–384 (section VII.H)

Chadan K. (1967): On a class of completely transparent nonlocal two-body potentials, *Nuovo Cimento* 47. A, 510–525 (section VIII.H, Foreword)

Chadan K. (1976): The number of bound states of singular oscillating potentials *Lett. Math. Phys.* 1, 281–287 (section II.H)

Chadan K. (1978): in *Applied Inverse Problems* (Sabatier P. C., ed.), Springer-Verlag, Berlin (sections II.7, III.7)

Chadan K. (1984): Le problème inverse en constante de couplage, *C. R. Acad. Sci. Paris*, *Sér. II* 299, 271–274 (section XVIII.4)

Chadan K., and Grosse H. (1983): New bounds on the number of bound states, *J. Phys. A* 16, 955–961 (section II.7)

Chadan K., and Grosse H. (1984): Le problème inverse en constante de couplage II, *C. R. Acad. Sci. Paris Sér. II* 299, 1305–1308 (section XVIII.4)

Chadan K., and Kobayashi R. (1986): Le problème inverse en constante de couplage III, *C. R. Acad. Sci. Paris. Sér. II* 303, 329–332 (section XVIII.4)

Chadan K., and Martin A. (1977): Inequalities on the number of bound states in oscillating potentials, *Commun. Math. Phys.* 53, 221–231 (section II.H)

Chadan K., and Martin A. (1979): Scattering theory and dispersion relations for a class of long-range oscillating potentials, *Comm. Math. Phys.* 70, 1–27 (section II.7)

Chadan K., and Montes A. (1968): Derivation of nonrelativistic sum rules from the causality condition of Wigner and Van Kampen, *J. Math. Phys.* 9, 1898–1914 (sections V.H, IX.9)

Chadan K., and Musette M. (1987): Le problème inverse dans la constante de couplage en intervalles infinis, *C. R. Acad. Sci. Paris Sér. II*, 305, 1409–1412 (section XVIII.4). See also preprint (Orsay, 1988)

Charap J. (1959): The field theoric definition of the nuclear potential I. *Nuovo Cimento* 14, 540–559 (section XIII.2)

Charap J., and Fubini (1960): The field theoric definition of the nuclear potential II. *Nuovo Cimento* 15, 73–86 (section XIII.2)

Cheney M. (1984): Inverse scattering in dimension two, *J. Math.* 25, 94–102 (section XIV.7)

Cheney M. (1985): Two-dimensional inverse scattering: Compactness of the generalized Marchenko operator, *J. Math. Phys.* 26, 743–752 (Foreword)

Cheney M., and Kristensson G. (1988): Three-dimensional inverse scattering: layer stripping formulas and ill-posedness results, Preprint of Royal Institute of Technology, TRITA-TET 87-4 (34 pages) (section XIV.7)

Cheney M., and Rose J. H. (1985): Three-dimensional inverse scattering: High-frequency analysis of Newton's Marchenko equation, *J. Math. Phys.* 26, 436–439 (section XIV.7)

Cheney M., Rose J. H., and De Facio B. (1986): Three-dimensional inverse scattering for the classical wave equation with variable speed, in *Rev. Prog. NDE*, Vol. 5 (in press) (D. O. Thompson and D. E. Chimenti, eds.), New York, pp. 1–5 (section XIV.7)

Cheney M., Rose J. H., and De Facio B. (1988): A fundamental integral equation of scattering theory, *SIAM J. Math. Anal.* in press (section XIV.7)

Child M. S. (1974): *Molecular Collision Theory* Academic Press, New York (section XVI.H)

Chudov L. A. (1949): The inverse Sturm–Liouville problem, *Nat. Sbornik* 25 (67), 451–456 (section III.H, Foreword)

Chudov L. A. (1956): A new variant of an inverse Sturm–Liouville problem on a finite interval, *Dokl. Akad. Nauk. SSSR* 109, 40–43 (section III.H, Foreword)

Claerbout J. F. (1968): Synthesis of a layered medium from its acoustic transmission response, *Geophys.* 33, 264–269 (section XIX.2)

Coddington E. A., and Levinson N. (1965): *Theory of Ordinary Differential Equations.* McGraw-Hill, New York (section II.H)

Coen S. (1981a): Inverse scattering of a layered and dispersionless dielectric half-space, Part 1: Reflection data from plane waves at normal incidence, *IEEE Trans. Antennas and Propagation* AP-29, 726–732 (section XIX.2)

Coen S. (1981b): On the elastic profiles of a layered medium from data, Part 1: Plane-wave sources, *J. Acoust. Soc. Amer.* 70, 172–175 (section XIX.2)

Coen S. (1981c): On the elastic profiles of a layered medium from reflection data, Part 2: Impulsive point source, *J. Acoust. Soc. Amer.* 70, 1473–1479 (section XIX.2)

Cohen A., and Kappeler T. (1985): Scattering and inverse scattering for steplike potentials in the Schrödinger equation, *Indiana Univ. Math. J.*, 34, 127–180 (section XVII.5)

Colin L. (Ed.) (1972): *Mathematics of Profile Inversion*: Proceedings of a Workshop held at Ames Research Center, Moffett Field, Calif., July 12–16, 1971, NASA Technical Memorandum: NASA TM X-62, 150. Lectures contained in this book are referred to separately under the name of their authors.

Colton D. and Kress R. (1986): *Integral Equation: Methods in Scattering Theory.* John Wiley & Sons, New York

Common A. K., and Martin A. (1987): Improved constraints on states bound by a class potentials, *J. Phys. A* 20, 4247–4255 (section IV.5)

Corbella O. D. (1970): Inverse scattering problem for Dirac particles: Explicit expressions for the values of the potentials and their derivative at the origin in terms of the scattering and bound-state data, *J. Math. Phys.* 11, 5, 1695–1713 (section VII.H, Foreword)

Corbella O. D. (1971): L-Wave inverse scattering problem, *J. Math. Phys.* 12, 1873–1882 (section VII.H, Foreword)

Corinaldesi E. (1954): Construction of potentials from phase shift and binding energies of relativistic equations, *Nuovo Cimento* 11, 5, 468–478 (section IX.5, Foreword)

Corinaldesi E.: Errata-Corrige—Construction of potentials from phase shift and binding energies of relativistic equations, *Nuovo Cimento* 12, 469 (section IX.5)

Cornille H. (1967): Connection between the Marchenko formalism and N/D equations: Regular interactions I. *J. Math. Phys.* 8, 11, 2268–2281 (section VII.H)

Cornille H. (1970a): Connection between Marchenko formalism and N/D equations: Regular interactions II. *J. Math. Phys.* 11, 1, 61–78 (section VII.H)

Cornille H. (1970b): Existence and uniqueness of crossing symmetric N/D-type equations corresponding to the Klein–Gordon equations, *J. Math. Phys.* 11, 1, 79–88 (section VII.H)

Cornille H. (1973): Some general results for the inverse scattering theory at fixed angular momentum or at fixed energy, Inst. Phys. Nucléaire, Division Physique Théorique IPNO/TH.72.8 (section XIII.4)

Cornille H. (1976): Differential equations satisfied by Fredholm determinants and application to the inversion formalism for parameter dependent potentials, *J. Math. Phys.* 17, 12, 2143–2158 (section IX.9)

Cornille H., and Drouffe J. M. (1974): Phase-shift ambiguities for spinless and $L_{max} \leqslant 4$ elastic scattering, *Nuovo Cimento* 20A, 3, 401–436 (section X.1)

Cornille H., and Martin A. (1962): *Nuovo Cimento* 26, 298–327 Propriétés analytiques de l'amplitude de diffusion de deux particules chargées interagissant par un potentiel du type de Yukawa (section VII.H)

Cornille H., and Martin A. (1972a): Asymptotic equality of cross-sections for line-reversed reactions, *Nuc. Phys.* B48, 104–116

Cornille H., and Martin A. (1972b): Constraints on the phase of scattering amplitudes due to positivity, *Nuc. Phys.* B49, 413–440.

Cornille H., and Rubinstein G. (1968): Threshold-behavior requirements for p-wave N/D equations, *J. Math. Phys.* 9, 1501–1516 (section VII.H)

Corones J. P., Davidson M. E., and Krueger R. J. (1983): Direct and inverse scattering in the time domain via invariant imbedding equations, *J. Acoust. Soc. Amer.* 74, 1535–1541 (section XVII.5)

Corones J. P., Davidson M. E., and Krueger R. J. (1985): Dissipative inverse problems in the time-domain, in *Inverse Methods in Electromagnetic Imaging* (W. M. Boerner et al., eds.), Reidel, Dordrecht, Part 1, pp. 121–130 (section XIX.2)

Corones J., and Krueger R. (1983): Obtaining scattering kernels using invariant imbedding, *J. Math. Anal. Appl.* 93, 393–495 (section XIX.2)

Corones J. P., Krueger R. J., and Weston V. H. (1984): Some recent results in inverse scattering theory, in *Inverse Problems of Acoustic and Elastic Waves* (F. Santosa et al., eds.), SIAM, Philadelphia, pp. 65–81 (section XVII.5)

Coudray C. (1977): The inverse problem: some applications of the Newton-Sabatier method. *Lett. Nuovo Cimento*, 19, 319–323 (section XIX.2)

Coudray C. (1979): Quelques aspects du problème inverse de la diffusion à énergie fixée, Thèse, Orsay (section XIX.2)

Coudray C., and Coz M. (1970): Generalized translation operators and the construction of potentials at fixed energy, *Ann. Phys.* 61, 2, 488–529 (sections XII.7, XV.2)

Coudray C., and Coz M. (1972): Construction of relativistic potentials when the energy is fixed, *J. Math. Phys.* 12, 1166–1178 (section XV.2)

Cox J. R. (1962): Construction of potentials from the many-channel S-matrix, PhD Thesis, Indiana University (section IX.4, Foreword)

Cox J. R. (1964): Many-channel Bargmann potentials, *J. Math. Phys.* 5, 8, 1065–1069 (section IX.4, Foreword)

Cox J. R. (1966): Many-channel Gel'fand–Levitan equations, *Ann. Phys.* 39, 2, 216–236 (section IX.4, Foreword)

Cox J. R. (1967): On the determination of the many-channel potential matrix and S-matrix from a single function, *J. Math. Phys.* 8, 12, 2327–2331 (section IX.4, Foreword)

Cox J. R. (1975): On angular momentum and channel coupling for a meromorphic many-channel S-matrix, *J. Math. Phys.* 16, 1410–1415 (section IX.4, Foreword)

Cox J. R., and Garcia H. R. (1975): Construction of a meromorphic many channel *p*-wave S-matrix, *J. Math. Phys.* 16, 1402–1409 (section IX.4, Foreword)

Cox J. R., and Thompson K. W. (1969): Some exact solutions of the Schrödinger equation at fixed energy, *Bull. Am. Phys. Soc.* 14, 579 (sections XII.6, XII.7, Foreword)

Cox J. R., and Thompson H. W. (1970a): On the inverse scattering problem at fixed energy for potentials having nonvanishing first moments, *J. Math. Phys.* 11, 3, 805–814 (sections XII.6, XII.7, XII.H, Foreword)

Cox J. R., and Thompson H. W. (1970b): Note on the uniqueness of the solution of an equation of interest in the inverse scattering problem, *J. Math. Phys.* 11, 3, 815–817 (sections XII.6, XII.H, Foreword)

Coz M. (1980): A direct study of a Marchenko fundamental equation with centripetal potential, *J. Math. Phys.* 22 (8) (section IX.H)

Coz M., and Coudray C. (1973): Existence of generalized translation operators from the Agranowitch–Marchenko transformation (Jost solution), *J. Math. Phys.* 14, 11, 1574–1578 (section IX.1)

Crawford F. H., and Jorgensen T., Jr. (1936): The band spectra of the hybrides of lithium: III Potential curves and isotope relations, *Phys. Rev.* 49, 745–752 (Foreword)

Cremers D. J., and Birkebak R. C. (1966): Application of the Abel integral equation to spectrographic data, *Appl. Opt.* 5, 6, 1057–1064 (section XVI.7)

Craig I. J. D., and Brown J. C. (1986): *Inverse Problems in Astronomy*, Adam Hilger, Bristol and Boston (Preface)

Crichton J. G. (1966): Phase-shift ambiguities for spin-independent scattering, *Nuovo Cimento* 45A, 1, 256–258 (section X.1)

Crum M. M. (1955): Associated Sturm–Liouville systems, *Quart. J. Math. Oxford* (2) 6, 121–127 (section IV.H)

Cuer M. (1976): Ambiguités semi-classiques, Thèse de doctorat de 3éme cycle, Phys. Math. Montpellier (sections XVI.2, XVI.6, XVI.H)

Cuer M. (1977): Some examples of transparent potentials in the classical approximation, *J. Math. Phys.* 18, 658–661 (section XVI.6)

Cuer M. (1979): Semi-classical analysis of optical model ambiguities, *Ann. Physics*, 120, 1–15 (section XVI.H)

Dahlberg B. E. J., and Trubowitz E. (1986): The inverse Sturm–Liouville III, *Comm. Pure Appl. Math.* 37, 255–267 (section XVIII.3)

Dammert O. (1983): Transmission through a system of potential barriers. I. Transmission coefficient, *J. Math. Phys.* 24, 2163–2170 (section XVI.H)

Darboux G. (1894): *Leçons sur la Théorie Générale des Surfaces*, Vol. III, Gauthier-Villars, Paris, pp. 438–444

Dautray R., and Lions L. (1986): *Analyse Mathématique et Calcul Numérique pour les Sciences et les Techniques*, Vols. 1, 2 and 3, Collection CEA, Masson, Paris (section XVIII.4)

De Alfaro V. (1958): On the inversion problem for a Klein–Gordon wave equation, *Nuovo Cimento* 4, 675–681 (section IX.5)

De Alfaro V., and Regge T. (1961): A discussion of the Noyes and Wong equation for potential scattering, *Nuovo Cimento* 20, 5, 956–962 (section VII.H)

De Alfaro V., and Regge T. (1965): *Potential Scattering*, John Wiley and Sons, New York (section II.H)

De Alfaro V., Regge T., and Rossetti C. (1962): *Nuovo Cimento* 26, 1029–1062. Dynamical equation and angular momentum (section VII.H)

De Facio B., and Brander O. (1988a): Coherent-state path-integral formulation of inverse scattering theory, *Rend. Circ. Mat. Palermo* in press (section XIV.7)

De Facio B., and Brander O. (1988b): Multidimensional $n \geq 3$ inverse scattering theory, in *Nonlinear Evolutions* (J. Leon, ed.), World Scientific, Singapore (section XVI.H)

De Facio B., and Moses H. E. (1980): The Gel'fand–Levitan equation can give simple examples of non-self-adjoint operators with complete eigenfunctions and spectral representations. I. Ghosts and resonance, *J. Math. Phys.* 21 (7), 1716–1723 (section XVII.5)

De Facio B., and Rose J. H. (1985): Inverse-scattering theory for the non-spherically-symmetric three-dimensional plasma wave equation, *Phys. Rev. A* 31, 897–902 (sections XIV.5, XIV.7)

De Facio B., and Rose J. H. (1986): A perturbation method for inverse scattering in three-dimensions based on the exact inverse scattering equations, *Review of Progress in Quantitative Nondestructive Evaluation* 5A, 345–353 (section XIV.7)

Degasperis A. (1970): On the inverse problem for the Klein–Gordon S-wave equation. *J. Math. Phys.* 11, 2, 551–567 (sections VII.H, IX.5, Foreword)

Degasperis A., and Sabatier P. C. (1987): Extension of the one-dimensional scattering theory, and ambiguities, *Inverse Problems* 3, 73–110 (sections XVII.2, 4, XVII.5).

Deift P. A. (1978): Applications of a commutation formula, *Duke Math. J.* 45, 267–311 (sections XVII.2, XVII.5, XVIII.4)

Deift P., and Trubowitz E. (1979): Inverse scattering on the line, *Comm. Pure Appl. Math.* XXXII, 121–251 (sections XVII.1, XVII.2, XVII.5)

Delsarte J. (1938): Sur certaines transformations fonctionnelles relatives aux équations linéaires aux dérivées partielles du second ordre, *C. R. Acad. Sci. Paris* 206, 1780–1782 (section XV.2)

Delsarte J., and Lions J. L. (1957): Transmutations d'opérateurs différentiels dans le domaine complexe, *Comment. Math. Helv.* 32, 113–127 (section XV.2)

Demkov Yu. N., Ostrovsky V. N., and Berezina N. B. (1971): Uniqueness of the Firsov inversion method and focusing potentials, *Soviet Phys. JETP* 33, 867–870 (section XVI.H)

Dikii C. A. (1958): Trace formulas for Sturm–Liouville differential operators, *Usp. Mat. Nauk* 13, 111–143; translated in *Amer. Math. Soc. Transl.* 18, 81–116 (section VII.H)

Di Salvo E., and Viano G. A. (1976): Uniqueness and stability in the inverse problem of scattering theory, Preprint—Istituto Nazionale Fisica Nucleare—Sezione di Genova (section XIII.2)

Dodd R. K., Eilbeck J. C., Gibbon J. D., and Morris H. C. (1982): *Solitons and Nonlinear Wave Equations*, Academic Press, New York (section XVII.5)

Dolveck-Guilpart B. (1988): Practical construction of potentials corresponding to exact rational coefficients, pp. 361–370, in *"Some Topics on Inverse Problems"* (P. C. Sabatier, ed.), World Scientific, Singapore (section XVII.5)

Drisko R. M., Satchler G. R., and Bassel R. H. (1963): Ambiguities in the optical potential for strongly absorbed projectiles, *Phys. Lett.* 5, 5, 347–350 (section XVI.3)

Dunford N., and Schwartz J. T. (1963): *Linear Operators*, Interscience Publishers (section II.H)

Dyson F. J. (1975): Photon noise and atmospheric noise in active optical systems, *Journ. Opt. Soc. Am.* 65, 5, 551–558 (section IX.H)

Dyson F. J. (1976a): Old and new approaches to the inverse scattering problem *in Studies in Mathematical Physics*, ed. by E. H. Lieb, B. Simon, and A. S. Wightman, Princeton (section IX.H)

Dyson F. J. (1976b): Fredholm determinants and inverse scattering problems, *Comm. Math. Phys.* 47, 2, 171–184 (section IX.H)

Eckaus N., and Van Harten A. (1981): *The Inverse Scattering Transformation and the Theory of Solitons, An Introduction*, North-Holland, Amsterdam (section XVII.5)

Eisberg R. M., and Porter C. E. (1961): Scattering of α-particles, *Rev. Mod. Phys.* 33, 190–230 (section XVI.4)

Engl H. W., and Groetsch C. W. (1987): *Inverse and Ill-Posed Problems*, Academic Press, New York (section XIX.2, Preface)

Enss V. (1978): Asymptotic completeness for quantum mechanical potential scattering, *Comm. Math. Phys.* 61, 285–291 (section XIV.7)

Erdelyi A. (1956): *Asymptotic Expansions*, Dover Publications (section XVI.H)

Erdelyi A. (1960): Asymptotic solutions of differential equations with transition points of singularities, *J. Math. Phys.* 1, 16–26 (section XVI.H)

Erdelyi A., Magnus W., Oberhettinger F., and Tricomi F. G. (1954): *Higher Transcendental Functions and Tables of Integral Transforms*,* McGraw-Hill

Faddeev, L. D. (1956): Uniqueness of solutions of the inverse scattering problem, Vestnik Leningrad University 11, 126–130 [*Math. Rev.* 18, 1, 259 (1957)] (section XVII.5)

Faddeev L. D. (1957): An expression for the trace of the difference between two singular differential operators of the Sturm–Liouville type, *Dokl. Akad. Nauk. SSSR* 115, 878–881 (section VII.H, Foreword)

* All the special functions that appear in this book are defined according to these tables.

Faddeev L. D. (1958): On the relation between S-matrix and potential for the one-dimensional Schrödinger operator, *Dokl. Akad. Nauk. SSSR* 121, 63–66: [*Math. Rev.* 20, 773 (1959)] (section XVII.5, Foreword)

Faddeev, L. D. (1959): The inverse problem in the quantum theory of scattering, *J. Math. Phys.* 4, 1, 72–104 (1963) translated from: *Usp. Mat. Nauk* 14 (sections II.H, IV.H, V.H, Foreword)

Faddeev L. D. (1964): Properties of the S-matrix of the one-dimensional Schrödinger equation, *Trudy Mat. Inst. Steklow* 73, 314–336 translated in *Amer. Math. Soc.* (2) 65, 139–166 (section XVII.H, Foreword)

Faddeev L. D. (1965): Increasing solutions of the Schrödinger equation, *Dokl. Acad. Nauk. SSSR* 165, 514–517 [*Soviet Phys. Dokl.* 10, 1033–1035] (section XIV.3, Foreword)

Faddeev L. D. (1966): Factorization of the S-matrix for the multidimensional Schrödinger operator, *Dokl. Akad. Nauk. SSSR* 167, 69–72 [*Sov. Phys. Dokl.* 11, 209–211] (Foreword)

Faddeev L. D. (1971): Three-dimensional inverse problem in the quantum theory scattering, Preprint Kiev ITP 71.106 E (sections XIV.1, XIV.3, XIV.H, Foreword)

Faddeev L. D. (1976): Inverse problem of quantum scattering, II, translated from Itagi, Nauki, Tekhesiki. *Sovrem. Probl. Mat.* 3, 93–180, Plenum, New York (sections XVII.2, XVII.5, XVIII.4)

Faddeev L. D., and Takhtadjian L. A. (1986): *Hamiltonian Operators and the Theory of Solitons*, Springer-Verlag, New York (section XVII.5)

Fawcett J. (1984): On the stability of inverse scattering problems, *Wave Motion* 6, 489–499 (section XIX.2)

Fedorjuk M. V. (1965): Asymptotic behavior in the one-dimensional scattering problem, *Dokl. Akad Nauk SSSR* 162, 676–678 (section XVII.5)

Fiedeldey H., Lipperheide R., Naidoo K., and Sofianos S. A. (1984): Semi-classical and quantal inversion of nuclear scattering at fixed energy, *Phys. Rev. C* 30, 434–440 (section XIX.2)

Fiedeldey H., Lipperheide R., and Sofianos S. (1982): Inverse problem of quantal potential scattering at fixed energy, Proceedings of the Workshop on Numerical Treatment of Inverse Problems for Differential and Integral Equations, Heidelberg (section XIX.2)

Fiedeldey H., Sofianos S. A., Allen L. J., and Lipperheide R. (1985): Equivalent local potential for non-alpha scattering, *Phys. Rev. C* 31, 2300–2302 (section XIX.2)

Fiedeldey H., Sofianos S. A., Allen L. J., and Lipperheide R. (1986): Determination of nonlocal potentials from the phase shifts, *Phys. Rev. C* 33, 1581–1587 (section VIII.5)

Firsova N. E. (1975a): *Mat. Zametki* 18, 831–843 [trans. *Math. Notes* 18, 1085–1091 (1975)] (Foreword)

Firsov O. B. (1953): Determination of forces acting between atoms with the use of the differential cross-section of elastic scattering, *Zh. Eksp. Theor. Fiz.* 24, 279–283 (section XVI.H, Foreword).

Firsova N. E. (1975b): Riemann surface of quasimomentum and scattering theory for the perturbed Hill operator, *Zap. Nauchn. Sem. Leningrad. Otdel. Mat. Inst. Steklov.* 51, 183–196 [trans. *J. Soviet Math.* 11, 487–497 (1979)] (Foreword)

Flügge S., and Vollmer G. (1971): Determination of potentials by WKB inversion of term formulae, *Z. Phys.* 248, 1, 1–6 (section XVI.H)

Ford K. W., Hill D. L., Wakano M., and Wheeler J. A. (1959a): Quantum effects near a barrier maximum, *Ann. Phys.* 7, 239–258 (section XVI.H)

Ford K. W., and Wheeler J. A. (1959b): Semiclassical description of scattering, *Ann. Phys.* 7, 259–286 and 287–322 (section XVI.4, XVI.H)

Frank W. M., Land D. J., and Spector R. M. (1971): Singular potentials, *Rev. Mod. Phys.* 43, 36–98 (section II.6)

Freeman M. P., and Katz S. (1960): Determination of the radial distribution of brightness in a cylindrical luminous medium with self-absorption, *J. Opt. Soc. Am.* 50, 8, 826–830 (section XVI.7)

Friedland S. (1977): Inverse eigenvalue problems, *Linear Algebra appl.* 17, 15–51 (section XIX.3)

Friedland S. (1979): The reconstruction of a symmetric matrix from the spectral data, *J. Math. Anal. Appl.* 71, 412–422 (section XIX.3)

Friedland S., Robbin J. W., and Sylvester J. H. (1984): On the crossing rule. *Comm. Pure Appl. Math.* 37, 19–37 (section XIX.3)

Friedman A. (1957): On the properties of a singular Sturm–Liouville equation determined by its spectral functions, *Michigan Math. J.* 4, 137–145 (sections III.H, V.H)

Froberg C. E. (1947): Calculation of the interaction between two particles from the asymptotic phase, *Phys. Rev.* 72, 519–520 (section X.1, Foreword)

Froberg C. E. (1948): Calculation of the potential from the asymptotic phase I. *Ark. Mat. Astron. Fys.* 34A, 28, 1–16 (section X.1, Foreword)

Froberg C. E. (1949): Calculation of the potential from the asymptotic phase II, *Ark. Mat. Astron. Fys.* 36A, 11, 1–55 (section X.1, Foreword)

Froberg C. E. (1951): On determination of proton–proton interaction from scattering experiments, *Ark. Fys.* 3, 1, 1–34 (section IX.1, Foreword)

Frobrich P., Lipperheide R., and Fiedeldey H. (1979): Long-range heavy-ion potential induced by multiple Coulomb excitation, *Phys. Rev. Lett.* 43, 1147–1150 (section XIX.2)

Fröman N., and Fröman P. O. (1965): *JWKB Approximation*, North Holland (section XIV.H)

Fulton T., and Newton R. G. (1956): Explicit noncentral potentials and wave functions for given s-matrices, *Nuovo Cimento*, 3, 677–717 (section IV.H, Foreword)

Funke H., and Zakhariev B. N. (1987): The quantum inverse problem for a special class of nonspherical potentials, in Sabatier (1987, pp. 99–115) (section XIX.2)

Gangal A. D., and Kupsch J. (1984): Determination of the scattering amplitude, *Comm. Math. Phys.* 93, 333–339 (section X.H)

Gardner C. S., Greene J. M., Kruskal M. D., and Miura R. M. (1967): Method for solving the Korteweg–de Vries equation, *Phys. Rev. Lett.* 19, 1095–1097 (section XVII.H, Foreword)

Gardner C. S., Greene J. M., Kruskal M. D., and Miura R. M. (1974): Korteweg-de Vries equation and generalizations VI: Methods for exact solutions, *Comm. on Pure and Appl. Math.* 27, 97–133 (section XVII.H)

Gasymov M. G. (1967): The inverse scattering problem for a system of Dirac equations of order 2n, *Sov. Phys. Dokl.* 11, 8, 676–678 (section IX.5, Foreword)

Gasymov M. G. (1968): The inverse scattering problem for a system of Dirac equations of order 2n, *Trudy Moscow Mat. Obsc.* 19, translated in *Trans. Moscow Math. Soc.* 19, 41–119 (section IX.5, Foreword)

Gasymov M. G., and Levitan B. M. (1966a): The inverse problem for a Dirac system. *Dokl. Akad. Nauk. SSSR* 167, 967–970 [*Sov. Math.* 7, 495–499] and see also *Dokl. Akad. Nauk. SSSR* 167, 1219–1222 (section IX.5, Foreword)

Gasymov M. G., and Levitan B. M. (1966b): Determination of Dirac's system from the scattering phase, *Doklady Akad. Nauk SSSR* 167, 543–547 (section IX.H)

Gazanhes C., Herauld J. P., and Stephanakis K. (1981): Reflection coefficient identification by means of correlation: application to a layered medium, *J. Acoust. Soc. Amer.* 69, 720–727 (section XIX.2)

Ge D. B. (1987): Numerical and approximate methods for 1D electromagnetic inverse scattering, in Sabatier (1987a, pp. 71–77) (section XIX.2)

Gel'fand I. M., and Levitan B. M. (1951a): On the determination of a differential equation by its spectral function, *Dokl. Akad. Nauk. USSR* 77, 557–560 (section III.H)

Gel'fand I. M., and Levitan B. M. (1951b): On the determination of a differential equation by its spectral function. *Izv. Akad. Nauk. SSR* ser. Math. 15, 309–360 translated in *Amer. Math. Soc. Transl.* ser. 2, 1, 253–304, (1955) (section III.H, Foreword)

Gerber R. B., and Karplus M. (1970): Determination of the phase of the scattering amplitude from the differential cross section, *Phys. Rev.* 1, 4, 998–1012 (section X.1, Foreword)

Gerber R. B., Buch V., Buck U., Maneke M., and Schleusener J. (1980): Direct inversion of rotationally inelastic cross sections: Determination of the asymptotic Ne–D2 potential, *Phys. Rev. Lett.* 44, 1397–1400 (section XIX.2)

Geronimo J. S. (1982): Scattering theory and matrix orthogonal polynomials on the real line, *Circuits Systems Signal Process* 1, 472–495 (sections XVII.5, XVIII.4)

Gersten A. (1969): Ambiguities of complex phase shift analysis, *Nucl. Phys.* B12, 537–548 (section X.1)

Gerver M., and Markushevich V. (1966): Determination of a seismic wave velocity from the travel time curve. *Geophys. J. R. Astron. Soc.* 11, 165–173 (section XVI.7)

Gerver M. L., and Markushevich V. (1967): On the characteristic properties of travel-time curves, *Geophys. J. R. Astron. Soc.* 13, 241–246 (section XVI.7)

Gladwell G. M. (1984): The inverse problem for a vibrating beam, *Proc. Roy. Soc. London Ser. A* 393, 277–295 (section XVIII.4)

Gladwell G. M. (1985): Qualitative properties of vibrating systems, *Proc. Roy. Soc. London Ser. A* 401, 299–315 (section XVII.5)

Gladwell G. M. (1986a): *Inverse Problems in Vibration*, Martinus Ninjhoff (section XVII.4, Preface)

Gladwell G. M. (1986b): Inverse problems in vibration, *Appl. Mech. Rev.* 39, 1013–1018 (section XVIII.4)

Glaser V., Grosse H., and Martin A. (1978): Bounds on the number of eigenvalues of the Schrödinger operator, *Comm. Math. Phys.* 59, 197–212 (section XVIII.4)

Glaser V., Martin A., Grosse H., and Thirring W. (1976): A family of optimal conditions for the absence of bound states in a potential: *in Studies in Mathematical Physics*, edited by E. H. Lieb, B. Simon, and A. S. Wightman, Princeton

Gokeler M. (1981): Quantum Gel'fand–Levitan method for the cubic Schrödinger field theory with attractive coupling, Thesis, Bonn University (section IX.H)

Goldberger M. L., Lewis H. W., and Watson K. M. (1963): Use of intensity correlations to determine the phase of a scattering amplitude, *Phys. Rev.* 132, 6, 2764–2787 (section X.1)

Goldberger M. L., Lewis H. W., and Watson K. M. (1966): Intensity—correlation spectroscopy, *Phys. Rev.* 142, 1, 25–32 (section X.1)

Goldberger M. L., and Watson K. M. (1964): Measurement of time correlations for quantum-mechanical systems, *Phys. Rev.* 134, 4B, 919–928 (section X.1)

Goldberger M. L., and Watson K. M. (1965a): Fluctuations with time of scattered-particle intensities, *Phys. Rev.* 137, 5B, 1396–1409 (section X.1)

Goldberger M. L., and Watson K. M. (1965b): Accuracy of measurement for counting and intensity-correlation experiments, *Phys. Rev.* 140, 2B, 500–509 (section X.1)

Goldman I. I., and Migdal A. B. (1954): The theory of scattering in the semiclassical approximation, *Sov. Phys.* JETP 1, 304 (section XVI.H)

Gopinath B. (1976): The solution to an inverse problem in stratified dielectric media, *J. Math. Phys.* 17, 1099–1104 (section XIX.2)

Gopinath B., and Sondhi M. M. (1970): Determination of the shape of the human vocal tract from acoustical measurements, *Bell Syst. Tech. J.* 49, 6, 1195–1214 (section XVII.4)

Gopinath B., and Sondhi M. M. (1971): Inversion of the telegraph equation and synthesis of nonuniform lines, *Proc. IEEE* 59, 3, 383–392 (section XIX.2)

Gorenflo R., and Kovetz Y. (1966): Solution of an Abel-type integral equation in the presence of noise by quadratic programming. *Numer. Math.* 8, 392–406 (section XVI.7)

Goupillaud P. (1961): An approach to inverse filtering of non-surface layer effects from seismic records, *Geophys.* 26, 754–760 (section XIX.2)

Gourdin M., and Martin A. (1957a): Détermination d'un potentiel séparable à partir des déphasages, cas où tan δ est une fraction rationnelle de l'énergie, *C. R. Acad. Sc. Paris*, 244, 1329–1330 (section VIII.H, Foreword)

Gourdin M., and Martin A. (1957b): Interaction non locale séparable et matrices de collision, *Nuovo Cimento* 6, 4, 757–779 (section VIII.H, Foreword)

Gourdin M., and Martin A. (1958): Exact determination of a phenomenological separable interaction, *Nuovo Cimento* 8, 5, 699–719 (section VIII.H, Foreword)

Gray S., and Symes W. (1985): Stability considerations for one-dimensional inverse problems, *Geophys. J. R. Astron. Soc.* 80, 149–163 (section XIX.2)

Green, (1837): On the motion of waves in a variable canal of small depth and width, *Cambridge Philos. Soc. Trans.* 6, 223–230 (section XVI.H)

Grosse H., and Martin A. (1979): Theory of the inverse problem for confining potentials, *Nucl. Phys. B* 148, 413–432 (section XVIII.4)

Grunbaum F. A. (1978): An inverse eigenvalue problem for random matrices, *SIAM J. Appl. Math.* 35, 268–273 (section XVIII.4)

Grunbaum F. A. (1985): The role of strict interlacing in inverse eigenvalue problems. Report Mathematics Berkeley (section XVIII.4)

Grunbaum F. A. (1986): An inverse spectral problem for band matrices" Report Mathematics Berkeley (section XVIII.4)

Gugushvili H. I., and Mentkovsky Y. L. (1972a): The free Riemann–Green's function of the inverse problem of scattering theory, *Nuovo Cimento* 10A, 2, 292 (section IX.1)

Gugushvili H. I., and Mentkovsky Y. L. (1972b): The inverse problem of scattering theory and Riemann's method, *Nuovo Cimento* 10A, 2, 277 (section IX.I)

Guillaume M. O., and Jaulent L. (1982): A Zakharov–Shabat inverse scattering problem with a polynomial spectral dependence in potential, *Lett. Math. Phys.* 6, 189–198 (section XVII.5)

Habashy T. M., and Mittra R. (1987a): On some inverse methods in electromagnetics, *J. Electromagnetic Waves Appl.* 1, 25–58 (section XIX.2)

Habashy T. M., and Mittra R. (1987b): On some inverse methods in electromagnetics, *J. Electromagnetic Waves Appl.* 1, 59–94 (section XIX.2)

Habiby-Asharafi F., and Mendel J. M. (1982): Estimation of parameters in lossless layered media systems, *IEEE Trans. Automat. Control.* AC-28, 1081–1091 (section XIX.2)

Hagin F. (1981): A stable approach to solving one-dimensional inverse problems, *SIAM J. Appl. Math.* 40, 439–453 (section XIX.2)

Hald O. H. (1972): On discrete and numerical inverse Sturm–Liouville problems, Upsala University, Department of Computer Sciences, Report 42 (section XIX.2)

Hald O. H. (1976): Inverse eigenvalue problems for Jacobi matrices, *Linear Algebra Appl.* 14, 63–85 (section XVIII.4)

Hald O. H. (1977a): Discrete inverse Sturm–Liouville problems, *Numer. Math.* 27, 249–256 (section XVIII.4)

Hald O. H. (1977b): Inverse eigenvalue problems for layered media, *Comm. Pure Appl. Math.* XXX, 69–94 (section XVIII.4)

Hald O. H. (1978a): The inverse Sturm–Liouville problem with symmetric potentials, *Acta Math.* 141, 263–291 (section XVIII.4)

Hald O. H. (1978b): The inverse Sturm–Liouville problem and the Rayleigh–Ritz method, *Math Comput.* 32, 687–705 (sections XVII.5, XVIII.4)

Hald O. H. (1980): Inverse eigenvalues for the mantle, *Geophys. J. R. Astr. Soc.* 62, 41–48 (section XIX.2)

Hald O. H. (1979): Numerical solution of the Gel'fand–Levitan equation *Linear Algebra Appl.* 28, 99–111 (section XIX.2)

Hald O. H. (1983): Inverse eigenvalue problems for the mantle, II *Geophys. J. R. Astr. Soc.* 72, 139–164 (sections XVIII.4, XIX.2)

Hald O. H. (1984): Discontinuous inverse eigenvalue problems, *Comm. Pure Appl. Math.* XXXVII, 539–577 (sections XVII.5, XVIII.4)

Hamilton E. L., Bucker H. P., and Whitney J. A. (1970): Velocities of compressional and shear waves in marine sediments determined in situ from a research submersible, *J. Geophys. Res.* 75, 4039–4049 (section XIX.2)

Hammer C. L., and Shrauner J. E. (1978): Path-integral representation for the *S*-matrix, *Phys. Rev. D* 118, 373–384 (section XVI.4)

Hanbury-Brown R., and Twiss R. Q. (1954): A new type of interferometer for use in radio astronomy, *Philos. Mag.* 45, 663–682 (section X.1)

Hanbury-Brown R., and Twiss R. Q. (1956): The question of correlation between photons in coherent light rays, *Nature* 178, 1447 (section X.1)

Hanbury-Brown R., and Twiss R. Q. (1957): *Proc. R. Soc. London* 242A, 300 (section X.1)

Heading J. (1962): *An Introduction to Phase Integral Method*, Methuen, London (section XVI.H)

Hecht W. (1972): An analytic approximation method for the one-dimensional Schrödinger equation, *J. Math. Phys.* 13, 1291–1295 (section XVII.5)

Hecht C. E., and Mayer J. E. (1957): Extension of the WKB equation, *Phys. Rev.* 106, 6, 1156–1160 (section XVI.H)

Hefter E. F. (1984a): Inverse methods and solitons in nonrelativistic quantum mechanics, *Acta Phys. A* 65, 377–393 (section XIX.2)

Hefter E. F. (1984b): Mass dependence of the real optical model potential for nucleon-nucleus scattering, *Phys. Lett. B* 141 (section XIX.2)

Hefter E. F., De Llano M., and Mitropolsky I. A. (1984b): Inverse methods and nuclear radii, *Phys. Rev. C* 30, 2042–2049 (section XIX.2)

Hefter E. F., and Gridnev K. A. (1984): *a + a* collisions via solitons, *Progr. Theoret. Phys.* 72, 549–562 (section XIX.2)

Heim D. S., and Sharpe C. B. (1967): The synthesis of nonuniform lines of finite length Part I, *IEEE Trans. on Circuit Theory* 14, 394–403 (section XVII.4)

Heisenberg W. (1943): Die beobachtbaren Grössen in der Theorie der Elementarteilchen, *Z. Phys.* 120, 513 and 673 (Foreword)

Heisenberg W. (1944): Die beobachtbaren Grössen in der Theorie der Elementarteilchen III., *Z. Phys.* 123, 93 (Foreword)

Heniger A., and Loeffel J. J. (1973): Une classe d'algorithmes pour le problème inverse de la diffusion, *Helv. Phys. Acta* 46, 462–464 (section XIII.3)

Helmberg G. (1969): *Introduction to Spectral Theory in Hilbert Spaces*, North-Holland, Amsterdam (section XVIII.4).

Henkin G. M., and Leiterer J. (1984): *Theory of Functions on Complex Manifolds*, Birkäuser, Basel (section XVII.5)

Henkin G. M., and Novikov R. G. (1984): *Theoret. and Math. Phys.* 61, 199–214 (section XVII.5)

Henkin G. M., and Novikov R. G. (1987): $\bar{\partial}$ methods and multidimensional inverse problems (in Russian), *Ivst. Math. USSR* 42, Transl. *Russian Math. Surveys* (section XIV.7)

Henkin G. M., and Novikov R. G. (1988a): A multidimensional inverse problem in quantum and acoustic scattering, *Inverse Problems* (sections XIV.7, XVII.5)

Henkin G. M., and Novikov R. G. (1988b): Solution of a multidimensional inverse scattering problem on the basis of generalized dispersion relations. *Dokl. Akad. Nauk SSSR* 292, 814–818 transl. in *Sov. Math. Dokl.* 35, 153–158 (sections XIV.4, XIV.7)

Henry J. H., Raju J. J., Wong Y. W., and Andres R. P. (1973): Abstracts of papers of VIII ICPEAC Beograd 35 (section XVI.5)

Herglotz V. G. (1907): Uber das Benndorfsche problem der Fortpflanzungages-chwindigkeit der Erdbebenstrablen, *Z. Phys.* 8, Jahrgang, no. 5, 145–147 (section XVI.7)

Herlitz S. I. (1963): A method for computing the emission distribution in cylindrical eight sources, *Ark. Fys.* 23, 49, 571–574 (section XVI.7)

Hermann G. T. et al. (1987): See Sabatier P. C. (1987c)

Hille E. (1969): *Lectures on Ordinary Differential Equations*, Addison Wesley, Reading (section II.H)

Hochstadt H. (1976): On the determination of the density of a vibrating string from spectral data, *Math. Anal. Appl.* 55, 673–685 (section XVIII.4)

Hochstadt H. (1977): On the well-posedness of the inverse Sturm–Liouville problems, *J. Differential Equations* 23, 402–413 (section XVIII.4)

Hodgson P. E. (1963): *The Optical Model of Elastic Scattering*, Clarendon Press, Oxford (section XIX.2)

Hodgson P. E. (1971): *Nuclear Reactions and Nuclear Structure*, Clarendon Press, Oxford (section XIX.2)

Holmberg B. (1952): A remark on the uniqueness of the potential determined from the asymptotic phase, *Nuovo Cimento* 9, 7, 597–604 (sections III.H, IV.H, Foreword)

Hooshyar M. A. (1971): On the inverse scattering problem at fixed energy for the tensor and spin-orbit potentials, *J. Math. Phys.* 12, 2243–2258 (section XV.1, Foreword)

Hooshyar M. A. (1972): The inverse scattering problems at fixed energy for L^2 dependent potentials, *J. Math. Phys.* 13, 1931–1933 (section XV.1, Foreword)

Hooshyar M. A. (1975): Construction of spin-orbit potentials from the phase shifts at fixed energy, *J. Math. Phys.* 16, 257–262 (section XV.1, Foreword)

Hooshyar M. A. (1980): Construction of $L2$ dependent potentials *J. Math. Phys.* 21, 1695–1697 (section XIX.2)

Hooshyar M. A., Nadeau R., and Razavy M. (1982): Determination of the neutron–nucleus optical potential from the angular dependence of the scattering amplitude, *Phys. Rev. C* 25, 1187–1193 (section XIX.2)

Hooshyar M. A., and Ravazy M. (1982a): A continued fraction approach to the inverse problem electrical conductivity, *Geophys. J. R. Astron. Soc.* 71, 127–138 (section XIX.2)

Hooshyar M. A., and Razavy M. (1982b): Phenomenological local nucleon–nucleon potentials obtained from the inverse scattering problem at a fixed energy, *Phys. Rev. C* 29, 20–28 (section XIX.2)

Hooshyar M. A., and Razavy M. (1983): A method for constructing wave velocity and density profiles from the angular dependence of the reflection coefficient, *J. Acoust. Soc. Amer.* 73, 19–23 (section XIX.2)

Hooshyar M. A., and Razavy M. (1981): A discrete method of constructing central and angular-momentum dependent potentials, *Canad. J. Phys.* 59, 1627–1634 (section XIX.2)

Hooshyar M. A., and Razavy M. (1986): The inverse problem of geomagnetic induction at a fixed frequency, National Science and Engineering Research Council of Canada (section XV.H)

Howard M. S. (1981): Inversion of the first-order equations governing sound propagation in a layered medium, Ph.D. Thesis, Indiana University (sections XVII.4, XVII.5)

Howard M. S. (1983): Inverse scattering for a layered acoustic medium using the first-order equations of motion, *Geophys.* 48, 163–170 (section XVII.4, XVII.5)

Hoyt F. C. (1939): The determination of force fields from scattering in the classical theory, *Phys. Rev.* 55, 664–665 (section XVI.H, Foreword)

Hruslov E. Ja. (1976): Asymptotics of the solution of the Cauchy problem for the Korteweg–deVries equations with initial data of step type, *Mat. Sbornik* 28 (141), 229–248 (section XVII.5)

Hylleraas E. A. (1948): Calculation of a perturbing central field of force from the elastic scattering phase shift, *Phys. Rev.* 74, 1, 48–51 (section III.H, Foreword)

Hylleraas, E. A. (1963a): On the inversion of eigenvalue problems, *Ann. Phys.* 25, 309 (section III.H)

Hylleraas E. A. (1963b): Resonance method in scattering theory, Institute for Theoretical Physics University of Oslo Report 18 (section III.H, Foreword)

Hylleraas E. A. (1963c): On the inversion of Eigen problems, Institute for Theoretical Physics University of Oslo Report 20 (section III.H, Foreword)

Hylleraas E. A. (1964): Determination of a perturbing potential from its scattering phase shift and bound state energy levels, *Nucl. Phys.* 57, 208–231 (section III.H, Foreword)

Inguva R., and Baker-Jarvis J. (1986): Principle of maximum entropy and inverse scattering problems, in *Maximum Entropy and Bayesian Methods in Applied Statistics* (J. Justice, ed.), Cambridge University Press, Cambridge, pp. 300–315 (section IX.H)

Ioannides A. A., and Mackintosh R. S. (1985): A method for S-matrix to potential inversion at fixed energy, I. Method, description and evaluation, *Nucl. Phys. A* 438, 354–380 (section XIX.2)

Isaacson E. L. and Trubowitz E. (1983): The inverse Sturm–Liouville problem I, *Comm. Pure Appl. Math.* 36, 767–783 (section XVIII.4)

Isaacson E. L., Mc Kean H. P., and Trubowitz E., (1984): The inverse Sturm–Liouville problem II, *Comm. Pure Appl. Math.* 37, 1–11

Itzykson C., and Martin A. (1973): Phase-shift ambiguities for analytic amplitudes, *Nuovo Cimento* 17A, 2, 245–287 (sections X.1, X.2, X.3, Forword)

Ivanov G. A., Pivovarchik V. N., and Popova A. M. (1985): The S-matrix asymptotic in the case of singular potentials at high complex energies, *J. Phys. A* 18, 265–270 (section III.H)

Jaggard D. L., and Olson K. E. (1985): Numerical reconstruction for dispersionless refractive profiles, *J. Opt. Soc. Amer A* 2, 1931–1936 (section XIX.2)

Jauch J. M. (1957): On the relation between scattering phase and bound states. *Helv. Phys. Acta* 30, 143 (section VIII.H)

Jauho P. (1950): On the unique determination of the nuclear potential between charged nucleons with the aid of scattering experiments, *Ann. Acad. Sci. Fenn. Ser.* A.I. 80, 7–44 (section IX.1)

Jaulent M. (1972): On an inverse scattering problem with an energy dependent potential, *Ann. Inst. Henri Poincaré* 17, 363–378 (section IX.8, Foreword)

Jaulent M. (1975): Sur le problème inverse de la diffusion pour l'équation de Schrödinger radiale avec un potentiel dépendant de l'énergie, *C. R. Acad. Sc. Paris* 280, 1467–1470 (section IX.8, Foreword)

Jaulent M. (1976): Inverse scattering problems in absorbing media, *J. Math. Phys.* 17, 7, 1351–1360 (section XVII.5, Foreword)

Jaulent M. (1982): The inverse scattering problem for LCRG transmission lines, *J. Math. Phys.* 23, 2286–2290 (section XVII.5)

Jaulent M., and Jean C. (1972): The inverse s-wave scattering problem for a class of potentials depending on energy, *Commun. Math. Phys.* 28, 177–220 (sections IX.8, XV.1)

Jaulent M., and Jean C. (1975): Le problème inverse pour l'équation de Schrödinger à une dimension avec un potentiel dépendant de l'énergie, *Cahiers de Mathématiques de Montpellier*, 7 (section XVII.H)

Jaulent M., and Jean C. (1976): The inverse problem for the one-dimensional Schrödinger equation with an energy-dependent potential I and II, *Ann. Inst. Henri Poincaré* 25, 2, 105–118 and 119–137 (section XVII.4)

Jaulent M., and Jean C. (1981a): A Schrödinger inverse scattering problem with a spectral dependence in the potential, *Lett. Math. Phys.* 6, 183–190 (section XVII.5)

Jaulent M., and Jean C. (1981b): Solution of a Schrödinger inverse scattering problem with a polynomial spectral dependence in the potential, *J. Math. Phys.* 23, 258–266 (section XVII.5)

Jaulent M., and Manna M. (1987): Solution of nonlinear equations by $\bar{\partial}$ analysis: the Kdv hierarchy, in *Inverse Problems, an Interdisciplinary Study* (P. C. Sabatier, ed.), Academic Press, New York, pp. 429–443 (section XVII.5). See also *idem Phys. Lett. A* 117, 62 (1986); *Inverse Problems* 2, L35 (1986); *Inverse Problems* 3, 13 (1987); *Europhys. Lett.* 2, 891 (1986).

Jaulent M., Manna M., Martinez-Alonso L. (1988): $\bar{\partial}$ equations in the theory of integrable systems, *Inverse Problems* 4, 123–150 (section XVII.5)

Jaulent M., and Miodek I. (1977): Connection between Zakharov–Shabat and Schrödinger-type inverse scattering transforms, *Lett. Nuovo Cimento* 20, 655–660 (section XVII.5)

Jean C. (1983): A completeness theorem relative to one-dimensional Schrödinger equations with energy-dependent potentials, *Ann. Inst. H. Poincaré* XXXVIII, 15–35 (section XVII.5)

Jean C., and Sabatier P. C. (1973): On an inverse scattering problem of quantum theory, *Nuovo Cimento* 18A, 1, 105–150 (section XIII.4, Foreword)

Jeffreys (1923): On certain approximate solutions of linear differential equations of the second order, *Proc. London Math. Soc.* 23, 428 (section XVI.H)

Jordan A. K., and Ahn S. (1979): Inverse scattering theory and profile reconstruction, *Proc. IEE-E* 126, 945 (section XVII.5)

Jordan A. K., and Lakshmanasamy S. (1987): Inverse scattering problem with linearly superposed reflection coefficients, in Sabatier (1987a, pp. 79–88) (section XIX.2)

Jorgens K. (1964): Spectral theory of second order ordinary differential operators, Lectures delivered at Aarhus Universitet—Matematik Institut (section II.H)

Jost R. (1956): Eine Bemerkung Uber der Zusammenhang von Streuphase und potential, *Helv. Phys. Acta* 29, 410–418 (section V.H, Foreword)

Jost R., and Kohn W. (1952a): Construction of a potential from a phase shift, *Phys. Rev.* 87, 6, 977–992 (section III.H, Foreword)

Jost R., and Kohn W. (1952b): Equivalent potentials, *Phys. Rev.* 88, 2, 382–385 (sections III.H, IV.H, VI.H, Foreword)

Jost R., and Kohn W. (1953): On the relation between phase shift energy levels and the potential, *Danske Vid. Selsk. Math. Fys. Medd.* 27, 9, 3–19 (sections III.H, IV.H, VI.H, Foreword)

Jost R., and Pais A. (1951): On the scattering of a particle by a static potential, *Phys. Rev.* 82, 6, 840–851 (section II.1)

Kac I. S., and Krein M. G. (1974a): On the spectral functions of the string, *Amer. Math. Soc. Transl.* 2, 103, 19–102 (section XVIII.4)

Kac I. S., and Krein M. G. (1974b): R-functions–analytic functions mapping the upper half-plane into itself, *Amer. Math. Soc. Transl.* 2, 103, 1–18 (section XVIII.4)

Kac M. (1966): Can one hear the shape of a drum? *Am. Math. Monthly* 73, no. 4, part II, 1–23 (section XVIII.4, Foreword)

Kalinkin B. N., Kochkina T. P., and Pustylnik B. I. (1963): The quasi-classical analysis of the elastic scattering of complex nuclei, *Acta Phys.* XXIV, 427–434 (section XVI.H)

Karlsson B. R. (1972): Off-shell extension of the partial wave transition amplitude, *Phys. Rev. D.* 6, 6, 1662–1669 (section IX 7)

Karlsson B. R. (1974): Inverse problem of scattering and off-shell continuation of the transition amplitude, *Phys. Rev. D.* 10, 6, 1985–1991 (section IX.7)

Karlsson B. R. (1978): Inverse method for off-shell continuation of the scattering amplitude in quantum mechanics, in *Applied Inverse Problems* (P. C. Sabatier, ed.), Springer-Verlag, New York (section XVII.2)

Kato T. (1976): *Perturbation Theory for Linear Operators*, Springer-Verlag, Heidelberg (section XIV.7)

Kay I. (1955): The inverse scattering problem, Research Report No. EM.74.NY, N.Y. University, Inst. Math. Sc. Electromagnetic Research (section XVII.H)

Kay I. (1959): On the determination of the free electron distribution of an ionized gas, Research Report No. EM.141, N.Y. University (section XVII.4)

Kay I. (1960): The inverse scattering problem when the reflection coefficient is a rational function, *Commun. Pure Appl. Math.* 13, 371–398 (section XVII.H)

Kay I. (1972): The inverse scattering problem for transmission lines in *Mathematics of Profile Inversion*, L. Colin ed. 6.2–6.17 (section XVII.H)

Kay I., and Moses H. E. (1955): The determination of the scattering potential from the spectral measure function I: Continuous spectrum, *Nuovo Cimento* 2, 5, 917–961 (section III.H, Foreword)

Kay I., and Moses H. E. (1956a): The determination of the scattering potential from the spectral measure function II: Point eigenvalues and proper eigenfunctions, *Nuovo Cimento* 3, 1, 66–84 (section III.H, Foreword)

Kay I., and Moses H. E. (1956b): The determination of the scattering potential from the spectral measure function III: Calculation of the scattering potential from the scattering operator for the one dimensional Schrödinger equation, *Nuovo Cimento* 3, 2, 276–304 (section XVII.H, Foreword)

Kay I., and Moses H. E. (1956c): Reflectionless transmission through dielectrics and scattering potentials, *J. Appl. Phys.* 27, 1503–1508 (section XVII.4, Foreword)

Kay I., and Moses H. E. (1957): The determination of the scattering potential from the spectral measure functions, IV "Pathological" scattering problems in one-dimension, *Il Nuovo Cimento Suppl.* (ser. 10) 5, 230–242 (section XVII.H, Foreword)

Kay I., and Moses H. E. (1961a): The determination of the scattering potential from the spectral measure function V. The Gel'fand–Levitan equation for the three-dimensional scattering problem, *Nuovo Cimento* 22, 689–705 (section XIV.H, Foreword)

Kay I., and Moses H. E. (1961b): A simple verification of the Gel'fand–Levitan equation for the three-dimensional scattering problem, *Comm. Pure Appl. Math.* 14, 435–445 (section XIV.H, Foreword)

Kay I., and Moses H. E. (1982): *Inverse Scattering Papers* (1955–1963), Math. Sci. Press (Massachussetts) (sections XVII.3, XVII.7)

Keller J. B. (1962): Determination of a potential from its energy levels and undetectability of quantization at high energy, *Am. J. Phys.* 30, 22–26 (section XVI.H)

Keller J. B., Kay I., and Shmoys J. (1956): Determination of the potential from scattering data, *Phys. Rev.* 102, 557–559 (sections XVI.7, XVI.H, Foreword)

Kermode M. W., Cooper S. G., and Allen L. J. (1986): A quantum mechanical inverse scattering method at fixed energy, *Inverse Problems* 2, 353–361 (section XIX.2)

Khristov E. Kh. (1981): On spectral properties of operators generating KdV-type equations, *J. Differential Equations* (section XVII.5)

Klaus M. (1977): On the bound state of Schrödinger operators in one dimension, *Ann. Phys.* 108, 288 (sections XVII.2, XVII.5)

Klaus M. (1979): A remark about weakly coupled one-dimensional Schrödinger operators, *Helv. Phys. Acta* 52, 223 (sections XVII.2, XVII.5)

Klaus M. (1981): Some remarks on double-wells in one and three dimensions, *Ann. Inst. H. Poincaré Sect. A* 34, 405–417 (sections XVII.2, XVII.5)

Klaus M. (1987): Low energy behavior of the scattering matrix for the Schrödinger equation on the line, Preprint, Virginia Polytechnic Institute and State University, 14 pages (sections XVII.2, XVII.5)

Klaus M., and Simon B. (1980): Coupling constant thresholds in nonrelativistic quantum mechanics. I. Short-range two body case, *Ann. Physics* 130, 251–281 (section XVII.5)

Klein O. (1932): Zur Berechnung von Potentialkurven für zweiatomige Moleküle mit Hilfe von Spektralternee, *Z. Phys.* 76, 226–235 (section XVI.H, Foreword)

Klepikov N. P. (1962): Removal of ambiguities in phase-shift analysis, *Sov. Phys. JETP* 14, 4, 846–851 (section X.1)

Klepikov N. P. (1964): On the completeness of the "complete scattering experiment," *Zh. Eksp. Theor. Fiz.* 47, 757–766 [*Sov. Phys. JETP* 20, 505] (Foreword)

Klingbeil (1972): Determination of interatomic potentials by the inversion of elastic differential cross-section data I. An inversion procedure, *J. Chem. Phys.* 56, 132 (section XVI.4)

Kohler W. (1986): Reflection from a one-dimensional, totally refracting random multi-layer, (section XVII.5)

Kolb P., Collino F., and Lailly P. (1986): Pre-stack inversion of a 1-medium, *Proc. IEE-E* 74, 499–508 (section XIX.2)

Konopelchenko B. G. (1987): *Nonlinear Integrable Equations*, Springer-Verlag, New York (section XVII.5)

Korolyova T. K., and Zhigunov V. P. (1983): On potential reconstruction from experimental data, Institute for High Energy Physics, Serpukhov preprint (section XIX.2)

Kowalski J., and Fry J. L. (1987): Tunneling in one-dimensional ideal barriers, (section XVII.5)

Kramers H. A. (1926): Wellenmdranik und halbzajlige quantisierung, *Z. Phys.* 39, 828–840 (section XVI.H)

Krappe H. J., and Lipperheide R. (1985): Advanced methods in the evaluation of nuclear scattering data (Proceedings, Berlin), Lecture Notes in Physics, Vol. 236, Springer-Verlag, New York (section XIX.2)

Krappe H. J., Massmann H., and Rossner H. H. (1987): Some nonlinear, ill-posed problems in nuclear physics from a practitioner's point of view, in Sabatier (1987a, pp. 129–140) (section XIX.2).

Krappe H. J., and Rossner H. H. (1985): The regularization method in heavy-ion optical potential analyses, in Krappe and Lipperheide (1985, pp. 242–248) (section XIX.2)

Krein M. G. (1974): Chebyshev–Markov inequalities in the theory of the spectral functions of the vibrating string, *Amer. Math. Soc. Transl.* (2) 103, 103–127 (section XVIII.4)

Krein M. G. (1951): Solution of the inverse Sturm–Liouville problem, *Dokl. Akad. Nauk SSSR* 76, 21–24 [*Math. Rev.* 12, 613] and Determination of the density of a nonhomogeneous symmetric cord by its frequency spectrum, *ibid.*, 345–348 [*Math. Rev.* 13 43] (section III.H, Foreword)

Krein M. G. (1953a): On the transfer function of a one-dimensional boundary problem of second order, *Dokl. Akad. Nauk. SSSR* 88, 405–408, [*Math. Rev.* 15, 316 (1954)] (sections III.H, IV.H, IX.2, XVII.H, Foreword)

Krein M. G. (1953b): On some case of effective determination of the density of an inhomogeneous cord from its spectral function, *Dokl. Akad. Nauk SSSR (N.S.)* 93, 617–620, [*Math. Rev.* 15, 796 (1954)] translated (see *Math. Rev.* 17, 740 (1956)) (sections III.H, IX.2, XVII.H, Foreword)

Krein M. G. (1954): On integral equations generating differential equations of second order, *Dokl. Akad. Nauk. SSSR* 97, 21–24, [*Math. Rev.* 16 (1955)] (sections III.H, IV.H, IX.2, Foreword)

Krein M. G. (1955): On determination of the potential of a particle from its S-function, *Dokl. Akad. Nauk. SSSR* 105, 433, [*Math, Rev.* 17, 1210 (1956)] (sections II.H, III.H, Foreword)

Krein M. G. (1956): On the theory of accelerants and s-matrices of canonical form, *Dokl. Akad. Nauk. SSSR* 111, 1167–80 (sections III.H, IV.H, IX.2)

Krein M. G. (1958): Integral equation on a half line with difference type kernels. *Usp. Math. Nauk.* 13, 3, [*Math. Rev.* 21, 1507 (1960)] (section II.H)

Kristensson G. (1986): The one-dimensional inverse scattering problem for an increasing potential, *J. Math. Phys.* 27, 804–815 (section XVIII.4)

Kristensson G., and Krueger R. J. (1986a): Direct and inverse scattering in the time domain for a dissipative wave equation. I. Scattering operators, *J. Math. Phys.* 27, 1667–1682 (section XVII.5)

Kristensson G., and Krueger R. J. (1986b): Direct and inverse scattering in the time domain for a dissipative wave equation. II. Simultaneous reconstruction of dissipation and phase velocity profiles, *J. Math. Phys.* 27, 1683–1693 (section XVII.5)

Kristensson G., and Krueger R. J. (1987): Direct and inverse scattering in the time domain for a dissipative wave equation. III. Scattering operators in the presence of a phase velocity mismatch, *J. Math. Phys.* 28, 360–370 (section XVII.5)

Kritikos H. N., Jaggard D. L., and Ge D. B. (1982): Numerical reconstruction of smooth dielectric profiles, *Proc. IEE-E,* 70, 295–297 (section XIX.2)

Kujawski E. (1972): Inverse problem for heavy-ion elastic scattering and the connection between parametrized phase-shift and optical potential models, *Phys. Rev. C* 6, 709–719 (section XIX.2)

Kujawski E. (1973): Computational approach to the inverse problem in the JWKB-approximation, *Phys. Rev. C* 8, 100–106 (sections XVI.H, XIX.2)

Kunetz G. (1961): Essai d'analyse de traces sismiques, *Geophys. Prospect.* 9, 317–341 (section XIX.2)

Kunetz G. (1963): Quelques exemples d'analyse d'enregistrements sismiques, *Geophys. Prospect.* 11, 409–422 (section XIX.2)

Kunetz G., and D'Erceville R. (1962): Sur certaines propriétés d'une onde acoustique plane de compression dans un milieu stratifié, *Ann. Géophys.* 18, 351–369 (section XIX.2)

Kunetz G., and Rocroi J. P. (1970): Traitement automatique des songages électriques, *Geophys. Prospect.* 18, 157–198 (section XIX.2)

Laddomada C., and Gui-Zhang Tu (1982): Backlund transformations for the Jaulent–Miodek equations, *Lett. Math. Phys.* 6, 453–462 (section XVII.5)

Lambert F., Corbella O., and Thome Z. D. (1975): Scattering at low energies and the inverse problem at fixed angular momentum, *Nucl. Phys.* B90, 267–284 (section VI.H)

Langer R. E. (1932): On the asymptotic solutions of differential equations with an application to the Bessel functions of large complex order, *Trans. Amer. Math. Soc.* 34, 447–480 (section XVI.H)

Langer R. E. (1934): The asymptotic solutions of certain linear ordinary differential equations of the second order, *Trans. Amer. Math. Soc.* 36, 90–106 (XVI.H)

Langer R. E. (1937): On the connection formulas and the solutions of the wave equation, *Phys. Rev.* 51, 669–676 (section XVI.H)

Langer R. E. (1949): On the connection formulas and the solutions of the wave equation, *Trans. Amer. Math. Soc.* 67, 461 (section XVI.H)

Lavine R. B., and Nachman A. I. (1987a): The Faddeev–Lippmann–Schwinger equation in multidimensional quantum inverse scattering, in *Inverse Problems: An*

Interdisciplinary Study (P. C. Sabatier, ed.), Academic Press, New York (sections XIV.4, XIV.7)

Lavine R. B., and Nachman A. I. (1987b): On the inverse scattering transform for the *n*-dimensional Schrödinger operator, in *Topics in Soliton Theory and Exactly Solvable Nonlinear Equations* (M. Ablowitz, ed.), World Scientific, Singapore (sections XIV.4, XIV.7).

Lax P. D. (1968): Integrals of nonlinear equations of evolution and solitary waves, *Comm. Pure and Appl. Math.* 21, 467–490 (section XVII.H)

Lax P. D., and Phillips, R. S. (1967): *Scattering Theory*, Academic Press, N.Y.; specially pp. 173 ff. (Foreword)

Legendre J. (1982): Problème de Schrödinger sur la ligne avec conditions dissymetriques et applications. Thèse de Phys. Théor., Université des Sciences et Techniques du Languedoc, Montpellier (section XVII.5)

Léon J. J. P. (1980): The inverse problem for a generalized Klein–Gordon system, *Lett. Nuovo Cimento* 29, 45–50 (section XVII.5)

Léon J. J. P. (1981): Inverse scattering connections, *J. Math. Phys.* 22, 965–969 (section XVII.5)

Léon J. J. P. (1988): Discontinuous soliton-like solution to the self-induced transparency equations, *Phys. Lett. A* 131, 79–84 (section XVII.5)

Lesselier D. (1978): Determination of index profiles by time-domain reflectometry, *J. Optics (Paris)* 9, 349–358 (section XIX.2)

Lesselier D. (1981): Diagnostic de milieux inhomogènes unidimensionnels par échographie électromagnétique, *Rev. CETHEDEC* 68, 1–42 (section XIX.2)

Lesselier D. (1981b): Optimization theory and time-domain inverse scattering, *Radio Sci.* 16, 1059–1063 (section XIX.2)

Lesselier D. (1982a): Optimization techniques and inverse problems: reconstruction of conductivity profiles in the time domain, *IEEE Trans. Antennas and Propagation*, AP-30, 59–65 (section XIX.2)

Lesselier D. (1982b): Optimization techniques and inverse problems: Probing of acoustic impedance profiles in time domain, *J. Acoust. Soc. Amer.* 72, 1276–1284 (section XIX.2)

Levi D., and Ragnisco O. (1982): Bäcklund transformation vs. the dressing method, *Lett. Nuovo Cimento* 33, 401–406 (section XVII.5)

Levi D., Ragnisco O., and Sym A. (1984): Dressing methods vs. classical Darboux transformation, *Nuovo Cimento* (section XVII.5)

Levin B. Y. (1956): Fourier and Laplace's types of transformation by means of solutions of differential equations of second order, *Dokl. Akad. Nauk. SSSR*, 106, 187–190, [*Math. Rev.* 18, 35 (1957)] (section V.H)

Levinson N. (1949a): Determination of the potential from the asymptotic phase, *Phys. Rev.* 75, 7, 1445 (section III.H)

Levinson N. (1949b): On the uniqueness of the potential in a Schrödinger equation for a given asymptotic phase, *Danske Vid. Selsk. Math. Fys. Medd.* 25, bd9, 1–29 (section III.H, Foreword)

Levinson N. (1953): Certain explicit relationship between phase shift and scattering potential, *Phys. Rev.* 89, 4, 755–757 (section III.H, Foreword)

Levitan B. M. (1949): The application of generalized displacement operators to linear differential equations of the second order, *Usp. Mat. Nauk.* 4, 3–112 (Translated by Atkinson, F. V.), *Amer. Math. Soc. Transl.* 59, Ser. I, 1–135 (section XV.1)

Levitan B. M. (1956): Certain questions in the spectral theory of self-adjoint differential operators, *Usp. Mat. Nauk.* 11, 117–144, translated in *Amer. Mat. Soc. Trans.* 18, 49–80 (section III.H, Foreword)

Levitan B. M. (1963): Determination of the Sturm–Liouville differential equation by two spectra, *Dokl. Akad. Nauk. SSSR* 150, 474–476 translated in *Soviet Math.* 4, 691–693 see also *Math. Rev.* 27, 379 (1964) (sections III.H, XVIII.4, Foreword)

Levitan B. M. (1964a): Determination of a Sturm–Liouville differential equation in terms of two spectra, *Izv. Akad. Nauk. SSSR* Ser. Math. 28, 63–78, [*Math. Rev.* 28, 3194] (sections III.H, XVIII.4, Foreword)

Levitan B. M. (1964b): *Generalized Translation Operators and Some of The Applications*, translated from the Russian by the Israel Program for Scientific Translations, Jerusalem—Published in the U.S.A. by D. Davey and Co., Inc., New York (section XV.2, Foreword)

Levitan B. M. (1973a): On the asymptotic behavior of the spectral function of a self-adjoint differential equation of the second order and expansion in eigenfunctions, *Amer. Math. Soc. Transl.* (2) 102, 191–229 (section XVIII.4)

Levitan B. M. (1973b): On the spectral function of the equation $y'' + (\lambda - q(x))y = 0$ *Amer. Math. Soc. Transl.* 102, 231–243 (section XVIII.4)

Levitan B. M. (1984): *Sturm–Liouville Inverse Problems* (in Russian), Nauka, Moscow (section XVIII.4)

Levitan B. M., and Sargsjan I. S. (1975): *Introduction to Spectral Theory*, American Mathematical Society, Providence, RI. (section XVIII.4)

Levitan B. M., and Gasymov M. G. (1964): Determination of a differential equation by two of its spectra, *Russian Math. Survey* 19, 1–63 (section III.H, Foreword)

Levitan B. M., and Sargsjan I. S. (1965): An extension of the solutions of a one-dimensional Dirac system, *Soviet Math.* 6, 1588–1591 (section IX.5)

Levy B. C. (1985): Layer by layer reconstruction methods for the earth resistivity from direct current measurements, *IEEE Trans. Geosci. Remote Sensing* GE-23, 841–850 (section XIX.2)

Liebenzon Z. L. (1966): An inverse problem of spectral analysis of ordinary differential operators of higher order, *Trans. Moscow Math. Soc.* 17, 78–163 (section III.H)

Lions J. L. (1958): Opérateurs de transmutation singuliers et équations d'Euler–Poisson–Darboux généralisées, (section XV.2)

Liouville J. (1837): Mémoire sur le développement des fonctions ou parties de fonctions en séries dont les divers termes sont assujettis à satisfaire à une même équation différentielle du second ordre contenant un paramètre variable, *J. de Math.* 2, 16–35 and 418–436 (section XVI.H)

Lipperheide R. (1984): A method for the determination of the optical potential from nuclear scattering, *Cahiers Math. Montpellier* 32, 178–210 (section XIX.2)

Lipperheide R., and Fiedeldey H. (1978): Inverse problem for potential scattering at fixed energy, *Z. Phys. A* 286, 45–56 (section XIX.2)

Lipperheide R., and Fiedeldey H. (1981): Inverse problem for potential scattering at fixed energy. II., *Z. Phys. A* 301, 81–89 (section XIX.2)

Lipperheide R., and Fiedeldey H. (1982): Potential inversion for scattering at fixed energy, *Phys. Rev. C* 26, 770–772 (section XIX.2)

Lipperheide R., Fiedeldey H., Haberzettl H., and Naidoo K. (1979): Determination of the potential for back-angle enhanced elastic heavy-ion scattering. Application to the scattering of ^{16}O on ^{28}Si, *Phys. Lett. B* 82, 39–42 (section XIX.2)

Li-Yi-Shen (1965): On an inverse eigenvalue problem for a second-order differential equation with boundary dependence on the parameter, *Acta Math. Sinica* 15, 74–80 (section XVIII.4)

Li-Yi-Shen (1981): One special inverse problem of the second-order differential equation on the whole real axis, *Chinese Ann. Math.* 2, 147–156 (section XVIII.4)

Ljance V. E. (1964): A differential operator with spectral singularities I and II, *Mat. Sb.* 64, 521–561 and *Mat. Sb.* 65, 47–103 translated in *Amer. Math. Soc. Transl.* (2) 60, 185–225 and 226–283 (section III.H)

Loeffel J. J. (1968): On an inverse problem in potential scattering theory, *Ann. Inst. Henri Poincaré* 8, 339–447 (sections XII.2, XVI.6, Foreword)

Lonngren K., and Scott A. C. (eds.) (1978): *Solitons in Action* Academic Press, New York (section XVII.H)

Loubatieres Y. (1985): Problème inverse à energie fixée de l'équation radiale de Schrödinger pour des potentiels de portée finie, Thèse de Doctorat, Université des Sciences et Techniques du Languedoc, Montpellier (sections XIII.4, XIII.6)

Ma S. T. (1946): Redundant zeros in the discrete energy spectra in Heisenberg's theory of characteristic matrix, *Phys. Rev.* 69, 668 (Foreword)

Ma S. T. (1947): On a general condition of Heisenberg for the S-matrix, *Phys. Rev.* 71, 195–200 and 210 (Foreword)

Mace D., and Lailly P. (1986): Solution of the VSP one-dimensional inverse problem, *Geophys. Prospect.* (section XIX.2)

Mackintosh R. S., and Ioannides A. A. (1985): Inversion as a means of understanding nuclear potentials, in Krappe and Lipperheide (1985, pp. 283–296), (section XIX.2)

Mal'cenko V. I. (1966): The inverse problem for the equations of quantum mechanics with energy-dependent potentials, *Ukr. Mat. Zh.* 18, 2, 126–129, translated in *Amer. Math. Soc. Transl.* (2), 75, 227–230 (section III.H, Foreword)

Maldonado C. D., Caron A. P., and Olsen H. N. (1965): New method for obtaining emission coefficients from emitted spectral intensities Part I. Circularly symmetric light sources, *J. Opt. Soc. Am.* 55, 10, 1247–1254 (section XVI.7)

Maldonado C. D., and Olsen H. N. (1966): New method for obtaining emission coefficients from emitted spectral intensities Part II. Assymetrical sources, *J. Opt. Soc. Am.* 56, 10, 1305–1313 (section XVI.7)

Malyarov V. V., Poplavsky I. V., and Popushoi N. N. (1975): Inverse problem of scattering theory in complex λ-plane in the presence of Coulomb field with allowance for absorption, *Soviet Phys. JETP* 41, 210–211 (section XIX.2). See also (same authors) *Soviet J. Nucl. Phys.* 22, 445 (1976); *Soviet J. Nucl. Phys.* 25, 38 (1977) (section XIX.2)

Marchenko V. A. (1950): Concerning the theory of a differential operator of the second order, *Dokl. Akad. Nauk. SSSR* (N.S.) 72, 457–460, [*Math. Rev.* 12, 183 (1951)] (section III.H, Foreword)

Marchenko V. A. (1952): Some problems in the theory of one-dimensional second-order differential operators Part I. *Trudy Mosk. Math. Obs.* 1, 327–420 (section III.4, Foreword)

Marchenko V. A. (1953): Some problems in the theory of one-dimensional second-order differential operators Part III. *Trudy Mosk. Math. Obs.* 2, 3–82 (section III.H, Foreword)

Marchenko V. A. (1955): The construction of the potential energy from the phases of the scattered waves, *Dokl. Akad. Nauk. SSSR* 104, 695–698, [*Math. Rev.* 17, 740 (1956)] (sections IV.H, V.H, Foreword)

Marchenko V. A. (1958): Expansion in eigenfunctions of non-self-adjoint singular differential operators of second order, *Amer. Math. Soc. Transl.* 2, 25, 77–130 (section XVIII.4)

Marchenko V. A. (1968): Stability in the inverse problem of scattering theory, *Mat. Sbornik* 77, 139–162 (sections XVII.2, XVII.5)

Marchenko V. A. (1986): *Sturm–Liouville Operators and Applications*, Birhäuser, Basel (section XVIII.4, Preface)

Martin A. (1958): On the validity of Levinson's theorem for nonlocal interactions, *Nuovo Cimento* 7, 5, 607–627 (section VIII.H)

Martin A. (1959): On the analytic properties of partial wave scattering amplitudes obtained from the Schrödinger equation, *Nuovo Cimento*, 14, 2, 403–425 (section VII.H)

Martin A. (1961): S-matrix, left-hand cut discontinuity and potential, *Nuovo Cimento* 19, 6, 1257–1265 (section VII.H)

Martin A. (1965): Analyticity in potential scattering: Progress in Elementary Particle and Cosmic Ray Physics, Vol. VIII, North Holland 1–66 (section VII.H)

Martin A. (1969): Construction of the scattering amplitude from the differential cross-sections, *Nuovo Cimento* 59A, 1, 131–152 (section X.1, Foreword)

Martin A. (1973): Reconstruction of scattering amplitude from differential cross-sections, Lectures given at the adriatic meeting on particle physics Rovjni (sections X.1, X.2, Foreword)

Martin A., and Sabatier P. C. (1977): Impedance, zero energy wave-function, and bound states, *J. Math. Phys.* 18, 1623–1626 (section XIV.7)

Martin A., and Targonsky G. Y. (1961): On the uniqueness of a potential fitting a scattering amplitude at a given energy, *Nuovo Cimento* 20, 6, 1182–1190 (section XIII.2, Foreword)

Maslov V. P. (1972): *Théorie des Perturbations et Méthodes Asymptotiques*, Dunod, Paris (section XIV.7)

Mason E. A. (1957): Scattering of low velocity molecular beams in gases, *J. Chem. Phys.* 26, 3, 667–677 (section XVI.4)

Mason E. A., and Monchick L. (1967): Methods for the determination of intermolecular forces, *Adv. Chem. Phys.* 12, 329–387, especially 351–354 (sections XVI.4, XVI.H, Foreword)

Matveev V. B. (1980): Darboux transformations and nonlinear equations, in *Problèmes Inverses Evolution Nonlinéaire* (P. C. Sabatier, ed.), Edition du CNRS Paris, pp. 247–266. See also Matveev V. B. (1979): *Lett. Math. Phys.* 3, 213; 3, 217; 3, 425 (section XVII.5)

May K. E., Munchow M., and Slcheid W. (1984): The modified Newton method for the solution of the inverse scattering problem with charged particles at fixed energy, *Phys. Lett. B* 141, 1–4 (section XI.H)

May K. E., and Scheid W. (1987): Inversion of elastic $^{12}C + ^{12}C$ phases with the modified Newton method, in Sabatier (1987a, pp. 117–127) (section XIX.2)

McAulay A. (1985): Pre-stack inversion with plane-layer point source modeling, *Geophys.* 50, 77–89 (section XIX.2)

McLaughlin J. R. (1986a): Analytical method for recovering coefficients in differential equations from spectral data, *SIAM Rev.* 28, 53–72 (section XVII.5)

McLaughlin J. R. (1987b): Inverse spectral theory using nodal points as data—a uniqueness result, (section XVII.5)

McLaughlin J. R., and Rundel W. (1987): A uniqueness theorem for an inverse Sturm–Liouville problem, *J. Math. Phys.* 28, 1471–1472 (section XVIII.4)

McWilliams B. (1979): Confining potentials for quarkonium and the inverse scattering problem, *Phys. Rev. D* 20, 1221–1228 (section XVIII.4)

Melin A. (1985): Operator methods for inverse scattering on the real line, *Comm. Partial Differential Equations* 10, 677–766 (section XVII.5)

Melin A. (1987a): Some problems in inverse scattering theory, University of Lund (sections XI.7, XVII.2, XVII.5)

Melin A. (1987b): Some mathematical problems in inverse potential scattering, Séminaire equations aux dérivées particlles 1986–87 (section XIV.H)

Melnikov V. N., Rudyak B. N., and Zakhariev B. N. (1977): A simple method for solving the scattering problem, Preprint, Dubna 1977 (section XIX.2)

Mendel J. M. (1986): Some modeling problems in reflection seismology, *IEEE ASSP Magazine* 1, 4–17 (section XIX.2)

Mendel J. M., and Habibi-Ashrafi F. (1980): A survey of approaches to solving inverse problems for lossless layered media systems, *IEEE Trans. Geosci. Remote Sensing* GE-18, 320–330 (section XIX.2)

Millane R. P., and Bates R. H. T. (1982): Inverse methods for branches ducts and transmission lines, *Proc. IEEE-E* (section XIX.2)

Miller W. H. (1968): Uniform semiclassical approximations for elastic scattering and eigenvalue problems, *J. Chem. Phys.* 48, 1, 464–467 (section XVI.H, Foreword)

Miller W. H. (1969): WKB solution of inversion problems for potential scattering, *J. Chem. Phys.* 51, 3631–3638 (sections XVI.5, XVI.H, Foreword)

Miller W. H. (1971): Additional WKB inversion relations for bound-state and scattering problem, *J. Chem. Phys.* 54, 10, 4174–4177 (sections XVI.4, XVI.H, Foreword)

Mills R. L., and Reading J. F. (1969): Inversion problem with separable potentials, *J. Math. Phys.* 10, 2, 321–331 (section VIII.H, Foreword)

Minerbo G. N., and Levy M. E. (1969): Inversion of Abel's integral equation by means of orthogonal polynomials, *SIAM J. Numer. Anal.* 6, 598–616 (section XVI.7)

Miodek I. (1976): Inverse problem for complex "r^a-analytic" potentials of finite range, *J. Math. Phys.* 17, 168–173 (section XV.1)

Mittra R., and Itoh T. (1975): A method for measuring the refractive index profile of thin-film waveguides, *IEEE Trans. Microwave Theory Tech.*, MTT-23, 176–177 (section XIX.2)

Miura R. (1976): The Korteweg–de Vries equation, a survey of results, *SIAM Rev.* 18, 412–459 (section XVII.5)

Morawetz C. S. (1975): Notes on time decay and scattering for some hyperbolic problems, *Regional Conference Series in Applied Mathematics*, No. 19, SIAM, Philadelphia (section XIV.7)

Morawetz C. S. (1981): A formulation for higher dimensional inverse problems for the wave equation, *Comput. Math. Appl.*, 7, 319–331 (sections XIV.5, XIV.7)

Morawetz C. S., and Kriegswann G. A. (1983): *SIAM J. Appl. Math.* 43, 844–854 (section XIV.7)

Morel P. (1978a): Diverses données spectrales pour le problème inverse discret de Sturm–Liouville, pp. 147–179; Algorithmes pour un problème discret de Sturm–Liouville, pp. 371–387, in *Applied Inverse Problems* (P. C. Sabatier, ed.), Springer-Verlag, New York (section XIX.2)

Morel P. (1978b): Problèmes inverses spectraux discrets, Thèse de Doctorat de Mathématiques, Bordeaux (section XIX.2)

Morse P. M. (1929): Diatomic molecules according to the wave mechanics II. Vibrational levels, *Phys. Rev.* 43, 57–67 (Foreword)

Moses H. E. (1956): Calculation of the scattering potential from reflection coefficients, *Phys. Rev.* 102, 2, 559–567 (sections III.H, XVII.H, Foreword)

Moses H. E. (1964): Generalizations of the Jost functions, *J. Math. Phys.* 5, 833–840 (section XVII.H)

Moses H. E (1978): A generalization of the inverse scattering problem for the one-dimensional Schrödinger equation and application to the Korteweg–de Vries equation. A variational principle, in *Solitons in Action* (K. Lonngren and A. C. Scott, eds.), Academic Press, New York (section XVII.5)

Moses H. E. (1978): A generalization of the direct and inverse problem for the radial Schrödinger equation, *Stud. Appl. Math.* 58, 187–207 (section IX.H)

Moses H. E. (1979a): Gel'fand–Levitan equations with comparison measures and comparison potentials, *J. Math. Phys.* 20, 2047–2053 (section XVII.5)

Moses H. E. (1979b): An example of the effect of rescaling of the reflection coefficient on the scattering potential for the one-dimensional Schrödinger equation, *Stud. Appl. Math.* 60, 177–181 (section XVII.5)

Moses H. E. (1979c): Jost solutions and Green's functions for the three-dimensional Schrödinger equation, *J. Math. Phys.* 20, 1151–1156 (section XIV.7)

Moses H. E. (1980): A kernel of Gel'fand–Levitan type for the three-dimensional Schrödinger equation, *J. Math. Phys.* 21, 83–89 (section XIV.7). Erratum: *J. Math. Phys.* 22, 3010 (1981)

Moses H. E. (1983): Example of two distinct potentials without point eigenvalues which have the same scattering operator with the reflection coefficient $c(0) = -1$, *Phys. Rev. A* 27, 2220–2221 (section XVII.5)

Moses H. E. (1984): The time-dependent inverse source problem for the acoustic and electromagnetic equations in the one- and three-dimensional cases, *J. Math. Phys.* 25, 1905–1923 (section XVII.5)

Moses H. E., and Prosser R. T. (1983): Phases of complex functions for the amplitudes of the functions and the amplitudes of the Fourier and Melin transforms, *J. Opt. Soc. Amer.* 73 (section X.7)

Moses H. E., and Tuan S. F. (1959): Potential with zero scattering phases, *Nuovo Cimento* 13, 198–206 (section VI.H, Foreword)

Mostafavi M., and Mittra R. (1972): Remote probing of inhomogeneous media using parameter optimization techniques, *Radio Sci.* 7, 1105–1111 (section XIX.2)

Mott N. F., and Massey H. S. N. (1965): *The Theory of Atomic Collisions*, third ed., Oxford (section I.H)

Munchow M., and Scheid W. (1980): Modification of the Newton method for the inverse-scattering problem at fixed energy, *Phys. Rev. Lett.* 44, 1299–1302 (section XIX.2)

Mushkhelishvili N. I. (1953): Singular Integral Equations, Noordhoff (section VIII.H)

Muzafarov V. M. (1986a): An inverse scattering problem for nonlocal potentials. I. The family of phase equivalent wave functions, *Modern Phys. Lett. A* 1, 449–454 (section VIII.H)

Muzafarov V. M. (1986b): An inverse scattering problem for nonlocal potentials. II. The family of phase equivalent potentials, *Mod. Phys. Lett A* 2, 177–182 (section VIII.H)

Muzafarov V. M. (1988): Inverse scattering problem for a family of phase-equivalent nonlocal potentials, *Inverse Problems* 4, 185–210 (Foreword, section VIII.5)

Nachman I., and Ablowitz M. J. (1984a): A multidimensional inverse-scattering method, *Stud. Appl. Math.* 71, 243–250 (section XIV.4)

Nachman I., and Ablowitz M. J. (1984b): Multidimensional inverse-scattering for first-order systems, *Stud. Appl. Math.* 71, 251–262 (sections XIV.4, XIV.7)

Nachman I., Sylvester J., and Uhlmann G. (1988): An n-dimensional Borg–Levinson theorem, (section XIV.7)

Nashed Z. (1981): Operator-theoretic and computational approaches to ill-posed problems with applications to antennas theory, *IEEE Trans. Antennas and Propagation* AP-29, 220–231 (section XIX.2)

Neuhaus M. G. (1955): About the determination of the asymptotic behavior of the function $q(x)$ on the basis of the spectral function of $y'' + q(x)y$, *Dokl. Akad. Nauk. SSSR* 102, 25–28 (section V.H)

Newell A. C. (1985): *Solitons in Mathematics and Physics*, SIAM, Philadelphia (section XVII.5)

Newton R. G. (1955): Connection between the S-matrix and the tensor force, *Phys. Rev.* 100, 1, 412–428 (section IX.3)

Newton R. G. (1956): Remarks on scattering theory, *Phys. Rev.* 101, 1588–1596 (sections III.H, IV.H, V.H, VII.H, Foreword)

Newton R. G. (1957a): Electron scattering by the deuteron, *Phys. Rev.* 105, 763–764 (section VI.H, Foreword)

Newton R. G. (1957b): Deuteron photodisintegration cross section, *Phys. Rev.* 107, 1025–1027 (section VI.H, Foreword)

Newton R. G. (1962): Construction of potentials from the phase shifts at fixed energy. *J. Math. Phys.* 3. 1, 75–82 (sections XI.H, XII.2, XII.H, XVI.6, Foreword)

Newton R. G. (1966): *Scattering Theory of Waves and Particles*, McGraw-Hill, New York (section XII.7)

Newton R. G. (1967): Connection between complex angular momenta and the inverse scattering problem at fixed energy, *J. Math. Phys.* 8, 8, 1566–1570 (section XII.7, Foreword)

Newton R. G. (1968): Determination of the amplitude from the differential cross section by unitarity, *J. Math. Phys.* 9, 12, 2050–2055 (section X.6, Foreword)

Newton R. G.* (1973): The Gel'fand–Levitan method in the inverse scattering problem, Proceedings of the NATO Advanced Study Institute on Scattering Theory, Denver Colorado, J. A. Levita and J. P. Marchand eds. (section XVI.1, Foreword)

Newton R. G. (1974): The three-dimensional inverse scattering problem in quantum mechanics, Indiana University, Bloomington (section XIV.1, XIV.3, XIV.H, Foreword)

Newton R. G. (1977): Noncentral potentials: The generalized Levinson theorem and the structure of the spectrum, *J. Math. Phys.* 18, 1348–1356 (section XIV.7)

Newton R. G. (1979): Inverse scattering as a Hilbert problem, Lecture at the International Symposium on Ill-posed Problems, University of Delaware, October 2–6, 1979 (section XIV.7)

Newton R. G. (1980a): Inverse scattering. I. One dimension, *J. Math. Phys.* 21. 493–505 (sections XIV.5, XIV.7)

Newton R. G. (1980b): Inverse scattering. II. Three dimensions, *J. Math. Phys.* 21, 1698–1715, and see op. cit. 22, 631 (sections XIV.5, XIV.7)

Newton R. G. (1980c): New result on the inverse scattering problem in three dimensions, *Phys. Rev. Lett.* 43, 541–542 (sections XIV.5, XIV.7)

Newton R. G. (1981a): Inverse scattering. III. Three dimensions (*continued*), *J. Math. Phys.* 22, 2191–2200, and see op. cit. 23, 693 (sections XIV.5, XIV.7)

Newton R. G. (1981b): Inversion of reflection data for layered media: A review of exact methods, *Geophys. J. R. Astron. Soc.* 65, 191–215 (section XVII.5)

Newton R. G. (1982a): *Scattering Theory of Waves and Particles*, 2nd ed., Springer-Verlag, New York (sections XIV.4, XIV.5, XIV.7)

Newton R. G. (1982b): Inverse scattering. IV. Three dimensions: Generalized Marchenko construction with bound states, *J. Math. Phys.* 23, 594–604 (sections XIV.4, XIV.5, XIV.7)

Newton R. G. (1982c): On a generalized Hilbert problem, *J. Math. Phys.* 23, 2257–2265 (sections XIV.5, XIV.7)

* Professor Newton ran across two material errors in the formula (4.91) of this paper, which should read:

$$2\,\mathrm{Re}\int_0^x dk\, A(\mathbf{k},\mathbf{k})e^{ikt} = -\sum_n K_n |c_n(\hat{k})|^2 e^{-K_n t} \quad (t>0)$$

Newton R. G. (1983a): The Marchenko and Gel'fand–Levitan methods in the inverse scattering problem in one and three dimensions, *Conference on Inverse Scattering: Theory and Application*" (J. B. Bednar et al., eds.), SIAM, Philadelphia, pp. 1–74 (section XVII.5)

Newton R. G. (1983b): Inverse scattering by a local impurity in a periodic potential in one dimension, *J. Math. Phys.* 24, 2152–2162 (section XVII.5)

Newton R. G. (1984a): An inverse spectral problem in three dimensions, *SIAM-AMS Proc.* 14, 81–90 (section XIV.5)

Newton R. G. (1984b): Remarks on inverse scattering in one dimension, *J. Math. Phys.* 25, 2991–2997 (section XVII.5)

Newton R. G. (1985a): Review of the Marchenko method for inverse scattering in three dimensions, (section XIV.5)

Newton R. G. (1985b): A Faddeev–Marchenko method for inverse scattering in three dimensions, *Inverse Problems* 1, 127–132 (section XIV.7)

Newton R. G. (1985c): Variational principles for inverse scattering, *Inverse Problems* 1, 371–380 (sections XIV.5, XIV.7)

Newton R. G. (1985d): Inverse scattering by a local impurity in a periodic potential in one dimension. II, *J. Math. Phys.* 26, 311–316 (Foreword, section XVII.5)

Newton R. G., and Fulton T. (1957): Phenomenological neutron–proton potentials, *Phys. Rev.* 107, 1103–1111 (section VI.H)

Newton R. G., and Jost R. (1955): The construction of potentials from the S-matrix for systems of differential equations, *Nuovo Cimento* 1, 4, 590–622 (section IX.3, Foreword)

Novikov R. G. (1986a): Reconstruction of a two-dimensional Schrödinger operator from the scattering amplitude for fixed energy, *Funktsional Anal. i Prilozhen.* 20, 90–91 [transl. *Functional Anal. Appl.* 20, 246–248 (1986)] (section XIV.7, Foreword)

Novikov R. G. (1986b): Construction of a two-dimensional Schrödinger operator with given scattering amplitude at fixed energy, *Teoret. i Mat. Fiz.* 66, 234–240 [transl. *Theoret. and Math. Phys.* 66, 154–158 (1986)] (section XIV.7)

Novikov R. G. (1987): Spectral problem for the equation $-\Delta\Psi + (V(x) - EU(x)\Psi = 0$, *Dokl. Akad. Math. SSSR* (section XIV.7)

Novikov R. G., and Henkin G. M. (1987a), Solution of a multidimensional inverse scattering problem on the basis of generalized dispersion relations, *Dokl. Akad. Nauk SSSR* 292, 814–818 [transl. *Soviet Math. Dokl.* 35, 153–157] (Foreword)

Novikov R. G., and Henkin G. M. (1987b): *Usp. Mat. Nauk* 42, 93–151 (in Russian) (section XIV.H)

Nussenzveig H. M. (1972): *Causality and Dispersion Relations*, Academic Press, New York (section XIV.6)

Olver F. W. J. (1961): Error bounds for the Liouville–Green (or WKB) approximation, *Proc. Cambridge Philos. Soc.* 57, 790–810 (section XVI.H)

Olver F. W. J. (1965a): Error analysis of phase-integral methods I. General theory for simple turning points, *J. Res. Nat. Bur. Stand.* 69B, 4, 271–290 (section XVI.H)

Olver F. W. J. (1965b): Error analysis of phase-integral methods II. Application to wave-penetration problems, *J. Res. Natl. Bur. Stand.* 69B, 4, 291–300 (section XVI.H)

Olver F. W. J. (1974): *Asymptotic Expansions and Special Functions*, Academic Press, New York (section XVI.H)

Oryu S. (1983): Separable expansion fit to the Reid soft-core fully off-shell t matrices: $^1S_0$ and $^3S_1 - {}^3D_1$ states, *Phys. Rev. C* 27, 2500–2514 (section VIII.H)

Osborn T. A., and Bollé D. (1977): An extended Levinson's theorem, *J. Math. Phys.* 18, 432–440 (section XIV.7)

Osborne A. R. (1983): The spectral transform: Methods for the Fourier analysis of nonlinear wave data, Lectures presented at the Workshop for Statics and Dynamics of Nonlinear Systems, Erice (section XIX.2)

Osborne T. A., and Bergamasco L. (1985): The small-amplitude limit of the spectral transform for the periodic Korteweg–de Vries equation, *Nuovo Cimento* 85, 229–240 (section XIX.2)

Osborne T. A., and Bergamasco L. (1986): The solitons of Zabusky and Kruskal revisited: Perspective in terms of the periodic spectral transform, *Physica D* 18, 26–46 (section XIX.2)

Osborne T. A., and Burch T. L. (1980): Internal solitons in the Andaman Sea, *Science*, 208, 451–460 (section XIX.2)

Osborne T. A., Provenzale A., and Bergamasco L. (1982a): On the Stokes wave in shallow water: Perspective in the context of spectral-transform theory, *Nuovo Cimento C* 5, 597–611 (section XIX.2)

Osborne T. A., Provenzale A., and Bergamasco L. (1982b): Nonlinear Fourier analysis of localized wave fields described by the Korteweg–de Vries equation, *Nuovo Cimento C* 5, 612–632 (section XIX.2)

Osborne T. A., Provenzale A., and Bergamasco L. (1982c): Theoretical and numerical methods for nonlinear Fourier analysis of shallow-water wave data, *Nuovo Cimento C* 5, 633–648 (section XIX.2)

Osborne T. A., Provenzale A., and Bergamasco L. (1983): The nonlinear Fourier analysis of internal solitons in the Andaman Sea, *Lett. Nuovo Cimento* 36, 593–599 (section XIX.2)

Paivarinta L., and Somersalo E. (1987): The uniqueness of the one-dimensional electromagnetic inversion with bounded potentials, *J. Math. Anal. Appl.* 127, 312–333 (section XIX.2)

Paley R., and Wiener N. (1934): Fourier transforms in the complex domain, American Mathematical Society (section XIII.3)

Panzone R. (1984): Sobre algunos problemas inversos y un teorema de Euler concerniente a la cuerda vibrante, *Anales de la Academia Nacional de Ciencias Exactas, Fisicas y Naturales, Buenos Aires* 36, 73–90 (section XVIII.4)

Paul A. K. (1967): Ionospheric electron-density profiles with continuous gradients and underlying ionization corrections I. The math. physical problem of real-height determination from ionograms, *Radio Sci.* 2 (New series) no. 10, 1127–1133 (section XVI.7)

Paul A. K., and Smith G. H. (1968): Generalization of Abel's solution for both magnetoionic components in the real-height problem, *Radio Sci.* 3. 2, 163–170 (section XVI.7)

Pauly H. D. (1973): *Physical Chemistry, an Advanced Treatise*, Vol. IV, Ch. IV, Academic Press, New York (section XVI.4)

Pauly H. D., and Toennies J. P. (1965): The study of intermolecular potentials with molecular beams and thermal energies, *Adv. At. Mol. Phys.* 1, 195 (sections XVI.4, XVI.H)

Pearson D. B. (1975): An example in potential scattering illustrating the breakdown of asymptotic completeness, *Comm. Math. Phys.* 40, 125–146 (section II.6)

Pechenik K. R., and Cohen J. M. (1981): Inverse scattering—Exact solution of the Gel'fand–Levitan equation, *J. Math. Phys.* 22, 1513–1516 (section XVII.5)

Pechenik K. R., and Cohen J. M. (1983a): Inverse scattering with coinciding-pole reflection coefficients, *J. Math. Phys.* 24, 115–118 (section XVII.5)

Pechenik K. R., and Cohen J. M. (1983b): Exact solutions to the valley in inverse scattering", *J. Math. Phys.* 24, 406–409 (section XVII.5)

Pelosi V. (1978): Survey of the phenomenological approach to the inverse problem in elementary particle scattering, in Sabatier (1978, pp. 330–350) (section XIX.5)

Percival J. C. (1962): Energy moments of scattering phase shifts, *Proc. Phys. Soc.* 80, 1290–1300 (section VII.H, Foreword)

Percival J. C., and Roberts M. J. (1963): Energy moments of scattering phase shifts II. Higher partial waves, *Proc. Phys. Soc.* 82, 519–529 (section VII.H, Foreword)

Peterson Bo., and Ström S. (1974): Matrix formulation of acoustic scattering from an arbitrary number of scatterers, *J. Acoust. Soc. Amer.* 56, 771–780 (section XIX.2)

Petras M. (1962): On properties of Jost functions and potentials with singularities, *Czech. J. Phys.* B12, 87–92 (section IX.9)

Phinney R. A. (1970): Reflection of acoustic waves from a continuously varying interfacial region, *Review of Geophysics and Space Physics*, 8, 3, 517–532 (section XVI.7)

Pietarinen (1974): Invited talk given at the Triangle Seminar Smolanice, Czechoslovakia (section X.6)

Pike E. R. (1964): On the related-equation method of asymptotic approximation (W.K.B. or A.A. method), *Q. J. Mech. Appl. Math.* XVII (section XVI.H)

Pivovarchik V. N., Poplavsky I. V., and Suzko A. A. (1984): Complete set of radial Schrödinger equation solutions with a linear-dependent potential, *Phys. Lett. A* 101, 72–74 (section XIX.2)

Pivovarchik V. N., Suzko A. A., and Zakhariev B. N. (1986): New exactly solved models with bound states above the scattering threshold, *Physica Scripta* 34, 101–105 (section XIX.2)

Plekhanov E. B., Suzko A. S., and Zakhariev B. N. (1982): Multichannel and multidimensional Bargmann potentials, *Ann. Physik* 39, 7, 313–396 (section VI.3)

Plessas W. (1986): Separable interactions and their role in few-body systems, in *Few Body Methods: Principles and Applications* (T. K. Lim, C. G. Bao, D. P. Hou and H. S. Huber, eds.), pp. 43–75 (sections VIII.H, XIX.3)

Pogorzelski W. (1966): *Integral Equations*, Pergamon (section I.H)

Portinari J. C. (1966): The one-dimensional inverse scattering problem, Ph.D. Thesis: M.I.T. Department of Electrical Engineering (section XVII.H)

Portinari J. C. (1967): Finite-range solutions to the one-dimensional inverse scattering problem, *Ann. Phys.* 45, 445–451 (section XVII.H)

Poschel J., and Trubowitz E. (1987): *Inverse Spectral Theory* Academic Press, New York (Preface, section XVIII.4)

Pöschl G., and Teller E. (1933): Bemerkungen zur Quantenmechanik des anharmonischen Oszillators, *Zeitschr. f. Physik*, 83, 143–151 (Foreword)

Povzner A. Y. A. (1948): On differential equations of Sturm–Liouville type on a half-axis, *Mat. Sb.* 23 (6.5) 3–53, [*Math. Rev.* 10, 299 (1949)] (section III.H)

Prats F., and Toll J. S. (1959): Construction of the Dirac Equation central potential from phase shifts and bound states, *Phys. Rev.* 113, 1, 363–370 (section IX.5, Foreword)

Pritchard D. E. (1972): Finding the potential directly from scattering data, *J. Chem. Phys.* 56, 8, 4206–4212 (section XVI.4)

Prosser R. T. (1969): Formal solutions of inverse scattering problems, *J. Math. Phys.* 10, 1819–1822 (section III.H, Foreword)

Prosser R. T. (1976): Formal solutions of inverse scattering problems, II, *J. Math. Phys.* 17, 1775–1779 (section III.H, Foreword)

Prosser R. T. (1980): Formal solutions of inverse scattering problems, III, *J. Math. Phys.* 21, 2648–2653 (section XIV.7, Foreword)

Prosser R. T. (1982): Formal solutions of inverse scattering problems IV, *J. Math. Phys.* 23, 2127–2130 (section XIV.7, Foreword)

Prosser R. T. (1983): Approximation methods and error estimates for inverse scattering problems, in *Conference on Inverse Scattering: Theory and Application* (J. B. Bednar et al., eds.), SIAM, Philadelphia (sections XVII.2, XVII.5)

Prosser R. T. (1984): On the solutions of the Gel'fand–Levitan equation, *J. Math. Phys.* 25, 1924–1929 (section XVII.5)

Qing Xu Bang and Shao Chang Qui (1985): *Lett. Math. Phys.* 9, 183–186 (section XVII.5)

Rajala J., and Riska D. O. (1981): On the inversion problem of magnetic ore prospecting, Report series in *Physics Helsinki* (section XIX.2)

Ramm A. G. (1965): Conditions under which scattering matrices are analytic, *Sov. Phys. Dokl.* 9, 645–647 (sections V. H, XVII.H)

Ramm A. G. (1987a): Characterization of the scattering data in multidimensional inverse scattering problem, in Sabatier (1987a, pp. 153–169) (section XIV.7)

Ramm A. G. (1987b): A method for solving the three-dimensional inverse-scattering problem, Preprint (6 pages). On a question of R. G. Newton, Preprint (6 pages) (section XIV.7)

Ramm A. G. (1988a): Inverse scattering on half-line, *J. Math. Anal. Appl.* 133, 543–572

Ramm A. G. (1988b): Recovery of the potential from fixed-energy scattering data, *Inverse Problems* 4, 877–886

Ramm A. G., and Weaver O. L. (1987): A characterization of the scattering data in the 3D inverse scattering problem, *Inverse Problems* 3, 149–152 (section XIV.7)

Lord Rayleigh J. W. S. (1877): *The Theory of Sound*, Dover Publications, New York, 1945 [Tom I, p. 216] (section III.H, Foreword)

Lord Rayleigh J. W. S. (1945): *The Theory of Sound* (1st ed., 1877), Dover, New York (section XVIII.4)

Razavy M. (1975): Determination of the wave velocity in an homogeneous medium from the reflection coefficient, *J. Acoust. Soc. Amer.* 58, 956–963 (section XIX.2)

Redmond P. J. (1964): Some remarks concerning a pathological matrix of interest in the inverse scattering problem, *J. Math. Phys.* 5, 11, 1547–1554 (section XII.H)

Rees A. L. G. (1947): The calculation of potential energy curves from bound spectro-scopic data, *Proc. Phys. Soc. London* 59, 998–1008 (section XVI.H, Foreword)

Regge T. (1959): Introduction to complex orbital moments, *Nuovo Cimento* 14, 5, 951–976 (sections XI.H, XIII.3, Foreword)

Reignier J. (1979): Remark on the inverse scattering problem at fixed energy, *Lett. Nuovo Cimento* 24, 139–143 (section XIII.4)

Reilly M. H., and Jordan A. K. (1981): The applicability of an inverse method for reconstruction of electron-density profiles, *IEEE Trans. Antennas and Propagation* AP-29, 245–252 (section XIX.2)

Remler E. A. (1971): Complex angular momentum analysis of atom scattering experiments, *Phys. Rev.* A3, 6, 1949–1954 (section XVI.4)

Rich V. G., Bobbio S. M., Champion R. L., and Doversfihe L. (1971): Inversion problem for ion atom differential elastic scattering, *Phys. Rev.* A4, 6, 2253–2260 (section XVI.4)

Richard V. (1985): The 1-D seismic inverse problem: Choice of a viable optimization method and well-posed questions for application to field data, *Proc. RCP 264*, Montpellier, 555–570 (section XIX.2)

Riska D. O. (1981): (section XIX.H)

Riska D. O. (1982a): On the electrodynamical inversion problem for vertically layered earth, *Comment. Phys.-Math.* 53, 1–23 (section XIX.2)

Riska D. O. (1982b): Iterative solution of the electrodynamical inversion problem for vertically varying earth, in *Proc. of R.C.P. 264* (P. C. Sabatier, ed.), *Cahiers Math. Montpellier* 28, 137–154 (section XIX.2)

Roberts M. J. (1963): Energy moments and cross sections for a Gaussian potential, *Proc. Phys. Soc. London* 82, 594–604 (section VII.H, Foreword)

Roberts M. J. (1964a): Negative energy moments and low energy approximations to phase shifts, *Proc. Phys. Soc. London* 83, 503–517 (section VII.H, Foreword)

Roberts M. J. (1964b): Alternative derivation of the moment relations involving scattering phase shifts, *Proc. Phys. Soc. London* 84, 825–826 (section VII.H, Foreword)

Roberts M. J. (1965): Moment relations for Schrödinger two-body cross sections, *Proc. Phys. Soc. London* 86, 683–691 (section VII.H, Foreword)

Rofe-Beketov F. S., and Hristov E. H. (1969): Some analytical questions and the inverse Sturm–Liouville problem for an equation with highly singular potential, *Sov. Math. Dokl.*, 10, 432–435 (sections II.6, III.6)

Rose J. H., Cheney M., and De Facio B. (1984): The connection between time- and frequency-domain three-dimensional inverse scattering methods, *J. Math. Phys.* 25, 2995–3004 (sections XIV.6, XIV.7)

Rose J. H., Cheney M., and De Facio B. (1985a): Physical basis of three-dimensional inverse scattering for the plasma wave equation, *J. Opt. Soc. Amer. A* 2 (section XIV.7)

Rose J. H., Cheney M., and De Facio B. (1985b): Three-dimensional inverse scattering: Plasma and variable velocity wave equations, *J. Math. Phys.* 26, 2803–2813 (section XI.H)

Rose J. H., Cheney M., and De Facio B. (1985c): Connection between time- and frequency-domain three-dimensional inverse problems for the Schrödinger equation, in *Review of Progress in Quantitative Nondestructive Evaluation*, Vol. 4A, (D. O. Thompson and D. E. Chimenti, eds.) pp. 535–541 (section XIV.6)

Rose J. H., Cheney M., and De Facio B. (1985d): Three-dimensional inverse scattering: Plasma and variable velocity wave equations, *J. Math. Phys.* 26, 2803–2813 (section XIV.7)

Rose J. H., Cheney M., and De Facio B (1986a): Determination of the wave field from scattering data, *Phys. Rev. Lett.* 57, 783–786 (section XIV.7)

Rose J. H., De Facio B., and Cheney M. (1988): A new integral wave equation which unifies the inverse and direct scattering problems, *Wave Motion* in press (section XIV.7)

Rosen N., and Morse P. M. (1932): On the vibrations of polyatomic molecules, *Phys. Rev.* 42, 210–217 (Foreword)

Rosen M., and Yennie D. R. (1965): A modified WKB approximation for phase shifts, *J. Math. Phys.* 5, 11, 1505–1515 (section XVI.H)

Rousselet B. (1976): *Problèmes Inverses de Valeurs Propres*, Springer-Verlag, New York (section XIX.3)

Rousselet B. (1982): Quelques résultats en optimisation de domaines Doctoral Thesis, Nice (section XIX.3)

Rudyak B. V., Suzko A. A., and Zakhariev B. N. (1984): Exactly solvable models (Crum–Krein transformations in the (2, E)-plane), *Physica Scripta*, 29, 515–517 (section XIX.2)

Rydberg R. (1931): Grasphische Darstellung einiger barden spektroskopischer Ergebnisse, *Z. Phys.* 73, 376–385 (section XVI.H)

Rydberg R. (1933): Über einige Potentialkurven der Quecksilberhydrids, *Z. Phys.* 80, 514–524 (section XVI.H, Foreword)

Sabatier P. C. (1965): On the asymptotic approximation for the elastic scattering by a potential I. Monotonic potential and uncritical range, *Nuovo Cimento* 37, 1180–1226 (section XVI.3). *Remark*: The other parts of this study appear in Sabatier 1966a

Sabatier P. C. (1966a): Le problème inverse à énergie fixée en mécanique quantique, Thèse de Doctorat d'Etat, Orsay, 222 p. (sections XII.7, XII.H, XVI.H, Foreword)

Sabatier P. C. (1966b): Asymptotic properties of the potentials in the inverse-scattering problem at fixed energy, *J. Math. Phys.* 7, 8, 1515–1531 (sections XII.2, XII.3, XII.4, XII.5, XII.7, XII.H, Foreword)

Sabatier P. C. (1966c): Analytic properties of a class of potentials and the corresponding Jost functions,* *J. Math. Phys.* 7, 11, 2079–2091 (sections XII.7, XII.H, Foreword)

Sabatier P. C. (1966d): Interpolation des fonctions d'onde dans le plan complexe du moment angulaire, *C. R. Acad. Sc. Paris* 263, 788–790 (section XII.8, Foreword)

Sabatier P. C. (1967a): General method for the inverse scattering problem at fixed energy, *J. Math. Phys.* 8, 4, 905–918 (sections XII.6, XII.7, XII.H, Foreword)

Sabatier P. C. (1967b): Interpolation formulas in the angular momentum plane,† *J. Math. Phys.* 8, 10, 1957–1972 (sections XI.H, XII.8, XII.9, Foreword)

Sabatier P. C. (1967c): Approximation des potentiels par les potentiels de classe C, *C. R. Acad. Sc. Paris* 265, 5–8 (section XIII.4, Foreword)

* Page 2081, 4 lines before 1.21 read "order 2" (and not "order at least equal to 2"). Page 2091, there is an overall sign mistake in the formulas A IV 5.8.

† Page 1961, 3 lines before (1.63), read $\lambda = -\mu$.

Sabatier P. C. (1968): Approach to scattering problems through interpolation formulas and application to spin-orbit potentials,* *J. Math. Phys.* 9, 8, 1241–1258 (section XV.1, Foreword)

Sabatier P. C. (1971): New tool for scattering studies, *J. Math. Phys.* 12, 7, 1303–1326 (section XIII.4, Foreword)

Sabatier P. C. (1972a): Complete solution of the inverse scattering problem at fixed energy,† *J. Math. Phys.* 13, 5, 675–699 (sections XII.4, XVI.6, Foreword)

Sabatier P. C. (1972b): Comparative evolution of the inverse problems, in *Mathematics of profile inversion*, L. Colin Ed., 9.2–9.34 (sections VI.H, XI.H, XVI.H, Foreword)

Sabatier P. C. (1973a): Ambiguities in the construction of potentials from phase-shifts, *Intern. J. of Quantum Chemistry Symp.* 7, 421–425 (section XVI.H, Foreword)

Sabatier P. C. (1973b): Discrete ambiguities and equivalent potentials,‡ *Phys. Rev.* A8, 589–601 (sections XVI.2, XVI.H, Foreword)

Sabatier P. C. (1974a): Construction of the scattering amplitude from the elastic cross section at a given energy, Invited lecture in the Summer School on Inverse problems organized by the American Mathematical Society in Los Angeles. *Cahiers mathématiques de Montpellier*, 4 (section X.1, Foreword)

Sabatier P. C. (1974b): Problème direct et inverse de diffraction d'une onde élastique par une zone perturbée sphérique, *C. R. Acad. Sci. Paris* 278, série B, 545–547 section VI.H, Foreword)

Sabatier P. C. (1974c): Problème direct et inverse de diffraction d'une onde élastique par une zone perturbée sphérique, Compléments, *C, R. Acad. Sci. Paris*, 278, série B, 603–606 (section VI.H, Foreword)

Sabatier P. C. (1977): On geophysical inverse problems and constraints, *J. Geophys.* 43, 115–137.

Sabatier P. C. (1978): *Applied Inverse Problems*, Lecture Notes in Physics, Vol. 85, Springer-Verlag, Berlin, (Section XIX.2)

Sabatier P. C. (1978a): Problème inverse des modes normaux d'une sphère élastique, *Revue CETHEDEC* 56, 145–151 (sections XVIII.2, XVIII.4)

Sabatier P. C. (1978b): Spectral and scattering inverse problems, *J. Math. Phys.* 19, 2410–425 (section XVII.2, XVII.4)

* In the formulas (2.72), (2.73), (2.74), of this paper, the term V_s (but not $\pm \lambda V_s$) should have a factor of 1/2. Remark also that in (2.71), the $(l = 0)$ term is in fact absent, and the sum should start at $l = 1$.

† There are in this reference some unpleasant misprints, in particular p. 679, 7th line after (2.11), read $\bar{k}(r) +$; p. 683, formula (2.93), read $w^{-1}(\sin w)u\Phi(u)$; p. 685, formula (3.2), read $\Phi_2^{(1)}(u)$; two lines after (3.2), read $L_{1 + 1/\varepsilon}(u)$; formula (3.4) read $(d/du)(\pi u)^{-1} \ldots$; p. 686, formula (3.28), read in the third line:

$$-\psi(u)[P_{l+1}(1 - 2u^2) - P_{l-1}(1 - 2u^2)]_\alpha^\mu + \int_\alpha^\beta \psi'(u)[P_{l+1}(1 - 2u^2) - P_{l-1}(1 - 2u^2)]du;$$

p. 697 formula A 99, read:

$$10r^{-5}[K(r, r)]^2 + 2r^{-2}V(r)K(r, r) + \cdots$$

and some (trivial) terms are lacking; in lines after (4.99), read $\partial/\partial r$, $\partial/\partial r$, $\partial/\partial r$, Cr, $r \leqslant r'$, [I (4.54.55)].

‡ In this paper, page 591, lines between formulas (2.2) and (2.3), and in (2.3) read $(\pi/2)g(t)$ instead of $g(t)$; page 592, in (2.8) and (2.9), read $(\pi/2)[g(t)]^2$; page 598, last line, replace "must be finite" by "must be finite for fixed λ^M."

Sabatier P. C. (1979a): On extremal solutions of Sturm–Liouville inverse problems, in *Inverse and Improperly Posed Problems in Differential Equations* (G. Anger, ed.), Akademi-Verlag, Berlin, pp. 223–232 (Sections XVIII.4, XIX.2). See also *idem*, Around the classical string problem, in *Nonlinear Evolution Equations and Dynamical Systems* (M. Boiti, F. Pempinelli, and G. Soliani, eds.), Springer-Verlag, New York, pp. 85–102 (1980)

Sabatier P. C. (1980): Problèmes inverse, évolution nonlineaire, *Proc. RCP 264*, Editions du C.N.R.S., Paris (section XVIII.2)

Sabatier P. C. (1983a): Application de la théorie de l'inversion, *Revue CETHEDEC* 76, 1–18. See also Sabatier P. C., Well-posed questions for ill-posed problems (invited paper to the 1983 joint meeting of the IEEE Geoscience and Remote Sensing Society and URSI/USNS, San Francisco) (section XIX.2)

Sabatier P. C. (1983b): Rational reflection coefficients and inverse scattering on the line. *Nuovo Cimento B* 78, 235–248. See also *idem*, Rational reflection coefficients in one-dimensional inverse scattering and applications, in Bednar et al. (1983, pp. 75–100); and *idem*, Critical analysis of the mathematical methods used in electromagnetic inverse theories (sections XVII.5, XIX.2)

Sabatier P. C. (1983c): Theoretical considerations for inverse scattering, *Radio Sci.* 18, 1–18 (section XIX.2)

Sabatier P. C. (1984): Well-posed questions and exploration of the space of parameters in linear and nonlinear inversion, in Santosa, Pao et al. (1984, pp. 82–103) (sections XVII.5, XIX.2)

Sabatier P. C. (1985a): Introduction to ill-posed aspects of nuclear scattering, in Krappe and Lipperheide (1985, pp. 1–19) (section XIX.2)

Sabatier P. C. (1985b): Algorithmic approaches to sets of good answers in inverse problems, in *Distributed Parameter Systems* (F. Kappel et al., eds.), pp. 312–323, Springer-Verlag, New York (section XVII.5)

Sabatier P. C. (1986): Reconstruction ambiguities of inverse scattering on the line, in *Inverse Problems* (J. R. Cannon and U. Hornung, eds.), International Series of Numerical Mathematics, Vol. 77, Birkhauser, Basel (section XVII.5)

Sabatier P. C. (1987a): Inverse Problems: An interdisciplinary study, *Advances in Electronics and Electron Physics*, Suppl. 19, Academic Press (section XIX.2)

Sabatier P. C. (1987b): A few geometrical features of inverse and ill-posed problems, in Engl and Groetsch (1987) (section XIX.2)

Sabatier P. C. (ed.) (1987c): Basic methods of tomography and inverse problems, A set of lectures by G. T. Herman, H. K. Tuy, K. J. Langenberg, and P. C. Sabatier. Malvern Physics Series, Adam Hilger, Bristol and Philadelphia (sections XVII.4, XIX.1, XIX.2)

Sabatier P. C. (1987d): Remark on the three-dimensional mixed impedance potential equation, *Inverse Problems* 3, L83–86 (section XVII.5)

Sabatier P. C. (1988a): For an impedance scattering theory, in *Non Linear Evolutions* (J. Léon, ed.), pp. 723–745, World Scientific, Singapore (section XVII.5)

Sabatier P. C. (ed.) (1988b): *Some Topics on Inverse Problems*, World Scientific, Singapore (section XVII.H)

Sabatier P. C. (1989): Three-dimensional Impedance Scattering Theory, in "Electro-Magnetic and Acoustic Scattering. Detection and Inverse Problem", (C. Bourrely, P. Chiapetta and B. Torresain, ed), World Scientific, Singapore (section XIV.6)

Sabatier P. C., and Dolveck-Guilpart B. (1988): On modeling discontinuous media. One-dimensional approximations, *J. Math. Phys.* (section XVII.5)

Sabatier P. C., and Quyen Van Phu F. (1971): Numerical computations in the inverse scattering problem at fixed energy, *Phys. Rev.* D4, 127–132 (section XIII.4, XVI.6)

Saito Y. (1982a): An inverse problem in potential theory and the inverse scattering problem, *J. Math. Kyoto Univ.* 22, 370–321 (sections XIV.5, XIV.7)

Saito Y. (1982b): Some properties of the scattering amplitude and the inverse scattering problem, *Osaka J. Math.* 19, 527–547 (sections XIV.5, XIV.7)

Saito Y. (1984): An asymptotic behavior of the S-matrix and the inverse scattering problem, *J. Math. Phys.* 25, 3105–3111 (sections XIV.5, XIV.7)

Saito Y. (1986): An approximation formula in the inverse scattering problem, *J. Math. Phys.* 27, 1145–1153 (section XIV.7)

Saito Y. (1987): Schrödinger operators with a nonspherical radiation condition, *Pacific J. Math.* 126, 331–359 (section XIV.7)

Sakmar I. A. (1978): Conditions on the uniqueness of the solution of the elastic unitarity equation, *J. Math. Phys.* 19, 2124–2125 (Foreword)

Sakmar I. A. (1979): An angle-dependent lower bound on the solution of the elastic unitarity integral and a new uniqueness condition resulting from it, *J. Math. Phys.* 20, 11–12 (section X.7)

Sakmar I. A. (1980): Variational calculation of a quantity which determines the uniqueness of the phase of the scattering amplitude, *Lett. Nuovo Cimento* 28, 118–120 (section X.7)

Sanders, W. A., and Mueller C. R. (1963): Direct determination of intermolecular potential parameters from scattering phase shifts, *J. Chem. Phys.* 39, 2572–2580 (Foreword)

Santosa F. (1982): Numerical scheme for the inversion of acoustical impedance profile based on the Gel'fand–Levitan method, *Geophys. J. R. Astron. Soc.* 70, 229–243 (section XIX.2)

Santosa F., Pao Y. S., Symes W. W., and Holland C. (1984): *Inverse Problems of Acoustic and Elastic Waves*, SIAM, Philadelphia (section XIX.2)

Santosa F., and Symes W. W. (1983): Inversion of impedance profile from band-limited data, 1–7 IGARSS '83 (section XIX.2)

Santosa F., and Symes W. W. (1987): Inversion of band-limited seismograms, *Inverse Problems* 3, 477–500 (section XIX.2)

Schaubert D. H., and Mittra R. (1977): A spectral domain method for remotely probing stratified media, *IEEE Trans. Antennas and Propagation* AP-25, 2, 261–265 (section XIX.2)

Schmid E. W. (1987): The problem of using energy-dependent nucleon–nucleon potentials in nuclear physics *Helv. Phys. Acta* 60, 394–397 (section XIX.2)

Schützer W., and Tiomno J. (1951): On the connection of the scattering and derivative matrices with causality, *Phys. Rev.* 83, 249–251 (section XIV.6)

Scott A. C., Chu F. Y. T., and McLaughlin D. W. (1973): The soliton: a new concept in applied science, *Proceedings of the IEEE* 61, 1443–1483 (section XVII.H, Foreword)

Segur H. (1973): The Korteweg–de Vries equation and water waves. Solutions of the equation, *J. Fluid Mech.* 59, 721–736 (section XVII.5)

Semjonov V. M., Gridnev K. A., Hefter E. F., and Subbotin V. B. (1987): Application of solitons to nuclear physics or the inverse mean field method (IMEFIM, Leningrad State University), 161–173 (section XIX.2)

Sengupta A., and Ganguly K. (1980): Inverse eigenvalue problem for a finite multiplying slab *J. Phys. A* 13, 2341–2352 (section XIX.2)

Sharpe C. B. (1963): Some properties of infinite line, *Q. Appl. Math.* 21, 337–342 (section XVII.4)

Sharpe C. B. (1977): An inverse problem associated with the discrete Schrödinger equation, *Siam J. Appl. Math.* 32, 2, 405–412 (section XVII.4)

Shmoys (1961): Proposed diagnostic method for cylindrical plasmas, *J. Appl. Phys.* 32, 689–695 (section XVII.4)

Sigmund P., and Llillemark U. (1974): Inversion or total cross-sections for repulsive ion atom scattering in the classical regime, *Physica* 71, 258–265 (section XVI.5)

Simon B. (1971): *Quantum Mechanics for Hamiltonians Defined as Quadratic Forms*, Princeton University Press, Princeton, NJ, pp. 94–131 (section XIV.7)

Simon B. (1975): The bound state of weakly coupled Schrödinger operators in one and two dimensions, *Ann. Physics* 97, 279–288 (section XIV.7)

Singh P. P., Malmin R. E., High M., and Devins D. W. (1969): Optical-model analysis of alpha particle scattering from Mg^{24}, *Phys. Rev. Lett.* 23, 19, 1124–1127 (sections XVI.3, XVI.H)

Singh P. P., and Schwandt P. (1972): Uniqueness in the radial dependence of the optical potential for alpha-nucleus scattering, *Phys. Lett.* 42B, 2, 181–185 (sections XVI.3, XVI.H)

Smith F. T. (1969): Elastic and inelastic atom–atom scattering in *Lectures in Theoretical Physics and Atomic Collision Processes*, ed. by Geltman et al. Gordon & Breach, New York, Vol. XIC p. 95 (section XVI.4)

Smith F. T., Marchi R. P., Aberth W., Lorents D. C., and Heinz O. (1967): Collision spectroscopy I. Analysis of the scattering of He^+ by Ne and Ar, *Phys. Rev.* 161, 1, 31–46 (section XVI.4)

Smith F. T., Marchi R. P., and Dedrich K. G. (1966): Impact expansions in classical and semi-classical scattering, *Phys. Rev.* 150, 79–92 (section XVI.4)

Sohin A. S. (1972): On one integral representation of Hankel functions, *Dokl. Akad. Nauk SSSR* (section III.7)

Somersalo E. (1987): Inverse scattering for standing wave solutions of the Schrödinger equation, *J. Math. Phys.* 28, 2416–2419 (section XIV.7)

Sondhi M. M. (1981a): The inverse problem for the vocal tract: some recent experimental and computational developments, in *Proceedings of RCP 264* (P. C. Sabatier, ed.), Cahiers Math. Montpellier, 28, 155–175 (section XIX.2)

Sondhi M. M. (1981b): Acoustical inverse problem for the cochlea, *J. Acoust. Soc. Amer.* 69, 500–504 (section XIX.2)

Sondhi M. M., and Gopinath B. (1972): Determination of the shape of a lossy vocal tract. Proceedings of the 7th international congress on acoustics Budapest, Hungary, Akademiai Kiado 1971 (section XVII.4)

Sondhi M. M., and Resnick J. R. (1982): Speech synthesis from measured vocal tract shapes, *J. Acoust. Soc. Amer.* (section XIX.2)

Stashevskaya V. V. (1953): On inverse problems of spectral analysis for a class of differential equations, *Dokl. Akad. Nauk. SSSR* 93, 409–411 (section III.H)

Stefanescu I. S. (1980): On the stable analytic continuation with rational functions, *J. Math. Phys.* 21, 175–188 (section X.7)

Stefanescu I. S. (1981): On the determination of πN phase shifts from isospin constraints and fixed t analyticity, *Acta Phys. Austriaca, Suppl.* XXIII, 695–701 (section X.7)

Stefanescu I. S. (1982): On the determination of πN phase shifts from isospin constraints and fixed t analyticity, *J. Math. Phys.* 23, 1190 (section X.7)

Stefanescu I. S. (1985): On the phase retrieval problem in two dimensions, *J. Math. Phys.* 26, 2141 (section X.7)

Stefanescu I. S. (1986): On the stable analytic continuation with a condition of uniform boundedness, *J. Math. Phys.* 27, 2657 (section X.7)

Stefanescu I. S. (1987): Zero trajectories and the problem of the reconstruction of the phase, *Fortschr. Phys.* 35, 573–673 (section X.7)

Steinmann O. (1957): Equivalent periodic potentials, *Helv. Phys. Acta* 30, 6, 515–520

Stickler D. C. (1983): Inverse scattering in a stratified medium, *J. Acoust. Soc. Amer.* 74, 994–1005 (section XIX.2)

Stickler D. C. (1986): Inverse scattering for stratified elastic media, *Wave Motion* 8, 101–112 (section XIX.2)

Stickler D. C., and Deift P. (1981): Inverse problem for a stratified ocean and bottom, *J. Acoust. Soc. Amer.* 70, 1723–1727 (section XIX.2)

Stokes (1857): On the discontinuity of arbitrary constants which appear in divergent developments, *Trans. Cambridge Phil. Soc.* 10, 106–128 (section XVI.H)

Strom S. (1974): T matrix for electromagnetic scattering from an arbitrary number of scatterers with continuously varying electromagnetic properties, *Phys. Rev. D* 10, 2685–2690 (section XIX.2)

Suzuki T. (1986): Gel'fand–Levitan's theory and related inverse problems, in *Inverse Problems* (I. R. Cannon and L. V. Hornung, eds.), Birkhaüser, Basel (section IX.H)

Swan P. (1967): Deduction of nonsingular potentials from scattering phase shifts II. Charged particles, *Nucl. Phys.* A90, 436–448 (section VI.H)

Swan P., and Pearce W. A. (1966): Deduction of potentials from scattering phase shifts I. Neutral particles, *Nucl. Phys.* 79, 77–107 (section VI.H, Foreword)

Symes W. W. (1979): Inverse boundary value problems and a theorem of Gel'fand and Levitan, *J. Math. Anal. Appl.* 71, 379–402 (section XVIII.4)

Symes W. W. (1981a): Invertibility of the trace operator and local energy decay for the string with BV coefficient, (section XIX.2)

Symes W. W. (1981b): Stable solution of the inverse reflection problem for a smoothly stratified elastic medium, *SIAM J. Math. Anal.* 12, 421–453 (section XVIII.4)

Symes W. W. (1983a): Impedance profile inversion via the first transport equation *J. Math. Anal. Appl.* 94, 435–453 (section XIX.5)

Symes W. W. (1983b): A trace theorem for solutions of the wave equation, and the remote determination of acoustic sources *Math. Methods Appl. Sci.* 5, 131–152 (section XIX.2)

Symes W. W. (1985): Scattering properties for the velocity inversion problem, in *Proceedings of the SEG/SIAM/SPE Meeting (Houston, Jan. 85)* (section XIX.2)

Symes W. W. (1986): On the relation between coefficient and boundary values for solutions of Webster's horn equation, *SIAM J. Math. Anal.* 17, 1400–1420 (section XVII.5)

Szu H. H., Carroll C. E., Yang C. C., and Ahn S. (1976): A new functional equation in the plasma inverse problem and its analytic properties, *J. Math. Phys.* 17, 1236–1247 (section XVII.5)

Tabakin F. (1969): Inverse scattering problem for separable potentials, *Phys. Rev.* 177, 1443–1451 (section VIII.H, Foreword)

Tabbara W. (1979): Reconstruction of permittivity profiles from a spectral analysis of the reflection coefficient, *IEEE Trans. Antennas and Propagation* AP-27, 2, 241–244 (section XIX.2)

Talenti G. (Ed.) (1986): *Inverse Problems*. Lecture Notes in Mathematics. Springer-Verlag, Berlin, Heidelberg, New York (Preface)

Thaker H. B., Quigg C., and Rosner J. L. (1978): Inverse scattering problem for Quarkonium systems, *Phys. Rev.* D 18, 274–286 and 287–295

Theis W. R. (1956): Explicit potentials for given scattering functions of charged particles with absorption, *Z. Naturforsch.* 11a, 889 (section IV.H, Foreword)

Thornton B. S. (1979): Inversion of the geophysical inverse problem for n layers with nonuniqueness reduced to n cases, *Geophys.* 44, No. 4, 801–819 (section XIX.2)

Tijhuis A. G. (1981): Iterative determination of permittivity and conductivity profiles of a dielectric slab in the time domain, *IEEE Trans. Antennas and Propagation* AP-29, 2, 239–245 (section XIX.2)

Tijhuis A. G. (1987): *Electromagnetic Inverse Profiling. Theory and Numerical Implementation.* VNU Science Press, Utrecht (section XIX.2)

Titchmarsh E. C. (1948): *Theory of Fourier Integrals*, Oxford University Press (section VIII.H)

Titchmarsh E. C. (1962): *Eigenfunction Expansions Part I*, Oxford University Press. 2nd ed. (section II.H)

Tortorella M. (1972): Existence and uniqueness questions for the unitary equation, *J. Math. Phys.* 13, 11, 1764–1767 (section X.1, Foreword)

Tortorella M. (1974): The unitarity equation for scattering in the absence of spherical symmetry, *J. Math. Phys.* 15, 6, 745–747; *J. Math. Phys.* 15, 7, 1153 (section X.1)

Tortorella M. (1975): Experimental uncertainties in the problem of the unitary equation *J. Math. Phys.* 16, 4, 823–828 (section X.1)

Tortorella M., and Leise J. A. (1976): The unitary equations for matrices and the problem of determining the scattering amplitude from the differential cross-section, *Siam J. on Appl. Math.* 30, 3, 407–420 (section X.1)

Tricomi F. G. (1957): *Integral Equations*, Interscience Publishers, New York (section XIII.3)

Trlifaj L. (1980): The Jost solutions for Yukawa potentials, *J. Math. Phys.* 21, 1724–1731 (section VII.H)

Trlifaj L. (1981): The radial Marchenko–Blazek integral equation for complete spectra, *Czech. J. Phys. B* 31, 355–366 (section V.H)

Trubowitz E. (1977): The inverse problem for periodic potentials, *Comm. Pure Appl. Math.* 30, 321–337 (section XVIII.4)

Turchin V. F. (1985): Statistical regularization, in Krappe and Lipperheide (1985) (section XIX.2)

Turchin V. F., Kozlov V. P., and Malkevitch M. S. (1971): *Sov. Phys. Usp.* 13, 681–703 (section XIX.2)

van Kampen N. G. (1951): Note on the analytic continuation of the S-matrix, *Philos. Mag.* 42, 851–855 (section V.H, Foreword)

van Kampen N. G. (1953): *S*-matrix and causality conditions, II. Nonrelativistic particles, *Phys. Rev.* 91, 1267–1276. See also *idem, Phil. Mag.* 7, 871; *Phys. Rev.* 89, 1072; *Physica* 20, 115; *Physica* 21, 127 (section XIV.6)

Verde M. (1959): The inversion problem in wave mechanics and dispersion relations, *Nucl. Phys.* 9, 255–266 (section IX.5, Foreword)

Viano G. A. (1969): A note on inverse problems in potential scattering. *Nuovo Cimento* 63A, 2, 581–597 (section XIII.2)

Vidberg H. J., and Riska D. O. (1983): The inverse scattering problem for reflection of electromagnetic dipole radiation from earth with vertical variation, (section XIX.2)

Vogelzang E., Yevick D., and Ferwerda H. A. (1983): A numerical procedure for solving the inverse scattering problem for stratified dielectric media, *Optics Comm.* 45, 376–379 (section XIX.2)

Volk V. Y. (1953): On inverse formulas for a differential equation with a singularity at $x = 0$, *Usp. Mat. Nauk (N.S.)* 8, 4 (56) 141–151 (section III.H)

Vollmer G. (1969): Inverse problem in atom–atom scattering in WKB approach. *Z. Phys.* 226, 423–431 (sections XVI.4, XVI.H)

Wadati M., and Kamijo T. (1974): On the extension of the inverse scattering method, *Progr. Theoret. Phys.* 52, 397–414. See also their lecture in Bullough and Caudrey (1980) (section XVII.5)

Ware J. A., and Aki K. (1968): Continuous and discrete inverse scattering problems in a stratified elastic medium. Part 1: Plane waves at normal incidences, *J. Acoust. Soc. Amer.* 45, 4, 911–921 (section XIX.2)

Weidelt P. (1972): The inverse problem of geomagnetic induction, *Z. Geophys.* 38, 257–289 (section VI.H)

Weis R., and Scharf G. (1971): The inverse problem of potential scattering according to the Klein–Gordon equation, *Helv. Phys. Acta* 44, 910–929 (section IX.5, Foreword)

Weis R., Stahel W., and Scharf C. (1972): The inverse problem of potential scattering according to the Dirac equation, *Nuclear Phys.* A183 337–351 (section IX.5, Foreword)

Weiss J. (1969): The $N_1(s, r)/D_1(s, r)$ approach and the standard equations of the inverse scattering problem, *Czech. J. Phys.* B19, 1578 (section VII.H)

Weiss J. (1970): Inverse scattering problem including arbitrary parameters, *Fyz. Cas.* 20, no. 1, 3 (section VII.H)

Weiss J. (1971a): Relativistic Jost functions and potentials, *Czech. J. Phys.* B21, 1019 (section VII.H, Foreword)

Weiss J. (1971b): The N/D method off the mass shell for the Klein–Gordon equation and the inversion problem, *Czech. J. Phys.* B21, 1032 (section VII.H)

Wentzel G. (1926): Eine verallgeneinerung der quantem-bedingungen für die Zwecke der Wellenmechanik, *Z. Phys.* 38, 518–529 (section XVI.H)

Weston V. H. (1987): Factorization of the wave equation in higher dimensions, *J. Math. Phys.* 28, 1061–1068 (section XIV.7)

Wheeler J. A. (1937): On the mathematical description of light nuclei by the method of resonating group structure, *Phys. Rev.* 52, 1107 (Foreword)

Wheeler J. A. (1955): Scattering and potential, *Phys. Rev.* 99, 630 (section XVI.H, Foreword)

Wheeler J. A. (1976): Semiclassical analysis illuminates the connection between potential and bound states and scattering. *Studies in Mathematical Physics*: Essays in honor of Valentine Bargmann, Princeton University (sections XVI.5, XVI.6, Foreword)

Wigner E. P. (1955a): On the development of the compound nucleus model, *Amer. J. Phys.* 23, 371–382 (Preface)

Wigner E. P. (1955b): Lower limit for the energy derivative of the scattering phase shift, *Phys. Rev.* 98, 145–147; see also *idem, Rev. Mexicana Fís.* 1, 91 (1952); *Amer. J. Phys.* 23, 271 (1955) (section XIV.6)

Wigner E. P. (1964): Dispersion relations and their connection with causality, *Course XXIX International School of Physics "Enrico Fermi"*, Academic Press, New York (section XIV.6)

Wigner E. P., and Von Neumann J. (1954): Significance of Lowener's theorem in the quantum theory of collisions, *Ann. Math.* 54, 418 (section XIV.6)

Wiechert E. (1907): Uber Erdbehenwellen I., *March. Ges. Wis. Gottingen Math. Phys. HI.*, 415–429 (section XVI.7)

Wildermuth K. (1949): Der analytische Zusammenhang zwischen den Streumatrix— Elementen und den discreten stationären Zustanden in der Heisenbergschen S-Matrix-Theorie, *Z. Phys.* 127, 85–91 (Foreword)

Willis C. (1985): Inverse Sturm–Liouville problems with two discontinuities, *Inverse Problems* 1, 263–289 (section XVIII.4)

Willis C. (1986): An inverse method using toroidal mode data, *Inverse Problems* 2, 111–130 (section XVIII.4)

Wohlers M. R. (1966): A realizability theory for smooth lossless transmission lines, *IEEE Trans. on Circuit Theory* 13, 356–363 (section XVII.6)

Wuenschel P. C. (1960): Seismogram synthesis including multiples and transmission coefficients, *Geophys.* 25, 106–129 (section XIX.2)

Yagle A. E. (1986): Differential and integral methods for multidimensional inverse scattering problems, *J. Math. Phys.* 27, 2584–2591 (section XIV.7)

Yagle A. E. (1987): Multidimensional inverse scattering: An orthogonalization formulation, *J. Math. Phys.* 28, 1481–1491 (section XIV.7).

Yagle A. E., and Levy B. C. (1984): Application of the Schur algorithm to the inverse problem for a layered acoustic medium, *J. Acoust. Soc. Amer.* 76, 301–308 (section XVII.5)

Reference List 493

Yagle A. E., and Levy B. C. (1985a): A fast algorithm solution of the inverse problem for a layered acoustic medium probed by spherical harmonic waves, *J. Acoust. Soc. Amer.* 78, 729–737 (section XIV.7)

Yagle A. E., and Levy B. C. (1985b): A layer-stripping solution of the inverse problem for a one-dimensional elastic medium, *Geophys.* 50, 425–433 (section XVII.5)

Yagle A. E., and Levy B. C. (1986): Layer-stripping solutions of multidimensional inverse scattering problems *J. Math. Phys.* 27, 1701–1710 (section XIV.7)

Yamaguchi Y. (1954): Two-Nucleon Problem when the Potential is non local but separable I. *Phys. Rev.* 95, 1628 (section VIII.H)

Yamaguchi Y. (1955): Id. II, *Phys. Rev.* 95, 1635 (section VIII.H)

Yennie D. R., Boos F. R., and Ravenhall D. G. (1965): Analytic distorted-wave approximation for high-energy electron scattering calculations, *Phys. Rev.* 137, 4B, 882–903 (section XVI.H)

Yosida K. (1978): *Functional Analysis* 5th ed., Springer-Verlag, New York (section XVII.5)

Youla D. C. (1964): Analysis and synthesis of arbitrarily terminated lossless non-uniform lines, *IEEE Trans. on Circuit Theory* 11, 363–372 (section XVII.4)

Zachary W. W. (1981): Povzner–Levitan representations for eigenfunctions of higher order differential equations, 1–8 (section XVIII.4)

Zachary W. W. (1984): An inverse spectral theory of Gel'fand–Levitan type for higher order differential operators, *Lett. Math. Phys.* 8, 403–411; Erratum (1985): *Lett. Math. Phys.* 9, 263 (sections XV.H, XVIII.4)

Zachary W. W. (1986): An inverse scattering formalism for higher-order differential operators, *J. Math. Anal. Appl.* 117, 449–495 (sections III.7, XV.H)

Zakhariev B. N. (1986): Three-body inverse scattering problem, Report of the Joint Institute for Nuclear Research, Dubna (section III.6)

Zakhariev B. N., and Funke H. (1986): Inverse scattering problem for deformed three-dimensional potentials, Report of the Joint Institute for Nuclear Research, Dubna (section XIX.2)

Zakhariev B. N., and Suzko A. A. (1985): Potential and quantum scattering: Direct and inverse problems (in Russian), Energo Atomisdat, Moscow. To be published in English by Springer-Verlag, New York (section XIX.2)

Zakharov V. E., and Shabat A. B. (1972): Theory of two-dimensional self-focusing and one-dimensional modulation of waves in nonlinear media, *Sov. Phys. JETP* 34, 62–69 (section XVII.5)

Zheng W. M. (1984): The Darboux transformation and solvable double-well potential models for Schrödinger equations, *J. Math. Phys.* 25, 88–90 (section XVII.5)

Zhigunov V. P. (1977): Statistical approach to the solution of inverse problem of scattering quantum theory. Serpukhov Preprint, IHEP 77–123 (section XIX.2)

Zhigunov V. P. (1980): Potential reconstruction at fixed energy. Serpukhov Report, IHEP 79–107 (section XIX.2)

Zirilli F. (1976): Absence of positive eigenvalues in a class of 2-body quantum systems with nonlocal interactions, *Nuovo Cimento A* 34, 385–389 (section VIII.5)

Subject Index